Sensitization of Cancer Cells
for Chemo/Immuno/Radio-therapy

CANCER DRUG DISCOVERY
AND DEVELOPMENT SERIES

BEVERLY A. TEICHER, PhD, SERIES EDITOR

For other titles published in the series, go to
www.springer.com/humana
select the subdiscipline
search for your title

Sensitization of Cancer Cells for Chemo/Immuno/Radio-therapy

Edited by

Benjamin Bonavida, PhD

University of California at Los Angeles, Los Angeles, CA, USA

 Humana Press

Editor
Benjamin Bonavida, PhD
University of California at Los Angeles
Los Angeles, CA, USA

Series Editor
Beverly A. Teicher, PhD
Department of Oncology Research
Genzyme Corporation
Framingham, MA, USA

ISBN: 978-1-934115-29-9   e-ISBN: 978-1-59745-474-2
DOI: 10.1007/978-1-59745-474-2

Library of Congress Control Number: 2008929553

Cover design: Provided by Raymond Bonavida

Printed on acid-free paper

9 8 7 6 5 4 3 2 1

springer.com

Preface

Cancer chemotherapy can be traced in the 1940s when mustine (the prototype nitrogen mustard anticancer chemotherapeutic drug) was injected into a patient with non-Hodgkin's lymphoma, resulting in a dramatic reduction in tumor masses [1]. Thereafter, we witnessed the discovery and the important application of several new drugs, such as methotrexate [2], 6-mercaptopurine (6-MP), vincristine (vinca alkaloid), and aminopterin (folic antagonists) [3]. Then the concept of combination chemotherapy was introduced in 1965 by James Holland, Emil Freireich, and Emil Frei, who administered methotrexate, vincristine, 6-mercaptopurine, and prednisone, together referred as the POMP regimen, which resulted in long-term remission in children with acute lymphoblastic leukemia (ALL). This combination approach was extended to the lymphomas by Vincent T. DeVita and George Canellos in the late 1960s, when it was found that nitrogen mustard, vincristine, procarbazine, and prednisone—known as the MOPP regimen—could cure patients with Hodgkin's and non-Hodgkin's lymphoma. Thereafter, new drugs were discovered, including taxanes, camptothecins, platinum-based agents, nitrosoureas, anthracyclines, and epipodophyllotoxins [4]. The successes of combination chemotherapy suggested that all cancers could be treated provided the correct combination of drugs at the correct doses and correct intervals were established. However, while chemotherapeutic drugs were effective with minimal knowledge of underlying mechanisms of action, new studies began to unravel the genetic nature of cancer and the development of targeted therapies.

Targeted therapies include monoclonal antibodies, cell-mediated immunotherapy, gene immunotherapy, and the development of inhibitors interfering with survival antiapoptotic signaling pathways in cancers. While these novel approaches have significantly improved the outcome of many cancer patients, there remains a major problem in the development of cancer resistance to conventional and novel cytotoxic therapies. Further, since most cytotoxic therapies mediate their activities by inducing programmed cell death, or apoptosis, tumor cells develop mechanisms to resist apoptosis and thus acquire a phenotype of cross-resistance to most cytotoxic stimuli. Therefore, there is an urgent need to unravel the underlying mechanisms of resistance at the biochemical and genetic levels and the development of agents that can reverse resistance, directly or in combination with other cytotoxics. The objective of this book is to select novel approaches developed to reverse tumor cell resistance to chemo/immuno/radio-therapy and the use of various sensitizing agents in combination with various cytotoxics [5]. This volume is by no means exhaustive of this subject matter, but primarily introduces several current approaches that have been developed by established investigators in the field. The volume is arbitrarily divided into several main topics, recognizing that the contents of several chapters in one topic can overlap with other topics.

There are several contributions on tumor cell sensitization based on approaches to target cell surface receptors and how such targeting agents sensitize tumor cells to apoptosis. Dr. Vollmers and colleagues describe the use of monoclonal antibodies as sensitizing agents to reverse epithelial cancers to apoptosis. Dr. Penichet and colleagues developed monoclonal antibodies directed against the overexpressed transferrin receptor on tumor cells. They also genetically engineered a fusion protein that was found to be cytotoxic and also sensitizes tumor cells to various chemotherapeutic drugs. Dr. Bonavida and colleagues discuss the FDA-approved chimeric anti-CD20 mAb, rituximab, and its ability to sensitize drug-resistant B-NHL to apoptosis by various chemotherapeutic drugs. They describe rituximab-mediated inhibition of several anti-apoptotic and constitutively activated signaling pathways and that are responsible for chemosensitization. Dr. Sakai and colleagues examine the role of the TRAIL death receptor, DR5, and its upregulation by various agents, leading to sensitization of TRAIL-resistant tumor cells to TRAIL-induced apoptosis. It is noteworthy that TRAIL and agonist DR4/DR5 mAbs are currently being tested in phases I and II clinical trials for various cancers. Dr. Murphy and colleagues used proteasome inhibitors to sensitize tumor cells to immune-mediated apoptosis.

Several contributors describe their findings by targeting constitutively activated cell survival pathways in cancer. Dr. Kerbel and colleagues describe the use of anti-angiogenic inhibitors as chemosensitizing agents, with particular emphasis on metastatic disease. Dr. McCubrey and colleagues describe the constitutively activated cell survival pathways, namely, the Raf/MEK/ERK and PI3/AKT pathways, and the use of cell membrane–permeable small-molecular-weight inhibitors that target these pathways and can be used as chemosensitizing agents. Drs. Rosato and Grant describe the use of histone deacetylase inhibitors in combination with other agents for the reversal of tumor cell resistance. Dr. Sorokin describes the role of eicosanoids in the regulation of tumor cell resistance to apoptosis and the various means to target these lipids in order to reverse chemoresistance.

There are several contributions that investigate targeting of transcription factors as sensitizing agents. Dr. Chatterjee and colleagues examine the relationship between the transcriptional regulation of survival pathways and inhibition of these pathways, and shifting the balance to reverse resistance.

They describe the roles of Raf kinase inhibitory protein (RKIP) as apoptotic and signal transducer and activator of transcription (STAT3) as antiapoptotic and describe the opposing effects of these two gene products. Dr. Gambari describes novel RNA-DNA–based strategies as chemosensitizing agents by targeting selected mRNAs with antisense oligonucleotides or small interfering RNAs (siRNA) or targeting transcription factors with decoy oligonucleotides. Drs. Maina and Domo examine the beneficial effect of combining inhibitors of *p53* as sensitizing agents when used in combination with conventional chemo- and radio-therapies to reverse resistance. Dr. Bonavida and colleagues examine the role of various inhibitors, such as nitric oxide (NO) donors, as sensitizing agents leading to inhibition of the transcription factors NF-κB and Yin Yang1 (YY1). Inhibition of these transcription factors upregulates death receptors (FAS, DR5) and sensitizes tumors cells to FAS ligand and TRAIL-induced apoptosis. Dr. Aggarwal and colleagues used several natural products that inhibit NF-κB and sensitize tumor cells to both chemotherapy and radiation.

Due to the fact that the apoptotic pathways are dysregulated in cancer, and primarily there is overexpression of antiapoptotic gene products or underexpression of apoptotic gene products, sensitizing agents that can regulate these gene products and interfere with apoptotic pathways may reverse resistance when used in combination with other cytotoxics. Several contributors used such approaches. Dr. Johnson examines the application of inhibitors of the Bcl-2 family as chemo- and radio-sensitizers. These studies were undertaken both in vitro and in vivo for their potential clinical application. Dr. Johnston and colleagues also used the strategy of interfering with the dysregulated apoptotic pathways in cancer and describe various means to interfere with antiapoptotic pathways by using, for example, antisense and siRNA as sensitizing agents. Dr. Li and colleagues discuss the use of peptides and peptide mimetics as sensitizing agents and their possible application in clinical trials as a new approach for cancer therapy. Dr. Mayo and colleagues discuss the utility of nonpeptide mimetics to sensitize tumor cells when used in combination with subtoxic doses of chemotherapy and radiation. Drs. Sarkar and Lee discuss the effects of combining isoflavones and conventional therapeutics. Isoflavones and derivatives exert many effects on cancer cells, such as regulating several survival pathways and apoptotic pathways. Drs. Schwenzer and Förster discuss antisense oligonucleotides and siRNA applications in therapy and their use in ongoing clinical trials.

The approach of tailored customizing therapy for individual cancer patients requires a thorough understanding of the genetic makeup of the patient and its cancer and the pharmacogenetics of drugs. Drs. Efferth and Wink discuss the pharmacogenetic approach to compare monogenetic disease with a more complex disease such as cancer. These studies open the way to design personalized custom-tailored therapy. Also, Drs. Stivala and her colleagues discuss the importance of how genetic abnormalities may influence the response to treatment. They also discuss current strategies to integrate pharmacogenetics into the development of anticancer drugs.

Clearly, this volume represents a broad overview of the field of cancer sensitization and introduces several novel approaches that can be used to reverse cancer resistance through the application of a variety of sensitizing agents. Readers are also encouraged to read several reviews on related topics. As editor, I wish to thank all of the contributors, without whom this book could not have been realized. In addition, I acknowledge the administrative and

technical support of Maggie Yang and Erica Keng for their diligent and professional input. Lastly, I wish to thank my wife and two sons for their unconditional support during the preparation of this volume.

Benjamin Bonavida, PhD

References

1. Goodman LS, Wintrobe MM, Dameshek W, et al. Nitrogen mustard therapy. Use of methyl-bis (beta-chloroethyl) amine hydrochloride and tris (beta-chloroethyl) amine hydrochloride for Hodgkin's disease, lymphosarcoma, leukemia, and certain allied and miscellaneous disorders. J Am Med Assoc 1946, 105:475–476. Reprinted in JAMA 1984, 251:2255–2261. PMID 6368885.
2. Li MC, Hertz R, Bergenstal DM. Therapy of choriocarcinoma and related trophoblastic tumors with folic acid and purine antagonists. N Engl J Med 1958, 259:66–74. PMID 1246307.
3. Farber S, Diamond LK, Mercer RD, et al. Temporary remissions in acute leukemia in children produced by folic antagonist, 4-aminopteroylglutamic acid (aminopterin). N Engl J Med 1948, 238:787–793.
4. Papac RJ. Origins of cancer therapy. Yale J Biol Med 2001, 74:391–398. PMID 11922186.
5. Ng C-P, Bonavida B. A new challenge to immunotherapy by tumors that are resistant to apoptosis: two complementary signals to overcome cross-resistance. Adv Cancer Res 2002, 85:145–174.

Contents

Part V   Sensitization Tailored to Individual Patients

Contributors

Stephen L. Abrams
Department of Microbiology & Immunology,
Brody School of Medicine at East Carolina
University, Greenville, NC

Bharat B. Aggarwal
Cytokine Research Laboratory, Department
of Experimental Therapeutics, and Department
of Radiation Oncology, The University of Texas
MD Anderson Cancer Center, Houston, TX

Sudharsana Rao Ande
Manitoba Institute of Cell Biology,
CancerCare Manitoba

Shantanu Banerji
Manitoba Institute of Cell Biology, CancerCare
Manitoba: Section of Haematology and Oncology,
Department of Internal Medicine, University
of Manitoba, Winnipeg, Canada

Versha Banerji
Manitoba Institute of Cell Biology, CancerCare
Manitoba: Section of Haematology and Oncology,
Department of Internal Medicine, University of
Manitoba, Winnipeg, Canada

Stravoula Baritaki
Department of Microbiology, Immunology
and Molecular Genetics, Jonsson Comprehensive
Cancer Center, David Geffen School of Medicine
at UCLA, University of California,
Los Angeles, CA

Jorg Basecke
Division of Hematology and Oncology,
Department of Medicine, Göttingen, Germany

Fred E. Bertrand
Department of Microbiology & Immunology,
Brody School of Medicine at East Carolina
University, Greenville, NC

Antonio Bonati
Department of Clinical Sciences, University
of Parma, Parma, Italy

Benjamin Bonavida
Department of Microbiology, Immunology, and
Molecular Genetics, Jonsson Comprehensive
Cancer Center, David Geffen School of
Medicine at UCLA, University of California
at Los Angeles, Los Angeles, CA

Stephanie Brändlein
Institute of Pathology, University of Würzburg,
Würzburg, Germany

Bibiana Bruni
Department of Biomedical Sciences,
University of Catania, Catania, Italy

Calogero Cannavò
Department of Biomedical Sciences,
University of Catania, Catania, Italy

Erika Cecchin
Experimental and Clinical Pharmacology Unit,
Centro di Riferimento Oncologico, National
Cancer Institute, Aviano, Italy

William H. Chappell
Department of Microbiology & Immunology,
Brody School of Medicine at East Carolina
University, Greenville, NC

Devasis Chatterjee
Departments of Medicine, Pathology and
Surgical Research, Rhode Island Hospital
and the Warren Alpert Medical School of
Brown University, Providence, RI

Y. Eugene Chin
Departments of Medicine, Pathology and
Surgical Research, Rhode Island Hospital and
the Warren Alpert Medical School of Brown
University, Providence, RI

Tracy R. Daniels
Division of Surgical Oncology, Department
of Surgery, David Geffen School of Medicine,
University of California at Los Angeles,
Los Angeles, CA

Ruud P.M. Dings
Department of Biochemistry, Molecular Biology
& Biophysics, University of Minnesota,
Minneapolis, MN

Rosanna Dono
Developmental Biology Institute of Marseille-
Luminy (IBDML) (CNRS—Univ. Méditerranée),
Marseille, France

Thomas Efferth
German Cancer Research Center,
Pharmaceutical Biology of Natural Products,
Heidelberg, Germany

Urban Emmenegger
Molecular and Cellular
Biology Research, Sunnybrook Health Sciences
Centre and Medical Biophysics, University
of Toronto, Toronto, Ontario, Canada

Yvonne Förster
Technical University Dresden, Department
of Biochemistry, Dresden, Germany

Giulio Francia
Molecular and Cellular Biology Research,
Sunnybrook Health Sciences Centre and
Medical Biophysics, University of Toronto,
Toronto, Ontario, Canada

Richard A. Franklin
Department of Microbiology & Immunology,
Brody School of Medicine at East Carolina
University, Greenville, NC

Alberto Fulvi
Department of Biomedical Sciences,
University of Catania, Catania, Italy

Roberto Gambari
Department of Biochemistry and Molecular
Biology, Molecular Biology Section,
Ferrara University, Ferrara, Italy

Steven Grant
Massey Cancer Center, Department of Medicine,
Virginia Commonwealth University, Richmond,
VA

William H.D. Hallett
Department of Microbiology and Immunology,
University of Nevada School of Medicine,
University of Nevada, Reno, Reno, NV

Gustavo Helguera
Division of Surgical Oncology, Department
of Surgery, David Geffen School of Medicine,
University of California at Los Angeles,
Los Angeles, CA

Sara Huerta-Yepez
Department of Microbiology, Immunology
and Molecular Genetics, Jonsson Comprehensive
Cancer Center, David Geffen School of Medicine
at UCLA, University of California,
Los Angeles, CA

Ganesh Jagetia
Cytokine Research Laboratory,
Department of Experimental Therapeutics,
and Department of Radiation Oncology,
The University of Texas MD Anderson
Cancer Center, Houston, TX

Ali R. Jazirehi
Department of Microbiology, Immunology and
Molecular Genetics, Jonsson Comprehensive
Cancer Center, David Geffen School of Medicine
at UCLA, University of California,
Los Angeles, CA

Daniel E. Johnson
Department of Medicine,
University of Pittsburgh, Pittsburgh, PA

Patrick G. Johnston
Drug Resistance Laboratory, Centre for Cancer
Research and Cell Biology,

Queen's University Belfast, Belfast,
N. Ireland

Robert S. Kerbel
Molecular and Cellular Biology Research,
Sunnybrook Health Sciences Centre and
Medical Biophysics, University of Toronto,
Toronto, Ontario, Canada

Mark Klein
Department of Biochemistry, Molecular Biology
& Biophysics, University of Minnesota,
Minneapolis, MN

Sunil Krishnan
Cytokine Research Laboratory, Department
of Experimental Therapeutics, and Department
of Radiation Oncology, The University of Texas
MD Anderson Cancer Center, Houston, TX

Brian D. Lehmann
Department of Clinical Sciences,
University of Parma, Parma, Italy

Minghui Li
Department of Microbiology and Immunology,
University of Nevada School of Medicine,
University of Nevada, Reno, Reno, NV

Yiwei Li
Department of Pathology, Karmanos
Cancer Institute, Wayne State University School
of Medicine, Detroit, MI

Massimo Libra
Department of Biomedical Sciences,
University of Catania, Catania, Italy

Daniel B. Longley
Drug Resistance Laboratory, Centre for
Cancer Research and Cell Biology,
Queen's University Belfast, Belfast, N. Ireland

Marek Los
Manitoba Institute of Cell Biology, CancerCare
Manitoba: Department of Human Anatomy and
Cell Biology, Department of Biochemistry and
Medical Genetics, University of Manitoba,
Winnipeg, Canada

Dale E. Ludwig
ImClone Systems, New York, NY

Paolo Lungi
Department of Clinical Sciences,
University of Parma, Parma, Italy

Subbareddy Maddika
Manitoba Institute of Cell Biology,
CancerCare Manitoba: Department of
Biochemistry and Medical Genetics,
University of Manitoba, Winnipeg, Canada

Flavio Maina
Developmental Biology Institute of Marseille-
Luminy (IBDML) (CNRS—Univ. Méditerranée),
Marseille, France

Alberto M. Martelli
Department of Human Anatomical Sciences,
University of Bologna, Bologna, Italy

Kevin H. Mayo
Department of Biochemistry,
Molecular Biology & Biophysics, University
of Minnesota, Minneapolis, MN

James A. McCubrey
Department of Microbiology & Immunology,
Brody School of Medicine at East Carolina
University, Greenville, NC

Melissa L. Midgett
Department of Microbiology & Immunology,
Brody School of Medicine at East Carolina
University, Greenville, NC

Michele Milella
Regina Elena Cancer Center, Rome, Italy

Negin Misaghian
Department of Microbiology & Immunology,
Brody School of Medicine at East Carolina
University, Greenville, NC

Milad Motarjemi
Department of Microbiology and Immunology,
University of Nevada School of Medicine,
University of Nevada, Reno, Reno, NV

William J. Murphy
Department of Microbiology and Immunology,
University of Nevada School of Medicine,
University of Nevada, Reno, Reno, NV

Isabel I. Neacato
David Geffen School of Medicine,
University of California at Los Angeles,
Los Angeles, CA

Neil W. Owens
Department of Chemistry, University of Manitoba,
Winnipeg, Canada

Manuel L. Penichet
Division of Surgical Oncology, Department of
Surgery, Department of Microbiology,
Immunology, and Molecular Genetics,
David Geffen School of Medicine,
University of California at Los Angeles,
Los Angeles, CA

Iran Rashedi
Manitoba Institute of Cell Biology,
CancerCare Manitoba: Department of Biochemis-
try and Medical Genetics, University of Manitoba,
Winnipeg, Canada

Murray B. Resnick
Departments of Medicine, Pathology and
Surgical Research, Rhode Island Hospital and
the Warren Alpert Medical School of Brown
University, Providence, RI

Roberto R. Rosato
Massey Cancer Center,
Department of Medicine, Virginia Commonwealth
University, Richmond, VA

Edmond Sabo
Departments of Medicine,
Pathology and Surgical Research, Rhode Island
Hospital and the Warren Alpert Medical School
of Brown University, Providence, RI

Toshiyuki Sakai
Department of Molecular-Targeting Cancer
Prevention, Graduate School of Medical Science,
Kyoto Prefectural University of Medicine,
Kyoto, Japan

Fazlul H. Sarkar
Department of Pathology, Karmanos Cancer
Institute, Wayne State University School of
Medicine, Detroit, MI

Frank Schweizer
Department of Chemistry, University of Manitoba,
Winnipeg, Canada

Bernd Schwenzer
Technical University Dresden, Department of
Biochemistry, Dresden, Germany

Andrey Sorokin
Department of Medicine, Medical College
of Wisconsin, Milwaukee, WI

Demetrios A. Spandidos
Department of Clinical Virology,
Faculty of Medicine, University of Crete,
Heraklion, Greece

Kristin M. Stadelman
Department of Microbiology & Immunology,
Brody School of Medicine at East Carolina
University, Greenville, NC

Linda S. Steelman
Department of Microbiology & Immunology,
Brody School of Medicine at East Carolina
University, Greenville, NC

Franca Stivala
Department of Biomedical Sciences,
University of Catania, Catania, Italy

Eriko Suzuki
Department of Applied Chemistry,
Keio University, Yokohama, Japan

Agostino Tafuri
Department of Hematology-Oncology,
La Sapienza, University of Rome,
Rome, Italy

Jackson R. Taylor
Department of Microbiology & Immunology,
Brody School of Medicine at East Carolina
University, Greenville, NC

David M. Terrian
Department of Anatomy & Cell Biology,
Brody School of Medicine at East
Carolina University, Greenville, NC

Giuseppe Toffoli
Experimental and Clinical
Pharmacology Unit, Centro di Riferimento
Oncologico, National Cancer Institute,
Aviano, Italy

Kazuo Umezawa
Department of Applied Chemistry,
Keio University, Yokohama, Japan

Mario I. Vega
Department of Microbiology,
Immunology and Molecular Genetics,
Jonsson Comprehensive Cancer Center,
David Geffen School of Medicine at UCLA,
University of California, Los Angeles, CA

H. Peter Vollmers
Institute of Pathology, University of Würzburg,
Würzburg, Germany

Timothy R. Wilson
Drug Resistance Laboratory, Centre for Cancer
Research and Cell Biology, Queen's University
Belfast, Belfast, N. Ireland

Michael Wink
Institute of Pharmacy and Molecular
Biotechnology, University of Heidelberg,
Germany

Ellis W.T. Wong
Department of Microbiology &
Immunology, Brody School of Medicine

at East Carolina University,
Greenville, NC

Kam C. Yeung
Department of Biochemistry and Molecular
Biology, Medical College of Ohio, Toledo, Ohio

Tatsushi Yoshida
Department of Molecular-Targeting
Cancer Prevention, Graduate School of
Medical Science, Kyoto Prefectural
University of Medicine,
Kyoto, Japan

Anne Zuse
Manitoba Institute of Cell Biology,
CancerCare Manitoba

COLOR PLATES

that were starting to fail (as shown in top panel **[A]** after around 100 days) Ld CTX plus trastuzumab therapies. Second-line regimens were as indicated, with the addition of bevacizumab causing a further growth delay of approximately 4 weeks. (Adapted from du Manoir JM, Francia G, Man S, et al. Strategies for delaying or treating in vivo acquired resistance to trastuzumab in human breast cancer xenografts. Clin Cancer Res 2006, 12:904–916.

COLOR PLATE 5   An example of how chronic combination oral metronomic low-dose CTX and UFT prolongs survival of mice with advanced metastatic disease. (From Munoz R, Man S, Shaked Y, et al. Highly efficacious nontoxic preclinical treatment for advanced metastatic breast cancer using combination oral UFT-cyclophosphamide metronomic chemotherapy. Cancer Res 2006, 66:3386–3391.) **A.** 231/LM2-4 human breast metastatic variant cells were orthotopically injected into the MFPs of 6- to 8-week-old CB17 SCID mice. When tumors reached volumes of approximately 200 mm^3, treatment with either vehicle control, or 15 mg/kg/day UFT by gavage, or 20 mg/kg per day CTX through the drinking water, or a combination of CTX and UFT treatments was initiated. Tumors were measured weekly and tumor volume was plotted accordingly. Arrow indicates time of initiation of treatment. **B.** 6-week-old CB-17 SCID mice were recipients of 231/LM2-4 transplanted cells. When tumors reached 400 mm^3 (which took approximately 3 weeks) primary tumors were surgically removed. Treatment with vehicle control, 15 mg/kg per day UFT by gavage, 20 mg/kg per day CTX through the drinking water, or the daily combination of metronomic UFT and CTX, were initiated 3 weeks after surgery on a daily non-stop basis. For example, in the experiment shown in **(B),** the duration of the therapy was 140 days, and was initiated on day 43, 3 weeks after surgery, with termination at day 183. Mice were monitored frequently according to the institutional guidelines. A Kaplan-Meier survival curve was plotted accordingly for all treated groups, as indicated in the figure. **A, B.** n = 7–9/group. NS = normal saline, Veh = vehicle control (0.1% HPMC). Note that effects on primary tumor **(A)** were minor and in no way predictive of the survival benefits seen with UFT and CTX on metastatic disease (recorded as survival)

COLOR PLATE 6   Metabolism of arachidonic acid and synthesis of major eicosanoids. Reactions catalyzed by cyclooxygenases are shown in the pink field, reactions catalyzed by lipoxygenases are shown in the green field and reactions catalyzed by CYP monooxidases are shown in the yellow field. Also shown are five major prostanoids, which are synthesized by prostaglandin synthases from PGH$_2$, and LTA$_4$ converted from product of lipoxygenase reaction, which is further converted to LTB$_4$ by LTA$_4$ hydrolase

COLOR PLATE 7   Structure of PNA-DNA-PNA chimeras targeting NF-kappaB–related proteins. The molecular structures and the sequence of the double-stranded PNA-DNA-PNA chimera mimicking NF-kappaB binding sites are modified from Borgatti et al. [116], Romanelli et al. [118], and Gambari et al. [109]

COLOR PLATE 8   Mechanism of tumor cell sensitization to Fas-L–induced apoptosis by IFN-γ: Pivotal role of NO. IFN-γ or other agents, such as TNF-a, IL-1, or LPS, upregulate NF-kB, which in turn regulates positively the transcription of NOSII. NOSII catalyses the biosynthesis of NO by L-arginine. NO can also be released in the cytosol by treatment of cells with an NO donor such as SNAP or DETANONOate. Free nitric oxide may react with O$_2$ (discontinuous line), resulting in the formation of reactive nitrogen species (RNS) such as (ONOO$^-$), which upregulate Fas and cause oxidative damage in protein and nucleic acids leading to apoptosis. Alternatively (continuous line), NO or NO$^+$ ion is capable of forming S-nitrosothiols resulting in S-nitrosylation of several proteins, including YY1, which acts as a repressor of Fas transcription. Thus, inducible levels of Fas by NO are able to overcome tumor resistance to Fas-L and sensitize them to Fas-L–mediated apoptosis

COLOR PLATE 9   Mechanism of tumor cell sensitization to TRAIL-induced apoptosis by NO. Treatment of several tumor cell lines with NO donors such as DETANONOate and TRAIL results in apoptosis and synergy is achieved. The synergy is the result of complementation in which each agent partially activates the apoptotic pathway and the combination results in apoptosis. The signal provided by NO partially inhibits NF-kB activity, and this leads to downregulation of antiapoptotic proteins of the Bcl-2 family such as Bcl-xL, and inhibition of cIAP family members (i.e.,

XIAP, cIAP-1, cIAP-2). In addition DETANONOate also partially activates the mitochondria and release of modest amounts of cytochrome C and Smac/DIABLO into the cytosol in the absence of caspase-9 activation. The NO-induced NF-kB suppression also inhibits the negative transcriptional regulator of DR5, YY1, resulting in DR5 upregulation. Thus, the combination treatment with TRAIL and DETANONOate results in significant activation of the mitochondria and release of high levels of cytochrome C and Smac/DIABLO, activation of caspases-9 and -3, promoting apoptosis. The role of Bcl-xL in the regulation of TRAIL apoptosis has been corroborated by the use of the chemical inhibitor 2MAM-A3 in several cell lines, which also sensitized the cells to apoptosis

COLOR PLATE 10   Model of the G6PD Aachen tetramer. This G6PD variant has originally been described by Kahn et al. [253]. The mutation in the G6PD Aachen variant has been determined by us previously [254]. A mutation 1089C>G results in a predicted amino acid change 363Asn>Lys. The 1089C>G point mutation is unique, but produces the identical amino acid change found in another variant of G6PD deficiency, G6PD Loma Linda. The 363Asn>Lys exchange in G6PD Loma Linda is caused by a 1089C>A mutation [255]. Using the available three-dimensional structure of the human G6PD tetrameric protein complex [256], the location of the point mutation of amino acid 363 in G6PD Aachen is found at the surface of a monomer in close proximity to $NADP^+$ and more than 20Å away from the glucose-6-phosphate binding site. This residue is probably involved in $NADP^+$ binding that in turn is required for tetramer stability [256]. Thus, Arg363 may be required to indirectly maintain the structural integrity of the functional unit. Replacing it with a positively charged Lys residue would lead to charge–charge repulsion between Lys363 and $NADP^+$, thus affecting $NADP^+$ binding and tetramer formation. The two pairs of dimer-forming monomers are colored in ice blue/blue and in orange/red. The mutation site (Asn363) and the cofactor $NADP^+$ are shown in van der Waals representation in purple and grey, respectively. The conserved eight-residue peptide RIDHYLGK corresponding to the substrate binding site is colored in yellow. The figure was prepared from the crystal structure 1QKI using the program VMD [257]

COLOR PLATE 11   Molecular modeling of SNP-based variants of MGMT. **A.** Wild-type structure of MGMT. Mutation sites are shown in van der Waals representation. Color codes of helices: red, N-terminal α-helices, blue, DNA recognition site with helix-turn-helix motif. The second DNA recognition site binds to the major groove of DNA, yellow, 3–10 helix with conserved Pro-Cys-His-Arg sequence. The active Cys145 is located here. Color code of loops: orange, Asn-hinge that joins the DNA recognition helix and the active site. It also provides 40% of the contact between N-terminal and C-terminal domains, white, DNA binding wing. O6-alkylguanine lies between the binding wing and the recognition helix. Color code of side chains shown in CPK representation: per atom, conserved active Pro-Cys-Arg sequence, pink, Zn^{2+} binding site (residues C5, C24, H29, H85). Cys5 is missing, since it is not resolved in the crystal structure 1QNT. **B.** Localization of amino acid change and overlap of structures of all mutants. Mutated side chains are shown in van der Waals representations. (Reprinted from Schwarzl SM, Smith JC, Kaina B, et al. Molecular modeling of O6-methylguanine-DNA methyltransferase mutant proteins encoded by single nucleotide polymorphisms. Int J Mol Med 2005, 16:553–557.)

Part I
Sensitization via Membrane-Bound Receptors

Chapter 1
Sensitization of Epithelial Cancer Cells with Human Monoclonal Antibodies

H. Peter Vollmers and Stephanie Brändlein

1 Natural IgM Antibodies

Natural IgM antibodies are part of the innate immunity and important components of first line defense mechanisms [1–3]. They are germ-line encoded, and are produced by a small subset of B lymphocytes, B1, or CD5+ cells [4–6]. The repertoire and reactivity pattern of natural antibodies is remarkably stable within each species and even between species. This genetic stability seems to be the result of an evolutionary selection process providing an inherent legacy of specificities capable of protecting against pathogen invasion, malignant cells, and other harmful alterations. Natural IgM antibodies have been shown to be involved in early recognition of external invaders such as bacteria and viruses [3, 7], but these natural IgM antibodies also seem to be involved in recognition and elimination of precancerous and cancerous lesions [8–12].

Although coded by a limited set of germ-line immunoglobulin genes, these IgM antibodies give a sufficient protection without additional mutational adaptation [13–18]. This is only possible, if IgM antibodies do not detect single structures, but instead bind to patterns of conservative structures, which are expressed by "non-self" structures. Furthermore, this recognition system guarantees that the innate immune response need not follow all mutational changes. However, natural antibodies are ideal antitumor weapons, because even if they have low affinity and show some oligo-reactivity, they have some unique properties that antibodies produced by xeno-immunization or phage display-

technique do not have: They are tumor specific by nature.

In an extensive investigation of the antitumor defense in humans, several tumor-specific IgM antibodies could be isolated [8, 19, 20]. All these human monoclonal antibodies analyzed so far have some typical features in common. They are pentameric molecules, coded by specific germ-line families and they are equipped mainly with lambda chains, in contrast to the majority of circulating antibodies [8]. These natural IgM antibodies preferentially bind to tumor-specific carbohydrate epitopes on post-transcriptionally modified cell surface antigens [8, 21, 22]. Another typical feature of the natural IgM antibodies is their ability to induce apoptosis in malignant cells in a death domain independent way [8, 12, 21, 23–26]. This is done for example by blocking of growth-factor receptors, cross-linking of modified anticomplement receptors or by increasing the intracellular level of neutral lipids [12, 21, 22, 26] (Fig. 1.1).

2 Growth Factor Receptors

Cancer cells are hyper-proliferating cells, with a huge need of growth factors [27, 28]. Epidermal growth factor receptors (EGF-Rs) are therefore often found overexpressed on a variety of malignant cells [29]. Antibodies binding to EGFRs have been used to block these receptors and to inhibit the binding of the specific growth factor ligands [30]. The effect of these antibodies can be

From: *Sensitization of Cancer Cells for Chemo/Immuno/Radio-therapy, 1st Edition.*
Edited by: Benjamin Bonavida © Humana Press, Totowa, NJ

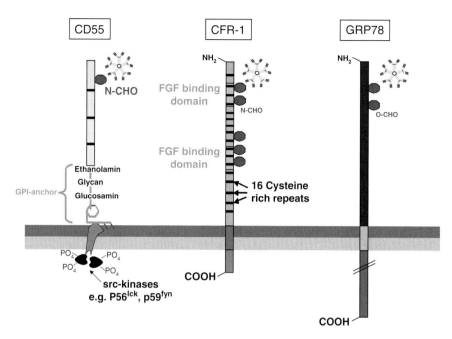

FIGURE 1.1. Tumor-specific post-transcriptionally modified cell surface antigens detected by natural IgM antibodies

enhanced in combinatorial approaches with conventional methods like radiation and chemotherapy and vice versa [31–35]. In breast cancer trials the monotherapeutic effect of trastuzumab (Herceptin) could clearly be enhanced in a synergistic manner when the antibody was used in combination with chemotherapy [32, 36–39]. Similar results were obtained for panitumumab (ABX-EGF) and cetuximab (Erbitux) in numerous cancer models, including lung, kidney, and colon [40].

The natural monoclonal IgM antibody PAM-1 was isolated from a patient with a stomach carcinoma and binds to a tumor-specific isoform of CFR-1 (Cysteine-rich fibroblastic growth factor receptor) [9, 22]. The CFR-1 is a 130-kDa integral membrane glycoprotein, homologous to CFR-1 (cysteine-rich fibroblast growth factor receptor), which has so far only been detected and described in Golgi of embryonic chicken cells and CHO cells [41]. The receptor is homologous to a rat protein, cloned as a Golgi-specific protein, designated MG160, which is involved in the processing and secretion of growth factors and was recently found in pancreatic cancer [42–46]. The human homologue, E-selectin ligand 1 (ESL-1), is a cytokine, expressed on myeloid and some lymphoma cells and is modulated by cell adhesion molecules that cause the binding of neutrophils to the endothelium [47, 48].

PAM-1 binds to a N-linked carbohydrate epitope which is specific for a post-transcriptionally modified isoform of CFR-1 (see Fig. 1.1) [22]. CFR-1/PAM-1 is expressed on almost all epithelial cancers of every type and origin, but not on healthy tissue [9, 22]. It is also found on precursor lesions such as *H. pylori*–induced gastritis, intestinal metaplasia and dysplasia of the stomach, ulcerative colitis–related dysplasia and adenomas of the colon, Barrett metaplasia and dysplasia of the esophagus, squamous cell metaplasia and dysplasia of the lung, cervical intraepithelial neoplasia I–III, ductal and lobular carcinoma in situ of the breast and prostate intraepithelial neoplasia (Fig. 1.2) [9, 10, 22]. PAM-1 inhibits tumor growth in vitro and in animal systems, by inducing apoptosis [25]. The effect is not dependent on the pentameric form and cross-linking, suggesting a similar mechanism as described for anti-EGFR antibodies [30].

3 Complement Decay Molecules

Decay acceleration factors (DAF) are surface receptors that protect host tissues from complement activation. They prevent cell damage by dissociating the classical and alternative pathway C3 convertases [49, 50]. CD55/DAF is such a complement

H&E   **IgM-control**   **PAM-1**

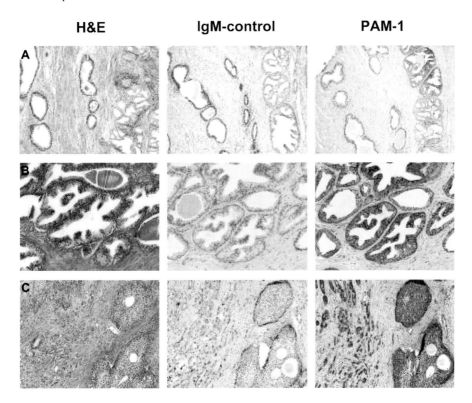

FIGURE 1.2. Immunohistochemical staining of antibody PAM-1 on different precursor lesions of prostate carcinoma. Paraffin sections were stained with hematoxylin-eosin, unspecific human IgM as a negative control and antibody PAM-1: **A.** Normal prostate tissue and low grade prostate intraepithelial neoplasia (PIN). **B.** High-grade PIN. **C.** Prostate adenocarcinoma and high grade PIN (original magnification, ×100) (*See Color Plates*)

regulatory protein, which is expressed in two different isoforms generated by differential splicing. While DAF-A is secreted from cells, DAF-B is linked to cells by a glycosylphosphatidylinositol (GPI-) anchor [51]. Both forms are further modified by different glycosylation patterns, resulting in sizes of 55–100 kDa molecular weight [52]. DAF-B is expressed on all cell types that can get in contact with the complement system [53].

Malignant cells often over-express DAF/CD55 molecules, to increase their protection level against complement attacks [49, 54]. Over-expression was found in, e.g., breast, colon, and stomach carcinoma [55–57]. This over-expression makes DAF a suitable target for cancer vaccines in the treatment of colon carcinomas [58]. Vaccination with a human anti-idiotype antibody that mimics DAF was used as an adjuvant treatment of colon carcinoma and resulted in an activation of a cellular antitumor response [59].

On the other hand, this over-expression of DAF and other complement inactivating molecules

limits therapeutical approaches which depend on the help of complement, like in antibody-dependent cellular cytotoxicity. However, if malignant cells over-express and/or modify protection molecules with cell-specific alterations, these new epitopes can be used as targets for therapeutical approaches. The human monoclonal antibody SC-1, which was isolated from a patient with a signet ring cell carcinoma of the stomach [24], reacts with a N-linked carbohydrate epitope present on an isoform of DAF-B (subsequently named DAF^{SC-1}) with a molecular weight of approximately 82 kDa (see Fig. 1.1) [21]. The antibody reacts with over 70% of all diffuse-type and intestinal-type gastric adenocarcinomas [60]. Clinical studies have shown that specific induction of regression and apoptosis can be induced in primary stomach cancers without any detected toxic cross-reactivity to normal tissue [61].

Binding of SC-1 induces specific apoptosis of stomach carcinoma cells both in vitro and in experimental in vivo systems (Fig. 1.3) [23, 62]. The effect depends on the cross-linking activity of the antibody which most likely inactivates the complement decay molecules. Shortly after binding, caspase-3 and -8 are activated resulting in cleavage of cytokeratin 18. Furthermore, there is a down-regulation of topoisomerase II, a short increase in the intracellular Ca^{2+}-concentration, which is not necessary for the apoptotic event

but seems to be in involved in the regulation of DAF^{SC-1} expression [21, 62].

4   Heat Shock Proteins

Stress or heat shock proteins (HSPs) are ubiquitous and highly conserved cytoprotective proteins [63]. They play an essential role in intracellular

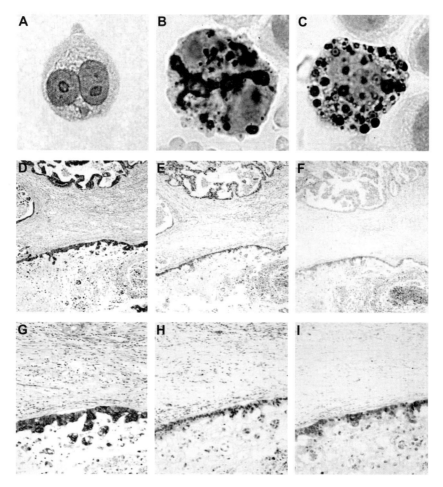

FIGURE 1.3. SC-1 induced apoptosis in vitro and in vivo. Cleavage of cytokeratin 18 in SC-1-treated apoptotic stomach carcinoma cells in vitro. Immunohistochemical staining of cytospin preparations reveals that 24 hours after induction of apoptosis, cleavage of cytokeratin 18 starts (**B**), and after 48 hours, apoptotic bodies are released from the cells (**C**). In (**A**), a nonapoptotic cell is shown. (Original magnification × 400) DNA fragmentation in SC-1-treated apoptotic stomach carcinoma cells in vivo. Apoptotic stomach carcinoma cells in a metastasised tumour of a 50-year-old patient after treatment with the antibody SC-1. The patient received a single dose of antibody SC-1 and the tumour specimen was investigated for SC-1 induced apoptosis using the Klenow FragEL DNA fragmentation Kit (Oncogene, Boston). **D, G.** Control antibody CK8, tumour cells are stained. **E, H.** Positive control, all cell nuclei are stained. **F, I.** only the nuclei of apoptotic tumor cells are stained and normal not malignant tissue is not affected. (Original magnification, ×100 (**D, E, F**)/×200 (**G, H, I**)) (*See Color Plates*)

"housekeeping" by assisting the correct folding of nascent and stress-accumulated misfolded proteins and preventing their aggregation [64, 65]. HSPs allow the cells to survive to otherwise lethal conditions and play an essential role in tumor growth. They promote autonomous cell proliferation and inhibit death pathways induced by therapeutical approaches. Their expression in malignant cells is closely associated with a poor prognosis and resistance to therapy [66].

GRP78, also referred to as BiP, is a member of the HSP70 family [67]. It is induced in a wide variety of cancer cells and cancer biopsy tissues [68], and contributes to tumor growth and drug resistance [69].

However, the discovery of GRP78 expression on the cell surface of cancer cells further leads to the development of new therapeutic approaches targeted against cancer [70]. The human monoclonal antibody SAM-6 was isolated from a gastric cancer patient [26]. In binding and functional studies it was found that the SAM-6. reactivity is restricted to malignant tissue [26, 71]. The binding of SAM-6 could be removed by glycosidase treatment of the target cell, indicating a carbohydrate epitope of the antibody [72]. This expression of post-transcriptionally modified carbo-epitopes seems to be a common feature for malignant cells and was already proven for a series of other tumor-specific human monoclonal IgM antibodies [8, 21, 22].

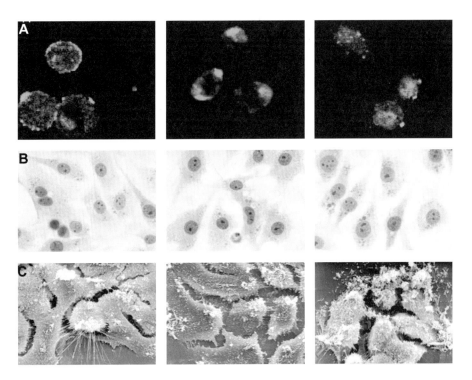

FIGURE 1.4. SAM-6 induced apoptosis: mode of action. Immunofluorescence of SAM-6 endocytosis. Pancreas carcinoma cells BXPC-3 were incubate with fluorochrome labeled SAM-6 antibody. After 30, 60, and 90 minutes cells were exposed on slides, fixed, and analyzed using confocal microscopy. 30 minutes, antibody binding; 60 minutes, "capping"; 90 minute antibody SAM-6 is completely internalized into the cell. Sudan III staining of neutral lipids in SAM-6 treated tumor cells. Pancreas carcinoma cells BXPC-3 were incubated with antibody SAM-6 antibody or for 2, 24, and 48 hours. An accumulation of red stained lipid droplets is visible in antibody SAM-6 treated tumor cells. Magnification ×200. Scanning electron microscopy of SAM-6 antibody-induced apoptosis. Stomach carcinoma cells 23132/87 were incubated with antibody SAM-6 for 2, 24 and 48 hours. Samples were proceeded for scanning electron microscopy and analyzed by ZEISS DSM 962. On the SAM-6 treated tumor cells apoptotic effects such as stress fibers, loss of cell-cell contacts, and clusters of apoptotic bodies are visible (*See Color Plates*)

The SAM-6 receptor is a tumor-specific isoform of GRP78, the epitope an O-linked carbohydrate (see Fig. 1.1) [72].

The antibody SAM-6 also binds to oxLDL and induces an excess of intracellular lipids, by over-feeding malignant cells with oxLDL via a receptor-mediated endocytosis (Fig. 1.4) [71]. The treated cells over-accumulate depots of cholesterol and triglyceride esters. Lipids are essential for normal and malignant cells during growth and differentiation. The turnover is strictly regulated because an uncontrolled uptake and accumulation is cytotoxic and can lead to lipo-apoptosis, lipoptosis [26]. This was shown in several animal studies and was also described for some inherited and acquired human diseases [73, 74]. When lipids over-accumulate in non-adipose tissue due to over-nutrition, fatty acids enter deleterious pathways such as ceramide production, and can cause apoptosis [73]. In mice and rats it was shown that lipotoxic cardiomyopathy is caused by accumulation of cardiotoxic lipids, which can induce the death of cardiac monocytes [75, 76]. Similar data on heart failure induced by lipid accumulation were obtained for humans by analyzing post mortem samples [77, 78].

The lipid over-accumulation induced by antibody SAM-6 is tumor-specific, nonmalignant cells neither bind the antibody nor harvest lipids after incubation with it. Shortly after internalization of the antibody-oxLDL–receptor complex and formation of lipid depots cytochrome c is released by mitochondria. Followed by this initiator-caspases 8 and 9 and effector-caspases 3 and 6 are activated and the apoptotic cascade starts (see Fig. 1.4) [72]. The interference with the lipid content in tumor cells by antibodies might be a novel avenue of cancer therapy.

5   Summary

Cancer cells respond like nontransformed cells to apoptotic signals, but they normally have a higher level of resistance. They use specific internal and external molecular changes, which make them less sensitive to death signals. On the other hand, malignant cells very often use their aberrant glycosylation machinery to modify carbohydrate residues on surface receptors. Antibodies that bind to this cancer-specific epitopes can be used to make cancer cells more sensitive to conventional therapeutical approaches.

References

1. Bohn J. Are natural antibodies involved in tumour defence? Immunol Lett 1999, 69(3):317–320.
2. Boes M. Role of natural and immune IgM antibodies in immune responses. Mol Immunol 2000, 37(18):1141–1149.
3. Ochsenbein AF, Fehr T, Lutz C, et al. Control of early viral and bacterial distribution and disease by natural antibodies. Science 1999, 286(5447):2156–2159.
4. Casali P, Notkins AL. CD5+ B lymphocytes, polyreactive antibodies and the human B-cell repertoire. Immunol Today 1989, 10(11):364–368.
5. Berland R, Wortis HH. Origins and functions of B-1 cells with notes on the role of CD5. Annu Rev Immunol 2002, 20:253–300.
6. Kantor AB, Merrill CE, Herzenberg LA, et al. An unbiased analysis of V(H)-D-J(H) sequences from B-1a, B-1b, and conventional B cells. J Immunol 1997, 158(3):1175–1186.
7. Ulvestad E, Kanestrom A, Sonsteby LJ, et al. Diagnostic and biological significance of anti-p41 IgM antibodies against Borrelia burgdorferi. Scand J Immunol 2001, 53(4):416–421.
8. Brandlein S, Pohle T, Ruoff N, et al. Natural IgM antibodies and immunosurveillance mechanisms against epithelial cancer cells in humans. Cancer Res 2003, 63(22):7995–8005.
9. Brandlein S, Beyer I, Eck M, et al. Cysteine-rich fibroblast growth factor receptor 1, a new marker for precancerous epithelial lesions defined by the human monoclonal antibody PAM-1. Cancer Res 2003;63(9):2052–2061.
10. Brandlein S, Eck M, Strobel P, et al. PAM-1, a natural human IgM antibody as new tool for detection of breast and prostate precursors. Hum Antibodies 2004, 13(4):97–104.
11. Vollmers HP, Brandlein S. The "early birds": natural IgM antibodies and immune surveillance. Histol Histopathol 2005, 20(3):927–937.
12. Vollmers HP, Brandlein S. Death by stress: natural IgM-induced apoptosis. Methods Find Exp Clin Pharmacol 2005, 27(3):185–191.
13. Schatz DG, Oettinger MA, Schlissel MS. V(D)J recombination: molecular biology and regulation. Annu Rev Immunol 1992, 10:359–383.
14. Lewis SM. The mechanism of V(D)J joining: lessons from molecular, immunological, and comparative analyses. Adv Immunol 1994, 56:27–150.
15. Gellert M. Recent advances in understanding V(D)J recombination. Adv Immunol 1997, 64:39–64.

16. Cedar H, Bergman Y. Developmental regulation of immune system gene rearrangement. Curr Opin Immunol 1999, 11(1):64–69.

17. Papavasiliou F, Jankovic M, Gong S, et al. Control of immunoglobulin gene rearrangements in developing B cells. Curr Opin Immunol 1997, 9(2):233–238.

18. Constantinescu A, Schlissel MS. Changes in locus-specific V(D)J recombinase activity induced by immunoglobulin gene products during B cell development. J Exp Med 1997, 185(4):609–620.

19. Vollmers HP, Brandlein S. Nature's best weapons to fight cancer. Revival of human monoclonal IgM antibodies. Hum Antibodies 2002, 11(4):131–142.

20. Brandlein S, Vollmers HP. Natural IgM antibodies, the ignored weapons in tumour immunity. Histol Histopathol 2004, 19(3):897–905.

21. Hensel F, Hermann R, Schubert C, et al. Characterization of glycosylphosphatidylinositol-linked molecule CD55/decay-accelerating factor as the receptor for antibody SC-1-induced apoptosis. Cancer Res 1999, 59(20):5299–5306.

22. Hensel F, Brandlein S, Eck M, et al. A novel proliferation-associated variant of CFR-1 defined by a human monoclonal antibody. Lab Invest 2001, 81(8):1097–1108.

23. Vollmers HP, Dammrich J, Ribbert H, et al. Apoptosis of stomach carcinoma cells induced by a human monoclonal antibody. Cancer 1995, 76(4):550–558.

24. Vollmers HP, O'Connor R, Muller J, et al. SC-1, a functional human monoclonal antibody against autologous stomach carcinoma cells. Cancer Res 1989, 49(9):2471–2476.

25. Brandlein S, Pohle T, Vollmers C, et al. CFR-1 receptor as target for tumor-specific apoptosis induced by the natural human monoclonal antibody PAM-1. Oncol Rep 2004, 11(4):777–784.

26. Pohle T, Brandlein S, Ruoff N, et al. Lipoptosis: tumor-specific cell death by antibody-induced intracellular lipid accumulation. Cancer Res 2004, 64(11):3900–3906.

27. Perona R. Cell signalling: growth factors and tyrosine kinase receptors. Clin Transl Oncol 2006, 8(2):77–82.

28. Guillemard V, Saragovi HU. Novel approaches for targeted cancer therapy. Curr Cancer Drug Targets 2004, 4(4):313–326.

29. Scaltriti M, Baselga J. The epidermal growth factor receptor pathway: a model for targeted therapy. Clin Cancer Res 2006, 12(18):5268–5272.

30. Vokes EE, Chu E. Anti-EGFR therapies: clinical experience in colorectal, lung, and head and neck cancers. Oncology (Williston Park) 2006, 20(5 Suppl 2):15–25.

31. Meyerhardt JA, Fuchs CS. Epidermal growth factor receptor inhibitors and colorectal cancer. Oncology (Williston Park) 2004, 18(14 Suppl 14):35–38.

32. Mendelsohn J, Baselga J. Epidermal growth factor receptor targeting in cancer. Semin Oncol 2006, 33(4):369–385.

33. Milas L, Raju U, Liao Z, et al. Targeting molecular determinants of tumor chemo-radioresistance. Semin Oncol 2005, 32(6 Suppl 9):S78–81.

34. Sartor CI. Mechanisms of disease: radiosensitization by epidermal growth factor receptor inhibitors. Nat Clin Pract Oncol 2004, 1(2):80–87.

35. Harari PM, Huang SM. Combining EGFR inhibitors with radiation or chemotherapy: will preclinical studies predict clinical results? Int J Radiat Oncol Biol Phys 2004, 58(3):976–983.

36. Yeon CH, Pegram MD. Anti-erbB-2 antibody trastuzumab in the treatment of HER2-amplified breast cancer. Invest New Drugs 2005, 23(5):391–409.

37. Tokunaga E, Oki E, Nishida K, et al. Trastuzumab and breast cancer: developments and current status. Int J Clin Oncol 2006, 11(3):199–208.

38. Kataoka A, Ishida M, Murakami S, et al. Sensitization of chemotherapy by anti-HER. Breast Cancer 2004, 11(2):105–115.

39. Mendelsohn J. Antibody-mediated EGF receptor blockade as an anticancer therapy: from the laboratory to the clinic. Cancer Immunol Immunother 2003, 52(5):342–346.

40. Cohenuram M, Saif MW. Panitumumab the first fully human monoclonal antibody: from the bench to the clinic. Anticancer Drugs 2007, 18(1):7–15.

41. Burrus LW, Zuber ME, Lueddecke BA, et al. Identification of a cysteine-rich receptor for fibroblast growth factors. Mol Cell Biol 1992, 12(12):5600–5609.

42. Gonatas JO, Chen YJ, Stieber A, et al. Truncations of the C-terminal cytoplasmic domain of MG160, a medial Golgi sialoglycoprotein, result in its partial transport to the plasma membrane and filopodia. J Cell Sci 1998, 111(Pt 2):249–260.

43. Gonatas JO, Mezitis SG, Stieber A, et al. MG-160. A novel sialoglycoprotein of the medial cisternae of the Golgi apparatus [published erratum appears in J Biol Chem 1989, 264(7):4264]. J Biol Chem 1989, 264(1):646–653.

44. Stieber A, Mourelatos Z, Chen YJ, et al. MG160, a membrane protein of the Golgi apparatus which is homologous to a fibroblast growth factor receptor and to a ligand for E-selectin, is found only in the Golgi apparatus and appears early in chicken embryo development. Exp Cell Res 1995, 219(2):562–570.

45. Zuber ME, Zhou Z, Burrus LW, et al. Cysteine-rich FGF receptor regulates intracellular FGF-1 and FGF-2 levels. J Cell Physiol 1997, 170(3):217–227.

46. Crnogorac-Jurcevic T, Efthimiou E, Capelli P, et al. Gene expression profiles of pancreatic cancer and stromal desmoplasia. Oncogene 2001, 20(50):7437–7446.

47. Steegmaier M, Borges E, Berger J, et al. The E-selectin-ligand ESL-1 is located in the Golgi as well as on microvilli on the cell surface. J Cell Sci 1997, 110(Pt 6):687–694.

48. Steegmaier M, Levinovitz A, Isenmann S, et al. The E-selectin-ligand ESL-1 is a variant of a receptor for fibroblast growth factor. Nature 1995, 373(6515):615–620.

49. Koretz K, Bruderlein S, Henne C, et al. Decay-accelerating factor (DAF, CD55) in normal colorectal mucosa, adenomas and carcinomas. Br J Cancer 1992, 66(5):810–814.

50. Cheung NK, Walter EI, Smith-Mensah WH, et al. Decay-accelerating factor protects human tumor cells from complement-mediated cytotoxicity in vitro. J Clin Invest 1988, 81(4):1122–1128.

51. Caras IW, Davitz MA, Rhee L, et al. Cloning of decay-accelerating factor suggests novel use of splicing to generate two proteins. Nature 1987, 325(6104):545–549.

52. Hara T, Matsumoto M, Fukumori Y, et al. A monoclonal antibody against human decay-accelerating factor (DAF, CD55), D17, which lacks reactivity with semen-DAF. Immunol Lett 1993, 37(2–3):145–152.

53. Lublin DM, Krsek-Staples J, Pangburn MK, et al. Biosynthesis and glycosylation of the human complement regulatory protein decay-accelerating factor. J Immunol 1986, 137(5):1629–1635.

54. Berstad AE, Brandtzaeg P. Expression of cell membrane complement regulatory glycoproteins along the normal and diseased human gastrointestinal tract. Gut 1998, 42(4):522–529.

55. Niehans GA, Cherwitz DL, Staley NA, et al. Human carcinomas variably express the complement inhibitory proteins CD46 (membrane cofactor protein), CD55 (decay-accelerating factor), and CD59 (protectin). Am J Pathol 1996, 149(1):129–142.

56. Spendlove I, Ramage JM, Bradley R, et al. Complement decay accelerating factor (DAF)/CD55 in cancer. Cancer Immunol Immunother 2006, 55(8):987–995.

57. Mikesch JH, Schier K, Roetger A, et al. The expression and action of decay-accelerating factor (CD55) in human malignancies and cancer therapy. Cell Oncol 2006, 28(5–6):223–232.

58. Spendlove I, Li L, Carmichael J, et al. Decay accelerating factor (CD55): a target for cancer vaccines? Cancer Res 1999, 59(10):2282–2286.

59. Durrant LG, Buckley DJ, Robins RA, et al. 105Ad7 cancer vaccine stimulates anti-tumour helper and cytotoxic T-cell responses in colorectal cancer patients but repeated immunisations are required to maintain these responses. Int J Cancer 2000, 85(1):87–92.

60. Vollmers HP, Eck M, Brändlein S, et al. SC-1, a human antibody for treatment of gastric carcinomas (GCs): results of a multi-ethnic expression study of the SC-1 antigen. In: American Society of Clinical Oncology Annual Proceedings of the 40th ASCO Meeting, 2004. New Orleans: American Society of Clinical Oncology, J Clin Oncol 2004.

61. Vollmers HP, Zimmermann U, Krenn V, et al. Adjuvant therapy for gastric adenocarcinoma with the apoptosis-inducing human monoclonal antibody SC-1: first clinical and histopathological results. Oncol Rep 1998, 5(3):549–552.

62. Hensel F, Hermann R, Brandlein S, et al. Regulation of the new coexpressed CD55 (decay-accelerating factor) receptor on stomach carcinoma cells involved in antibody SC-1-induced apoptosis. Lab Invest 2001, 81(11):1553–1563.

63. Macario AJ, Conway de Macario E. Sick chaperones, cellular stress, and disease. N Engl J Med 2005, 353(14):1489–1501.

64. Ma Y, Hendershot LM. The role of the unfolded protein response in tumour development: friend or foe? Nat Rev Cancer 2004, 4(12):966–977.

65. Bukau B, Weissman J, Horwich A. Molecular chaperones and protein quality control. Cell 2006, 125(3): 443–451.

66. Calderwood SK, Khaleque MA, Sawyer DB, et al. Heat shock proteins in cancer: chaperones of tumorigenesis. Trends Biochem Sci 2006, 31(3):164–172.

67. Hendershot LM. The ER function BiP is a master regulator of ER function. Mt Sinai J Med 2004, 71(5):289–297.

68. Li J, Lee AS. Stress induction of GRP78/BiP and its role in cancer. Curr Mol Med 2006, 6(1):45–54.

69. Tsutsumi S, Namba T, Tanaka KI, et al. Celecoxib upregulates endoplasmic reticulum chaperones that inhibit celecoxib-induced apoptosis in human gastric cells. Oncogene 2006, 25(7):1018–1029.

70. Delpino A, Piselli P, Vismara D, et al. Cell surface localization of the 78 kD glucose regulated protein (GRP 78) induced by thapsigargin. Mol Membr Biol 1998, 15(1):21–26.

71. Brändlein S, Rauschert N, Rasche N, et al. The human IgM antibody SAM-6 induces tumor-specific apoptosis with oxidized LDL. Mol Cancer Ther 2007, 6(1).

72. Rauschert N, Brändlein S, Holzinger E, Hensel F, Müller-Hermelink HK and Vollmers HP. A new tumor-specific variant of GRP78 as target for antibody-based therapy. Lab Invest 2008, 88:375–386.

73. Unger RH. Lipotoxic diseases. Annu Rev Med 2002, 53:319–336.

74. de Vries JE, Vork MM, Roemen TH, et al. Saturated but not mono-unsaturated fatty acids induce apoptotic

cell death in neonatal rat ventricular myocytes. J Lipid Res 1997, 38(7):1384–1394.

75. Chiu HC, Kovacs A, Ford DA, et al. A novel mouse model of lipotoxic cardiomyopathy. J Clin Invest 2001, 107(7):813–822.

76. Zhou YT, Grayburn P, Karim A, et al. Lipotoxic heart disease in obese rats: implications for human obesity. Proc Natl Acad Sci U S A 2000, 97(4):1784–1789.

77. Kang PM, Izumo S. Apoptosis and heart failure: a critical review of the literature. Circ Res 2000, 86(11):1107–1113.

78. Narula J, Pandey P, Arbustini E, et al. Apoptosis in heart failure: release of cytochrome c from mitochondria and activation of caspase-3 in human cardiomyopathy. Proc Natl Acad Sci U S A 1999, 96(14):8144–8149.

Chapter 2
Targeting the Transferrin Receptor to Overcome Resistance to Anti-Cancer Agents

Tracy R. Daniels, Isabel I. Neacato, Gustavo Helguera, and Manuel L. Penichet

1 The Transferrin Receptor

The transferrin receptor (TfR) plays an important role in iron uptake and delivery [1]. The primary role of the TfR is to internalize iron through the binding of its natural ligand, the transferrin (Tf) protein that carries iron through the circulation. Iron is necessary for various cell processes such as respiration, metabolism, DNA synthesis and the proper functioning of various heme and nonheme proteins that require iron as a cofactor [1]. In addition, the TfR seems to be important for other processes such as cell growth and proliferation [1].

The TfR, a 180-kDa homodimeric glycoprotein, is a type II transmembrane receptor that has three important domains for its function (Fig. 2.1). It is composed of a C-terminal domain, also known as the ectodomain, a transmembrane region, and an N-terminal domain that is on the cytosolic side of the membrane. The ectodomain is important for binding to Tf for the internalization of iron. Two TfR genes have been identified, TfR1 and TfR2. Furthermore, the TfR2 gene produces two transcripts, α and β, that are produced by alternative splicing. TfR2α shows similarity with TfR1 in that they exhibit a 45% similarity and 66% homology in their ectodomain. However, the cytoplasmic domains of the two proteins demonstrate no similarity [1]. The TfR2β transcript lacks the transmembrane and cytoplasmic domains and its

function remains unknown. TfR1 and TfR2α differ in cell surface expression and gene regulation. The TfR1 is ubiquitously expressed on normal cells at low levels. Increased TfR1 expression is observed on cells with a high proliferation rate, including cancer cells. TfR2 expression is limited to hepatocytes and enterocytes in the small intestine [1]. TfR2 expression has been found in some human cell lines such as B and myeloid cell lines as well as some cell lines derived from solid tumors [1]. TfR1 is post-transcriptionally regulated directly by intracellular iron levels as compared with TfR2 that is not. TfR2 is thought to be primarily regulated by the cell cycle and iron-bound Tf [1]. Thus both receptors differ considerably in expression and regulation indicating different roles in iron delivery. In addition, TfR1 has a 25-fold greater affinity for Tf relative to TfR2, indicating the main role of TfR1 in iron homeostasis [1].

Tf is an 80 kDa monomeric glycoprotein composed of two lobes; an N and C lobe that are separated by a short spacer sequence (reviewed in [1]). Each lobe is capable of binding one iron molecule. The number of iron molecules bound to Tf has an important effect on the affinity of Tf for the TfR. At physiological conditions, holo Tf or diferric Tf (two iron) has the greatest affinity followed by monoferric (one iron), while apo-Tf (no iron) has the lowest affinity for the receptor [1]. Thus, iron uptake by the cell is mediated mostly through the

From: *Sensitization of Cancer Cells for Chemo/Immuno/Radio-therapy, 1st Edition.*
Edited by: Benjamin Bonavida © Humana Press, Totowa, NJ

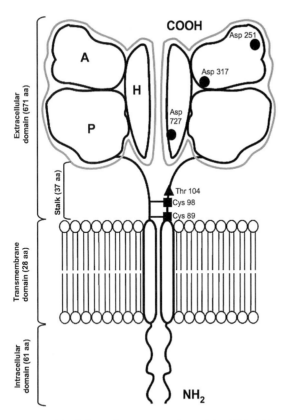

FIGURE 2.1. Schematic representation of the TfR. This receptor is a type II receptor found on the cell surface as a homodimer consisting of two monomers linked by disulfide bridges at cysteines 89 and 98 (■). The TfR contains an intracellular domain, a transmembrane domain, and a large extracellular domain. There is an O-linked glycosylation site at threonine 104 (▲) and three N-linked glycosylation sites on asparagine residues 251, 317, and 727 (●). The extracellular domain of the TfR consists of three subdomains: apical (A), helical (H) and protease-like domain (P). (Daniels et al. *Clinical Immunology* 2006, (121):144–158. Copyright 2006 Reprinted with the kind permission of Elsevier USA.)

interaction of diferric Tf and the TfR. The formation of a complex between diferric Tf and the TfR allows iron internalization into cells through a receptor-mediated endocytosis pathway (Fig. 2.2). This complex is internalized in a clathrin-coated pit into the cell and delivered into endosomes. Protons are pumped into the endosome causing an acidic change in the pH environment. This causes a conformational change in Tf that results in the release of iron. Iron can then be transported out of the endosome into the cytosol through a divalent metal

transporter (DMT1). The Tf/TfR complex remains inside the endosome until it is brought to the cell surface where apo-Tf dissociates from the TfR and is then free to circulate and bind free iron.

Many studies have used the TfR as a target for the delivery of various therapeutic agents (reviewed in [2]). The high expression of the TfR in cancer cells (that can be 10 to 100-fold greater than normal cells), its cell surface accessibility, and constitutive recycling pathway make this receptor an attractive target for immunotherapy. Importantly, either Tf or anti-TfR antibodies can mediate delivery of molecules by TfR targeting. The following is a discussion of the various strategies that have utilized targeting of the TfR (summarized in Fig. 2.3 and Table 2.1) to overcome cancer cell resistance to therapy or to provide the first hit in the "two hit signal" model to sensitize resistant cells to chemotherapeutic agents as combination treatment strategies. Both strategies are of great importance in treating patients whose cancers have developed resistance to common therapies and have thus developed more aggressive malignancies.

2 Tf Conjugates to Overcome Chemoresistance

Doxorubicin (Adriamycin®) (ADR) is an anthracycline chemotherapeutic drug used to treat a variety of cancers. ADR blocks DNA synthesis along with the activity of topoisomerase II, an enzyme that helps to relax the coil and extend the DNA molecule prior to DNA synthesis or RNA transcription. When used as a single treatment modality, ADR often exhibits devastating side effects including cardiotoxicity, myelosuppression, nephrotoxicity, and extravasation [3]. Systemic drug toxicity is often attributed to quick diffusion throughout the body resulting in a homogeneous tissue distribution [4]. The potential benefits of ADR treatment may also be blocked by the development of drug resistant cancer cells. ADR resistance can be attributed to many molecular events. Includes the overexpression of the multi-drug resistance (MRP) gene that codes for an active drug efflux pump P-glycoprotein on the cell surface that decreases cellular accumulation of the drug [5, 6]. ADR resistance may also be attributed to the impaired ability of drug trafficking or altered intracellular distribution within the cell

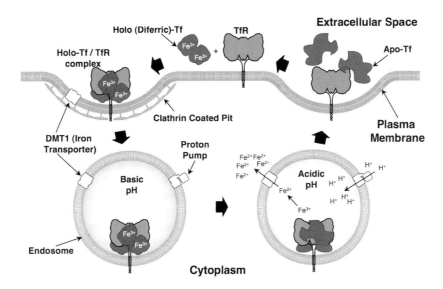

FIGURE 2.2. *(See Color Plates.)* Cellular uptake of iron through the Tf system via receptor-mediated endocytosis. The TfR consists of a dimer on the surface of the cell. Each receptor monomer binds one Tf molecule that consists of two lobes (the N and C lobes). Each lobe of Tf binds one iron molecule and thus two diferric Tf molecules bind to the receptor with high affinity. Endocytosis of the diferric Tf/TfR complex occurs via clathrin-coated pits and the complex is delivered into endosomes. Protons are pumped into the endosome resulting in a decrease in pH that stimulates a conformational change in Tf and its subsequent release of iron. The iron is then transported out of the endosome into the cytosol by the divalent metal transporter 1 (DMT1). Apo-Tf remains bound to the TfR while in the endosome and is only released once the complex reaches the cell surface. (Daniels et al. *Clinical Immunology* 2006, (121):144–158. Copyright 2006 Reprinted with the kind permission of Elsevier USA.)

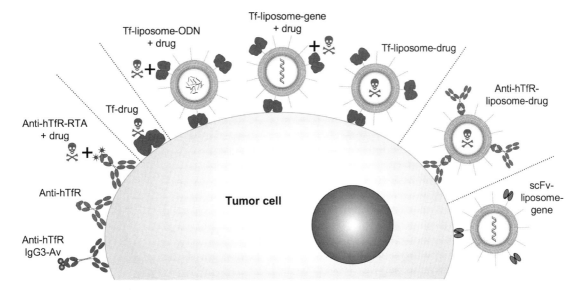

FIGURE 2.3. *(See Color Plates.)* Strategies for targeting the TfR to overcome resistance to chemotherapy. The TfR can be targeted by direct interaction with conjugates of its ligand Tf as well as by monoclonal antibodies or antibody conjugates that target the extracellular domain of the TfR. Targeting the TfR has been utilized to deliver chemotherapeutic drugs, protein toxins, and liposomes containing either drugs, antisense oligonucleotides, or genes into malignant cells in order to sensitize resistant cells to chemotherapy. Adapted from (Daniels et al. *Clinical Immunology* 2006, (121):144–158. Copyright 2006 Reprinted with the kind permission of Elsevier USA.)

TABLE 2.1. Targeting the TfR to overcome resistance or to sensitize cancer cells to chemotherapy.

Conjugate	Cell model	Comments	References
Tf conjugates			
Tf-ADR	K562 erythroleukemia 3-fold	3-fold Increase in cytotoxicity compared with free ADR;	11
		Restored IC$_{50}$ of ADR-resistant cells to that of sensitive cells	12
	HL-60 leukemia 10-fold	10-fold Increase in sensitivity in ADR-resistant cells	11, 13
	Lovo colorectal carcinoma Hep2 laryngeal carcinoma	5- to 10-fold increase in sensitivity in ADR-resistant cells	13
	H-MESO-1 mesothelioma	5- to 10-fold increase in sensitivity in ADR-resistant cells; Prolonged the life span in a xenograft model	13
	KB cervical cancer	Decreased IC$_{50}$ 4-40 fold (depending on the cell line)	14
Tf-GN	CCRF-CEM T-lymphoma	Cytotoxicity enhanced by Bortezomib	16
Tf-GN-ADR	MCF-7/ADR	Decreased IC$_{50}$ 100-fold	17
Tf-Liposome conjugates			
Tf-Liposome-ADR	MXT-2B metastatic mammary cancer 2-fold	2-fold Increase in cytotoxicity compared with nontargeted liposome; 2.4-fold increase compared with ADR alone	18
Tf-liposome-cisplatin	MKN45 Gastric cancer	Increase in survival rate in a xenograft model	20
Tf-liposome-hαFR ODN	MDA-MB-435 Breast cancer	Increased cytotoxicity 10-fold compared with ADR alone; Sensitized cells to ADR treatment	21
Tf-liposome-Bcl-2 ODN	K562 Erythroleukemia	Decreased Bcl-2 expression 10-fold compared with ODN alone; Sensitized cells to daunorubicin (10-fold increase in cytotoxicity)	25
	MCF-7 Breast cancer	Increase in apoptosis, cytochrome c release, and caspase-8 activation	26
Tf-liposome-p53	JSQ-3 Head and neck cancer	Sensitization to radiotherapy (dependent on level of restored p53 expression); Complete tumor regression and no recurrences for up to 6 months in a xenograft model	29
Antibody-liposome conjugates			
OKT9-liposome-ADR	K562/ADM	Murine anti-hTfR IgG1; Decreased IC$_{50}$ 3.5 fold compared with free ADR; Restored intracellular levels of ADR in resistant cells	19
ScFv-liposome-p53	MDA-MB-435 Breast cancer DU145 Prostate cancer	Increased p53 expression 10-fold	28
	JSQ-3 Head and neck cancer	Systemic delivery increased p53 expression in subcutaneous tumors in a xenograft model	28
	435/LCC6 Breast cancer	Combination treatment with docetaxel prolonged survival in a xenograft model	28

Antibody or antibody conjugates			
A24 + chemotherapy	Activated PBMC from normal or acute T-cell leukemia/lymphoma	Murine IgG2b anti-hTfR; proapoptotic by itself; Competes with Tf and blocks iron uptake; Decreases TfR surface expression; Enhanced cytotoxicity observed with combined treatment with chemotherapy	30–31
7D3-RTA + ADR	H-MESO-1 Mesothelioma	Murine IgG1 anti-hTfR; no affects when used alone; 7D3-RTA increased cytotoxicity; 7D3-RTA+ADR no additive effects observed in vitro, but enhanced median survival in a xenograft model	33
454A12-RTA + rhIFN	OVCAR-3 Ovarian cancer	Murine IgG1 anti-hTfR; no affects when used alone; No effect of rhIFN alone; control median survival = 41 days; 454A12-RTA increased survival to 89 days; 454A12-RTA+rhIFN increased survival to >120 days; Tumor burden dependent; Chemotherapy decreased tumor burden prior to treatment	35
Antibody fusion proteins			
Anti-rat TfR IgG3-Av	Y3-Ag 1.2.3 rat myeloma C58 (NT) D.1.G.OVAR.1 rat T-cell lymphoma	proapoptotic alone; Delivers active biotinylated agents into tumor cells	36, 40
Anti-human TfR IgG3-Av	Malignant B-cell lines (ARH-77 and IM-9) and primary CD138+ cells isolated from patients (one diagnosed with multiple myeloma and one with plasma cell leukemia)	Induced rapid degradation of TfR; Cross-linking of TfR is important for cytotoxicity; Induced iron-dependent apoptosis	36, 37

rather than its accumulation inside the cell [7, 8]. It has been shown that chemotherapeutic drugs are distributed throughout the cytoplasm and nuclei of sensitive cells. In resistant cells, the localization of these drugs changes and they are only found within discrete cytoplasmic organelles [9, 10]. Thus, drug localization within the tumor cell is important for therapeutic efficacy.

In an attempt to overcome these problems, delivery of ADR into the tumor through targeted therapy via the TfR has been widely studied (reviewed in [2]). Conjugation of Tf to ADR has been effective in overcoming drug resistance in a variety of malignant cell lines. In fact, ADR resistant sublines of K562 human erythroleukemic cells and HL-60 human promyelocytic leukemia cells exhibited enhanced sensitivity to Tf-ADR relative to the use of ADR alone. Resistant K562 and HL-60 cells were 3- and 10-fold more sensitive to the conjugate than ADR alone, respectively [11]. Another study demonstrated that the K562 parental cell line had an IC_{50} of 3.2 nM with ADR alone compared with 20 nM for the resistant subline [12]. Treatment of the resistant subline with Tf-ADR induced potent cytotoxic effects with IC_{50} of 3.6 nM, which was comparable to that of the K562 parental (sensitive) cell line.

Other studies with the resistant sublines of Lovo (human colon carcinoma), H-MESO-1 (human mesothelioma), Hep2 (human laryngeal carcinoma cells), and HL-60 (human promyelocytic leukemia) also exhibited cytotoxicity with Tf-ADR treatment [13]. More importantly, the conjugate showed greater potency toward resistant cell lines relative to sensitive cell lines. In ADR resistant cells a 5- to 10-fold increase in cytotoxicity was observed with Tf-ADR treatment relative to ADR alone. ADR-sensitive cells exhibited a smaller increase in cytotoxicity when treated with Tf-ADR (4- to 5-fold more toxic than free ADR) [13]. Multidrug resistant (MDR) KB (human cervical cancer) also exhibited significant cytotoxicity to the Tf-ADR conjugate [14]. Toxicity of the Tf-ADR conjugate was observed in MDR KB cell lines; KB-8-5 (partially MDR) and the highly MDR cells KB-C1 and KB-V1. Treatment with the conjugate in all KB MDR cells exhibited greater cytotoxicity and a lower IC_{50} than ADR treatment alone. In vivo studies with nude mice bearing the H-MESO-1 cell line prolonged the life span of the treated mice

compared with ADR alone or unlinked ADR and Tf [13]. Treatment with the conjugate increased the life span of mice from 39% in mice treated with ADR alone, to 69% in mice treated with the Tf-ADR conjugate. Even though there were no long-term survivors, tumor burden was significantly decreased in conjugate treated mice relative to ADR treatment alone [13].

A Tf conjugate consisting of the anti-neoplastic metallo-drug gallium nitrate (GN) has also be targeted by conjugates of Tf in order to sensitize cancer cells to common therapeutics. GN shares chemical characteristics with iron and binds free Tf in human blood (reviewed in [15]). The Tf-GN complexes enter cells through TfR-mediated endocytosis. In this way, GN blocks cellular iron uptake and induces a lethal iron deprivation within the cell. GN also has direct effects on cells, such as inhibition of DNA polymerases, membrane tyrosine phosphatase activity, magnesium-dependent ATPase, and tubulin polymerization. However, the extent to which these mechanisms contribute to the anti-cancer effects of GN are not clear. The final result of GN treatment is the induction of apoptosis through a mitochondrial-mediated pathway. GN alone is commonly used for the treatment of non-Hodgkin's lymphoma [15]. However, this activity was recently shown to be enhanced by the proteosome inhibitor Bortezomib in the CCRF-CEM human T-lymphoma cell line suggesting the efficacy of the combination treatment with the Tf-GN conjugate [16]. Furthermore, a Tf-GN-ADR conjugate showed cytotoxicity in an MDR cell line MCF-7/ADR (human breast cancer cell line resistant to ADR) [17]. This study indicated that the MCF-7/ADR was 1,000-fold more resistant than the parental cell line MCF-7 to free ADR treatment. Tf-GN-ADR exhibited the same inhibitory effect as ADR alone in MCF-7. However, in the resistant cell line MCF-7/ADR, treatment with the Tf-GN-ADR conjugate decreased the IC_{50} by 100-fold [17]. Another conjugate consisting of Tf, iron, and ADR (Tf-Fe-ADR) also demonstrated enhanced cytotoxicity, but only by 10-fold. Tf-GN-ADR was also compared with Tf-GN alone and was found to be 500- and 3,000-fold more cytotoxic to MCF-7 and MCF-7/ADR, respectively. This study indicates that the Tf-GN-ADR conjugates reversed ADR resistance that was accompanied by a decrease in the MRP expression

[17]. Drug resistance in cancer patients introduces various complications in treating relapsed and disseminated malignancies. These studies indicate that the Tf-ADR or Tf-GN-ADR conjugates can be effectively used as sensitizing agents to induce cytotoxic effects in ADR-resistant or MDR cell lines. Furthermore, these conjugates have the potential to be utilized to treat patients who have developed MDR for the effective treatment of cancer malignancies.

3 TfR Targeted Liposomal Delivery of Chemotherapeutic Drugs

The delivery of ADR through the targeting of the TfR by a Tf-liposomal system may also be a successful therapeutic approach. Efficacy of this system has been shown in vitro in MXT-B2, a human metastatic mammary carcinoma cell line [18]. The Tf conjugated liposomes encapsulating ADR enhanced cytotoxicity about 2-fold compared with a nontargeted liposome carrying ADR and 2.4-fold compared with free ADR in MXT-B2 cells. The anti-human TfR (hTfR) antibody OKT9 (IgG1) has also been used to deliver liposomes encapsulating ADR into the resistant human erythroleukemic cell line K562/ADM [19]. In this cell line, the targeted liposomal delivery of ADR decreased the IC_{50} 3.5-fold compared with free ADR. Interestingly, the IC_{50} of the targeted liposome was similar to free ADR in the parental, ADR-sensitive K562 cell line. This study also showed that the intracellular level of ADR in the K562/ADM (resistant) cells rapidly decreased (due to drug efflux) after free ADR treatment and was significantly lower (15- to 45-fold) compared with the parental K562 cell line, which maintained high levels of intracellular ADR. However, resistant cells treated with the OKT9-ADR–liposomes maintained similar levels of intracellular ADR compared with sensitive cells. This suggests that delivery of ADR through the TfR results in decreased ADR efflux, resulting in higher intracellular accumulation and cytotoxic effects. How the TfR is involved in the blockage of ADR efflux remains elusive; however, the route

of delivery through TfR mediated endocytosis is expected to play a key role in altering the trafficking of ADR so that it is retained within the resistant tumor cell [7, 8].

The efficient delivery of cisplatin, a platinum-containing chemotherapeutic drug that forms cytotoxic adducts with the DNA, into cells has also been achieved through Tf-liposomal therapy [20]. This system utilizes polyethylene glycol (PEG) to avoid uptake by the reticuloendothelial system and thus increases the circulation time of the liposomes. Tf is then covalently linked to PEG to form a liposome that targets the TfR on tumor cells. This system efficiently delivered cisplatin in peritoneal dissemination xenografts of MKN45, a human gastric cancer cell line, in BALB/cA-JcI-nu nude mice. An increase in the survival rate was observed in mice treated with this targeted liposomal system (40% of mice survived after 60 days of treatment) as compared with control animals (21-day survival), PEG liposomes (50-day survival), and cisplatin alone (45-day survival). More importantly, all survivors treated with the targeted liposomes were tumor free after the 60-day period. Tf-liposome internalization into cancer cells was greater than nontargeted liposomes or PEG alone, indicating the importance of targeting the TfR. Also uptake by the spleen and liver was significantly lower than nontargeted liposomes further demonstrating the specific delivery of cisplatin to tumor cells.

4 TfR Targeted Liposomal Gene Therapy to Sensitize Cells to Chemotherapy

Targeted liposomes can also be used for gene therapy purposes to sensitize cells to chemotherapy. This strategy can be used to deliver antisense oligodeoxyribonucleotides (ODN) to cells in order to knock down expression of tumor-promoting proteins. In addition, this technique can also be used to delivery entire genes into cells in order to restore expression of antitumor proteins whose expression has been lost during malignant transformation. Both possibilities are further discussed in the following sections.

4.1   Human α Folate Receptor Oligodeoxyribonucleotide Delivery

The efficacy of Tf-liposomes to deliver antisense nucleotides complementary to the human α folate receptor gene sequence (hαFR) through the TfR has been demonstrated [21]. The hαFR is a surface glycoprotein that has high affinity for folic acid and reduced folates that promotes cell proliferation. This receptor is highly upregulated on cancer cells [22] and thus promotes the growth of the tumor. Tf-mediated delivery of ODN to knock down hαFR expression in a panel of human breast carcinoma cell lines significantly decreased cell survival [21]. It was also demonstrated that targeted delivery of the hαFR ODN sensitizes MDA-MB-435 human breast cancer cells to ADR treatment. In fact, cell death was increased by 5-fold compared with delivery of the sense ODN control and 10-fold compared with ADR alone treated cells.

4.2   Bcl-2 Oligodeoxyribonucleotide Delivery

The upregulation of Bcl-2, an anti-apoptotic protein in various malignant cells, is often associated with the development of chemoresistance [23], including resistance to daunorubicin (Cerubidine®) an anthracycline topoisomerase inhibitor that blocks DNA synthesis and RNA transcription [24]. In order to block this up-regulation and overcome chemoresistance, Bcl-2 ODN have been delivered into tumor cells via Tf conjugated liposomes [25]. Efficient uptake of these liposomes in K562 human erythroleukemia cells was observed compared with nontargeted liposomes and could be blocked by the addition of free Tf. Bcl-2 expression was downregulated 2-fold compared with the nontargeted liposomes and 10-fold relative to the Bcl-2 ODN alone.

Tf targeted delivery of Bcl-2 ODN sensitized cells to chemotherapy as demonstrated by a 10-fold increase in toxicity observed with the combination treatment compared with nontargeted liposomal delivery or daunorubicin alone. Tf-liposomes have also been utilized to deliver Bcl-2 ODN into breast cancer cells as a chemosensitizing agent. Synergistic effects were observed in MCF-7 breast cancer cells with the combination treatment of Tf-Bcl-2 ODN liposomes and cisplatin [26]. Cells

containing the p53 tumor suppressor protein (see following section) as well as those lacking p53 expression were found to be sensitive to the combination treatment [26]. Thus, this treatment strategy may be beneficial irrespective of the p53 status of the cell. These studies demonstrate the efficacy of using Tf targeted liposomes as chemosensitizing agents to overcome resistance to single treatment strategies.

4.3   Wild-Type p53 Gene Delivery

The tumor suppressor protein p53 is a transcription factor that is activated upon DNA damage and has been termed the "guardian of the genome" [27]. Activation of p53 leads to cell cycle arrest, DNA repair, and or even cell death if the DNA damage is too extensive. p53 is mutated in many types of malignancies and this loss of function results in genomic instability and impaired apoptosis. The reintroduction of the wild-type p53 gene into malignant cells has been a common goal for many gene therapy strategies.

Single chain antibodies (scFv) targeting the TfR have also been used as delivery vehicles complexed to liposomes for targeted gene delivery. One such anti-TfR scFv contains the single chain variable region of the murine IgG1 anti-hTfR 5E9 antibody conjugated to the surface of liposomes. This immunolipoplex successfully delivered the p53 tumor suppressor gene both in vitro and in vivo [28]. In vitro p53 gene delivery was increased 10-fold in MDA-MB-435 human breast cancer cells and DU145 human prostate cancer cells. In a mouse xenograft model of the human squamous cell carcinoma of the head and neck using the JSQ-3 (containing mutated p53) cell line, systemic delivery of the 5E9 scFv-targeted liposomes efficiently delivered the p53 gene to tumor cells that were implanted subcutaneously and also resulted in high expression of the wild-type p53 protein [28]. Much lower p53 expression was observed in mice treated with a nontargeted liposomes containing the p53 gene. A xenograft model of breast cancer metastasis (using the 435/LCC6 cell line derived from mouse ascites of breast cancer cell line MDA-MB-435) was used to determine the in vivo efficacy of the 5E9 scFv targeted liposomes to sensitize cells to treatment with the microtubule stabilizer docetaxel [28]. No antitumor efficacy was observed with docetaxel treatment alone, which suggests that this tumor model is highly chemoresistant. Systemic administration of 5E9-liposomes containing the p53 gene in combination with docetaxel significantly prolonged survival (60–80% long term survival) compared with either agent alone. Importantly, no general toxicity was observed in any of the treated mice. The antitumor activity was equivalent to Tf-liposomes containing the p53 gene tested in the same study. Altogether this study demonstrates that delivery of wild-type p53 to tumor cells successfully sensitizes them to chemotherapy and demonstrated antitumor effects at lower chemotherapeutic doses. This strategy could also potentially decrease the severity of docetaxel-induced side effects. Ultimately this study demonstrates that targeting of the TfR is beneficial for the systemic delivery and restorative expression of wild type p53 as a chemosensitizing agent.

In vitro studies have shown that high levels of wild-type p53 expression can also be obtained using liposomes conjugated to Tf [29]. These studies were performed in the human head and neck cell line JSQ-3 that was derived from a radioresistant tumor of the nasal vestibule. Restoration of p53 expression in these cells sensitized them to radiation treatment, which was dependent on the level of wild-type p53 expression. In vivo efficacy was also demonstrated in a xenograft mouse model [29]. JSQ-3 tumors were established subcutaneously in athymic nude mice. After 7–10 days of tumor growth, mice were treated systemically with Tf-p53–liposomes twice a week for a total of 5 injections. Radiotherapy at the tumor site began 48 hours after the first injection of the Tf-p53–liposomes and was given every 48 hours for a total of 10 treatments. This treatment strategy resulted in complete tumor regression and blocked recurrence for up to 6 months.

5   Anti-TfR antibodies and Their Conjugates For Sensitization of Tumors to Chemotherapy

Another strategy to overcome the resistance to common anticancer therapies is to use anti-human TfR (hTfR) antibodies alone (if they interfere with TfR function) or conjugates made with these anti-bodies in combination treatment strategies with chemotherapeutic drugs. For example, the murine anti-hTfR IgG2b antibody A24 has been used for this purpose. A24 competes with Tf for TfR binding that leads to a block in iron uptake, impaired cycling of the TfR, and decreased surface expression of the receptor [30, 31]. A24 alone has antiproliferative and proapoptotic against peripheral blood mononuclear cells (PBMC) activated by phytohemagglutinin (PHA)/IL-2 or cocultured with activated dendritic cells. PBMC activated T cells isolated from both normal donor and patients with adult T-cell leukemia/lymphoma (ATL) were sensitive to the effects of the A24 antibody. Enhanced cytotoxic effects were observed when tumor cells isolated from a patient with chronic ATL were treated with a combination of the A24 antibody and chemotherapy (either the nucleoside reverse transcriptase inhibitor zidovudine (AZT), the cytokine interferon-α, or the topoisomerase II inhibitor VP-16) [31].

Some anti-hTfR antibodies do not have cytotoxic effects when used as a single treatment. For example, the murine anti-hTfR IgG1 antibody 7D3 has no cytotoxic effects when used alone. However, when used as a component of an immunotoxin (a cell-specific ligand linked to a plant or bacterial toxin or modified toxin subunit) [32] its antitumor effects can be greatly enhanced [33]. The ricin toxin, derived from the seed of the *Ricinus communis* plant, is a ribosomal inactivating protein that is composed of two chains connected by a disulfide bond [34]. The A chain of ricin (RTA) contains the N-glycosidase enzyme that blocks ribosomal activity, while the B chain (RTB) is responsible for binding to the cell surface. RTA has no cytotoxic effects by itself because it can not enter the cells and has been used for the development of potent hTfR targeted immunotoxins (reviewed in [2]). 7D3 when chemically conjugated to RTA demonstrated significant antitumor effects in H-MESO-1, a human mesothelioma cell line [33]. These effects of the 7D3-RTA conjugate have been observed both in vitro and in vivo in a xenograft mouse model. In vitro analysis of the combination treatment of 7D3-RTA and ADR showed no synergistic effects of the combination treatment. However, in vivo studies demonstrated that this combination treatment enhanced the median survival to 31 days compared with 22 days for ADR alone and 23 days

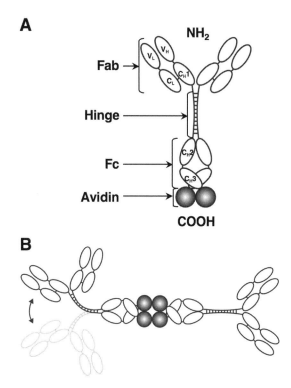

A

NH₂

Fab →

Hinge —

Fc —

Avidin —

COOH

B

FIGURE 2.4. Schematic representation of the structure of anti-TfR IgG3-Av. A) This fusion protein consists of a murine/human chimeric antibody specific for either rTfR or hTfR. Each fusion protein has one avidin molecule genetically fused to each C_H3 domains of the human IgG3 Fc region. B) FPLC analysis suggest that anti-TfR IgG3-Av exists as a dimer in solution [36]. The extended hinge region of the IgG3 molecule allows for increased flexibility of the fusion protein.

binant human interferon-α (rhuIFN). Treatment with rhuIFN alone demonstrated no antitumor effects. Mice treated with 454A12-RTA alone showed an increased median survival time (89 days) compared with the saline control group (41 days). Combination treatment of 454A12-RTA and rhuIFN further increased the survival time to more than 120 days. However, if treatment was initiated when the tumor burden was higher (15 days after tumor injection) no additive effect was observed with the 454A12-RTA and rhuIFN treatment, indicating that tumor burden plays a role in determining antitumor efficacy of the combination treatment. The pretreatment of animals with large tumor burdens with chemotherapy, either cyclophosphamide (a nitrogen mustard that cross-links DNA) or cisplatin, to reduce tumor volume prior to treatment significantly enhanced the antitumor effects of 454A12-RTA and rhuIFN combined therapy. After chemotherapy, mice treated with rhuIFN alone showed a significant increase in survival (89 days) compared with control mice (77 days). 454A12-RTA alone increased survival to 129 days while the combination of 454A12-RTA and rhuIFN increased survival to 162 days [35]. This study clearly demonstrates the benefit of using multicomponent treatment strategies as cancer therapeutics.

6   A Universal Delivery System for Targeted Cancer Therapy

We have previously constructed two chimeric antibody fusion proteins that target either the rat TfR (rTfR) or the human TfR (hTfR) [36, 37]. Both antibodies contain human IgG3 constant regions and have chicken avidin (Av) genetically fused to the C_H3 domains (Fig. 2.4). Anti-rTfR IgG3-Av contains the variable regions of the murine monoclonal antibody OX26 while anti-hTfR IgG3-Av is composed of the variable regions of the murine antibody 128.1. These antibody fusion proteins are the first anti-TfR of its kind, since IgG1 and IgG2 antibodies have been more commonly used for therapeutic purposes. The rational behind utilizing IgG3 is due to the presence of an extended hinge that may facilitate simultaneous binding to antigen and Fc receptors to maximize its therapeutic potential. Both antibodies were developed to function as universal vectors for the delivery

for the 7D3-RTA conjugate alone [33]. The addition of a third component, the sodium ionophore monensin, further increased survival to 41 days. Thus, the combination of 7D3-RTA with chemotherapy demonstrates synergistic antitumor effects that could be used for the treatment of solid tumors in humans.

The murine anti-hTfR IgG1 (454A12) antibody alone also is not an effective treatment. However, an 454A12-RTA conjugate has shown synergistic effects with standard therapeutic regimes. The efficacy of the conjugated to RTA was studied in a human ovarian carcinoma xenograft model [35]. Ten days after intraperitoneal injection of OVCAR-3 cells, mice were treated with the 454A12-RTA immunotoxin alone or in combination with recom-

of biotinylated agents into malignant cells that overexpress TfR. Unexpectedly, we have found that both of these antibodies alone are capable of inducing significant antiproliferative/proapoptotic effects in malignant cells, including cells that have developed resistance to various therapeutic drugs. Dimerization of these fusion proteins was shown to be important for this cytotoxicity. FPLC analysis showed that the anti-rTfR IgG3-Av exists as a dimer in solution [36]. Avidin is a tetramer comprised of four noncovalent monomers. Since each anti-hTfR IgG3-Av molecule contains two avidin moieties, dimerization may be caused by the formation of the tetrameric form of avidin (see Fig. 2.4). This suggests that cross-linking of the TfR due to the tetravalency of anti-hTfR IgG3-Av

is responsible, at least in part, for the cytotoxicity of the fusion protein. However, we can rule out the possibility that avidin (a positively charged molecule with heparin-binding ability) [38], the extended hinge region of human IgG3 [39], and/or the binding of the Fc fragment of IgG3 to Fc receptors may contribute to cross-linking induced by anti-hTfR IgG3-Av.

The efficacy of anti-rTfR IgG3-Av as a universal vector for targeted delivery of biotinylated agents has been demonstrated [36]. The avidin moiety of this fusion protein forms strong noncovalent interactions with biotinylated molecules such as glucose oxidase and β-galactosidase. Studies of this antibody fusion protein complexed to biotinylated-FITC indicate that the complex is internalized through

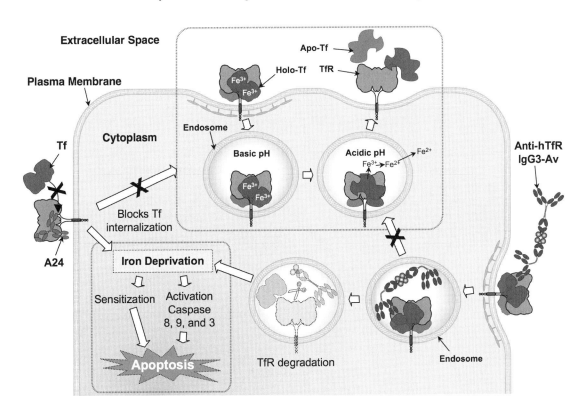

FIGURE 2.5. Schematic representation of the proposed mechanism of cytotoxicity of anti-hTfR antibodies. Under normal conditions (top box), diferric transferrin binds the TfR on the surface of the cell and is internalized via a clathrin-coated pit into an endosome. Due to the decrease in pH, iron dissociates from Tf and is pumped out of the endosome. The Tf/TfR complex traffics back to the cell surface where Tf dissociates from its receptor. Treatment with the A24 anti-TfR antibody (left side of the figure) blocks the binding of Tf and thus leads to iron deprivation, apoptosis and/or chemosensitization. Anti-hTfR IgG3-Av treatment (right side) disrupts the normal cycling pathway leading to the degradation of the TfR. Iron deprivation results are lethal to some malignant B cells. In cells that are resistant to anti-hTfR IgG3-Av cytotoxicity, the treatment could be used to sensitize cells to chemotherapy due to induction of iron starvation. A24 has also been shown to disrupt TfR cycling leading to a reduction of cell surface TfR expression. This may also contribute to the cytotoxic activity of A24. (See Color Plates)

receptor-mediated endocytosis [36]. The antibody fusion protein was also found to effectively deliver large biotinylated proteins (β-galactosidase, 464 kDa) into malignant cells that retained activity in the intracellular environment. The anti-rTfR IgG3-Av exhibited antiproliferative/proapoptotic effects in rat cell lines of hematopoietic origin that have developed resistance to therapeutic agents. Anti-rTfR IgG3-Av exhibited these effects in Y3-Ag 1.2.3, a rat myeloma cell line resistant to 8-azaguanine (a purine analog interferes with purine biosynthesis), and C58 (NT) D.1.G.OVAR.1, a rat T-cell lymphoma resistant to 6-thioguanine (an antimetabolite cancer drug that interferes with purine biosynthesis) and ouabain (a poisonous glycoside that blocks the sodium potassium ATPase). These studies indicate that anti-rat IgG3-Av has dual activity through direct cytotoxic activity and the delivery of active biotinylated molecules.

Anti-rTfR IgG3-Av has also been used for antibody-directed enzyme prodrug therapy (ADEPT) [40]. ADEPT works in a two-step manner, the first is the administration of the antibody fusion protein conjugated to a prodrug processing enzyme in order to deliver the enzyme to the tumor cell. In the second step, a prodrug is administered in a nontoxic form following antibody/enzyme treatment and is metabolized by the tumor-localized enzyme. This form of therapy works to potentially eliminate nonspecific effects in normal cells that are caused by the use of chemotherapeutic agents alone. This form of therapy also has potential for eliminating minimal residual disease associated with treating solid tumors. As previously discussed, anti-rTfR IgG3-Av has been shown to efficiently deliver biotinylated proteins into cells, however limitations associated with binding variability can occur due to the presence of various biotinylated sites within the protein. Fusion of P67, the carboxy-terminal domain of human propionyl-CoA carboxylase α subunit, to a protein can eliminate this variability since P67 allows biotinylation at only one residue (lysine-669) by human or E. coli biotin protein ligases. In these studies, P67 was genetically fused to FCU1, a genetically engineered chimeric protein consisting of cytosine deaminase (CD) and uracil phosphoribosyltransferase (UPRT). CD converts the nonactive and nontoxic prodrug 5-fluorocytosine (5-FC) to a highly toxic 5-fluorouracil (5-FU). UPRT can then convert 5-FU to the toxic metabolite 5-fluorouridine 5′-monophosphate (5-FUMP). P67 was also fused to an E. coli enzyme PNP (purine nucleoside phosphorylase). PNP works by cleaving the prodrug 2-fluoro-2′-dexoyadenosine (F-dAdo) to produce the highly toxic 2-fluoroadenine (F-Ade). These two fusion proteins were then conjugated to anti-rTfR IgG3-Av after monobiotinylation of P67. Efficient monobiotinylation of P67 (>95%) was achieved and the prodrug processing enzymes retained their activity in the antibody fusion protein/enzyme complex. The antigen binding capability of the antibody was also preserved allowing successful targeting of the tumor cells. Anti-rTfR-Av effectively localized FCU1 and PNP to tumor cells so that the pro-dugs were metabolized into their cytotoxic forms. The prodrugs alone had cytotoxic effects at high concentrations, however, cytotoxicity increased by 50-fold when the cells were pretreated with either anti-rTfR-Av/P67-FCU1 or anti-rTfR-Av/P67-PNP in Y3-Ag1.2.3 rat myeloma cells [40]. These cells were found to be more susceptible to F-dAdo treatment. No cytotoxic effects were observed for a nontargeted complex. Importantly, this study supports the use of this antibody fusion protein as a delivery system for biotinylated therapeutic agents into tumor cells and its possible use as a sensitizing agent in resistant tumor cells.

Like its rat counterpart, the anti-hTfR IgG3-Av fusion protein has also demonstrated intrinsic cytotoxicity in human malignant cells. This was indicated by studies in the human erythroleukemia cell line K562, a panel of human malignant B-cell lines as well as malignant cells isolated from patients with multiple myeloma and plasma cell leukemia [36, 37]. ARH-77 and IM-9, Epstein-Barr virus (EBV) transformed lymphoblastoid cell lines, were found to be most sensitive to anti-hTfR IgG3-Av treatment [37]. Different levels of sensitivity to anti-hTfR IgG3-Av were observed for other malignant B-cell lines. More importantly, the fusion protein exhibited anti-proliferative effects in 8226/DOX40, a doxorubicin resistant variant of 8226/S, indicating that this antibody was capable of inducing cytotoxic effects in chemoresistant cells.

The mechanism of anti-hTfR IgG3-Av cytotoxicity has recently been explored (see Fig. 2.5) [37]. Upon treatment with anti-hTfR IgG3-Av, the TfR is rapidly degraded. Confocal microscopy indicated that the TfR colocalized with an intracellular compartment expressing LAMP-1 within 15 minutes of treatment. This indicated that trafficking is altered and that the TfR is directed to a lysosomal compartment. Cells treated with anti-hTfR IgG3-Av indicated the presence of small TfR fragments that appeared to be the result of proteolytic cleavage, suggesting that degradation may take place in or prior to the LAMP-1 intracellular compartment. Degradation of the TfR was partially blocked by a cysteine protease inhibitor indicating that a papain-like cysteine protease is involved in the degradation of the N-terminus of the receptor. Other protease inhibitor treatments tested did not completely block degradation, suggesting other protease families may also be involved. Treatment with chloroquine or ammonium chloride showed that the degradation of the receptor was not sensitive to changes in the pH of the lysosome, indicating that the proteases involved in TfR degradation were still active and functional under these conditions.

Iron is important for dividing cells and its absence has been shown to cause mitochondrial-mediated cell death [41]. In two highly sensitive human B-cell lines (IM-9 and ARH-77), anti-hTfR IgG3-Av causes iron deprivation in cells leading to mitochondria depolarization and activation of caspase 9, 8, and 3 [37]. These cytotoxic effects are blocked by iron supplementation indicating the role of iron deprivation in inducing such effects. Treatment of anti-hTfR IgG3-Av and a pan-caspase inhibitor, Z-VAD-FMK, only partially blocked cell death, suggesting cell death is not solely dependent on the caspase activation pathway, but that other mechanisms may be involved. In addition, cells treated with the antibody fusion protein and Z-VAD-FMK still exhibited loss of mitochondrial potential and eventually died. Thus, anti-hTfR IgG3-Av induced lethal iron deprivation that can lead to mitochondrial-mediated apoptosis and chemosensitization

These studies support the use of this fusion protein as an effective therapeutic agent that can induce significant cytotoxic effects in addition to its ability to deliver biotinylated molecules into cancer cells. Anti-hTfR IgG3-Av may also be used in combination treatment strategies with nonbiotinylated agents (such as chemotherapeutic drugs) to sensitize resistant cells to chemotherapy. This antibody fusion protein has the potential to be used as a universal delivery vector for the treatment of human malignancies and as an effective sensitizing agent for the "two-hit signal" model.

7   Conclusion

Upregulation of the TfR on the surface of malignant cells and its constitutive internalization pathway make this an attractive molecule for targeted cancer therapy. The main advantage of targeting the TfR is the versatility of both the targeting strategy itself as well as the agent to be delivered. The TfR can be targeted by its ligand Tf, antibodies, or antibody fragments specific for its extracellular domain. In addition, delivery of a therapeutic agent that targets a defect in cancer cells, like the restoration of wild-type p53 expression, further enhances tumor specificity without affecting normal cells. In these two ways, targeting the TfR would help to minimize the nonspecific effects of the treatment by more adequately targeting the malignant cells. Furthermore, it can be successfully used for the treatment of chemoresistant tumors and has served to sensitize cancer cells to cell death induced by chemotherapy, either by directly inducing iron starvation or by the delivery of agents the make the tumor cells more susceptible to chemotherapy. Thus, the TfR can potentially open new avenues of combination therapies for the treatment of human cancers.

References

1. Daniels TR, Delgado T, Rodriguez JA, et al. The transferrin receptor part I. Biology and targeting with cytotoxic antibodies for the treatment of cancer. Clin Immunol 2006, 121:144–158.
2. Daniels TR, Delgado T, Helguera G, et al. The transferrin receptor part II: targeted delivery of therapeutic agents into cancer cells. Clin Immunol 2006, 121:159–176.
3. Myers CE, Chabner BA. Anthracyclins. In: Frogg MH. Goodman M, et al. (eds.) Cancer chemotherapy: principles and practice. Philadelphia: Lippincott, 1990.
4. Takakura Y, Hashida M. Macromolecular drug carrier systems in cancer chemotherapy: macromolecular prodrugs. Crit Rev Oncol Hematol 1994, 18:207–231.
5. Gottesman MM, Pastan I. Biochemistry of multidrug resistance mediated by the multidrug transporter. Annu Rev Biochem 1993, 62:385–427.
6. Bellamy WT. P-glycoproteins and multidrug resistance. Annu Rev Pharmacol Toxicol 1996, 36:161–183.
7. Schuurhuis GJ, Broxterman HJ, Cervantes A, et al. Quantitative determination of factors contributing to doxorubicin resistance in multidrug-resistant cells. J Natl Cancer Inst 1989, 81:1887–1892.

8. Larsen AK, Escargueil AE, Skladanowski A. Resistance mechanisms associated with altered intracellular distribution of anti-cancer agents. Pharmacol Ther 2000, 85:217–229.

9. Coley HM, Amos WB, Twentyman PR, et al. Examination by laser scanning confocal fluorescence imaging microscopy of the subcellular localisation of anthracyclines in parent and multidrug resistant cell lines. Br J Cancer 1993, 67:1316–1323.

10. Weaver JL, Pine PS, Aszalos A, et al. Laser scanning and confocal microscopy of daunorubicin, doxorubicin, and rhodamine 123 in multidrug-resistant cells Exp Cell Res 1991, 196:323–329.

11. Fritzer M, Barabas K, Szuts V, et al. Cytotoxicity of a transferrin-Adriamycin conjugate to anthracycline-resistant cells. Int J Cancer 1992, 52:619–623.

12. Hatano T, Ohkawa K, Matsuda M. Cytotoxic effect of the protein-doxorubicin conjugates on the multidrug-resistant human myelogenous leukemia cell line, K562, in vitro. Tumour Biol 1993, 14:288–294.

13. Singh M, Atwal H, Micetich R. Transferrin directed delivery of Adriamycin to human cells. Anti-cancer Res 1998, 18:1423–1427.

14. Fritzer M, Szekeres T, Szuts V, Het al. Cytotoxic effects of a doxorubicin-transferrin conjugate in multidrug-resistant KB cells. Biochem Pharmacol 1996, 51:489–493.

15. Chitambar CR. Gallium compounds as antineoplastic agents. Curr Opin Oncol 2004, 16:547–552.

16. Chitambar CR, Wereley JP, Matsuyama S. Gallium-induced cell death in lymphoma: role of transferrin receptor cycling, involvement of Bax and the mitochondria, and effects of proteasome inhibition. Mol Cancer Ther 2006, 5:2834–2843.

17. Wang F, Jiang X, Yang DC, et al. Doxorubicin-gallium-transferrin conjugate overcomes multidrug resistance: evidence for drug accumulation in the nucleus of drug resistant MCF-7/ADR cells. Anticancer Res 2000, 20:799–808.

18. Lopez-Barcons LA, Polo D, Llorens A, et al. Targeted Adriamycin delivery to MXT-B2 metastatic mammary carcinoma cells by transferrin liposomes: effect of Adriamycin ADR-to-lipid ratio. Oncol Rep 2005, 14:1337–1343.

19. Suzuki S, Inoue K, Hongoh A, et al. Modulation of doxorubicin resistance in a doxorubicin-resistant human leukaemia cell by an immunoliposome targeting transferring receptor. Br J Cancer 1997, 76:83–89.

20. Iinuma H, Maruyama K, Okinaga K, et al. Intracellular targeting therapy of cisplatin-encapsulated transferrin-polyethylene glycol liposome on peritoneal dissemination of gastric cancer. Int J Cancer 2002, 99:130–137.

21. Jhaveri MS, Rait AS, Chung KN, et al. Antisense oligonucleotides targeted to the human alpha folate receptor inhibit breast cancer cell growth and sensitize the cells to doxorubicin treatment. Mol Cancer Ther 2004, 3:1505–1512.

22. Ross JF, Chaudhuri PK, Ratnam M. Differential regulation of folate receptor isoforms in normal and malignant tissues in vivo and in established cell lines. Physiologic and clinical implications. Cancer 1994, 73:2432–2443.

23. Adams JM, Cory S. The Bcl-2 apoptotic switch in cancer development and therapy. Oncogene 2007, 26:1324–1337.

24. Kostanova-Poliakova D, Sabova L. Anti-apoptotic proteins-targets for chemosensitization of tumor cells and cancer treatment. Neoplasma 2005, 52:441–449.

25. Chiu SJ, Liu S, Perrotti D, et al. Efficient delivery of a Bcl-2-specific antisense oligodeoxyribonucleotide (G3139) via transferrin receptor-targeted liposomes. J Control Rel 2006, 112:199–207.

26. Basma H, El-Refaey H, Sgagias MK, et al. BCL-2 antisense and cisplatin combination treatment of MCF-7 breast cancer cells with or without functional p53. J Biomed Sci 2005, 12:999–1011.

27. Smith ND, Rubenstein JN, Eggener SE, et al. The p53 tumor suppressor gene and nuclear protein: basic science review and relevance in the management of bladder cancer. J Urol 2003, 169:1219–1228.

28. Xu L, Tang WH, Huang CC, et al. Systemic p53 gene therapy of cancer with immunolipoplexes targeted by anti-transferrin receptor scFv. Mol Med 2001, 7:723–734.

29. Xu L, Pirollo KF, Tang WH, et al. Transferrin-liposome-mediated systemic p53 gene therapy in combination with radiation results in regression of human head and neck cancer xenografts. Hum Gene Ther 1999, 10:2941–2952.

30. Moura IC, Lepelletier Y, Arnulf B, et al. A neutralizing monoclonal antibody (mAb A24) directed against the transferrin receptor induces apoptosis of tumor T lymphocytes from ATL patients. Blood 2004, 103:1838–1845.

31. Moura IC, Centelles MN, Arcos-Fajardo M, et al. Identification of the transferrin receptor as a novel immunoglobulin (Ig)A1 receptor and its enhanced expression on mesangial cells in IgA nephropathy. J Exp Med 2001, 194:417–425.

32. Frankel AE. Increased sophistication of immunotoxins. Clin Cancer Res 2002, 8:942–924.

33. Griffin TW, Stocl M, Collins J, et al. Combined antitumor therapy with the chemotherapeutic drug doxorubicin and an anti-transferrin receptor immunotoxin: in vitro and in vivo studies. J Immunother 1992, 11:12–18.

34. Stirpe F, Barbieri L, Battelli MG, et al. Ribosome-inactivating proteins from plants: present status and future prospects. Biotechnology (N Y) 1992, 10:405–412.

35. Pearson JW, Hedrick E, Fogler WE, et al. Enhanced therapeutic efficacy against an ovarian tumor xenograft of immunotoxins used in conjunction with recombinant alpha-interferon. Cancer Res 1990, 50:6379–6388.

36. Ng PP, Dela Cruz JS, Sorour DN, et al. An anti-transferrin receptor-avidin fusion protein exhibits both strong proapoptotic activity and the ability to deliver various molecules into cancer cells. Proc Natl Acad Sci U S A 2002, 99:10706–10711.

37. Ng PP, Helguera G, Daniels TR, et al. Molecular events contributing to cell death in malignant human hematopoietic cells elicited by an IgG3-avidin fusion protein targeting the transferrin receptor. Blood 2006, 108:2745–2754.

38. Kett WC, Osmond RI, Moe L, et al. Avidin is a heparin-binding protein. Affinity, specificity and structural analysis. Biochim Biophys Acta 2003, 1620:225–234.

39. Phillips ML, Tao MH, Morrison SL, et al. Human/mouse chimeric monoclonal antibodies with human IgG1, IgG2, IgG3 and IgG4 constant domains: electron microscopic and hydrodynamic characterization. Mol Immunol 1994, 31:1201–1210.

40. Asai T, Trinh R, Ng PP, et al. A human biotin acceptor domain allows site-specific conjugation of an enzyme to an antibody-avidin fusion protein for targeted drug delivery. Biomol Eng 2005, 21:145–155.

41. Greene BT, Thorburn J, Willingham MC, Aet al. Activation of caspase pathways during iron chelator-mediated apoptosis. J Biol Chem 2002, 277:25568–25575.

Chapter 3
Chemo-Immunosensitization of Resistant B-NHL as a Result of Rituximab (anti-CD20 mAb)-Mediated Inhibition of Cell Survival Signaling Pathways

Benjamin Bonavida, Ali R. Jazirehi, Mario I. Vega, Sara Huerta-Yepez, Kazuo Umezawa, and Eriko Suzuki

1 Introduction

Non-Hodgkin's lymphoma (NHL) consists of a large group of neoplasms of the immune system and represents a heterogeneous group of diseases that share the expansion of both B and T cells. The majority of NHL are of B-cell origin and greater than 95% express CD20 [1]. The incidence of NHL has been rising steadily for a number of years and currently NHL is the fifth most common cancer. Combination chemotherapy with CHOP (cyclophosphamide, doxorubicin, vincristine, and prednisone) is the standard treatment option for aggressive NHL [2]. However, only 40% of patients with aggressive NHL are cured. Treatment of patients with NHL with monoclonal antibody has provided an alternative approach. The majority of B-NHL expresses CD20 on the cell surface and this receptor has been exploited as a therapeutic target. CD20 is a tetra-membrane–spanning protein (molecular weight 33–37 kDa) located on chromosome 11q12-q13.1. CD20 expression on the cell surface is present as multimers. Li et al. [3] reported that more than 80% of CD20 in Ramos cells appears to be localized as microvilli. It is resident in lipid raft domains of the plasma membrane [4].

Rituximab (a chimeric anti-CD20 monoclonal antibody), the first FDA-approved antibody for cancer, was initially used for the treatment of follicular lymphoma (FL) and low-grade NHL but its use has expanded in combination with chemotherapy for other B-cell malignancies [5–7]. Various mechanisms have been proposed for the in vivo effects of rituximab, such as complement-dependent cytotoxicity (CDC) [8], antibody-dependent cellular cytotoxicity (ADCC) [8, 9], and direct apoptotic effects [10]. However, these mechanisms do not explain rituximab-mediated inhibition of cell growth as well as patients' failure to respond to rituximab treatment while harboring CD20-expressing tumors and the previously mentioned effector mechanisms are still operational. We reported the novel observation that rituximab treatment of B-NHL cell lines sensitized the drug resistant B-NHL cell lines to apoptosis by a variety of chemotherapeutic drugs [11]. The underlying molecular mechanisms of sensitization are briefly described in the following.

The possible interaction of rituximab-treated tumors with host effector cells, as an additional mechanism of its anti-tumor effect, has not been explored with the exception of the Fc effector functions in CDC and ADCC. Our findings established

From: *Sensitization of Cancer Cells for Chemo/Immuno/Radio-therapy, 1st Edition.*
Edited by: Benjamin Bonavida © Humana Press, Totowa, NJ

the role of rituximab-mediated sensitization of B-NHL cells to death ligands (Fas-L, TRAIL)–induced apoptosis via inhibition of NF-kB and YY1 and upregulation of death receptors. The underlying molecular mechanism of immunosensitization is described in the following.

2 Rituximab-Mediated Inhibition of Constitutively Activated Cell Survival Anti-Apoptotic Signaling Pathways: Pivotal Role in Chemosensitization of Drug-Resistant B-NHL

It has been proposed that CD20 may serve as a membrane channel implicated in the regulation of calcium flux. CD20 was co-isolated with tyrosine and serine/threonine kinases on incubation of cells with anti-CD20 antibody, implicating CD20 in transmembrane signaling pathways [12]. There is no direct interaction of the cytoplasmic membrane with signaling, as signaling results from association of CD20 with raft signaling platforms [13]. Rituximab induces re-distribution of CD20 to lipid rafts without modifying the overall context in lipid surfaces and signaling proteins. It leads to decreased Lyn activity, depending on the presence of CBP/PAG. The raft aliquots contain increasing amounts of CD20, so the total protein kinase activity, as measured by levels of phosphorylated Lyn and CBP/PAG, decreases without changes in nonphosphorylated proteins [14]. We have also found decreased phosphorylated Lyn in NHL cell lines upon treatment with rituximab, resulting in downstream inhibition of various survival signaling pathways [15]. Bezombes et al. [16] reported that rituximab treatment leads to a rapid and transient increase in acid sphingomyelinase activity concomitant with cellular ceramide generation in raft microdomains. Growth inhibition by rituximab was suggested to be mediated through a ceramide-triggered signaling pathway leading to induction of cell cycle–dependent kinase inhibitors such as p27 through a mechanism dependent on mitogen-activated protein kinase (MAPK) [12]. Initial studies with rituximab treatment of several B-NHL cell lines demonstrated that it sensitized drug-resistant

tumor cells to the induction of apoptosis by various chemotherapeutic drugs [11]. Based on this observation, we systematically investigated the underlying molecular mechanisms responsible for rituximab-mediated chemosensitization. The findings were concomitant with rituximab-mediated inhibition of the anti-apoptotic gene products such as Bcl-2/Bcl-$_{xL}$. Therefore, we examined the effect of rituximab on various pathways that may regulate the transcription/expression/activity of these gene products. Thus far, we have identified the following pathways, namely, p38 MAPK, NF-kB, ERK1/2, and Akt pathways largely responsible for the regulation of Bcl-2/Bcl-$_{xL}$. Each of these signaling pathways affected by rituximab is described briefly in the following.

2.1 Rituximab-Mediated Inhibition of Tumor-Derived Resistance Factor(s): Role of the p38 MAPK Signaling Pathway in Resistance

It has been reported that the secretion of certain cytokines by tumor cells renders them resistant to the cytotoxic effects of chemotherapeutic drugs [17, 18]. We hypothesized that the chemosensitization of NHL cell lines might be due to rituximab-mediated inhibition of these tumor-derived resistance factors. Using as a model the diffuse large B-cell lymphoma (DLBCL) AIDS-derived B-NHL cell line 2F7, which secretes cytokines such as TNFα and IL-10, we have found that rituximab selectively disrupted the IL-10 autocrine/paracrine loop and this disruption correlated with selective downregulation of Bcl-2 expression and chemosensitization. Rituximab-mediated inhibition of IL-10 secretion resulted in downregulation of the constitutive activity of signal transducer and activator of transcription 3 (STAT3) in those cells (through IL-10–IL-10R interaction), and STAT3 inhibition resulted in inhibition of Bcl-2 transcription and expression. The direct role by which rituximab induced inhibition of STAT3 activity and Bcl-2 expression was corroborated by the use of a STAT3 inhibitor, piceatannol (shown to inhibit the JAK1/Tyk-2–dependent STAT3 and STAT5 signaling pathways) [19].

Rituximab inhibited the constitutive activity of Lyn and p38 MAPK activities, resulting in inhibition of IL-10 transcription via inhibition of Sp1. The role of p38 MAPK in the regulation of IL-10 was corroborated by using specific pharmacologic inhibitors of the p38 MAPK pathway and implicating the roles of Src kinases and NF-κB. Rituximab-mediated inhibition of MAPK activity and IL-10 transcription correlated with the inhibition of both STAT3 activity and Bcl-2 expression and resulted in drug-induced apoptosis. Treatment with *cis*-diamminedichloroplatinum (CDDP, cisplatin) induced the generation of mitochondrial reactive oxygen species, specifically intracellular peroxides. The combination of rituximab and CDDP acted synergistically to induce apoptosis through mitochondria-mediated apoptotic events (type II) [20]. Many studies have reported that IL-10 is increased in the serum of many patients with NHL and that this increase correlates with a lower rate of survival [21]. These studies introduce a novel therapeutic strategy aiming at interfering with IL-10 synthesis and secretion via inhibitors of the p38 MAPK/STAT3 pathways.

2.2 Rituximab-Mediated Inhibition of the Raf-1/MEK1/2/ERK1/2/AP-1 Pathway: Pivotal Role in Inhibition of Bcl-2/Bcl-$_{xL}$ and Chemosensitization

The studies performed in the DLBCL 2F7 cell line following rituximab treatment did not address whether the same effects are observed in the FL and NHL cell lines. Studies in Ramos and Daudi cells revealed that rituximab selectively inhibited Bcl-$_{xL}$ expression, leading to chemosensitization [22]. Bcl-$_{xL}$ expression is under the transcriptional regulation of AP-1. The ERK1/2 pathway is constitutively activated in Ramos and Daudi cells; its inhibition by rituximab sensitizes the cells to drug-induced apoptosis. The phosphorylation-dependent state of Raf-1/MEK1/2/ERK1/2 was significantly decreased 3 to 6 hours after rituximab treatment, concomitant with inhibition of MEK1/2 kinase activity. The role of the ERK1/2 pathway in the regulation of Bcl-$_{xL}$ and chemosensitization was corroborated by the use of specific pharmacologic inhibitors of the pathway, which also sensitized the cells to drug-induced apoptosis. Bcl-$_{xL}$ is abundantly expressed in lymphoma [23] and protects the cells from apoptosis induced by DNA-damaging agents. We further

analyzed mechanisms that underlie rituximab-mediated inhibition of the ERK1/2 pathway. Raf-1 kinase inhibitor protein (RKIP) is recently identified as a negative regulator of the ERK1/2 pathway [24, 25]. Rituximab upregulates the expression of RKIP and facilitates the association of RKIP and Raf-1 [15]. These findings unraveled a novel mechanism by which rituximab inhibits the ERK1/2 pathway and selectively downregulates Bcl-$_{xL}$ expression. Consequently, this study identified the components of the ERK1/2 pathway, Bcl-$_{xL}$ and RKIP, as potential targets for therapeutic intervention and might provide a molecular basis for the use of inhibitors of the ERK1/2 pathway in combination with chemotherapeutic drugs in the treatment of resistant B-NHL.

2.3 Rituximab-Mediated Inhibition of the NF-*k*B Signaling Pathway: Pivotal Role in the Inhibition of Bcl-$_{xL}$ Transcription, Expression, and Chemosensitization

Previous studies have reported that NF-κB regulates Bcl-$_{xL}$ gene expression [26, 27]. Therefore, we examined whether rituximab modified the constitutive activation of the NF-κB pathway in Ramos and Daudi cell lines. Rituximab decreases the phosphorylation of NF-κB–inducing kinase (NIK), IκB kinase (IKK), and IκB-α diminishes IKK activity, and decreases NF-κB DNA-binding activity. In addition, rituximab significantly upregulated RKIP expression, thus interrupting the NF-κB signaling pathway concomitant with Bcl-$_{xL}$ downregulation. The induction of RKIP expression augmented its physical association with endogenous NIK, IKK, and transforming growth factor β–activated kinase 1 (TAK1), resulting in decreased activity of the NF-κB pathway and diminishing NF-κB DNA-binding activity; this result confirms the role of RKIP inhibition of NF-κB [28]. The role of NF-κB inhibition in downregulation of Bcl-$_{xL}$ and chemosensitization was corroborated by the use of various pharmacologic inhibitors of this pathway. These studies identified the components of the NF-κB pathway and RKIP as potential targets for therapeutic intervention. The findings also provide a novel approach for the use of rituximab or pharmacologic inhibitors of NF-κB in combination with subtoxic concentrations of chemotherapeutic drugs in the treatment of NHL refractory to rituximab and to other drugs [7].

2.4 Rituximab-Mediated Inhibition of the Akt/PI3K Pathway: Pivotal Role in the Inhibition of Phospho-Bad and Augmentation of Associations of Bad with Bcl-$_{xL}$

The Akt pathway has been reported to regulate Bcl-$_{xL}$ activity and expression [29]. Rituximab treatment of Ramos cells inhibited the PI3K/Akt pathway and resulted in inhibition of phospho PI3K, PDK-1, and Akt with no effect on nonphosphorylated proteins. In addition, inhibition of phospho-Bad by rituximab augmented the association of Bad with Bcl-$_{xL}$ to form complexes. Formation of these complexes changed the ratio of cytoplasmic Bad/Bcl-xL in favor of Bad, thus potentiating mitochondrial membrane depolarization and activation of caspases 9, 7, and 3 when used in combination with chemotherapeutic drugs. The role of the Akt/PI3K pathway in the regulation of chemoresistance was corroborated by the use of the Akt inhibitor Ly-294002 and by transfection with small interfering RNA (siRNA) Akt [30]. These findings implicated the Akt pathway in the regulation of drug-resistance in B-NHL, whereas its modification by rituximab sensitized the cells to drug-induced apoptosis. These findings identified the Akt pathway as a target for therapeutic intervention.

Inhibition of the above signaling pathways by rituximab resulted in inhibition of anti-apoptotic gene product (Bcl-xL) and reversal of drug resistance. The signaling pathways modified by rituximab identified novel potential therapeutic targets, whose intervention can mimic rituximab-mediated chemosensitizing effects. It is important to note that the findings achieved in the laboratory with various cell lines cannot be generalized to patients' lymphoma of different origins and clinical behavior. Also, the data cannot be extrapolated to B-cell diseases for which there are no cell lines, such as follicular lymphoma. Clearly, if validations are obtained, sensitizing agents such as those reported by us have potential for use in combination with chemotherapeutic drugs, thus mimicking rituximab and leading to reversal of drug resistance.

3 Rituximab-Mediated Inhibition of Constitutively Activated and Overexpressed NF-kB and YY1 Repressor Activities: Pivotal Role in Sensitization to Cytotoxic Ligands (Fas-L, TRAIL) via Upregulation of Corresponding Death Receptors

In studies performed with tumor cells other than B-NHL, we have found that inhibition of the NF-κB pathway sensitizes tumor cells to Fas ligand–induced apoptosis concomitant with upregulation of Fas expression. We have further demonstrated that the upregulation of Fas expression was the result of inhibition of the transcription repressor Yin Yang 1 (YY1) that is under the regulation of NF-κB [31]. We have also reported that inhibition of the transcription repressor Yin Yang 1 (YY1) by S-nitrosation [32] or siRNA YY1 [33] resulted in upregulation of Fas expression and sensitization to Fas ligand (FasL)–induced apoptosis in ovarian and prostate carcinoma cell lines. YY1 negatively regulates Fas transcription via its association to the silencer region of the Fas promoter. Three elements potentially responsive to YY1 were found to cluster in a very close proximity within the Fas promoter silencer region between −1619 and −1533 base pairs relative to the transcription initiation site. Deletion of the silencer region of the Fas promoter in a reporter assay resulted in augmentation of Fas expression of the tumor cells [31]. Since rituximab also inhibited the NF-kB pathway as described, we examined whether rituximab also inhibits YY1 and sensitizes the cells to Fas-L and TRAIL-induced apoptosis.

3.1 Rituximab Sensitizes B-NHL Cell Lines to Fas-L–Induced Apoptosis: Pivotal Role of YY1 Inhibition and Up-regulation of Fas Expression

Treatment with rituximab of B-NHL cell lines resistant to FasL or CH-11 agonist monoclonal antibody resulted in significant inhibition of YY1 expression and activity, upregulation of Fas expression, and sensitization to CH-11–induced apoptosis.

Fas expression was upregulated as early as 6 hours after rituximab treatment, as determined by flow cytometry for surface expression, reverse transcriptase polymerase chain reaction for transcription (RT-PCR), and Western blot for total protein. Rituximab-induced inhibition of YY1 expression was determined by Western blot and its DNA activity by electrophoretic mobility shift assay (EMSA). The involvement of NF-κB and YY1 in the regulation of Fas expression was corroborated by using Ramos cells with a dominant active inhibitor of NF-κB and by silencing YY1 with YY1 siRNA, respectively. The role of rituximab-mediated inhibition of the p38 MAPK, NF-κB, and YY1 pathways in the regulation of Fas and sensitization to CH-11–induced apoptosis was validated by the use of specific pharmacologic inhibitors of these pathways, all of which resulted in sensitization to CH-11–induced apoptosis. Treatment with rituximab alone had no effect on the activation of caspases or on the mitochondria. Likewise, treatment with CH-11 resulted in modest activation of caspases 8 and 9, which correlated with moderate induction of apoptosis. However, treatment with a combination of rituximab and CH-11 resulted in mitochondrial depolarization, release of cytochrome C and the mitochondrial protein Smac/DIABLO, activation of caspases 9 and 3, and poly(ADP-ribose) polymerase (PARP) cleavage, suggesting the involvement of the type II mitochondrial apoptotic pathway [34]. The activation of the mitochondrial pathway by the combination of rituximab and CH-11 may be the result of the inhibition of the anti-apoptotic gene products Bcl-2 and Bcl-$_{xL}$ by rituximab both of which will decrease mitochondrial integrity, resulting in mitochondrial outer membrane collapse and the cytosolic release of apoptotic proteins (cytochrome C and Smac/DIABLO).

Chemoresistance and Fas resistance in NHL cell lines are commonly regulated by the constitutive activation of NF-κB. We have demonstrated, however, that chemoresistance is regulated by Bcl-$_{xL}$, whereas Fas resistance is regulated by YY1. The rituximab-mediated inhibition of NF-κB activity inhibits Bcl-$_{xL}$ expression, leading to chemosensitization, and also inhibits YY1, resulting in sensitization to CH-11–induced apoptosis. These differentially regulated mechanisms for chemoresistance and immune resistance were identified from

findings using biologically engineered cell lines and specific pharmacologic inhibitors. Treatment with specific inhibitors for NF-κB sensitized NHL cells to both drug-induced and CH-11–induced apoptosis. The role of Bcl-$_{xL}$ expression in the regulation of drug resistance—but not Fas resistance—was demonstrated by the failure of rituximab to sensitize Ramos cells that overexpressed Bcl-$_{xL}$ to drug-induced apoptosis, although rituximab induced sensitization to CH-11–induced apoptosis in the same cells. These findings clearly establish distinct regulatory mechanisms modulated by rituximab in NHL cells downstream of NF-κB for sensitization to Fas and drug-induced apoptosis. This finding may have clinical implications: Tumors that overexpress Bcl-$_{xL}$ and are refractory to treatment with chemotherapeutic drugs, alone or in combination with rituximab, may still be sensitive to killing by rituximab in combination with immunotherapy [35].

3.2 Rituximab Sensitizes B-NHL Cells to TRAIL-Induced Apoptosis: Pivotal Role of YY1 Inhibition and Upregulation of DR5 Expression

There is great interest to date in the therapeutic application of TRAIL or agonist monoclonal antibodies directed against TRAIL death receptors, DR4 and DR5, in the treatment of patients who are refractory to current conventional therapeutics. In contrast to Fas ligand, TRAIL is not toxic to normal tissues and it has been shown to exhibit significant tumoricidal activity against a variety of human cancer cell lines, both in vitro and in vivo [36]. Therefore, we have examined, based on our findings with rituximab-mediated sensitization to Fas ligand, whether rituximab can also sensitize TRAIL-resistant B-NHL cells to TRAIL-induced apoptosis. TRAIL induces apoptosis in tumor cells by binding to death receptors TRAIL R1/DR4 and TRAIL R2/DR5 [37]. While TRAIL is a promising drug, however, the majority of tumor cells are resistant to apoptosis induced by TRAIL [38]. Resistance to TRAIL appears to result from the dysregulation of various gene products that are involved in the regulation of the apoptotic signaling pathways. These may involve the low expression of death receptors, overexpression of decoy receptors, hyperactivation of survival pathways, inhibition of apoptotic genes, defects in caspase activation, etc. [39]. Therefore, it is important that TRAIL-resistant tumor cells are sensitized for TRAIL-induced apoptosis. Reported studies have demonstrated that chemotherapeutic drugs used at low concentrations can sensitize drug-resistant and TRAIL-resistant tumor cells to TRAIL-induced apoptosis and that sensitization was concomitant with upregulation of DR5 [40–43]. These findings implied that the sensitizing drugs may be regulating a transcription repressor leading to upregulation of DR5 transcription. Yoshida et al. [44] have examined the DR5 promoter and demonstrated that the transcription factor SP1 positively regulates DR5 transcription. We have also identified in the DR5 promoter a putative DNA-binding site that can negatively regulate DR5 transcription and expression. We have recently established in prostate cancer cells that YY1 negatively regulates DR5 transcription and inhibition of YY1 by drugs resulted in upregulation of DR5 expression and sensitization to TRAIL [45]. These findings are reminiscent of the role YY1 plays as a transcription repressor for Fas transcription as previously described [31] and are further demonstrated in rituximab-mediated sensitization to Fas ligand apoptosis [35, 46].

The preceding findings implicating YY1 in the regulation of DR5 and TRAIL sensitivity prompted us to study whether rituximab can sensitize B-NHL cells to TRAIL-induced apoptosis. We have already reported that rituximab inhibits YY1 expression and activity in B-NHL cells; thus, inhibition of YY1 by rituximab may mimic drug-induced inhibition of YY1, resulting in upregulation of DR5 and sensitization to TRAIL. This hypothesis was tested in B-NHL cells lines following treatment with rituximab and TRAIL. Treatment of Ramos, 2F7, Daudi, and Raji B-NHL cell lines with rituximab and thereafter treated with TRAIL resulted in significant sensitization to TRAIL-induced apoptosis, and synergy was achieved. We also demonstrate that treatment with rituximab significantly upregulated the expression of DR5. The inhibition of YY1 by rituximab coincided with upregulation of DR5 and sensitization to TRAIL (Vega et al., unpublished).

It is possible that rituximab in vivo may exert a new mechanism of action, namely, the sensitization of tumor cells to host FasL and TRAIL-induced apoptosis expressed by cytotoxic immune cells (NK, CD8[+] T cells). The failure of certain patients to respond to rituximab may be due, in part, to the failure to sensitize the tumor cells to host immune–mediated apoptosis. Such failure may be due to mutations, as about 20% of B-cell lymphomas derived from post–germinal center B cells have been found to carry mutations or defects in the expression level of Fas. Alternatively, the cells may not be responding to rituximab-mediated inhibition of the survival pathways and thus cannot be sensitized to FasL-mediated apoptosis. Likewise, for TRAIL, it is possible that failure to respond to TRAIL may be due to poor expression of TRAIL on effector cells or the inability of rituximab to upregulate DR5 transcription and expression.

4 Development of Resistance to Rituximab Treatment: Reversal of Resistance by Pharmacologic Inhibitors Targeting the Constitutively Hyperactivated/ Survival Anti-apoptotic Signaling Pathways

Although rituximab used as monotherapy or in combination with chemotherapy has significantly improved the treatment of patients with NHL, there remains the problem of patients initially unresponsive to rituximab and a subset of patients who are unresponsive to further treatment. The mechanisms of unresponsiveness are not clear. It has been postulated that CD20 downregulation [47], loss of CD20 [48], and circulating CD20 [49] may be responsible for resistance. Based on our findings of rituximab-mediated inhibition of cell signaling, we hypothesized that the development of resistance may emanate from the failure of rituximab to signal the cells effectively, as well as from the development of hyperactivated survival signaling pathways and upregulation of anti-apoptotic gene products. To test this hypothesis, we generated in vitro rituximab-resistant (RR) clones from Ramos, Daudi, and 2F7 cells by culturing the cells in the presence of increasing concentrations of rituximab for several weeks and multiple cycles of limited dilutions. Representative clones (2F7RR1, Ramos RR1, and

Daudi RR1) were examined for their response to rituximab as compared with wild-type cells [50]. These clones expressed some loss of CD20 on the cell surface, were not responsive to complement-dependent cytotoxicity or antibody-dependent cellular cytotoxicity, were not growth inhibited by rituximab, and did not undergo apoptosis after cross-linking of rituximab with a secondary anti–human IgG. Preliminary findings demonstrate that in wild-type Daudi cells, rituximab induces a rapid and transient increase in A-SMase activity paralleled with cellular ceramide generation in lipid rafts. In addition, rituximab treatment externalizes both ceramide and A-SMase, which co-localizes with the CD20 receptor (Bezombes et al, unpublished). In the resistant clones, however, rituximab-induced A-SMase translocation and ceramide generation at the cell surface was reduced [7]. These findings suggest that the failure of rituximab to mediate cell signaling may be due to failure of CD20 migration to lipid rafts and the initiation of cell signaling. Further studies revealed that rituximab failed to inhibit the p38 MAPK, ERK1/2, and NF-κB signaling pathways. In addition, the clones showed hyperactivation of these pathways, with overexpression of Bcl-2 and Bcl-$_{xL}$. Compared with the response seen in wild-type cells, the clones showed cross-resistance to high concentrations of chemotherapeutic drugs intended to induce apoptosis [50]. We then examined whether the RR clones that failed to be chemosensitized by rituximab may be reversed to sensitivity. We found with wild-type cells that sensitizing agents that interfere with the signaling pathways modified by rituximab produced significant chemosensitization, comparable to the sensitization induced by rituximab. Treatment of RR clones with inhibitors of the NF-κB pathway (e.g., DHMEQ, bortezomib, Bay 11-7085), the ERK1/2 pathway (e.g., PD098059), the p38 MAPK pathway (e.g., SB203580), and Bcl-2 (e.g., 2MAM-A3) sensitized the resistant clones to various chemotherapeutic drugs (e.g., paclitaxel, vincristine, VP-16 [etoposide], CDDP, and doxorubicin). These findings revealed that resistance of NHL cells to rituximab, alone or in combination with cytotoxic therapeutics, may be amenable to treatment by combining sensitizing agents with low doses of conventional chemotherapy or immunotherapy. In addition, these findings identified several potential new targets for the generation of

a new class of inhibitors against these targets for intervention to reverse resistance.

5   Concluding Remarks

The current therapeutic use of rituximab, alone or in combination with chemotherapy, in the treatment of NHL, has made a significant impact on previously untreated and relapsed or refractory patients. When combined with chemotherapy, it showed a synergy in vivo and is now standard treatment for many B-cell lymphomas. However, the challenge of further improving the response rate and prolonging overall survival remains, along with the need to effectively treat unresponsive patients. One approach to these challenges is the development of new therapeutics, based on the underlying biologic and molecular mechanisms responsible for rituximab responsiveness and unresponsiveness. Rituximab has shown success in the treatment of several types of lymphoma, but the in vivo mechanisms of action are not fully understood. We propose several mechanisms of potential clinical significance. It is clear that rituximab interacts with CD20, leading to cell surface redistribution, modifying several interacting proteins and kinases in lipid rafts and triggering cell signaling. Our findings identified several intracellular survival signaling pathways modified by rituximab, leading to inhibition of various kinases, transcription factors, and anti-apoptotic gene products, all of which participate in the chemosensitizing and immunosensitizing effects induced by rituximab, as well as in the establishment of synergistic cytotoxicity. We have also discovered that rituximab sensitizes cells to both FasL- and TRAIL-induced apoptosis, a mechanism that may play a significant role in vivo (Fig. 3.1). In model systems studied with rituximab-resistant clones, the mechanism of rituximab resistance was shown to be due to the failure of rituximab to signal the resistant cells, as well as the development of hyperactivated survival anti-apoptotic pathways in the cells. The resistant clones, however, could be chemosensitized by pharmacologic inhibitors and reverse resistance. These studies have provided new insights into the identification at the molecular level of potential biomarkers of prognostic significance that may a priori identify responsive versus unresponsive patients.

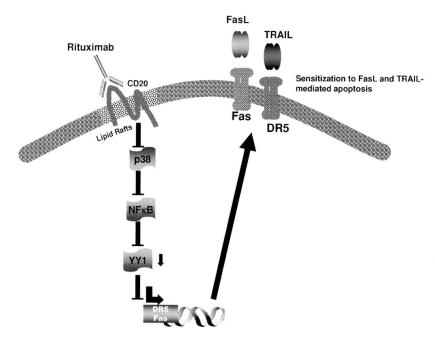

FIGURE 3.1. This figure schematically represents rituximab-mediated immunosensitization of B-NHL to both FasL-and TRAIL-induced apoptosis. Via rituximab-mediated inhibition of the Nf-kB and YY1 transcriptional activities, Fas and DR5 are upregulated and the cells are sensitized to FasL and TRAIL-induced apoptosis

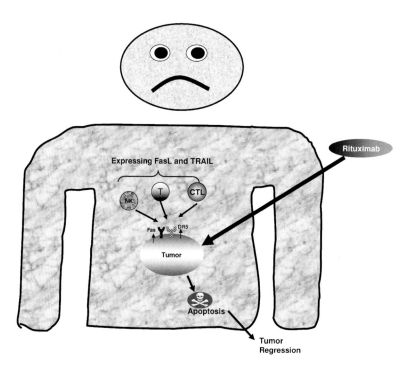

FIGURE 3.2. This figure schematically represents a cancer patient with B-NHL who is treated with rituximab. This treatment upregulates the death receptors Fas and DR5 on the tumor cells, and the cells become sensitive to host effector cells expressing FasL and TRAIL

They also identify new therapeutic approaches to treat unresponsive patients through the administration of targeted sensitizing agents that modify resistant factors to lower the resistance threshold and allow successful treatment in combination with low doses of conventional chemotherapy or immunotherapy. In addition, based on rituximab-induced immuno-sensitization, we propose that treatment with rituximab and immunomodulating agents that upregulate the expression of death ligands on host effector cells may participate in tumor rejections (Fig. 3.2).

6 Summary

Rituximab (chimeric anti-CD20 mAb) has been used successfully in the treatment of B- non Hodgkin's lymphoma (NHL) as single agent or in combination with chemotherapeutic drugs. The combination treatment improved significantly the response rate when compared with treatment with single agent. The *in vivo* mechanism by which rituximab exerts its effects has been proposed to be mediated by ADCC, CDC and apoptosis. However, the intracellular molecular signaling mediated by rituximab-CD20 interaction and responsible for the synergy achieved by combination treatment with drugs has not been examined. Further, the mechanism underlying the development of resistance to rituximab treatment is not clear. Our laboratory has systemically investigated rituximab-mediated cell signaling using B-NHL cell lines as models. Our studies demonstrated that rituximab inhibits several survival pathways such as Raf-1/MEK1/2/ERK1/2, NF-κB, p38 MAPK and Akt, all of which resulted in inhibition of pivotal anti-apoptotic gene products (Bcl-2 and Bcl-xL) and led to chemosensitization of drug-resistance tumor cells. Rituximab-induced inhibition of the ERK 1/2 and NF-κB pathways was concomitant with the induction of Raf-1 Kinase Inhibitor Protein (RKIP), which negatively regulates the activity of those pathways. In addition, since the NF-κB pathway negatively regulates tumor cell sensitivity to TNF family of death ligands (TNF-α, Fas-L, TRAIL), we examined the effect of rituximab treatment on immunosensitization of B-NHL to death ligand apoptosis. Our findings revealed for the first time that rituximab sensitizes drug/FasL/TRAIL-resistant B-NHL cells to FasL-

and TRAIL-mediated apoptosis. Sensitization by rituximab was the result of rituximab-mediated inhibition of NF-κB activity and inhibition of the transcription repressor Yin Yang 1 (YY1). YY1 was shown to negatively regulate the expression of Fas and TRAIL death receptor DR5. Various lines of evidence implicated the direct relationship between YY1 in the negative regulation of Fas and DR5 with resistance. Hence, inhibition of YY1 results in upregulation of Fas and DR5 and sensitizes the cells to Fas-L and TRAIL-induced apoptosis, respectively. The potential mechanism for the development of rituximab resistance was also investigated by generating a model of *in vitro* rituximab-resistant B-NHL clones that served as model for examination. The clones exhibited hyperactivated cell survival pathways and overexpression of anti-apoptotic gene products which could not be chemo-immunosensitized by rituximab. Pharmacological inhibitors of the survival signaling pathways, however, reverse drug/immune resistance in both wild-type cells and resistant clones. Overall, the present findings identify several novel intracellular pathways modified by rituximab that participate in the sensitization of NHL cells to both chemotherapy and immunotherapy. In addition, several therapeutic targets have been identified whose modification by pharmacological inhibitors reverse resistance. These targets are of significant clinical relevance and can be used for prognosis, diagnosis and development of novel treatment strategies for patients with NHL.

Acknowledgments: The author acknowledges the Ann C. Rosenfield Fund under the direction of David Leveton, the Fogarty International Center Fellowship D3TW00132, and the Jonnson Comprehensive Cancer Center. The assistance of Maggie Yang, Alina Katsman, and Tania Golkar is acknowledged in the preparation of this manuscript.

References

1. Ngan BY, Picker LJ, Medeiros LJ, et al. Immunophenotypic diagnosis of non-Hodgkin's lymphoma in paraffin sections. Co-expression of L60 (Leu-22) and L26 antigens correlates with malignant histologic findings. Am J Clin Pathol 1989, 91: 579–583.

2. Fisher RI, Gaynor ER, Dahlberg S, et al. Comparison of a standard regimen (CHOP) with three intensive chemotherapy regimens for advanced non-Hodgkin's lymphoma. N Engl J Med 1993, 328:1002–1006.

3. Li Z, David G, Hung KW, et al. Cdk5/p35 phosphorylates mSds3 and regulates mSds3-mediated repression of transcription. J Biol Chem 2004, 279:54438–54444.

4. Cragg MS, Walshe CA, Ivanov AO, et al. The biology of CD20 and its potential as a target for mAb therapy. Curr Dir Autoimmun 2005, 8:140–174.

5. Hainsworth JD. Prolonging remission with rituximab maintenance therapy. Semin Oncol 2004, 31:17–21.

6. Coiffier B. First-line treatment of follicular lymphoma in the era of monoclonal antibodies. Clin Adv Hematol Oncol 2005, 3:484–491, 505.

7. Jazirehi AR, Bonavida B. Cellular and molecular signal transduction pathways modulated by rituximab (Rituxan, anti-CD20 mAb) in non-Hodgkin's lymphoma: implications in chemosensitization and therapeutic intervention. Oncogene 2005, 24:2121–2143.

8. Reff ME, Carner K, Chambers KS, et al. Depletion of B cells in vivo by a chimeric mouse human monoclonal antibody to CD20. Blood 1994, 83:435–445.

9. Kennedy AD, Solga MD, Schuman TA, et al. An anti-C3b(i) mAb enhances complement activation, C3b(i) deposition, and killing of CD20+ cells by rituximab. Blood 2003, 101:1071–1079.

10. Shan D, Ledbetter JA, Press OW. Signaling events involved in anti-CD20-induced apoptosis of malignant human B cells. Cancer Immunol Immunother 2000, 48:673–683.

11. Demidem A, Salahuddin Z, Lam T, et al. Sensitization of AIDS related non-Hodgkin's B lymphoma cell lines to cytotoxic drugs/toxins by interferon-γ. Int J Oncol 1996, 8:461–468.

12. Deans JP, Li H, Polyak MJ. CD20-mediated apoptosis: signalling through lipid rafts. Immunology 2002, 107:176–182.

13. Deans JP, Robbins SM, Polyak MJ, et al. Rapid redistribution of CD20 to a low density detergent-insoluble membrane compartment. J Biol Chem 1998, 273:344–348.

14. Semac I, Palomba C, Kulangara K, et al. Anti-CD20 therapeutic antibody rituximab modifies the functional organization of rafts/microdomains of B lymphoma cells. Cancer Res 2003, 63:534–540.

15. Jazirehi AR, Vega MI, Chatterjee D, Goodglick L, Bonavida B. Inhibition of the Raf-MEK1/2-ERK1/2 signaling pathway, Bcl-xL downregulation, and chemosensitization of non-Hodgkin's lymphoma B cells by Rituximab. Cancer Res 2004, 64: 7117–7126.

16. Bezombes C, Grazide S, Garret C, et al. Rituximab antiproliferative effect in B-lymphoma cells is associated with acid-sphingomyelinase activation in raft microdomains. Blood 2004, 104:1166–1173.

17. Borsellino N, Belldegrun A, Bonavida B. Endogenous interleukin 6 is a resistance factor for cis-diamminedichloroplatinum and etoposide-mediated cytotoxicity of human prostate carcinoma cell lines. Cancer Res 1995, 55:4633–4639.

18. Mizutani Y, Bonavida B, Nio Y, et al. Overcoming TNF-alpha and drug resistance of human renal cell carcinoma cells by treatment with pentoxifylline in combination with TNF-alpha or drugs: the role of TNF-alpha mRNA downregulation in tumor cell sensitization. J Urol 1994, 151:1697–1702.

19. Su L. Distinct mechanisms of STAT phosphorylation via the interferon-alpha/beta receptor. J Biol Chem 2000, 275:12661–12666.

20. Alas S, Ng CP, Bonavida B. Rituximab modifies the cisplatin-mitochondrial signaling pathway, resulting in apoptosis in cisplatin-resistant non-Hodgkin's lymphoma. Clin Cancer Res 2002, 8:836–845.

21. Blay JY, Burdin N, Rousset F, et al. Serum interleukin-10 in non-Hodgkin's lymphoma: a prognostic factor. Blood 1993, 82:2169–2174.

22. Jazirehi AR, Gan XH, De Vos S, et al. Rituximab (anti-CD20) selectively modifies Bcl-xL and apoptosis protease activating factor-1 (Apaf-1) expression and sensitizes human non-Hodgkin's lymphoma B cell lines to paclitaxel-induced apoptosis. Mol Cancer Ther 2003, 2:1183–1193.

23. Xerri L, Parc P, Brousset P, et al. Predominant expression of the long isoform of Bcl-x (Bcl-xL) in human lymphomas. Br J Haematol 1996, 92:900–906.

24. Yeung K, Seitz T, Li S, et al. Suppression of Raf-1 kinase activity and MAP kinase signalling by RKIP. Nature 1999, 401:173–177.

25. Yeung K, Janosch P, McFerran B, et al. Mechanism of suppression of the Raf/MEK/extracellular signal-regulated kinase pathway by the raf kinase inhibitor protein. Mol Cell Biol 2000, 20:3079–3085.

26. Ghosh S, Karin M. Missing pieces in the NF-kappaB puzzle. Cell 2002, 109(Suppl):S81–96.

27. Dixit V, Mak TW. NF-kappaB signaling. Many roads lead to madrid. Cell 2002, 111:615–619.

28. Yeung KC, Rose DW, Dhillon AS, et al. Raf kinase inhibitor protein interacts with NF-kappaB-inducing kinase and TAK1 and inhibits NF-kappaB activation. Mol Cell Biol 2001, 21:7207–7217.

29. Vivanco I, Sawyers CL. The phosphatidylinositol 3-Kinase AKT pathway in human cancer. Nat Rev Cancer 2002, 2:489–501.

30. Suzuki E, Umezawa K, Bonavida B. Rituximab inhibits the constitutively activated PI3K-Akt pathway in B-NHL cell lines: involvement in chemosensitization to drug-induced apoptosis. Oncogene 2007, 26:6184–6193.

31. Garban HJ, Bonavida B. Nitric oxide inhibits the transcription repressor Yin-Yang 1 binding activity at the silencer region of the Fas promoter: a pivotal role for nitric oxide in the upregulation of Fas gene expression in human tumor cells. J Immunol 2001, 167:75–81.

32. Hongo F, Garban H, Huerta-Yepez S, et al. Inhibition of the transcription factor Yin Yang 1 activity by S-nitrosation. Biochem Biophys Res Commun 2005, 336:692–701.

33. Huerta-Yepez S, Vega M, Jazirehi A, et al. Nitric oxide sensitizes prostate carcinoma cell lines to TRAIL-mediated apoptosis via inactivation of NF-kappa B and inhibition of Bcl-xl expression. Oncogene 2004, 23:4993–5003.

34. Barnhart BC, Alappat EC, Peter ME. The CD95 type I/type II model. Semin Immunol 2003, 15:185–193.

35. Vega MI, Jazirehi AR, Huerta-Yepez S, et al. Rituximab-induced inhibition of YY1 and Bcl-xL expression in Ramos non-Hodgkin's lymphoma cell line via inhibition of NF-kappa B activity: role of YY1 and Bcl-xL in Fas resistance and chemoresistance, respectively. J Immunol 2005, 175:2174–2183.

36. Ashkenazi A, Pai RC, Fong S, et al. Safety and antitumor activity of recombinant soluble Apo2 ligand. J Clin Invest 1999, 104:155–162.

37. Sheridan JP, Marsters SA, Pitti RM, et al. Control of TRAIL-induced apoptosis by a family of signaling and decoy receptors. Science 1997, 277:818–821.

38. Shankar S, Srivastava RK. Enhancement of therapeutic potential of TRAIL by cancer chemotherapy and irradiation: mechanisms and clinical implications. Drug Resist Updat 2004, 7:139–156.

39. Zhang L, Fang B. Mechanisms of resistance to TRAIL-induced apoptosis in cancer. Cancer Gene Ther 2005, 12:228–237.

40. Jazirehi AR, Ng CP, Gan XH, Schiller G, Bonavida B. Adriamycin sensitizes the Adriamycin-resistant 8226/Dox40 human multiple myeloma cells to Apo2L/tumor necrosis factor-related apoptosis-inducing ligand-mediated (TRAIL) apoptosis. Clin Cancer Res 2001, 7:3874–3883.

41. Sheikh MS, Burns TF, Huang Y, et al. p53-dependent and -independent regulation of the death receptor KILLER/DR5 gene expression in response to genotoxic stress and tumor necrosis factor alpha. Cancer Res 1998, 58:1593–1598.

42. Nagane M, Pan G, Weddle JJ, et al. Increased death receptor 5 expression by chemotherapeutic agents in human gliomas causes synergistic cytotoxicity with tumor necrosis factor-related apoptosis-inducing ligand in vitro and in vivo. Cancer Res 2000, 60:847–853.

43. Shankar S, Chen X, Srivastava RK. Effects of sequential treatments with chemotherapeutic drugs followed by TRAIL on prostate cancer in vitro and in vivo. Prostate 2005, 62:165–186.

44. Yoshida T, Maeda A, Tani N, et al. Promoter structure and transcription initiation sites of the human death receptor 5/TRAIL-R2 gene. FEBS Lett 2001, 507:381–385.

45. Baritaki S, Huerta-Yepez S, Sakai T, et al. Chemotherapeutic drugs sensitize cancer cells to TRAIL-mediated apoptosis: upregulation of DR5 and inhibition of Yin Yang 1. Mol Cancer Ther 2007, 6:1387–1399.

46. Vega MI, Huerta-Yepez S, Jazirehi AR, et al. Rituximab (chimeric anti-CD20) sensitizes B-NHL cell lines to Fas-induced apoptosis. Oncogene 2005, 24:8114–8127.

47. Kennedy AD, Beum PV, Solga MD, et al. Rituximab infusion promotes rapid complement depletion and acute CD20 loss in chronic lymphocytic leukemia. J Immunol 2004, 172:3280–3288.

48. Haidar JH, Shamseddine A, Salem Z, et al. Loss of CD20 expression in relapsed lymphomas after rituximab therapy. Eur J Haematol 2003, 70:330–332.

49. Manshouri T, Do KA, Wang X, et al. Circulating CD20 is detectable in the plasma of patients with chronic lymphocytic leukemia and is of prognostic significance. Blood 2003, 101:2507–2513.

50. Jazirehi AR, Vega MI, Bonavida B. Development of rituximab-resistant lymphoma clones with altered cell signaling and cross-resistance to chemotherapy. Cancer Res 2007, 67:1270–1281.

Chapter 4
Agents that Regulate DR5 and Sensitivity to TRAIL

Tatsushi Yoshida and Toshiyuki Sakai

1 TRAIL-DR5 Pathway

Tumor necrosis factor–related apoptosis-inducing ligand (TRAIL) is a type II transmembrane protein and a member of the TNF gene superfamily, which was identified and cloned based on sequence homology to Fas and TNF [1, 2]. TRAIL interacts with two pro-apoptotic death receptors, DR5 and DR4 (also called TRAIL-R2 and TRAIL-R1), and two decoy receptors, DcR1 and DcR2 (also called TRAIL-R3 and TRAIL-R4) [3]. DR5 and DR4 contain an intracellular death domain. As shown in Fig. 4.1, interaction of TRAIL causes trimerization of the receptors, and the activated pro-apoptotic receptors bind to an adapter protein called the Fas-associated death domain (FADD) between death domains of receptors and FADD. FADD interacts with pro-caspase-8 or -10 and forms death-inducing signaling complex (DISC), which leads to the cleavage and activation of pro-caspase-8 or -10. In type I cells, the activated caspase-8/-10 triggers apoptosis via the activation of effector caspase-3 [4]; however, in type II cells, an amplification loop through the mitochondrial pathway is required for apoptosis. Activated caspase-8 cleaves Bid, and the truncated Bid (tBid) releases cytochrome C and Smac from mitochondria to cytosol via Bax and Bak. Bcl-2 and Bcl-xL are anti-apoptotic factors inhibiting cytochrome C release. The released cytochrome C forms a complex including Apaf-1 and caspase-9, and leads to the activation of caspase-9, which consequently activates caspase-3

and induces apoptosis. IAP, XIAP, and survivin block apoptosis by inhibiting the activities of caspase-3 and -9. On the other hand, Smac released from mitochondria inhibits the action of these members of the IAP family and acts as a pro-apoptotic protein. In contrast, DcR1 does not contain an intracellular death domain and DcR2 contains a truncated death domain. These two decoy receptors are unable to transmit death signaling and act as anti-apoptotic receptors by competing with DR5 and DR4 for TRAIL interaction.

TRAIL induces apoptosis selectively in cancer cells in vitro and in vivo, with little or no toxicity in normal cells [5–7]; therefore, TRAIL is one of the most promising new candidates for cancer therapeutics, although many types of cancer cells remain resistant to TRAIL [8]. Thus, to overcome resistance to TRAIL and to use TRAIL as a more powerful tool for cancer treatment, agents that can sensitize cancer cells to TRAIL play an important role. As one strategy to resolve this issue, many DR5 inducers have been discovered.

1.1 Agents That Regulate DR5 and Sensitivity to TRAIL

Many agents have been reported to upregulate DR5 and sensitize malignant tumor cells to TRAIL-induced apoptosis. Here we classified the agents as shown in Fig. 4.2 and described as follows: (1) conventional anti-tumor agents used clinically; (2) new-age anti-tumor agents recently approved as

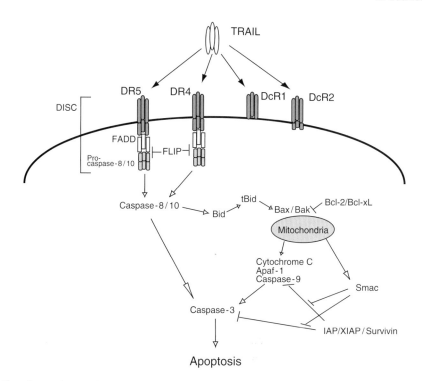

FIGURE 4.1. Signaling pathway of TRAIL-induced apoptosis. TRAIL interacts with pro-apoptotic receptors (DR5 and DR4) and decoy receptors (DcR1 and DcR2). Interaction of TRAIL with DR5 or DR4 forms a complex, called DISC, containing FADD and pro-caspase-8/-10, and causes activation of caspase-8/-10. In type I cells, activated caspase-8/-10 cleaves and activates caspase-3, leading to apoptosis. In type II cells, activated caspase-8/-10 changes Bid to tBid. The tBid releases cytochrome C and Smac from mitochondria via Bax/Bak. Cytochrome C forms a complex with Apaf-1 and caspase-9, leading to caspase-9 activation. Activated caspase-9 cleaves and activates caspase-3, leading to apoptosis. FLIP inhibits apoptosis by prevention of DISC formation. Bcl-2/BclxL inhibits Bax/Bak activity and blocks apoptosis. IAP, XIAP, and survivin inhibit caspase-3 and -9 activities and block apoptosis. Smac inhibits the actions of IAP, XIAP and survivin and advances apoptosis

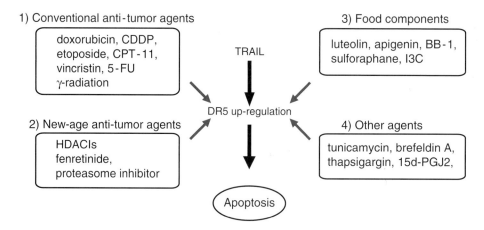

FIGURE 4.2. Agents that upregulate DR5 and sensitize TRAIL-induced apoptosis. DR5 is a receptor of TRAIL. Interaction of TRAIL with DR5 causes apoptosis. Many agents that upregulate DR5 have been reported. In this chapter, we categorized these agents into four classes: (1) conventional anti-tumor agents, (2) new-age anti-tumor agents, (3) food components, and (4) other agents

anti-tumor agents or undergoing clinical studies; (3) food components reported to prevent tumor growth in vitro or in vivo; and (4) other agents used in basic research and promising as anti-tumor agents in the future.

1.1.1 Conventional Anti-tumor Agents

Wu et al. explored doxorubicin-induced transcripts and found that killer/DR5 was upregulated [9]. DR5 was upregulated by doxorubicin through the tumor-suppressor p53 gene-dependent pathway. Other DNA-damaging agents such as etoposide, 5-fluorouracil, or ionizing radiation also induced DR5 expression. Nagane et al. hypothesized that conventional DNA-damaging chemotherapy might enhance the cytotoxicity of TRAIL, since DNA damage upregulates DR5 [10]. *Cis*-diamminedichloroplatinum (II) (CDDP) and etoposide, which are commonly used in human glioma treatment, upregulated DR5, and sensitized TRAIL-induced cytotoxicity in various types of glioma cells. Furthermore, in vivo administration of TRAIL with CDDP suppressed tumor formation and increased the survival rate in nude mice bearing human glioma xenografts.

Baitaki et al. have shown that chemotherapeutic agents such as CDDP, etoposide, doxorubicin and vincristine upregulate DR5 expression and sensitize TRAIL-induced apoptosis in prostate cancer and melanoma cells in a p53-independent manner [11]. These agents inhibited transcription factor YY1 expression and its DNA-binding activity. Furthermore, YY1 siRNA upregulates DR5 and sensitizes TRAIL-induced apoptosis, suggesting that these chemotherapeutic agents sensitize tumor cells to TRAIL concurrently with DR5 upregulation through the inhibition of YY1 expression.

Naka et al. have demonstrated that CPT-11 (irinotecan hydrochloride) in combination with TRAIL enhanced the anti-tumor effect in SCID mice engrafted with colon tumors taken from fresh surgical specimens [12]. Tumors treated with CPT-11 showed increased membrane expression of DR5, suggesting that CPT-11 sensitized TRAIL-induced apoptosis by upregulation of DR5.

The nontagged-recombinant form of TRAIL does not induce apoptosis in most normal human cells; however, the histidine-tagged form of TRAIL exhibits cytotoxicity to normal human hepatocytes [13]. These results raise a question as to whether chemotherapeutic agent-mediated sensitization of TRAIL causes cytotoxicity against normal hepatocytes. Ganten et al. studied this issue and found that TRAIL in combination with 5-FU, irinotecan, or etoposide seems to be safe for primary human hepatocytes, although the combination of TRAIL with high-dose CDDP exhibits cytotoxicity [14].

1.1.2 New-Age Anti-tumor Agents

Histone deacetylase inhibitors (HDACIs) are novel anti-tumor agents that induce cell cycle arrest [15] and apoptosis in malignant tumor cells [16]. HDACIs upregulate the transcription of certain genes via the inhibition of HDAC and changes in chromatin structure [16]. Derivatives of butyric acid, suberoylanilide hydroxamic acid (SAHA), torichostatin A (TSA), and valproic acid are known to be HDACIs. SAHA, also called vorinostat or Zolinza, was approved by the US Food and Drug Administration (FDA) for the treatment of cutaneous T-cell lymphoma patients in 2006. Nakata et al. found that SAHA, TSA, and sodium butyrate are able to upregulate DR5 through transcription in T-cell leukemia Jurkat cells [17]. HDACIs sensitized TRAIL-induced apoptosis and the combination strongly cleaved caspase-3, -8, -9, and -10 and Bid in a dose-dependent manner. In contrast, HDACIs neither upregulate DR5 nor sensitize TRAIL-induced apoptosis in normal human lymphocytes of the peripheral blood mononuclear cell (PBMC) fraction, suggesting that a combination of HDACIs and TRAIL might be useful for cancer therapeutics. It has been reported that HDACIs also sensitizes TRAIL-induced apoptosis accompanied with DR5 upregulation in renal [18], bladder [19], breast [20, 21] and colon [22] cancer, and leukemia [23].

Fenretinide (N-[4-Hydroxyphenyl]retinamide, 4HPR) is a semisynthetic retinoid that has been examined in clinical trials, including a phase III study, as a cancer chemopreventive agent in a variety of malignant tumors [24]. Fenretinide induces apoptosis through a mitochondrial pathway in cell lines derived from a variety of malignant tumors [25]. Kouhara et al. examined the effect of fenretinide on the death-receptor pathway and found DR5 upregulation by fenretinide [26]. Kouhara et al. hypothesized that fenretinide can sensitize TRAIL-induced apoptosis due to DR5 upregulation. Fenretinide enhanced TRAIL-induced apoptosis in four kinds of colon cancer cells. Fenretinide also upregulated DR5 at the transcriptional level through a transcription factor, CHOP.

Bortezomib, also known as PS-341 or Velcade, is a proteasome inhibitor and has recently approved by the FDA for the treatment of multiple myeloma [27]. Johnson et al. found that bortezomib overcomes the resistance of prostate, colon and bladder cancer cells to TRAIL-induced apoptosis and increases the protein level of DR5 associated with ubiquitination [28]. The combination of bortezomib and TRAIL efficiently induces cell death in Bcl-xL-overexpressed prostate cancer and Bax-deficient colon cancer cells. He et al. have also shown that another proteasome inhibitor, MG132, upregulates DR5 and effectively cooperates with TRAIL to induce apoptosis in both Bax-proficient and -deficient colon cancer cells [29]. Interestingly, MG132 increases DR5 mRNA as well as protein. Yoshida et al. examined the underlying mechanism of DR5 mRNA induction by MG132 and found that a transcription factor, CHOP, is responsible for DR5 induction via direct binding to the DR5 gene promoter [30].

1.1.3   Food Components

Luteolin is a naturally occurring flavone contained in vegetables such as broccoli, celery, perilla leaf and seeds, and carrots. Luteolin possesses a potent preventive effect against skin papilloma in mice by topical application [31]. Horinaka et al. studied the mechanism of apoptosis induced by luteolin in cervical cancer HeLa cells and found that it partly depends on DR5 upregulation [32]. Horinaka et al. hypothesized that luteolin can sensitize TRAIL-induced apoptosis because luteolin increased the quantity of DR5 protein [33]. Indeed, luteolin sensitized HeLa cells to TRAIL-induced apoptosis. Moreover, the sensitized effect of luteolin on TRAIL-induced apoptosis was reduced when DR5 siRNA blocked the DR5 expression induced by luteolin. Interestingly, luteolin did not upregulate DR5 expression nor sensitize TRAIL-induced apoptosis in PBMC derived from normal human cells, suggesting that the combination of luteolin and TRAIL may kill cancer cells but not normal cells.

Apigenin is also a kind of flavone contained in Chinese cabbage, bell pepper, garlic, celery, and guava. The typical administration of apigenin

reduced tumor formation in the skin induced by chemical carcinogens or UV in mice [34, 35]. Moreover, IP administration of apigenin inhibits melanoma growth and metastasis in vivo [36]; thus, apigenin has anti-tumor potential. Horinaka et al. found that apigenin increased DR5 protein in Jurkat T-cell leukemia cells [37]; however, apigenin only slightly increased DR5 mRNA and did not increase DR5 promoter activity. Since apigenin is reported to act as a proteosome inhibitor, it suggests that apigenin stabilized DR5 protein by proteasome inhibition. Apigenin also sensitized Jurkat cells to TRAIL-induced apoptosis. Apigenin induced DR5 protein and enhanced TRAIL sensitivity in prostate cancer DU145 and colon cancer DLD-1 cells as well as Jurkat cells but not in normal human PBMC, suggesting that the combined treatment of apigenin and TRAIL is effective against various types of malignancies.

Hasegawa et al. carried out screening to find natural products that can overcome TRAIL resistance in adult T-cell leukemia/lymphoma (ATLL) [38]. Dihydroflavonol extracted from *Blumea balsamifera* (BB-1) exhibited striking synergism with TRAIL in ATLL. BB-1 increased DR5 expression through transcription and DR5 siRNA inhibited the synergism between BB-1 and TRAIL, indicating that the combined effect is due to DR5 upregulation.

Sulforaphane is a naturally occurring isothiocyanate produced from vegetables such as broccoli, and effectively prevents chemically induced carcinogenesis in rats [39, 40]. Furthermore, oral administration of sulforaphane drastically inhibits the growth of human prostate cancer xenografts [41], and IV injection inhibits the growth of human breast cancer xenografts in mice [42]. Matsui et al. studied the effect of sulforaphane on human osteosarcoma Saos2 or MG63 cells [43], in which it very weakly induced apoptosis. TRAIL also weakly induced apoptosis in both cells; however, sulforaphane enhanced TRAIL-induced apoptosis and a combination of sulforaphane and TRAIL strongly induced apoptosis. They found that sulforaphane induced DR5 expression and that the sensitizing effect of sulforaphane on TRAIL-induced apoptosis partly depends on DR5 upregulation by sulforaphane using DR5 siRNA. Kim et al. also found that sulforaphane upregulates DR5 and sensitizes TRAIL-induced apoptosis in hepatoma cells [44]. They investigated the mechanism by which sulforaphane upregulates DR5, showing upregulation at a transcriptional level through the Sp1-binding site of DR5 promoter, and transactivation is mediated by the generation of reactive oxygen species.

Indole-3-carbiol (I3C) is a common phytochemical present in vegetables including cabbage, broccoli, cauliflower and kale. Jeon et al. reported that I3C induces DR5 at both transcriptional and translational levels, and augments TRAIL-induced apoptosis in prostate cancer cells [45].

1.1.4 Other Agents

Tunicamycin, a naturally occurring antibiotic, blocks the first step of the lipid-linked saccharide pathway and inhibits the N-linked oligosaccharide formation of glycoproteins [46]. Shiraishi et al. searched for a new strategy to overcome the resistance of hormone-refractory prostate cancer cells and found that tunicamycin acts as a potent enhancer of TRAIL-induced apoptosis in prostate cancer cells [47]. As a single agent, tunicamycin or TRAIL only moderately induced apoptosis in prostate cancer PC-3 cells, although tunicamycin strongly enhanced TRAIL-induced apoptosis. Shiraishi et al. investigated the expression of death receptor–related genes following tunicamycin treatment to elucidate the mechanism of tunicamycin-mediated enhancement of TRAIL-induced apoptosis, and found that tunicamycin specifically upregulated DR5 mRNA. Shiraishi et al. searched for the factor responsible for DR5 upregulation by tunicamycin and found that a transcription factor, CHOP, upregulates DR5 through a CHOP-binding site on the DR5 promoter. Tunicamycin also increased CHOP protein in prostate cancer cells. It has been reported that tunicamycin induces endoplasmic reticulum stress (ER stress) following the inhibition of N-linked glycosylation. Other ER stress inducers, thapsigargin, and brefeldin A, also upregulate DR5 and sensitize TRAIL-induced apoptosis via CHOP [48–50], suggesting that tunicamycin upregulates DR5 and acts as an enhancer of TRAIL owing to the induction of ER stress.

Prostaglandins are naturally occurring cyclic 20-carbon fatty acids synthesized from arachidonate released from membrane phospholipids. 15-Deoxy-$\Delta^{12,14}$-prostaglandin J_2 (15d-PGJ$_2$) inhibits the cell proliferation of malignant tumor cells such as breast [51], colon [52], and gastric [53] cancer

cells, and neuroblastoma cells [54]. Nakata et al. examined the effect of 15d-PGJ$_2$ on the expression of death receptors and found that 15d-PGJ$_2$ upregulates DR5 in prostate cancer PC-3 and T-cell leukemia Jurkat cells [55]. 15d-PGJ$_2$ is a natural endogenous ligand of a nuclear receptor, peroxisome proliferator-activated receptor γ (PPARγ); however, PPARγ is not responsible for DR5 upregulation by 15d-PGJ$_2$. The stability of DR5 mRNA was increased by 15d-PGJ$_2$. Indeed, 15d-PGJ$_2$ sensitized PC-3 and Jurkat cells to TRAIL-induced apoptosis.

1.2 Other Systems Sensitizing TRAIL

DR5 upregulation is one of the mechanisms sensitizing tumor cells to TRAIL-induced apoptosis. The regulation of factors related to the pathway on TRAIL-mediated apoptosis, as shown in Fig. 4.1, also sensitizes TRAIL action. PPARγ modulators induced the ubiquitination and proteasome-dependent degradation of FLIP, and sensitized tumor but not normal cells to apoptosis induction by TRAIL [56]. Triterpenoids downregulate FLIP and sensitize acute myelogenous leukemia to TRAIL [57]. A nitric oxide donor, DETANONOate, sensitizes TRAIL-induced apoptosis via NF-kB inhibition and downregulation of the NF-kB downstream gene, Bcl-xL, in prostate cancer cells [58]. Downregulation of XIAP and Survivin also sensitizes cancer cells to TRAIL [59–61]. Furthermore, Smac peptide and Smac mimic small molecules sensitize various tumor cells to TRAIL [62, 63].

1.3 Future Direction

Recently, agonistic antibodies against DR5, which can substitute for the TRAIL function, have been developed [64–66]. Conventional anti-tumor agents that can upregulate DR5 also sensitize tumor cells to DR5-agonistic antibodies as well as recombinant TRAIL.

TRAIL is an endogenous cytokine secreted in response to signals such as interferon (IFN), retinoic acid (RA), and bacillus Calmette-Guérin (BCG) [67–70]. Tecchio et al. have demonstrated that INFα stimulates the expression of TRAIL, releases a soluble form of TRAIL in both human neutrophils and monocytes, and enhances apoptosis in human leukemic cells [67]. Clarke et al. reported

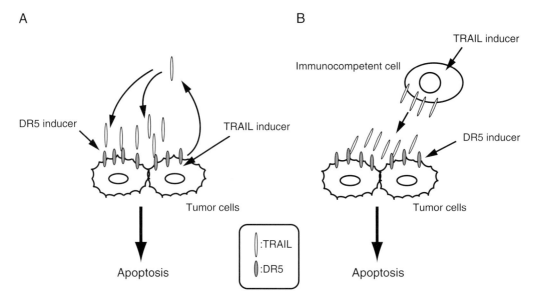

FIGURE 4.3. Future direction of the study of DR5 inducer and TRAIL. TRAIL, as well as DR5, is induced by various agents, since TRAIL is an endogenous cytokine. We demonstrate two models here. **A.** Agents that can induce TRAIL act on tumor cells and the stimulated tumor cells secrete a soluble form of TRAIL. Secreted TRAIL interacts with DR5 induced by other agents in an autocrine or paracrine manner, and consequently induces apoptosis in tumor cells. **B.** Agents that can induce TRAIL act on immunocompetent cells that release a soluble form of TRAIL. Secreted TRAIL interacts with DR5 induced by other agents on tumor cells, and causes apoptosis in tumor cells

that IFN-γ and 9-cis-RA treatment induces TRAIL expression in breast cancer cells and kills other target cancer cells but not normal cells via paracrine action [68]. Ludwig et al. found that BCG, a potent immunostimulant used in bladder cancer treatment, induces TRAIL expression [69]. Moreover, Kemp et al. have shown that BCG induces the release of soluble TRAIL from neutrophils, and that both the BCG-stimulated neutrophils and the culture medium kill RT-4 bladder target cells [70].

Therefore, instead of exogenous treatment of recombinant TRAIL or antibodies against DR5, low-molecular-weight compounds that can induce TRAIL expression or secretion may have a synergistic effect with DR5 inducers for tumor treatment. As shown in Fig. 4.3A, we demonstrate one model in which the agents directly affect tumor cells and induce TRAIL, and then secreted TRAIL acts on tumor cells in an autocrine or paracrine manner. TRAIL-inducing agents in combination with DR5 inducer may effectively augment apoptosis in tumor cells. In another model, shown in Fig. 4.3B, the agents affect immunocompetent cells and induce TRAIL secretion from these cells.

Secreted TRAIL acts on tumor cells in which DR5 is upregulated by other agents, and the combination of TRAIL and DR5 inducers may effectively kill tumor cells.

In conclusion, we believe that this new cancer therapeutic strategy using combinations of different agents resulting in synergistic effects is very promising. We termed the strategy combination-oriented molecular-targeting cancer therapy or prevention.

References

1. Wiley SR, Schooley K, Smolak PJ, et al. Identification and characterization of a new member of the TNF family that induces apoptosis. Immunity 1995, 3:673–682.
2. Pitti RM, Marsters SA, Ruppert S, et al. Induction of apoptosis by Apo-2 ligand, a new member of the tumor necrosis factor cytokine family. J Biol Chem 1996, 271:12687–12690.
3. Ashkenazi A. Targeting death and decoy receptors of the tumour-necrosis factor superfamily. Nat Rev Cancer 2002, 2:420–430.
4. Wang S, El-Deiry WS. TRAIL and apoptosis induction by TNF-family death receptors. Oncogene 2003, 22:8628–8633.

5. Walczak H, Miller RE, Ariail K, et al. Tumoricidal activity of tumor necrosis factor-related apoptosis-inducing ligand *in vivo*. Nat Med 1999, 5:157–163.

6. Ashkenazi A, Pai RC, Fong S, et al. Safety and anti-tumor activity of recombinant soluble Apo2 ligand. J Clin Invest 1999, 104:155–162.

7. Lawrence D, Shahrokh Z, Marsters S, et al. Differential hepatocyte toxicity of recombinant Apo2L/TRAIL versions. Nat Med 2001, 7:383–385.

8. Zhang L, Fang B. Mechanisms of resistance to TRAIL-induced apoptosis in cancer. Cancer Gene Ther 2005, 12:228–237.

9. Wu GS, Burns TF, McDonald ER 3rd, et al. KILLER/DR5 is a DNA damage-inducible p53-regulated death receptor gene. Nat Genet 1997, 17:141–143.

10. Nagane M, Pan G, Weddle JJ, et al. Increased death receptor 5 expression by chemotherapeutic agents in human gliomas causes synergistic cytotoxicity with tumor necrosis factor-related apoptosis-inducing ligand *in vitro* and *in vivo*. Cancer Res 2000, 60:847–853.

11. Baritaki S, Huerta-Yepez S, Sakai T, et al. Chemo-therapeutic drugs sensitize cancer cells to TRAIL-mediated apoptosis: upregulation of DR5 and inhibition of YY1. Mol Cancer Ther 2005, 175:2174–2183.

12. Naka T, Sugamura K, Hylander BL, et al. Effects of tumor necrosis factor-related apoptosis-inducing ligand alone and in combination with chemotherapeutic agents on patients' colon tumors grown in SCID mice. Cancer Res 2002, 62:5800–5806.

13. Jo M, Kim TH, Seol DW, et al. Apoptosis induced in normal human hepatocytes by tumor necrosis factor-related apoptosis-inducing ligand. Nat Med 2000, 6:564–567.

14. Ganten TM, Koschny R, Sykora J, et al. Preclinical differentiation between apparently safe and potentially hepatotoxic applications of TRAIL either alone or in combination with chemotherapeutic drugs. Clin Cancer Res 2006, 12:2640–2646.

15. Yoshida M, Beppu T. Reversible arrest of proliferation of rat 3Y1 fibroblasts in both the G1 and G2 phases by trichosanthin A. Exp Cell Res 1988, 177:122–131.

16. Marks PA, Richon VM, Rifkind RA. Histone deacetylase inhibitors: inducers of differentiation or apoptosis of transformed cells. J Natl Cancer Inst 2000, 92:1210–1216.

17. Nakata S, Yoshida T, Horinaka M, et al. Histone deacetylase inhibitors upregulate death receptor 5/TRAIL-R2, and sensitize apoptosis induced by TRAIL/APO2-L in human malignant tumor cells. Oncogene 2004, 23:6261–6271.

18. VanOosten RL, Earel JK Jr, Griffith TS. Enhancement of Ad5-TRAIL cytotoxicity against renal cell carcinoma with histone deacetylase inhibitors. Cancer Gene Ther 2006, 13:628–632.

19. Earel JK Jr, Van Oosten RL, Griffith TS. Histone deacetylase inhibitors modulate the sensitivity of tumor necrosis factor-related apoptosis-inducing lig-and-resistant bladder tumor cells. Cancer Res 2006, 66:499–507.

20. Singh TR, Shankar S, Srivastava RK. HDAC inhibitors enhance the apoptosis-inducing potential of TRAIL in breast carcinoma. Oncogene 2005, 24:4609–4623.

21. Butler LM, Liapis V, Bouralexis S, et al. The histone deacetylase inhibitor, suberoylanilide hydroxamic acid, overcomes resistance of human breast cancer cells to Apo2L/TRAIL. Int J Cancer 2006, 119: 944–954.

22. Kim YH, Park JW, Lee JY, et al. Sodium butyrate sensitizes TRAIL-mediated apoptosis by induction of transcription from the DR5 gene promoter through Sp1 sites in colon cancer cells. Carcinogenesis 2004, 25:1813–1820.

23. Guo F, Sigua C, Tao J, et al. Cotreatment with his-tone deacetylase inhibitor LAQ824 enhances Apo-2L/tumor necrosis factor-related apoptosis inducing ligand-induced death inducing signaling complex activity and apoptosis of human acute leukemia cells. Cancer Res 2004, 64:2580–2589.

24. Malone W, Perloff M, Crowell J, et al. Fenretinide: a prototype cancer prevention drug. Expert Opin Invest Drugs 2003, 12:1829–1842.

25. Wu JM, Di Pietrantonio AM, Hsieh TC. Mechanism of fenretinide (4-HPR)-induced cell death. Apoptosis 2001, 6:377–688.

26. Kouhara J, Yoshida T, Nakata S, et al. Fenretinide upregulates DR5/TRAIL-R2 expression via the induction of the transcription factor CHOP and combined treatment with fenretinide and TRAIL induces synergistic apoptosis in colon cancer cell lines. Int J Oncol 2008, 68:1180–1186.

27. Richardson PG, Barlogie B, Berenson J, et al. A phase 2 study of bortezomib in relapsed, refractory myeloma. N Engl J Med 2003, 348:2609–2617.

28. Johnson TR, Stone K, Nikrad M, et al. The protea-some inhibitor PS-341 overcomes TRAIL resistance in Bax and caspase 9-negative or Bcl-xL overex-pressing cells. Oncogene 2003, 22:4953–4963.

29. He Q, Huang Y, Sheikh MS. Proteasome inhibitor MG132 upregulates death receptor 5 and cooper-ates with Apo2L/TRAIL to induce apoptosis in Bax-proficient and -deficient cells. Oncogene 2004, 23:2554–2558.

30. Yoshida T, Shiraishi T, Nakata S, et al. Proteasome inhibitor MG132 induces death receptor 5 through CCAAT/enhancer-binding protein homologous protein. Cancer Res. 2005, 65:5662–5667.

31. Ueda H, Yamazaki C, Yamazaki M. Inhibitory effect of Perilla leaf extract and luteolin on mouse skin tumor promotion. Biol Pharm Bull 2003, 26: 560–563.

32. Horinaka M, Yoshida T, Shiraishi T, et al. Luteolin induces apoptosis via death receptor 5 upregulation in human malignant tumor cells. Oncogene 2005, 24:7180–7189.

33. Horinaka M, Yoshida T, Shiraishi T, et al. The combination of TRAIL and luteolin enhances apoptosis in human cervical cancer HeLa cells. Biochem Biophys Res Commun 2005, 333:833–838.

34. Wei H, Tye L, Bresnick E, et al. Inhibitory effect of apigenin, a plant flavonoid, on epidermal ornithine decarboxylase and skin tumor promotion in mice. Cancer Res 1990, 50:499–502.

35. Birt DF, Mitchell D, Gold B, et al. Inhibition of ultraviolet light induced skin carcinogenesis in SKH-1 mice by apigenin, a plant flavonoid. Anticancer Res 1997, 17:85–91.

36. Caltagirone S, Rossi C, Poggi A, et al. Flavonoids apigenin and quercetin inhibit melanoma growth and metastatic potential. Int J Cancer 2000, 87:595–600.

37. Horinaka M, Yoshida T, Shiraishi T, et al. The dietary flavonoid apigenin sensitizes malignant tumor cells to tumor necrosis factor-related apoptosis-inducing ligand. Mol Cancer Ther 2006, 5:945–951.

38. Hasegawa H, Yamada Y, Komiyama K, et al. Dihydroflavonol BB-1, an extract of natural plant Blumea balsamifera, abrogates TRAIL resistance in leukemia cells. Blood 2006, 107:679–688.

39. Zhang Y, Kensler TW, Cho CG, et al. Anticarcinogenic activities of sulforaphane and structurally related synthetic norbornyl isothiocyanates. Proc Natl Acad Sci U S A 1994, 91:3147–3150.

40. Chung FL, Conaway CC, Rao CV, et al. Chemoprevention of colonic aberrant crypt foci in Fischer rats by sulforaphane and phenethyl isothiocyanate. Carcinogenesis 2000, 21:2287–2291.

41. Singh AV, Xiao D, Lew KL, et al. Sulforaphane induces caspase-mediated apoptosis in cultured PC-3 human prostate cancer cells and retards growth of PC-3 xenografts in vivo. Carcinogenesis 2004, 25:83–90.

42. Jackson SJ, Singletary KW. Sulforaphane: a naturally occurring mammary carcinoma mitotic inhibitor, which disrupts tubulin polymerization. Carcinogenesis 2004, 25:219–227.

43. Matsui TA, Sowa Y, Yoshida T, et al. Sulforaphane enhances TRAIL-induced apoptosis through the induction of DR5 expression in human osteosarcoma cells. Carcinogenesis 2006, 27:1768–1777.

44. Kim H, Kim EH, Eom YW, et al. Sulforaphane sensitizes tumor necrosis factor-related apoptosis-inducing ligand (TRAIL)-resistant hepatoma cells to TRAIL-induced apoptosis through reactive oxygen species-mediated upregulation of DR5. Cancer Res 2006, 66:1740–1750.

45. Jeon KI, Rih JK, Kim HJ, et al. Pretreatment of indole-3-carbinol augments TRAIL-induced apoptosis in a prostate cancer cell line, LNCaP. FEBS Lett 2003, 544:246–251.

46. Elbein AD. Inhibitors of the biosynthesis and processing of N-linked oligosaccharide chains. Annu Rev Biochem 1987, 56:497–534.

47. Shiraishi T, Yoshida T, Nakata S, et al. Tunicamycin enhances tumor necrosis factor-related apoptosis-inducing ligand-induced apoptosis in human prostate cancer cells. Cancer Res. 2005, 65:6364–6370.

48. He Q, Lee DI, Rong R, et al. Endoplasmic reticulum calcium pool depletion-induced apoptosis is coupled with activation of the death receptor 5 pathway. Oncogene 2002, 21:2623–2633.

49. Yamaguchi H, Bhalla K, Wang HG. Bax plays a pivotal role in thapsigargin-induced apoptosis of human colon cancer HCT116 cells by controlling Smac/Diablo and Omi/HtrA2 release from mitochondria. Cancer Res 2003, 63:1483–1489.

50. Yamaguchi H, Wang HG. CHOP is involved in endoplasmic reticulum stress-induced apoptosis by enhancing DR5 expression in human carcinoma cells. J Biol Chem 2004, 279:45495–45502.

51. Clay CE, Namen AM, Atsumi G, et al. Influence of J series prostaglandins on apoptosis and tumorigenesis of breast cancer cells. Carcinogenesis 1999, 20:1905–1911.

52. Shimada T, Kojima K, Yoshiura K, et al. Characteristics of the peroxisome proliferator activated receptor γ (PPARγ) ligand induced apoptosis in colon cancer cells. Gut 2002, 50:658–664.

53. Liu JD, Lin SY, Ho YS, et al. Involvement of c-jun N-terminal kinase activation in 15-deoxy-$\Delta^{12,14}$-prostaglandin J_2-and prostaglandin A1-induced apoptosis in AGS gastric epithelial cells. Mol Carcinog 2003, 37:16–24.

54. Kim EJ, Park KS, Chung SY, et al. Peroxisome proliferator-activated receptor-γ activator 15-deoxy-$\Delta^{12,14}$-prostaglandin J_2 inhibits neuroblastoma cell growth through induction of apoptosis: association with extracellular signal-regulated kinase signal pathway. J Pharmacol Exp Ther 2003, 307:505–517.

55. Nakata S, Yoshida T, Shiraishi T, et al. 15-Deoxy-$\Delta^{12,14}$-prostaglandin J_2 induces death receptor 5 expression through mRNA stabilization independently of PPARγ and potentiates TRAIL-induced apoptosis. Mol Cancer Ther 2006, 5:1827–1835.

56. Kim Y, Suh N, Sporn M, et al. An inducible pathway for degradation of FLIP protein sensitizes tumor cells to TRAIL-induced apoptosis. J Biol Chem 2002, 277:22320–22329.

57. Suh WS, Kim YS, Schimmer AD, et al. Synthetic triterpenoids activate a pathway for apoptosis in AML cells involving downregulation of FLIP and sensitization to TRAIL. Leukemia 2003, 17: 2122–2129.

58. Huerta-Yepez S, Vega M, Jazirehi A, et al. Nitric oxide sensitizes prostate carcinoma cell lines to TRAIL-mediated apoptosis via inactivation of NF-kappa B and inhibition of Bcl-xl expression. Oncogene 2004, 23:4993–5003.

59. Lee TJ, Jung EM, Lee JT, et al. Mithramycin A sensitizes cancer cells to TRAIL-mediated apoptosis by downregulation of XIAP gene promoter through Sp1 sites. Mol Cancer Ther 2006, 5:2737–2746.

60. Kim EH, Kim SU, Choi KS. Rottlerin sensitizes glioma cells to TRAIL-induced apoptosis by inhibition of Cdc2 and the subsequent downregulation of survivin and XIAP. Oncogene 2005, 24:838–849.

61. Lu M, Kwan T, Yu C, et al. Peroxisome proliferator-activated receptor γ agonists promote TRAIL-induced apoptosis by reducing survivin levels via cyclin D3 repression and cell cycle arrest. J Biol Chem 2005, 280:6742–6751.

62. Li L, Thomas RM, Suzuki H, et al. A small molecule Smac mimic potentiates TRAIL- and TNFα-mediated cell death. Science 2004, 305:1471–1474.

63. Fulda S, Wick W, Weller M, et al. Smac agonists sensitize for Apo2L/TRAIL- or anticancer drug-induced apoptosis and induce regression of malignant glioma *in vivo*. Nat Med 2002, 8:808–815.

64. Ichikawa K, Liu W, Zhao L, et al. Tumoricidal activity of a novel anti-human DR5 monoclonal antibody without hepatocyte cytotoxicity. Nat Med 2001, 7:954–960.

65. Georgakis GV, Li Y, Humphreys R, et al. Activity of selective fully human agonistic antibodies to the TRAIL death receptors TRAIL-R1 and TRAIL-R2 in primary and cultured lymphoma cells: induction of apoptosis and enhancement of doxorubicin- and bortezomib-induced cell death. Br J Haematol 2005, 130:501–510.

66. Marini P, Denzinger S, Schiller D, et al. Combined treatment of colorectal tumours with agonistic TRAIL receptor antibodies HGS-ETR1 and HGS-ETR2 and radiotherapy: enhanced effects *in vitro* and dose-dependent growth delay *in vivo*. Oncogene 2006, 25:5145–5154.

67. Tecchio C, Huber V, Scapini P, et al. IFNα-stimulated neutrophils and monocytes release a soluble form of TNF-related apoptosis-inducing ligand (TRAIL/Apo-2 ligand) displaying apoptotic activity on leukemic cells. Blood 2004, 103:3837–3844.

68. Clarke N, Jimenez-Lara AM, Voltz E, et al. Tumor suppressor IRF-1 mediates retinoid and interferon anticancer signaling to death ligand TRAIL. EMBO J 2004, 23:3051–3060.

69. Ludwig AT, Moore JM, Luo Y, et al. Tumor necrosis factor-related apoptosis-inducing ligand: a novel mechanism for Bacillus Calmette-Guérin-induced antitumor activity. Cancer Res 2004, 64:3386–3390.

70. Kemp TJ, Ludwig AT, Earel JK, et al. Neutrophil stimulation with *Mycobacterium bovis* bacillus Calmette-Guérin (BCG) results in the release of functional/soluble TRAIL Apo-2L. Blood 2005, 106:3474–3482.

Chapter 5
Proteasome Inhibition: Potential for Sensitization of Immune Effector Mechanisms in Cancer

Milad Motarjemi, William H.D. Hallett, Minghui Li, and William J. Murphy

1 Introduction

The proteasome is a critical component in the natural degradation of eukaryotic cellular proteins [1–4]. The 26S proteasome, a large intracellular adenosine 5′-triphosphate-dependent protease, is critical to the proper functioning and survival of the cell [2–6], making it a valuable and novel target in cancer therapy. The proteasome plays a key regulatory function in many vital cellular processes, including managing damaged, misfolded, or oxidized proteins [7]. Inhibition of the proteasome may cause various disruptions in the cellular microenvironment resulting in an arrest of cell cycling and triggering intrinsic apoptotic pathways [8–13]. Bortezomib (Velcade®, formerly PS-341), the first drug to enter clinical trials as a proteasome inhibitor, was granted approval for use as treatment for patients with relapsed multiple myeloma (MM) [1, 5, 14]. The hastened approval of bortezomib came from the US FDA in May of 2003 after phase II clinical trials demonstrated about one in three patients with advanced MM showed complete response to treatment [15, 16]. The therapeutic responses to bortezomib in preclinical and clinical studies reinforce the role of the proteasome as a useful target in cancer therapy.

Proteasome-mediated degradation of cellular proteins occurs as a result of a highly regulated, tri-enzymatic process of ubiquitination [10, 17]. Proteins that are to be degraded are marked with an addition of ubiquitin chains [18]. The C-terminal glycine residue of ubiquitin covalently binds specific lysine residues on targeted proteins. The affinity of ubiquitination is enzymatically regulated, creating an affinity-control mechanism for substrate selection [17]. This is followed by identification of the ubiquitinated protein by the 19S regulatory component of the proteasome complex, and then finally degradation of the selected protein in the 20S catalytic central portion of the proteasome [1, 9, 13, 17, 19–21]. Although most protein degradations are targeted by ubiquitination, some proteins may also be degraded in an ubiquitin-independent process. Proteins with lesser structural stability that are degraded as a part of the normal cellular life cycle, and proteins that rely on the proteasome as a normal post-translational regulatory component are subject to ubiquitin-independent degradation [22]. Also, damaged or misfolded proteins are subject to ubiquitin-independent, 19S-independent degradation under various conditions of cellular stress [22]. The timely degradation of cellular proteins is crucial to many functions, including transcriptional regulation, and various cellular stress responses [8–10, 13].

The processing of proteins for major histocompatibility complex (MHC) class I molecule presentation has also been shown to be another key function of the proteasome. Peptide-aldehydes used to inhibit the function of the 20S component reduced the degradation of ubiquitinated and non-ubiquitinated proteins as well as reduced the cell's ability to generate peptides presented on

MHC class I molecules [23]. However, inhibitors such as bortezomib bind only the chymotryptic catalytic subunit of the proteasome core, possibly allowing some peptide processing for MHC class I molecules due to incomplete inhibition of proteasome activity. In the presence of some proteasome inhibitors, presentation of certain antigens was not affected or was even enhanced [24–28]. Other pathways for peptide generation such as IFN-γ stimulated cytosolic proteases have been implicated in maintaining a level of MHC class I expression in the presence of proteasome inhibition, indicating a proteasome-independent component also plays a role in antigen presentation [25, 26].

The multifarious nature of proteasome function allows for many different potential mechanisms of action in cancer therapy. Although the therapeutic character of proteasome inhibition is most likely due to a complex composite of cellular effects, transcriptional regulation of nuclear factor-kappaB (NF-κB) appears to be a major contributor to the overall beneficial effects of proteasome inhibition. This pathway controls the activation of NF-κB, which is a transcriptional factor that plays a critical role in cellular survival [29]. The Rel/NF-κB family of proteins are dimeric transcription factors that bind a common DNA sequence motif [30–33]. NF-κB is normally bound to the NF-κB inhibitor (I-κB), which is a cytoplasmic retention protein (32). Stimulation of the NF-κB inhibitor kinase (IKK) causes the phosphorylation event of I-κB. This induces the ubiquitination and degradation of the I-κB, allowing NF-κB to translocate into the nucleus and bind transcription sequences in the DNA [34, 35]. NF-κB has the ability to counterbalance apoptotic signals by activating genes involved in inflammatory responses, such as those encoding hematopoietic growth factors, chemokines, and leukocyte adhesion molecules [31].

Following proteasome inhibition, many other proteins also contribute to the transition from normal cell cycling to arrest and apoptosis. Cyclin-dependent kinase inhibitors, p21 and p27, experience elevated levels resulting in cell cycle arrest in G_1-S following proteasome inhibition. The cell will also increase levels of cyclin A, B, D, and E simultaneously to the elevated level of cyclin-dependent kinase inhibitors; this contradictory signaling is also thought to promote the initiation of apoptosis [5, 17]. Tumor protein 53 (p53), a transcriptional

factor that regulates cell cycling and functions as a tumor suppressor, experiences elevated levels following proteasome inhibition, resulting in the upregulation of p21 and Bcl-2–associated X protein (Bax). Therefore, the increase in the pro-apoptotic Bax protein overcomes the effects of anti-apoptotic Bcl-2 expression inducing apoptosis [5, 17]. The activation of c-Jun N-terminal kinase, which results in the activation of caspase-8 and -3, and the accumulation of damaged, oxidized, or misfolded proteins also contribute to the overall transition to an apoptotic state [35]. It is apparent that proteasome inhibition has the potential to affect a variety of cellular systems, of which show different levels of importance to different cell lines in both human and murine (Fig. 5.1).

2 Proteasome Inhibition in Immune-Mediated Disorders

Proteasome inhibition has shown to affect multiple immune and inflammatory cellular responses both in vitro and in vivo. The role of the proteasome in regulating the activation of NF-κB and the importance of the transcription factor in promoting inflammation has put proteasome inhibition on stage as a potential therapy for a wide range of immune disorders. Bortezomib has shown preclinical efficacy in treatment of a range of disorders from psoriasis and arthritis to autoimmune encephalomyelitis and graft-versus-host disease (GVHD), a potentially lethal complication of allogeneic hematopoietic stem cell transplantation [11, 36–40]. The administration of bortezomib at the time of allogeneic bone marrow transplantation demonstrates its capacity to both prevent the onset of acute GVHD and maintain the vital graft-versus-tumor effect in a murine model [37, 41, 42]. Interestingly, delayed treatment (3+ days after transplant) showed marked increases in GVHD-dependent gastrointestinal toxicity, which was strongly correlated with a TNFα-receptor upregulation, and increases in TNFα, IL-1β, and IL-6 in serum [41]. The role of NF-κB with bortezomib in GVHD treatment is still unclear. The NF-κB inhibitor, PS-1145, was also successful in preventing GVHD, but did not produce the toxicities when administered later after BMT, indicating different mechanisms of action for the two drugs [42]. However, like PS-1145,

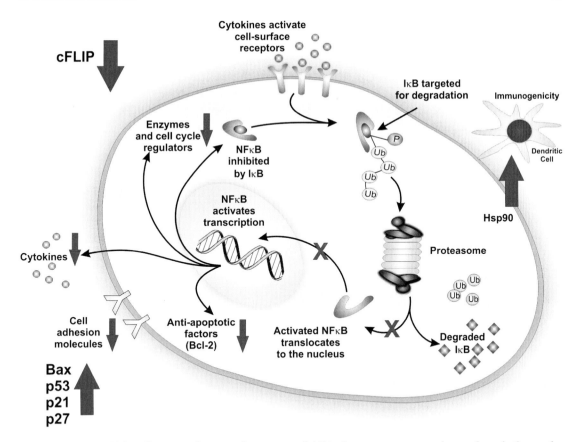

FIGURE 5.1. Intracellular effects post-bortezomib treatment. Inhibited proteasome can no longer degrade the regulatory component of NF-kB, therefore downregulating anti-apoptotic proteins (Bcl-2), cytokine expression, and the regulatory enzymes required for cell cycling. Bortezomib has also been shown to upregulate the expression of Hsp90, which is critical to enhanced immunogenicity via cell-to-cell contact with DCs. Decreased cFLIP expression is also strongly correlated with the increased sensitivity to death receptor ligands, such as TRAIL

bortezomib has been shown to effectively reduce the proliferative capability and cause a direct apoptotic effect of T-lymphocytes in vitro [43].

Lymphoproliferative tumors, such as myeloma, have been shown to have prosurvival type interactions with dendritic cells (DCs) that promote clonogenic growth [44]. Cell-mediated immunity is regulated through a vital interface of antigen and T-lymphocyte by the DCs, which specialize in antigen presentation that stimulates the proliferation of T-lymphocytes [44–47]. Bortezomib has been shown to interfere with tumor-DC interactions [45]. The dose-dependent induction of apoptosis in both myeloid and plasmacytoid DCs indicates that bortezomib's disruptive effects in tumor-DC interaction may be a contributing factor to the overall therapeutic effects in the treatment of GVHD in myeloma [48, 49]. Further studies will be necessary

to judge the efficacy of proteasome inhibition as an effective treatment for GVHD as well as other immune-mediated disorders or solid cancers.

A recent study indicates that bortezomib is able to increase the immunogenic character of human myeloma through the exposure of heat shock protein 90 (hsp90) on apoptotic tumor cells [50]. The ingestion of myeloma cells by the DCs after treatment with bortezomib led to induced antitumor immunity. This induced immunity could not be reproduced with gamma-irradiation or steroid treatment. Combination treatment of bortezomib with geldanamycin, an hsp90 inhibitor, increased tumor cell apoptosis, but abolished all immunogenic response [50]. The current model suggest that heat shock proteins play a crucial role in immunogenic response and that cell-to-cell contact of DCs and apoptotic tumor cells is mediated through the upregulation of hsp90 on the cell surface [51, 52]. What is most notable is the occurrence of enhanced anti-tumor immune surveillance following drug-induced apoptosis.

3 Use of Proteasome Inhibitors to Create Cellular Sensitization to Death Pathways

Because of the crucial role of the proteasome in affecting apoptosis, drugs such as bortezomib allow for the exploration of synergy between proteasome inhibition and other apoptosis-inducing agents for cases of drug resistance and lack of cellular sensitivity to therapeutic intervention.

The feasibility of using bortezomib as a single agent on solid malignancies has been investigated in various phase I and II studies and found to be limited in certain cases due to dose-limiting toxicities of diarrhea and neuropathy [53, 54]. However, bortezomib has been shown to downregulate the expression of PI3K-Akt to undetectable levels through the induction of PTEN expression [55]. The downregulation of Akt has proven to sensitize cells to radiation [56, 57]. A head and neck cancer cell line (SQ20B) was effectively sensitized using bortezomib, lowering cellular survival by more than 40%. Further research will be needed to evaluate the efficacy in in vivo models, as timing and dosing may be critical factors.

The death receptor ligands, Fas ligand (FasL), tumor necrosis factor alpha (TNFα) and tumor necrosis factor-related apoptosis-inducing ligand (TRAIL) have considerable potential in use as cancer therapeutics [58]. Development of treatments with TNFα and another member of the TNF superfamily, FasL, were limited due to the nonselective nature and extreme toxicity associated with the agents [59]. TRAIL, on the other hand, exhibits a level of selectivity. Normal, non-transformed cells show significant resistance to TRAIL [60]. This selective toxicity is the source of much of the interest for the use of TRAIL as a novel cancer therapy.

Numerous cell lines, especially malignant tumors, show resistance to apoptosis inducing death ligands due to various reasons ranging from repeated exposure resistance to dysfunction of death receptors due to mutation. Dysfunctions of the Fas-associated

death domain (FADD) or caspase-8 could also contribute to death ligand resistance [61]. Two human pancreatic cancer cell lines (Panc-1 and HS766T) show resistance to TRAIL due to lower levels of expression of the TRAIL receptors, DR4 and DR5. When treated with bortezomib, responsiveness to TRAIL was restored in both cell lines [62]. The effect of bortezomib was pinpointed with a small interfering RNA construct targeted at p65 subunit of NF-κB and also with PS-1145, a chemical inhibitor of I-κB kinase (IKK). NF-κB was shown to be an inhibitory agent of TRAIL-induced apoptosis and therapy with NF-κB inhibitors in combination with death receptor ligands shows synergy capable of sensitizing cancer lines to apoptosis [62, 63].

Overexpression of cellular FADD-like interleukin-1 β-converting enzyme-inhibitory protein (cFLIP) has also been strongly correlated with many TRAIL-resistant cancer lines. Treatment with bortezomib showed marked decreases in cFLIP expression in murine acute myeloid leukemia (C1498) and murine renal cancer (Renca) cell lines [59, 62, 64, 65]. In combination assays, bortezomib and recombinant TRAIL showed increased induction of apoptosis than either agent alone. Additionally, C1498 was selectively purged from normal bone marrow after treatment with bortezomib and TRAIL [65]. When the purged bone marrow was introduced into syngeneic murine recipients, the combination treatment allowed for tumor-free survival. It has also been shown that down-regulation of cFLIP sensitizes human lung, breast and ovarian cancer cells to TRAIL-induced apoptosis [66], indicating the future potential of proteasome inhibition used in combination with other novel therapeutics.

Mantle cell lymphoma (MCL) cells, which have a natural resistance to intrinsically signaled apoptosis [67, 68], appear to have a susceptibility to TRAIL induced apoptosis, regardless of receptor levels, Bcl-2 family, or caspase-regulator expression, and was associated with NF-κB inhibition and decreased expression of cFLIP [69]. Proteasome inhibition with bortezomib, which has shown marked decreases in cFLIP expression [59, 62, 64, 65], was able to create synergistic effect in MCL cells with low cFLIP expression [59, 70], but was less effective in high cFLIP expressing MCL cell lines such as JVM-2 [69, 71]. The intracellular accumulation of cFLIP post-bortezomib treatment of high cFLIP expressing cell lines could be the major obstacle to total synergistic effect. Selective inhibition of the I-κB kinase with BMS-345541 allowed complete TRAIL-induced apoptosis of resistant and sensitive MCL cells. Interestingly, a decrease in DR5 expression caused by selective I-κB kinase inhibition did not affect the sensitivity to TRAIL-induced apoptosis [69]. MCL cell incubation with recombinant TRAIL, which preferentially binds DR5 rather than DR4 [72], did not activate cell death even in the presence of BMS-345541. Selective inhibition of I-κB kinase in combination with TRAIL or other anti-DR4 antibodies could show promising models for further investigation.

The combined effect of proteasome inhibition and TRAIL-induced apoptosis appears to show cell-line specific modalities, including various different pre–cell-death regulatory responses. In TRAIL-resistant, proteasome inhibition-resistant RKO colon carcinoma cells, DNA fragmentation, mitochondrial depolarization, and upregulation of caspase-3 enzyme activities (DEVDase) were only observed under combined treatments [73]. Levels of XIAP, survivin, Bcl-2, and Bcl-X_L, all anti-apoptotic proteins regulated by NF-κB, were not affected. The release of apoptosis-inducing factor was also not observed. Mitochondrial depolarization causing the release of cytochrome c [74], HtrA2/Omi, and Smac/DIABLO followed only the combined proteasome inhibition and TRAIL treatment. In RKO cells, treatment with TRAIL alone only managed partial activation of caspase-3, where caspase-3 was not cleaved into active components—indicated by the absence of p17, low DEVDase activity, and no DNA fragmentation [73]. Where TRAIL alone did not release any mitochondrial products, treatment with only a proteasome inhibitor, epoxomicin, caused the release of cytochrome c and HtrA2/Omi without activation of caspase-3 or DNA fragmentation. Only the combination of epoxomicin and TRAIL were able to induce the release of Smac/DIABLO. The release of Smac/DIABLO closely correlates with the activation of caspase-3 (p20 to p17), increased DEVDase activity, and DNA fragmentation [73] (Fig. 5.2). Smac/DIABLO functions to counter the caspase-inhibitory properties of the IAP family,

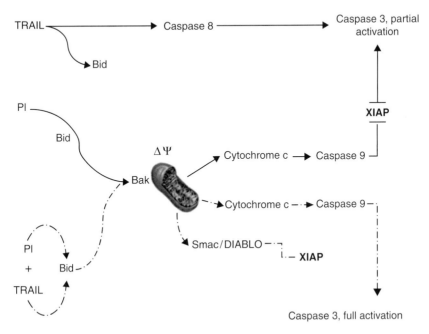

FIGURE 5.2. TRAIL and proteasome inhibition combined treatment allows full activation of caspase-3. In TRAIL resistant, and proteasome inhibition resistant cell lines, only partial activation of caspase-3 is possible. When treatments are combined, full activation of caspase-9 and caspase-3 is achieved through the release of anti-IAP factor, Smac/DIABLO

especially XIAP [75]. Smac/DIABLO is capable of dissociating IAP proteins from their binding sites on procaspase-3 and -9 [76, 77]. In cases in which IAP proteins are caspase-bound, both intrinsic and extrinsic apoptotic pathways are blocked. This model could possibly explain why caspase-9 (–) [78] and Apaf-1 (–) [79] cells are sensitive to TRAIL/proteasome inhibition–combined treatment.

The mechanism of the selective release of Smac/DIABLO is not understood, but bortezomib and other proteasome inhibitors have proved to be able to upregulate the levels of pro-apoptotic Bik and Bim via decreased proteasomal degradation causing intracellular accumulation [79–81]. siRNA downregulation of Bik shows that it is a primary cytotoxic effect of bortezomib. Bik translocates to the mitochondria, causing a permeability shift and the release of cytochrome c, which in turn activates caspase-9 and other effector caspases. The current model suggests that TRAIL-induced apoptosis involving the activation of Bid and partial activation of caspase-3 function synergistically with bortezomib-activated Bak and Bik pathways to cause the release of Smac/DIABLO from the mitochondria for enhanced apoptotic signaling [82, 83].

The use of proteasome inhibition to sensitize tumor cells to death may be best utilized in the context of adoptive transfer of natural killer (NK) cells following hematopoietic stem cell transplantation (HSCT) for patients with hematologic cancers. NK cells are members of the innate immune response who play a role in the defense against viral pathogens, parasites, and neoplastic cells [84, 85]. NK cells have multiple mechanisms of cytotoxicity, including cytolytic mechanisms such as perforin and granzymes, as well as death ligand molecules including TNFα, FasL, and TRAIL. Because NK cells do not initiate GVHD, they may be transferred into patients undergoing transplant in high numbers with relatively low risk of toxicity. NK cells also produce abundant granulocyte-colony stimulating factor (G-CSF), which promotes myeloid recovery following transplant [86]. These combined characteristics of NK cells make the use of these cells in combination with bortezomib an attractive therapy for patients undergoing transplants for neoplastic malignancies. Preliminary data from our laboratory indicate that bortezomib does indeed sensitize tumor cells to NK cell-mediated killing.

4 Bortezomib and CDDO

2-Cyano-3, 12-dioxooleana-1, 9(11)-diene-28-oic acid (CDDO), a synthetic triterpenoid, emerged as a novel cancer therapeutic [87–94]. CDDO, and new, more potent derivatives (CDDO-Me and CDDO-IM) show promising anticancer properties of which the mechanisms of the action are still unclear [94]. Cytotoxic effects of the CDDO family include a broad range of effects that inhibit prosurvival mechanisms and/or cause the altered function of regulatory proteins:

- Modulated NF-κB activity and NF-κB gene expression. Suppression of NF-κB directed through the inhibition of IκBα kinase activation, which results in the inhibition of IκBα phosphorylation, degradation, p65 phosphorylation, and p65 translocation. Apparent suppression of NF-κB also resulted in the down-regulation of NF-κB dependent survival gene expression [94].
- Peroxisome proliferator-activated receptor-γ (PPAR$_\gamma$) ligand. Transactivation of PPAR$_\gamma$ by CDDO caused cell cycle arrest in G$_1$-S and G$_2$-M, apoptosis and cell growth inhibition in a time- and dose-dependent manner [91]. Growth inhibition was correlated with the upregulation of caveolin-1 [87].
- Direct permeabilization of the inner mitochondrial membrane. CDDO-Me induced dysfunction of the mitochondria causing the rapid depletion of mitochondrial glutathione, loss of cardiolipin, and inhibition in mitochondrial respiration, all contributing to the activation of the intrinsic apoptotic pathway [93].
- Disruption of microtubule polymerization. CDDO selectively binds to β-tubulin both in vivo and in vitro, causing inhibition of polymerization contributing to the initiation of the intrinsic apoptotic pathway in M-phase [88].

Because of the success of CDDO in apoptotic induction in leukemia cells [89, 94], further investigation with CDDO-IM, the more potent derivative of the CDDO family, was conducted—ultimately showing promising results in the initiation of apoptosis in MM cells that were previously resistant to conventional treatments without causing decreased viability of normal cells [95]. CDDO-IM also decreased IL-6 secretion caused by MM cell adhesion to bone marrow stromal cells [95].

Apoptosis induction as a result of CDDO-IM was not effected by IL-6 or insulin growth factor-1. The combined use of CDDO-IM and proteasome inhibition has shown very promising results. Low-dose therapies with CDDO-IM and bortezomib have shown an ability to overcome the cytoprotective effects of prosurvival proteins such as NF-κB, Bcl2, and Hsp27, and drug resistance in patients with MM [95].

With a new drive towards molecular targeting therapeutics for cancer therapy, the role of bortezomib was vital to the validation of the proteasome as a relevant target. Bortezomib, which was the first anti-MM drug approved for use in many years, showed unexpected efficacy due to the presumed lack of specificity of proteasome inhibiting compounds. Although bortezomib is currently available to patients with MM who have had at least one prior therapy, the precise mechanisms of action are still unclear. Further studies, which will elucidate such mechanisms, will allow for further optimization and potential combination approaches with either chemotherapeutic- or immune-based therapies. Bortezomib is also currently available for use in combination with other therapeutics in clinical studies. The use of proteasome inhibition to create cell-sensitization is an exciting new development that could bolster the effectiveness of traditional cancer drugs for patients with resistant cancers or developed resistances. Cell sensitization used in conjunction with novel therapeutics such as TRAIL or CDDO may result in increased efficacy. Future studies aim to continue the evaluation of proteasome inhibition as a possible therapy for immune-mediated disorders, such as GVHD, and current assessments are gauging the safety and efficacy of bortezomib for use in lymphomas and a range of other cancers.

Acknowledgments: This publication was funded all or in part by National Institutes of Health grant R01CA95327-02 and American Cancer Society grant RSG-020169.

The authors would like to thank Lisbeth Welniak for critically reviewing the manuscript and Ruth Gault for creating the figures.

References

1. Adams J. Development of the proteasome inhibitor PS-341. Oncologist 2002, 7:9–16.

2. Adams J. Proteasome inhibition in cancer: development of PS-341. Semin Oncol 2001, 28:613–619.

3. Adams J. Proteasome inhibitors as new anticancer drugs. Curr Opin Oncol 2002, 14:628–634.

4. Ciechanover A. The ubiquitin-proteasome proteolytic pathway. Cell 1994, 79:13–21.

5. Adams J, Palombella VJ, Sausville EA, et al. Proteasome inhibitors: a novel class of potent and effective antitumor agents. Cancer Res 1999, 59:2615–2622.

6. Montagut C, Rovira A, Albanell J. The proteasome: a novel target for anticancer therapy. Clin Transl Oncol 2006, 8:313–317.

7. Wang J, Maldonado MA. The ubiquitin-proteasome system and its role in inflammatory and autoimmune diseases. Cell Mol Immunol 2006, 3:255–261.

8. Adams J. Preclinical and clinical evaluation of proteasome inhibitor PS-341 for the treatment of cancer. Curr Opin Chem Biol 2002, 6:493–500.

9. Adams J, Palombella VJ, Elliott PJ. Proteasome inhibition: a new strategy in cancer treatment. Invest New Drugs 2000, 18:109–121.

10. Hershko A, Ciechanover A. The ubiquitin system. Annu Rev Biochem 1998, 67:425–479.

11. Palombella VJ, Conner EM, Fuseler JW, et al. Role of the proteasome and NF-kappaB in streptococcal cell wall-induced polyarthritis. Proc Natl Acad Sci U S A 1998, 95:15671–15676.

12. Palombella VJ, Rando OJ, Goldberg AL, et al. The ubiquitin-proteasome pathway is required for processing the NF-kappa B1 precursor protein and the activation of NF-kappa B. Cell 1994, 78:773–785.

13. Varshavsky A. The ubiquitin system. Trends Biochem Sci 1997, 22:383–387.

14. Cheson BD. Hematologic malignancies: new developments and future treatments. Semin Oncol 2002, 29:33–45.

15. Anonymous. FDA approves Velcade for multiple myeloma treatment. FDA News 2003.

16. Richardson PG, Barlogie B, Berenson J, et al. A phase 2 study of bortezomib in relapsed, refractory myeloma. N Engl J Med 2003, 348:2609–2617.

17. Almond JB, Cohen GM. The proteasome: a novel target for cancer chemotherapy. Leukemia 2002, 16:433–443.

18. Kloetzel PM. Antigen processing by the proteasome. Nat Rev Mol Cell Biol 2001, 2:179–187.

19. Coux O, Tanaka K, Goldberg AL. Structure and functions of the 20S and 26S proteasomes. Annu Rev Biochem 1996, 65:801–847.

20. Craiu A, Gaczynska M, Akopian T, et al. Lactacystin and clasto-lactacystin beta-lactone modify multiple proteasome beta-subunits and inhibit intracellular protein degradation and major histocompatibility complex class I antigen presentation. J Biol Chem 1997, 272:13437–13445.

21. Orlowski M. The multicatalytic proteinase complex, a major extralysosomal proteolytic system. Biochemistry 1990, 29:10289–10297.
22. Shringarpure R, Grune T, Mehlhase J, et al. Ubiquitin conjugation is not required for the degradation of oxidized proteins by proteasome. J Biol Chem 2003, 278:311–318.
23. Rock KL, Gramm C, Rothstein L, et al. Inhibitors of the proteasome block the degradation of most cell proteins and the generation of peptides presented on MHC class I molecules. Cell 1994, 78:761–771.
24. Anton LC, Snyder HL, Bennink JR, et al. Dissociation of proteasomal degradation of biosynthesized viral proteins from generation of MHC class I-associated antigenic peptides. J Immunol 1998, 160:4859–4868.
25. Luckey CJ, King GM, Marto JA, et al. Proteasomes can either generate or destroy MHC class I epitopes: evidence for nonproteasomal epitope generation in the cytosol. J Immunol 1998, 161:112–121.
26. Vinitsky A, Anton LC, Snyder HL, et al. The generation of MHC class I-associated peptides is only partially inhibited by proteasome inhibitors: involvement of nonproteasomal cytosolic proteases in antigen processing? J Immunol 1997, 159:554–564.
27. Yellen-Shaw AJ, Wherry EJ, Dubois GC, et al. Point mutation flanking a CTL epitope ablates in vitro and in vivo recognition of a full-length viral protein. J Immunol 1997, 158:3227–3234.
28. Wherry EJ, Golovina TN, Morrison SE, et al. Re-evaluating the generation of a "proteasome-independent" MHC class I-restricted CD8 T cell epitope. J Immunol 2006, 176:2249–2261.
29. Dutta J, Fan Y, Gupta N, et al. Current insights into the regulation of programmed cell death by NF-kappaB. Oncogene 2006, 25:6800–6816.
30. Gilmore TD, Koedood M, Piffat KA, et al. Rel/NF-kappaB/IkappaB proteins and cancer. Oncogene 1996, 13:1367–1378.
31. Hideshima T, Chauhan D, Richardson P, et al. NF-kappa B as a therapeutic target in multiple myeloma. J Biol Chem 2002, 277:16639–16647.
32. Karin M. How NF-kappaB is activated: the role of the IkappaB kinase (IKK) complex. Oncogene 1999, 18:6867–6874.
33. Karin M, Ben-Neriah Y. Phosphorylation meets ubiquitination: the control of NF-[kappa]B activity. Annu Rev Immunol 2000, 18:621–663.
34. Karin M, Delhase M. The I kappa B kinase (IKK) and NF-kappa B: key elements of proinflammatory signalling. Semin Immunol 2000, 12:85–98.
35. Mitsiades N, Mitsiades CS, Poulaki V, et al. Biologic sequelae of nuclear factor-kappaB blockade in multiple myeloma: therapeutic applications. Blood 2002, 99:4079–4086.
36. Luo H, Wu Y, Qi S, et al. A proteasome inhibitor effectively prevents mouse heart allograft rejection. Transplantation 2001, 72:196–202.
37. Sun K, Welniak LA, Panoskaltsis-Mortari A, et al. Inhibition of acute graft-versus-host disease with retention of graft-versus-tumor effects by the proteasome inhibitor bortezomib. Proc Natl Acad Sci USA 2004, 101:8120–8125.
38. Vanderlugt CL, Rahbe SM, Elliott PJ, et al. Treatment of established relapsing experimental autoimmune encephalomyelitis with the proteasome inhibitor PS-519. J Autoimmun 2000, 14:205–211.
39. Wu T, Sozen H, Luo B, et al. Rapamycin and T cell costimulatory blockade as post-transplant treatment promote fully MHC-mismatched allogeneic bone marrow engraftment under irradiation-free conditioning therapy. Bone Marrow Transplant 2002, 29:949–956.
40. Zollner TM, Podda M, Pien C, et al. Proteasome inhibition reduces superantigen-mediated T cell activation and the severity of psoriasis in a SCID-hu model. J Clin Invest 2002, 109:671–679.
41. Sun K, Wilkins DE, Anver MR, et al. Differential effects of proteasome inhibition by bortezomib on murine acute graft-versus-host disease (GVHD): delayed administration of bortezomib results in increased GVHD-dependent gastrointestinal toxicity. Blood 2005, 106:3293–3299.
42. Vodanovic-Jankovic S, Hari P, Jacobs P, et al. NF-kappaB as a target for the prevention of graft-versus-host disease: comparative efficacy of bortezomib and PS-1145. Blood 2006, 107:827–834.
43. Kaufmann SH, Vaux DL. Alterations in the apoptotic machinery and their potential role in anticancer drug resistance. Oncogene 2003, 22:7414–7430.
44. Sallusto F, Lanzavecchia A. The instructive role of dendritic cells on T-cell responses. Arthritis Res 2002, 4(Suppl 3):S127–132.
45. Kukreja A, Hutchinson A, Dhodapkar K, et al. Enhancement of clonogenicity of human multiple myeloma by dendritic cells. J Exp Med 2006, 203:1859–1865.
46. Lanzavecchia A, Sallusto F. Antigen decoding by T lymphocytes: from synapses to fate determination. Nat Immunol 2001, 2:487–492.
47. Lanzavecchia A, Sallusto F. Regulation of T cell immunity by dendritic cells. Cell 2001, 106:263–266.
48. Chen L, Arora M, Yarlagadda M, et al. Distinct responses of lung and spleen dendritic cells to the TLR9 agonist CpG oligodeoxynucleotide. J Immunol 2006, 177:2373–2383.
49. Chen X, Reed-Loisel LM, Karlsson L, et al. H2-O expression in primary dendritic cells. J Immunol 2006, 176:3548–3556.

50. Spisek R, Charalambous A, Mazumder A, et al. Bortezomib enhances dendritic cell (DC) mediated induction of immunity to human myeloma via exposure of cell surface heat shock protein 90 on dying tumor cells: therapeutic implications. Blood 2007, 109(11):4839–4845.

51. Dai J, Liu B, Caudill MM, et al. Cell surface expression of heat shock protein gp96 enhances cross-presentation of cellular antigens and the generation of tumor-specific T cell memory. Cancer Immunol 2003, 3:1.

52. Liu B, Dai J, Zheng H, et al. Cell surface expression of an endoplasmic reticulum resident heat shock protein gp96 triggers MyD88-dependent systemic autoimmune diseases. Proc Natl Acad Sci U S A 2003, 100:15824–15829.

53. Aghajanian C, Soignet S, Dizon DS, et al. A phase I trial of the novel proteasome inhibitor PS341 in advanced solid tumor malignancies. Clin Cancer Res 2002, 8:2505–2511.

54. Ludwig H, Khayat D, Giaccone G, et al. Proteasome inhibition and its clinical prospects in the treatment of hematologic and solid malignancies. Cancer 2005, 104:1794–1807.

55. Weber CN, Cerniglia CJ, Maity A, et al. Bortezomib sensitizes human head and neck carcinoma cells SQ20B to radiation. Cancer Biol Ther 2007, 6.

56. Anai S, Goodison S, Shiverick K, et al. Combination of PTEN gene therapy and radiation inhibits the growth of human prostate cancer xenografts. Hum Gene Ther 2006, 17:975–984.

57. Ng KT, Man K, Ho JW, et al. Marked suppression of tumor growth by FTY720 in a rat liver tumor model: the significance of down-regulation of cell survival Akt pathway. Int J Oncol 2007, 30:375–380.

58. Wajant H, Gerspach J, Pfizenmaier K. Tumor therapeutics by design: targeting and activation of death receptors. Cytokine Growth Factor Rev 2005, 16:55–76.

59. Sayers TJ, Murphy WJ. Combining proteasome inhibition with TNF-related apoptosis-inducing ligand (Apo2L/TRAIL) for cancer therapy. Cancer Immunol Immunother 2006, 55:76–84.

60. Walczak H, Miller RE, Ariail K, et al. Tumoricidal activity of tumor necrosis factor-related apoptosis-inducing ligand in vivo. Nat Med 1999, 5:157–163.

61. Zhang L, Fang B. Mechanisms of resistance to TRAIL-induced apoptosis in cancer. Cancer Gene Ther 2005, 12:228–237.

62. Khanbolooki S, Nawrocki ST, Arumugam T, et al. Nuclear factor-kappaB maintains TRAIL resistance in human pancreatic cancer cells. Mol Cancer Ther 2006, 5:2251–2260.

63. Kamat AM, Sethi G, Aggarwal BB. Curcumin potentiates the apoptotic effects of chemotherapeutic agents and cytokines through down-regulation of nuclear factor-{kappa}B and nuclear factor-{kappa}B-regulated gene products in IFN-{alpha}-sensitive and IFN-{alpha}-resistant human bladder cancer cells. Mol Cancer Ther 2007, 6:1022–1030.

64. Brooks AD, Ramirez T, Toh U, et al. The proteasome inhibitor bortezomib (Velcade) sensitizes some human tumor cells to Apo2L/TRAIL-mediated apoptosis. Ann N Y Acad Sci 2005, 1059:160–167.

65. Sayers TJ, Brooks AD, Koh CY, et al. The proteasome inhibitor PS-341 sensitizes neoplastic cells to TRAIL-mediated apoptosis by reducing levels of c-FLIP. Blood 2003, 102:303–310.

66. Clarke P, Tyler KL. Down-regulation of cFLIP following reovirus infection sensitizes human ovarian cancer cells to TRAIL-induced apoptosis. Apoptosis 2007, 12:211–223.

67. Brody J, Advani R. Treatment of mantle cell lymphoma: current approach and future directions. Crit Rev Oncol/Hematol 2006, 58:257–265.

68. Ogura M. [Recent progress in the therapeutic strategy for follicular lymphoma and mantle cell lymphoma]. [Rinsho ketsueki] Japanese J Clin Hematol 2006, 47:495–512.

69. Roue G, Perez-Galan P, Lopez-Guerra M, et al. Selective inhibition of I{kappa}B kinase sensitizes mantle cell lymphoma B cells to TRAIL by decreasing cellular FLIP level. J Immunol 2007, 178:1923–1930.

70. An J, Sun YP, Adams J, et al. Drug interactions between the proteasome inhibitor bortezomib and cytotoxic chemotherapy, tumor necrosis factor (TNF) alpha, and TNF-related apoptosis-inducing ligand in prostate cancer. Clin Cancer Res 2003, 9:4537–4545.

71. Pham LV, Tamayo AT, Yoshimura LC, et al. Inhibition of constitutive NF-kappa B activation in mantle cell lymphoma B cells leads to induction of cell cycle arrest and apoptosis. J Immunol 2003, 171:88–95.

72. Kelley RF, Totpal K, Lindstrom SH, et al. Receptor-selective mutants of apoptosis-inducing ligand 2/tumor necrosis factor-related apoptosis-inducing ligand reveal a greater contribution of death receptor (DR) 5 than DR4 to apoptosis signaling. J Biol Chem 2005, 280:2205–2212.

73. Nagy K, Szekely-Szuts K, Izeradjene K, et al. Proteasome inhibitors sensitize colon carcinoma cells to TRAIL-induced apoptosis via enhanced release of Smac/DIABLO from the mitochondria. Pathol Oncol Res 2006, 12:133–142.

74. Liu X, Kim CN, Yang J, et al. Induction of apoptotic program in cell-free extracts: requirement for dATP and cytochrome c. Cell 1996, 86:147–157.

75. Verhagen AM, Ekert PG, Pakusch M, et al. Identification of DIABLO, a mammalian protein that promotes apoptosis by binding to and antagonizing IAP proteins. Cell 2000, 102:43–53.

76. Du C, Fang M, Li Y, et al. Smac, a mitochondrial protein that promotes cytochrome c-dependent caspase activation by eliminating IAP inhibition. Cell 2000, 102:33–42.

77. Petak I, Vernes R, Szucs KS, et al. A caspase-8-independent component in TRAIL/Apo-2L-induced cell death in human rhabdomyosarcoma cells. Cell Death Diff 2003, 10:729–739.

78. Johnson TR, Stone K, Nikrad M, et al. The proteasome inhibitor PS-341 overcomes TRAIL resistance in Bax and caspase 9-negative or Bcl-xL overexpressing cells. Oncogene 2003, 22:4953–4963.

79. Nikrad M, Johnson T, Puthalalath H, et al. The proteasome inhibitor bortezomib sensitizes cells to killing by death receptor ligand TRAIL via BH3-only proteins Bik and Bim. Mol Cancer Ther 2005, 4:443–449.

80. Zhu H, Guo W, Zhang L, et al. Proteasome inhibitors-mediated TRAIL resensitization and Bik accumulation. Cancer Biol Ther 2005, 4:781–786.

81. Zhu H, Zhang L, Dong F, et al. Bik/NBK accumulation correlates with apoptosis-induction by bortezomib (PS-341, Velcade) and other proteasome inhibitors. Oncogene 2005, 24:4993–4999.

82. Li H, Zhu H, Xu CJ, et al. Cleavage of BID by caspase 8 mediates the mitochondrial damage in the Fas pathway of apoptosis. Cell 1998, 94:491–501.

83. Adrain C, Creagh EM, Martin SJ. Apoptosis-associated release of Smac/DIABLO from mitochondria requires active caspases and is blocked by Bcl-2. EMBO J 2001, 20:6627–6636.

84. Cudkowicz G, Bennett M. Peculiar immunobiology of bone marrow allografts. II. Rejection of parental grafts by resistant F 1 hybrid mice. J Exp Med 1971, 134:1513–1528.

85. Hallett WH, Murphy WJ. Positive and negative regulation of Natural Killer cells: therapeutic implications. Sem Cancer Biol 2006, 16:367–382.

86. Murphy WJ, Keller JR, Harrison CL, et al. Interleukin-2-activated natural killer cells can support hematopoiesis in vitro and promote marrow engraftment in vivo. Blood 1992, 80:670–677.

87. Chintharlapalli S, Papineni S, Konopleva M, et al. 2-Cyano-3,12-dioxoolean-1,9-dien-28-oic acid and related compounds inhibit growth of colon cancer cells through peroxisome proliferator-activated receptor gamma-dependent and -independent pathways. Mol Pharmacol 2005, 68:119–128.

88. Couch RD, Ganem NJ, Zhou M, et al. 2-cyano-3,12-dioxooleana-1,9(11)-diene-28-oic acid disrupts microtubule polymerization: a possible mechanism contributing to apoptosis. Mol Pharmacol 2006, 69:1158–1165.

89. Konopleva M, Contractor R, Kurinna SM, et al. The novel triterpenoid CDDO-Me suppresses MAPK pathways and promotes p38 activation in acute myeloid leukemia cells. Leukemia 2005, 19:1350–1354.

90. Konopleva M, Zhang W, Shi YX, et al. Synthetic triterpenoid 2-cyano-3,12-dioxooleana-1,9-dien-28-oic acid induces growth arrest in HER2-overexpressing breast cancer cells. Mol Cancer Ther 2006, 5:317–328.

91. Lapillonne H, Konopleva M, Tsao T, et al. Activation of peroxisome proliferator-activated receptor gamma by a novel synthetic triterpenoid 2-cyano-3,12-dioxooleana-1,9-dien-28-oic acid induces growth arrest and apoptosis in breast cancer cells. Cancer Res 2003, 63:5926–5939.

92. Pedersen IM, Kitada S, Schimmer A, et al. The triterpenoid CDDO induces apoptosis in refractory CLL B cells. Blood 2002, 100:2965–2972.

93. Samudio I, Konopleva M, Pelicano H, et al. A novel mechanism of action of methyl-2-cyano-3,12 dioxoolean-1,9 diene-28-oate: direct permeabilization of the inner mitochondrial membrane to inhibit electron transport and induce apoptosis. Mol Pharmacol 2006, 69:1182–1193.

94. Shishodia S, Sethi G, Konopleva M, et al. A synthetic triterpenoid, CDDO-Me, inhibits IkappaBalpha kinase and enhances apoptosis induced by TNF and chemotherapeutic agents through down-regulation of expression of nuclear factor kappaB-regulated gene products in human leukemic cells. Clin Cancer Res 2006, 12:1828–1838.

95. Chauhan D, Li G, Podar K, et al. The bortezomib/proteasome inhibitor PS-341 and triterpenoid CDDO-Im induce synergistic anti-multiple myeloma (MM) activity and overcome bortezomib resistance. Blood 2004, 103:3158–3166.

Part II
Sensitization via Inhibition of Cell Survival Pathways (Excluding Apoptotic Signaling Pathways)

Chapter 6
Angiogenesis Inhibitors as Enabling Agents for the Chemotherapeutic Treatment of Metastatic Disease

Giulio Francia, Urban Emmenegger, and Robert S. Kerbel

1 Introduction

The strategy of preventing the formation of new blood vessels by tumors—tumor angiogenesis—achieved a milestone in cancer therapeutics in 2003, with the announcement of a positive clinical trial involving the antiangiogenic agent bevacizumab (Avastin®). This phase III trial demonstrated a survival benefit of bevacizumab (which targets vascular endothelial growth factor, or VEGF) combined with IFL chemotherapy (irinotecan, 5-fluorouracil, and leucovorin), compared with chemotherapy alone, in patients with metastatic colorectal cancer [1]. The study was the culmination of decades of research which began in earnest in 1971 when Judah Folkman first proposed the hypothesis of targeting tumor angiogenesis as a novel approach to the treatment of cancer [2]. However, not only did the study by Hurwitz et al. trial reveal, for the first time, a survival advantage in patients receiving an angiogenesis inhibitor incorporated in a standard chemotherapy regimen, it also set the stage for angiogenesis inhibitors as a new class of chemosensitizers. This represents a paradox, since intuitively one would predict that tumor associated blood vessels would facilitate the delivery and perfusion of chemotherapeutics to a solid tumor mass, and therefore inhibition of angiogenesis would do the opposite, i.e., reduce the delivery of chemotherapy thus rendering it less effective. Yet in spite of this, as early as 1992 Teicher et al. had already described results from preclinical studies showing that inhibiting angiogenesis enhanced the therapeutic benefit of chemotherapy—and presumably radiotherapy [3]. This turned out to be a seminal observation since some antiangiogenic drugs such as bevacizumab that target a single angiogenic factor/pathway, appear to have minimal or no antitumor activity in advanced stage metastatic cancers. Instead the benefit of using such a drug appears to be when it is used as part of a treatment "cocktail" involving other drugs, especially other chemotherapeutics [4]. While addition of bevacizumab to chemotherapy can improve clinical benefit, the effects remain modest. Improvements can be expected by a better understanding of the mechanisms underlying antiangiogenic drug mediated chemosensitization. This chapter describes the evolution of theory and practice of inhibitors of angiogenesis in cancer therapy, as well as their application as chemosensitizers both in the laboratory and the clinical setting, with particular emphasis on metastatic disease. For relevant areas not covered by this chapter, such as the radiosensitizing effects of antiangiogenic drugs, the reader is referred to a recent and exhaustive review [5].

From: *Sensitization of Cancer Cells for Chemo/Immuno/Radio-therapy, 1st Edition.*
Edited by: Benjamin Bonavida © Humana Press, Totowa, NJ

2   Historical Perspective

Following the original postulation by Folkman in 1971, that targeting tumor angiogenesis could provide an effective antitumor growth strategy [2], an intensive search began to identify angiogenesis promoting factors (Fig. 6.1). These would not only permit a better understanding of the mechanisms by which a tumor stimulates its own blood supply, but targeting such factors would permit a proof-of-concept test of the basic hypothesis. Probably the most significant milestone in this field was the discovery of VEGF (originally termed vascular permeability factor, or VPF, by Dvorak's group), by Ferrara's group [6–7]. Subsequently, massive efforts were directed at molecularly describing, targeting and neutralizing this pro-angiogenic molecule, culminating (see Fig. 6.1) in the generation of the humanized anti-(human) VEGF antibody bevacizumab [7] and later other types of VEGF or VEGF receptor targeting drugs. During the development of this second milestone, two very significant discoveries were reported. The first was the description by Teicher and colleagues of the benefit of combining angiogenesis inhibitors, such as TNP-470, a fumagillin analogue, with conventional chemotherapy regimens [3]. The second was a hypothesis proposing that endothelial cells, which lack the genetic

instability that drives mutation and heterogeneity of tumor populations, should not be able to develop a drug-resistant phenotype to cytotoxic agents [8], which target endothelial cells in the growing tumor neovasculature. Therefore the combination of angiogenesis inhibitors and chemotherapy would not only enhance the cytotoxic effects of the latter, but also could conceivably delay or prevent the appearance of drug-resistant tumors. By way of example, VEGF can function as a potent and specific survival factor for angiogenic endothelial cells, which could limit the ability of chemotherapy to damage or destroy activated endothelium. Neutralizing VEGF activity would therefore enhance the antiangiogenic effects of chemotherapy [9–11]. We are presently entering a period in which these ideas are being validated and (accordingly) refined, and thus raising new questions, including: is the combination therapy concept "universal," i.e., will angiogenesis inhibitors sensitize all tumor types to chemotherapy? Will angiogenesis inhibitors sensitize all currently known chemotherapy drugs or regimens, or will sensitization be largely restricted to certain classes of cytotoxic drugs? In what clinical setting (e.g., (neo)adjuvant or advanced disease stage) are such combination treatments likely to deliver the greatest therapeutic benefit? What are the underlying mechanisms?

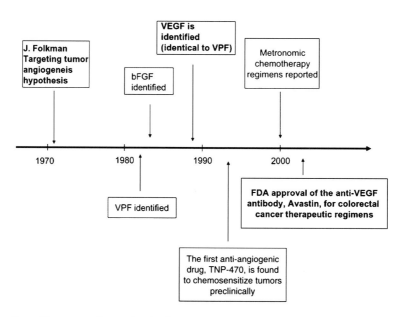

FIGURE 6.1. Timeline of important discoveries leading up to FDA approval of Avastin

3 Preclinical Studies

A number of antibodies and small molecule inhibitors are currently employed in preclinical studies of inhibition of angiogenesis, some of which (i.e., those discussed in this chapter) as listed in Table 6.1. Aside from bevacizumab and metronomic chemotherapy, which are further detailed in the following sections notable small molecule receptor tyrosine kinase inhibitors (RTKIs) include sunitinib (S011248) and sorafenib (BAY 43-9006), which were recently approved as monotherapies for kidney cancer [12, 13]. In addition, as an alternative to neutralizing the VEGF ligand, there are also various drugs that block particular VEGF receptors. For example, DC101 is an antibody directed against mouse VEGF receptor 2, which has shown excellent chemosensitizing properties in a number of preclinical studies [14], including experimental metastasis models [15, 16]. The corresponding anti-human-VEGF receptor 2 antibody, termed IMC1121b, is currently being tested in clinical trials.

3.1 Bevacizumab Therapy Experiments in Preclinical Tumor Models, Including Metastatic Models

Studies on the impact of targeting VEGF in preclinical models have employed either bevacizumab or the progenitor mouse antibody A4.6.1 (from which

the humanized monoclonal antibody bevacizumab was generated) [17]. Single-agent administration was reported to inhibit primary human tumor xenograft growth of at least 13 different tumor types, including prostate, breast and colon carcinoma, glioblastoma, and neuroblastoma. Addition of chemotherapy (e.g., paclitaxel, doxorubicin, or topotecan) to bevacizumab administration was found to produce additive or synergistic tumor inhibition in models of human ovarian, prostate and breast carcinoma grown in nude mice [17]. Following A4.6.1 administration to mice bearing experimental human colorectal cancer xenografts, a 10-fold reduction in the number of liver metastases was observed, compared with controls [18]. Similar results were observed with experimental metastasis models of prostate cancer [19] and Wilms' tumor [20].

One further application of bevacizumab has been for delaying the emergence of drug resistance to certain agents [21]. Thus orthotopically implanted MDA231H2N, a MDA-MB-231 human breast cancer variant genetically modified to overexpress Her-2, was found initially to respond well to trastuzumab (Herceptin®) plus cyclophosphamide chemotherapy, although eventually resistance developed (Fig. 6.2) [21]. Analysis of tissue preparations from the drug-resistant tumors revealed elevated VEGF mRNA expression compared with the control-treated tumors, tentatively suggesting a possible role between VEGF overexpression and the development of resistance, concordant

TABLE 6.1. A list of examples, by no means exhaustive, of agents with anti-angiogenic activity currently under investigation.

Agent	Description	Target	Status
Bevacizumab	Humanized monoclonal antibody	human VEGF-A	FDA Approved
IMC-C1121b	Humanized monoclonal antibody	human VEGFR-2	Phase I/II
DC101	Monoclonal antibody	Mouse VEGFR-2	
Metronomic chemotherpay	Continuous low-dose of chemotherapy	Endothelial cells, EPCs	Phase II/III
SU11248/Sunitinib	Small molecule inhibitor of various RTKs	VEGFR-2, PDGFR, FLT-3, c-Kit	FDA Approved
BAY-43-9006/ Sorafenib	Small molecule inhibitor of various RTKs	Raf Kinase, VEGFR-2, PDGFR, FLT-3, c-Kit	FDA Approved
PTK787	Small molecule inhibitor of various RTKs	VEGFR-1, -2, -3, PDGFR, c-Kit	Phase III
IMC-225 (Erbitux)[1]	Humanized monoclonal antibody	EGFR	FDA Approved
Herceptin (Trastuzumab)[1]	Humanized monoclonal antibody	ERB-2 (HER-2/neu)	FDA Approved

For further reference see: Ferrara et al. Nat Rev Drug Disc 2004, Jain et al. Nat Clin Pract Oncol 2006, as well as Kerbel and Folkman Nat Rev Cancer 2002.

[1] These antibodies inhibit the oncogene-mediated overexpression of pro-angiogenic factors such as VEGF, thus 'indirectly' inhibiting tumor angiogenesis. EPC- Endothelial Progenitor cells, RTK- Receptor Tyrosine Kinase.

with previous similar findings by Viloria-Petit et al., who studied acquired resistance to antibodies targeting EGF receptors [22]. Further evidence for this hypothesis was obtained by the addition of bevacizumab treatment when tumors were beginning to relapse while on trastuzumab plus cyclophosphamide combination therapy, which then resulted in a further growth delay compared with tumors maintained on the original therapy alone (see Fig. 6.2). This result has two important ramifications. First, it serves as further evidence (and another example) of a chemosensitizing role of bevacizumab. Second, it challenges the current dogma that tumors expressing little or no VEGF

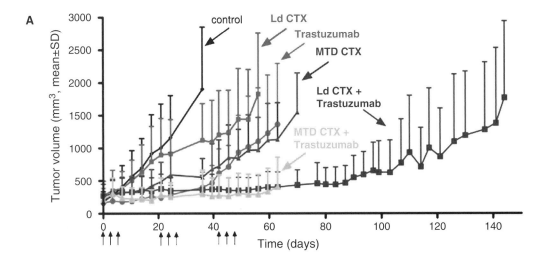

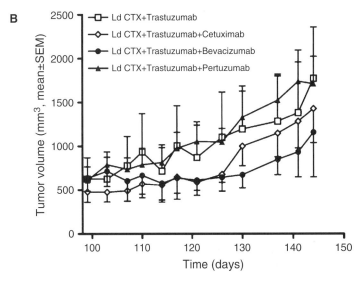

FIGURE 6.2. In vivo treatment of Her-2 positive MDA231-H2N human mammary tumors grown orthotopically in female SCID mice. **A.** Antitumor effects of low-dose metronomic cyclophosphamide (Ld CTX), maximum tolerated dose cyclophosphamide (MTD CTX), or trastuzumab alone, and combination regimens using low dose metronomic cyclophosphamide plus trastuzumab, or trastuzumab in combination with MTD cyclophosphamide. Arrows indicate time of MTD CTX dosing. **B.** Addition of second line therapies to tumors of MDA231-H2N that were starting to fail (as shown in top panel [A] after around 100 days) Ld CTX plus trastuzumab therapies. Second-line regimens were as indicated, with the addition of bevacizumab causing a further growth delay of approximately 4 weeks. (Adapted from du Manoir JM, Francia G, Man S, et al. Strategies for delaying or treating in vivo acquired resistance to trastuzumab in human breast cancer xenografts. Clin Cancer Res 2006, 12:904–916) (*See Color Plates*)

should not be considered for treatment with bevacizumab. Indeed, when such tumors start to fail conventional chemotherapy-based regimens, they may well concomitantly upregulate VEGF, making them good de novo potential targets for bevacizumab. It is of interest that Orlando et al. [23] recently reported that the combination of trastuzumab with metronomic cyclophosphamide was found to be effective in a study involving 22 patients with Her-2 positive metastatic breast cancer.

3.2 Metronomic (Antiangiogenic) Chemotherapy in Preclinical Models and Metastasis Therapy Experiments

Metronomic chemotherapy is a term coined to describe the continuous or frequent repetitive low-dose administration of chemotherapeutic agents with no prolonged breaks, as originally described by Browder et al. [24] and Klement et al. [9]. These groups reported the antiangiogenic and antitumor effects of low-dose cyclophosphamide [24] or vinblastine [9] in different preclinical models. Fundamentally important was an experiment conducted by Browder et al. showing that tumors made resistant to maximum tolerated dose (MTD) cyclophosphamide in vivo by repeated exposure to cycles of MTD cyclophosphamide would respond to metronomically administered cyclophosphamide. This result indicated that the antitumor mechanisms of metronomic dosing were distinct from MTD dosing of the same drug. A number of groups have since independently reproduced and validated this therapeutic strategy [14].

The metronomic concept is relatively new and, as with many new concepts, the initial investigative steps are of major influence but carry a high level of empiricism. One notable example is the effect of metronomic cyclophosphamide plus UFT (tegafur-uracil) in a model of advanced metastatic breast cancer in mice, which resulted in a very significant survival increase [25]. The decision to test UFT was primarily based on the use of this drug in a metronomic-like setting in the clinic, coupled with enthusiasm over encouraging clinical results, e.g., in the adjuvant therapy of resected non-small cell lung cancer where UFT was given daily at low dose every day for 2 years [26]. Of considerable interest, the preclinical application in the metastasis model (Fig. 6.3) [25] revealed an increase in survival that

was not observed when the same tumor line was grown orthotopically at the primary site.

3.2.1 Future Challenges Regarding Metronomic Regimens

The results by Munoz et al., as well as similar results previously reported [27], provide compelling evidence of the potential benefits of metronomic chemotherapy combinations. Furthermore, since a majority of preclinical studies employ primary tumor models (which are seldom grown orthotopically, as SC injections tend to be the preference due to ease of injection and subsequent tumor monitoring), the results from such studies are scrutinized long before most attempts are made evaluating the same strategy on metastatic models. Consequently, it is highly unlikely that a conventional research approach would have revealed the effect of the cyclophosphamide plus UFT combination on this metastasis model. Most likely it would have been discarded following disappointing results on primary tumors. Yet it is precisely metastatic disease, especially advanced metastasis, which is one of the major hurdles of cancer treatment. The report by Munoz et al. serves as an impetus for undertaking additional studies employing models of advanced metastatic disease.

It should also be noted that one concern with current preclinical studies is the use of tumor size reduction as a measure of the effectiveness of a therapeutic regimen, which may not be the most effective way to assess the impact of an angiogenesis inhibitor or treatment regimen. Similarly there is a lack of biomarkers that can be used to assess the correct dosing and biological activity of angiogenesis inhibitors although some small recent inroads have been made in these areas. For example, plasma levels of VEGF have been shown to be indicative of the optimal dosing of anti-VEGF receptor 2 targeting antibodies [28] and measurements of the numbers of circulating endothelial progenitor cells have been used to determine the optimal biologic dose of antiangiogenic agents [29].

So why are some metronomic-based regimens effective anti-angiogenic strategies at some anatomic sites, but not others? Are tumors growing at one site (e.g., in the lungs) more susceptible than others to combination therapies involving angiogenesis inhibitors? Would substituting metronomic

A Primary Tumor

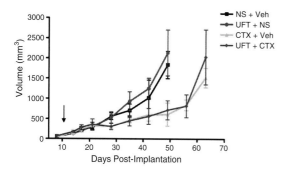

B Metastatic Disease

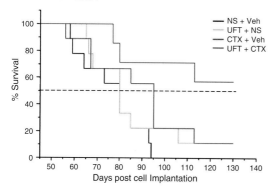

FIGURE 6.3. An example of how chronic combination oral metronomic low-dose CTX and UFT prolongs survival of mice with advanced metastatic disease. (From Munoz R, Man S, Shaked Y, et al. Highly efficacious nontoxic preclinical treatment for advanced metastatic breast cancer using combination oral UFT-cyclophosphamide metronomic chemotherapy. Cancer Res 2006, 66:3386–3391.) **A.** 231/LM2-4 human breast metastatic variant cells were orthotopically injected into the 6- to 8-week-old CB17 SCID mice. When tumors reached volumes of approximately 200 mm³, treatment with either vehicle control, or 15 mg/kg/day UFT by gavage, or 20 mg/kg per day CTX through the drinking water, or a combination of CTX and UFT treatments was initiated. Tumors were measured weekly and tumor volume was plotted accordingly. Arrow indicates time of initiation of treatment. **B.** 6-week-old CB-17 SCID mice were recipients of 231/LM2-4 transplanted cells. When tumors reached 400 mm³ (which took approximately 3 weeks) primary tumors were surgically removed. Treatment with vehicle control, 15 mg/kg per day UFT by gavage, 20 mg/kg per day CTX through the drinking water, or the daily combination of metronomic UFT and CTX, were initiated 3 weeks after surgery on a daily non-stop basis. For example, in the experiment shown in **(B)**, the duration of the therapy was 140 days, and was initiated on day 43, 3 weeks after surgery, with termination at day 183. Mice were monitored frequently

cyclophosphamide with VEGF neutralizing antibodies (as a means of delivering an antiangiogenic effect) still produce greater survival when treating preclinical advanced breast cancer metastasis, combined with UFT? Or do individual antiangiogenic strategies show distinct spectra of activity toward different tumors? Obtaining answers to these questions, as well as effective use of appropriate preclinical metastasis models and biomarkers to monitor dosing and response, will allow for a clearer definition of the extent to which antiangiogenic agents may be combined with chemotherapy.

3.3 Proposed Mechanisms of Action for Bevacizumab and Metronomic Chemotherapy

3.3.1 Metronomic Chemotherapy

Initial studies by Browder et al. and by Klement et al. suggested that metronomic chemotherapy led to the induction of endothelial cell death [9, 24]. These studies were followed by others using a 6-day in vitro exposure protocol of very low concentrations of various chemotherapeutic agents on different human normal and tumor cell lines [30]. This revealed an exquisite sensitivity of human umbilical vein endothelial cells, but not fibroblasts or tumor cells to such therapy. A subsequent microarray analysis screening of in vitro grown endothelial cells subjected to 6-day in vitro metronomic treatment revealed the upregulation of thrombospondin-1 (TSP-1) mRNA compared with control [31]. Consistent with these findings, metronomic cyclophosphamide was found to be less effective in TSP-1 null mice (in which endothelial cells cannot respond to the metronomic regimen by increasing TSP-1 expression) compared with wild-type mice bearing the same tumor model. Thus

according to the institutional guidelines. A Kaplan-Meier survival curve was plotted accordingly for all treated groups, as indicated in the figure. **A, B.** n = 7–9/group. NS = normal saline, Veh = vehicle control (0.1% HPMC). Note that effects on primary tumor (**A**) were minor and in no way predictive of the survival benefits seen with UFT and CTX on metastatic disease (recorded as survival) (*See Color Plates*)

upregulation of TSP-1, a well-known endogenous inhibitor of angiogenesis, was the first proposed molecular mechanism to explain the antiangiogenic basis of metronomic chemotherapy, a hypothesis confirmed by others [32]. In this respect there are two important factors to note. First, the results with TSP-1 null mice clearly showed that metronomic cyclophosphamide retained significant effectiveness in the TSP-1 null background, suggesting the existence of additional mechanisms. Furthermore recent reports (discussed later in this chapter) have described other mechanisms, including impairment of bone marrow–derived circulating endothelial cells, the selective targeting of tumor stem cells, and boosting the immune system by targeting T regulatory cells [33–35].

3.3.2 Bevacizumab

The target of bevacizumab is VEGF, which effectively negates ambiguity over its mechanisms of action, especially in preclinical models, in which the only target is human VEGF secreted by a human tumor xenograft. However, although the dogma is that neutralizing tumor secreted VEGF prevents endothelial cell growth, thus inhibiting angiogenesis, there is a growing body of evidence pointing to tumor cells expressing functional receptors to VEGF as well [36]. These tumors cells could therefore presumably be targeted by bevacizumab. This could be difficult to assess preclinically as the host (mouse) VEGF could also stimulate tumor-bound VEGF receptors without interference by bevacizumab. Notwithstanding such arguments, some studies suggest that the central mechanism of chemosensitization by bevacizumab is by sequestration of the survival factor function that VEGF has for the endothelial cell, which after chemotherapy fails to recover, and this in turn leads to inhibition of tumor angiogenesis.

3.4 Proposed Mechanisms to Explain the Effectiveness of Angiogenesis Inhibitors as Chemosensitizing Agents

Following the clinical trial success and approval of bevacizumab, a number of hypotheses to explain the chemosensitizing mode of action of inhibitors of angiogenesis were put forward [4]. These are "vessel normalization," inhibition of the rate and extent of tumor repopulation, augmentation of the anti-vascular effects of chemotherapy, thus preventing or suppressing chemotherapy induced vasculogenesis and targeting the vascular tumor stem cell niche. The four hypotheses are described in the following with a cautionary note that they need not be mutually exclusive.

3.4.1 Induction of "Vessel Normalization"

This hypothesis, proposed by R. Jain and colleagues, states that angiogenesis inhibitors can bring about a transient phase during which some new vessels within a tumor mass are pruned and eradicated, thus leading to a more efficient blood flow through parts of the tumor. Also reduction in vascular permeability would occur, thus reducing interstitial fluid pressure. Antiangiogenic therapy would therefore transiently "normalize" the chaotic and dysfunctional tumor vascular network. This effect would result in a better delivery of chemotherapy to the tumor bed and therefore produce an overall chemosensitizing effect [37, 38]. This hypothesis also suggests that chemotherapy ought to be administered within a certain period of time following the delivery of angiogenesis inhibitors, to coincide with the period of maximized blood flow through the tumor—otherwise known as a "normalization" window. It is yet unclear how such a window would be ascertained in a patient with multiple metastatic lesions, each one of which would have a distinct microenvironment. On the other hand, the attractiveness of this hypothesis lies in the fact that it can be essentially tested via time-dosed schedules, and it is therefore likely that ongoing clinical trials (differing in timing of administration) will soon begin to provide data to better attest to the validity and/or prevalence of this concept.

3.4.2 Blunting Tumor Repopulation

Even when a tumor mass responds to chemotherapy with massive tumor cell death and tumor shrinkage, this is usually followed by rapid tumor regrowth and repopulation [39]. The rebound, which occurs in the rest periods between cycles of treatment, can in some cases be sufficiently robust as to ultimately negate any initial benefits of the chemotherapy. It has been hypothesized [4, 39] that this tumor repopulation would be dependent on the tumor vascular network to provide nutrients and oxygen required for tumor cell growth. Therefore inhibitors of angiogenesis may eliminate one of the crucial mechanisms necessary for the rapid regrowth after cytotoxic chemotherapy [40]. It should be noted that this hypothesis is also relevant to the emerging theory of tumor stem cells in cancer growth—and regrowth after cancer therapy—as discussed in the following.

3.4.3 Augmentation of the Antivascular Effects of Chemotherapy by Targeting Circulating Bone Marrow–Derived Endothelial Progenitor Cells

One intriguing question for tumor angiogenesis researchers has been: Why is there little evidence that MTD chemotherapy has an antiangiogenic effect? After all, such regimens aggressively target rapidly dividing cells, which should include tumor-stimulated endothelial cells. One reason may be that VEGF acts as an essential "survival" factor for endothelial cells following chemotherapy, and bevacizumab negates this protective effect, thus rendering the dependent tumor vasculature sensitive to therapy [10]. Another hypothesis that has recently emerged is that bone marrow–derived circulating endothelial progenitor cells (CEPs) are rapidly released from the marrow compartment following MTD chemotherapy, and circulate to the tumor vasculature where they substitute, repopulate, and repair the damaged endothelium, thus maintaining tumor angiogenesis intact [33, 41, 42]. Such a pro-vasculogenic "rebound" process has been reported in mice when treated with a so-called vascular disrupting agent (VDA) such as combretastatin or a second-generation derivative called OXi-4503 [41]. The CEPs are rapidly mobilized by drug treatment and home to the viable rim of tumor tissue, which typically remains after treatment with this type of drug, where they facilitate tumor cell repopulation. So this type of drug and tumor response is a paradigm for a therapy that can cause significant tumor cell kill that is followed by rapid tumor regrowth. By combining an antiangiogenic drug with VDA treatment, the acute CEP mobilization and tumor homing process can be blocked, thus amplifying the efficacy of the VDA treatment [41]. Since MTD chemotherapy can also cause an acute mobilization of CEPs [33], a similar process

of tumor homing and boosting tumor angiogenesis may occur. If so, blocking it with an antiangiogenic drug would slow down tumor repopulation during the drug-free break periods [42]. Thus cytotoxic chemotherapy may have antithetical effects on tumor angiogenesis: (1) suppressing it by direct targeting/killing of activated endothelial cells in the growing tumor neovasculature, and (2) stimulating it by mobilizing CEPs, but concurrent treatment with an antiangiogenic drug such as an antibody to VEGF receptor 2 can amplify the first effect and block the second so that the overall potential of some conventional cytotoxic chemotherapy regimens to mediate an antiangiogenic effect that contributes to the antitumor efficacy is enhanced.

3.4.4 Targeting the Vascular Cancer Stem Cell Niche in Order to Enhance the Activity of Cytotoxic Therapy Toward Cancer Stem Cells

Of relevant note is a distinct emerging hypothesis that vascular endothelial cells may contribute to the creation of a niche that promotes the existence of tumor subpopulations with stem cell–like properties [43]. These "tumor stem cells" are thought to be resistant to chemotherapy and could contribute to the rapid repopulation of a tumor following treatment. Thus drugs such as bevacizumab may interfere with the vascular niche, leading (indirectly) to the eradication of tumor subpopulations with stem cell–like properties and therefore prevent the tumor repopulation phenomenon described in the preceding. Such a selective targeting of tumor stem cell–like populations was recently described with metronomic cyclophosphamide therapy, but crucially not when the same drug was administered in an MTD fashion, using a mouse glioma model [34].

3.5 Toxicity Associated with Combinations of Angiogenesis Inhibitors and Chemotherapy

As novel antiangiogenic strategies are tested for their ability to render chemotherapy more effective, the toxicity of particular combinations will inevitably emerge. For example, in a clinical trial testing gemcitabine plus cisplatin in combination with the small molecule VEGF receptor 2 inhibitor

SU5416, 42% of patients had thromboembolic or other vascular complications [44]. A subsequent preclinical study showed that the TFP/TPFI ratio (expression of tissue factor, a pro-coagulatory factor, over that of tissue factor pathway inhibitor) in human umbilical vein endothelial cells (HUVECs) exposed in vitro to different drugs was strikingly high in the drug combination that caused notable toxicity in the aforementioned trial [45]. Curiously in this assay angiogenic drugs such as bevacizumab showed a negligible effect on the TFP/TPFI ratio, confirming the low toxicity profile of angiogenesis inhibitors when used as single agents. This may be a good example of "translational" research, in which clinical observations lead to preclinical model studies that suggest a means to overcome the problem (in this case of high toxicity). It is likely that more assays like the TFP/TPFI ratio will have to be developed. This is in part due to the fact that inhibitors of angiogenesis do not exhibit the classical toxic behavior of most chemotherapeutic drugs, for which the optimal biological dose is typically the MTD [14]. Therefore single-agent clinical trials (or preclinical studies for that matter) could fail to reveal any serious toxicity profile of a new drug, which, unfortunately, would appear in a phase II and III trial when combined with chemotherapy or other agents.

With regard to low-dose metronomic cyclophosphamide as an antiangiogenic strategy, it is worthwhile to point to the minimal toxicity associated with this regimen. Preclinical studies have shown that, similar to low-dose vinblastine, the metronomic cyclophosphamide protocol does not exhibit any overt/serious gross toxicity, even after 6 months or more of continuous administration [9, 21]. An exhaustive preclinical study on the toxicity profile of low-dose metronomic cyclophosphamide identified only sustained lymphopenia as a potentially serious side effect [46]. However, it is unlikely that such regimens will be applied on their own, but rather they will be combined with other anticancer treatment strategies. As reported these can give rise to unexpected complications. On the other hand it is also notable that metronomic chemotherapy has been combined with a number of other different anticancer agents (e.g., DC101 [9], tirapazamine [46], trastuzumab [21, 23]) and thus far no severe toxicity has been noted.

3.6 Complex Therapeutic Combination Regimens

In addition to many promising novel combinations of angiogenesis inhibitors plus chemotherapy, there is the interesting possibility of additionally incorporating a third therapeutic concept or class of anticancer agents. For example, Pietras and Hanahan reasoned that since pericyte coverage may contribute to the tumor endothelium being resistant to MTD chemotherapy, targeting the pericyte population would improve the efficacy of metronomic chemotherapy [47]. Through a series of elegant experiments in the transgenic RipTag2 model of pancreatic islet carcinoma they devised a "chemo-switch" regimen (Fig. 6.4), which resulted in significantly extended survival compared with controls. The treatment protocol consists of an upfront short course of MTD cyclophosphamide followed by a "switch" to a metronomic cyclophosphamide dosing in combination with sunitinib to both target pericytes and endothelial cells.

Another example arises from the consideration that chronic angiogenesis inhibition ought to reduce nutrient and oxygen delivery to a tumor, with increased hypoxia being the expected eventual outcome. Under such conditions adding a drug that targets hypoxic tumor cells should increase the therapeutic efficacy. In a preclinical study, using PC-3 human prostate xenografts in nude mice, this was indeed found to be the case. Low-dose metronomic cyclophosphamide was employed as an antiangiogenic strategy that was combined with tirapazamine [48]. Tirapazamine is a "hypoxic cell cytotoxin" (i.e., a DNA-damaging drug activated under conditions of hypoxia), and this therapy combination resulted in a significant tumor growth delay. Therefore it will be important to test, once efficacious combinations such as low-dose cyclophosphamide plus UFT or "chemo-switch" regimens have been identified, if these approaches can be further enhanced by agents such as tirapazamine. This could set the stage for a new era of complex, yet rationally designed, therapeutic strategies.

4 Clinical Aspects

4.1 Single-Agent Antiangiogenic Therapy

The response rates of single-agent antiangiogenic drugs in most clinical trials of advanced cancer are generally low, with the exception of certain

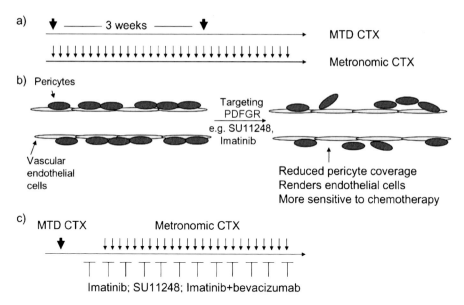

FIGURE 6.4. Evolution of the "chemo-switch" therapy strategy. **A.** Compared with MTD CTX, shown as a CTX bolus dose followed by a 3-week rest period, metronomic dosing is defined by the absence of rest periods and continuous administration of lower CTX doses (daily administration is shown here). **B.** Pietras and Hannahan conceived a strategy to try and improve the sensitivity of endothelial cells to chemotherapy by inhibiting pericyte coverage of the tumor endothelium using PDGFR targeting drugs, such as sunitinib. **C.** This approach eventually resulted in the "chemo-switch" strategy, where an upfront (debulking) MTD CTX is given followed by a metronomic scheduling combined with vascular endothelial and pericyte targeting drugs

RTKIs (e.g. sunitinib) in specific indications (e.g. renal cell cancer). This explains why many such compounds, e.g. bevacizumab, are preferentially being tested clinically in conjunction with conventional chemotherapy. In 2004, the monoclonal anti-VEGF antibody bevacizumab became the first US FDA-approved antiangiogenic agent. Bevacizumab was initially approved for use in combination with intravenous 5-fluorouracil-based chemotherapy for first-line treatment of patients with metastatic carcinoma of the colon or rectum. Other indications soon followed, i.e., use in combination with oxaliplatin and intravenous 5-fluorouracil-based chemotherapy for second-line treatment of colorectal cancer [49], and in combination with carboplatin and paclitaxel for the first-line treatment of patients with unresectable, locally advanced, recurrent or metastatic non-squamous non-small cell lung cancer [50]. Currently, more than 30 compounds with antiangiogenic properties are evaluated in clinical trials (http://www.cancer.gov/clinicaltrials/developments/anti-angio-table).

As far as the development of bevacizumab is concerned, five phase II studies were initiated in 1998, including three studies involving single-agent bevacizumab for androgen-independent prostate, metastatic breast, and renal cell cancer [7]. The results of these three single-agent bevacizumab studies are summarized in Table 6.2, including two studies that were later initiated. Objective responses are generally seen in less than 10% of patients, with ovarian cancer being the exception. Even in the case of renal cell carcinoma, a strongly VEGF-dependent disease, the response rate does not exceed 10%.

Objective responses to the antiangiogenic multi-targeted RTKIs sunitinib (FDA approved for renal cell carcinoma and gastrointestinal stromal tumors) [12] and sorafenib (FDA approved for renal cell cancer) [13] are similarly low with the exception being the treatment of renal cell carcinoma with sunitinib. In fact, a response rate of 31% was observed in a phase III trial of previously untreated patients with metastatic renal cell carcinoma receiving sunitinib [51]. The true clinical benefit of sunitinib in this and another phase III study involving patients with gastrointestinal stromal tumors after imatinib failure [52], however, is better reflected by unprecedented high stable disease rates and meaningful prolongation of time to progression. Furthermore, both sunitinib and sorafenib show very promising clinical activity in many other tumor types, such as breast, non-small cell lung, prostate, and thyroid cancer, as well as hepatocellular carcinoma [53, 54].

4.2 Combination Therapies

4.2.1 Bevacizumab Combined with Conventional Chemotherapy

Two of the aforementioned five initial phase II trials involved the combination of bevacizumab with standard chemotherapy regimens for metastatic colorectal and non-small cell lung cancer (Table 6.3), resulting in the most encouraging efficacy results that then were further explored in phase III trials [7]. By combining IFL chemotherapy with bevacizumab given at a dose of 5 mg/kg every 2 weeks, Hurwitz et al. demonstrated a significantly increased overall response rate (44.8% versus 34.8%) and median survival time (20.3 versus 15.6 months) in the first-line treatment of metastatic colorectal cancer compared with IFL chemotherapy only [1]. Similarly, adding bevacizumab at a dose of 10 mg/kg every other week to FOLFOX4 (i.e., oxaliplatin, 5-fluorouracil, and leucovorin) chemotherapy resulted in improved progression-free and overall survival as well as an objective response rate in previously treated advanced colorectal cancer [49]. By applying the same principle, combination therapies of bevacizumab with paclitaxel-carboplatin or weekly (3 out of 4 weeks) paclitaxel were shown to be superior to chemotherapy alone in the treatment of recurrent or advanced (non-squamous) non-small cell lung cancer [50] and metastatic breast cancer [55], respectively (see Table 6.3).

This paradigm, however, seems not to be universally applicable. In fact, a randomized phase III trial comparing capecitabine with or without bevacizumab in patients with extensively pretreated metastatic breast cancer did not result in improved progression-free or overall survival despite an increased response rate of 19.8% (versus 9.1% in the capecitabine-only arm) [56]. There are several explanations why the primary endpoint of this study, i.e., improved progression-free survival, was not met. First, the postulated chemosensitizing effect of bevacizumab toward capecitabine might only be weak or absent, although the increased response rate suggests otherwise. Second, resistance to capecitabine might have developed very rapidly despite combination with bevacizumab in this very advanced patient population. The latter might also account for the low response rate (1%) and short progression-free survival (3.5 months) observed in colorectal cancer patients administered bevacizumab and fluorouracil/leucovorin following progression after both irinotecan- and oxaliplatin-based regimens [57]. This study and the experience

TABLE 6.2. Activity of single-agent bevacizumab.

	Objective RR (%)	SD (%)	PFS/TTP (median, m)	Comments
Androgen-independent prostate cancer (Reese 01)	0	-	4	Phase II, n = 15 BEV 10 mg/kg q2w
Metastatic breast cancer (Cobleigh 2003)	6.7	16	5.6	Phase I/II, n = 75 BEV 3, 10 or 20 mg/kg q2w
Renal cell carcinoma (Yang 2003)	10	-	4.8	Phase II, n = 39 BEV 10 mg/kg q2w, (3 mg/kg dose: objective RR 0%)
Recurrent NHL (Stopeck 2005)	5	20	5	Phase II, n = 46 BEV 10 mg/kg q2w
Recurrent ovarian cancer (Burger 2005)	18	55	-	Phase II, n = 62 BEV 15 mg/kg q3w
Colorectal cancer (Giantonio 2007)	3.3	-	2.7	Phase III three-arm trial, BEV arm n = 243, 2nd line, 10 mg/kg q2w

RR = response rate, PFS = progression free survival, TTP = time to progression, OS = overall survival, BEV = bevacizumab, NHL = Non-Hodgkin's lymphoma, m = months, w = weeks.

TABLE 3. Clinical trials of bevacizumab combined with conventional chemotherapy.

Endpoint	Hurwitz et al. mCRC 1st line		Giantonio et al. mCRC 2nd line		Sandler et al. NSCLC 1st line		Miller et al. ABC 1st line	
	IFL + BEV	IFL	FOLFOX4 + BEV	FOLFOX4	P-C + BEV	P-C	P + BEV	P
Median overall survival (months)	20.3	15.6	12.9	10.8	12.3	10.3	11.4	6.11
One-year survival rate (%)	4.3	63.4	-	-	51	44	-	-
Median progression-free survival (months)	10.6	6.2	7.3	4.7	6.2	4.5	-	-
Overall response rate (%)	44.8	34.8	22.7	8.6	15	35	29.9	13.8

mCRC = metastatic colorectal cancer, NSCLC = non-small cell lung cancer, ABC = advanced breast cancer, IFL = irinotecan-fluorouracil-leucovorin, FOLFOX4 = oxaliplatin-fluorouracil-leucovorin, P-C = paclitaxel-carboplatin , P = paclitaxel, BEV = bevacizumab.

in breast cancer suggest that earlier use of combined antiangiogenic and cytotoxic therapy might be more promising. However, even when used in early stages, fewer than half of colorectal, non-small cell lung and breast cancer patients benefit from adding bevacizumab to standard chemotherapy. Moreover, notoriously chemoresistant tumors such as pancreatic cancer seem not to be amenable at all to the chemosensitizing effects of bevacizumab. In fact, the recent interim results of a phase III clinical trial involving gemcitabine with and without bevacizumab for the treatment of advanced pancreatic cancer did not reveal a benefit of the combination (CALGB 80303, http://www.calgb.org).

4.2.2 Targeted Antiangiogenic Drugs Combined with Low-Dose Metronomic Chemotherapy

Based on the preclinical evidence reported above suggesting synergistic activity, not only conventional, MTD but also low-dose metronomic chemotherapy protocols are currently tested in combination with a variety of compounds featuring antiangiogenic properties, including the aromatase inhibitor letrozole [58], cyclooxygenase-2 inhibitors [59–61], thalidomide [59, 62], and bevacizumab [63, 64]. Three randomized phase II studies are of particular interest. Bottini et al. showed in elderly breast cancer patients receiving neoadjuvant letrozole alone or combined with metronomic cyclophos-

phamide an increase in the overall response rate from 71.9% in the letrozole only arm to 87.7% in the combination arm, without adding significant toxicity [58]. Similarly, the combination of metronomic cyclophosphamide and methotrexate with 10 mg/kg of bevacizumab q2weeks resulted in increased partial response (29% versus 10%) and stable disease rates (41% versus 38%) compared with metronomic chemotherapy alone in stage IV breast cancer patients that had received ≤1 prior chemotherapy regimen for metastatic disease [63]. The median time to progression in the combination group was 5.5 months compared with 2.0 months in the patients receiving metronomic cyclophosphamide and methotrexate only. On the other hand, thalidomide, a more promiscuous compound with antiangiogenic activity among others, seems not to be beneficial when coadministered with metronomic cyclophosphamide and methotrexate in advanced metastatic breast cancer patients [62].

4.2.3 Antiangiogenic RTKIs Combined with Conventional Chemotherapy

Whereas adding bevacizumab to various conventional chemotherapy regimens is beneficial in tumors as diverse as colorectal, breast, and non-small cell lung cancer, it still remains to be seen whether similar or even better results can be achieved with antiangiogenic RTKIs. For example, the combination of vatalanib (PTK787) with

FOLFOX4 chemotherapy for untreated (CONFIRM 1 study) or previously treated (CONFIRM 2 study) metastatic colorectal cancer did not result in a significant change in overall survival despite improved progression-free survival [38]. However, poor prognosis patients with high serum LDH levels might derive a significant benefit, as suggested by subgroup analyses. Mature results of clinical trials combining conventional chemotherapy with the RTKIs sunitinib and sorafenib are eagerly awaited. It will be interesting to see whether the safety profile in such trials is as favorable as been seen in combination regimens using bevacizumab, and whether improved efficacy benefit is observed.

4.3 Future Challenges

Although combinations of bevacizumab with standard chemotherapy improve outcome parameter in various tumor types, the percentage of objective responses stays below 50% (see Table 6.2), leaving ample room for improvement.

4.3.1 Dosing

The dosing of antiangiogenic drugs remains very empirical in the absence of validated pharmacodynamic surrogate markers [65]. For instance, with respect to bevacizumab, there is no consistent dose–response relationship across various clinical trials. As an example, both 5 mg/kg [1] and 10 mg/kg [49] doses given q2weeks have been successfully applied in the treatment of colorectal cancer. In a randomized phase II trial, however, the 5 mg/kg dose of bevacizumab was superior to 10 mg/kg when combined with 5-fluorouracil [66]. A u-shaped dose–response relationship might apply for at least some antiangiogenic drugs [67–69]. Conceivably, high doses could lead to extensive vessel rarefaction and reduced intratumoral cytotoxic drug deposition [70]. Given the complex biology of a given tumor and its vasculature, individual dosing might be necessary to optimize clinical outcome [71].

4.3.2 Timing

The optimal timing of cytotoxic and antiangiogenic drug administration might be especially an issue when using combinations involving RTKIs given their shorter biological half-life compared with antiangiogenic antibodies. It remains to be seen whether continuous or intermittent administration of such agents results in the best therapeutic index in combination protocols. Ongoing clinical trials are also assessing the use of antiangiogenic agents as a maintenance therapy strategy after initial tumor debulking. Finally, it needs to be defined whether there is a role for continued antiangiogenic therapy beyond progression despite combined cytotoxic and antiangiogenic treatment.

4.3.3 Tumor Stage

Theoretical considerations and (pre)clinical data suggest that the antitumor effects of most antiangiogenic agents are more pronounced in early disease stages when the tumor load is low [72, 73], or if the tumor mass can be reduced either before antiangiogenic therapy is initiated [47] or by concomitant MTD chemotherapy. It will therefore be interesting to see whether the impact of antiangiogenic agents combined with cytotoxic therapy is even more pronounced in the adjuvant setting.

4.3.4 Type of Cytotoxic Drug(s) Used in Combination Regimens

It remains to be seen whether the postulated chemosensitizing properties of antiangiogenic agents are similarly pronounced across the various groups of cytotoxic agents, or whether certain chemotherapeutics or combination regimens are more likely to result in beneficial effects when used in combination with antiangiogenic compounds.

5 Turning The Current Dogma Upside Down

As discussed, the objective response rates in clinical trials using single-agent antiangiogenic compounds such as bevacizumab are generally very low (see Table 6.2). This contrasts with commonly more pronounced prolongation of survival rates seen in preclinical xenograft studies [17]. This discrepancy could derive at least in part from phenotypic differences of the (murine) vasculature of rapidly growing ectopic xenografts compared with human tumors, which generally develop over prolonged periods of time. Mature pericyte-covered vessels have been shown to be more resistant to

VEGF blockade than immature vessels lacking proper pericyte coverage as commonly found in tumors [74, 75].

Most cytotoxic drugs exert antiangiogenic effects [76]. If given in a conventional MTD manner however, this antiangiogenic activity is followed by rapid repair processes in the drug-free break period, mediated by local angiogenesis and the influx of circulating pro-angiogenic/vasculogenic cells, including CEPs, mobilized from the bone marrow [24, 33]. This nascent vasculature is likely very susceptible, i.e., sensitized, to the effects of antiangiogenic agents. The beneficial effects of such combinations might therefore not only derive from the chemosensitizing effects of antiangiogenic agents, but also from the "antiangiogenesis-sensitizing" properties of cytotoxic drugs (Fig. 6.5).

Chemotherapy-induced tumor cell expression of angiogenic factors such as VEGF may be the major driving force behind these repair processes [77–81]. Both endothelial and tumor cell survival following chemotherapy administration might depend on this cytotoxic stress-induced expression changes [10, 36]. In addition, the "cytokine storm" following cytotoxic therapy involves growth factors such as G-CSF and M-CSF, known to accelerate neovascularization [82–86]. Similarly, erythropoietin stimulates the mobilization of endothelial progenitors cells [87]. Such cytokines could therefore result in deleterious pro-angiogenic activity if uncontested, yet also sensitize the growing vasculature to the effects of antiangiogenic agents. Intriguingly, growth factors such as G-CSF and erythropoietin are commonly used as part of the supportive care in the context of MTD chemotherapy.

If chemotherapy modifies the physiology of tumors in a way that would sensitize them to the effects of antiangiogenic agents, one might wonder whether such changes could be obtained with more specificity and fewer side effects than usually associated with MTD chemotherapy. In fact, and as previously mentioned, our group recently showed that VDAs induce an acute mobilization of endothelial progenitor cells, which home to the remaining viable tumor rim [41]. Blunting of this response with the VEGF receptor 2 blocking antibody DC101 in mice results in better tumor control [41].

6 Conclusion

The field of angiogenesis is seemingly set for an explosive expansion and outreach into combinations with other areas of anticancer treatment, particularly

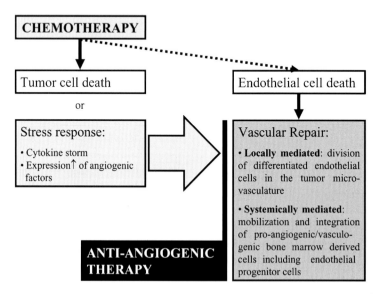

FIGURE 6.5. Cytotoxic therapy targets the tumor vasculature, but is also at the basis of repair processes in the drug-free break period, e.g., via stimulating tumor cells to express various angiogenic factors and cytokines. The repair-related nascent neovasculature is likely very susceptible to the effects of antiangiogenic agents. As such, chemotherapy might "antiangiogenesis-sensitize" tumors

chemotherapy. However, a cautionary note is also called for. New targets and drugs remain to be validated from the bench side, where there is already a confounding plethora of candidates, to the clinic. Some of these drugs, such as bevacizumab, sunitinib, and sorafenib, are already available for standard treatment, and a large number of others are currently being tested in clinical trials. Inevitably as these agents are tested for their efficacy in combination with standard care, so will their relative ability to serve as one kind or another chemosensitizer (e.g., to alkylating agents or to taxanes). This will allow for a better definition of inhibition of angiogenesis as a form of cancer treatment in its own right, or (alternatively) as a general potentiator, or enabler of certain established chemotherapy regimens and perhaps other types of therapy as well.

Acknowledgments. The authors' research summarized in this review was primarily supported by grants from the Canadian Health Institutes, the National Cancer Institute of Canada, and the National Cancer Institute (NIH), USA to RSK as well as sponsored research agreements with Taiho Pharmaceuticals, Japan, and ImClone Systems Inc., New York/USA. Dr. Kerbel is a recipient of a Tier 1 Canada Research Chair in Tumor Biology, Angiogenesis and Antiangogenic Therapy.

References

1. Hurwitz H, Fehrenbacher L, Novotny W, et al. Bevacizumab plus irinotecan, fluorouracil, and leucovorin for metastatic colorectal cancer. N Engl J Med 2004, 350:2335–42.
2. Folkman J. Tumor angiogenesis: therapeutic implications. N Engl J Med 1971, 285:1182–1186.
3. Teicher BA, Holden SA, Ara G, et al. Potentiation of cytotoxic cancer therapies by TNP-470 alone and with other anti-angiogenic agents. Int J Cancer 1994, 57:920–925.
4. Kerbel RS. Antiangiogenic therapy: a universal chemosensitization strategy for cancer? Science 2006, 312:1171–1175.
5. Nieder C, Wiedenmann N, Andratschke N, et al. Current status of angiogenesis inhibitors combined with radiation therapy. Cancer Treat Rev 2006, 32:348–364.
6. Senger DR, Galli SJ, Dvorak AM, et al. Tumor cells secrete a vascular permeability factor that promotes accumulation of ascites fluid. Science 1983, 219:983–985.
7. Ferrara N, Hillan KJ, Gerber HP, et al. Discovery and development of bevacizumab, an anti-VEGF antibody for treating cancer. Nat Rev Drug Discov 2004, 3:391–400.
8. Kerbel RS. Inhibition of tumor angiogenesis as a strategy to circumvent acquired resistance to anticancer therapeutic agents. Bioessays 1991, 13:31–36.
9. Klement G, Baruchel S, Rak J, et al. Continuous low-dose therapy with vinblastine and VEGF receptor-2 antibody induces sustained tumor regression without overt toxicity. J Clin Invest 2000, 105:R15–24.
10. Tran J, Master Z, Yu JL, et al. A role for survivin in chemoresistance of endothelial cells mediated by VEGF. Proc Natl Acad Sci USA 2002, 99:4349–4354.
11. Sweeney CJ, Miller KD, Sissons SE, et al. The antiangiogenic property of docetaxel is synergistic with a recombinant humanized monoclonal antibody against vascular endothelial growth factor or 2-methoxyestradiol but antagonized by endothelial growth factors. Cancer Res 2001, 61:3369–3372.
12. Goodman VL, Rock EP, Dagher R, et al. Approval summary: sunitinib for the treatment of imatinib refractory or intolerant gastrointestinal stromal tumors and advanced renal cell carcinoma. Clin Cancer Res 2007, 13:1367–1373.
13. Kane RC, Farrell AT, Saber H, et al. Sorafenib for the treatment of advanced renal cell carcinoma. Clin Cancer Res 2006, 12:7271–7278.
14. Kerbel RS, Kamen BA. The anti-angiogenic basis of metronomic chemotherapy. Nat Rev Cancer 2004, 4:423–436.
15. Inoue K, Slaton JW, Davis DW, et al. Treatment of human metastatic transitional cell carcinoma of the bladder in a murine model with the anti-vascular endothelial growth factor receptor monoclonal antibody DC101 and paclitaxel. Clin Cancer Res 2000, 6:2635–2643.
16. Bruns CJ, Liu W, Davis DW, et al. Vascular endothelial growth factor is an in vivo survival factor for tumor endothelium in a murine model of colorectal carcinoma liver metastases. Cancer 2000, 89:488–499.
17. Gerber HP, Ferrara N. Pharmacology and pharmacodynamics of bevacizumab as monotherapy or in combination with cytotoxic therapy in preclinical studies. Cancer Res 2005, 65:671–680.
18. Warren RS, Yuan H, Matli MR, et al. Regulation by vascular endothelial growth factor of human colon cancer tumorigenesis in a mouse model of experimental liver metastasis. J Clin Invest 1995, 95:1789–1797.
19. Melnyk O, Zimmerman M, Kim KJ, et al. Neutralizing anti-vascular endothelial growth factor antibody

inhibits further growth of established prostate cancer and metastases in a pre-clinical model. J Urol 1999, 161:960–963.

20. Rowe DH, Huang J, Kayton ML, et al. Anti-VEGF antibody suppresses primary tumor growth and metastasis in an experimental model of Wilms' tumor. J Pediatr Surg 2000, 35:30–32, discussion 32–33.

21. du Manoir JM, Francia G, Man S, et al. Strategies for delaying or treating in vivo acquired resistance to trastuzumab in human breast cancer xenografts. Clin Cancer Res 2006, 12:904–916.

22. Viloria-Petit A, Crombet T, Jothy S, et al. Acquired resistance to the antitumor effect of epidermal growth factor receptor-blocking antibodies in vivo: a role for altered tumor angiogenesis. Cancer Res 2001, 61:5090–5101.

23. Orlando L, Cardillo A, Ghisini R, et al. Trastuzumab in combination with metronomic cyclophosphamide and methotrexate in patients with HER-2 positive metastatic breast cancer. BMC Cancer 2006, 6:225.

24. Browder T, Butterfield CE, Kraling BM, et al. Antiangiogenic scheduling of chemotherapy improves efficacy against experimental drug-resistant cancer. Cancer Res 2000, 60:1878–1886.

25. Munoz R, Man S, Shaked Y, et al. Highly efficacious nontoxic preclinical treatment for advanced metastatic breast cancer using combination oral UFT-cyclophosphamide metronomic chemotherapy. Cancer Res 2006, 66:3386–3391.

26. Kato H, Ichinose Y, Ohta M, et al. A randomized trial of adjuvant chemotherapy with uracil-tegafur for adenocarcinoma of the lung. N Engl J Med 2004, 350:1713–1721.

27. Kerbel RS. Human tumor xenografts as predictive preclinical models for anticancer drug activity in humans: better than commonly perceived—but they can be improved. Cancer Biol Ther 2003, 2: S134–139.

28. Bocci G, Man S, Green SK, et al. Increased plasma vascular endothelial growth factor (VEGF) as a surrogate marker for optimal therapeutic dosing of VEGF receptor-2 monoclonal antibodies. Cancer Res 2004, 64:6616–6625.

29. Shaked Y, Emmenegger U, Man S, et al. Optimal biologic dose of metronomic chemotherapy regimens is associated with maximum antiangiogenic activity. Blood 2005, 106:3058–3061.

30. Bocci G, Nicolaou KC, Kerbel RS. Protracted low-dose effects on human endothelial cell proliferation and survival in vitro reveal a selective antiangiogenic window for various chemotherapeutic drugs. Cancer Res 2002, 62:6938–6943.

31. Bocci G, Francia G, Man S, et al. Thrombospondin 1, a mediator of the antiangiogenic effects of low-dose metronomic chemotherapy. Proc Natl Acad Sci U S A 2003, 100:12917–12922.

32. Hamano Y, Sugimoto H, Soubasakos MA, et al. Thrombospondin-1 associated with tumor microenvironment contributes to low-dose cyclophosphamide-mediated endothelial cell apoptosis and tumor growth suppression. Cancer Res 2004, 64:1570–1574.

33. Bertolini F, Paul S, Mancuso P, et al. Maximum tolerable dose and low-dose metronomic chemotherapy have opposite effects on the mobilization and viability of circulating endothelial progenitor cells. Cancer Res 2003, 63:4342–4346.

34. Folkins C, Man S, Xu P, et al. Anticancer therapies combining antiangiogenic and tumor cell cytotoxic effects reduce the tumor stem-like cell fraction in glioma xenograft tumors. Cancer Res 2007, 67:3560–3564.

35. Ghiringhelli F, Menard C, Puig PE, et al. Metronomic cyclophosphamide regimen selectively depletes CD4(+)CD25 (+) regulatory T cells and restores T and NK effector functions in end stage cancer patients. Cancer Immunol Immunother 2007, 56:641–648.

36. Wey JS, Stoeltzing O, Ellis LM. Vascular endothelial growth factor receptors: expression and function in solid tumors. Clin Adv Hematol Oncol 2004, 2:37–45.

37. Jain RK. Normalization of tumor vasculature: an emerging concept in antiangiogenic therapy. Science 2005, 307:58–62.

38. Jain RK, Duda DG, Clark JW, et al. Lessons from phase III clinical trials on anti-VEGF therapy for cancer. Nat Clin Pract Oncol 2006, 3:24–40.

39. Kim JJ, Tannock IF. Repopulation of cancer cells during therapy: an important cause of treatment failure. Nat Rev Cancer 2005, 5:516–525.

40. Hudis CA. Clinical implications of antiangiogenic therapies. Oncology (Williston Park) 2005, 19:26–31.

41. Shaked Y, Ciarrocchi A, Franco M, et al. Therapy-induced acute recruitment of circulating endothelial progenitor cells to tumors. Science 2006, 313: 1785–1787.

42. Shaked Y, Kerbel RS. Antiangiogenic strategies on defense: blocking rebound by the tumor vasculature after chemotherapy. Cancer Res 2007, in press.

43. Calabrese C, Poppleton H, Kocak M, et al. A perivascular niche for brain tumor stem cells. Cancer Cell 2007, 11:69–82.

44. Kuenen BC, Rosen L, Smit EF, et al. Dose-finding and pharmacokinetic study of cisplatin, gemcitabine, and SU5416 in patients with solid tumors. J Clin Oncol 2002, 20:1657–1667.

45. Ma L, Francia G, Viloria-Petit A, et al. In vitro procoagulant activity induced in endothelial cells by chemotherapy and antiangiogenic drug combinations: modulation by lower-dose chemotherapy. Cancer Res 2005, 65:5365–5373.

46. Emmenegger U, Man S, Shaked Y, et al. A comparative analysis of low-dose metronomic cyclophosphamide reveals absent or low-grade toxicity on tissues highly sensitive to the toxic effects of maximum tolerated dose regimens. Cancer Res 2004, 64:3994–4000.

47. Pietras K, Hanahan D. A multitargeted, metronomic, and maximum-tolerated dose "chemo-switch" regimen is antiangiogenic, producing objective responses and survival benefit in a mouse model of cancer. J Clin Oncol 2005, 23:939–952.

48. Emmenegger U, Morton GC, Francia G, et al. Low-dose metronomic daily cyclophosphamide and weekly tirapazamine: a well-tolerated combination regimen with enhanced efficacy that exploits tumor hypoxia. Cancer Res 2006, 66:1664–1674.

49. Giantonio BJ, Catalano PJ, Meropol NJ, et al. Bevacizumab in combination with oxaliplatin, fluorouracil, and leucovorin (FOLFOX4) for previously treated metastatic colorectal cancer: results from the Eastern Cooperative Oncology Group Study E3200. J Clin Oncol 2007, 25:1539–1544.

50. Sandler A, Gray R, Perry MC, et al. Paclitaxel-carboplatin alone or with bevacizumab for non-small-cell lung cancer. N Engl J Med 2006, 355:2542–2550.

51. Motzer RJ, Hutson TE, Tomczak P, et al. Sunitinib versus interferon alfa in metastatic renal-cell carcinoma. N Engl J Med 2007, 356:115–124.

52. Demetri GD, van Oosterom AT, Garrett CR, et al. Efficacy and safety of sunitinib in patients with advanced gastrointestinal stromal tumour after failure of imatinib: a randomised controlled trial. Lancet 2006, 368:1329–1338.

53. Chow LQ, Eckhardt SG. Sunitinib: from rational design to clinical efficacy. J Clin Oncol 2007, 25:884–896.

54. Wilhelm S, Carter C, Lynch M, et al. Discovery and development of sorafenib: a multikinase inhibitor for treating cancer. Nat Rev Drug Discov 2006, 5:835–844.

55. Miller KD, Wang M, Gralow J, et al. A randomized phase III trial of paclitaxel versus paclitaxel and bevacizumab as first-line therapy for locally recurrent or metastatic breast cancer: a trial coordinated by the Eastern Cooperative Oncology Group (E2100). San Antonio Breast Cancer Symposium, 2005. Breast Cancer Res Treat. 2005, 95(suppl 1): abstract 3.

56. Miller KD, Chap LI, Holmes FA, et al. Randomized phase III trial of capecitabine compared with bevacizumab plus capecitabine in patients with previously treated metastatic breast cancer. J Clin Oncol 2005, 23:792–799.

57. Chen HX, Mooney M, Boron M, et al. Phase II multicenter trial of bevacizumab plus fluorouracil and leucovorin in patients with advanced refractory colorectal cancer: an NCI Treatment Referral Center Trial TRC-0301. J Clin Oncol 2006, 24:3354–3360.

58. Bottini A, Generali D, Brizzi MP, et al. Randomized phase II trial of letrozole and letrozole plus low-dose metronomic oral cyclophosphamide as primary systemic treatment in elderly breast cancer patients. J Clin Oncol 2006, 24:3623–3628.

59. Kieran MW, Turner CD, Rubin JB, et al. A feasibility trial of antiangiogenic (metronomic) chemotherapy in pediatric patients with recurrent or progressive cancer. J Pediatr Hematol Oncol 2005, 27:573-81.

60. Buckstein R, Kerbel RS, Shaked Y, et al. High-Dose celecoxib and metronomic "low-dose" cyclophosphamide is an effective and safe therapy in patients with relapsed and refractory aggressive histology non-Hodgkin's lymphoma. Clin Cancer Res 2006, 12:5190–198.

61. Young SD, Whissell M, Noble JC, et al. Phase II clinical trial results involving treatment with low-dose daily oral cyclophosphamide, weekly vinblastine, and rofecoxib in patients with advanced solid tumors. Clin Cancer Res 2006, 12:3092–3098.

62. Colleoni M, Orlando L, Sanna G, et al. Metronomic low-dose oral cyclophosphamide and methotrexate plus or minus thalidomide in metastatic breast cancer: antitumor activity and biological effects. Ann Oncol 2006, 17:232–238.

63. Burstein HJ, Spigel D, Kindsvogel K, et al. Metronomic chemotherapy with and without bevacizumab for advanced breast cancer: a randomized phase II study. San Antonio Breast Cancer Symposium, 2005. Breast Cancer Res Treat. 2005,95(suppl 1): abstract 4.

64. Garcia AA, Oza AM, Hirte H, et al. Interim report of a phase II clinical trial of bevacizumab (Bev) and low dose metronomic oral cyclophosphamide (mCTX) in recurrent ovarian (OC) and primary peritoneal carcinoma: A California Cancer Consortium Trial. ASCO Annual Meeting, 2005. J Clin Oncol 2005, 23(16s): abstract 5000.

65. Jubb AM, Oates AJ, Holden S, et al. Predicting benefit from anti-angiogenic agents in malignancy. Nat Rev Cancer 2006, 6:626–635.

66. Kabbinavar F, Hurwitz HI, Fehrenbacher L, et al. Phase II, randomized trial comparing bevacizumab plus fluorouracil (FU)/leucovorin (LV) with FU/LV alone in patients with metastatic colorectal cancer. J Clin Oncol 2003, 21:60–65.

67. Slaton JW, Perrotte P, Inoue K, et al. Interferon-alpha-mediated down-regulation of angiogenesis-related genes and therapy of bladder cancer are dependent on optimization of biological dose and schedule. Clin Cancer Res 1999, 5:2726–2734.

68. Kisker O, Becker CM, Prox D, et al. Continuous administration of endostatin by intraperitoneally

implanted osmotic pump improves the efficacy and potency of therapy in a mouse xenograft tumor model. Cancer Res 2001, 61:7669–7674.

69. Panigrahy D, Singer S, Shen LQ, et al. PPARgamma ligands inhibit primary tumor growth and metastasis by inhibiting angiogenesis. J Clin Invest 2002, 110:923–932.

70. Bergsland E, Dickler MN. Maximizing the potential of bevacizumab in cancer treatment. Oncologist 2004, 9(Suppl 1):36–42.

71. Emmenegger U, Kerbel RS. A dynamic de-escalating dosing strategy to determine the optimal biological dose for antiangiogenic drugs. Clin Cancer Res 2005, 11:7589–7592.

72. Bergers G, Javaherian K, Lo KM, et al. Effects of angiogenesis inhibitors on multistage carcinogenesis in mice. Science 1999, 284:808–812.

73. Kerbel R, Folkman J. Clinical translation of angiogenesis inhibitors. Nat Rev Cancer 2002, 2:727–739.

74. Benjamin LE, Golijanin D, Itin A, et al. Selective ablation of immature blood vessels in established human tumors follows vascular endothelial growth factor withdrawal. J Clin Invest 1999, 103:159–165.

75. Eberhard A, Kahlert S, Goede V, et al. Heterogeneity of angiogenesis and blood vessel maturation in human tumors: implications for antiangiogenic tumor therapies. Cancer Res 2000, 60:1388–1393.

76. Miller KD, Sweeney CJ, Sledge GW Jr. Redefining the target: chemotherapeutics as antiangiogenics. J Clin Oncol 2001, 19:1195–1206.

77. Lev DC, Ruiz M, Mills L, et al. Dacarbazine causes transcriptional up-regulation of interleukin 8 and vascular endothelial growth factor in melanoma cells: a possible escape mechanism from chemotherapy. Mol Cancer Ther 2003, 2:753–763.

78. Riedel F, Gotte K, Goessler U, et al. Targeting chemotherapy-induced VEGF up-regulation by VEGF antisense oligonucleotides in HNSCC cell lines. Anticancer Res 2004, 24:2179–2183.

79. Miyahara Y, Yoshida S, Motoyama S, et al. Effect of cis-diamine dichloroplatinum on vascular endothelial growth factor expression in uterine cervical carcinoma. Eur J Gynaecol Oncol 2004, 25:33–39.

80. Wild R, Dings RP, Subramanian I, et al. Carboplatin selectively induces the VEGF stress response in endothelial cells: potentiation of antitumor activity by combination treatment with antibody to VEGF. Int J Cancer 2004, 110:343–351.

81. Trieu V, Ran S, Desai N. Investigation of chemotherapy-induced tumor angiogenesis: Rationale for combination of nab-paclitaxel with anti-VEGF therapy, AACR Annual Meeting, 2007, Abstract 4636.

82. Kawano Y, Takaue Y, Saito S, et al. Granulocyte colony-stimulating factor (CSF), macrophage-CSF, granulocyte-macrophage CSF, interleukin-3, and interleukin-6 levels in sera from children undergoing blood stem cell autografts. Blood 1993, 81:856–860.

83. Chen YM, Whang-Peng J, Liu JM, et al. Serum cytokine level fluctuations in chemotherapy-induced myelosuppression. Jpn J Clin Oncol 1996, 26:18–23.

84. Chen YM, Whang-Peng J, Liu JM, et al. Elevation of serum interleukin-6 levels before peak of serum granulocyte colony-stimulating factor level in chemotherapy-induced myelosuppressive patients. J Immunother Emphasis Tumor Immunol 1995, 17:249–254.

85. Minamino K, Adachi Y, Okigaki M, et al. Macrophage colony-stimulating factor (M-CSF), as well as granulocyte colony-stimulating factor (G-CSF), accelerates neovascularization. Stem Cells 2005, 23:347–354.

86. Bracci L, Moschella F, Sestili P, et al. Cyclophosphamide enhances the antitumor efficacy of adoptively transferred immune cells through the induction of cytokine expression, B-cell and T-cell homeostatic proliferation, and specific tumor infiltration. Clin Cancer Res 2007, 13:644–653.

87. Heeschen C, Aicher A, Lehmann R, et al. Erythropoietin is a potent physiologic stimulus for endothelial progenitor cell mobilization. Blood 2003, 102:1340–1346.

88. Reese DM, Fratesi P, Corry M, et al. A phase II trial of humanized anti-vascular endothelial growth factor antibody for the treatment of androgen-independent prostate cancer. Prostate J 2001, 3:65–70.

89. Cobleigh MA, Langmuir VK, Sledge GW, et al. A phase I/II dose-escalation trial of bevacizumab in previously treated metastatic breast cancer. Semin Oncol 2003, 30:117–124.

90. Yang JC, Haworth L, Sherry RM, et al. A randomized trial of bevacizumab, an anti-vascular endothelial growth factor antibody, for metastatic renal cancer. N Engl J Med 2003, 349:427–434.

91. Stopeck AT, Bellamy W, Unger J, et al. Phase II trial of single agent bevacizumab (Avastin) in patients with relapsed, aggressive non-Hodgkin's lymphoma (NHL): Southwest Oncology Group Study S0108. J Clin Oncol 2005, 23s:Abstract 6592.

92. Burger RA, Sill M, Monk BJ, et al. Phase II trial of bevacizumab in persistent or recurrent epithelial ovarian cancer (EOC) or primary peritoneal cancer (PPC): a Gynecologic Oncology Group (GOG) study. J Clin Oncol 2005, 23s:Abstract 5009.

Chapter 7
Targeting Survival Cascades Induced by Activation of Ras/Raf/MEK/ERK and PI3K/Akt Pathways to Sensitize Cancer Cells to Therapy

James A. McCubrey, Richard A. Franklin, Fred E. Bertrand, Jackson R. Taylor, William H. Chappell, Melissa L. Sokolosky, Ellis W.T. Wong, Stephen L. Abrams, Kristin M. Stadelman, Negin Misaghian, Dale L. Ludwig, Jorg Basecke, Massimo Libra, Franca Stivala, Michele Milella, Agostino Tafuri, Alberto M. Martelli, Paolo Lunghi, Antonio Bonati, David M. Terrian, Brian D. Lehmann, and Linda S. Steelman

1 Overview of the Ras/Raf/MEK/ERK Pathway

The Ras/Raf/MEK/ERK pathway is activated by many growth factors and cytokines, which are important in driving proliferation and preventing apoptosis [1–5]. An overview of the effects of Ras/Raf/MEK/ERK and PI3K/Akt pathways on downstream signaling pathways leading to growth and the prevention of apoptosis is presented in Fig. 7.1. After receptor ligation, Shc, a Src homology (SH)-2 (SH2)-domain containing protein, becomes associated with the c-terminus of the growth factor receptor [6–8]. Shc recruits the GTP-exchange complex Grb2/Sos resulting in the loading of membrane bound Ras with GTP [9, 10]. Ras:GTP then recruits Raf to the membrane, where it becomes activated, likely via a Src-family tyrosine (Y) kinase [11–13]. Raf is responsible for phosphorylation of the mitogen associated/extracellular regulated kinase-1 (MEK1) [14–16]. MEK1 phosphorylates extracellular regulated kinases 1 and 2 (ERKs 1 and 2) on specific threonine (T) and Y residues [14–16]. Activated ERK1 and ERK2 serine (S)/T kinases phosphorylate and activate a variety of substrates including p90^{Rsk1}[17–23]. p90^{Rsk1} can activate the cAMP response element binding protein (CREB) transcription factor (20). Moreover, ERK can translocate to the nucleus and phosphorylate additional transcription factors such as Elk1, CREB, and Fos, which bind promoters of many genes, including growth factor and cytokine genes important in stimulating the growth and survival of multiple cell types [24–35]. The Raf/MEK/ERK pathway can also modulate the activity of many proteins involved in apoptosis, including Bcl-2, Bad, Bim, Mcl-1, caspase 9, and survivin [36–45]. Raf-1 has many roles that are independent of MEK and ERK, and many of these non-MEK/ERK functions are involved in the prevention of apoptosis [4].

Recently Raf-1 was shown to interact with mammalian sterile 20-like kinase (MST-2) and prevent its dimerization and activation [46]. MST-2 is a kinase, which is activated by pro-apoptotic agents

From: *Sensitization of Cancer Cells for Chemo/Immuno/Radio-therapy, 1st Edition.*
Edited by: Benjamin Bonavida © Humana Press, Totowa, NJ

Interactions Between Raf/MEK/ERK and PI3K/Akt Pathways

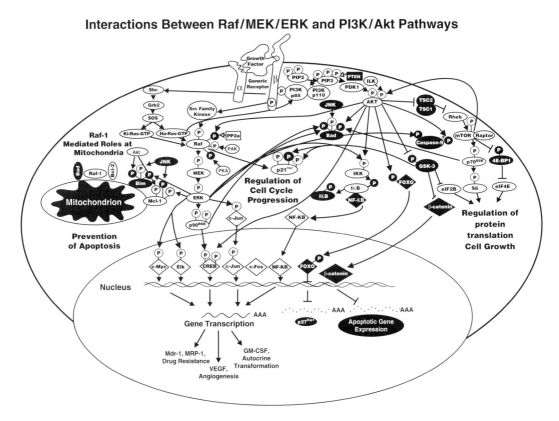

FIGURE 7.1. Interactions between Raf/MEK/ERK and PI3K/Akt pathways. The Raf/MEK/ERK and PI3K pathways are regulated by Ras as well as various kinases. Many kinases serve to phosphorylate S/T and Y residues on Raf. Some of these phosphorylation events serve to enhance Raf activity (shown by a black P in a white circle), whereas others serve to inhibit Raf activity (shown by a white P in a black circle. Moreover there are phosphatases such as PP2A, which remove phosphates on certain regulatory residues. The downstream transcription factors regulated by this pathway are indicated in diamond shaped outlines

such as staurosporine and Fas ligand. Raf-1, but not B-Raf, binds MST-2. Depletion of MST-2 from Raf-1$^{-/-}$ cells abrogated sensitivity to apoptosis. Overexpression of MST-2 increased sensitivity to apoptosis. It was proposed that Raf-1 might control MST-2 by sequestering it into an inactive complex. This complex of Raf-1:MST-2 is independent of MEK and downstream ERK. Raf-1 can also interact with the apoptotic signal kinase (ASK1) to inhibit apoptosis [47]. ASK1 is a general mediator of apoptosis and it is induced in response to a variety of cytotoxic stresses including tumor necrosis factor (TNF), Fas and reactive oxygen species (ROS). ASK1 appears to be involved in the activation of the JNK and p38 MAP kinases. This

is another interaction of Raf-1 that is independent of MEK and ERK.

2 Overview of the PI3K/Akt Pathway

Growth factor/cytokine receptor ligation also leads to rapid activation of phosphatidylinositol 3-kinase (PI3K) [48–50]. PI3K consists of an 85-kDa regulatory subunit, which contains SH2 and SH3 domains, and a 110-kDa catalytic subunit [48–50]. Cytokine stimulation often creates a PI3K binding site on the cytokine receptor. The p85 subunit SH2 domain associates with this site [48–50]. The p85 subunit is then phosphorylated, which leads to activation of the p110 catalytic subunit. Activated PI3K phosphorylates the membrane lipid phosphatidylinositol (4,5)-bisphosphate [PtdIns(4,5)P_2] to phosphatidylinositol (3,4,5)-tris-phosphate [PtdIns(3,4,5)P_3], which activates PI3K-dependent kinase (PDK1). PDK1 then phosphorylates Akt at threonine 308 (T308). A second kinase phosphorylates Akt on serine 473 (S473) [51–55].

Akt can transduce an anti-apoptotic signal by phosphorylating downstream target proteins involved in the regulation of cell growth [e.g., glycogen synthase kinase-3β (GSK-3β), ASK1, Bim, Bad, MDM-2, p21^{Cip1}, X-linked inhibitor of apoptosis (XIAP), and the Foxo3a transcription factor] [55–66]. Phosphorylated Foxo3a loses its ability to induce Fas, p27^{Kip1}, Bim, Noxa, and Puma gene transcription [67, 68]. Akt also phosphorylates I-κK, which subsequently phosphorylates I-κB, resulting in its ubiquitination and subsequent degradation in proteosomes [69–76]. Disassociation of I-κB from NF-κB enables NF-κB to translocate into the nucleus to promote gene expression. The PI3K/Akt pathway can also phosphorylate and activate CREB, which regulates transcription of anti-apoptotic genes including Mcl-1 and Bcl-2 [77–79]. The PI3K pathway also results in activation of the mammalian target of rapamycin (mTOR) and ribosomal protein kinases such as p70S6K [80–87]. It is worth noting that Akt can cause the activation of specific substrates (e.g., IκKα and CREB) or may mediate the inactivation of other proteins (e.g., Raf-1, B-Raf [by the Akt related kinase SGK], p21^{Cip-1}, Bim, Bad, procaspase-9, FOXO3a, and GSK-3β).

The PI3K pathway is negatively regulated by phosphatases. PTEN (phosphatase and tensin homologue deleted on chromosome 10) is primarily a lipid phosphatase that removes the 3-phosphate from the PI3K lipid product PtdIns (3,4,5)P_3 to produce PtdIns (4,5)P_2, which prevents Akt activation [50, 88–92]. PTEN is also a protein phosphatase [92–94]. Two other phosphatases, SHIP-1 and SHIP-2, remove the 5-phosphate from PtdIns(3,4,5)P_3 to produce PtdIns(3,4)P_2 [95–99]. Mutations in these phosphatases, which eliminate their activity, can lead to tumor progression. Consequently, the genes encoding these phosphatases are referred to as anti-oncogenes or tumor suppressor genes.

3 Interactions Between PI3K/Akt and Raf/MEK/ERK Pathways That Regulate Apoptosis

Akt can phosphorylate Raf-1 and B-Raf and lead to their inactivation [100–103]. Akt can also activate Raf-1 through a Ras-independent but protein kinase C (PKC)-dependent mechanism, which results in the prevention of apoptosis [104]. Suppression of apoptosis in some cells by Raf and MEK requires PI3K-dependent signals [105–110]. An overview of the effects of these pathways on the prevention of apoptosis is presented in Fig. 7.2.

Both PI3K/Akt and Raf/MEK/ERK pathways contribute to the transcriptional regulation of Bcl-2 family members as they can regulate CREB phosphorylation. CREB binds the Mcl-1 and Bcl-2 promoter regions [79, 111, 112]. Moreover, both pathways phosphorylate pro-apoptotic Bcl2 homology-3 (BH3) only domain protein Bad, which prevents its apoptotic effects and it becomes cytoplasmically localized [113–116]. Another MAP kinase, Jun N-terminal kinase (JNK) phosphorylates 14-3-3 proteins, which results in their disassociation from cytoplasmically localized Bad. Bad then translocates to the mitochondrion [117]. When Bad associates with Bcl-2 or Bcl-X$_L$, Bad promotes apoptosis by preventing Bcl-2 or Bcl-X$_L$ from interacting with Bax [118–124]. Bad is phosphorylated in most acute myelogenous leukemia (AML) specimens suggesting that inhibition of Bad phosphorylation may be therapeutically important in AML [125]. In contrast, the anti-apoptotic Mcl-1 protein is not reported to interact with Bad [119]. An overview of the effects of the PI3K/Akt and Raf/MEK/ERK pathways on Bad phosphorylation and the prevention of apoptosis is presented in Fig. 7.3.

Both the Raf/MEK/ERK and PI3K/Akt pathways can phosphorylate the BH3 only domain protein Bim [40, 62]. When Bim is phosphorylated by ERK and Akt, it is targeted for ubiquitination and degradation in the proteosome [45]. Mcl-1 can bind Bim which prevents the activation and mitochondrial translocation of Bax [44, 45]. In contrast, JNK can phosphorylate Bim at S65, which enhances its ability to induce Bax activation and hence stimulates apoptosis [120]. Mcl-1 can also bind pro-apoptotic Bak [119]. The Mcl-1:Bak interaction

Growth Factor Mediated Signal Transduction
Pathways and Prevention of Apoptosis

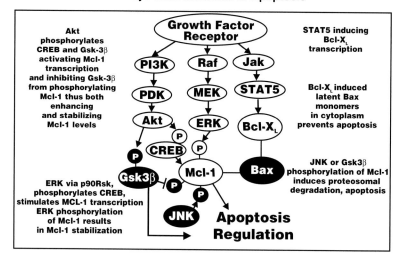

FIGURE 7.2. Growth factor mediated signal transduction pathways and prevention of apoptosis. Growth factors can induce multiple signal transduction pathways which can effect the expression of apoptotic molecules by transcriptional and post-transcriptional mechanisms

Effects of Raf/MEK/ERK, PI3K/Akt and JNK Pathways
on Bad Phosphorylation and Apoptosis

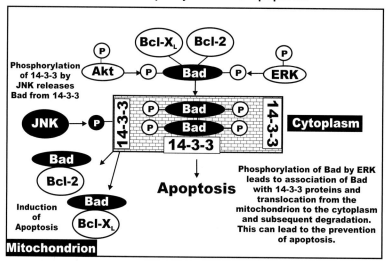

FIGURE 7.3. Effects of Raf/MEK/ERK and PI3K/Akt and JNK Pathways on Bad Phosphorylation and the Induction of Apoptosis. All three of these pathways can phosphorylate Bad on different residues, which affect its activity and interactions with Bcl-2, Bcl-X_L, and 14-3-3 proteins. Phosphorylation events mediated by Raf/MEK/ERK and PI3K/Akt result in Bad being associated with 14-3-3 proteins and translocation from the mitochondrion to the cytoplasm. Bcl-2 and Bcl-X_L remain associated with Bax and Bak, which prevent their activation and lead to suppression of apoptosis. In contrast, phosphorylation of Bad by JNK results in its dissociation from 14-3-3 proteins and Bad localizes to the mitochondrion and binds Bcl-2 and Bcl-X_L. Bax and Bak are then able to induce apoptosis

can be disrupted by the binding of the BH3-only domain Noxa protein which results in Mcl-1 being ubiquitinated and degraded in the proteosome [121]. Bak can then form active dimers and induce apoptosis. Unlike Bcl-2 and Bcl-X$_L$, the half-life of the Mcl-1 protein is short due to the amino terminal PEST sequence and its expression is regulated by both transcriptional and post-translational mechanisms [126]. Certain chemotherapeutic drugs such as taxol will induce Mcl-1 phosphorylation at different sites than those phosphorylated by ERK (T163) [44]. Oxidative stress can activate JNK, which induces the phosphorylation of Mcl-1 on S121 and T163 [127]. Cytokine deprivation of certain hematopoietic cells induces GSK-3β, which in turn induces the phosphorylation of Mcl-1 at S159 that results in its ubiquitination and degradation [128]. Akt phosphorylates GSK-3β suppressing its phosphorylation of Mcl-1. Altering the levels and phosphorylation state of Mcl-1 plays important roles in the regulation of apoptosis. A diagram depicting the effects of signaling pathways on Bim and Mcl-1 and apoptosis is presented in Fig. 7.4.

The expression of BH3-only domain Puma and Noxa proteins are under the control of the PI3K/Akt pathway [129]. Noxa interacts specifically with Mcl-1 but not with Bcl-2 or Bcl-X$_L$ [119]. Bak associates with Mcl-1 and Bcl-X$_L$ but not Bcl-2. Upon induction of Puma and Noxa by p53, Puma and Noxa displace Mcl-1 from Bak and Bak is able to oligomerize and induce apoptosis. This may lead to Mcl-1 degradation and apoptosis. The Raf/MEK/ERK pathway increases Mcl-1 protein levels and stability. This may lead to an increase in Mcl-1 associated with Noxa and Puma and a decrease in free Bak levels and less apoptosis.

Human caspase 9 was originally thought to be phosphorylated by Akt, but the murine caspase 9 lacks the Akt consensus phosphorylation site [17]. Caspase 9 is phosphorylated by the Raf/MEK/ERK pathway at T125, which inhibits activation of the caspase cascade [18]. Mcl-1 is a substrate for

Effects of Raf/MEK/ERK, PI3K/Akt, JNK and p53 Pathways on Noxa and Puma and the Induction of Apoptosis

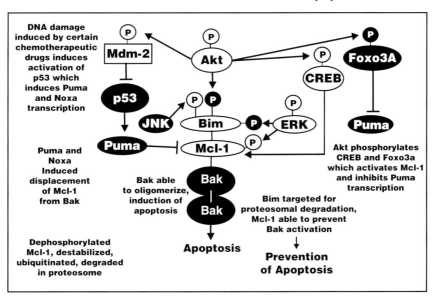

FIGURE 7.4. Effects of Raf/MEK/ERK and PI3K/Akt and JNK pathways on Bim phosphorylation and the induction of apoptosis. All three of these pathways can phosphorylate Bim on different residues that affect its activity and interactions with Mcl-1 and Bax and Bak. Phosphorylation events mediated by Raf/MEK/ERK and PI3K/Akt result in the prevention of Bax and Bak activation and lead to Bim being targeted to proteosome ubiquitination and degradation. In contrast phosphorylation of Bim by JNK results in its dissociation of Bim:Mcl-1 heterodimers, Mcl-1 is targeted to the proteosome, ubiquitination, and degradation and Bim-mediated activation of Bax and Bak

activated caspase 3; thus, decreased caspase 9 activation by ERK phosphorylation will reduce caspase 3 activation and Mcl-1 cleavage and apoptosis will be suppressed. An overview of the effects of the PI3K/Akt and Raf/MEK/ERK cascades caspase 9 phosphorylation and the prevention of apoptosis is presented in Fig. 7.5.

Many cytokines and growth factors can also induce the Jak/STAT pathway, which regulates the transcription of Bcl-X$_L$ [122]. Bcl-X$_L$ can prevent the formation of Bax:Bax homodimers [123]. Hence the Raf/MEK/ERK, PI3K/Akt, Jak/STAT and JNK pathways regulate many molecules involved in prevention of apoptosis. Dysregulation of these pathways may lead to malignant transformation.

4 Roles of the Ras/Raf/MEK/ERK Pathway in Neoplasia

Both the Raf/MEK/ERK and PI3K/Akt pathways can be activated by mutations/amplifications of upstream growth factor receptors. An illustration

Effects of Raf / MEK / ERK and PI3K / Akt on Caspase 9 Phosphorylation and Prevention of Apoptosis

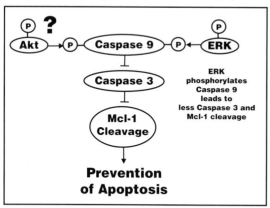

FIGURE 7.5. Effects of Raf/MEK/ERK and PI3K/Akt pathways on caspase-9 phosphorylation and the prevention of apoptosis. The Raf/MEK/ERK pathway phosphorylates caspase-9, which prevents activation of caspase-3. With less caspase-9 activation, less caspase-3 will be activated and more full-length Mcl-1 will be present. The ability of Akt to phosphorylate caspase-9 is controversial, as the Akt consensus phosphorylation site is present in human but not mouse caspase-9

of some of the receptors, kinases and phosphatases mutated/amplified in human cancer is presented in Fig. 7.6. Cancer remains the second leading cause of death in the United States despite recent advances in treatment of patients with anti-neoplastic drugs.

Mutations that lead to the expression of constitutively active Ras proteins have been observed in approximately 30% of human cancers [130, 131]. These are often point mutations which alter key residues that affect Ras activity, although amplification of Ras is also detected in some tumors. Mutations that result in increased Ras activity also perturb the Raf/MEK/ERK kinase cascade. *B-Raf* has been reported to be mutated in approximately 7% of all cancers [132]. However, this frequency may change as more and diverse tumors are examined for *B-Raf* mutation. Recent studies indicated that mutated alleles of *Raf-1* are present in therapy induced acute myelogenous leukemia (t-AML) [133]. These leukemias arose after chemotherapeutic treatment of breast cancer patients. The mutated *Raf-1* genes detected were transmitted in the germ line; thus, they are not a spontaneous mutation in the leukemia but may be associated with the susceptibility to induction of t-AML in these Austrian breast cancer patients.

For many years, the *Raf* oncogenes were not thought to be frequently mutated in human cancer and more attention to abnormal activation of this pathway was dedicated to *Ras* mutations, which can regulate both the Raf/MEK/ERK and PI3K/Akt pathways. However, it was shown recently that *B-Raf* is frequently mutated in melanoma (27–70%), papillary thyroid cancer (36–53%), colorectal cancer (5–22%) and ovarian cancer (30%) [132, 134–136]. The reasons for mutation at *B-Raf* and not *Raf-1* or *A-Raf* in melanoma patients are not entirely clear. Based on the mechanism of activation of *B-Raf*, it may be easier to select for *B-Raf* than either *Raf-1* or *A-Raf* mutations. Due to the amino acids present in certain key regulatory sites in the different *Raf* isoforms, activation of *B-Raf* would require one genetic mutation, whereas activation of either *Raf-1* or *A-Raf* needs two genetic events. It was proposed recently that the structure of B-Raf, Raf-1, and A-Raf may dictate the ability of activating mutations to occur at these molecules, which can permit the selection of oncogenic forms [132, 136, 137]. These predictions have arisen

Sites of Mutation which can Result in
Activation of Raf/MEK/ERK and PI3K/Akt Cascades

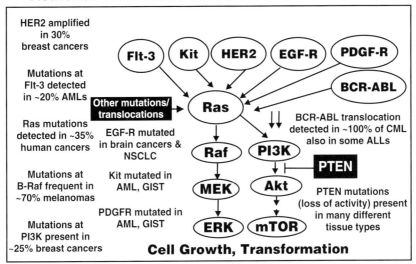

FIGURE 7.6. Sites of mutation that can result in activation of the Raf/MEK/ERK and PI3K/Akt pathways. Mutations have been detected in EGF-R, Flt-3, Kit, PDGF-R, PI3K, PTEN, Ras, and B-Raf. The BCR-ABL chromosomal translocation is present in virtually all CMLs and some ALLs. Many of these mutations and chromosomal translocations result in activation of the Raf/MEK/ERK and PI3K/Akt cascades

from determining the crystal structure of B-Raf [137]. Like many enzymes, B-Raf is proposed to have small and large lobes, which are separated, by a catalytic cleft. The structural and catalytic domains of B-Raf and the importance of the size and positioning of the small lobe may be critical in its ability to be stabilized by certain activating mutations. In contrast, the precise substitutions in A-Raf and Raf-1 are not predicted to result in small lobe stabilization thus preventing the selection of mutations at *A-Raf* and *Raf-1*, which would result in activated oncogenes [137]. Raf-1 has been known for years to interact with heat shock protein 90 (Hsp90). Hsp90 may stabilize activated Raf-1, B-Raf, and A-Raf. The role that Hsp90 plays in selection of activated *Raf* mutations is highly speculative yet very intriguing. The effects of drugs that target Hsp90 are discussed later.

The most common *B-Raf* mutation is a change at nucleotide 600, which converts a valine to a glutamic acid (V600E) [132]. This *B-Raf* mutation accounts for over 90% of the *B-Raf* mutations found in melanoma and thyroid cancer. It has been

proposed that *B-Raf* mutations may occur in certain cells that express high levels of B-Raf due to hormonal stimulation. Certain hormonal signaling events elevate intracellular cAMP levels, which result in B-Raf activation, leading to proliferation. Melanocytes and thyrocytes are two such cell types, which have elevated B-Raf expression as they are often stimulated by the appropriate hormones [138]. Moreover, it is now thought that B-Raf is the most important kinase in the Raf/MEK/ERK cascade [132]; thus, mutation at *B-Raf* activates downstream MEK and ERK. In some models wild-type and mutant B-Raf activates Raf-1, which in turn activates MEK and ERK [132, 139, 140].

In some cells, *B-Raf* mutations are believed to be initiating events, but are not sufficient for full-blown neoplastic transformation [141, 142]. Moreover, there appear to be cases in which certain *B-Raf* mutations (V600E) and *Ras* mutations are not permitted in the transformation process, as they might result in hyperactivation of Raf/MEK/ERK signaling and expression, which leads to cell cycle arrest [134]. In contrast, there are other situations, which depend on the particular *B-Raf* mutation and require both B-Raf and Ras mutations for transformation. The *B-Raf* mutations in these cases result in weaker levels of B-Raf activity [134, 142].

Different *B-Raf* mutations have been mapped to various regions of the B-Raf protein. Some of the other *B-Raf* mutations are believed to result in B-Raf molecules with impaired B-Raf activity, which must signal through Raf-1 [132, 140]. Heterodimerization between B-Raf and Raf-1 may allow the impaired B-Raf to activate Raf-1. Other mutations, such as D593V, may activate alternative signal transduction pathways [132].

It has been reported that a high frequency of AML and acute lymphocytic leukemias (ALL) (>50%) display constitutive activation of the Raf/MEK/ERK pathway in absence of any obvious genetic mutation [143, 144]. While there may be some unidentified mutation at one component of the pathway or a phosphatase that regulates the activity of the pathway, the genetic nature of constitutive activation of the Raf/MEK/ERK pathway is unknown. Elevated expression of ERK in AMLs and ALLs is associated with a poor prognosis [144]. Raf and potentially more effective MEK inhibitors may prove useful in the treatment of a large percentage of AML and ALLs.

Recently it was shown that there can be a genetic basis for the sensitivity of non-small cell lung cancers to epidermal growth factor receptor (EGFR) inhibitors [145–150]. In addition, some melanoma cells carrying B-Raf mutations are sensitive to MEK inhibitors while cells lacking these B-Raf mutations are resistant [151]. We have shown that introduction of activated EGFR mutants into hematopoietic cells renders them sensitive to EGFR inhibitors [145, 152, 153]. Furthermore, introduction of activated Ras, Raf, MEK genes into hematopoietic cells makes them sensitive to MEK inhibitors [154–158].

but is not believed to be frequently mutated in leukemia [167–171]. Mutations and hemizygous deletions of PTEN have been detected in AML and NHL [172–178]. Increased Akt expression is linked with tumor progression and drug/hormonal resistance [179–182]. SHIP mutations have been detected in AML (183, 184). Thus, there are many possible mechanisms which can lead to elevated Akt levels.

The relationship between dysregulated PI3K activity and the onset of cancer is well documented. The PI3K is the predominant growth factor–activated pathway in LNCaP human prostate carcinoma cells [185, 186]. Other reports directly implicate PI3K activity in a variety of human tumors, including breast cancer [187], lung cancer [188], melanomas [189], and leukemia [190], among others. Activated Akt can affect the expression and regulation of the responses of hormone receptors and hence lead to ineffectiveness of hormone ablation therapies [191–193].

Activated Akt has been reported to be detected in over 50% of primary AML samples and detection of activated Akt is associated with a poor prognosis [144]. Furthermore, the Akt pathway has been shown to be involved in regulation of multidrug resistance protein-1 (MRP-1) and drug resistance in AML [194–197]. Taken together, these data endorse the substantial role that PI3K signaling plays in oncogenesis and drug resistance. Moreover, targeted inhibition of the central components of this pathway appears to be an excellent choice for future therapeutic approaches. It has been observed that overexpression of both the Raf/MEK/ERK and PI3K/Akt pathways in AML is associated with a worse prognosis than overexpression of a single pathway [144]. Thus the development of inhibitors that target both pathways or the formulation of combinations of inhibitors may prove effective in the treatment of certain cancers.

5 Roles of the PI3K/Akt Pathway in Neoplasia

Some Ras mutations can result in PI3K/Akt activation [159–165]. Mutations at the p85 subunit of PI3K have been detected in Hodgkin's lymphoma cells [166]. The p110 subunit of PI3K is frequently mutated (~25%) in breast and some other cancers

6 Signaling Pathways and Hematopoietic Cancer

Approximately 42,000 people in the United States die each year from leukemia and lymphomas, which represents 10% of all cancer deaths. While vastly improved therapeutic approaches have been developed for CML and childhood ALL, therapy for adult AML and CLL have not yet yielded significant improvements. Approximately 11,000 Americans will be diagnosed with AML this year, and about 75% will die eventually from this disease. While improvements in the outcomes have been observed with young AML patients over the past 40 years, progress in the treatment of older AML patients has not been as significant [198]. Fifty to 75% of adult AML patients achieve complete remission with combination chemotherapy, which consists of the deoxycytidine analogue cytarabine and an anthracycline antibiotic (doxorubicin, daunorubicin, idarubicin, or the anthracenedione mitoxantrone, which inhibit the enzyme topoisomerase IIa). However, this treatment is not always effective, as only approximately 25% of these patients enjoy long-term survival [198]. The incidence of AML increases with age, 1.2 cases per 100,000 at age 30 and greater than 20 cases per 100,000 at age 80 [198]. Unfortunately, survival decreases with age. As the average life span of Americans increases due to improvements in health care and life styles, AML will be an increasing problem in American health care.

While approximately 50% of AML cases have genetic aberrations which can be identified (e.g., deletions such as 5q-, translocations such as t(8;21) AML-ETO, or duplications such as Flt-3 internal tandem duplication [ITD]), the other 50% do not have currently identifiable genetic mutations (199). Unlike chronic myeloid leukemia (CML) where the BCR-ABL translocation is present in virtually all patients and the majority of patients are sensitive to Imatinib, which inhibits the BCR-ABL oncoprotein, treatment with a targeted "upstream" inhibitor (e.g., Flt-3 inhibitor) would be ineffective in many AML cases.

Mutations of upstream receptors such as Flt-3 (20–30%), Kit (7–17% of AMLs), Fms (12% of MDS), and granulocyte colony stimulating factor receptor (G-CSF-R) have been documented in AML and may cause Ras/Raf/MEK/ERK pathway activation (see Fig. 7.6) [200–205]. Furthermore overexpression of vascular endothelial growth factor receptors (VEGF-R) has been observed in AML, which could result in activation of this pathway [206].

Upregulation of the Ras/Raf/MEK/ERK and PI3K/Akt pathways and phosphorylation of the downstream target Bad are observed frequently in AML patient specimens and associated with a poorer prognosis than patients lacking these changes [36, 125, 207–210]. Aberrant expression of multiple signaling pathways is associated with a worse prognosis [207]. Flt-3 ITD mutations are present in 20–30% of AMLs, and these patients have a poorer prognosis than patients lacking these mutations [211]. Flt-3 and other mutations may result in activation of the Ras/Raf/MEK/ERK and PI3K/Akt pathways [212–219].

7 Targeted Therapy in AML

While treatment of some subsets of AML, such as acute promyelocytic leukemia (APL), have shown great success with retinoids and arsenic trioxide, a significant problem in the remainder of AML patients is that most chemotherapy does not ultimately work and eventually the patients relapse and succumb to the disease [198]. Another continuing problem in AML therapy is the emergence of drug resistance [220–222]. Unlike the success observed with Imatinib and Dasatinib in treatment of CML [223], similar successes have not been observed in AML. Flt-3 inhibitors have been developed, but only approximately 20% of AMLs have mutations at Flt-3, which render them somewhat sensitive to Flt-3 inhibitor monotherapy [224, 225].

In some AML patients, MDM2 is overexpressed, which enhances the tumorgenicity and the loss of apoptotic processes resulting in a poor prognosis. When MDM2 ubiquitionates WT p53, it becomes targeted for degradation in the proteosome. p53 controls the transcription of many genes and is critical for the fate of the cells. Many of these targets are involved in p53-dependent apoptotic processes (e.g., Puma and Noxo). Recently, Andreeff and his group have observed that the small molecular weight, membrane permeable, Nutlins bind and inhibit MDM2. This results in increased levels of WT p53 and the induction of apoptosis and cell cycle arrest. Therefore, by using this treatment, the activities of p53 is enhanced and tumor growth is suppressed. More studies need to be performed, but according to this group, Nutlins are a novel therapeutic approach in the treatment of AML and other chemorefractory conditions [226]. Thus, targeting MDM2 with Nutlins may be a therapeutic approach in cancers that have WT p53.

8 Signaling Pathways and Breast Cancer

Breast cancer is among the most common forms of cancer, with over 210,000 new cases diagnosed in the United States each year and unfortunately, approximately 40,000 women will die from breast cancer this year. Among women in the United States, it affects about 1 in 7 and is the second most frequent cause of cancer death [227]. Although much progress has been achieved in breast cancer treatment, metastatic breast cancer remains a generally incurable and fatal disease, as 50% of patients die from the disease. Cytotoxic drug treatment is an important weapon against cancer; however, cancerous cells frequently develop drug resistance to these agents.

Breast cancer originates from genetic causes. Mutated or amplified genes are either inherited or occur sporadically. Hereditary breast cancer accounts for only about 10% of all breast cancer cases and generally results from lack of a tumor suppressor gene as opposed to gain of an oncogene. Approximately 45% of hereditary breast cancer is attributable to mutations in breast cancer associated gene-1 (BRCA1) and an additional 45% is attributable to mutation in BRCA2 (227). Other tumor suppressor genes implicated in hereditary breast cancer include p53 and PTEN [228]. The p53 tumor suppressor is a transcription factor involved in cell cycle regulation and DNA damage repair. Germline p53 mutation is present in approximately 50% of patients with Li-Fraumeni syndrome (LFS), which is a multicancer familial syndrome that includes adrenocortical carcinoma, brain tumors, leukemia, and osteosarcomas in addition to early onset breast cancer. Breast cancer attributable to germ-line p53 mutation in the absence of LFS is rare. Germ-line PTEN mutation is present in approximately 80% of patients with Cowden's syndrome [228, 229]. This disease, which is also known as multiple hamartoma syndrome, is another familial syndrome that includes many different types of cancer conditions, including early onset breast cancer. Mutations have been reported to occur at PTEN in breast cancer in varying frequencies (5–21%) [230–234]. Loss of heterozygosity (LOH) is probably more common (30%) [233]. PTEN promoter methylation leads to low PTEN expression. In one study, 26% of primary breast cancers had low PTEN levels, which correlated with lymph node metastases and poor prognoses [235, 236]. Mutations at certain residues of PTEN, which are associated with Cowden's disease, affect the ubiquitination of PTEN and prevent nuclear translocation. These mutations leave the phosphatase activity intact [237, 238]. Inhibition of PTEN activity leads to centromere breakage and chromosome instability. Thus PTEN has diverse activities. Disruption of PTEN activity by various genetic mechanisms could have vast effects on different processes affecting the sensitivity of breast cancers to various therapeutic approaches.

Sporadic breast cancer accounts for the remaining 90% of all breast cancer cases. The PI3K p110 catalytic subunit is mutated in approximately 25% of breast cancer specimens and these mutations frequently result in activation of its kinase activity [167–170, 230, 239]. Somatic mutation of p53 is associated with many cancers and exists in approximately 20% of sporadic breast cancer cases. In contrast, somatic mutation of BCRA1 or BCRA2 is rare in breast cancer patients. Another important cause of sporadic breast cancer is amplification/overexpression of HER2, which occurs in approximately 30% of breast cancer. This gene encodes human epidermal growth factor (EGF) receptor-2 (HER2 a.k.a., c-ErbB-2), which is a receptor tyrosine kinase (RTK). Expression and activity of downstream signal transduction cascades, such as the Raf/MEK/ERK and PI3K/Akt pathways, change as a result of these mutations. ERK and Akt are frequently activated in breast cancer specimens [230, 240].

Association of genes that regulate signal transduction pathways with breast cancer implies an important role of these pathways in this disease. Perhaps the best example of this is association of HER2 gene amplification with breast cancer. While a normal breast cell possesses 20,000–50,000 HER2 molecules, amplification of this gene can increase levels of HER2 up to two million molecules per cell [241, 242]. Overexpression of HER2 in breast cancers is linked to comedo forms of ductal carcinoma in situ (DCIS) and occurs in approximately 90% of these cases. HER2 overexpression will lead to increased expression of both the Raf/MEK/ERK and PI3K/Akt pathways.

We have observed that ERK is activated after tamoxifen treatment of the MCF-7 breast cancer cell line. This may be important with regard to the sensitivity of breast cancer cells to MEK inhibitors. ERK can phosphorylate and contribute to the inactivation of the tuberous sclerosis complex (TSC-2). Akt can also phosphorylate TSC-2, at a different residue, which leads to its inactivation. This leads to mTOR activation. Inhibition of TSC-2 phosphorylation by MEK and PI3K/Akt inhibitors may make cells more sensitive to chemotherapy and hormonal therapy.

Furthermore, activated Akt is often upregulated in breast cancer cells and its overexpression is associated with a poor prognosis. However, this may actually render the breast cancer cells sensitive to Akt as well as downstream mTOR inhibitors. The formation of the rapamycin sensitive mTORC1 complex (consisting of mTOR, regulatory-associated protein of mTOR) (Raptor and mLST8) in the breast cancer cells that overexpress activated Akt may be different than in drug-sensitive breast cells that do not overexpress activated Akt. In cells which express activated Akt, Akt should phosphorylate TSC-2 resulting in its inactivation. The mTORC1 complex is formed and downstream p70S6K and 4E-BP1 are phosphorylated, allowing the disassociation of eIF-4E, ribosome biogenesis, and protein synthesis. In contrast, in the absence of Akt and ERK activation, this complex should not be formed. Rapamycin targets this complex; hence, the cells that constitutively express activated Akt cells are more sensitive to rapamycin than the breast cancer cells that do not overexpress activated Akt. In the cells that do not constitutively express activated Akt, this complex should be transiently assembled after growth factor treatment. In contrast, the assembly of the rapamycin-insensitive mTORC2 complex (consisting of rapamycin-insensitive companion of mTOR (Rictor), mTOR, mLST8) should be lower in the cells that constitutively express activated Akt than in those cells that do not, as there is equilibrium between the mTORC1 and mTORC2 complexes. The significance of these biochemical complex signaling events is that in drug-resistant breast cancer, cells that overexpress activated Akt, or lack expression of PTEN, they have an Achilles hell with regard

to therapeutic intervention in that they are highly sensitive to rapamycin treatment.

9 Aberrant Regulation of Apoptosis May Contribute to Breast Cancer and Subsequent Drug Resistance

Cell death following cytotoxic drug treatment is generally apoptotic as opposed to necrotic. Many chemotherapeutic drugs induce apoptosis by activating the intrinsic cell death pathway, which involves cytochrome c release and activation of the apoptosome-catalyzed caspase cascade. During apoptosis, activation of caspase family cysteine proteases occurs. Although various cytotoxic drugs differ in their mechanism of action, each ultimately relies upon built-in apoptotic machinery to elicit cell death [228, 229, 242–245]. Caspase family cysteine proteases are responsible for proteolytic cleavage of cellular proteins carboxy terminal to aspartate residues.

In a study involving 46 breast cancer patients, 75% lacked caspase-3 mRNA transcripts and protein expression [246]. The MCF-7 cell line has a mutation in caspase-3 and is deficient in certain aspects of apoptosis. In this respect, MCF-7 cells are a stringent model to investigate breast cancer apoptosis [247]. Caspase-9 can be activated in MCF-7 cells, which can result in the sequential activation of caspases-7 and -6 [248]. Caspase-7 is activated by the apoptosome complex and forms a XIAP-caspase-7 complex. This XIAP–caspase-7 complex is more stable in MCF-7 cells due to the absence of functional caspase-3. Furthermore, the PI3K/Akt pathway can phosphorylate XIAP, which prevents its ubiquitination and subsequent proteosomal degradation [249]. ERK activity maintains XIAP levels; however, the mechanism by which this occurs is unknown. Resistance to chemotherapeutic drugs induced by the Raf/MEK/ERK and PI3K/Akt pathways may be due in part to prolonged XIAP expression which prevents caspase-7 from exerting its effects on apoptosis [249, 250]. ERK phosphorylates caspase-9, which inhibits its activation. Negative regulation of caspases by ERK represents a mechanism by which Raf/MEK/ERK pathway activation prevents apoptosis.

10   Signaling Pathways and in Prostate Cancer

Increased expression of the Ras/Raf/MEK/ERK pathway has been associated with advanced prostate cancer, hormonal independence, and a poor prognosis [251–255]. The mechanisms of activation of this cascade in prostate cancer have not been well established, as mutations at Ras, Raf, MEK, or ERK have not been frequently reported in prostate cancer; however, amplification of Ras has been detected in some prostate cancers. Alternatively, the Ras/Raf/MEK/ERK pathway could be induced by autocrine- and paracrine-acting growth factors in prostate cancer cells.

Interestingly, we observed that introduction of activated Raf genes (Raf-1 and B-Raf) did not increase the chemoresistance of prostate cancer cells, while introduction of activated Akt genes did increase their chemoresistance [256, 257]. Some prostate cancer cell lines such as LNCaP and PC3 cells have PTEN mutations and express high levels of active Akt, and express low levels of active Raf/MEK/ERK pathway members. Hormonal-independent advanced prostate cancer cell lines such as DU145 and PC3 tend to express low levels of activated Raf, MEK, and ERK. Thus while it is logical to believe that activation of the Ras/Raf/MEK/ERK cascade could contribute to prostate cancer progression and hormonal independence, it may not be that simple, as some of these established prostate cancer lines express low levels of Raf/MEK/ERK.

Recently, a role for Raf kinase inhibitory protein (RKIP) in prostate cancer was hypothesized. Certain advanced prostate cancers express lower amounts of RKIP than less malignant prostate cancer specimens [258, 259]. Inhibition of RKIP expression makes certain prostate cancer cells more metastatic [258]. The mechanism responsible for this increase in metastasis is believed to be due to the enhanced activity of the Raf/MEK/ERK signaling pathways. RKIP is not thought to alter the tumorigenic properties of prostate cancer cells; rather, it is thought to be a suppressor of metastasis and may function by decreasing vascular invasion [258]. RKIP associates with centrosomes and kinetochores and regulates the spindle checkpoint [260]. The Raf/MEK/ERK cascade regulates RKIP, the aurora B kinase, and the spindle checkpoint. Small changes in Raf/MEK/ERK activity can affect the fidelity of the cell cycle.

The role of the Raf/MEK/ERK pathway in prostate cancer remains controversial. The studies with RKIP suggest that increasing Raf activity, by inhibition of RKIP after phosphorylation by PKC, is somehow linked with metastasis in prostate cancer [259]. However, the Raf/MEK/ERK pathway may be shut off in advanced prostate cancer due to the deletion of the PTEN gene, which normally regulates Akt activity by counterbalancing PI3K activity [261]. In some but not all cell lineages, Akt may inhibit Raf-1 activity by phosphorylation of Raf-1 on S259.

The expression of the Raf/MEK/ERK pathway may be decreased in some prostate cancer cell lines isolated from advanced prostate cancer patients by the deletion or inactivation of p53. p53 could influence the expression and activation of the Raf/MEK/ERK pathway by multiple mechanisms. For example, p53 could alter the expression of phosphatases that regulate the activity of Raf/MEK/ERK or stimulate the transcription of the Raf/MEK/ERK genes [262]. After DNA damage, p53 may activate phosphatases that serve to fine tune the Raf/MEK/ERK cascade. In contrast, after growth factor stimulation, p53 may induce map kinase phosphatase (MKP1) or other phosphatases that alter activity of the Raf/MEK/ERK cascade. Some events associated with phosphatases involve the removal of a phosphate group, which results in inactivation of the protein; in contrast, some dephosphorylation events (e.g., removal of the phosphate from S259) of Raf-1 by PP2A contributed to activation of Raf-1.

Heparin-binding epidermal growth factor (hbEGF) has been observed to be a transcriptional target of p53. p53 can transactivate hbEGF, which in turn activates both the Raf/MEK/ERK and PI3K/Akt pathways, suggesting that p53 can activate survival pathways, which affect the balance between life and death in response to genotoxic stresses [262, 263]. Raf activates hbEGF expression in some cell types, and the effects of this induction on p53 activity are poorly characterized. Furthermore, Akt can phosphorylate MDM-2 increases cytoplasmic to nuclear shuttling and decreases p53 activity.

Clearly, the interactions between p53, and the Raf/MEK/ERK and PI3K/Akt are very complicated and likely feed back upon one another.

We have observed low expression of the Raf/MEK/ERK pathway in certain prostate cancer cells (LNCaP and PC3), which have overexpression of activated Akt due to deletion or mutation of the PTEN phosphatase [256, 257]. Although the Raf/MEK/ERK genes are present in these cells and Raf/MEK/ERK can be induced upon treatment with certain chemotherapeutic drugs, their expression under normal growth conditions is very low. This may indicate that elevated Akt expression shuts off the Raf/MEK/ERK pathway, which may normally induce cell cycle arrest or senescence of prostate cells. The situation becomes even more complicated in advanced prostate cancer cells that have lost functional p53 activity. The expression of the Raf/MEK/ERK cascade is detected at lower levels in some representative cell lines (PC3 and DU145) than in prostate cancer lines that have both functional p53 and PTEN (22Rv1). Introduction of WT p53 into PC3 and DU145 cells increases both the sensitivity of the cells to chemotherapeutic drugs and expression and activation of the Raf/MEK/ERK cascade. Thus, therapies aimed at increasing Raf/MEK/ERK activation might be more effective in the treatment of certain prostate cancers, as they may induce terminal differentiation, senescence, or cell cycle arrest of the cells [264–266]. In contrast, therapies aimed at decreasing Raf/MEK/ERK activity are likely to be more appropriate for the treatment of hematopoietic and breast malignancies in which overexpression of this cascade is associated with proliferation and drug resistance.

11   Therapeutic Targeting of the Raf/MEK/ERK, PI3K/Akt, and Apoptotic Pathways

Small molecule inhibitors such as Imatinib have proved effective in the treatment of CML and certain other cancers that proliferate in response to BCR-ABL, platelet-derived growth factor receptor (PDGF-R), and c-Kit [228–230, 267–269], such as gastrointestinal stromal tumors (GIST). As stated, lung carcinomas that have mutations in EGFR are sensitive to EGFR inhibitors (6). Raf and MEK inhibitors have been developed and some are in clinical trials [228–230, 269, 270]. We have determined that a consequence of doxorubicin treatment of breast, hematopoietic, and prostate cancer cell lines is the induction of ERK [4]. Eliminating this deleterious side effect of these therapies may enhance their ability to kill drug-resistant cancer. PI3K, PDK, Akt, and mTOR inhibitors have been developed. mTOR inhibitors have been used for many years as immunosuppressive drugs in kidney transplant patients. A side effect of mTOR inhibitors is the inhibition of a negative feed back pathway, which results in Akt activation [271]. Bcl-2 inhibitors have been developed that suppress Bcl-2 and Bcl-X$_L$ but not Mcl-1 [272]. MDM-2 inhibitors have been developed that enhance WT p53 stability and activity [273]. These inhibitors may augment the effects of chemo-, radio- and hormonal-therapy of breast cancers. A diagram illustrating the sites of these and other inhibitors is presented in Fig. 7.7.

Raf inhibitors have been developed and some are being evaluated in clinical trials [274–278]. Certain Raf inhibitors have been developed that are small molecule competitive inhibitors of the ATP-binding site of Raf protein. These inhibitors (e.g., L-779,450, ZM 336372, Bay 43-9006, aka Sorafenib) bind the Raf kinase domain and therefore prevent its activity. Some Raf inhibitors may affect a single Raf isoform (e.g., Raf-1); others may affect Raf proteins, which are more similar (Raf-1 and A-Raf); still other pan Raf inhibitors may affect all three Raf proteins (Raf-1, A-Raf, and B-Raf). We have observed that the L-779,450 inhibitor suppresses the effects of A-Raf and Raf-1 more than the effects of B-Raf [274]. Like many Raf inhibitors, L-779,450 is not specific for Raf; it also inhibits the closely related p38^{MAPK}. Likewise, Sorafenib inhibits other kinases besides Raf (e.g., VEGF-II receptor, PDGF-R, Kit, Flt-3, Fms) and is more appropriately referred to as a multi-kinase inhibitor. Knowledge of the particular Raf gene mutated or overexpressed in certain tumors may provide critical information regarding how to treat the patient, as some cancers that overexpress a particular Raf gene may be more sensitive to inhibition by agents that target that particular Raf protein. Inhibition of certain Raf proteins might prove beneficial, whereas suppression of other Raf proteins under certain circumstances might prove

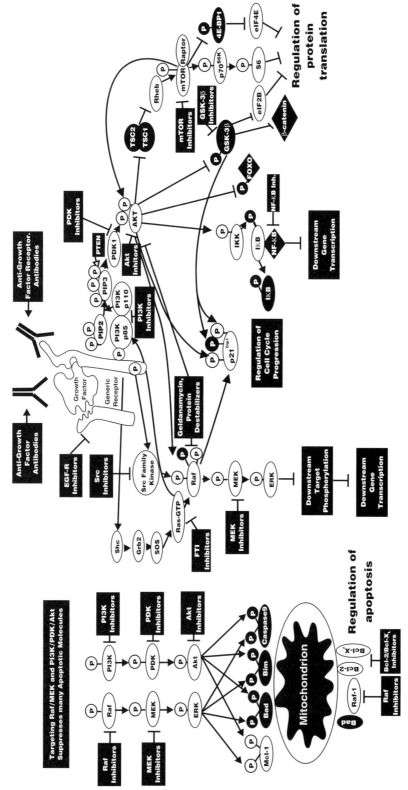

FIGURE 7.7. Targeting signal transduction pathways. Potential sites of action of small molecular weight inhibitors and cytotoxic antibodies are indicated. In some cases inhibitors will suppress growth, apoptotic and cell cycle regulatory pathways. This diagram serves to illustrate the concept that targeting Raf/MEK/ERK and PI3K/Akt can have dramatic effects on many growth regulatory molecules. Proteins inactivated by S/T phosphorylation induced by the PI3K/Akt pathway are shown in black circles with white P in a black circle. ALL, acute lymphocytic leukemia; AML, acute myeloid leukemia; GIST, gastrointestinal stromal tumor; NSCLC, non-small cell lung cancer

detrimental. Thu, the development of unique and broad-spectrum Raf inhibitors may prove useful in human cancer therapy.

Chaperonin proteins such as 14-3-3 and Hsp90 regulate Raf activity [279]. Raf activity is regulated by dimerization. These biochemical properties result in Raf activity being sensitive to drugs that block protein:protein interactions such as geldanamycin [280]. Geldanamycin and its 17-allyamino-17-demethoxygeldanamycin (17-AAG) analogue are nonspecific Raf inhibitors, as they also affect the activity of many proteins that are stabilized by interaction with Hsp90. Geldanamycin and 17-AAG are currently in clinical trials [281]. We often think of a single Raf protein carrying out its biochemical activity. However, Raf isoforms dimerize with themselves and other Raf isoforms to become active. Drugs such as coumermycin, which inhibit Raf dimerization, and others, such as geldanamycin, which prevent interaction of Raf with Hsp90 and 14-3-3 proteins suppress Raf activity. Geldanamycin has also been shown to be effective in suppressing the growth of NSCLC cells that are gefitinib- and erlotinib-resistant due to a second mutation in the EGFR [282]. Furthermore, 17-AAG potentiated the effects of paclitaxel in ovarian cancer lines that expressed high levels of activated Akt [283].

An alternative approach to targeting Raf is to prevent Raf activation by targeting kinases (e.g., Src, PKC, PKA, PAK, or Akt) and phosphatases (e.g., PP2A) involved in Raf activation. Some Src kinase inhibitors such as Dastinib would be predicted to inhibit Raf activation as it should suppress Raf-1 activation by Src. It is worth noting that some of these kinases normally inhibit Raf activation (Akt and PKA). A major limitation of this approach would be that these kinases and phosphatases could result in activation or inactivation of other proteins and would have other effects on cell physiology.

Currently it is believed that MEK1 is not frequently mutated in human cancer. However, aberrant expression of MEK1 is observed in many different cancers due to the activation of the Raf/MEK/ERK pathway by upstream kinases (e.g., BCR-ABL) and growth factor receptors (e.g., EGFR) as well as other unknown mechanisms. Specific inhibitors to MEK have been developed (PD98059, U0126, PD184352 [aka, CI1040], PD-0325901, array MEK inhibitors

[ARRY-142886], and others) [270]. The successful development of MEK inhibitors may be due to the relatively few phosphorylation sites on MEK involved in activation/inactivation. Furthermore, effective targeting of MEK1,2 is highly specific as ERK1,2 are the only downstream targets. An advantage of targeting the Raf/MEK/ERK cascade is that it can be targeted without knowledge of the precise genetic mutation, which results in its aberrant activation. This is important, as the nature of the critical mutation(s), which leads to the malignant growth of at least 50% of AMLs and other cancers, is not currently known. An advantage of targeting MEK is that Raf/MEK/ERK pathway is a convergence point where a number of upstream signaling pathways can be blocked with the inhibition of a single kinase (MEK).

To our knowledge, no small molecular weight ERK inhibitors have been developed yet; however, inhibitors to ERK could prove very useful as ERK can phosphorylate many targets (Rsk, c-Myc, Elk, and at least 150 more). There are at least two ERK molecules regulated by the Raf/MEK/ERK cascade, ERK1 and ERK2. Little is known about the different in vivo targets of ERK1 and ERK2. However ERK2 has been postulated to have pro-proliferative effects while ERK1 has anti-proliferative effects [285]. Development of specific inhibitors to ERK1 and ERK2 might eventually prove useful in the treatment of certain diseases.

12 Combination Therapies to Enhance Toxicity

An approach that we have been investigating recently is to determine whether inhibition of two signal transduction pathways is a more effective means to induce apoptosis than inhibiting a single signal transduction pathway. We have observed that inhibition of the Raf/MEK/ERK and PI3K/Akt pathways is usually a more effective means to induce apoptosis, and synergy between the two inhibitors is often observed. Many transformed cells have elevated Raf/MEK/ERK and/or PI3K/Akt signaling. These two pathways play prominent roles in the promotion of growth and the prevention of apoptosis. The PI3K/Akt pathway may be inhibited with PI3K (LY294002, PX-866),

PDK1 (OSU-03012, Celecoxib), Akt (A-443654) inhibitors, or downstream mTOR inhibitors such as rapamycin and modified rapamycins (CCI-779 and RAD001). Initially mTOR inhibitors showed much promise as PTEN is often deleted in various tumors. However, it has been recently determined that the mTOR pathway has a complicated feed back loop that actually involves suppression of Akt; hence, mTOR inhibitors would be predicted to activate Akt in some cells. Recent evidence has highlighted that mTOR can also be activated by Raf/MEK/ERK [197, 286]. This may well be another relevant cross-talk between the Ras/Raf/MEK/ERK and the PI3K/Akt pathways and might offer a further rationale for treatments combining drugs that inhibit both signaling networks. The effects of the combination of mTOR and Sorafenib are being evaluated in clinical trials to treat melanoma [287]. The effects of combining EGFR and mTOR or mTOR and MEK inhibitors on cell cycle progression in the induction of apoptosis in kidney cells were examined and synergistic effects were observed [288]. The effects of EGFR and MEK inhibitors were enhanced by the addition of rapamycin, which resulted in enhanced G_1 arrest. Similar experiments have been performed on NSCLC with gefitinib and either the MEK inhibitor U0126 or the farnesyl transferase inhibitor SCH66336 [289].

In some cases, the precise gene responsible for driving proliferation of the malignant cell is known (e.g., BCR-ABL in CML, EGFR in some cases of non-small cell lung cancer, FLT-3 in some AMLs and B-Raf in melanoma). However, even in the previously listed cancers, there may be additional genes that are also critical for malignant transformation. Treatment of some of these diseases with specific kinase inhibitors is often effective; however, resistance to the inhibitors may develop due to further mutations in aberrant kinases that often prevent the signal transduction inhibitor from inhibiting the altered kinase. In these novel "drug-resistant" cases, additional therapeutic approaches are necessary. In some of these cases it may be possible to inhibit the drug-resistant cells with novel inhibitors that will suppress the resistant oncoprotein or combinations of MEK and PI3K/Akt inhibitors. We have observed that Imatinib-resistant hematopoietic cells (which have mutated BCR-ABL kinase) are sensitive to MEK inhibitors. This result is not

surprising, as a Src inhibitor (Dasatinib) is being used to inhibit Imatinib resistant cells as they often have overexpression of an activated Src family kinase, such as Lyn, which likely acts by inducing the Raf/MEK/ERK cascade.

Classical chemotherapy often remains the most used anticancer therapy for many different types of cancer treatment. Drugs such as doxorubicin and taxol are effective in the treatment of many cancers, even though in some cases drug resistance does develop after prolonged treatment. Doxorubicin and taxol target cellular events such as DNA replication and cell division, which are downstream of the targets of signal transduction pathway inhibitors. Thus, by combining classical chemotherapy with targeted therapy, it may be possible to enhance toxicity while lowering the effective concentrations of classical chemotherapeutics necessary for effective elimination of the particular tumor.

We have investigated the effects of combining classical chemotherapy with signal transduction inhibitors in suppressing the growth of hematopoietic cells that grow in response to activated Raf and Akt or ErbB. Treatment of transformed cells with MEK or PI3K/Akt inhibitors with either doxorubicin or taxol resulted in a synergistic response documenting the effectiveness of classical chemotherapy with targeted therapy.

13 Combining Signal Transduction Inhibitors with Antibody-, Hormonal-, and Chemotherapeutic-Based Therapies

Recent studies have indicated that the effectiveness of certain antibody-based therapies (e.g., Herceptin, which targets HER2) may be greatly enhanced by inclusion of mTOR inhibitors. These observations have been seen in both preclinical studies performed in tissue culture and xenograft models and are being further evaluated in phase II clinical trials [290]. The cytotoxic effects of Herceptin can also be improved by addition of an inhibitor such as Lapatinib, which targets both EGFR and HER2 [291].

The effects of combining PI3K and mTOR inhibitors with the chemotherapeutic drug fludarabine have been examined in human leukemia cell lines

[292]. Combination of fludarabine and either PI3K or mTOR inhibitors resulted in increased apoptosis compared with what was observed after fludarabine treatment alone.

Rapamycin exerted synergistic effects when combined with paclitaxel, carboplatin, and vinorelbine in certain responsive breast cancer lines in vitro [293]. Rapamycin combined with paclitaxel resulted in a significant reduction in tumor volume in xenograft models when rapamycin-sensitive tumors were examined. mTOR inhibitors also increased the chemosensitivity of cervical cancer cells to paclitaxel [294]. The effects of rapamycin on the sensitivity to paclitaxel are dependent on functional glycogen synthase kinase 3β (Gsk-3β) as rapamycin induced toxicity in Gsk-3β WT cells but not in Gsk-3β null cells [295].

Combinations of rapamycin and the cell cycle check point kinase (Chk1) inhibitor UCN-01 also resulted in a synergistic induction of apoptosis in human leukemic cells that was regulated by the Raf/MEK/ERK, Akt and JNK signal transduction pathways [296]. Coadministration of UCN-01 and rapamycin reduced the levels of Mcl-1, Bcl-X_L, cyclin D1, and p34^{cdc2}. Similar studies were performed with the farnesyltransferase inhibitor L744832 and UCN-01, which also revealed a synergistic interaction in terms of the induction of apoptosis and interruption of both Akt and MEK/ERK pathways and activation of SEK1/JNK [297]. L744832 blocked the induction of ERK normally stimulated by UCN-01.

Novel PI3K inhibitors have been developed. PWT-458 is a novel pegylated-17-hydroxywortmannin, which inhibits PI3K and has been shown to suppress glioma and NSCLC and renal cell carcinoma in xenograft models [298]. PWT-458 augmented the anticancer effects of paclitaxel and pegylated rapamycin in certain xenograft models.

The PI3K inhibitor LY294002 has been shown to block drug export from drug-resistant colon carcinoma cells overexpressing MRP-1 [299]. Furthermore, combining the PI3K inhibitor with doxorubicin resulted in enhanced apoptosis, whereas combining doxorubicin with the MEK inhibitor did not.

Perifosine is an oral bioactive novel alkylphospholipid that inhibits Akt. Perifosine enhanced dexamethasome-, doxorubicin-, melphalan-, and bortezomib-induced multiple myeloma cytotoxicity [300]. Furthermore perifosine synergistically increased the effects of etoposide on the induction of apoptosis in human T-ALL cells [301]. Additional Akt inhibitors have been developed. An Akt inhibitor developed by Abbott (A-443654) agumented the effectiveness of paclitaxel and rapamycin in suppressing tumor growth in xenograft models [302]. Treatment of cells with this Akt inhibitor resulted in increased detection of activated Akt. Similar events are also observed with certain MEK inhibitors, as incubation of cells with some MEK inhibitors results in increased levels of activated MEK but suppressed levels of activated ERK. Two problems associated with the Abbott Akt inhibitor is increased toxicity and glucose secretion. Unfortunately, there are also toxicity problems with the PI3K inhibitor LY294002 and pharmacologic problems with some of the MEK inhibitors (CI1040), preventing their usage in human cancer patients.

Suntinib was developed as a selective inhibitor of VEGFR. However, it has since been shown to have multiple targets. Suntinib sensitizes ovarian cancer cells to cisplatin via suppression of nucleotide excision repair activity by inhibiting the expression of G_1 cell cycle checkpoint regulators (p53, p21^{Cip-1}, p27^{Kip-1}, and MDM2) [303]. The chemosensitizing effects of Suntinib may be mediated by inhibiting G_1 checkpoint control and upregulating the apoptotic response to cisplatin.

Multitargeted kinase inhibitors such as Sorafenib and Sunitib are being combined with an antibody (Bevacizumab) that targets the VEGF and evaluated in clinical trials [287]. Bevacizumab is also being combined with Erlotinib, an EGFR inhibitor, in a phase II clinical trial for renal cell carcinoma patients. Furthermore, Beracizumab and mTOR inhibitors are being combined in clinical trials for renal cell carcinoma and melanoma patients [304].

14 Enhancing the Effects of Ras/Raf/MEK/ERK Pathway Inhibitors by Combination Therapy

Although the precise targets of farnesyltransferase inhibitors remain controversial, the farnesyltransferase inhibitor R115777 (Zarnestra) was shown to result in disease stabilization in 64% of multiple myeloma patients in a phase II clincal trial [305].

Furthermore R115777 was found to synergize with paclitaxel and docetaxel but not with doxorubicin, 5-flurouracil, cisplatin, melphalan, mitoxantrone, and dexamethasone.

A side effect of some chemotherapeutic drugs such as taxol is the induction of the Raf/MEK/ERK pathway. Activation of this pathway can under certain circumstances promote proliferation and prevent apoptosis. Combining taxol treatment with MEK inhibitors has been observed to synergistically enhance apoptosis and inhibit tumor growth [306, 307]. The synergistic effects of paclitaxel and MEK inhibitors are complex and not fully elucidated but may be mediated in part by inhibiton of Bad phosphorylation at S112 by ERK [308].

Moreover, the cytotoxic effects of combinations of MEK inhibitors and paclitaxel may be specific for cells of certain origins and may depend on the levels of endogenous activated MEK/ERK present in those cells. Some studies with NSCLC cells which constitutively expressed activated MEK/ERK revealed no increase in paclitaxel-induced apoptosis upon treatment with the MEK inhibitor [309]. In contrast, addition of a dominant negative MEK gene to these cells potentiated paclitaxel induced apoptosis.

MEK inhibitors have also been observed to affect cisplatin resistance in squamous cell carcinoma implicating the Raf/MEK/ERK pathway in their drug resistance [310]. In neuroblastoma cells, cisplatin induced apoptosis was associated with an increase in p53 and Bax proteins. Activated ERK1,2 levels also increased earlier in these cells upon cisplatin treatment. Culture of these cells with MEK inhibitors blocked apoptotic cell death which prevented the cisplatin induced accumulation of p53 and Bax [311].

MEK inhibitors have also been observed to synergize with arsenic trioxide (ATO) to induce apoptosis in acute promyelocytic leukemia (APL) and AML cells [312, 313]. The p53-related gene p73 is a molecular target of the combined therapy. ATO modulates the expression of the p73 gene by inducing the proapoptotic and antiproliferative 73 isoforms. p53 Requires p63 and p73 for the induction of apoptosis in response to DNA-damaging drugs. p73 exists as multiple transactivation competent (TA) proapoptotic and antiproliferative p73 COOH-terminal splicing isoforms (α, β, γ, δ, ε, ζ) of which the two major forms are p73α and p73β.

Dominant-negative (ΔN) p73 variants are expressed from a second promoter. These DN ΔNp73 variants lack the amino-terminal transactivation domain, act as trans-repressors of p53- and p73-dependent transcription, and have antiapoptotic and pro-proliferative potential. Treatment of APL cells with the PD184352 MEK inhibitor reduced the level of ΔNp73 and decreased the ATO-mediated upregulation of ΔNp73, thus causing an increase in the TA/ΔNp73 ratio of dual-treated cells. High doses of ATO induced p53 accumulation in 11 of 21 patients. Combined treatment resulted in the induction of the proapoptotic p53/p73 target gene *p53AIP1* (p53-regulated apoptosis-inducing protein: (1) and greatly enhanced the apoptosis of treated cells [313]. Thus this study documented the effectiveness of combining ATO with MEK inhibitors in the treatment of APL and identified the molecular mechanism responsible for the observed synergism.

MEK inhibitors have been observed to synergize with UCN-01 and induce apoptosis in multiple myeloma cells [314]. Part of the synergy may be due to UCN-01 inducing ERK activation which is suppressed by the MEK inhibitor.

It should be pointed out that the combination of MEK inhibitors and a chemotherapeutic drug may not always result in a positive interaction and in some cases combination therapy results in an antagonistic response. For example, combining MEK inhibitors with betulinic acid, a drug lethal for melanoma cells, antagonized the effects that betulinic acid normally has on apoptosis [315]. Furthermore, the precise timing of the addition of two drugs is important as they may differentially affect cell cycle progression; therefore, one drug may need to be added before the other for a synergynistic response to be observed and perhaps to prevent an antagonistic response [301].

15   Role of the Raf/MEK/ERK Pathway in Drug Resistance to Reactive Oxygen Intermediate– Inducing Cancer Treatments

Many cancer therapies induce the generation of oxygen radicals within cells. These therapies include treatments such as, chemotherapeutic drugs, irradiation, and newer treatments such as photodynamic therapy (PDT). Doxorubicin, one of the most effective chemotherapeutic drugs against a wide variety of cancers, works via two main mechanisms to exert antitumor effects and toxicity. Doxorubicin intercalates in the DNA and interferes with DNA polymerase by disrupting helicase activity [316]. It also induces the production of free radicals and oxidative stress, which are involved in its antitumor effects [317, 318]. The generation of oxygen radicals is important for the therapeutic effectiveness of doxorubicin because scavenging reactive oxygen intermediates results in decreased cell killing to this drug [319].

The initial reactive oxygen species generated as a consequence of ionization radiation is OH⁻, which is short-lived and only diffuses about 4 nm before reacting. Secondary reactive oxygen species, produced in response to ionizing radiation include O^{2-} and H_2O_2. Studies with fluorescent dyes demonstrated generation of reactive oxygen species within cells within 15 minutes after irradiation [320]. Similar to doxorubicin the generation of oxygen radicals is important for the therapeutic effectiveness of radiation therapy because scavenging reactive oxygen intermediates results in decreased cell killing in response to radiation [321].

PDT is a three-component treatment that is used for the treatment of cancer [322]. PDT requires a photosensitizer, molecular oxygen and a laser of a wavelength matching the absorption spectrum of the photosensitizer (porphyrins and porphyrin-related compounds). When a porphyrin molecule absorbs light, it can transform an oxygen molecule to an activated state. Similar to doxorubicin and irradiation, PDT also requires the production of oxygen radicals to mediate some of its antitumor effects [323]. Thus, three well known cancer treatments result in the generation of reactive oxygen intermediates. These same three treatments have also been shown to lead to activation of ERK1,2 [324–327].

The Raf/MEK/ERK signaling pathway can play an adaptive role in protecting cells from oxidative stress [328]. In a nonmalignant murine alveolar epithelial cell line, blocking MEK activation using the MEK inhibitor U0126 prevents hypoxia-induced Nrf2 up-regulation [328]. Deletion of ASK1 protects cells from oxidant-induced cell death but not death receptor-induced apoptosis

[329]. Conversely, hydrogen peroxide is capable of inducing apoptosis in cardiomyocytes which can be increased in MEKK1 negative cells [330]. Deletion of ASK1 protects against hydrogen peroxide–induced apoptosis in fibroblasts and also prevents prolonged p38 activation, suggesting an apoptotic role for p38 in response to oxidative stress [331]. Ras activation and subsequent signaling via Rho can also activate this pathway, as does ligation of the TNF receptor [332–335]. Redox activation of ERK5/BMK1 exhibits an anti-apoptotic effect [336]. U0126 and PD98059 are also reported to inhibit the activity of MEK5, the MAPKK involved in ERK5/BMK1 activation [337–339]. Suzuki et al. found that these inhibitors decreased PC12 cell viability in response to hydrogen peroxide treatment. This decrease in cell viability occurred when the ERK5/BMK1 protein was completely downregulated using siRNA, suggesting that the effects of U0126 and PD98059 were mediated in part via ERK5/BMK1 pathway [336]. These data indicate the potential for both the ERK1/2 and ERK5/BMK pathways to promote treatment resistance to currently used reactive oxygen intermediate–inducing cancer treatments.

16   Conclusions

Over the past 25 years, there has been much progress in elucidating the involvement of the Ras/Raf/MEK/ERK and PI3K/PTEN/Akt cascades in promoting normal cell growth, regulating apoptosis, as well as the etiology of human neoplasia and the induction of chemotherapeutic drug resistance. From initial seminal studies that elucidated the oncogenes present in avian and murine oncogenes, we learned that ErbB, Ras, Src, Abl, Raf, PI3K, Akt, Jun, Fos, Ets, and NF-κB (Rel) were originally cellular genes which were captured by retroviruses. Biochemical studies continue to elucidate the roles that these cellular and viral oncogenes have in cellular transformation. We have learned that many of these oncogenes are connected to the Ras/Raf/MEK/ERK and PI3K/PTEN/PDK/Akt pathways and either feed into this pathway (e.g., BCR-ABL, ErbB) or are downstream targets, which regulate gene expression (e.g., Jun, Fos, Ets, and NF-κB).

The Ras/Raf/MEK/ERK pathway has what often appears to be conflicting roles in cellular proliferation, differentiation, and the prevention of apoptosis. Classical studies have indicated that Ras/Raf/MEK/ERK can promote proliferation and malignant transformation in part due to the stimulation of cell growth and at the same time the prevention of apoptosis. Furthermore, an often overlooked aspect of Raf/MEK/ERK is its effects on cytokine and growth factor gene transcription that can stimulate proliferation. The latest "hot" area of the Ras/Raf/MEK/ERK pathway is the discovery of mutation of the B-Raf gene in human cancer, which can promote proliferation and transformation [132]. However, it should be remembered that only a few years ago, hyperactivation of B-Raf and Raf-1 was proposed to promote cell cycle arrest [4]. Thus, it is probably fine-tuning of these mutations, which dictates whether there is cell cycle arrest or malignant transformation.

Initially it was thought that Raf-1 was the most important Raf isoform. Raf-1 was the earliest studied Raf isoform and homologous genes are present in both murine and avian transforming retroviruses. Originally it was shown that Raf-1 was ubiquitously expressed, indicating a more general and important role, whereas B-Raf and A-Raf had more limited patterns of expression. However, it is now believed that not only B-Raf is the more important activator of the Raf/MEK/ERK cascade and in some cases, activation of Raf-1 may require B-Raf. However, Raf-1 rears its head again in the cancer field by the recent discovery that there are mutant Raf-1 alleles in certain therapy induced t-AMLs that are transmitted in a Mendelian fashion [133]. The role of A-Raf remains poorly defined yet it is an interesting isoform. It is the weakest Raf kinase, yet it can stimulate cell cycle progression and proliferation without having the negative effects on cell proliferation that B-Raf and Raf-1 can exert [337, 339].

Activation of the Raf proteins is very complex as there are many phosphorylation sites on Raf. Phosphorylation at different sites can lead to either activation or inactivation. Clearly there are many kinases and phosphatases that regulate Raf activity and the state of phosphorylation will determine whether Raf is active or inactive. While the kinases involved in regulation in Raf/MEK/ERK have been extensively studied, there is only very limited knowledge of the specific phosphatases involved in these regulatory events.

Raf-1 has many roles, which are apparently independent of downstream MEK/ERK. Some of these functions occur at the mitochondria and are intimately associated with the prevention of apoptosis. Raf-1 may function as a scaffolding molecule to inhibit the activity of kinases that promote apoptosis.

The Raf/MEK/ERK pathway is both positively [Hsp90, kinase suppressor of Ras (KSR), MEK partner-1 (MP-1)] and negatively (RKIP, 14-3-3) regulated by association with scaffolding proteins. The expression of some of the scaffolding proteins is altered in human cancer (e.g., RKIP) in some cases. Some of these scaffolding proteins (e.g., Hsp90) are being evaluated as potential therapeutic targets (geldanamycin). Potential roles of Hsp90 in stabilizing activated forms of Raf are intriguing and may allow the evolution of activated mutant forms of Raf.

The Raf/MEK/ERK pathway is intimately linked with the PI3K/PTEN/Akt pathway. Ras can regulate both pathways. Furthermore, in some cell types, Raf activity is negatively regulated by Akt indicating a cross-talk between the two pathways. Both pathways may result in the phosphorylation of many downstream targets and impose a role in the regulation of cell survival and proliferation. These pathways phosphorylate many key proteins involved in apoptosis (Bad, Bim, Mcl-1, Caspase 9, Ask-1, and others) which serve to alter their activities and subcellular localization. The phosphorylation events mediated by Raf/MEK/ERK and PI3K/Akt pathways are associated with the prevention of apoptosis. In contrast, another MAPK, JNK also phosphorylates many of these molecules, and these phosphorylation events often have opposite effects as those elicited by the Raf/MEK/ERK and PI3K/Akt pathways. Interestingly, Ras and Raf mutations may not always result in similar outcomes. For example, a Ras mutation would be predicted to activate both the Raf/MEK/ERK and PI3K/Akt pathways. Activation of PI3K/Akt could result in the suppression of Raf/MEK/ERK. However, mutation at either B-Raf or Raf-1 would result in only activation of Raf/MEK/ERK.

Although we often think of phosphorylation of these molecules as being associated with the prevention of apoptosis and the induction of gene transcription, this view is oversimplified. For example, in certain situations such as in advanced prostate cancer, the Raf/MEK/ERK pathway may be inhibited; hence, the phosphorylation of Bad and CREB normally mediated by the Raf/MEK/ERK cascade, which is associated with the prevention of apoptosis, will be inhibited. Likewise, it is important to remember that phosphorylation at certain protein residues will result in enhanced activity, whereas phosphorylation at different residues will result in decreased activity. For example, phosphorylation of Bim by JNK is associated with the promotion of apoptosis while phosphorylation of Bim by Raf/MEK/ERK or PI3K/Akt pathways is associated with the prevention of apoptosis.

A consequence of diverse cancer therapies (chemotherapy, radiation therapy, photodynamic therapy) is the induction of the Raf/MEK/ERK pathway, which may in some cases provide a survival function. The mechanism of induction of these pathways may be in part in response to the ROS generated by the different therapies. Thus in some cases it may be appropriate to combine these conventional therapies with small molecular weight inhibitors which target the Raf/MEK/ERK pathway.

Although it has been known for many years that the Raf/MEK/ERK pathway can effect cell cycle arrest, differentiation and senescence, these are probably some of the least studied research areas in the field due to the often cell lineage–specific effects that must be evaluated in each cell type. An intriguing aspect of human cancer therapy is that in some cases stimulation of the Raf/MEK/ERK pathway may be desired to promote terminal differentiation, whereas in other types of malignant cancer cells that proliferate in response to Raf/MEK/ERK activity, inhibition of the Raf/MEK/ERK pathway may be desired to suppress proliferation. Thus, we must be flexible in dealing with the Raf/MEK/ERK pathway. As we learn more, our conceptions continue to change.

Acknowledgments. JAM, RAF, WHC, and LSS have been supported in part by a grant from the NIH (R01098195). MM, AT, AMM, PL, and AB were supported in part from grants from Associazione Italiana Ricerca sul Cancro (AIRC regional grants).

References

1. Steelman LS, Pohnert SC, Shelton JG, et al. JAK/STAT, Raf/MEK/ERK, PI3K/Akt and BCR-ABL in cell cycle progression and leukemogenesis. Leukemia 2004, 18:189–218.

2. Lee JT Jr, McCubrey JA. Targeting the Raf kinase cascade in cancer therapy—novel molecular targets and therapeutic strategies. Expert Opin Ther Targets 2002, 6:659–678.

3. Blalock WL, Weinstein-Oppenheimer C, Chang F, et al. Signal transduction, cell cycle regulatory, and anti-apoptotic pathways regulated by IL-3 in hematopoietic cells: possible sites for intervention with anti-neoplastic drugs. Leukemia 1999, 13:1109–1166.

4. McCubrey JA, Steelman LS, Chappell WH, et al. 2007. Roles of the Raf/MEK/ERK pathway in cell growth, malignant transformation and drug resistance. Biochem Biophys Acta, in press.

5. Kim D, Dan HC, Park S, et al. AKT/PKB signaling mechanisms in cancer and chemoresistance. Front Biosci 2005, 10:975–987.

6. Matsuguchi T, Salgia R, Hallek M, et al. Shc phosphorylation in myeloid cells is regulated by granulocyte macrophage colony-stimulating factor, interleukin-3, and steel factor and is constitutively increased by p210BCR/ABL. J Biol Chem 1994, 269:5016–5021.

7. Inhorn RC, Carlesso N, Durstin M, et al. Identification of a viability domain in the granulocyte/macrophage colony-stimulating factor receptor beta-chain involving tyrosine-750. Proc Natl Acad Sci U S A 1995, 92:8665–8669.

8. Okuda K, Foster R, Griffin JD. Signaling domains of the beta c chain of the GM-CSF/IL-3/IL-5 receptor. Ann N Y Acad Sci 1999, 872:305–313.

9. Tauchi T, Boswell HS, Leibowitz D, et al. 1994. Coupling between p210bcr-abl and Shc and Grb2 adaptor proteins in hematopoietic cells permits growth factor receptor-independent link to ras activation pathway. J Exp Med 1994, 179:167–175.

10. Lanfrancone L, Pelicci G, Brizzi MF, et al. Overexpression of Shc proteins potentiates the proliferative response to the granulocyte-macrophage colony-stimulating factor and recruitment of Grb2/Sos and Grb2/p140 complexes to the beta receptor subunit. Oncogene 1995, 10:907–917.

11. Karin M, Minden A, Lin A, et al. Differential activation of ERK and JNK mitogen-activated protein kinases by Raf-1 and MEKK. Science 1994, 266:1719–1723.

12. Lange-Carter CA, Johnson GL. Ras-dependent growth factor regulation of MEK kinase in PC12 cells. Science 1994, 265:1458–1461.

13. Marais R, Light Y, Paterson HF, et al. Ras recruits Raf-1 to the plasma membrane for activation by tyrosine phosphorylation. EMBO J 1995, 14:3136–3145.

14. Marais R, Light Y, Paterson HF, et al. Differential regulation of Raf-1, A-Raf, and B-Raf by oncogenic ras and tyrosine kinases. J Biol Chem 1997, 272:4378–4383.

15. Mason CS, Springer CJ, Cooper RG, et al. Serine and tyrosine phosphorylations cooperate in Raf-1, but not B-Raf activation. EMBO J 1999, 18:2137–2148.

16. Xu S, Robbins D, Frost J, et al. MEKK1 phosphorylates MEK1 and MEK2 but does not cause activation of mitogen-activated protein kinase. Proc Nat Acad Sci U S A 1995, 92:6808–6812.

17. Cardone MH, Roy N, Stennicke HR, et al. Regulation of cell death protease caspase-9 by phosphorylation. Science 1998, 282:1318–1321.

18. Allan LA, Morrice N, Brady S, et al. Inhibition of caspase-9 by phosphorylation at Thr125 by ERK MAP kinase. Nat Cell Biol 2003, 5:647–654.

19. Davis RJ, Derijard B, Raingeaud J, et al. Independent human MAP kinase signal transduction pathways defined by MEK and MKK isoforms. Science 1995, 267:682–685.

20. Xing J, Ginty DD, Greenberg ME. Coupling of the Ras-MAPK pathway to gene activation by Rsk2, a growth factor regulated CREB kinase. Science 1996, 273:959–963.

21. Coutant A, Rescan C, Gilot D, et al. PI3K-FRAP/mTOR pathway is critical for hepatocyte proliferation whereas MEK/ERK supports both proliferation and survival. Hepatology 2002, 36:1079–1088.

22. Iijima Y, Laser M, Shiraishi H, et al. c-Raf/MEK/ERK pathway controls protein kinase C-mediated p70S6K activation in adult cardiac muscle cells. J Biol Chem 2002, 277:23065–23075.

23. Blalock WL, Navolanic PM, Steelman LS, et al. Requirement for the PI3K/Akt pathway in MEK1-mediated growth and prevention of apoptosis: identification of an Achilles heel in leukemia. Leukemia 2003, 17:1058–1067.

24. Deng T, Karin M. c-Fos transcriptional activity stimulated by H-Ras-activated protein kinase distinct from JNK and ERK. Nature 1994, 371:171–175.

25. Davis RJ. Transcriptional regulation by MAP kinases. Mol Reprod Dev 1995, 4:459–467.

26. Robinson MJ, Stippec SA, Goldsmith E, et al. A constitutively active and nuclear form of the MAP kinase ERK2 is sufficient for neurite outgrowth and cell transformation. Curr Biol 1998, 21:1141–1150.

27. Aplin AE, Stewart SA, Assoian RK, et al. Integrin-mediated adhesion regulates ERK nuclear translocation and phosphorylation of Elk-1. J Cell Biol 2001, 153:273–282.

28. McCubrey JA, May WS, Duronio V, et al. Serine/threonine phosphorylation in cytokine signal transduction. Leukemia 2000, 14:9–21.

29. Tresini M, Lorenzini A, Frisoni L, et al. Lack of Elk-1 phosphorylation and dysregulation of the extracellular regulated kinase signaling pathway in senescent human fibroblast. Exp Cell Res 2001, 269:287–300.

30. Eblen ST, Catling AD, Assanah MC, et al. Biochemical and biological functions of the N-terminal, noncatalytic domain of extracellular signal-regulated kinase 2. Mol Cell Biol 2001, 21:249–259.

31. Adachi T, Kar S, Wang M, et al. Transient and sustained ERK phosphorylation and nuclear translocation in growth control. J Cell Physiol 2002, 192:151–159.

32. Wang CY, Bassuk AG, Boise LH, et al. Activation of the granulocyte-macrophage colony-stimulating factor promoter in T cells requires cooperative binding of Elf-1 and AP-1 transcription factors. Mol Cell Biol 1994, 14:1153–1159.

33. Thomas RS, Tymms MJ, McKinlay LH, et al. ETS1, NFkappaB and AP1 synergistically transactivate the human GM-CSF promoter. Oncogene 1997, 23:2845–2855.

34. Ponti C, Gibellini D, Boin F, et al. Role of CREB transcription factor in c-fos activation in natural killer cells. Eur J Immunol 2002, 32:3358–3365.

35. Fry TJ, Mackall CL. Interleukin-7: from bench to clinic. Blood 2002, 99:3892–3904.

36. Deng X, Kornblau SM, Ruvolo PP, et al. Regulation of Bcl2 phosphorylation and potential significance for leukemic cell chemoresistance. J Natl Cancer Inst Monogr 2001, 28:30–37.

37. Carter BZ, Milella M, Tsao T, et al. Regulation and targeting of antiapoptotic XIAP in acute myeloid leukemia. Leukemia 2003, 17:2081–2089.

38. Jia W, Yu C, Rahmani M, et al. Synergistic anti-leukemic interactions between 17-AAG and UCN-01 involve interruption of RAF/MEK- and AKT-related pathways. Blood 2003, 102:1824–1832.

39. Troppmair J, Rapp UR. Raf and the road to cell survival: a tale of bad spells, ring bearers and detours. Biochem Pharmacol 2003, 66:1341–1345.

40. Harada H, Quearry B, Ruiz-Vela A, et al. Survival factor-induced extracellular signal-regulated kinase phosphorylates BIM, inhibiting its association with BAX and proapoptotic activity. Proc Natl Acad Sci U S A 2004, 101:15313–15317.

41. Marani M, Hancock D, Lopes R, et al. Role of Bim in the survival pathway induced by Raf in epithelial cells. Oncogene 2004, 23:2431–2441.

42. Ley R, Balmanno K, Hadfield K, et al. Activation of the ERK1/2 signaling pathway promotes phosphorylation and proteasome-dependent degradation

of the BH3-only protein, Bim. J Biol Chem 2003, 278:18811–18816.

43. Weston CR, Balmanno K, Chalmers C, et al. Activation of ERK1/2 by deltaRaf-1:ER* represses Bim expression independently of the JNK or PI3K pathways. Oncogene 2003, 22:1281–1293.

44. Domina AM, Vrana J, Gregory MA, et al. MCL1 is phosphorylated in the PEST region and stabilized upon ERK activation in viable cells, and at additional sites with cytotoxic okadaic acid or taxol. Oncogene 2004, 23:5301–5315.

45. Gélinas C, White E. BH3-only proteins in control: specificity regulates MCL-1 and BAK-mediated apoptosis. Genes Dev 2006, 19:1263–1268.

46. O'Neill E, Rushworth L, Baccarini M, et al. Role of the kinase MST2 in suppression of apoptosis by the proto-oncogene Raf. Science 2004, 306:2267–2270.

47. O'Neill EE, Matallanas D, Kolch W. Mammalian sterile 20-like kinases in tumor suppression: an emerging pathway. Cancer Res 2005, 65:5485–5487.

48. Drexler HG. Expression of the FLT-3 receptor and response to FLT3 ligand by leukemic cells. Leukemia 1996, 10:588–599.

49. Rao P, Mufson RA. A membrane proximal domain of the human interleukin-3 receptor beta c subunit that signals DNA synthesis in NIH 3T3 cells 1995 specifically binds a complex of Src and Janus family tyrosine kinases and phosphatidylinositol 3-kinase. J Biol Chem 1995, 270:6886–6893.

50. Chang F, Lee JT, Navolanic PM, et al. Involvement of PI3K/Akt pathway in cell cycle progression, apoptosis, and neoplastic transformation: a target for cancer chemotherapy. Leukemia 2003, 17:590–603.

51. Troussard AA, Mawji NM, Ong C, et al. Conditional knock-out of integrin-linked kinase demonstrates an essential role in protein kinase B/Akt activation. J Biol Chem 2003, 278:22374–22378.

52. Xu Z, Ma DZ, Wang LY, et al. Transforming growth factor-beta1 stimulated protein kinase B serine-473 and focal adhesion kinase tyrosine phosphorylation dependent on cell adhesion in human hepatocellular carcinoma SMMC-7721 cells. Biochem Biophys Res Commun 2003, 312:388–396.

53. Persad S, Dedhar S. The role of integrin-linked kinase (ILK) in cancer progression. Cancer Metastasis Rev 2003, 22:375–384.

54. Kumar AS, Naruszewicz I, Wang P, et al. ILKAP regulates ILK signaling and inhibits anchorage-independent growth. Oncogene 2004, 23:3454–3461.

55. Songyang Z, Baltimore D, Cantley LC, et al. Interleukin 3-dependent survival by the Akt protein kinase. Proc Natl Acad Sci U S A 1997, 94:11345–11350.

56. Scheid MP, Duronio V. Dissociation of cytokine-induced phosphorylation of Bad and activation of PKB/akt: involvement of MEK upstream of Bad phosphorylation. Proc Natl Acad Sci U S A 1998, 95:7439–7444.

57. del Peso L, Gonzalez-Garcia M, Page C, et al. Interleukin-3-induced phosphorylation of BAD through the protein kinase Akt. Science 1997, 278:687–689.

58. Nakae J, Park BC, Accili D. Insulin stimulates phosphorylation of the forkhead transcription factor FKHR on serine 253 through a Wortmannin-sensitive pathway. J Biol Chem 1999, 274:15982–15985.

59. Brunet A, Bonni A, Zigmond MJ, et al. Akt promotes cell survival by phosphorylating and inhibiting a forkhead transcription factor. Cell 1999, 96:857–868.

60. Medema RH, Kops GJ, Bos JL, et al. Forkhead transcription factors mediate cell-cycle regulation by Ras and PKB through p27kip1. Nature 2000, 404:782–787.

61. Dijkers PF, Medema RH, Pals C, et al. Forkhead transcription factor FKHR-L1 modulates cytokine-dependent transcriptional regulation of p27Kip1. Mol Cell Biol 2000, 20:9138–9148.

62. Qi XJ, Wildey GM, Howe PH. Evidence that Ser87 of BimEL is phosphorylated by Akt and regulates BimEL apoptotic function. J Biol Chem 2006, 281:813–823.

63. Mayo LD, Donner DB. A phosphatidylinositol 3-kinase/Akt pathway promotes translocation of Mdm2 from the cytoplasm to the nucleus. Proc Natl Acad Sci U S A 2001, 98:10983–10985.

64. Gottlieb TM, Leal JF, Seger R, et al. Cross-talk between Akt, p53 and Mdm2: possible implications for the regulation of apoptosis. Oncogene 2002, 21:1299–1303.

65. Zhou BP, Hung MC. Novel targets of Akt, p21(Cip1. WAF1), and MDM2. Semin Oncol 2002, 29:62–70.

66. Dan HC, Sun M, Kaneko S, et al. Akt phosphorylation and stabilization of X-linked inhibitor of apoptosis protein (XIAP). J Biol Chem 2004, 279:5405–5412.

67. You H, Pellegrini M, Tsuchihara K, et al. FOXO3a-dependent regulation of Puma in response to cytokine/growth factor withdrawal. J Exp Med 2006, 203:1657–1663.

68. Obexer P, Geiger K, Ambros PF, et al. FKHRL1-mediated expression of Noxa and Bim induces apoptosis via the mitochondria in neuroblastoma cells. Cell Death Diff 2007, 14:534–547.

69. Ozes ON, Mayo LD, Gustin JA, et al. NF-kappaB activation by tumor necrosis factor requires the Akt serine-threonine kinase. Nature 1999, 401:82–85.

70. Romashkova JA, Makarov SS. NF-kappaB is a target of Akt in anti-apoptotic PDGF signaling. Nature 1999, 401:86–90.

71. Madrid LV, Wang CY, Guttridge DC, et al. Akt suppresses apoptosis by stimulating the transactivation potential of the Rel A/p65 subunit of NF-kappaB. Mol Cell Biol 2000, 20:1626–1638.

72. Howe CJ, LaHair MM, Maxwell JA, et al. Participation of the calcium/calmodulin-dependent kinases in hydrogen peroxide-induced Ikappa B phosphorylation in human T lymphocytes. J Biol Chem 2002, 277:30469–30476.

73. Howe CJ, LaHair MM, McCubrey JA, et al. Redox regulation of the CaM-kinases. J Biol Chem 2004, 279:44573–44581.

74. Hu MC, Lee DF, Xia W, et al. IkappaB kinase promotes tumorigenesis through inhibition of forkhead FOXO3a. Cell 2004, 117:225–237.

75. Mayo MW, Baldwin AS. The transcription factor NF-kappaB: control of oncogenesis and cancer therapy resistance. Biochim Biophys Acta 2000, 1470: M55–62.

76. Shishodia S, Aggarwal BB. Nuclear factor-kappaB activation mediates cellular transformation, proliferation, invasion angiogenesis and metastasis of cancer. Cancer Treat Res 2004, 119:139–173.

77. Du K, Montminy M. CREB is a regulatory target for the protein kinase Akt/PKB. J. Biol Chem 1998, 273:32377–32379.

78. Arcinas M, Heckman CA, Mehew JW, et al. Molecular mechanisms of transcriptional control of bcl-2 and c-myc in follicular and transformed lymphoma. Cancer Res 2001, 61:5202–5206.

79. Wang JM, Chao JR, Chen W, et al. The antiapoptotic gene Mcl-1 is upregulated by the phosphatidylinositol 3-kinase/Akt signaling pathway through a transcription factor complex containing CREB. Mol Cell Biol 1999, 19:6195–6206.

80. Mahalingam M, Templeton DJ. Constitutive activation of S6 kinase by deletion of amino-terminal autoinhibitory and rapamycin sensitivity domains. Mol Cell Biol 1996, 16:405–413.

81. Dufner A, Anjelkovic M, Burgering BMT, et al. Protein kinase B localization and activation and eukaryotic translational initiation factor 4E-binding protein phosphorylation. Mol Cell Biol 1999, 19:4525–4534.

82. Romanelli A, Martin KA, Toker A, et al. p70 S6 Kinase is regulated by protein kinase Cζ and participates in a phosphoinositide 3-kinase-regulated signaling complex. Mol Cell Biol 1999, 19:2921–2928.

83. Harada H, Andersen JS, Mann M, et al. P70S6 kinase signals cell survival as well as growth, inactivating the pro-apoptotic molecule Bad. Proc Natl Acad Sci U S A 2001, 98:9666–9670.

84. Edinger AL, Thompson CB. An activated mTOR mutant supports growth factor-independent, nutrient-dependent cell survival. Oncogene 2004, 23:5654–5663.

85. Panwalkar A, Verstovsek S, Giles FJ. Mammalian target of rapamycin inhibition as therapy for hematologic malignancies. Cancer 2004, 100:657–666.

86. Jonassen AK, Mjos OD, Sack MN. p70S6 kinase is a functional target of insulin activated Akt cell-survival. Biochem Biophys Res Commun 2004, 315:160–165.

87. Ma L, Chen Z, Erdjument-Bromage H, et al. Phosphorylation and functional inactivation of TSC2 by Erk implications for tuberous sclerosis and cancer pathogenesis. Cell 2005, 121:179–193.

88. Shaw RJ, Cantley LC. Ras, PI(3)K and mTOR signaling controls tumour cell growth. Nature 2006, 441:424–430.

89. Steck PA, Pershouse MA, Jasser SA, et al. Identification of a candidate tumour suppressor gene, MMAC1, at chromosome 10q23.3 that is mutated in multiple advanced cancers. Nat Genet 1997, 15:356–362.

90. Li DM, Sun H. TEP1, encoded by a candidate tumor suppressor locus, is a novel protein tyrosine phosphatase regulated by transforming growth factor beta. Cancer Res 1997, 57:2124–2129.

91. Li J, Yen C, Liaw D, et al. PTEN, a putative protein tyrosine phosphatase gene mutated in human brain, breast, and prostate cancer. Science 1997, 275:1943–1947.

92. Steelman LS, Bertrand FE, McCubrey JA. The complexity of PTEN: mutation, marker and potential target for therapeutic intervention. Expert Opinion Ther Targets 2004, 8:537–550.

93. Mahimainathan L, Choudhury GG. Inactivation of platelet-derived growth factor receptor by the tumor suppressor PTEN provides a novel mechanism of action of the phosphatase. J Biol Chem 2004, 279:15258–15268.

94. Raftopoulou M, Etienne-Manneville S, Self A, et al. Regulation of cell migration by the C2 domain of the tumor suppressor PTEN. Science 2004, 303:1179–1181.

95. Damen JE, Liu L, Rosten P, et al. The 145-kDa protein induced to associate with Shc by multiple cytokines is an inositol tetraphosphate and phosphatidylinositol 3,4,5-triposphate 5-phosphatase. Proc Natl Acad Sci U S A 1996, 93:1689–1693.

96. Kavanaugh WM, Pot DA, Chin SM, et al. Multiple forms of an inositol polyphosphate 5-phosphaatase from signaling complexes with Shc and Grb2. Curr Biol 1996, 6:438–445.

97. Lioubin MN, Algate PA, Tsai S, et al. P150Ship, a signal transduction molecule with inositol polyphosphate-5-phosphatase activity. Gene Dev 1996, 10:1084–1095.

98. Taylor V, Wong M, Brandts C, et al. 5 phospholipid phosphatase SHIP-2 causes protein kinase B inactivation and cell cycle arrest in glioblastoma cells. Mol Cell Biol 2000, 20:6860–6871.

99. Muraille E, Pesesse X, Kuntz C, et al. Distribution of the src-homology-2-domain-containing inositol 5-phosphatase SHIP-2 in both non-haemopoietic and haemopoietic cells and possible involvement in SHIP-2 in negative signaling of B-cells. Biochem J 1999, 342:697–705.

100. Rommel C, Clarke BA, Zimmermann S, et al. Differentiation stage-specific inhibition of the Raf-MEK-ERK pathway by Akt. Science 1999, 286:1738–1741.

101. Zimmermann S, Moelling K. Phosphorylation and regulation of Raf by Akt (Protein Kinase B). Science 1999, 286:1741–1744.

102. Guan K, Figueroa C, Brtva TR, et al. Negative regulation of the serine/threonine kinase B-Raf by Akt. J Biol Chem 2000, 275:27354–27359.

103. Zhang BH, Tang E, Zhu T, et al. Serum and glucocorticoid-inducible kinase SGK phosphorylates and negatively regulates B-Raf. J Biol Chem 2001, 276:31620–31626.

104. Majewski M, Nieborowska-Skorska M, Salomoni P, et al. Activion of mitochondrial Raf-1 is involved in the anti-apoptotic effects of Akt. Cancer Res 1999, 59:2815–2819.

105. McCubrey JA, Steelman LS, Blalock WL, et al. Synergistic effects of PI3K/Akt on abrogation of cytokine-dependency induced by oncogenic Raf. Adv Enzyme Regl 2001, 41:289–323.

106. McCubrey JA, Lee JT, Steelman LS, et al. Interactions between the PI3K and Raf signaling pathways can result in the transformation of hematopoietic cells. Cancer Detect Prev 2001, 25:375–393.

107. Gelfanov VM, Burgess GS, Litz-Jackson S, et al. Transformation of interleukin-3-dependent cells without participation of Stat5/Bcl-xL: cooperation of akt with raf/erk leads to p65 nuclear factor κB-mediated antiapoptosis involving c-IAP2. Blood 2001, 15:2508–2517.

108. von Gise A, Lorenz P, Wellbrock C, et al. Apoptosis suppression by Raf-1 and MEK1 requires MEK and phosphatidylinositol 3-kinase dependent signals. Mol Cell Biol 2001, 21:2324–2336.

109. Shelton JG, Steelman LS, Lee JT, et al. Effects of the RAF/MEK/ERK and PI3K/AKT signal transduction pathways on the abrogation of cytokine-dependence and prevention of apoptosis in hematopoietic cells. Oncogene 2003, 22:2478–2492.

110. Shelton JG, Blalock WL, White ER, et al. Ability of the activated PI3K/Akt oncoproteins to synergize with MEK1 and induce cell cycle progression and abrogate the cytokine-dependence of hematopoietic cells. Cell Cycle 2004, 3:503–512.

111. Yang E, Zha J, Jockel J, et al. Bad, a hetro-dimeric partner for Bcl-xL and Bcl-2 displaces Bax and promotes cell death. Cell 1995, 80:285–291.

112. Pugazhenthi S, Nesterova A, Sable C, et al. Akt/protein kinase B up-regulates Bcl-2 expression through cAMP-response element-binding protein. J Biol Chem 2000, 275:10761–10766.

113. Pugazhenthi S, Miller E, Sable C, et al. Insulin-like growth factor-I induces Bcl-2 promoter through the transcription factor c-AMP-response element-binding protein. J Biol Chem 1999, 274:27529–27535.

114. Bonni A, Brunet A, West AE, et al. Cell survival promoted by the Ras-MAPK signaling pathway by transcription-dependent and independent mechanisms. Science 1999, 286:1358–1362.

115. Datta SR, Dudek, H, Tao X, et al. Akt phosphorylation of BAD couples survival signals to the cell-intrinsic death machinery. Cell 1997, 91:231–241.

116. Harada H, Becknell B, Wilm M, et al. Phosphorylation and inactivation of BAD by mitochondria-anchored protein kinase A. Mol Cell 1999, 3:413–422.

117. Sunayama J, Tsuruta F, Masuyama N, et al. JNK antagonizes Akt-mediated survival signals by phosphorylating 14-3-3. J Cell Biol 2005, 170:295–304.

118. She QB, Solit DB, Ye Q, et al. The BAD protein integrates survival signaling by EGFR/MAPK and PI3K/Akt kinase pathways in PTEN-deficient tumor cells. Cancer Cell 2005, 8:297.

119. Chen L, Willis SN, Wei A, et al. Differential targeting of prosurvival Bcl-2 proteins by their BH3-only ligands allows complementary apoptotic function. Mol Cell 2005, 17:393–403.

120. Putcha GV, Le S, Frank S, et al. JNK-mediated BIM phosphorylation potentiates BAX-dependent apoptosis. Neuron 2003, 38:899–914.

121. Willis SN, Chen L, Dewson G, et al. Proapoptotic Bak is sequestered by Mcl-1 and Bcl-xL, but not Bcl-2, until displaced by BH3-only proteins. Gene Dev 2005, 19:1294–1305.

122. Nosaka T, Kawashima T, Misawa K, et al. STAT5 as a molecular regulator of proliferation, differentiation and apoptosis in hematopoietic cells. EMBO J 1999, 18:4754–4765.

123. Wang K, Gross A, Waksman G, et al. Mutagenesis of the BH3 domain of BAX identifies residues critical for dimerization and killing. Mol Cell Biol 1998, 18:6083–6089.

124. Ernst P, Fisher JK, Avery W, et al. Definative hematopoiesis requires the mixed-lineage gene. Dev Cell 2004, 6:437–443.

125. Zhao S, Konopleva M, Cabreira-Hansen M, et al. Inhibition of phosphatidylinositol 3-kinase dephosphorylates

BAD and promotes apoptosis in myeloid leukemias. Leukemia 2004, 18:267–275.

126. Vrana JA, Cleaveland ES, Eastman A, et al. Inducer-and cell type-specific regulation of antia-poptotic MCL-1 in myeloid leukemia and multiple myeloma cells exposed to differentiation-inducing or microtubule-disrupting agents. Apoptosis 2006, 11:1275–1288.

127. Inoshita S, Takeda K, Hatai T, et al. Phosphorylation and inactivation of myeloid cell leukemia 1 by JNK in response to oxidative stress. J Biol Chem 2002, 277:43730–43734.

128. Maurer U, Charvet C, Wagman AS, et al. Glycogen synthase kinase-3 regulates mitochondrial outer membrane permeabilization and apoptosis by desta-bilization of MCL-1. Mol Cell 2006, 21:749–760.

129. Yu J, Zhang L. The transcriptional targets of p53 in apoptosis control. Biochem Biophys Res Commun 2005, 331:851–858.

130. Flotho C, Valcamonica S, Mach-Pascual S, et al. RAS mutations and clonality analysis in children with juvenile myelomonocytic leukemia (JMML). Leukemia 1999, 13:32–37.

131. Stirewalt DL, Kopecky KJ, Meshinchi S, et al. FLT3, RAS, and TP53 mutations in elderly patients with acute myeloid leukemia. Blood 2001, 97:3589–3595.

132. Garnett MJ, Marais R. Guilty as charged: B-Raf is a human oncogene. Cancer Cell 2004, 6:313–319.

133. Zebisch A, Staber PB, Delavar A, et al. Two trans-forming C-RAF germ-line mutations identified in patients with therapy-related acute myeloiid leuke-mia. Cancer Res 2006, 166:3401–3408.

134. Davies H, Bignell GR, Cox C, et al. Mutations of the BRAF gene in human cancer. Nature 2002, 417:949–954.

135. Libra L, Malaponte G, Navolanic PM, et al. Analysis of BRAF mutation in primary and meta-static melanoma. Cell Cycle 2006, 4:968–970.

136. Fransen K, Klinntenas M, Osterstrom A, et al. Mutation analysis of the B-Raf, A-Raf and Raf-1 genes in human colorectal adenocarcinomas. Carcinogenesis 2004, 25:527–533.

137. Wan PT, Garnett MJ, Ros SM, et al. Mechanism of activation of the Raf-MEK signaling pathway by oncogenic mutations of B-Raf. Cell 2004, 116:856–867.

138. Busca R, Abbe P, Mantous F, et al. Ras mediates the cAMP-dependent activation of extracellular signal-regulated in melanocytes. EMBO J 2000, 19:2900–2910.

139. Rushworth LK, Hindley AD, O'Neil E, et al. Regulation and role of Raf-1.B-Raf heterodimeriza-tion. Mol Cell Biol 2006, 26:2262–2272.

140. Garnett MJ, Rana S, Paterson H, et al. Wild-type and mutant B-RAF activate C-RAF through distinct mechanisms involving heterodimerization. Mol Cell 2005, 20:963–969.

141. Rajagopalan H, Bordelli A, Lengauer C, et al. Tumorigenesis: Raf/Ras oncogenes and mismatch-repair status. Nature 2002, 418:934.

142. Yuen ST, Davies H, Chan TL, et al. Similarity of the phenotypic patterns associated with B-Raf and KRAS mutations in colorectal neoplasia. Cancer Res 2002, 62:6451–6455.

143. Ricciardi MR, McQueen T, Chism D, et al. Quantitative single cell determination of ERK phos-phorylation and regulation in relapsed and refrac-tory primary acute myeloid leukemia, Leukemia 2005, 29:1543–1549.

144. Kornblau SM, Womble M, Qiu YH, et al. Simultaneous activation of multiple signal transduc-tion pathways confers poor prognosis in acute myel-ogenous leukemia. Blood 2006, 108:2358–2365.

145. Shelton JG, Steelman LS, Abrams SL, et al. The epidermal growth factor receptor as a target for ther-apeutic intervention in numerous cancers—What's genetics got to do with it? Expert Opinion Ther Targets 2005, 9:1009–1030.

146. Lynch TJ, Bell DW, Sordella R, et al. Activating mutations in the epidermal growth factor receptor underlying responsiveness of non-small cancer to gefitinib. N Engl J Med 2004, 350:2129–2139.

147. Sordella R, Bell DW, Haber DA, et al. Gefitinib-sensitizing EGFR mutations in lung cancer activate anti-apoptotic pathways. Science 2004, 305:1163–1167.

148. Sequist LV, Haber DA, Lynch TJ. Epidermal growth factor receptor mutations in non-small cell lung cancer, predicting clinical response to kinase inhibi-tors. Clin Cancer Res 2005, 11:5668–5670.

149. Pao W, Miller V, Zakowski M, et al. EGF receptor mutations are common in lung cancers from "never smokers" are associated with sensitivity of tumors to gefitinib and erlotinib. Proc Natl Acad Sci U S A 2004, 101:13306–13311.

150. Pao W, Wang TY, Riely GJ, et al. KRAS mutations and primary resistance of lung adenocarcinomas to gefitinib or erlotinib. PLoS Med 2005, 2:57–61.

151. Solit DB, Garraway LA, Pratilas CA, et al. BRaf mutation predicts sensitivity to MEK inhibition. Nature 2006, 439:358–362.

152. Shelton JG, Steelman LS, Abrams SL, et al. Conditional EGFR promotes cell cycle progression and prevention of apoptosis in the absence of auto-crine cytokines. Cell Cycle 2005, 4:822–830.

153. McCubrey JA, Shelton JG, Steelman LS, et al. Conditionally active v-ErbB:ER transforms NIH-3T3

cells and converts human and mouse cells to cytokine-independence. Oncogene 2004, 23:7810–7820.

154. Konopleva M, Shi Y, Steelman LS, et al. Development of a conditional in vivo model to evaluate the efficacy of small molecule inhibitors for the treatment of Raf-transformed hematopoietic cells. Cancer Res 2005, 65:9962–9970.

155. Blalock WL, Pearce M, Steelman LS, et al. A conditionally-active form of MEK1 results in autocrine transformation of human and mouse hematopoeitic cells. Oncogene 2000, 19:526–536.

156. Hoyle PE, Moye PW, Steelman LS, et al. Differential abilities of the Raf family of protein kinases to abrogate cytokine-dependency and prevent apoptosis in murine hematopoietic cells by a MEK1-dependent mechanism. Leukemia 2000, 14:642–656.

157. McCubrey JA, Steelman LS, Hoyle PE, et al. Differential abilities of activated Raf oncoproteins to abrogate cytokine-dependency, prevent apoptosis and induce autocrine growth factor synthesis in human hematopoietic cells. Leukemia 1998, 12:1903–1929.

158. Shelton JG, Blalock WL, White ER, et al. Ability of the activated PI3K/Akt oncoproteins to synergize with MEK1 and induce cell cycle progression and abrogate the cytokine-dependence of hematopoietic cells. Cell Cycle 2004, 3:503–512.

159. Kubota Y, Ohnishi H, Kitanaka A, et al. Constitutive activation of PI3K is involved in the spontaneous proliferation of primary acute myeloid leukemia cells: direct evidence of PI3K activation. Leukemia 2004, 18:1438–1440.

160. Cuni S, Perez-Aciego P, Perez-Chacon G, et al. A sustained activation of PI3K/NF-kappaB pathway is critical for the survival of chronic lymphocytic leukemia B cells. Leukemia 2004, 18:1391–1400.

161. Rodriguez-Viciana P, Warne PH, Dhand R, et al. Phosphatidylinositol-3-OH kinase as a direct target of Ras. Nature 1994, 370:527–532.

162. Hu L, Shi Y, Hsu JH, et al. Downstream effectors of oncogenic ras in multiple myeloma cells. Blood 2003, 101:3126–3135.

163. Gire V, Marshall C, Wynford-Thomas D. PI-3-kinase is an essential anti-apoptotic effector in the proliferative response of primary human epithelial cells to mutant RAS. Oncogene 2000, 19:2269–2276.

164. Sun H, King AJ, Diaz HB, et al. Regulation of the protein kinase Raf-1 by oncogenic Ras through phosphatidylinositol 3-kinase, Cdc42/Rac and Pak. Curr Biol 2000, 10:281–284.

165. Ninomiya Y, Kato K, Takahashi A, et al. K-Ras and H-Ras activation promote distinct consequences on endometrial cell survival. Cancer Res 2004, 64:2759–2765.

166. Jucker M, Sudel K, Horn S, et al. Expression of a mutated form of the p85alpha regulatory subunit of phosphatidylinositol 3-kinase in a Hodgkin's lymphoma-derived cell line (CO). Leukemia 2002, 16:894–901.

167. Engleman JA, Luo J, Canley LC. The evolution phosphatidylinositol 3-kinases as regulators of growth and metabolism, Nat Rev Genet 2006, 7:606–619.

168. Vogt PK, Bader AG, Kang S. Phosphoinositide 3-kinase: from viral oncoprotein to drug target. Virology 2006, 344:131–138.

169. Bader AG, Kang S, Zhao L, et al. Oncogenic PI3K deregulates transcription and translation. Nat Rev Cancer 2005, 5:921–929.

170. Kang S, Bader AG, Zhao L, et al. Mutated PI 3-kinases: cancer targets on a silver platter. Cell Cycle 2005, 4:578–581.

171. Müller CI, Miller CW, Hofman W-K, et al. Rare mutations of the PIK3CA gene in malignancies of the hematopoietic system as well as endometrium, ovary, prostate and osteosarcomas, and discovery of a PIK3CA pseudogene. Leuk Res 2007, 31:27–32.

172. Leslie NR, Gray A, Pass I, et al. Analysis of the cellular functions of PTEN using catalytic domain and C-terminal mutations: differential effects of C-terminal deletion on signaling pathways downstream of phosphoinositide 3-kinase. Biochem J 2000, 346:827–833.

173. Dahia PL, Aquiar RC, Alberta J, et al. PTEN is inversely correlated with the cell survival factor Akt/PKB and is inactivated via multiple mechanism in haematological malignancies. Hum Mol Gen 1999, 8:185–193.

174. Sakai A, Thieblemont C, Wellmann A, et al. PTEN gene alterations in lymphoid neoplasms. Blood 1998, 92:3410–3415.

175. Aggerholm A, Gronbaek K, Guldberg P, et al. Mutational analysis of the tumour suppressor gene MMAC1/PTEN in malignant myeloid disorders. Eur J Haematol 2000, 65:109–113.

176. Nakahara Y, Nagai H, Kinoshita T, et al. Mutational analysis of the PTEN/MMAC1 gene in non-Hodgkin's lymphoma. Leukemia 1998, 12:1277–1280.

177. Butler MP, Wang SI, Chaganti RS, et al. Analysis of PTEN mutations and deletions in B-cell non-Hodgkin's lymphomas. Genes Chrom Cancer 1999, 24:322–327.

178. Herranz M, Urioste M, Santos J, et al. Allelic losses and genetic instabilities of PTEN and p73 in non-Hodgkin lymphomas. Leukemia 2000, 14:1325–1327.

179. Graff JR, Konicek BW, McNulty AM, et al. Increased Akt activity contributes to prostate cancer progression by dramatically accelerating prostate tumor growth and diminishing p27Kip-1 expression. J Biol Chem 2000, 275:24500–24505.

180. Staal SP. Molecular cloning of the Akt oncogene and its human homologues AKT1 and AKT2: amplification of AKT1 in a primary human gastric adenocarcinoma. Proc Natl Acad Sci U S A 1987, 84:5034–5037.

181. Cheng JQ, Godwin AK, Bellacosa A, et al. AKT2, a putative oncogene encoding a member of a subfamily of protein-serine/threonine kinases, is amplified in human ovarian carcinomas. Proc Natl Acad Sci U S A 1992, 89:9267–9271.

182. Cheng JQ, Ruggeri B, Klein WM, et al. Amplification of AKT2 in human pancreatic cells and inhibiton of AKT2 expression and tumorigenicity by anti-sense RNA. Proc Natl Acad Sci U S A 1996, 93:3636–3641.

183. Luo JM, Yoshida H, Komura S, et al. Possible dominant-negative mutation of the SHIP gene in acute myeloid leukemia. Leukemia 2003, 17:1–8.

184. Luo JM, Liu ZL, Hao HL, et al. Mutation analysis of SHIP gene in acute leukemia. Zhongguo Shi Yan Xue Ye Xue Za Zhi 2004, 12:420–426.

185. Hietakangas V, Cohen SM. Re-evaluating Akt regulation: role of TOR complex 2 in tissue growth. Genes Dev 2007, 21:632–637.

186. Lin J, Adam RM, Santiestevan E, et al. The phosphatidylinositol 3 -kinase pathway is a dominant growth factor-activated cell survival pathway in LNCaP human prostate carcinoma cells. Cancer Res 1999, 59:2891–2897.

187. Fry MJ. Phosphoinositide 3-kinase signalling in breast cancer: how big a role might it play? Breast Cancer Res 2001, 3:304–312.

188. Lin X, Bohle AS, Dohrmann P, et al. Overexpression of phosphatidylinositol 3-kinase in human lung cancer. Langenbecks Arch Surg 2001, 386:293–301.

189. Krasilnikov M, Adler V, Fuchs SY, et al. Contribution of phosphatidylinositol 3-kinase to radiation resistance in human melanoma cells. Mol Carcinog 1999, 24:64–69.

190. Martinez-Lorenzo MJ, Anel A, Monleon I, et al. Tyrosine phosphorylation of the p85 subunit of phosphatidylinositol 3-kinase correlates with high proliferation rates in sublines derived from the Jurkat leukemia. Int J Biochem Cell Biol 2000, 32:435–445.

191. Shou J, Massarweh S, Osborne CK, et al. Mechanisms of tamoxifen resistance: increased estrogen receptor-HER2/neu cross-talk in ER/HER2-positive breast cancer. J Natl Cancer Inst 2004, 96:926–935.

192. Frogne T, Jepsen JS, Larsen SS, et al. Antiestrogen-resistant human breast cancer cells require activated protein kinase B/Akt for growth. Endocr Rel Cancer 2005, 12:599–614.

193. Kirkegaard T, Witton CJ, McGlynn LM, et al. AKT activation predicts outcome in breast cancer patients treated with tamoxifen. J Pathol 2005, 207:139–146.

194. Nyakern M, Tazzari PL, Finelli C, et al. Frequent elevation of Akt kinase phosphorylation in blood marrow and peripheral blood mononuclear cells from high-risk myelodysplastic syndrome patients. Leukemia 2006, 20:230–238.

195. Mantovani I, Cappellini A, Tazzari PL, et al. Caspase-dependent cleavage of 170-kDa P-glycoprotein during apoptosis of human T-lymphoblastoid CEM cells. J Cell Physiol 2006, 207:836–844.

196. Nyakern M, Cappellini A, Mantovani J, et al. Synergistic induction of apoptosis in yhuman leukemia T cells by the Akt inhibitor perifosine and etoposide through activation of intrinsic and Fas-mediated extrinsic cell death pathways. Mol Cancer Ther 2006, 5:1559–1570.

197. Martelli AM, Nyakern M, Tabellini G, et al. Phosphoinositide 3-kinase/Akt signaling pathway and its therapeutical implications for human acute myelid leukemia. Leukemia 2006, 20:911–928.

198. Tallman MS, Gillliland DG, Rowe JM. Drug therapy for acute myeloid leukemia. Blood 2005, 106:1154–1163.

199. Vey N, Mozzinconacci MJ, Groulet-Martinec A, et al. Identification of new classes amoung acute myelogenous leukaemias with normal karyotype using gene expression profiling. Oncogene 2004, 23:9381–9391.

200. Kiyoi H, Towatari M, Yokota S, et al. Internal tandem duplication of the FLT3 gene is a novel modality of elongation mutation which causes constitutive activation of the product. Leukemia 1998, 12:1333–1337.

201. Shimada A, Taki T, Tabuchi K, et al. KIT mutations, and not FLT3 internal tandem duplication, are strongly associated with a poor prognosis in pediatric acute myeloid leukemia with t(8, 21): a study of the Japanese Childhood AML Cooperative Study Group. Blood 2006, 107:1806–1809.

202. Christiansen DH, Andersen MK, Desta F, et al. Mutations of genes in the receptor tyrosine kinase (RTK)RAS-BRAF signal transduction pathway in therapy-related myelodysplasia and acute myeloid leukemia. Leukemia 2005, 19:2232–2240.

203. Padua RA, Guinn BA, Al-Sabah AI, et al. RAS, FMS and p53 mutations and poor clinical outcome in myelodysplasias: a 10-year follow-up. Leukemia 1998, 12:887–892.

204. Dong F, Brynes RK, Tidow N, et al. Mutations in the gene for granulocyte colony-stimulating-factor receptor in patients with acute myeloid leukemia preceeded by severe congenital neutropenia. N Engl J Med 1995, 333:487–493.

205. Dong F, Dale DC, Bonilla MA, et al. Mutations in the granulocyte colony-stimulating factor receptor gene in patients with severe congenital neutropenia. Leukemia 1997, 11:120–125.

206. Hiramatsu A, Miwa H, Shikami M, et al. Disease-specific expression of VEGF and its receptors in AML cells, possible autocrine pathway of VEGF/type-1 receptor of VEGF in t(15, 17) and AML in VEGF/type2 receptor of VEGF in t(8, 21) AML. Leuk Lymph 2006, 47:89–95.

207. Gregorj C, Ricciardi MR, Petrucci MT, et al. ERK1/2 phosphorylation is an independent predictor of complete remission in newly diagnosed adult acute lymphoblastic leukemia. Blood 2007, 109:5473–5476.

208. Milella M, Konopleva M, Precupanu CM, et al. MEK blockade converts AML differentiating response to retinoids into extensive apoptosis. Blood 2007, 109:2121–2129.

209. Milella M, Kornblau SM, Estrov Z, et al. Therapeutic targeting of the MEK/MAPK signal transduction module in acute myeloid leukemia. J Clin Invest 2001, 108:851–859.

210. Staber PB, Linkesch W, Zauner D, et al. Common alterations in gene expression and increased proliferation in recurrent acute myeloid leukemia. Oncogene 2004, 29:894–904.

211. Stone RM, O'Donnell MR, Sekeres MA. Acute myeloid leukemia. Hematology Am Soc Hematol Educ Program 2004, 1:98–117.

212. Birkenkamp KU, Geugien M, Schepers H, et al. Constitutive NF-kappaB DNA-binding activity in AML is frequently mediated by a Ras/PI3-K/PKB-dependent pathway. Leukemia 2004, 18:103–112.

213. Stirewalt DL, Radich JP. The role of FLT3 in haematopoietic malignancies. Nat Rev Cancer 2003, 3:650–665.

214. Meshinchi S, Stirewalt DL, Alonzo TA, et al. Activating mutations of RTK/ras signal transduction pathway in pediatric acute myeloid leukemia. Blood 2003, 102:1474–1479.

215. Yokota S, Nakao M, Horiike S, et al. Mutational analysis of the N-ras gene in acute lymphoblastic leukemia: a study of 125 Japanese pediatric cases. Int J Hematol 1998, 67:379–387.

216. Hoelzer D, Gokbuget N. Recent approaches in acute lymphoblastic leukemia in adults. Crit Rev Oncol Hematol 2000, 36:49–58.

217. Pui CH, Evans WE. Genetic abnormalities and drug resistance in acute lymphoblastic leukemia. Adv Exp Med Biol 1999, 457:383–389.

218. Tazzari PL, Cappellini A, Ricci F, et al. Multidrug resistance-associated protein 1 expression is under the control of the phosphoinositide 3 kinase/Akt signal transduction network in human acute myelogenous leukemia blasts. Leukemia 2007, 21:427–438.

219. McKearn JP, McCubrey JA, Fagg B. Enrichment of hematopoietic precursor cells and cloning of multipotential B lymphocyte precursors. Proc. Natl. Acad. Sci U S A 1985, 85:7414–7418.

220. Teodori E, Dei S, Martelli C, et al. The functions and structure of ABC transporters: implications for the design of new inhibitors of Pgp and MRP1 to control multidrug resistance (MDR). Curr Drug Targets 2006, 7:893–909.

221. Polgar O, Bates SE. ABC transporters in the balance: is there a role in multidrug resistance? Biochem Soc Trans 2005, 33:241–245.

222. Ross DD. Modulation of drug resistance transporters as a strategy for treating myelodysplastic syndrome. Review. Bailliere's Best Practice in Clin Haematol 2004, 17:641–651.

223. Talpaz M, Shah NP, Kantarjian H, et al. Dasatinib in imatinib-resistant Philadelphia chromosome-positive leukemias. N Engl J Med 2006, 354:2531–2541.

224. Markovic A, MacKenzie KL, Lock RB. FLT-3: a new focus in the understanding of acute leukemia. Int J Biochem Cell Biol 2005, 37:1168–1172.

225. Stone RM, DeAngelo DJ, Klimek V, et al. Patients with acute myeloid leukemia and an activating mutation in FLT3 respond to a small-molecule FLT3 tyrosine kinase inhibitor, PKC412. Blood 2005, 105:54–60.

226. Kojima K, Konopleva M, Samudio IJ, et al. MDM2 antagonists induce p53-dependent apoptosis in AML: implications for leukemia therapy. Blood 2005, 106:3150–3159.

227. Centers for Disease Control and Prevention. Cancer: Symptoms of breast cancer. Atlanta: Centers for Disease Control and Prevention, 2006, http://www.cdc.gov/cancer/breast/basic_info/symptoms.htm.

228. Steelman LS, Bertrand FE, McCubrey JA. The complexity of PTEN: mutation, marker and potential target for therapeutic intervention. Expert Opinion Ther Targets 2004, 8:537–550.

229. Navolanic PM, Steelman LS, McCubrey JA. EGFR family signaling and its association with breast cancer development and resistance to chemotherapy. Int J Oncol 2003, 22:237–252.

230. Hollestelle A, Elstrodt F, Nagel JHA, et al. Phosphatidylinositol-3-OH kinase or Ras pathway mutations in human breast cancer cell lines. Mol Cancer Res 2007, 5:195–201.

231. Feilotter HE, Coulon V, McVeigh JL, et al. Analysis of the 10q23 chromosomal region and the PTEN

gene in human sporadic breast carcinoma. Br J Cancer 1999, 79:718–723.

232. Rhei E, Kang L, Bogomolniy F, et al. Mutation analysis of the putative tumor suppressor gene PTEN/MMAC1 in primary breast carcinomas. Cancer Res 1997, 57:3657–3659.

233. Singh B, Ittmann MM, Krolewski JJ. Sporadic breast cancers exhibit loss of heterozygosity on chromosome segment 10q23 close to the Cowden disease locus. Genes Chromosome Cancer 1998, 21:166–171.

234. DeGraffenried LA, Fulcher L, Friedrichs WE, et al. Reduced PTEN expression in breast cancer cells confers susceptibility to inhibitors of the PI3 kinase/Akt pathway. Ann Oncol 2004, 15:1510–1516.

235. Garcia JM, Silva J, Pena C, et al. Promoter methylation of the PTEN gene is a common molecular change in breast cancer. Genes Chromosomes Cancer 2004, 41:117–124.

236. Tsutsui S, Inoue H, Yasuda K, et al. Reduced expression of PTEN protein and its prognostic implications in invasive ductal carcinoma of the breast. Oncology 2005, 68:398–404.

237. Trotman LC, Wang X, Alimonti A, et al. Ubiquitination regulates PTEN nuclear import and tumor suppression. Cell 2007, 128:141–156.

238. Shen WH, Balajee AS, Wang J, et al. Essential role for nuclear PTEN in maintaining chromosomal integrity. Cell 2007, 128:157–170.

239. Tokunaga E, Kimura Y, Mashino K, et al. Activation of PI3K/Akt signaling and hormone resistance in breast cancer. Breast Cancer 2006, 13:137–144.

240. Greenman C, Stephens P, Smith R, et al. Patterns of somatic mutation in human cancer genomes. Nature 2007, 446:153–158.

241. Shelton JG, Steelman LS, Abrams SL, et al. The epidermal growth factor receptor as a target for therapeutic intervention in numerous cancers—what's genetics got to do with it? Expert Opinion Ther Targets 2005, 9:1009–1030.

242. Chang F, Steelman LS, Shelton JG, et al. Regulation of cell cycle progression and apoptosis by the Ras/Raf/MEK/ERK pathway. Int J Oncol 2003, 22:469–480.

243. Harada H, Andersen JS, Mann M, et al. p70S6 kinase signals cell survival as well as growth, inactivating the pro-apoptotic molecule Bad. Proc Natl Acad Sci U S A 2001, 98:9666–9670.

244. Martelli AM, Tazzari PL, Evangelisti C, et al. Targeting the phosphatidylinositol 3-kinase/Akt/mammalian target of rapamycin module for acute myelogenous leukemia therapy: from bench to bedside. Current Medicinal Chem 2007, 14:2009–2023.

245. Deng X, Xiao L, Lang W, et al. Novel role for JNK as a stress-activated Bcl2 kinase. J Biol Chem 2001, 276:23681–23688.

246. Devarajan E, Sahin AA, Chen JS, et al. Down-regulation of caspase 3 in breast cancer: a possible mechanism for chemoresistance. Oncogene 2002, 21:8843–8851.

247. Liang Y, Yan C, Schor NF. Apoptosis in the absence of caspase 3. Oncogene 2001, 20:6570–6578.

248. Twiddy D, Gohen GM, MacFarlane M, et al. Casape-7 is directly activated by the ~700-kDa apoptosome complex and is released as a stable XIAP-caspase-7 ~200-kDa complex. J Biol Chem 2006, 281:3876–3888.

249. Dan HC, Sun M, Naneko S, et al. Akt phosphorylation and stabilization of X-linked inhibitor of apoptosis protein (XIAP). J Biol Chem 2004, 279:5405–5412.

250. Gardai SJ, Whitlock BB, Xiao YQ, et al. Oxidants inhibit ERK/MAPK and prevent its ability to delay neutrophil apoptosis downstream of mitochondrial changes and at the level of XIAP, J Biol Chem 2004, 279:44695–44703.

251. Gioeli D, Mandell JW, Petroni GR, et al. Activation of mitogen-activated protein kinase associated with prostate cancer progression. Cancer Res 1999, 59:279–284.

252. Abreu-Martin MY, Chari A, Palladino AA, et al. Mitogen-activated protein kinase kinase kinase 1 activates androgen receptor-dependent transcription and apoptosis in prostate cancer. Mol Cell Biol 1999, 19:5143–5154.

253. Weber MJ, Gioeli D. Ras signaling in prostate cancer progression. J Cell Biochem 2004, 91:13–25.

254. Bakin RE, Gioeli D, Sikes RA, et al. Constitutive activation of the Ras/Mitogen-activated protein kinase signaling pathway promotes androgen hypersensitivity in LNCaP prostate cancer cells. Cancer Res 20003, 63:1981–1989.

255. Mukherjee R, Bartlett JMS, Krishna NS, et al. Raf-1 expression may influence progression to androgen insensitive prostate cancer. Prostate 2005, 64:101–107.

256. Lee JT, Steelman LS, McCubrey JA. Modulation of Raf/MEK/ERK pathway in prostate cancer drug resistance. Int J Oncol 2005, 26:1637–1645.

257. Lehmann BD, McCubrey JA, Jefferson HS, et al. A dominant role for p53-dependent cellular senescence in radiosensitization of human prostate cancer cells. Cell Cycle 2007, 6:595–605.

258. Fu Z, Smith PC, Zhang L, et al. Effects of raf kinase inhibitor protein on suppression of prostate cancer metastasis. JNCI 2003, 95:878–889.

259. Keller ET, Fu Z, Yeung K, et al. Raf kinase inhibitor protein: a prostate cancer metastasis suppressor gene. Cancer Lett 2004, 207:131–137.

260. Eves EM, Shapiro P, Naik K, et al. Raf kinase inhibitory protein regulates aurora B kinase and the spindle checkpoint. Mol Cell 2006, 23:561–574.

261. Cully M, You H, Levine AJ, et al. Beyond PTEN mutations: the PI3K pathway as an integrator of multiple inputs during tumorigenesis. Nat Rev Cancer 2006, 6:184–192.

262. Wu G. The functuional interactions between p53 and MAPK signaling pathways. Cancer Biol Ther 2004, 3:156–161.

263. Fang L, Li G, Liu G, et al. P53 induction of heparin-binding EGF-like growth factor counteracts p53 growth suppression through activation of MAPK and PI3K/Akt signaling cascades. EMBO J 2001, 20:1931–1939.

264. Blagosklonny MV, Prabhu NS, El-Deiry WS. Defects in p21WAF1/CIP1, Rb and c-myc signaling in phorbol ester-resistant cancer cells. Cancer Res 1997, 57:320–325.

265. Blagosklonny MV. The mitogen-activated protein kinase pathway mediates growth arrest or E1A-dependent apoptosis in SKBR3 human breast cancer cells. Int J Cancer 1998, 78:511–517.

266. Ravi RK, McMahon M, Yangang Z, et al. Raf-1-induced cell cycle arrest in LNCap human prostate cancer cells. J Cell Biochem 1999, 72:458–469.

267. Lee JT Jr, McCubrey JA. The Raf/MEK/ERK signal transduction cascade as a target for chemotherapeutic intervention. Leukemia 2002, 16:486–507.

268. Barnes G, Bulusu VR, Hardwick RH, et al. A review of the surgicial management of metastatic gastrointestinal stromal tumours (GISTs) on imatinib mesylate (Glivectrade mark). Int J Surg 2005, 3:206–212.

269. Hochhaus A, McCubrey JA, Killmann NMB. Spotlight imatinib: a model for signal transduction inhibitors. Leukemia 2002, 16:1205–1206.

270. Tortora G, Bianco R, Daniele G, et al. Overcoming resistance to molecularly targeted anticancer therapies: rational drug combinations based on EGFR and MAPK inhibition for solid tumors and hematological malignancies. Curr Drug Des, 10:81–100.

271. Sun SY, Rosenberg LM, Wang X, et al. Activation of Akt and eIF4E survival pathways by rapamycin-mediated mammalian target of rapamycin inhibition. Cancer Research 2005, 65:7052–7058.

272. Konopleva M, Contractor R, Tsao T, et al. Mechanisms of apoptosis sensitivity and resistance to the BH3 mimetic ABT-737 in acute myeloid leukemia. Cancer Cell 2006, 10:375–388.

273. Kojima K, Konopleva M, McQueen T, et al. MDM2 inhibitor Nutlin 3a induces p53-mediated apoptosis by transcription-dependent and transcription-independent mechanisms and may overcome Atm-mediated resistance to fludarabine in chronic lymphocytic leukemia. Blood 2006, 108:993–1000.

274. Shelton JG, Moye PW, Steelman LS, et al. Differential effects of kinase cascade inhibitors on neoplastic and cytokine-mediated cell proliferation. Leukemia 2003, 17:1765–1782.

275. Heimbrook DC, Huber HE, Stirdivant SM, et al. Identification of potent, selective kinase inhibitors of Raf. Proc Amer Assoc Cancer Res Annu Meeting 1998, 39:558.

276. Hall-Jackson CA, Eyers PA, Cohen P, et al. Paradoxical activation of Raf by a novel Raf inhibitor. Chem Biol 1999, 6:559–568.

278. Lyons JF, Wilhelm S, Hibner B, et al. Discovery of a novel Raf kinase inhibitor. Endocrine-Related Cancer 2001, 8:219–225.

279. Blagosklonny MV. Hsp-90-associated oncoproteins: multiple targets of geldanamycin and its analogs. Leukemia 2002, 16:455–462.

280. Lee JT, Steelman LS, McCubrey JA. PI3K Activation leads to MRP1 expression and subsequent chemoresistance in advanced prostate cancer cells. Cancer Res 2004, 64:8397–8404.

281. Workman P. Altered states: selectively drugging the Hsp90 cancer chaperone. Trends Mol Med 2004, 10:47–51.

282. Shimamura T, Lowell AM, Engelman JA, et al. Epidermal growth factor receptors harboring kinase domain mutations associate with the heat shock protein 90 chaperone and are destabilized following exposure to geldanamycins. Cancer Res 2005, 65:6401–6408.

283. Sain N, Krishnan B, Ormerod MG, et al. Potentiation of paclitaxel activity by the HSP90 inhibitor 17-allylamino-17-demethoxygeldanamycin in human ovarian carcinoma cell lines with high levels of activated AKT. Mol Cancer Ther 2006, 5:1197–1208.

284. Neckers L. Hsp90inhibitors as novel cancer chemotherapeutic agents. Trends Mol Med 2002, 8: S55–61.

285. Mazzucchelli C, Vantaggiato C, Ciamei A, et al. Knockout of ERK1 MAP kinase enhances synaptic plasticity in the striatum and facilitates striatal-mediated learning and memory. Neuron 2002, 34:807–820.

286. Rolfe M, McLeod LE, Pratt PF, et al. Activation of protein synthesis in cardiomyocytes by the hypertrophic agent phenylephrine requires the activation of ERK and involves phosphorylation of tuberous

sclerosis complex 2 (TSC2). Biochem J 2005, 388:973–984.

287. Sosman JA, Puzanov I, Atkins MB. Opportunities and obstacles to combination targeted therapy in renal cell cancer. Clin Cancer Res 2007, 13:764s–769s.

288. Costa LJ, Gemmill RM, Drabkin HA. Upstream signaling inhibition enhances rapamycin effect on growth of kidney cancer cells. Urology 2007, 69:596–602.

289. Janmaat ML, Rodriguez JA, Gallegos-Ruiz M, et al. Enhanced cytotoxicity induced by gefitinib and specific inhibitors of the Ras or phosphatidyl inositol-3 kinase pathways in non-small cell lung cancer cells. Int J Cancer 2006, 118:209–214.

290. Wang LH, Chan JLK, Li W. Rapamycin together with herceptin significantly increased anti-tumor efficacy compared to either alone in ErbB2 over expressing breast cancer cells. Int J Can 2007, 121, 2911–2918.

291. Konecny GE, Pegram MD, Venkatesan N, et al. Activity of the dual kinase inhibitor lapatinib (GW572016) against HER-2-overexpressing and trastuzumab-treated breast cancer cells. Cancer Res 2006, 66:1630–1639.

292. Yu C, Mao X, Li WX. Inhibition of the PI3K pathway sensitizes fludarabine-induced apoptosis in human leukemic cells through an inactivation of MAPK-dependent pathway. Biochem Biophys Res Commun 2005, 331:391–397.

293. Mondesire WH, Jian W, Zhang H, et al. Targeting mammalian target of rapamycin synergistically enhances chemotherapy-induced cytotoxicity in breast cancer cells. Clin Cancer Res 2004, 10:7031–7042.

294. Faried LS, Kanuma T, Nakazato T, et al. Inhibition of the mammalian target of Rapamycin (mTOR) by Rapamycin increases chemosensitivity of CaSki cells to paclitaxel. Eur J Can 2006, 42:934–947.

295. Dong J, Peng J, Zhang H, et al. Role of glycogen synthase kinase 3beta in rapamycin-mediated cell cycle regulation and chemosensitivity. Cancer Res 2005, 65:1961–1972.

296. Hahn M, Li W, Yu C, et al. Rapamycin and UCN-01 synergistically induce apoptosis in human leukemia cells through a process that is regulated by the Raf-1/MEK/ERK, Akt, and JNK signal transduction pathways. Mol Cancer Ther 2005, 4:457–470.

297. Dai Y, Rahmani M, Pei XY, et al. Farnesyltransferase inhibitors interact synergistically with the Chk1 inhibitor UCN-01 to induce apoptosis in human leukemia cells through interruption of both Akt and MEK/ERK pathways and activation of SEK1/JNK. Blood 2005, 105:1706–1716.

298. Yu K, Lucas J, Zhu T, et al. PWT-458, a novel pegylated-17-hydroxywortmannin, inhibits phosphatidylinositol 3-kinase signaling and suppresses growth of solid tumors. Cancer Biol Ther 2005, 4:538–545.

299. Abdul-Ghani R, Serra V, Gyorffy B, et al. The PI3K inhibitor LY294002 blocks drug export from resistant colon carcinoma cells overexpressing MRP-1. Oncogene 2006, 25, 1743–1752.

300. Hideshima T, Catley L, Yasui H, et al. Perifosine, an oral bioactive novel alkylphospholipid, inhibits Akt and induces in vitro and in vivo cytotoxicity in human multiple myeloma cells. Blood 2006, 107:4053–4062.

301. Tazzari PL, Tabellini G, Bortul R, et al. The insulin-like growth factor-I receptor kinase inhibitor NVP-AEW541 induces apoptosis in acute myeloid leukemia cells exhibiting autocrine insulin-like growth factor-I secretion. Leukemia 2007, 21:886–896.

302. Luo Y, Shoemaker AR, Liu X, et al. Potent and selective inhibitors of Akt kinases slow the progress of tumors in vivo. Mol Cancer Ther 2005, 4:977–986.

303. Zhong X, Li X, Wang G, et al. Mechanisms underlying the synergistic effect of SU5416 and cisplatin on cytotoxicity in human ovarian tumor cells. Int J Oncol 2004, 25:445–451.

304. Sosman JA, Puzanov I. Molecular targets in melanoma from angiogenesis to apoptosis. Clin Cancer Res 2006, 12:2376s–283s.

305. Zhu K, Gerbino E, Beaupre DM, et al. Farnesyltransferase inhibitor R115777 (Zarnestra, Tipifarnib) synergizes with paclitaxel to induce apoptosis and mitotic arrest and to inhibit tumor growth of multiple myeloma cells. Blood 2005, 105:4759–4766.

306. MacKeigan JP, Clements CM, Lich JD, et al. Proteomic profiling drug-induced apoptosis in non-small cell lung carcinoma: identification of RS/DJ-1 and RhoGDIalpha. Cancer Res 2003, 63:6928–6934.

307. McDaid HM, Lopez-Barcons L, Grossman A, et al. Enhancement of the therapeutic efficacy of taxol by the mitogen-activated protein kinase kinase inhibitor CI-1040 in nude mice bearing human heterotransplants. Cancer Res 2005, 65:2854–2860.

308. Mabuchi S, Ohmichi M, Kimura A, et al. Inhibition of phosphorylation of BAD and Raf-1 by Akt sensitizes human ovarian cancer cells to paclitaxel. J Biol Chem 2002, 277:33490–33500.

309. Brognard J, Dennis PA. Variable apoptotic response of NSCLC cells to inhibition of the MEK/ERK pathway by small molecules or dominant negative mutants. Cell Death Diff 2002, 9:893–904.

310. Aoki K, Ogawa T, Ito Y, et al. Cisplatin activates survival signals in UM-SCC-23 squamous cell carcinoma and these signal pathways are amplified in cisplatin-resistant squamous cell carcinoma. Oncol Rep 2004, 11:375–379.

311. Park SA, Park HJ, Lee BI, et al. Bcl-2 blocks cisplatin-induced apoptosis by suppression of ERK-mediated p53 accumulation in B104 cells. Brain Res Mol Brain Res 2001, 93:18–26.

312. Lunghi P, Tabilio A, Lo-Coco F, et al. Arsenic trioxide (ATO) and MEK1 inhibition synergize to induce apoptosis in acute promyelocytic leukemia cells. Leukemia 2005, 19:234–244.

313. Lunghi P, Costanzo A, Salvatore L, et al. MEK1 inhibition sensitizes primary acute myelogenous leukemia to arsenic trioxide-induced apoptosis. Blood 2006, 107:4549–4553.

314. Dai Y, Landowski TH, Rosen ST, et al. Combined treatment with the checkpoint abrogator UCN-01 and MEK1/2 inhibitors potently induces apoptosis in drug-sensitive and -resistant myeloma cells through an IL-6-independent mechanism. Blood 2002, 100:3333–3343.

315. Rieber M, Rieber MS. Signalling responses linked to betulinic acid-induced apoptosis are antagonized by MEK inhibitor U0126 in adherent or 3D spheroid melanoma irrespective of p53 status. Int J Cancer 2006, 118:1135–1143.

316. McCubrey J, LaHair M, Franklin RA. Reactive oxygen species-induced activation of the MAP kinase signaling pathway. Antiox Redox Signal 2006, 8:1745–1748.

317. Fornari FA, Randolph JK, Yalowich JC, et al. Interference by doxorubicin with DNA unwinding in MCF-7 breast tumor cells. Mol Pharmacol 1994, 45:649–656.

318. Gewirtz DA. A critical evaluation of the mechanisms of action proposed for the antitumor effects of the anthracycline antibiotics Adriamycin and daunorubicin. Biochem Pharmacol 1999, 57:727–741.

319. Singal PK, Li T, Kumar D, et al. Adriamycin-induced heart failure: mechanism and modulation. Mol Cell Biochem 2000, 207:77–86.

320. Friesen C, Fulda S, Debatin KM. Induction of CD95 ligand and apoptosis by doxorubicin is modulated by the redox state in chemosensitive- and drug-resistant tumor cells. Cell Death Diff 1999, 6:471–480.

321. Narayanan PK, Goodwin EH, Lehnert BE. Alpha particles initiate biological production of superoxide anions and hydrogen peroxide in human cells. Cancer Res 1997, 57:3963–3971.

322. Kobayashi D, Tokino T, Watanabe N. Contribution of caspase-3 differs by p53 status in apoptosis induced by X-irradiation. Jpn J Cancer Res 2001, 92:475–481.

323. Oleinick NL, Evans HH. The photobiology of photodynamic therapy: cellular targets and mechanisms. Radiat Res 1998, 150:S146–156.

324. Matroule JY, Carthy CM, Granville DJ, et al. Mechanism of colon cancer cell apoptosis mediated by pyropheophorbide-a methylester photosensitization. Oncogene 2001, 20:4070–4084.

325. Tong Z, Singh G, Rainbow AJ. Sustained activation of the extracellular signal-regulated kinase pathway protects cells from photofrin-mediated photodynamic therapy. Cancer Res 2002, 62:5528–5535.

326. Dent P, Yacoub A, Contessa J, et al. Stress and radiation-induced activation of multiple intracellular signaling pathways. Radiat Res 2003b, 159:283–300.

327. Dent P, Yacoub A, Fisher PB, et al. MAPK pathways in radiation responses. Oncogene 2003a, 22:5885–5896.

328. Abdelmohsen K, von Montfort C, Stuhlmann D, et al. Doxorubicin induces EGF receptor-dependent downregulation of gap junctional intercellular communication in rat liver epithelial cells. Biol Chem 2005, 386:217–223.

329. Papaiahgari S, Kleeberger SR, Cho HY, et al. NADPH oxidase and ERK signaling regulates hyperoxia-induced Nrf2-ARE transcriptional response in pulmonary epithelial cells. J Biol Chem 2004, 279:42302–42312.

330. Matsuzawa A, Nishitoh H, Tobiume K, et al. Physiological roles of ASK1-mediated signal transduction in oxidative stress- and endoplasmic reticulum stress-induced apoptosis: advanced findings from ASK1 knockout mice. Antioxid Redox Signal 2002, 4:415–425.

331. Minamino T, Yujiri T, Papst PJ, et al. MEKK1 suppresses oxidative stress-induced apoptosis of embryonic stem cell-derived cardiac myocytes. Proc Natl Acad Sci U S A 1999, 96:15127–15132.

332. Tobiume K, Matsuzawa A, Takahashi T, et al. ASK1 is required for sustained activations of JNK/p38 MAP kinases and apoptosis. EMBO Rep 2001, 2:222–228.

333. Modur V, Zimmerman G, Prescott S, et al. Endothelial cell inflammatory responses to tumor necrosis factor alpha. Ceramide-dependent and -independent mitogen-activated protein kinase cascades. J Biol Chem 1996, 271:13094–13102.

334. Teramoto H, Coso OA, Miyata H, et al. Signaling from the small GTP-binding proteins Rac1 and Cdc42 to the c-Jun N-terminal kinase/stress-activated protein kinase pathway. A role for mixed lineage kinase 3/protein-tyrosine kinase 1, a novel member of the mixed lineage kinase family. J Biol Chem 1996, 271:27225–27228.

335. Ichijo H, Nishida E, Irie K, et al. Induction of apoptosis by ASK1, a mammalian MAPKKK that activates SAPK/JNK and p38 signaling pathways. Science 1997, 275:90–94.

336. Suzaki Y, Yoshizumi M, Kagami S, et al. Hydrogen peroxide stimulates c-Src-mediated big mitogen-activated protein kinase 1 (BMK1) and the MEF2C signaling pathway in PC12 cells: potential role in cell survival following oxidative insults. J Biol Chem 2002, 277:9614–9621.

337. Touyz RM, Yao G, Viel E, et al. Angiotensin II and endothelin-1 regulate MAP kinases through different redox-dependent mechanisms in human vascular smooth muscle cells. J Hypertens 2004, 22:1141–1149.

338. Kamakura S, Moriguchi T, Nishida E. Activation of the protein kinase ERK5/BMK1 by receptor tyrosine kinases. Identification and characterization of a signaling pathway to the nucleus. J Biol Chem 1999, 274:26563–26571.

339. Cavanaugh JE, Ham J, Hetman M, et al. Differential regulation of mitogen-activated protein kinases ERK1/2 and ERK5 by neurotrophins, neuronal activity, and cAMP in neurons. J Neurosci 2001, 21:434–443.

Chapter 8
Histone Deacetylase Inhibitors and Anticancer Activity

Roberto R. Rosato and Steven Grant

1 Introduction

Structural alterations of DNA and their corresponding genetic defects are recognized as some of the fundamental characteristics of cancer, whereas gene mutations, deletions, and chromosomal alterations appear to play a critical role in the etiology of the disease. These epigenetic modifications have become the target of novel experimental therapeutic approaches representing a component of a more generalized concept in which the ultimate goal is to address the specific molecular alteration responsible for transformation. The notion of molecularly targeted therapies have radically changed the way cancer therapeutics is currently envisioned [1]. These compounds, in contradistinction to conventional cytotoxic drugs, which exert a more generally disruptive effect on key cellular structures or processes (e.g., DNA replication, formation of the mitotic spindle, etc.), are specifically designed to target pathways directly implicated in neoplastic transformation. There has been substantial progress in the development of drugs targeting epigenetic regulatory processes, and one of the areas that has become the subject of considerable attention is the development of HDAC inhibitors (HDACIs). The present chapter focuses on the analysis of this group of compounds as anticancer therapeutic agents.

2 Histone Deacetylases: HDAC Inhibitors

In eukaryotic cells, histone proteins assemble the vast quantity of genomic DNA into a manageable size and structure that can be easily accommodated by the nucleus. This structure is comprised of nucleosomes, which constitute the fundamental subunit of chromatin [2]. The nucleosome is organized around an octamer consisting of four corehistones (H2A, H2B, H3, and H4) [3]. In order to render DNA accessible to the actions of transcription factors, the chromatin must be remodeled [4]. The architecture of chromatin can be reversibly modified by the action of ATP-dependent chromatin-remodeling complexes that modify nucleosome positions and disrupt DNA–histone interactions [5]. In addition, histones can be altered post-translationally by multiple processes, including acetylation, methylation, phosphorylation, ribosylation, and sumoylation, each of which influences the likelihood of gene transcription [6]. Moreover, and adding further to the complexity, alterations in histone phosphorylation can modify acetylation-related events [6]. Of these processes, acetylation has received the most attention to date [7, 8]. Histone acetylation involves a group of evolutionary conserved lysine amino acids located at the N-termini of the histone tail and is reciprocally regulated by

From: *Sensitization of Cancer Cells for Chemo/Immuno/Radio-therapy, 1st Edition.*
Edited by: Benjamin Bonavida © Humana Press, Totowa, NJ

histone acetylases (HATs) and histone deacetylases (HDACs) [8]. In general, acetylation of positively charged lysine residues results in charge neutralization, diminished binding of histones to the negatively charged DNA backbone, and relaxation of the chromatin structure, which allows a variety of factors (e.g., co-activators) to gain access to the transcriptional machinery, thereby promoting gene expression [9]. Conversely, deacetylation of histones favors chromatin compaction, which is generally associated with gene repression, although depending upon the cellular context, acetylation of histones can also be associated with downregulation of various genes [10].

Histone acetylation is controlled by two classes of enzymes, histone acetyltransferases (HATs) and histone deacetylases (HDACs) [9, 11]. HATs are subdivided into three groups based on the presence of conserved motifs: the GNAT family, the MYST family, and the p300/CBP family [12]. On the other hand, HDACs are grouped, based on their structural homology with distinct yeast HDACs, into two classes: a- the classical, Zn^{2+}-dependent HDAC family consisting of classes I (HDAC1, -2, -3, and -8), that is related to yeast Rpd3 deacetylase; -IIa/b (HDAC4, -5, -6, -7, -9, and -10), related to yeast Hda1; and -IV (HDAC11); and b-, the recently discovered, Zn^{2+}-independent, NAD^+-dependent class III SIR2 family of HDACs (Sirt1, -2, -3, -4, -5, -6, and -7) [8, 11, 13, 14]. Class I HDACs are expressed in a variety of tissues and localize primarily in the nucleus, whereas class II HDACs display a more restricted tissue distribution and can shuttle between the nucleus and cytoplasm, suggesting type-specific functions and cellular substrates [8, 14–16].

Various classes of HDACIs have been developed, and based upon their structural and chemical characteristics, can be subdivided into five distinct groups: (1) hydroxamic acids (e.g., SAHA, pyroxamide, TSA, oxamflatin, CHAPs; NVP-LAQ824; BL1521; NVP-LBH589); (2) short-chain fatty acids (e.g., sodium butyrate (SB), AN-9; phenylbutyrate (PB); phenylacetate (PA); valproic acid); (3) synthetic benzamide derivatives (e.g., MS-275 and CI-994); (4) cyclic tetrapeptides (e.g., depsipeptide, trapoxin, apicidin, CHAPs); and (5) others (depudecin, MGCD-0103). The concentrations at which these HDACIs exhibit biological activity differ markedly, and range from mM concentrations in the case of sodium butyrate [17, 18] to nM concentrations in the case of NVP-LAQ824 or NVP-LBH589 [19–22].

Hydroxamic acid derivatives consist of a wide variety of compounds, some of which, e.g., the antifungal agent trichostatin A (TSA), were among the first agents shown to exhibit HDAC inhibitory activity [23]. The chemical structure of these agents revealed that the hydroxamic acid group inhibits HDACs by both coordinating a Zn^{+2} cation situated at the base of the HDAC tubular catalytic site and by contacting active-site residues [24]. Several newer generation hydroxamic acid derivatives have been synthesized and are currently at various stages of clinical evaluation. Examples of these compounds include pyroxamine, SAHA (Vorinostat), LAQ824, BL1521 and LBH589 [16, 22, 25–28]. The newly developed hydroxamic HDACIs are prime examples of rationally developed molecules whose design is based on structural analysis of the catalytic HDAC sites [27]. Recently, SAHA (vorinostat, Zolinza®; Merck) was approved by the FDA in the United States for treatment of cutaneous T-cell lymphoma, representing the first HDAC inhibitor to reach such approval [29]. Vorinostat has also shown evidence of activity in phase I trials in patients with refractory AML and MDS [30]. Short-chain fatty acids (SCFA) [e.g., sodium butyrate (SB), AN-9; phenylbutyrate (PB); phenylacetate (PA); valproic acid] have been the subject of interest since the earliest investigations of HDACIs. The prototypical SCFA SB functions as a non-competitive inhibitor of HDACs, but its limited potency and specificity have limited its application as a therapeutic agent [31–33]. Several SB derivatives and precursors have been synthesized and screened, including PA, PB, AN-9, and valproic acid, all of which have undergone clinical evaluation, particularly VA, which has been in clinical use for over two decades as an anticonvulsant [34–37].

The synthetic benzamide derivative group of HDACIs is characterized chemically by the presence of a benzamide ring [38]. Two agents, CI-994 and MS-275, have shown a broad spectrum of activity in preclinical murine and human tumor xenografts model systems in the micromolar range [39–42]. Several clinical trials have been conducted with both compounds administered alone or in some cases in combination with radiation or established cytotoxic agents [39, 43–45]. The cyclic tetrapeptides (e.g., depsipeptide, trapoxin, apicidin) [38, 46–48] represent the most structurally complex class of HDACIs. Recently, several new molecules have emerged from this class including tetrapeptides containing various zinc binding functional groups including trifluoroethyl and pentafluoroethyl ketones [49], retrohydroxamic acids [50], or sulfhydryl moieties [51]. Depudecin is a fungal metabolite derived from *Alternaria brassicicola* that induces hyperacetylation of histones in a dose-dependent manner and inhibits 50% of purified HDAC1 at ~50 µM [52]. MGCD0103, a non-hydroxamate small molecule HDAC inhibitor, selectively targets certain specific class I HDAC isoforms at submicromolar IC50s in vitro and induces hyperacetylation of histones in vivo. It is currently undergoing phase I/II clinical trials and has recently shown preliminary evidence of activity in patients with refractory AML [53, 54].

3 HDAC Inhibitors in Cancer Therapy: Molecular Mechanisms

Some of the most relevant molecular events involved in the anticancer activity of HDACIs are summarized here, with an emphasis on the new insights into the mechanism of action by which HDACIs induce cell death. While HDACIs are modulators the chromatin structure and it might thereby be expected that they primarily act by altering gene expression, accumulating evidence indicates that such an action may represent only one of a large number of additional mechanisms. In addition to reprogramming neoplastic cells, it has become clear that HDACIs can also act by inducing apoptosis in those cells, either indirectly by altering the expression of survival-related genes, or perhaps more directly by triggering biochemical changes in the cell unrelated to gene expression. Several mechanisms have been implicated in HDACI-mediated anticancer activity. These include effects on the expression of genes (epigenetic alterations), generation of reactive oxygen species (ROS) and pro-apoptotic lipid second messenger ceramide, modulation of the extrinsic, receptor-mediated apoptotic pathway, acetylation of protein chaperones (e.g., Hsp90), disruption of cell cycle check-

points, and effects on NF-κB activity. Moreover, in most cases, HDACIs-induced lethality may involve several of these factors. A summary of these most relevant mechanisms is presented in the following.

3.1 HDACIs and Epigenetics

HDACIs can induce tumor cell maturation, particularly in leukemia cells [55], and it is assumed that this reflects re-expression of differentiation-related genes that have been repressed during the course of leukemogenesis. Gene expression in eukaryotes is a highly regulated process [56]. Regulation of gene transcription by epigenetic mechanisms includes DNA methylation, post-translational histone modifications, as well as RNA-associated silencing [57, 58]. Contrary to expectations, initial analyses of the transcriptional response to HDACIs utilizing either differential display, which involved a limited number of genes [59], or DNA microarray approaches, revealed changes in only 2–10% of the expressed genes, respectively. Furthermore, a similar number of genes were up- and downregulated [10, 60–62]. Notably, exposure to HDACIs has been shown to restore silenced gene expression. For example, in colon and breast cell lines exposed to SAHA, the expression of a subset of genes was re-established, rendering them competent to undergo growth arrest, differentiation, and/or apoptosis [61]. Moreover, several of the reported effects that HDACIs exert on the cell cycle could be attributed, at least in part, to the changes induced in key cell cycle-related genes [59, 62]. Of these, the most common HDACI-induced gene is *CDKN1A*, which encodes for the cyclin-dependent kinase inhibitor p21[WAF1/CIP1] [27, 63, 64]. Other genes include *CDKN2A* (which encodes p16[INK4]); the putative tumor suppressor gelsolin, which is involved in control of cell shape and tumor invasiveness [46, 47]; histone H2B; cyclin E and thioredoxin-binding protein 2 (TBP2); α-tubulin; and finally, genes linked to thymidylate synthetase and CTP synthase, both active in DNA synthesis, which were downregulated [62], as were others including cyclins D1 and A, ErbB2, importin β, and VEGF [65].

Tumor suppressor genes such as INK4A, BCRA1, E-cadherin among others [66–68], are frequently silenced by mechanisms that include extensive hypermethylation or histone hypoacetylation [58].

Tumors exhibiting such a phenotype appear particularly appropriate candidates for such epigenetic strategies. Moreover, therapies combining both classes of drugs, i.e., DNA-demethylating agents and HDACIs, can reduce concentrations needed to achieve therapeutic activity, resulting in marked antitumor synergism [69–71]. Notably, clinical trials combining HDACIs with hypomethylating agents have shown promising activity in patients with refractory AML [72]. Attempts have also been made to employ HDACIs in order to circumvent the antiapoptotic and pro-survival effects of oncogenic fusion proteins. For example, in leukemia cells, the PML-RAR and AML-ETO fusion proteins have been shown to act as co-repressors through formation of complexes with other proteins, including HDACs [73, 74]. Notably, a strategy combining pharmacologic concentrations of all-trans-retinoic acid (ATRA), which can overcome transcriptional repression by the PML-RAR protein, with the HDACI butyrate resulted in a marked increase in activity in acute promyelocytic leukemia (APL) cells in vitro [75]. Furthermore, administration of phenylbutryate overcame ATRA resistance in one patient with APL [76]. However, evidence of the generalizability of this strategy is currently lacking. Nevertheless, the concept of combining HDACIs with other classes of agents directed specifically against co-repressor complexes deserves further investigation.

3.2 HDAC Inhibitors and Generation of Reactive Oxygen Species

It is increasingly recognized that reactive oxygen species (ROS) play a key role in diverse pathologic states, including cancer [77]. Diverse neoplastic cells display marked perturbations in oxidative status, and it has been reported that antioxidants can inhibit tumor cell proliferation [78]. Furthermore, ROS generation has been viewed as a tumor suppressive mechanism through induction of cancer cell apoptosis and senescence [79–81]. Significantly, ROS generation has been identified as a critical event in cell death induction by multiple HDACIs [40, 82–86]. For example, Vorinostat was found to trigger cell death in human T cell CEM-CCRF cells via the mitochondria-mediated cell death pathway through a mechanism involving ROS generation and cytochrome

c release [83]. Similarly, increased levels of ROS were observed in Bcr/Abl$^+$ K562 cells undergoing apoptosis in response to Vorinostat [85]. The central role of Vorionostat-induced ROS generation to lethality was supported by the finding that co-administration of free-radical scavengers dramatically reduced Vorinostat cytotoxicity in both CEM-CCRF and K562 cells [83, 85]. A similar critical role for ROS generation in HDACI-induced lethality has been observed in the case of both MS-275 and LAQ824 [40, 86]. Studies with MS-275 revealed that cell death induced by this compound stemmed from the pronounced early generation of ROS, event that was followed by mitochondrial damage, activation of caspases, and degradation of diverse proteins, including p21$^{WAF1/CIP1}$, p27^{KIP}, Bcl-2, and pRb [40]. In contrast, when MS-275 was administered in vitro to primary nonproliferating B-cell chronic lymphocytic leukemia (CLL) cells, it induced selective toxicity and generated a detectable increase in ROS only after 15 hours of drug exposure [87]. Moreover, in these cells, MS-275-induced ROS generation was caspase-dependent in that it could be blocked by co-incubating CLL cells with a pan-caspase inhibitor [87]. Similarly, recent investigations showed also that LAQ824-induced ROS generation, mitochondrial injury, caspase-dependent Mcl-1 downregulation ceramide generation, and apoptosis observed in human leukemia cells were blocked free-radical scavengers, providing clear evidence of the functional role that ROS plays in LAQ824-mediated lethality [86]. ROS generation has also been shown to represent a primary mechanism underlying TSA-induced apoptosis in CD4$^+$ T cells that, when exposed to TSA, displayed increased peroxide and superoxide production; furthermore, pretreatment with free-radical scavengers or mitochondrial respiration inhibitors significantly reduced TSA-induced apoptosis [82].

The metabolic fate of ROS depends upon the function of various cellular antioxidant defense systems, of which the glutathione (GSH) redox cycle and redox sensitive proteins, which include glutathione S-transferase (GST) and thioredoxin, are of primary importance [88–90]. In this context, HDACIs have been shown to modulate redox status by perturbing both the GSH redox cycle [84, 91] and the expression of thioredoxin or thioredoxin-associated proteins [61, 92–94]. For example, recent evidence indicates that apoptosis induced by exposure of MCF-7 breast cancer cells to sodium butyrate is associated with a dose-dependent depletion of intracellular GSH associated with a very marked increase in the activities of enzymes of GSH consumption, including glutathione peroxidase, -reductase, –S-transferase, and catalase [84]. Moreover, sodium butyrate–induced GSH depletion may account for the sensitization observed in VCREMS cells (i.e., vincristine-resistant MCF-7) to doxorubicin, a pro-oxidant agent, through potentiation of oxidative stress and cell lethality [91].

Thioredoxin (TRX), together with glutaredoxin, represent protein disulfide reductases, which function as hydrogen donors for many protein targets, including ribonucleotide reductases and TRX peroxidases, as well as transcription factors such as NF-κB, estrogen, and glucocorticoid receptors [95]. Increased levels of TRX occur in a number of human cancers that may contribute to the resistance observed in some cell types or cancers to therapy by scavenging ROS generated by the drugs. In this regard, recent investigations have found that exposure of LNCaP prostate cells to Vorinostat was associated with both a decrease in the levels of TRX and increased expression of TBP2 (thioredoxin-binding protein-2), a protein that binds to and blocks TRX reducing activity [61]. Moreover, in a study investigating the possible basis for HDACI-mediated antitumor selectivity, a comparison was made between the effects of HDACI on TRX levels in normal cells versus their transformed counterparts. It was observed that both Vorinostat and MS-275 selectively increased the expression of TRX in normal but not in transformed cells, resulting in selective oxidative damage in the latter [92]. This finding may provide a possible theoretical basis for the selectivity of HDACIs in sparing normal cells while triggering lethal oxidative injury in their neoplastic counterparts.

3.3 Effects of HDACIs on the Extrinsic, Receptor-Mediated Apoptotic Pathway

Recent attention has focused on the HDACI-mediated modulation of a family of genes closely related to the extrinsic, receptor-mediated apoptotic pathway, including death receptors-4 and -5, Fas/FasL, TRAIL, and FLIP [21, 96–101]. Moreover, the ability of HDACIs to regulate expression of these genes

has been invoked as one of the factors that might account for the putative antitumor selectivity of HDACIs [101, 102]. Previous reports demonstrated that exposure to the hybrid polar HDAC inhibitor CBHA induced apoptosis in several neuroblastoma cell lines, and that this process was dependent upon the synthesis of de novo proteins linked to the expression of Fas/FasL [97]. Subsequently, apicidin, a cyclic tetrapeptide HDACI, was also shown to induce apoptosis in human leukemia cells (HL60) through a mechanism involving de novo protein synthesis and transient increased expression of both Fas and Fas ligand [96]. More recently, several other HDACIs including TSA, SAHA and sodium butyrate have been shown to modulate Fas/FasL expression levels and lethality as well [103–105]. Various HDACIs, including LAQ824, SAHA, MS-275, TSA, and sodium butyrate, act at least in part, by regulating the expression of components of the TRAIL pathway including TRAIL itself and TRAIL death receptors -4 and -5 [21, 100, 106, 107]. For example, exposure of human leukemia cells (i.e., Jurkat and SKW6.4) to LAQ824 was shown to induce the expression of TRAIL, DR4 and DR5 at both the protein and mRNA levels [21]. Simultaneously, levels of the antiapoptotic proteins FLIP, Bcl-2, and members of the IAP family were downregulated concomitantly with activation of caspases-9 and -3, suggesting that treatment with LAQ824, while inducing apoptosis through the intrinsic, mitochondrial pathway, also primes the cell death program by activating the extrinsic, receptor-mediated pathway [21]. Indeed, exposure of cells to LAQ824 in conjunction with TRAIL markedly enhanced DISC (death-inducing signaling complex) assembly and resulted in a very pronounced synergistic proapoptotic response [21]. A similar phenomenon has been described in CLL primary cells in which depsipeptide sensitized resistant primary CLL cells to TRAIL-induced apoptosis by facilitating formation of the DISC [106]. Other reports have also shown synergistic HDACI/TRAIL interactions in both human leukemia cells [108] and breast cancer cells [107].

As noted, HDACI-mediated modulation of the TRAIL pathway has been postulated to represent one of the factors contributing to the antitumor selectivity displayed by this class of compounds. Recently, in vivo and in vitro studies involving mice bearing either PML-RAR- or AML1-ETO-induced leukemia provided additional support for this concept [101]. In myeloid leukemias, both the PML-RAR and AML1-ETO fusion proteins form stable co-repressor complexes with HDACs to regulate target genes and to induce transformation of hematopoietic progenitors [109, 110]. In vivo treatment of PML-RAR mice with the HDACI valproic acid (VPA) significantly increased survival of these animals as did ATRA (all-trans-retinoic acid), which is a clinically useful APL-specific drug. However, in contrast to ATRA which primarily induced blast cell differentiation in the peripheral blood, bone marrow, and spleen, exposure to VPA resulted in extensive apoptosis of leukemic cells in the same sites without evidence of differentiation [101]. Furthermore, the proapoptotic effects of VPA were only observed in animals bearing leukemia and did not affect the control animals. Significantly, the specificity of VPA-induced apoptosis in leukemia cells was found to be dependent upon activation of the death receptor pathway, i.e., by upregulation of the expression of TRAIL, DR5, Fas, and FasL [101]. Importantly, modulation of the extrinsic pathway was not perturbed in normal hematopoietic progenitors, reinforcing the concept of the selective sensitivity of neoplastic cells to HDACIs.

Other studies have also addressed the basis for the tumor-selective action of HDACIs in leukemia cells and the relationship between HDACI action, p21$^{WAF1/CIP1}$-mediated differentiation, and the involvement of extrinsic apoptotic pathway. Nebbioso et al. [102] demonstrated that U937 cells exposed to HDACIs including MS-275, Vorinostat, and VPA displayed a marked increase in levels of expression of TRAIL, DR4, and DR5. Treatment with MS-275 significantly enhanced survival of mice harboring tumors generated by IP injection of U937 cells, but had no effect in mice inoculated with U937 cells genetically modified to knock down expression of either the TRAIL or p21 p21$^{WAF1/CIP1}$ genes. Similarly, HDACI-mediated regulation of TRAIL expression and its role in drug-induced lethality was observed in blasts from AML patients. Importantly, normal CD34$^+$ cells remained insensitive to this treatment despite displaying higher basal levels of TRAIL than the corresponding AML blasts [102]. Collectively, these findings provide new insights into the basis by which HDACIs might exert selective toxicity toward neoplastic cells, as well as the possible role

of the intrinsic apoptotic pathway in this phenomenon. They also supply a theoretical foundation for the development of novel combination strategies involving HDACIs and TRAIL for the treatment of various malignancies [21, 107, 108].

3.4 HDAC Inhibitors and NF-κB Activity

NF-κB transcriptional activity has been shown to play a key role in HDACI-induced differentiation and/or cell death [111–114]. A variety of recent reports have highlighted the complex relationship that exists between acetylation/deacetylation events and NF-κB activity. NF-κB consists of a family of transcription factors that control the expression of a variety of genes involved in immune responses and inflammation [115, 116]. In addition, NF-κB has been implicated in multiple other cellular functions such as growth regulation, protection against oxidative stress, and cell survival [115, 117–121]. Sustained NF-κB activation has been reported in several tumor types [122, 123] and may also play a critical role in the cytoprotective response of transformed cells to both radiotherapy and chemotherapy [121]. NF-κB transcriptionally activates a number of promoters including that of IkBα, its own inhibitor, as well as various antiapoptotic proteins, including Bcl-2, XIAP, FLIP and Bfl-1/A1 [124–128]. HDACIs have been found to activate NF-κB [129, 130], and recent studies have demonstrated a requirement for a functionally intact NF-κB pathway in the ability of the HDACI sodium butyrate to induce G$_1$ arrest and differentiation [111]. For example, human leukemia cells stably transfected with an IkBα "super-repressor," which lacks phosphorylation sites necessary for proteasomal degradation, exhibited impaired sodium butyrate-induced maturation and a reciprocal increase in apoptosis. Significantly, IkBα "super-repressor" cells also displayed impaired butyrate-mediated induction of p21 $^{WAF1/CIP1}$, a phenomenon that may be responsible for the failure of SB to promote G$_1$ arrest and differentiation in these cells [111, 131, 132]. Such results are in agreement with those performed in leukemia cells examining the role of NF-κB in synergistic interactions between HDACI and the CDK inhibitor flavopiridol [130]. In this study, exposure of human leukemia cells to either sodium butyrate or SAHA resulted in a marked increase in NF-κB DNA binding; however, co-administration of flavopiridol, a potent cyclin-dependent kinase inhibitor [133], transcriptional repressor [134, 135], and inhibitor of NF-κB [136], blocked HDACI-mediated NF-κB activation [130]. This process was accompanied with downregulation of XIAP and p21[WAFI/CIP1], both well-known NF-κB target genes [137–139].

As HDACI-induced NF-κB activation may limit the lethality these drugs show in certain cellular models [112], therapeutic strategies combining HDACIs with agents that either inhibit HDACI-mediated NF-κB activation114 or suppress proteasome-mediated IκBα degradation [140–142] have been proposed. Thus, marked synergistic lethality was observed in NSCLC cells co-exposed to Vorinostat and the soluble NF-κB inhibitor BAY-11-7085 [114]. Similar effects were observed when these cells were treated with the small-molecule proteasome inhibitor bortezomib followed by sodium butyrate [140]. A regimen combining HDACIs and proteasome inhibitors was also investigated in human leukemia Bcr/Abl+ cells [141] and human multiple myeloma cells [142]. In both cases, bortezomib inhibited HDACIs-induced NF-κB activation that, among other events, resulted in a marked increase in generation of ROS. Together, these observations highlight the role that the NF-κB pathway plays in regulating HDACI lethality, and importantly, how disabling of the NF-κB pathway might enhance the anti-tumor efficacy of HDACIs.

3.5 Generation of the Pro-apoptotic Lipid Second Messenger Ceramide

Many sphingolipid-regulated functions have important links to cancer initiation, progression, and response to anticancer treatments. In particular, ceramide appears to be intimately involved in the regulation of cancer cell growth, differentiation, senescence, and apoptosis [143, 144]. Moreover, the dynamic balance between the opposing effects of ceramide and the antiapoptotic sphingosine metabolite sphingosine-1-phosphate (S1P) is also emerging as a key regulator in determining cell fate [145, 146]. The pro-apoptotic role of ceramide has been extensively investigated as a mechanism of cytotoxicity of a broad variety of drugs including Fas,

TNFα [144, 147], and chemotherapeutic agents daunorubicin [148], etoposide [149], camptothecin [150], and fludarabine [151].

The role that ceramide may play in HDACI-induced lethality has been relatively unexplored. However, generation of ceramide has been invoked to explain the induction of apoptosis by butyric acid against human T cells [152] and LAQ824 in human leukemia U937 cells [86]. In addition, in Jurkat cells, recent studies showed that the HDACI MS-275 markedly increased the ability of fludarabine to trigger ceramide production, thereby contributing to the lethality of the MS-275/fludarabine regimen. Similarly, synergistic interactions between HDACIs and the alkyl-lysophospholipid perifosine in human leukemia cells (U937) were shown to be accompanied by a dramatic increase in ceramide formation [153]. Significantly, both ceramide production as well as lethality of this combination were attenuated by desipramine, an inhibitor of the acid sphingomyelinase pathway.

3.6 HDACIs and Modulation of Heat Shock Proteins

Earlier studies investigating TSA-mediated regulation of genes involved in the control of apoptosis revealed that TSA modulated mRNA levels of several members of the heat shock protein (Hsp) family [154, 155]. Heat shock proteins represent a group of molecules whose synthesis is activated under conditions of heat-related stress [156, 157]. Their physiologic role involves functioning as molecular chaperones that promote the refolding of misfolded proteins or their elimination under conditions in which irreversible damage occurs [156, 157]. Increased expression of these proteins enhances survival in tissues damaged by a variety of stressors. This property may represent a mechanism by which malignant cells, which in general display higher levels of chaperone proteins than their normal counterparts, maintain homeostasis under adverse conditions, permitting tumor cells to adapt to mutated signaling molecules that might otherwise be lethal [158–162]. Mammalian cells express six major Hsp families that are grouped according to their molecular weights: Hsp104, Hsp90, Hsp70, Hsp60, Hsp40, and the small Hsps (20–25 kDa) [156, 163]. Each member of the Hsp family has a distinct set of functions within the cell, and localize

to the cytosol, endoplasmic reticulum, and mitochondria, where they exert their chaperone activity
[163, 164]. Evidence that depsipeptide (FR901228)
was able to substantially increase the acetylation
of Hsp90 without affecting the expression levels
of the protein provided some insights on the role
of Hsp proteins in mechanisms of HDACI-induced
lethality [165]. Acetylation of Hsp90 by depsipeptide was associated with disruption of interactions
with client molecules critically involved in cell
survival decisions, including mutant p53 and Raf-1
[165]. In both cases, dissociation of Hsp90-p53
and/or Hsp90-Raf-1 was accompanied by the formation of complexes with Hsp70, a phenomenon
known to target oncoproteins for degradation by
the proteasome pathway [166, 167], and that pre-
ceded the depletion of both mutant p53 and Raf-1
[165]. A similar mechanism appeared to contribute
to depsipeptide-mediated depletion of ErbB1 and
ErbB2 in NSCLC cells [165]. Acetylation of Hsp90
has also been observed in human breast cancer
cells exposed to LAQ824 [168], where acetylated
Hsp90 appeared impaired in its capacity to act as
a chaperone for client proteins such as Her-2 and
Raf-1, which subsequently formed a complex with
Hsp70 prior to their proteasomal degradation [168].
Interestingly, given previous evidence that down-
regulation of Her-2 and its downstream signaling
targets sensitize cancer cells to chemotherapy, the
effects of LAQ824 were examined in combination
with a variety of agents including trastuzumab,
Taxotere, gemcitabine, and epothilone B [168]. In
all cases, co-administration of LAQ824 significantly
increased trastuzumab-, Taxotere-, gemcitabine-,
and epothilone B–induced apoptosis, events associated with a marked attenuation of Her-2, Raf-1, and
AKT levels [168]. Similar synergistic interactions
were observed in the case of malignant human Bcr-
Abl$^+$ CML-BC leukemia cells exposed to LAQ824
[169]. As in the case of human breast cancer cells,
Hsp90 became progressively acetylated over time
in K562 cells, a phenomenon that was associated
with the time- and dose-dependent disruption of
Hsp90-Bcr-Abl chaperone activity, increased Bcr-
Abl-Hsp70 association and a marked downregulation of the protein via the ubiquitin/proteasome
pathway [169]. Based upon these findings, new
strategies have been developed in which HDACIs
are combined with agents known to inhibit Hsp90
[22, 170, 171] or, alternatively, employed to exploit

HDACI-mediated effects on Hsp90 in order to
potentiate the lethal actions of other targeted agents
[169]. For example, the HDACIs Vorinostat and
sodium butyrate, when co-administered with the
Hsp90 antagonist 17-allylamino 17-demethoxy-
geldanamycin (17-AAG), synergistically induced
mitochondrial damage and apoptosis in human
leukemia cells including Bcr-Abl$^+$ cells [170, 171].
Similar interactions were recently reported with a
novel cinnamic hydroxamic acid analog HDACI,
LBH589, that were evaluated in both Bcr-Abl$^+$
K562 cells and acute leukemia MV4-11 cells
expressing the activating length mutation of FLT-3
[22]. In both cell types, LBH589 induced a marked
dose-dependent apoptotic response associated with
LBH589-mediated depletion of Bcr-Abl+ and
FLT-3 in K562 and MV4-11 cells, respectively [22].
As noted in the case of other HDACIs, LBH589
induced also acetylation of Hsp90 inhibiting its
chaperone activity and promoting the polyubiquit-
ylation and proteasomal degradation of Bcr-Abl
and FLT-3. Furthermore, the effects of LBH528 on
Bcr-Abl and FLT-3 degradation were potentiated by
co-exposing K562 and MV4-11 cells to the Hsp90
inhibitor 17-AAG, which also resulted in increased
levels of cell death [22]. Importantly, the combination of LBH529/17-AAG displayed limited toxicity
toward CD34$^+$ normal bone marrow progenitor
cells. Currently, LBH529 is being evaluated in
Phase I clinical studies [172]. Finally, the ability of
pan-HDAC inhibitors to function as tubulin acetylase inhibitors and mimic the actions of tubacin
[173], has recently been emphasized. Such agents
can acetylate and disrupt the function of perinuclear
aggresomes, which, in conjunction with interference with Hsp90 and proteasome activity, dramatically increases apoptosis in tumor cells [173].

3.7   HDACI-Mediated Effects
on the DNA Repair Pathway

A series of recent studies have highlighted a novel
mechanism through which HDACIs may affect
cancer cells that involves the DNA repair activity
[174–177]. For example, in human leukemia cells
it was observed that administration of HDACIs
not only trigger DNA damage responses but also
actual damage [177]. Increase in H2AX and ATM
phosphorylation, both markers of DNA damage,
appeared shortly after HDACI administration,

events that were followed by induction of apoptosis [177]. Moreover, HDACIs-mediated modulation of the DNA repair system played an important role in increasing the sensitivity of a variety of cancer cells to ionizing radiation where DNA damage and repair of double-strand breaks are critically involved [174–176].

3.8 HDACIs and Disruption of the G2M Checkpoint

One interesting characteristic of HDACIs is that they appear to be selectively cytotoxic to tumor cells, whereas normal cells seem to be relatively resistant to killing by this class of drugs [178, 179]. Several mechanisms have been invoked to explain this selectivity, including selective generation of ROS and induction of death receptors in transformed cells, as discussed. Another possible mechanism that may account for this specificity is related to cell cycle checkpoints, which are frequently defective in cancer cells [180, 181]. An intact checkpoint protects drug-resistant cells, whereas drug-sensitive cells exhibit checkpoint defects [178]. Activation of this checkpoint is only observed when drug-resistant cells are treated with high concentrations of HDACIs that would ordinarily be cytotoxic for drug-sensitive cells [179, 182]. Treatment of drug-sensitive cell lines with high concentrations of HDACIs kills a proportion of cells (20–50%) but also induces G1 arrest in the surviving population. These growth-arrested cells ultimately die, although the onset of cell death is delayed compared with cells that do not arrest in G1, suggesting that HDACIs may be able to kill nonproliferating as well as proliferating cells [178, 179, 183]. In fact, recent reports have shown that HDACIs are equally effective against both proliferating and growth-arrested tumor cells, emphasizing the potential of these agents for the treatment of not only rapidly proliferating tumors, but also those with a low mitotic index [183].

3.9 HDACIs and E2F1-Mediated Induction of Bim

The retinoblastoma (Rb)/E2F pathway is critically important in regulating both cell proliferation and apoptosis [184, 185]. This pathway links cellular proliferation control to apoptosis as a fail-safe mechanism to protect cells from aberrant oncogenic transformation. In fact, the loss of the Rb tumor suppressor gene, which may result in deregulated transcriptional E2F activity, has been reported in many human tumors [186]. However, E2F1 may also function as a tumor suppressor gene by inducing apoptosis. Indeed, deregulated E2F1 can trigger apoptosis by regulating the expression of proapoptotic genes including p14/p19ARF, p73, Apaf-1, caspase-3 and the BH3-only proteins PUMA, Noxa, Bim, and Hrk/DP5 [187, 188]. In this context it may be relevant that HDACIs target the Rb/E2F1 pathway related to apoptosis by activating the expression of pro-apoptotic Bim through an E2F1-dependent mechanism that also involves ASK1 induction [189, 190]. This mechanism has also been postulated as one that may contribute to the selectivity of transformed cells to HDACIs compared to their normal counterparts.

4 Summary and Conclusions

Although the present chapter has presented some of the most relevant mechanisms involved in HDACIs-mediated lethality separately, these are in all likelihood associated and are not mutually exclusive. Consequently, it is probable that more than one mechanism is responsible of the HDACI-mediated effects. For example, when human leukemia cells were exposed to MS-275 or LAQ-824, increased levels of ROS were detected. HDACI-mediated ROS generation was determined to have important roles in lethality [40, 86]. At the same time, induction of $p21^{WAF1/CIP1}$ expression, followed by its caspase-mediated degradation, played important roles in this process. Similarly, simultaneous degradation and/or transcriptional downregulation of anti-apoptotic proteins such as XIAP or Mcl-1 also represented critical events, all of which contributed to maximize anti-cancer activity [40, 86].

An expanding number of combination therapies have been rationally designed based in the mechanisms described above, and several of them are currently under clinical evaluation [26]. For example, based in observations that HDACIs-meditated lethality was markedly increased if the expression of $p21^{WAF1/CIP1}$ was impaired [18], coadministration of HDACIs (i.e., Vorinostat and sodium butyrate) with the potent transcriptional inhibitor flavopiridol was tested and developed [131, 132, 139, 191]. As anticipated, these two agents administered together at subtoxic concentrations displayed pronounced synergistic activities, and this regimen is currently being evaluated in a phase I clinical trial in patients with refractory AML and MDS. Similarly, based in the observation that HDACIs may increase death receptors 4 and 5 (DR4, DR5), rational combinations with TRAIL, an activator of these receptors have been investigated [21, 101, 102]. An additional example is the combination of HDACIs with agents that block HDACIs-induced NF-κB activity, thereby minimizing the protective, antiapoptotic effects that activation of this pathway exerts [141, 192–194]. Other strategies include the administration of HDACIs with Hsp90 inhibitor 17-AAG [22, 170] or ionizing radiation [174–176].

In summary, although it is widely recognized that as a group HDACIs modify gene expression and in so doing initiate events that lead to tumor cell differentiation and cell death, it is becoming increasingly apparent that other HDACI actions, including those unrelated to histone acetylation per se, may contribute to these events. More importantly, a growing understanding of the mechanisms of action involved in HDACIs anticancer activity has allowed the development of new rationally designed molecular targeted therapeutics with very promising outcomes. Whether these approaches will lead to improved antitumor activity and selectivity is likely to be answered in the next few years.

Acknowledgments. This work was supported by awards CA63753, CA61774, and CA93738 from the NIH, DAMD-17-03-1-0209 from the Department of Defense, The V Foundation, and award 6059-06 from the Leukemia and Lymphoma Society of America.

References

1. O'Dwyer ME, Mauro MJ, Druker BJ. STI571 as a targeted therapy for CML. Cancer Invest 2003, 21:429–438.
2. Chakravarthy S, Park YJ, Chodaparambil J, et al. Structure and dynamic properties of nucleosome core particles. FEBS Lett 2005, 579:895–898.
3. Strahl BD, Allis CD. The language of covalent histone modifications. Nature 2000, 403:41–45.
4. Cress WD, Seto E. Histone deacetylases, transcriptional control, and cancer. J Cell Physiol 2000, 184:1–16.
5. Peterson CL. Chromatin remodeling enzymes: taming the machines. Third in review series on chromatin dynamics. EMBO Rep 2002, 3:319–322.
6. Khan AU, Krishnamurthy S. Histone modifications as key regulators of transcription. Front Biosci 2005, 10:866–872.
7. Gray SG, Teh BT. Histone acetylation/deacetylation and cancer: an "open" and "shut" case? Curr Mol Med 2001, 1:401–429.
8. De Ruijter AJ, Van Gennip AH, Caron HN, et al. Histone deacetylases: characterisation of the classical HDAC family. Biochem J 2003, 370:737–749.
9. Gregory PD, Wagner K, Horz W. Histone acetylation and chromatin remodeling. Exp Cell Res 2001, 265:195–202.
10. Peart MJ, Smyth GK, van Laar RK, et al. Identification and functional significance of genes regulated by structurally different histone deacetylase inhibitors. Proc Natl Acad Sci U S A 2005, 102:3697–3702.
11. Gray SG, Ekstrom TJ. The human histone deacetylase family. Exp Cell Res 2001, 262:75–83.
12. Roth SY, Denu JM, Allis CD. Histone acetyltransferases. Annu Rev Biochem 2001, 70:81–120.
13. Blander G, Guarente L. The Sir2 family of protein deacetylases. Annu Rev Biochem 2004, 73:417–435.
14. Yang XJ, Gregoire S. Class II histone deacetylases: from sequence to function, regulation, and clinical implication. Mol Cell Biol 2005, 25:2873–2884.
15. Kao HY, Verdel A, Tsai CC, et al. Mechanism for nucleocytoplasmic shuttling of histone deacetylase 7. J Biol Chem 2001, 276:47496–47507.
16. Drummond DC, Noble CO, Kirpotin DB, et al. Clinical development of histone deacetylase inhibitors as anticancer agents. Annu Rev Pharmacol Toxicol 2004, 45:495–528.
17. Newmark HL, Young CW. Butyrate and phenylacetate as differentiating agents: practical problems and opportunities. J Cell Biochem Suppl 1995, 22:247–253.
18. Rosato RR, Wang Z, Gopalkrishnan RV, et al. Evidence of a functional role for the cyclin-dependent kinase-inhibitor p21WAF1/CIP1/MDA6 in promoting differentiation and preventing mitochondrial dysfunction and apoptosis induced by sodium butyrate in human myelomonocytic leukemia cells (U937). Int J Oncol 2001, 19:181–191.
19. Weisberg E, Catley L, Kujawa J, et al. Histone deacetylase inhibitor NVP–LAQ824 has significant activity against myeloid leukemia cells in vitro and in vivo. Leukemia 2004, 18:1951–1963.

20. Romanski A, Bacic B, Bug G, et al. Use of a novel histone deacetylase inhibitor to induce apoptosis in cell lines of acute lymphoblastic leukemia. Haematologica 2004, 89:419–426.

21. Guo F, Sigua C, Tao J, et al. Cotreatment with histone deacetylase inhibitor LAQ824 enhances Apo-2L/tumor necrosis factor-related apoptosis inducing ligand-induced death inducing signaling complex activity and apoptosis of human acute leukemia cells. Cancer Res 2004, 64:2580–2589.

22. George P, Bali P, Annavarapu S, et al. Combination of the histone deacetylase inhibitor LBH589 and the hsp90 inhibitor 17-AAG is highly active against human CML-BC cells and AML cells with activating mutation of FLT-3. Blood 2005, 105:1768–1776.

23. Yoshida M, Kijima M, Akita M, et al. Potent and specific inhibition of mammalian histone deacetylase both in vivo and in vitro by trichostatin A. J Biol Chem 1990, 265:17174–17179.

24. Finnin MS, Donigian JR, Cohen A, et al. Structures of a histone deacetylase homologue bound to the TSA and SAHA inhibitors. Nature 1999, 401: 188–193.

25. Marks PA, Miller T, Richon VM. Histone deacetylases. Curr Opin Pharmacol 2003, 3:344–351.

26. Rosato RR, Grant S. Histone deacetylase inhibitors in clinical development. Expert Opin Investig Drugs 2004, 13:21–38.

27. De Ruijter AJ, Kemp S, Kramer G, et al. The novel histone deacetylase inhibitor BL1521 inhibits proliferation and induces apoptosis in neuroblastoma cells. Biochem Pharmacol 2004, 68:1279–1288.

28. Villar-Garea A, Esteller M. Histone deacetylase inhibitors: understanding a new wave of anticancer agents. Int J Cancer 2004, 112:171–178.

29. Grant S, Easley C, Kirkpatrick P. Vorinostat. Nat Rev Drug Discov 2007, 6:21–22.

30. Garcia-Manero G, Issa JP, Cortes J, et al. Phase I study of oral suberoylanilide hydroxamic acid (SAHA), a histone deacetylase inhibitor, in patients (pts) with advanced leukemias or myelodysplastic syndromes (MDS). J Clin Oncol 2004, 22(14S):3027.

31. Candido EP, Reeves R, Davie JR. Sodium butyrate inhibits histone deacetylation in cultured cells. Cell 1978, 14:105–113.

32. Cousens LS, Gallwitz D, Alberts BM. Different accessibilities in chromatin to histone acetylase. J Biol Chem 1979, 254:1716–1723.

33. Kruh J. Effects of sodium butyrate, a new pharmacological agent, on cells in culture. Mol Cell Biochem 1982, 42:65–82.

34. Nudelman A, Gnizi E, Katz Y, et al. Prodrugs of butyric acid. Novel derivatives possessing increased aqueous solubility and potential for treating cancer and blood diseases. Eur J Med Chem 2001, 36: 63–74.

35. Nudelman A, Rephaeli A. Novel mutual prodrug of retinoic and butyric acids with enhanced anticancer activity. J Med Chem 2000, 43:2962–2966.

36. Reid T, Valone F, Lipera W, et al. Phase II trial of the histone deacetylase inhibitor pivaloyloxymethyl butyrate (Pivanex, AN-9) in advanced non-small cell lung cancer. Lung Cancer 2004, 45:381–386.

37. Witt O, Schweigerer L, Driever PH, et al. Valproic acid treatment of glioblastoma multiforme in a child. Pediatr Blood Cancer 2004, 43:181.

38. Kelly WK, O'Connor OA, Marks PA. Histone deacetylase inhibitors: from target to clinical trials. Expert Opin Investig Drugs 2002, 11:1695–1713.

39. Prakash S, Foster BJ, Meyer M, et al. Chronic oral administration of CI-994: a phase 1 study. Invest New Drugs 2001, 19:1–11.

40. Rosato RR, Almenara JA, Grant S. The histone deacetylase inhibitor MS-275 promotes differentiation or apoptosis in human leukemia cells through a process regulated by generation of reactive oxygen species and induction of p21CIP1/WAF1 1. Cancer Res 2003, 63:3637–3645.

41. LoRusso PM, Demchik L, Foster B, et al. Preclinical antitumor activity of CI-994. Invest New Drugs 1996, 14:349–356.

42. Graziano MJ, Spoon TA, Cockrell EA, et al. Induction of apoptosis in rat peripheral blood lymphocytes by the anticancer drug CI-994 (Acetyldinaline)(*). J Biomed Biotechnol 2001, 1:52–61.

43. Undevia SD, Kindler HL, Janisch L, et al. A phase I study of the oral combination of CI-994, a putative histone deacetylase inhibitor, and capecitabine. Ann Oncol 2004, 15:1705–1711.

44. Nemunaitis JJ, Orr D, Eager R, et al. Phase I study of oral CI-994 in combination with gemcitabine in treatment of patients with advanced cancer. Cancer J 2003, 9:58–66.

45. Camphausen K, Scott T, Sproull M, et al. Enhancement of xenograft tumor radiosensitivity by the histone deacetylase inhibitor MS-275 and correlation with histone hyperacetylation. Clin Cancer Res 2004, 10:6066–6071.

46. Marks PA, Rifkind RA, Richon VM, et al. Histone deacetylases and cancer: causes and therapies. Nat Rev Cancer 2001, 1:194–202.

47. Wang C, Fu M, Mani S, et al. Histone acetylation and the cell-cycle in cancer. Front Biosci 2001, 6:D610–D629.

48. Acharya MR, Figg WD. Histone deacetylase inhibitor enhances the anti-leukemic activity of an established nucleoside analogue. Cancer Biol Ther 2004, 3:719–720.

49. Jose B, Oniki Y, Kato T, et al. Novel histone deacety-lase inhibitors: cyclic tetrapeptide with trifluorome-thyl and pentafluoroethyl ketones. Bioorg Med Chem Lett 2004, 14:5343–5346.

50. Nishino N, Yoshikawa D, Watanabe LA, et al. Synthesis and histone deacetylase inhibitory activity of cyclic tetrapeptides containing a retrohydroxam-ate as zinc ligand. Bioorg Med Chem Lett 2004, 14:2427–2431.

51. Nishino N, Jose B, Okamura S, et al. Cyclic tetrapep-tides bearing a sulfhydryl group potently inhibit histone deacetylases. Org Lett 2003, 5:5079–5082.

52. Kwon HJ, Owa T, Hassig CA, et al. Depudecin induces morphological reversion of transformed fibroblasts via the inhibition of histone deacetylase. Proc Natl Acad Sci U S A 1998, 95:3356–3361.

53. Gelmon K, Tolcher A, Carducci M, et al. Phase I tri-als of the oral histone deacetylase (HDAC) inhibitor MGCD0103 given either daily of 3x weekly for 14 days every 3 weeks in patients (pts) with advanced solid tumors. J Clin Oncol 2005 ASCO Annual Meeting Proceedings 2005, 23[16S].

54. Garcia-Manero G, Minden MD, Estrov Z, et al. Clinical activity and safety of the histone deacety-lase inhibitor MGCD0103: results of a phase I study in patients with leukemia or myelodysplastic syn-dromes (MDS). J Clin Oncol 2006, 24(18S):6500.

55. Melnick A, Licht JD. Histone deacetylases as thera-peutic targets in hematologic malignancies. Curr Opin Hematol 2002, 9:322–332.

56. Wade PA. Transcriptional control at regulatory checkpoints by histone deacetylases: molecular con-nections between cancer and chromatin. Hum Mol Genet 2001, 10:693–698.

57. Jenuwein T, Allis CD. Translating the histone code. Science 2001, 293:1074–1080.

58. Egger G, Liang G, Aparicio A, et al. Epigenetics in human disease and prospects for epigenetic therapy. Nature 2004, 429:457–463.

59. Van Lint C, Emiliani S, Verdin E. The expression of a small fraction of cellular genes is changed in response to histone hyperacetylation. Gene Expr 1996, 5:245–253.

60. Lee H, Lee S, Baek M, et al. Expression profile analysis of trichostatin A in human gastric cancer cells. Biotech Lett 2002, 24:377–381.

61. Butler LM, Zhou X, Xu WS, et al. The histone deacetylase inhibitor SAHA arrests cancer cell growth, up-regulates thioredoxin-binding protein-2, and down-regulates thioredoxin. Proc Natl Acad Sci U S A 2002, 99:11700–11705.

62. Glaser KB, Staver MJ, Waring JF, et al. Gene expression profiling of multiple histone deacetylase (HDAC) inhibitors: defining a common gene set produced by HDAC inhibition in T24 and MDA car-cinoma cell lines. Mol Cancer Ther 2003, 2:151–163.

63. Li H, Wu X. Histone deacetylase inhibitor, Trichostatin A, activates p21WAF1/CIP1 expression through down-regulation of c-myc and release of the repression of c-myc from the promoter in human cervical cancer cells. Biochem Biophys Res Commun 2004, 324:860–867.

64. Qian DZ, Wang X, Kachhap SK, et al. The his-tone deacetylase inhibitor NVP-LAQ824 inhibits angiogenesis and has a greater antitumor effect in combination with the vascular endothelial growth factor receptor tyrosine kinase inhibitor PTK787/ ZK222584. Cancer Res 2004, 64:6626–6634.

65. Marks PA, Richon VM, Miller T, et al. Histone deacetylase inhibitors. Adv Cancer Res 2004, 91:137–168.

66. Jones PA, Baylin SB. The fundamental role of epigenetic events in cancer. Nat Rev Genet 2002, 3:415–428.

67. Johnstone RW, Licht JD. Histone deacetylase inhibi-tors in cancer therapy: is transcription the primary target? Cancer Cell 2003, 4:13–18.

68. Esteller M, Cordon-Cardo C, Corn PG, et al. p14ARF silencing by promoter hypermethylation mediates abnormal intracellular localization of MDM2. Cancer Res 2001, 61:2816–2821.

69. Cameron EE, Bachman KE, Myohanen S, et al. Synergy of demethylation and histone deacetylase inhibition in the re-expression of genes silenced in cancer. Nat Genet 1999, 21:103–107.

70. Zhu WG, Otterson GA. The interaction of histone deacetylase inhibitors and DNA methyltransferase inhibitors in the treatment of human cancer cells. Curr Med Chem Anti -Canc Agents 2003, 3:187–199.

71. Aparicio A, Weber JS. Review of the clinical experi-ence with 5-azacytidine and 5-aza-2′-deoxycytidine in solid tumors. Curr Opin Investig Drugs 2002, 3:627–633.

72. Gore SD, Baylin S, Sugar E, et al. Combined DNA methyltransferase and histone deacetylase inhibition in the treatment of myeloid neoplasms. Cancer Res 2006, 66:6361–6369.

73. Insinga A, Pelicci PG, Inucci S. Leukemia-associated fusion proteins. Multiple mechanisms of action to drive cell transformation. Cell Cycle 2005, 4:67–69.

74. He LZ, Tolentino T, Grayson P, et al. Histone deacetylase inhibitors induce remission in transgenic models of therapy-resistant acute promyelocytic leukemia. J Clin Invest 2001, 108:1321–1330.

75. Jing Y, Xia L, Waxman S. Targeted removal of PML-RARalpha protein is required prior to inhibition of histone deacetylase for overcoming all-trans retinoic acid differentiation resistance in acute promyelocytic leukemia. Blood 2002, 100:1008–1013.

76. Zhou DC, Kim SH, Ding W, et al. Frequent mutations in the ligand–binding domain of PML-RARalpha after multiple relapses of acute promyelocytic leukemia: analysis for functional relationship to response to all-trans retinoic acid and histone deacetylase inhibitors in vitro and in vivo. Blood 2002, 99:1356–1363.

77. Behrend L, Henderson G, Zwacka RM. Reactive oxygen species in oncogenic transformation. Biochem Soc Trans 2003, 31:1441–1444.

78. Irani K, Xia Y, Zweier JL, et al. Mitogenic signaling mediated by oxidants in Ras-transformed fibroblasts. Science 1997, 275:1649–1652.

79. Chung YM, Bae YS, Lee SY. Molecular ordering of ROS production, mitochondrial changes, and caspase activation during sodium salicylate-induced apoptosis. Free Radic Biol Med 2003, 34:434–442.

80. Curtin JF, Donovan M, Cotter TG. Regulation and measurement of oxidative stress in apoptosis. J Immunol Meth 2002, 265:49–72.

81. Chen QM, Bartholomew JC, Campisi J, et al. Molecular analysis of H2O2-induced senescent-like growth arrest in normal human fibroblasts: p53 and Rb control G1 arrest but not cell replication. Biochem J 1998, 332(Pt 1):43–50.

82. Moreira JM, Scheipers P, Sorensen P. The histone deacetylase inhibitor Trichostatin A modulates CD4+ T cell responses. BMC Cancer 2003, 3:30–47.

83. Ruefli AA, Ausserlechner MJ, Bernhard D, et al. The histone deacetylase inhibitor and chemotherapeutic agent suberoylanilide hydroxamic acid (SAHA) induces a cell-death pathway characterized by cleavage of Bid and production of reactive oxygen species. Proc Natl Acad Sci U S A 2001, 98:10833–10838.

84. Louis M, Rosato RR, Brault L, et al. The histone deacetylase inhibitor sodium butyrate induces breast cancer cell apoptosis through diverse cytotoxic actions including glutathione depletion and oxidative stress. Int J Oncol 2004, 25:1701–1711.

85. Yu C, Subler M, Rahmani M, et al. Induction of apoptosis in BCR/ABL+ cells by histone deacetylase inhibitors involves reciprocal effects on the RAF/MEK/ERK and JNK pathways. Cancer Biol Ther 2003, 2:544–551.

86. Rosato RR, Maggio SC, Almenara JA, et al. The histone deacetylase inhibitor LAQ-824 induces human leukemia cell death through a process involving XIAP down-regulation, oxidative injury, and the acid sphingomyelinase-dependent generation of ceramide. Mol Pharmacol 2006, 69:216–225.

87. Lucas DM, Davis ME, Parthun MR, et al. The histone deacetylase inhibitor MS-275 induces caspase-dependent apoptosis in B-cell chronic lymphocytic leukemia cells. Leukemia 2004, 18:1207–1214.

88. Fernandez-Checa JC. Redox regulation and signaling lipids in mitochondrial apoptosis. Biochem Biophys Res Commun 2003, 304:471–479.

89. Powis G, Montfort WR. Properties and biological activities of thioredoxins. Annu Rev Pharmacol Toxicol 2001, 41:261–295.

90. Fernandez-Checa JC, Kaplowitz N, Garcia-Ruiz C, et al. GSH transport in mitochondria: defense against TNF-induced oxidative stress and alcohol-induced defect. Am J Physiol 1997, 273:G7–17.

91. Louis M, Rosato RR, Battaglia E, et al. Modulation of sensitivity to doxorubicin by the histone deacetylase inhibitor sodium butyrate in breast cancer cells. Int J Oncol 2005, 26:1569–1574.

92. Ungerstedt JS, Sowa Y, Xu WS, et al. Role of thioredoxin in the response of normal and transformed cells to histone deacetylase inhibitors. Proc Natl Acad Sci U S A 2005, 102:673–678.

93. Marks PA. Thioredoxin in cancer—role of histone deacetylase inhibitors. Semin Cancer Biol 2006, 16:436–443.

94. Xu W, Ngo L, Perez G, et al. Intrinsic apoptotic and thioredoxin pathways in human prostate cancer cell response to histone deacetylase inhibitor. Proc Natl Acad Sci U S A 2006, 103:15540–15545.

95. Arner ES, Holmgren A. Physiological functions of thioredoxin and thioredoxin reductase. Eur J Biochem 2000, 267:6102–6109.

96. Kwon SH, Ahn SH, Kim YK, et al. Apicidin, a histone deacetylase inhibitor, induces apoptosis and Fas/Fas ligand expression in human acute promyelocytic leukemia cells. J Biol Chem 2002, 277:2073–2080.

97. Glick RD, Swendeman SL, Coffey DC, et al. Hybrid polar histone deacetylase inhibitor induces apoptosis and CD95/CD95 ligand expression in human neuroblastoma. Cancer Res 1999, 59:4392–4399.

98. Kim YH, Park JW, Lee JY, et al. Sodium butyrate sensitizes TRAIL-mediated apoptosis by induction of transcription from the DR5 gene promoter through Sp1 sites in colon cancer cells. Carcinogenesis 2004, 25:1813–1820.

99. Watanabe K, Okamoto K, Yonehara S. Sensitization of osteosarcoma cells to death receptor–mediated apoptosis by HDAC inhibitors through downregulation of cellular FLIP. Cell Death Diff 2005, 12:10–18.

100. Nakata S, Yoshida T, Horinaka M, et al. Histone deacetylase inhibitors upregulate death receptor 5/TRAIL-R2 and sensitize apoptosis induced by TRAIL/APO2-L in human malignant tumor cells. Oncogene 2004, 23:6261–6271.

101. Insinga A, Monestiroli S, Ronzoni S, et al. Inhibitors of histone deacetylases induce tumor-selective

apoptosis through activation of the death receptor pathway. Nat Med 2005, 11:71–76.

102. Nebbioso A, Clarke N, Voltz E, et al. Tumor-selective action of HDAC inhibitors involves TRAIL induction in acute myeloid leukemia cells. Nat Med 2005, 11:77–84.

103. Daehn IS, Varelias A, Rayner TE. Sodium butyrate induced keratinocyte apoptosis. Apoptosis 2006, 11:1379–1390.

104. Emanuele S, Lauricella M, Carlisi D, et al. SAHA induces apoptosis in hepatoma cells and synergistically interacts with the proteasome inhibitor Bortezomib. Apoptosis 2007.

105. Kim HR, Kim EJ, Yang SH, et al. Trichostatin A induces apoptosis in lung cancer cells via simultaneous activation of the death receptor-mediated and mitochondrial pathway? Exp Mol Med 2006, 38:616–624.

106. Inoue S, MacFarlane M, Harper N, et al. Histone deacetylase inhibitors potentiate TNF-related apoptosis-inducing ligand (TRAIL)-induced apoptosis in lymphoid malignancies. Cell Death Differ 2004, 11(Suppl 2):S193–S206.

107. Chopin V, Slomianny C, Hondermarck H, et al. Synergistic induction of apoptosis in breast cancer cells by cotreatment with butyrate and TNF-alpha, TRAIL, or anti-Fas agonist antibody involves enhancement of death receptors' signaling and requires P21(waf1). Exp Cell Res 2004, 298: 560–573.

108. Rosato RR, Almenara JA, Dai Y, et al. Simultaneous activation of the intrinsic and extrinsic pathways by histone deacetylase (HDAC) inhibitors and tumor necrosis factor-related apoptosis-inducing ligand (TRAIL) synergistically induces mitochondrial damage and apoptosis in human leukemia cells. Mol Cancer Ther 2003, 2:1273–1284.

109. Sirulnik A, Melnick A, Zelent A, et al. Molecular pathogenesis of acute promyelocytic leukaemia and APL variants. Best Pract Res Clin Haematol 2003, 16:387–408.

110. Melnick A, Carlile GW, McConnell MJ, et al. AML-1/ETO fusion protein is a dominant negative inhibitor of transcriptional repression by the promyelocytic leukemia zinc finger protein. Blood 2000, 96:3939–3947.

111. Dai Y, Rahmani M, Grant S. An Intact NF-kappaB Pathway is required for histone deacetylase inhibitor-induced G1 arrest and maturation in U937 human myeloid leukemia cells. Cell Cycle 2003, 2:467–472.

112. Mayo MW, Denlinger CE, Broad RM, et al. Ineffectiveness of histone deacetylase inhibitors to induce apoptosis involves the transcriptional activation of NF-kappa B through the Akt pathway. J Biol Chem 2003, 278:18980–18989.

113. Catley L, Weisberg E, Tai YT, et al. NVP-LAQ824 is a potent novel histone deacetylase inhibitor with significant activity against multiple myeloma. Blood 2003, 102:2615–2622.

114. Rundall BK, Denlinger CE, Jones DR. Combined histone deacetylase and NF-kappaB inhibition sensitizes non-small cell lung cancer to cell death. Surgery 2004, 136:416–425.

115. Chen LF, Greene WC. Shaping the nuclear action of NF-kappaB. Nat Rev Mol Cell Biol 2004, 5:392–401.

116. Ghosh S, Karin M. Missing pieces in the NF-kappaB puzzle. Cell 2002, 109 Suppl:S81–S96.

117. Karin M, Lin A. NF-kappaB at the crossroads of life and death. Nat Immunol 2002, 3:221–227.

118. Lezoualc'h F, Sagara Y, Holsboer F, et al. High constitutive NF-kappaB activity mediates resistance to oxidative stress in neuronal cells. J Neurosci 1998, 18:3224–3232.

119. Storz P, Toker A. Protein kinase D mediates a stress-induced NF-kappaB activation and survival pathway. EMBO J 2003, 22:109–120.

120. Quivy V, Van Lint C. Regulation at multiple levels of NF-kappaB-mediated transactivation by protein acetylation. Biochem Pharmacol 2004, 68: 1221–1229.

121. Orlowski RZ, Baldwin AS, Jr. NF-kappaB as a therapeutic target in cancer. Trends Mol Med 2002, 8:385–389.

122. French LE, Tschopp J. The TRAIL to selective tumor death. Nat Med 1999, 5:146–147.

123. Baldwin AS. Control of oncogenesis and cancer therapy resistance by the transcription factor NF-kappaB. J Clin Invest 2001, 107:241–246.

124. Stehlik C, de Martin R, Kumabashiri I, et al. Nuclear factor (NF)-kappaB-regulated X-chromosome-linked iap gene expression protects endothelial cells from tumor necrosis factor alpha-induced apoptosis. J Exp Med 1998, 188:211–216.

125. Grumont RJ, Rourke IJ, Gerondakis S. Rel-dependent induction of A1 transcription is required to protect B cells from antigen receptor ligation-induced apoptosis. Genes Dev 1999, 13:400–411.

126. Zong WX, Edelstein LC, Chen C, et al. The prosurvival Bcl-2 homolog Bfl-1/A1 is a direct transcriptional target of NF-kappaB that blocks TNFalpha-induced apoptosis. Genes Dev 1999, 13:382–387.

127. Kreuz S, Siegmund D, Scheurich P, et al. NF-kappaB inducers upregulate cFLIP, a cycloheximide-sensitive inhibitor of death receptor signaling. Mol Cell Biol 2001, 21:3964–3973.

128. Catz SD, Johnson JL. Transcriptional regulation of bcl-2 by nuclear factor kappa B and its significance in prostate cancer. Oncogene 2001, 20:7342–7351.

129. Suzuki M, Shinohara F, Sato K, et al. Interleukin-1beta converting enzyme subfamily inhibitors prevent induction of CD86 molecules by butyrate through a CREB-dependent mechanism in HL60 cells. Immunology 2003, 108:375–383.

130. Gao N, Dai Y, Rahmani M, et al. Contribution of disruption of the nuclear factor-kappaB pathway to induction of apoptosis in human leukemia cells by histone deacetylase inhibitors and flavopiridol. Mol Pharmacol 2004, 66:956–963.

131. Rosato RR, Almenara JA, Cartee L, et al. The cyclin-dependent kinase inhibitor flavopiridol disrupts sodium butyrate-induced p21WAF1/CIP1 expression and maturation while reciprocally potentiating apoptosis in human leukemia cells. Mol Cancer Ther 2002, 1:253–266.

132. Almenara J, Rosato R, Grant S. Synergistic induction of mitochondrial damage and apoptosis in human leukemia cells by flavopiridol and the histone deacetylase inhibitor suberoylanilide hydroxamic acid (SAHA). Leukemia 2002, 16:1331–1343.

133. De AW Jr, Mueller-Dieckmann HJ, Schulze-Gahmen U, et al. Structural basis for specificity and potency of a flavonoid inhibitor of human CDK2, a cell cycle kinase. Proc Natl Acad Sci U S A 1996, 93:2735–2740.

134. Chao SH, Fujinaga K, Marion JE, et al. Flavopiridol inhibits P-TEFb and blocks HIV-1 replication. J Biol Chem 2000, 275:28345–28348.

135. Chao SH, Price DH. Flavopiridol inactivates P-TEFb and blocks most RNA polymerase II transcription in vivo. J Biol Chem 2001, 276:31793–31799.

136. Takada Y, Aggarwal BB. Flavopiridol inhibits NF-kappaB activation induced by various carcinogens and inflammatory agents through inhibition of IkappaBalpha kinase and p65 phosphorylation: abrogation of cyclin D1, cyclooxygenase-2, and matrix metalloprotease-9. J Biol Chem 2004, 279:4750–4759.

137. Savickiene J, Treigyte G, Pivoriunas A, et al. Sp1 and NF-{kappa}B transcription factor activity in the regulation of the p21 and FasL promoters during promyelocytic leukemia cell monocytic differentiation and its associated apoptosis. Ann N Y Acad Sci 2004, 1030:569–577.

138. Tang G, Minemoto Y, Dibling B, et al. Inhibition of JNK activation through NF-kappaB target genes. Nature 2001, 414:313–317.

139. Rosato RR, Almenara JA, Kolla SS, et al. Mechanism and functional role of XIAP and Mcl-1 down-regulation

in flavopiridol/vorinostat antileukemic interactions. Mol Cancer Ther 2007, 6:692–702.

140. Denlinger CE, Keller MD, Mayo MW, et al. Combined proteasome and histone deacetylase inhibition in non-small cell lung cancer. J Thorac Cardiovasc Surg 2004, 127:1078–1086.

141. Yu C, Rahmani M, Conrad D, et al. The proteasome inhibitor bortezomib interacts synergistically with histone deacetylase inhibitors to induce apoptosis in Bcr/Abl+ cells sensitive and resistant to STI571. Blood 2003, 102:3765–3774.

142. Pei XY, Dai Y, Grant S. Synergistic induction of oxidative injury and apoptosis in human multiple myeloma cells by the proteasome inhibitor bortezomib and histone deacetylase inhibitors. Clin Cancer Res 2004, 10:3839–3852.

143. Ogretmen B, Hannun YA. Biologically active sphingolipids in cancer pathogenesis and treatment. Nat Rev Cancer 2004, 4:604–616.

144. Bektas M, Spiegel S. Glycosphingolipids and cell death. Glycoconj J 2004, 20:39–47.

145. Payne SG, Milstien S, Spiegel S. Sphingosine-1-phosphate: dual messenger functions. FEBS Lett 2002, 531:54–57.

146. Stunff HL, Milstien S, Spiegel S. Generation and metabolism of bioactive sphingosine-1-phosphate. J Cell Biochem 2004, 92:882–899.

147. Ding WX, Yin XM. Dissection of the multiple mechanisms of TNF-alpha-induced apoptosis in liver injury. J Cell Mol Med 2004, 8:445–454.

148. Bose R, Verheij M, Haimovitz-Friedman A, et al. Ceramide synthase mediates daunorubicin-induced apoptosis: an alternative mechanism for generating death signals. Cell 1995, 82:405–414.

149. Perry DK, Carton J, Shah AK, et al. Serine palmitoyltransferase regulates de novo ceramide generation during etoposide-induced apoptosis. J Biol Chem 2000, 275:9078–9084.

150. Chauvier D, Morjani H, Manfait M. Ceramide involvement in homocamptothecin- and camptothecin-induced cytotoxicity and apoptosis in colon HT29 cells. Int J Oncol 2002, 20:855–863.

151. Biswal SS, Datta K, Acquaah-Mensah GK, et al. Changes in ceramide and sphingomyelin following fludarabine treatment of human chronic B-cell leukemia cells. Toxicology 2000, 154:45–53.

152. Kurita-Ochiai T, Amano S, Fukushima K, et al. Cellular events involved in butyric acid-induced T cell apoptosis. J Immunol 2003, 171:3576–3584.

153. Rahmani M, Reese E, Dai Y, et al. Coadministration of histone deacetylase inhibitors and perifosine synergistically induces apoptosis in human leukemia cells through Akt and ERK1/2 inactivation and the

generation of ceramide and reactive oxygen species. Cancer Res 2005, 65:2422–2432.

154. Eickhoff B, Ruller S, Laue T, et al. Trichostatin A modulates expression of p21waf1/cip1, Bcl-xL, ID1, ID2, ID3, CRAB2, GATA-2, hsp86 and TFIID/TAFII31 mRNA in human lung adenocarcinoma cells. Biol Chem 2000, 381:107–112.

155. Eickhoff B, Germeroth L, Stahl C, et al. Trichostatin A-mediated regulation of gene expression and protein kinase activities: reprogramming tumor cells for ribotoxic stress-induced apoptosis. Biol Chem 2000, 381:1127–1132.

156. Sreedhar AS, Csermely P. Heat shock proteins in the regulation of apoptosis: new strategies in tumor therapy: a comprehensive review. Pharmacol Ther 2004, 101:227–257.

157. Blagosklonny MV. Hsp-90-associated oncoproteins: multiple targets of geldanamycin and its analogs. Leukemia 2002, 16:455–462.

158. Kimura E, Enns RE, Alcaraz JE, et al. Correlation of the survival of ovarian cancer patients with mRNA expression of the 60-kD heat-shock protein HSP-60. J Clin Oncol 1993, 11:891–898.

159. Ciocca DR, Clark GM, Tandon AK, et al. Heat shock protein hsp70 in patients with axillary lymph node–negative breast cancer: prognostic implications. J Natl Cancer Inst 1993, 85:570–574.

160. Santarosa M, Favaro D, Quaia M, et al. Expression of heat shock protein 72 in renal cell carcinoma: possible role and prognostic implications in cancer patients. Eur J Cancer 1997, 33:873–877.

161. Takayama S, Reed JC, Homma S. Heat-shock proteins as regulators of apoptosis. Oncogene 2003, 22:9041–9047.

162. Bagatell R, Whitesell L. Altered Hsp90 function in cancer: a unique therapeutic opportunity. Mol Cancer Ther 2004, 3:1021–1030.

163. Creagh EM, Sheehan D, Cotter TG. Heat shock proteins—modulators of apoptosis in tumour cells. Leukemia 2000, 14:1161–1173.

164. Beliakoff J, Whitesell L. Hsp90: an emerging target for breast cancer therapy. Anticancer Drugs 2004, 15:651–662.

165. Yu X, Guo ZS, Marcu MG, et al. Modulation of p53, ErbB1, ErbB2, and Raf-1 expression in lung cancer cells by depsipeptide FR901228. J Natl Cancer Inst 2002, 94:504–513.

166. Neckers L, Schulte TW, Mimnaugh E. Geldanamycin as a potential anti-cancer agent: its molecular target and biochemical activity. Invest New Drugs 1999, 17:361–373.

167. Citri A, Alroy I, Lavi S, et al. Drug-induced ubiquitylation and degradation of ErbB receptor tyrosine kinases: implications for cancer therapy. EMBO J 2002, 21:2407–2417.

168. Fuino L, Bali P, Wittmann S, et al. Histone deacetylase inhibitor LAQ824 down-regulates Her-2 and sensitizes human breast cancer cells to trastuzumab, taxotere, gemcitabine, and epothilone B. Mol Cancer Ther 2003, 2:971–984.

169. Nimmanapalli R, Fuino L, Bali P, et al. Histone deacetylase inhibitor LAQ824 both lowers expression and promotes proteasomal degradation of Bcr-Abl and induces apoptosis of imatinib mesylate-sensitive or -refractory chronic myelogenous leukemia-blast crisis cells. Cancer Res 2003, 63:5126–5135.

170. Rahmani M, Yu C, Dai Y, et al. Co-administration of the heat shock protein 90 antagonist 17-AAG with SAHA or sodium butyrate synergistically induces apoptosis in human leukemia cells. Cancer Res 2003, 63:8420–8427.

171. Rahmani M, Reese E, Dai Y, et al. Co–treatment with SAHA and 17-AAG synergistically induces apoptosis in Bcr-Abl+ cells sensitive and resistant to STI-571 in association with down-regulation of Bcr-Abl, abrogation of STAT5 activity, and Bax conformational change. Mol Pharmacol 2005, 67:1166–1176.

172. Phase I pharmacokinetic (PK) and pharmacodynamic (PD) study of LBH589A: a novel histone deacetylase inhibitor. 04, 2004.

173. Hideshima T, Bradner JE, Wong J, et al. Small-molecule inhibition of proteasome and aggresome function induces synergistic antitumor activity in multiple myeloma. Proc Natl Acad Sci U S A 2005, 102:8567–8572.

174. Zhang Y, Adachi M, Zou H, et al. Histone deacetylase inhibitors enhance phosphorylation of histone H2AX after ionizing radiation. Int J Radiat Oncol Biol Phys 2006, 65:859–866.

175. Munshi A, Kurland JF, Nishikawa T, et al. Histone deacetylase inhibitors radiosensitize human melanoma cells by suppressing DNA repair activity. Clin Cancer Res 2005, 11:4912–4922.

176. Munshi A, Tanaka T, Hobbs ML, et al. Vorinostat, a histone deacetylase inhibitor, enhances the response of human tumor cells to ionizing radiation through prolongation of gamma-H2AX foci. Mol Cancer Ther 2006, 5:1967–1974.

177. Gaymes TJ, Padua RA, Pla M, et al. Histone deacetylase inhibitors (HDI) cause DNA damage in leukemia cells: a mechanism for leukemia-specific HDI-dependent apoptosis? Mol Cancer Res 2006, 4:563–573.

178. Qiu L, Burgess A, Fairlie DP, et al. Histone deacetylase inhibitors trigger a G2 checkpoint in normal cells that is defective in tumor cells. Mol Biol Cell 2000, 11:2069–2083.

179. Burgess AJ, Pavey S, Warrener R, et al. Up-regulation of p21(WAF1/CIP1) by histone deacetylase

inhibitors reduces their cytotoxicity. Mol Pharmacol 2001, 60:828–837.

180. Elledge SJ. Cell cycle checkpoints: preventing an identity crisis. Science 1996, 274:1664–1672.

181. Warrener R, Beamish H, Burgess A, et al. Tumor cell-selective cytotoxicity by targeting cell cycle checkpoints. FASEB J 2003, 17:1550–1552.

182. Richon VM, Sandhoff TW, Rifkind RA, et al. Histone deacetylase inhibitor selectively induces p21WAF1 expression and gene-associated histone acetylation. Proc Natl Acad Sci U S A 2000, 97:10014–10019.

183. Burgess A, Ruefli A, Beamish H, et al. Histone deacetylase inhibitors specifically kill nonproliferating tumour cells. Oncogene 2004, 23:6693–6701.

184. DeGregori J. The genetics of the E2F family of transcription factors: shared functions and unique roles. Biochim Biophys Acta 2002, 1602:131–150.

185. Ginsberg D. E2F1 pathways to apoptosis. FEBS Lett 2002, 529:122–125.

186. Sherr CJ, McCormick F. The RB and p53 pathways in cancer. Cancer Cell 2002, 2:103–112.

187. Matsumura I, Tanaka H, Kanakura Y. E2F1 and c-Myc in cell growth and death. Cell Cycle 2003, 2:333–338.

188. Hershko T, Ginsberg D. Up-regulation of Bcl-2 homology 3 (BH3)-only proteins by E2F1 mediates apoptosis. J Biol Chem 2004, 279:8627–8634.

189. Zhao Y, Tan J, Zhuang L, et al. Inhibitors of histone deacetylases target the Rb-E2F1 pathway for apoptosis induction through activation of proapop-totic protein Bim. Proc Natl Acad Sci U S A 2005, 102:16090–16095.

190. Tan J, Zhuang L, Jiang X, et al. Apoptosis signal-regulating kinase 1 is a direct target of E2F1 and contributes to histone deacetylase inhibitor-induced apoptosis through positive feedback regulation of E2F1 apoptotic activity. J Biol Chem 2006, 281:10508–10515.

191. Rosato RR, Almenara JA, Yu C, et al. Evidence of a functional role for p21WAF1/CIP1 down-regulation in synergistic antileukemic interactions between the histone deacetylase inhibitor sodium butyrate and flavopiridol. Mol Pharmacol 2004, 65:571–581.

192. Dai Y, Rahmani M, Dent P, et al. Blockade of histone deacetylase inhibitor-induced RelA/p65 acetylation and NF-kappaB activation potentiates apoptosis in leukemia cells through a process mediated by oxidative damage, XIAP downregulation, and c-Jun N-terminal kinase 1 activation. Mol Cell Biol 2005, 25:5429–5444.

193. Duan J, Friedman J, Nottingham L, et al. Nuclear factor-kappaB p65 small interfering RNA or proteasome inhibitor bortezomib sensitizes head and neck squamous cell carcinomas to classic histone deacetylase inhibitors and novel histone deacetylase inhibitor PXD101. Mol Cancer Ther 2007, 6:37–50.

194. Bai J, Demirjian A, Sui J, et al. Histone deacetylase inhibitor trichostatin A and proteasome inhibitor PS-341 synergistically induce apoptosis in pancreatic cancer cells. Biochem Biophys Res Commun 2006, 348:1245–1253.

Chapter 9
Eicosanoids and Resistance of Cancer Cells to Chemotherapeutic Agents

Andrey Sorokin

1 Introduction

Apoptosis, or programmed cell death, is a normal physiologic process of cell deletion in embryonic development as well as in the maintenance of tissue and organ homeostasis. Inappropriate induction of apoptosis has been associated with organ injury, whereas failure to undergo apoptosis is known to participate in cell overgrowth and the development of many malignancies [1, 2]. The ability of tumor cells to avoid apoptosis determines their resistance to many chemotherapeutic cancer therapies and prevents their removal from the pool of proliferating cells. Eicosanoids, lipid mediators made by body cells, may make them resistant to apoptosis and treatment by standard chemotherapeutic agents. Eicosanoids are collectively known as molecules derived from 20-carbon polyunsaturated essential fatty acids, such as arachidonic acid. Three major classes of eicosanoids are prostanoids, products of cyclooxygenases, leukotrienes, products of lipoxygenases, and EETs and HETEs, products of CYP mono-oxygenases (Fig. 9.1). The formation and release of particular eicosanoids depends on the cellular presence and activity of these three major classes of enzymes, which are in control of arachidonic acid metabolism. Eicosanoids can be viewed as hormone-like messengers, able to act in autocrine and paracrine fashion to confer, among other things, cellular resistance to apoptosis. Resistance to apoptotic cell death is a characteristic feature of most if not all cancer cells and is one of the reasons for the failure of chemotherapeutic strategies. The aim of this chapter is to explore present knowledge about the potential contribution of eicosanoids to resistance of cancer cells to chemotherapeutic agents.

2 Cellular Synthesis of Eicosanoids

2.1 Prostanoids

Arachidonic acid, released from membrane glycerophospholipids by phospholipases, is catalyzed by cyclooxygenases (also known as prostaglandin H synthases) to the cyclic endoperoxidase, prostaglandin G2 (PGG_2), via an intermediate radical. PGG_2 is further converted to prostaglandin H2 (PGH_2) by a peroxidase reaction (see Fig. 9.1). These two steps occur at two distinct active sites of cyclooxygenase molecule [3, 4]. The cyclooxygenase active site is an L-shaped hydrophobic channel with active-site Tyr-385 directly involved in catalysis and other active-site residues controlling arachidonic acid positioning to make certain that PGG_2 is produced, not hydroperoxide side products [5]. Both radical abstraction by a tyrosyl radical and combined radical/carbocationic models have been proposed for this reaction, but a combined radical/carbocation mechanism seems to be less likely [6]. Generation of tyrosyl radical at Tyr-385 at cyclooxygenase active site is a consequence

From: *Sensitization of Cancer Cells for Chemo/Immuno/Radio-therapy, 1st Edition.*
Edited by: Benjamin Bonavida © Humana Press, Totowa, NJ

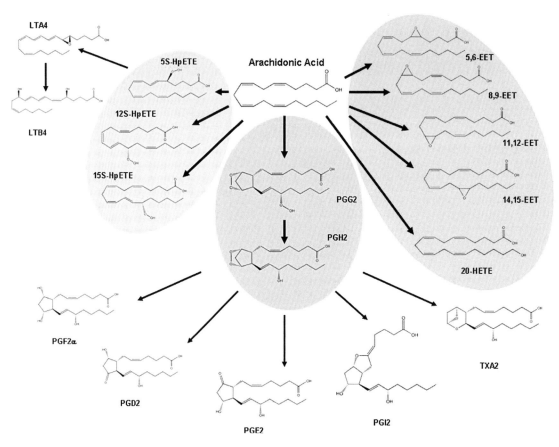

FIGURE 9.1. Metabolism of arachidonic acid and synthesis of major eicosanoids. Reactions catalyzed by cyclooxygenases are shown in the pink field, reactions catalyzed by lipoxygenases are shown in the green field and reactions catalyzed by CYP monooxidases are shown in the yellow field. Also shown are five major prostanoids, which are synthesized by prostaglandin synthases from PGH_2, and LTA_4 converted from product of lipoxygenase reaction, which is further converted to LTB_4 by LTA_4 hydrolase (*See Color Plates*)

of oxidation of the heme group at the peroxidase active site by a hydroperoxide. The peroxidase site activity catalyzes the two-electron reduction of the hydroperoxide bond of PGG_2 to yield the PGH_2 and site-directed mutagenesis indicated that the conserved cationic pocket is involved in enzyme-substrate binding [7]. Cyclooxygenases function as homodimers and each monomer contains its own cyclooxygenase and peroxidase active sites. It was recently shown, that while enzyme monomers comprising a dimer are identical in the resting enzyme, they differ from one another during catalysis: The nonfunctioning subunit provides structural support enabling its partner monomer to catalyze the cyclooxygenase reaction [8]. PGH_2 is a relatively unstable intermediate and is rapidly

converted to distinct prostanoids by corresponding terminal prostaglandin synthases [9]. Five major active prostanoids produced in vivo are $PGF_{2\alpha}$, PGD_2, PGE_2, prostacyclin (PGI_2) and thromboxane (TXA_2) (see Fig. 9.1). J-series prostaglandins including PGJ2, $\Delta12$-PGJ2, and 15-deoxy-$\Delta12$, 14- PGJ2 (15d-PGJ2) are naturally occurring metabolites of PGD2.

Two isoforms of cyclooxygenases have been characterized: cyclooxygenase 1 (Cox-1) and cyclooxygenase 2 (Cox-2) [10]. Cox-1 is expressed constitutively in most tissues, whereas Cox-2 is the inducible form of the enzyme that is produced upon stimulation with growth factors and cytokines (e.g., at sites of inflammation) [10, 11]. Notably, constitutive expression of Cox-2 is observed in restricted subpopulations of cells [12]. Renal cortical COX-2 expression was localized to the macula densa of the juxtaglomerular apparatus and to adjacent epithelial cells of the cortical thick ascending limb of Henle [13]. Since enforced activation of all three of the major mammalian MAPK leads to the induction of Cox-2 mRNA and protein [14], the constitutive activation of Cox-2 in these cells could be explained by constant exposure of macula densa cells to varying levels of luminal salt concentrations and stress-inducing variability in osmolarity [15]. In general, expression of Cox-2 mRNA is regulated by several transcription factors including the cAMP response element binding protein (CREB), nuclear factor kappaB (NFkB), and the CCAAT-enhancer binding protein (C/EBP) [16]. Remarkably, tumor progression is often accompanied by increased Cox-2 expression and selective Cox-2 inhibitors protect against the formation of multiple tumor types in experimental animals [17]. Increased expression of Cox-2 in cancer cells can be partially caused by constitutively active signaling cascades triggered by activating mutations in signaling molecules, which occur in carcinogenesis. Cox-2–mediated resistance to apoptosis of cancer cells is among several mechanisms of Cox-2 related tumor promotion [18, 19]. Since anticancer drugs typically possess pro-apoptotic properties and their efficiency is linked to ability to induce apoptotic cell death in cancer cells [20, 21], Cox-2 expression antagonizes anticancer treatment making cells resistant to apoptosis and therefore prevents the success of therapy.

Traditional nonsteroidal anti-inflammatory drugs (NSAIDs) inhibit both isoforms of the enzyme but have their anti-inflammatory, analgesic, and anti-pyretic effects due to inhibition of Cox-2. NSAIDs have recently attracted attention as an adjunct to cancer chemotherapy [22–24]. Individuals regularly taking aspirin or other NSAIDs, pharmacologic COX inhibitors, have a reduced incidence of colon cancer than those that do not [25]. These drugs though have undesirable side effects such as gastrointestinal ulceration, bleeding, and platelet dysfunctions due to inhibition of Cox-1. Since a new class of Cox-2 selective inhibitors (COXIBs) preferentially inhibits the Cox-2 enzyme thereby reducing side effects, these inhibitors have emerged as important therapeutic tools for treatment of pain and arthritis [26]. The initial enthusiasm about Cox-2 selective inhibitors has diminished recently because of reports suggesting an increased cardio-vascular risk associated with their use [27, 28] and an increasing body of evidence of their ability to act independently of their effect upon Cox-2 [29].

Nevertheless, treatment of HCA-7 cells (human colon cancer cells expressing Cox-2 constitutively) with a highly selective Cox-2 inhibitor, results in the inhibition of growth of the tumor cells and increases the number of apoptotic cells. These effects could be reversed by administration of the Cox-2 product, PGE_2 [30]. These studies and many similar studies provide the rationales that inhibition of Cox-2 may enhance the success of chemotherapy for cancer [31].

2.2 Leukotrienes

Leukotrienes are generated as a result of stereospecific insertion of molecular oxygen at the carbons 5, 8, 12, or 15 of the arachidonic acid backbone to produce 5-, 8-, 12-, and 15-HETES. The carbon numbering system of the arachidonic acid used for nomenclature of leukotrienes and products of cytochrome P450 enzymes is shown in Fig. 9.2. Generation of leukotrienes is catalyzed by non-heme, iron-containing dioxygenases, termed lipoxygenases (5-LOX, 8-LOX, 12-LOX, and 15-LOX, correspondingly). The human genome contains six functional LOX genes, which encode five typical lipoxygenases 5-LOX, 12S-LOX, 12R-LOX, 15-LOX-1, 15-LOX-2, and one atypical eLOX-3, which functions as a hydroperoxide isomerase (epoxyalcohol synthase) by using the product of 12R-LOX as the preferred substrate [32, 33]. 5-LOX

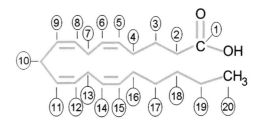

Arachidonic Acid

FIGURE 9.2. The carbon numbering system of the arachidonic acid. Circled numbers indicate positions of carbon atoms in the carbon backbone of AA, which are used for nomenclature of leukotrienes and products of CYP monooxygenases

acts with the help of 5-lipoxygenase activating protein (FLAP) and translocates to nuclear membrane to convert arachidonic acid to 5-S-HETE, which in turn is converted to LTA$_4$ and then to LTB$_4$ by LTA$_4$ hydrolase [34]. 5-LOX, 8-LOX, and 12-LOX are considered to be pro-carcinogenic, whereas LOX-15 has shown anticarcinogenic activity [34]. The data on the contribution of LOXs to apoptosis and tumorigenesis are limited, although the change in 15-LOX-1 expression was shown to contribute to colorectal tumorigenesis [35]. Transcriptional regulation of 15-LOX-1 is combinatorial and more than one factor is required to reverse 15-LOX-1 suppression in cancer cells [36]. Both isoforms of 15-lipoxygenase (15-LOX-1 and 15-LOX-2) have a tumor-suppressing role in breast cancer [37].

2.3 EETs and HETEs

Arachidonic acid (AA) is also metabolized via cytochrome P450 (CYP) enzymes to 19- and 20-hydroxyeicosatetraenoic acids (19- and 20-HETE) and 7-, 10-, 12-, 13-, 15-, 16-, 17-, and 18-HETEs in addition to epoxyeicosatrienoic acids (EETs) and dihydroxyeicosatetraenoic acids (DiHETEs) [38]. Cytochrome P450 enzymes of 4A family (CYP4A) catalyze omega-hydroxylation of AA to form 20-hydroxyeicosatetraenoic acid (20-HETE). 20-HETE stimulates mitogenic and angiogenic responses in vitro and in vivo [39]. Investigation of the role of 20-HETE and other HETEs and EETs in regulation of cancer cell growth is in its very early stages. In 2007, the first paper appeared that showed that the urinary concentration of 20-HETE is elevated in patients with prostatic hypertrophy and prostate cancer [40].

There are 18 cytochrome P450 gene families in mammals and many of these enzymes participate in the detoxication of drugs and environmental pollutants, sometimes generating reactive oxygenated intermediates, which can damage DNA or protein [41]. Analysis of their contribution to environmental carcinogenesis is beyond the scope of this chapter.

3 Signaling by Eicosanoids

Newly synthesized prostanoids are known to cross the membrane twice: They are released from the cytoplasm to the extracellular space and later act as local hormones in the vicinity of their production site and are again internalized prior to inactivation by oxidation (see Fig. 9.3). The efflux is either maintained by simple diffusion or facilitated by several prostaglandin carriers—transporters, which control energy-dependent prostaglandin transport across the plasma membrane [42]. Extracellular prostanoids carry out their biological function via activation of seven transmembrane domain G-protein–coupled receptors (GPCR), of which eight types and subtypes (FP, DP, IP, TP and EP$_{1-4}$) have been described [43]. The rank order of affinity of prostanoid ligands to each receptor is known and roles of individual receptors were revealed in individual mice knockdown systems [44]. The mouse FP receptor only binds PGF$_{2\alpha}$ with high affinity, IP receptor binds prostacyclin analogues, and thromboxane binds TP receptor. Correspondingly, mouse DP receptor binds PGD$_2$, PGD$_2$ can also bind and signal via chemoattractant receptor named CRTH2 (chemoattractant receptor homologous molecule expressed on Th2 cells), a seven-transmembrane G-protein–coupled receptor selectively expressed in Th2 cells, T-cytotoxic type 2 cells, eosinophils, and basophils [45]. DP receptor and CRTH2 receptor are termed DP1 and DP2 receptors. Four EP receptors all bind PGE$_2$, although with different affinity. The EP$_1$ receptor is coupled with the G$_q$ protein and thus activates phospholipase C and induces mobilization of intracellular Ca^{2+}. The EP$_2$ and EP$_4$ receptors are coupled with the G$_s$ protein; accordingly, they signal through elevation of intracellular cAMP levels and activation of protein kinase A. The EP$_3$ receptor is coupled with the G$_i$ protein and causes the reduction of intracellular cAMP levels. Although the major COX-2 derived prostaglandins implicated in oncogenesis are PGE$_2$ and TxA$_2$ [46], effects of Cox-2 selective inhibitors upon cancer cell proliferation and apoptosis not necessarily correlate with PGE$_2$ production [47], indicating that either other prostanoids play the crucial role in regulation of cancer cell proliferation or inhibitors have non-specific off-target effect. In addition to exerting their actions via G-protein–coupled receptors, eicosanoids have been shown to regulate cell functions through activation of peroxisome proliferator-activated receptors (PPAR), the superfamily of nuclear receptors that function as ligand-activated transcription factors [48]. Out of three PPAR isoforms (PPAR-α, PPAR-β/δ, and PPAR-γ), PPAR-γ was described

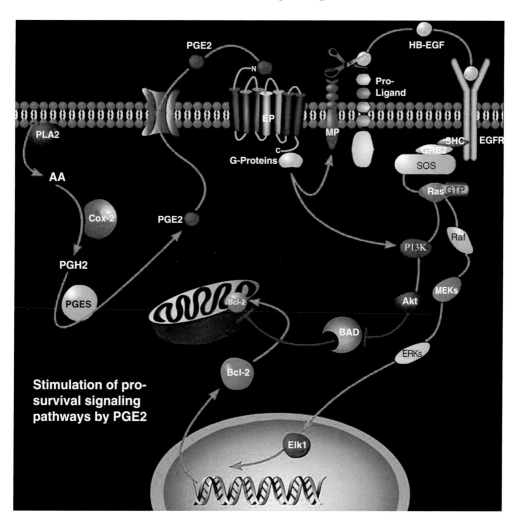

FIGURE 9.3. Stimulation of pro-survival signaling pathways by PGE_2. Arachidonic Acid (AA), released by phospholipase 2 (PLA2), is metabolized by cyclooxygenases to PGH_2 and further converted to PGE_2 by prostaglandin E synthase (PGES) prior to transporting out of the cell by prostaglandin transporter (PGT). PGE_2 binds to one of its G-protein–coupled receptors and triggers the transactivation of EGFR via metalloproteinase-dependent release of EGFR ligand from its transmembrane precursor. Activation of members of EGFR family results in recruitment of complex Shc-Grb2-Sos to close proximity of small G protein Ras and stimulation of ERK and PI-3K pro-survival pathways. ERK-mediated gene expression results in expression of anti-apoptotic Bcl-2 proteins, whereas PI-3K/Akt signaling promotes phosphorylation and inactivation of BAD (pro-apoptotic Bcl-2 protein) preventing its inhibitory action towards Bcl-2 in mitochondria. PI-3K pathways is also stimulated by PGE_2 receptors independent of EGFR transactivation

as an intracellular target of recently discovered prostaglandin, the cyclopentanone 15-deoxy-$\Delta^{12,14}$-PGJ_2 (15d-PGJ_2) [49].

There are four G-protein–coupled seven transmembrane domain receptors for leukotrienes, of which two mediate LTB4 actions (termed BLT1 and BLT2), and two bind LTD4 and LTC4 (termed cysLT1-R and cysLT2-R) [50].

Despite detailed studies characterizing multiple cellular actions of EETs and HETEs, transmembrane receptors for these products of CYP monooxygenases have never been identified [38]. One can argue that since these easily diffusible lipid messengers are capable of penetrating cells without aid from transmembrane receptors or transporters, their real receptors are the intracellular signaling molecules that have been described as their targets. The mechanism of action of HETEs and EETs remains uncertain. 20-HETE activates PKC and ERK cascades in renal arterioles [51]. One possibility is that 20-HETE may directly bind to, and activate, small G proteins from the Ras family of GTPases or raf proteins in the MAPK pathway, because these proteins are known to be activated by AA [51]. It must be taken into consideration, however, that AA also stimulates tyrosine phosphorylation of the adaptor protein Shc and its association with Grb2–Sos complex, signaling events preceding activation of Ras [52]. Previously reported inhibitory effect of 20-HETE upon Na$^+$ transport [53] is caused, at least in part, by inhibiting Na$^+$-K$^+$-ATPase and raising [Na$^+$] [54]. The question whether Na$^+$-K$^+$-ATPase or PKC serve as intracellular receptors for products of CYP4A enzymatic activity requires further investigation. The elucidation of direct intracellular targets of EETs and HETEs is necessary for understanding their role in cell metabolism and regulation of cell growth. It can not be excluded, however, that although transmembrane receptors for EETs and HETEs have escaped investigators' attention so far, future studies will uncover the novel class of receptor molecules responsible for specific binding of HETEs and EETs.

In summary, cellular actions of eicosanoids are subject to regulation at multiple levels: (1) availability of arachidonic acid, general precursor for all eicosanoids; (2) expression of particular arachidonic acid metabolizing enzyme (cyclooxygenases, lipoxygenases, and CYP monooxygenases); (3) expression and activity of distinct synthases or hydrolases that further modify primary product of arachidonic acid metabolism; and (4) availability of corresponding specific receptors.

4 Biological Function of Eicosanoids

Eicosanoids play an important role in regulation of a wide variety of physiologic and pathophysiologic processes, including among others cardiovascular homeostasis [39, 55], acute and chronic renal failure [56–58] and gastrointestinal inflammation [58, 59]. Detailed discussion of the role of eicosanoids in the regulation of basic physiologic processes

within the human body is beyond the scope of this chapter; however, we discuss briefly the role of eicosanoids in inflammation, immunity, and apoptosis since these functions seem relevant for the role of eicosanoids in cancer.

4.1 Inflammation

Inflammation is a type of tightly regulated unspecific immune response and is a basic way an organism reacts to infection or injury, leading to removal of the offending factor. The acute phase of inflammation, which lasts for only a few days, can be a part of the defense response, but chronic inflammation can lead to multiple diseases, including cancer [60]. During inflammation increased rates of production of arachidonic acid–derived eicosanoids occur and increased levels of these eicosanoids are observed in blood and tissues from patients with acute and chronic inflammatory conditions [61]. Cox-2 is induced in cells at the site of inflammation and cyclooxygenase products, such as PGE$_2$, are traditionally considered to act as pro-inflammatory substances. LTB4, produced by neutrophils, monocytes and macrophages, increases vascular permeability, serves as a potent chemotactic agent for leucocytes, and enhances generation of reactive oxygen species and production of inflammatory cytokines [61]. Recently, however, cyclooxygenase and lipoxygenase pathways were also shown to be essential for successful resolution of inflammation, the process that involves neutralizing of inflammatory stimulus and restoration of tissue structure and function [62]. The PGD$_2$-CRTH2 system plays a significant role in chronic allergic skin inflammation [45].

4.2 Immunity

Prostaglandins are well known immunomodulators capable of modulating the function of dendritic cells, which are essential for the initiation of immune responses by capturing, processing, and presenting antigens to T cells [63]. The overall impact of action of cyclooxygenase metabolites in immunity is determined by the profile of prostanoid synthesis and the repertoire of prostanoid receptors that is expressed by a particular immune cell population [63, 64]. PGE$_2$-EP4 signaling promotes the migration and maturation of Langerhans cells,

TABLE 9.1. Involvement of arachidonic acid pathways in common cancer types.

Common cancer type enzymes	Cyclooxygenases	Lipoxygenases	CYP4A
Bladder cancer	[197, 198]	[199, 200]	
Breast cancer	[18, 201, 202]	[37, 196, 203]	
Colon and rectal cancer	[94, 204, 205]	[34, 206]	
Endometrial cancer	[207, 208]		
Kidney cancer	[209–211]	[212]	
Leukemia	[213, 214]	[215–217]	
Lung cancer	[19, 218]	[219]	
Melanoma	[220, 221]	[222, 223]	
Non-Hodgkin's lymphoma	[224, 225]		
Pancreatic cancer	[181, 226]	[226, 227]	
Prostate cancer	[85, 86, 91]	[93, 228, 229]	[40]
Skin cancer	[230, 231]	[232]	
Thyroid cancer	[233, 234]		
Gliomas and meningiomas	[235]	[235]	[84]

a type of dendritic cells, and facilitates initiation of skin immune responses [64]. On the contrary, PGD_2 was shown to be produced in the skin and via activation of the D prostanoid receptor 1 (DP1) inhibits Langerhans cell trafficking [65]. TXA_2-TP signaling modulate peripheral immune response in the lymph node [64]. Tumor growth is favored by PGE_2-induced reduction in immunity [66].

4.3 Apoptosis

Cox-2 has emerged as a key player in the regulation of apoptosis in many types of cells, including intestinal epithelial cells [67], macrophages [68], renal medullary interstitial cells [69], embryonic stem cells [70], prostate cancer cell lines [71, 72], esophageal adenocarcinoma cells [73], mouse and pancreatic cancer cells [74]. The enzymatic activity of Cox-2 contributes to inhibition of trophic withdrawal apoptosis in PC12 rat pheochromocytoma cells [75]. Several mechanisms proposed to explain the anti-apoptotic effect of Cox-2 will be discussed in the section Stimulation of Pro-survival Signaling Pathways. Among most frequently suggested are modulation of expression of the anti-apoptotic protein Bcl-2 [67, 76] and regulation of Akt activation [77, 78].

Leukotrienes were reported to have both pro-survival [79, 80], and pro-apoptotic [81, 82] activity. Overexpression of 15-LOX was associated with increased thymic apoptosis [80] and LTD4, derived from arachidonic acid via the 5-LOX pathway, induced the apoptosis of B-1F cells,

established from estrogen-responsive mouse Leydig cell tumor [83]. Pro-survival effects of leukotrienes can be at least partially explained by induced expression of Cox-2 with corresponding increase of PGE_2 release [79].

Involvement of 19- and 20-HETEs in regulation of apoptosis remains largely unexplored. The observation that chronic administration of a selective inhibitor of the formation of 20-HETE reduced proliferation of rat gliosarcoma cells and caused three- to fourfold increase in the apoptotic index [84] is probably the first indication, that 20-HETE signaling is involved in resistance of cancer cells to apoptosis.

5 Arachidonic Acid Pathways in Oncogenesis

5.1 Involvement of Eicosanoids in Common Cancer Types

An involvement of arachidonic acid pathways in different cancer types is well documented. Table 9.1 contains selected references to studies and recent reviews suggesting participation of particular arachidonic acid metabolizing enzymes in common cancer types and two most common types of human brain tumor, gliomas and meningiomas. There are multiple studies addressing the role of cyclooxygenases and lipoxygenases; in particular, the types of cancer and references included in Table 9.1 should serve as a starting point for those interested in the topic. It is impossible to cite all important works in this chapter.

In contrast to cyclooxygenase and lipoxygenase pathways the evidence of CYP4A enzymes involved in carcinogenesis is largely missing. There are two probable reasons of the deficiency of the current knowledge with regard to involvement of CYP4A enzymes in cancer: (1) until recently selective inhibitors of CYP4A enzymes were not available, situation which has changed now [38]; and (2) enzymes in the cytochrome P450 families involved in eicosanoid production (CYP1, CYP2, CYP3, and CYP4 gene families) show some redundancy and overlapping substrate specificity, making a knockout approach rather challenging [41].

The role of eicosanoids in oncogenesis is often associated with control of apoptosis, promotion of invasiveness of cancer cells, and angiogenesis, increasing blood flow to the tumor. Involvements of eicosanoids in the colorectal, breast, and prostate cancer have received particular attention.

Inhibition of arachidonic acid release from phospholipids by PLA2 inhibitor 4-bromophenacyl bromide (4-BPB) prevents formation of all eicosanoids and inhibits prostate cancer cell invasiveness [81]. The word *prostaglandin* derives from prostate gland. Products of Cox-2-mediated metabolism of arachidonic acid were traditionally associated with functioning of the prostate gland, and the involvement of Cox-2 overexpression in prostate cancer has been suggested by multiple studies [71, 85, 86]. Inhibition of cyclooxygenases (but not lipoxygenases) had effect similar to effect of PLA2 inhibitor and, remarkably, inhibition was reversed by addition of PGE_2 [81]. Nevertheless, PGE_2 alone was not able to enhance invasiveness of prostate cancer cells [81].

Increased levels of prostaglandins, products of Cox enzymatic activity, have been found in malignant human prostate tumors and mouse prostate cancers [71]. Furthermore, increased synthesis of prostaglandins was shown to be associated with poor disease prognosis in humans [87]. Inhibitors of Cox-2 reduced human prostate tumor cell invasiveness and triggered apoptosis [77, 81, 88]. Comparison of prostanoid release from human prostate cancer cells with different degrees of invasiveness revealed threefold higher production of PGE_2 and twofold higher production of PGD_2 in cells with higher invasiveness [89]. Cox-2 expression was induced in a canine model of spontaneous

prostatic adenocarcinoma [90] and elevated levels of Cox-2 expression were demonstrated in cultured prostate cancer cells [76]. Notably, Cox-2 mRNA and protein are overexpressed in 83% of human prostate tumor samples [91]. Thus, the involvement of Cox-2 in prostate cancer is compelling.

The concentrations of 20-HETE and 12-HETE in urine samples of patients with prostatic diseases were significantly elevated compared with healthy men [40]. The source of HETEs in urine appears to be the prostatic gland, since its removal resulted in a radical decrease in urinary eicosanoids to levels of normal volunteers. However, in this study no correlation was found between the concentrations of these eicosanoids and either PSA level or grade of tumor [40].

5.2   Trials for the Prevention of Colorectal Adenomas and Cox-2 Inhibitors

Cox-2 involvement in colorectal tumor development is well demonstrated [92, 93]. Since overexpression of Cox-2 has been observed in colorectal cancers and adenomas in humans [94] and Cox-2 knockout mice demonstrated dramatic reduction of the number and size of the intestinal polyps [95], it was important to test the effect of Cox-2 inhibition as a chemopreventive intervention [96]. Analysis of two trials The Adenoma Prevention with Celecoxib (APC) and Prevention of Spontaneous Adenomatous Polyps (PreSAP) reached the conclusion that celecoxib is an effective agent for the prevention of colorectal adenomas but, because of potential cardiovascular events, cannot be routinely recommended for this indication [97, 98]. It appears that another selective Cox-2 inhibitor rofecoxib has similar limitations. Two recent clinical studies, aiming to reveal whether rofecoxib (Vioxx) and celecoxib (Celebrex) reduce the risk of colorectal cancer, showed that these drugs increased the probability of serious cardiovascular events, including death, when compared with the control group that took a placebo [99, 100]. The Adenomatous Polyp PRevention On Vioxx (APPROVe) trial carried out by Merck Research Laboratories established that 3.5% of participants taking rofecoxib versus 2.0% in the placebo group developed serious cardiovascular event [99]. The APC trial organized by National Cancer Institute and Pfizer Corporation revealed that 3.4% of participants taking 400 mg of celecoxib twice a day had died of a cardiovascular event or had experienced a heart attack, stroke, or heart failure; of those who took 200 mg of celecoxib twice a day 2.3% had one of these outcomes, compared with 1% of those who took a placebo twice a day for 3 years [100]. Although the absolute numbers of patients experiencing cardiovascular events was small, both trials were halted and Merck announced that it was voluntarily withdrawing rofecoxib from the market worldwide. Nevertheless, it is important to emphasize that analysis of celecoxib trial results showed that those participants taking celecoxib had fewer new adenomas and fewer new advanced adenomas than those on placebo. Celecoxib remains on the market for the treatment of chronic pain conditions and as an accessory to usual care for people with familial adenomatous polyposis. The APPROVe trial was a randomized, double-blind, placebo-controlled trial of the efficacy of oral rofecoxib, 25 mg/day, to prevent colorectal adenomas and it clearly showed that rofecoxib reduces the 3-year risk of colorectal adenomas among patients with an adenoma history [56]. An increased risk of clinically relevant upper gastrointestinal events when compared with placebo was associated with the use of rofecoxib [101]. The Pfizer-sponsored PreSAP trial was a randomized, placebo-controlled, double-blind study of the COX-2 inhibitor celecoxib given daily in a single 400-mg dose and it showed that celecoxib reduced the occurrence of colorectal adenomas within 3 years after polypectomy [96]. Thus, both rofecoxib and celecoxib are effective agents for the prevention of colorectal adenomas, but because of potential cardiovascular and gastrointestinal events, cannot be routinely recommended for chemoprevention [97]. Nonetheless, NIH supports a number of clinical trials examining the potential benefits and risks of using celecoxib to prevent and treat various cancers (Table 9.2). In practically all listed trials, efficiency of Cox-2 inhibitors is examined in combination with some chemotherapeutic agent. The negative features of Cox-2 inhibitors need to be weighed against their benefits as triggers of apoptosis in cancer cells. Due to the statistically significant increased risk of complications, Cox-2 inhibitors are not used as chemopreventive agents, but their benefits may be significant when using them in combination with

chemotherapeutic drugs to induce apoptosis of cancer cells or prevent tumor growth.

6 Mechanisms of Eicosanoid-Mediated Resistance to Chemotherapeutic Agents

6.1 Resistance to Apoptosis Is One of the Reasons for the Failure of Chemotherapy

The therapeutic objective of cancer treatment is to initiate selective death of cancer cells, but resistance of cancer cells to death-triggering chemotherapeutic agents is a concern for many types of cancer [102]. Resistance to apoptosis is one of the reasons for the failure of chemotherapeutic strategies of hormone refractory prostate cancer [103]. The clinical improvement in patients with metastatic prostate cancer following androgen ablation therapy is finite, typically lasting only 15 months [103–105]. The treatment of patients with relapsing cancer is usually disappointing, with chemotherapy having little or no effect due to high resistance of metastatic prostate cancer cells to apoptosis [22, 103]. The molecular mechanisms of the acquired resistance of prostate cancer cells to apoptosis are unknown and are the focus of current investigation by multiple scientific teams.

Anticancer therapy using combined proapoptotic therapy with inhibition of antiapoptotic factors; therefore, increasing chemotherapy sensitivity through the induction of apoptosis is a promising combination treatment strategy that is being evaluated in multiple clinical trials [106].

What are potential mechanisms by which eicosanoids can protect cancer cells from apoptosis induced by chemotherapeutic agents? Eicosanoid-dependent actions protecting cancer cells from chemotherapeutic agents include:

- Promoting proliferation of cancer cells [31, 107]
- Making cancer cells resistant to targeted growth factor receptor therapy by cross-activating the receptor signaling pathway downstream components [108]
- Inducing expression of membrane transporter proteins responsible for drug resistance phenotype [109]
- Stimulating pro-survival pathways, such as expression of Bcl-2 or Akt activation [110]

TABLE 9.2. NIH supported clinical trials examining potential benefits and risks of Cox-2 inhibitor Celecoxib in preventing and treating cancers.

Treatment	Second drug	Clinical trial ID	Condition
Celecoxib	Docetaxel	NCT00274898	Lung cancer
	Multiple drugs 1	NCT00357500	Young patients with relapsed or progressive cancer
	Unspecified	NCT00135018	Breast cancer
	Multiple drugs 2	NCT00300729	Non-small cell lung cancer
	Multiple drugs 3	NCT00061893	Ewing's family of tumors
	Capecitabine	NCT00305643	Breast/colorectal cancer
	Multiple drugs 4	NCT00268476	Prostate cancer
	Radio ter. and 5-FU	NCT00188565	Colorectal cancer
	EPO906	NCT00159484	Colon/colorectal cancer
	Taxotere	NCT00215345	Prostate cancer
	Erlotinib	NCT00314262	Precancerous conditions
	Erlotinib	NCT00088959	Lung cancer
	None	NCT00104767	Lung cancer
	None	NCT00099047	Multiple myeloma and cell neoplasm
Plasma			
	None	NCT00081263	Cervical cancer
	None	NCT00055978	Lung cancer
	None	NCT00393016	Colorectal cancer
	None	NCT00088972	Breast cancer
	Carboplatin/Paclitaxel	NCT00062179	Lung cancer
	None	NCT00104767	Lung cancer
	Eflornithine	NCT00033371	Colorectal cancer
	None	NCT00099047	Multiple myeloma and cell neoplasm
Plasma			
	Exemestane	NCT00085072	Breast cancer
	Temozolomide	NCT00112502	Brain tumors
	Erlotinib	NCT00314262	Precancerous conditions
	Capecitabine/Irinotecan	NCT00230399	Colorectal cancer
	Decitabine/Romidepsin	NCT00041158	Esophageal cancer and lung cancer

Clinical trial ID is the clinicaltrials.gov identifier.
Listed are NIH supported clinical trials active at the time of this writing.
Multiple drugs 1: etoposide, cyclophosphamide, thalidomide, celecoxib, and fenofibrate. Multiple drugs 2: either carboplatin + gemcitabine or carboplatin + vinorelbine. Multiple drugs 3: cyclophosphamide, doxorubicin, etoposide, filgrastim, ifosfamide, vinblastine, vincristine. Multiple drugs 4: zoledronate, docetaxel, prednisolone.

6.2 Contribution of Eicosanoid-Independent Effects of Cox-2

Cox-2 was reported to contribute to tumorigenesis and the malignant phenotype of tumor cells by inhibiting apoptosis, increasing angiogenesis and invasiveness, by modulation of inflammation/immunosuppression and conversion of procarcinogens to carcinogens [111]. Not all of these effects are modulated via production of eicosanoids. Both peroxidase and cyclooxygenase half-cycles of Cox-2–catalyzed reaction can be involved in oxidation of neuro-catechols and oxygen radical production [112]. In addition, antitumor effects of selective Cox-2 inhibitors are sometimes mediated through Cox-2 independent mechanisms and are not restricted to Cox-2 overexpressing tumors [113].

6.3 Eicosanoids Promote Tumor Growth via Modulation of Non-Cancer Cells

Without a doubt, some eicosanoid-mediated effects promoting tumor growth result from modulation of not only tumor cells, but also surrounding normal cells. Strong Cox-2 and VEGF expression is highly correlated with increased tumor microvascular density; new vessels proliferate in areas of the tumor that express Cox-2 [114].

There are a small number of studies in which commercial analogues of prostanoids were shown to have either a beneficial effect on survival of normal cells after exposure to chemotherapeutic agent, or caused unfavorable conditions for tumor growth. Beraprost (PGI_2 analogue) enhanced drug delivery to solid tumor and also reduced the blood supply to tumor tissues [115]. Iloprost (PGI_2 analogue) counteracted and reversed the damaging effect of chemotherapeutic agent 5-Fluorouracil on would-healing of colonic anastomoses in rats [116]. Misoprostol (PGE_1 analogue), a potent cytotoxic and oncogenic radioprotector, significantly inhibited tumor volumes in nude mice inoculated with human colon cancer cell lines and had an inhibitory effect upon early tumor growth in rats with 1,2-dimethyl-hydrazine–induced colon cancer [117].

6.4   Eicosanoids Promote Proliferation and Invasiveness of Cancer Cells

Multiple studies have demonstrated a close link between proliferation of cancer cells and production of either prostanoids or leukotrienes [31, 46, 80, 93, 118–120]. It is known that prostaglandins, particularly PGE_2, have a pro-proliferative effect on a wide variety of non-cancerous cells, including gastrointestinal epithelial cells [121], glomerular epithelial cells [122], glandular epithelial cells [123], and glomerular mesangial cells [124]. The basis of pro-proliferative action of prostanoids in normal and cancer cells may be not much different, but because of the enhanced production of prostanoids in cancer cells they exert their pro-proliferative effect predominantly in cancer cells and their close proximity. Prostanoids are also capable of enhancing activation of growth factor receptors in cells [122].

There is much less known about the contribution of products of CYP4A enzymes HETEs and EETs to the proliferation of cancer cells. It seems important that 20-HETE promotes proliferation of several forms of cancer cells in vitro and that inhibitors of the formation of 20-HETE inhibit the growth of cancer cells in vitro and the growth of brain tumors in rats in vivo [84, 125]. It was also shown that inhibitors of 20-HETE block angiogenesis induced by cancer cells in vivo [125]. These results provide a compelling reason to explore the role of 20-HETE and the potential of inhibitors of this pathway in the treatment of different types of cancer.

Among numerous mechanisms proposed to govern pro-proliferative effects of eicosanoids in cancer cells, the cross-talk between eicosanoid signaling pathways and pathways activated by growth factors or pharmacologic ligands recently attracted most of attention. The cross-talk between the nuclear transcription factor PPARδ, stimulated by its agonist GW501516, and prostaglandin signaling pathways (PPARδ-mediated enhancement of Cox-2 gene expression and resulting PGE_2-induced feedback activation of PPARδ through phosphorylation of $cPLA_{2\alpha}$) has been proposed to regulate human hepatocellular carcinoma cell growth [126].

Eicosanoids are also known to increase motility and invasiveness of cancer cells and induce local immunosuppression contributing to tumor development [93, 127]. Heregulin β1-mediated increased invasiveness of colon cancer cells is augmented at least partially by induction of PGE_2 [128].

6.5   Cross-activation of Receptor Signaling Pathway Downstream Components

Uncontrolled proliferation of cancer cells is often caused by aberrant signaling of receptor tyrosine kinases and particularly multiple members of the epidermal growth factor receptor (EGFR) family [129]. Tyrosine kinase activity is indispensable for EGFR signaling. One promising anticancer therapy is to block EGFR signaling in cancer cells by treatment with either monoclonal antibodies reacting with extracellular ligand binding domain of EGFR, such as cetuximab (Erbitux), or with small synthetic selective EGFR tyrosine kinase inhibitors, such as gefitinib and erlotinib [130]. In both cases tyrosine kinase activity of EGFR is inhibited and, accordingly, downstream signaling pathways controlling proliferation of cancer cells are blocked. Unfortunately, the efficiency of these inhibitors in vivo is significantly compromised by incomplete inhibition of transphosphorylation of the kinase inactive members of EGFR family, which then drive pro-survival signaling pathways [131]. It appears that PGE_2 signaling may confer cells resistance to targeted growth factor receptor therapy by cross-activation of the receptor signaling pathway downstream components [108]. Furthermore, Cox-2 overexpression was shown to induce EGFR activation via PGE_2-mediated activation of EP_1

receptor in human hepatocellular carcinoma cells [132]. PGE$_2$-induced transactivation of EGFR was demonstrated in gastric cancer cells [133], colon cancer cells [134, 135], cholangiocarcinoma cells via EP$_1$ receptor [119], neonatal ventricular myocytes via EP$_4$ receptor [136], and endometrial adenocarcinoma cells via EP$_2$ receptor [137].

Acting via their respective GPCRs, how can prostanoids transactivate EGFR? A recently proposed mechanism of transactivation of a number of receptor tyrosine kinases by various GPCRs involves processing of transmembrane growth factor precursors by metalloproteases of the ADAM (a disintegrin and metalloprotease) family of zinc-dependent proteases [138]. Activation of ADAM metalloproteases, probably mediated by induction of the serine/threonine kinase PKC (protein kinase C) [138], results in shedding processes that lead to the release of EGF-like ligands (HB-EGF, TGF-α, and amphiregulin) from their transmembrane precursors. Consequently, the released ligand binds to EGFR and triggers the standard EGFR-mediated signaling cascade (Fig. 9.3). PGE$_2$-induced EGFR transactivation in colon cancer cells involves signaling transduced via TGF-α, likely released by c-Src–activated ADAM metalloproteases [135]. So far the complete identity of the signaling elements in between the eicosanoid receptors and the ADAM proteases has remained elusive. Nevertheless, the combination of celecoxib with targeted growth factor receptor therapy is considered for the therapy of different types of cancers. In a phase I trial the optimal biological dose (OBD) of 600 mg bid celecoxib when combined with erlotinib was determined in therapy of advanced non-small cell lung cancer [139]. PGE$_2$ also promotes human cholangiocarcinoma cell growth and invasion through EP$_1$ receptor-mediated activation of the epidermal growth factor receptor and Akt [119].

Alternative intracellular mechanisms of transactivation of growth factor receptors by eicosanoids, which do not include activation of metalloproteases, have been described. In colorectal carcinoma cells, PGE$_2$ induced activation of EGFR via an intracellular Src-mediated event, but not through the release of an extracellular epidermal growth factor–like ligand [140]. It appears that PGE$_2$ induces the association of EP$_4$/β-arrestin 1/c-Src signaling complex resulting in the transactivation of the EGFR and downstream Akt signaling [141]. The intracellular mechanism of transactivation of EGFR by eicosanoids provides additional rationale for combining targeted EGFR therapy with inhibitors of eicosanoid signaling, since ligand-independent activation of EGFR is insensitive to treatment with monoclonal antibodies reacting with the extracellular ligand binding domain of EGFR. An additional level of regulation of prostaglandin signaling is provided by prostaglandin dehydrogenase (PGDH), an enzyme responsible for biochemical inactivation of prostaglandins. PGDH is downregulated in colorectal cancer, serves a tumor suppressor function and provides a possible Cox-2–independent way to target PGE$_2$ to inhibit cancer progression [142].

Recent data suggest an interplay of Cox-2 and Wnt signaling cascades in colorectal cancer [143]. Gα subunits of G proteins coupled with prostanoid receptors and Akt, stimulated as a result of EGFR transactivation, cooperate to cause the accumulation of β-catenin (member of canonical Wnt pathway) in the nucleus and cause the corresponding transcriptional activation [143]. Noteworthy crosstalk between Cox-2 and Wnt pathways offers novel targets for the development of anticancer drugs.

6.6 Expression of ABC Transporters

The ATP-binding cassette (ABC) family of transporter molecules, which require hydrolysis of ATP to run the transport mechanism, contains over 200 members, of which 48 are of human origin [144]. Expression of such ABC transporters as multidrug resistance protein 1 (MDR1, also termed P-glycoprotein and ABCB1), multidrug resistance (-associated) transporter 1 (MRP1 or ABCC1), and ABCG2 contributes to multidrug resistance phenotype which has been a major obstacle to the effective treatment of cancer [145, 146]. The cellular overproduction of multidrug transporter MDR1 confers upon cancer cells the ability to resist lethal doses of certain cytotoxic drugs by actively pumping the drugs out of the cells and thus reducing their cytotoxicity [147, 148]. MDR1 mediates active extrusion of such unrelated compounds as vinca alkaloids, anthracyclines, steroids, cyclosporines, and miscellaneous hydrophobic organic cations from the cell [149]. To enhance the anticancer efficacy of chemotherapeutic drugs against multidrug-resistant cancers it is important to either suppress the drug efflux mediated by ABC transporters or inhibit their transcriptional upregulation [150]. Even though the search for selective and potent inhibitors of MDR transporters is being actively conducted by the pharmaceutical industry [151], the metabolism of these inhibitors may have considerable implications for their effect on drug transport and their potential for treatment of drug resistance and enhancement of the effectiveness of the treatment of patients with drug-resistant cancer [152]. Pharmacologic intervention in upregulation of ABC transporters is an attractive alternative approach to increase effectiveness of anticancer drugs.

Resistance against etoposide, doxorubicin (also termed Adriamycin) and vincristine was reversed in human prostate cancer cell lines with leukotriene D4 antagonists MK-571 and zafirlukast. Notably, the effect of LD4 antagonists was associated with inhibition of multidrug resistance-associated protein 1 (MRP1) function [153].

It is generally accepted that regulation of MDR1 function is maintained via the regulation of MDR1 expression and prostanoids have emerged as a class of molecules capable of triggering cellular MDR1 overproduction [109]. There are several compelling data which link prostanoids with regulation of MDR1 expression: (1) products of Cox-2 enzymatic activity (PGE_2 and $PGF_{2\alpha}$) added directly to the culture medium of primary rat hepatocytes upregulated intrinsic mdr1b mRNA expression and MDR1-dependent transport activity, whereas structurally different cyclooxygenase inhibitors (indomethacin, meloxicam, NS398) mediated inhibition of EGF-induced MDR1 mRNA overexpression [154]; (2) adenovirus-mediated transfer of Cox-2 cDNA into glomerular mesangial cells resulted in increased expression of MDR1 mRNA and functional protein [155]; and (3) immunohistochemical analysis of human breast tumor specimen and human ovarian carcinomas revealed a strong correlation between expression of Cox-2 and MDR1 [156, 157]. A higher overall MDR1 immunoreactivity score is typical for Cox-2 positive breast cancer cases [158].

A selective Cox-2 inhibitor reduced MDR1 expression and function in medullary thyroid carcinoma, a highly chemoresistant malignant neoplasia deriving from parafollicular C cells, by a mechanism involving cyclooxygenase products other than

PGE_2[159]. In summary, an unfavorable prognostic and predictive significance of immunohistochemical estimation of Cox-2 and MDR1 was established in a number of human cancers advocating evaluation of coxibs as a supporting treatment in chemotherapy of chemoresistant cancers.

One of the MDR1 substrates is Adriamycin, an anthracycline chemotherapeutic agent with severe cardiotoxicity and nephrotoxicity, linked to its pro-apoptotic properties [160]. Our data suggest that induction of endogenous Cox-2 by cytokines results in stimulation of expression of endogenous MDR1 [161]. Because of the ability of Cox-2 inhibitors to attenuate this stimulation, it appears that stimulation depends upon Cox-2 enzymatic activity. It is of note, that MDR1 is a vital regulator of cell survival [162–165] and cells induced to overexpress MDR1 were reported to maintain resistance to cell death induced by such death stimuli as FasL, glucocorticoid hormones, tumor necrosis factor, and ultraviolet irradiation [162, 166]. Accordingly, Cox-2 expression and activity could be protective against apoptosis induced by these agents via upregulation of MDR1. Our recent data suggest that Cox-2 expression rescues kidney cells from apoptosis induced by Adriamycin due to increased cellular overproduction of MDR1 [161]. The mechanisms by which Cox-2, lipoxygenases and CYP4A enzymes protect cells from chemotherapeutic agents-induced apoptosis remain to be fully elucidated and are likely to consist of several anti-apoptotic components.

6.7 Stimulation of Pro-survival Signaling Pathways

Given that the ability of selective Cox-2 inhibitors to facilitate action of anticancer drugs and combat inflammation is accompanied by undesirable cardiotoxicity, it is important to uncover the molecular mechanisms of anti-apoptotic action of Cox-2 both in transformed and normal mammalian cells. Identification and targeting of alterations in apoptotic machinery, which enables cancer cells to escape chemotherapeutic agent-mediated apoptosis, might be a successful alternative approach to deal with chemotherapy-resistant tumors [102].

A number of mechanisms have been proposed to explain the anti-apoptotic effect of Cox-2 in cancer [167]. Proposed mechanisms include: (1) depletion

of arachidonic acid, which prevents the activation of neutral sphingomyelinase and, accordingly,
production of ceramide [168]; (2) regulation of
expression of the anti-apoptotic protein Bcl-2 [67,
76]; and (3) regulation of pro-survival phosphatidylinositol 3-kinase-Akt signaling pathway [77,
78]. All three proposed mechanisms can without
doubt play a significant role in protecting mammalian cells from apoptosis induced by a wide variety
of agents (Fig. 9.3). Nevertheless, none of them can
completely explain the Cox-2–mediated resistance
to apoptosis of cancer cells in all cases. For example, Cox-2–mediated resistance of LNCaP prostate
cancer cells to apoptosis is likely to be mediated by
alternative mechanisms for the following reasons:
(1) treatment of LNCaP cells with neutral or acidic
sphingomyelinase or addition of cell permeable
analogues of endogenous ceramide failed to induce
apoptosis although caused inhibition of cell proliferation [169]; (2) a survival signaling pathway
independent of phosphatidylinositol 3'-kinase and
Akt kinase was found in LNCaP cells [170]; and
(3) apoptosis of cancer induced by Cox-2 inhibitor
celecoxib was independent of Bcl-2 in LNCaP and
PC3 cells [77].

The phosphatidylinositol 3'-kinase (PI-3K)/Akt
signaling pathway is a key pro-survival signaling
cascade [171]. Within 5 minutes of AA addition to
prostate cancer cells PGE_2 expression was elevated
and phosphatidylinositol 3-kinase (PI-3K) significantly activated followed by activation of Akt at 30
minutes [172]. Eicosanoids, products of cyclooxygenases and lipoxygenases, activate PI-3K/Akt
pathway in bladder cancer cells [173], human lung
adenocarcinoma [78], human pancreatic cancer
cells [174]. Additional proof of regulation of
PI-3K/Akt by prostanoids comes from data with
selective Cox-2 inhibitors. Treatment of cancer
cells with selective Cox-2 inhibitors often results
in significant reduction of Akt phosphorylation
and signaling accompanied by increased apoptosis [175]. However, these data must be examined
taking into consideration that members of PI-3K/
Akt signaling cascade were shown to be Cox-2-
independent molecular targets of these inhibitors
[176]. Reversion of the inhibitor effect by addition
of exogenous prostanoids, a reliable confirmation of specificity of observed inhibitor effect is
only possible when the nature of Cox-2–derived
active eicosanoid product is known. In contrast,

data obtained with either enforced expression of
Cox-2 or with treatment of cells with exogenous
prostanoids tend to provide unequivocal evidence
for the Cox-2–dependent activation of pro-survival
pathways.

In our laboratory we used adenovirus mediated transfer of Cox-2 cDNA into prostate cancer
cells prior to analysis of sensitivity of cancer
cells to apoptotic agents. Since Cox-2 enzymatic
activity can induce transcriptional regulation of
MDR1 [161] and promote multidrug resistant
phenotype [109], we aimed to evaluate effect of
Cox-2 expression upon activity of an anticancer
agent that is not a substrate of MDR1. A natural
alkaloid sanguinarine derived from *Sanguinaria
canadensis* exhibits antimicrobial, antitumor
[177], antiangiogenic [178], and anti-inflammatory [179] properties and is not a substrate of
MDR1 [180]. Sanguinarine caused apoptosis of
immortalized human keratinocytes [181], human
epidermoid carcinoma [182], and human prostate
carcinoma cells [26]. Like many other antitumor
drugs, sanguinarine acts by interfering with the
function of DNA, inducing apoptosis; in contrast to many chemotherapeutic agents, it is also
effective against multidrug resistance [180]. For
this reason sanguinarine could be developed as
an anticancer drug against highly chemoresistant
malignant tumors [182]. Treatment of prostate
cancer cells with sanguinarine results in apoptotic
cells death accompanied with increased generation
of nitric oxide (NO) and superoxide radicals [183].
Although NO possesses both anti- and pro-apoptotic properties, NO and superoxide are known
to rapidly react to yield peroxynitrite, which
exhibits a wide array of tissue damaging effects,
ranging from DNA damage to protein nitration.
Overexpression of Cox-2 protected prostate cancer
cells from sanguinarine triggered apoptosis [183].
The protective effect of adenovirus-mediated Cox-
2 expression was nullified by exposure of cells
to the selective Cox-2 inhibitor celecoxib. It is
important that along with preventing apoptosis,
Cox-2 overexpression also inhibited NO formation in LNCaP cells, as detected by measurement
of the level of nitrites using Sievers Nitric Oxide
Analyzer or by detection of NO formation with
NO sensor using flow cytometry. Previously, the
ability of Cox-2 to counteract NO-mediated apoptotic cell death was demonstrated in two other cell

systems: in rat pheochromocytoma PC12 cells in which Cox-2 acted via modulation of expression of pro-survival gene capable of inhibiting production of NO [184], and in RAW 264.7 macrophages via regulation of cellular susceptibility toward NO [185]. Using rat pheochromocytoma PC12 cells engineered to stably express IPTG-inducible Cox-2 we demonstrated that trophic withdrawal apoptosis was markedly reduced in cells expressing Cox-2 when compared with control cells as assessed by annexin V staining, DNA laddering, caspase-3 activation assay, and Hoechst 33258 staining [75, 184]. Our studies suggest that generation of NO by neuronal nitric oxide synthase (NOS) is partially responsible for SAPK activation and apoptosis in PC12 cells following NGF withdrawal and that Cox-2–dependent overexpression of protein inhibitor of NOS (PIN, also termed dynein light chain, DLC) is the mechanism of cell protection from NO-mediated SAPK activation and apoptosis [184, 186].

PGE$_1$ (the pharmaceutical name Alprostadil) had anti-apoptotic effect on cultured neonatal rat cardiomyocytes [187]. PGE$_1$ increased effect of cisplatin intraperitoneal chemotherapy in rat peritoneal carcinomatosis model by decreasing the number of apoptotic renal cells and reducing the chance of cisplatin-induced renal failure [188]. In this study combination of PGE$_1$ with intraperitoneal cisplatin chemotherapy increased resistance to apoptosis of non-tumor cells and suggested a therapeutic benefit for patients with peritoneal carcinomatosis.

PGE$_2$ inhibits programmed cell death caused by selective Cox-2 inhibitors and induced Bcl-2 expression in human colon cancer (HCA-7) cells [30]. The Bcl-2 (B-cell CLL/lymphoma-2) family of proteins has been implicated as critical regulators in the intrinsic pathway of apoptosis and contains both anti-apoptotic and pro-apoptotic members. Multidomain anti-apoptotic members (Bcl-2, Bcl-xL, Bcl-w, Mcl-1, A1/Bfl-1, Boo/Diva, and NR-13) contain different combinations of four α-helices Bcl-2 homology domains (BH1-4). Protein–protein interactions mediated by these domains are essential for their pro-survival signaling [189]. Multidomain pro-apoptotic members (Bax, Bak, Bok/Mtd) contain BH1-3 domains and undergo a conformational change following an apoptotic stimuli and prior to their interaction with anti-apoptotic members of Bcl-2 family

[190]. BH3 only pro-apoptotic Bcl-2 proteins (Bid, Bad, Bim, Bik, Blk, Hrk, BNIP3, Nix, BMF, Noxa, and Puma) mainly act through inhibition of Bcl-2/Bcl-xL and activation of Bak and Bax [189, 190]. The "BH3-only" group of BCL-2 family was termed this way to emphasize that they have homology only within the minimal death domain, the BH3 amphipathic α helix 3. In cancer cells there is an apparent downregulation of apoptotic signaling, accompanied by increased expression of pro-apoptotic Bcl-2 proteins and decrease or inactivation of pro-apoptotic members of Bcl-2 family. It is likely that this balance is at least partially under control of eicosanoids. The bcl-2 gene expression increases during development of esophageal adenocarcinoma, and correlation was found with increase in Cox-2 gene expression [191]. PGE$_2$ induced Bcl-2 expression in squamous cell carcinoma [192] and epithelial ovarian cancer cells [193].

In spite of the different mechanisms by which anticancer drugs camptothecin, etoposide, staurosporine, 2-chloro-2'-deoxyadenosine, and nimesulide induce apoptosis, they all cause Bax activation and subcellular translocations from the cytosol to organellar membranes in human colon adenocarcinoma cells [194]. PGE$_2$ derivatives by signaling through the E prostaglandin receptor EP$_2$ were able to suppress Bax translocation to the mitochondrial membrane and prevent radiation-induced epithelial apoptosis [195].

7   Concluding Remarks

Resistance of cancer cells to chemotherapeutic agents, the major obstacle for successful chemotherapy, is the result of multiple processes regulating availability of anticancer drugs, survival of cancer cells, and status of surrounding normal cells. Eicosanoids, lipid mediators, play an active role in conferring the resistance of cancer cells to anticancer drugs. Combination therapy strategies that aim to inhibit pro-survival pathways in cancer cells and at the same time prevent of the overproduction of eicosanoids are most likely to succeed. These combination strategies are currently being tested in a number of ongoing clinical trials. Even though prostanoids, products of cyclooxygenase activity, are the most potent regulators of cancer cell survival among eicosanoids, leukotriene receptor antagonists are worth assessing for a therapeutic role in treatment of inflammatory conditions, including cancer [196]. Inhibitors of HETES and EETs generation and signaling are largely untested in major types of cancer and could provide effective alternative strategies for cancer treatment.

Acknowledgments. This work was supported by NIH grants DK 041684 and HL 022563.

References

1. Hengartner MO. The biochemistry of apoptosis. Nature 2000, 407:770–776.
2. Kam PC, Ferch NI. Apoptosis: mechanisms and clinical implications. Anaesthesia 2000, 55:1081–1093.
3. Marnett LJ, Rowlinson SW, Goodwin DC, et al. Arachidonic acid oxygenation by COX-1 and COX-2. Mechanisms of catalysis and inhibition. J Biol Chem 1999, 274:22903–22906.
4. Smith WL, DeWitt DL, Garavito RM. Cyclooxygenases: structural, cellular, and molecular biology. Annu Rev Biochem 2000, 69:145–182.
5. Thuresson ED, Lakkides KM, Rieke CJ, et al. Prostaglandin endoperoxide H synthase-1: the functions of cyclooxygenase active site residues in the binding, positioning, and oxygenation of arachidonic acid. J Biol Chem 2001, 276:10347–10357.
6. Silva PJ, Fernandes PA, Ramos MJ. A theoretical study of radical-only and combined radical/carbocationic mechanisms of arachidonic acid cyclooxygenation by prostaglandin H synthase. Theor Chem Acc 2007, 110:345–351.
7. Chubb AJ, Fitzgerald DJ, Nolan KB, et al. The productive conformation of prostaglandin G2 at the peroxidase site of prostaglandin endoperoxide H synthase: docking, molecular dynamics, and site-directed mutagenesis studies. Biochemistry 2006, 45:811–820.
8. Yuan C, Rieke CJ, Rimon G,. Partnering between monomers of cyclooxygenase-2 homodimers. Proc Natl Acad Sci U S A 2006, 103:6142–6147.
9. Helliwell RJ, Adams LF, Mitchell MD. Prostaglandin synthases: recent developments and a novel hypothesis. Prostaglandins Leukot Essent Fatty Acids 2004, 70:101–113.
10. Afek A, Zurgil N, Bar-Dayan Y, et al. Overexpression of 15-lipoxygenase in the vascular endothelium is associated with increased thymic apoptosis in LDL receptor-deficient mice. Pathobiology 2004, 71:261–266.
11. DuBois RN, Abramson SB, Crofford L. Cyclooxygenase in biology and disease. FASEB J 1998, 12:1063–1073.
12. Harris RC, Breyer MD. Physiological regulation of cyclooxygenase-2 in the kidney.
13. Harris RC, McKanna JA, Akai Y, et al. Cyclooxygenase-2 is associated with the macula densa of rat kidney and increases with salt restriction. J Clin Invest 1994, 94:2504–2510.
14. McGinty A, Foschi M, Chang YW, et al. Induction of prostaglandin endoperoxide synthase 2 by mitogen-activated protein kinase cascades. Biochem J 2000, 352(Pt 2):419–424.
15. Bell PD, Lapointe JY, Peti-Peterdi J. Macula densa cell signaling. Annu Rev Physiol 2003, 65:481–500.
16. Tsatsanis C, Androulidaki A, Venihaki M, et al. Signalling networks regulating cyclooxygenase-2. Int J Biochem Cell Biol 2006, 38:1654–1661.
17. Dannenberg AJ, Lippman SM, Mann JR, et al. Cyclooxygenase-2 and epidermal growth factor receptor: pharmacologic targets for chemoprevention. J Clin Oncol 2005, 23:254–266.
18. Arun B, Goss P. The role of COX-2 inhibition in breast cancer treatment and prevention. Semin Oncol 2004, 31:22–29.
19. Riedl K, Krysan K, Pold M, et al. Multifaceted roles of cyclooxygenase-2 in lung cancer. Drug Resist Updat 2004, 7:169–184.
20. Jendrossek V, Handrick R. Membrane targeted anticancer drugs: potent inducers of apoptosis and putative radiosensitisers. Curr Med Chem Anti-Canc Agents 2003, 3:343–353.
21. Kawanishi S, Hiraku Y. Amplification of anticancer drug-induced DNA damage and apoptosis by DNA-binding compounds. Curr Med Chem Anti-Canc Agents 2004, 4:415–419.
22. Moore BC, Simmons DL. COX-2 inhibition, apoptosis, and chemoprevention by nonsteroidal anti-inflammatory drugs. Curr Med Chem 2000, 7:1131–1144.
23. Subbaramaiah K, Zakim D, Weksler BB, et al. Inhibition of cyclooxygenase: a novel approach to cancer prevention. Proc Soc Exp Biol Med 1997, 216:201–210.
24. Thun MJ, Henley SJ, Patrono C. Nonsteroidal anti-inflammatory drugs as anticancer agents: mechanistic, pharmacologic, and clinical issues. J Natl Cancer Inst 2002, 94:252–266.
25. Avis I, Hong SH, Martinez A, et al. Five-lipoxygenase inhibitors can mediate apoptosis in human breast cancer cell lines through complex eicosanoid interactions. FASEB J 2001, 15:2007–2009.
26. Adhami VM, Aziz MH, Reagan-Shaw SR, et al. Sanguinarine causes cell cycle blockade and apoptosis of human prostate carcinoma cells via modulation of cyclin kinase inhibitor-cyclin-cyclin-dependent kinase machinery. Mol Cancer Ther 2004, 3:933–940.

27. Fitzgerald GA. Coxibs and cardiovascular disease. N Engl J Med 2004, 351:1709–1711.

28. Furberg CD, Psaty BM, Fitzgerald GA. Parecoxib, valdecoxib, and cardiovascular risk. Circulation 2005, 111:249.

29. Hanif R, Pittas A, Feng Y, et al. Effects of nonsteroidal anti-inflammatory drugs on proliferation and on induction of apoptosis in colon cancer cells by a prostaglandin-independent pathway. Biochem Pharmacol 1996, 52:237–245.

30. Sheng H, Shao J, Morrow JD, et al. Modulation of apoptosis and Bcl-2 expression by prostaglandin E2 in human colon cancer cells. Cancer Res 1998, 58:362–366.

31. Bortuzzo C, Hanif R, Kashfi K, et al. The effect of leukotrienes B and selected HETEs on the proliferation of colon cancer cells. Biochim Biophys Acta 1996, 1300:240–246.

32. Funk CD, Chen XS, Johnson EN, et al. Lipoxygenase genes and their targeted disruption. Prostaglandins Other Lipid Mediat 2002, 68–69:303–312.

33. Yu Z, Schneider C, Boeglin WE, et al. The lipoxygenase gene ALOXE3 implicated in skin differentiation encodes a hydroperoxide isomerase. Proc Natl Acad Sci U S A 2003, 100:9162–9167.

34. Shureiqi I, Lippman SM. Lipoxygenase modulation to reverse carcinogenesis. Cancer Res 2001, 61:6307–6312.

35. Shureiqi I, Wu Y, Chen D, et al. The critical role of 15-lipoxygenase-1 in colorectal epithelial cell terminal differentiation and tumorigenesis. Cancer Res 2005, 65:11486–11492.

36. Shureiqi I, Zuo X, Broaddus R, et al. 2006. The transcription factor GATA-6 is overexpressed in vivo and contributes to silencing 15-LOX-1 in vitro in human colon cancer. FASEB J 2006, 21:743–753.

37. Jiang WG, Watkins G, Douglas-Jones A, et al. Reduction of isoforms of 15-lipoxygenase (15-LOX)-1 and 15-LOX-2 in human breast cancer. Prostaglandins Leuk Essent Fatty Acids 2006, 74:235–245.

38. Roman RJ. P-450 metabolites of arachidonic acid in the control of cardiovascular function. Physiol Rev 2002, 82:131–185.

39. Miyata N, Roman RJ. Role of 20-hydroxyeicosatetraenoic acid (20-HETE) in vascular system. J Smooth Muscle Res 2005, 41:175–193.

40. Nithipatikom K, Isbell MA, See WA, et al. Elevated 12- and 20-hydroxyeicosatetraenoic acid in urine of patients with prostatic diseases. Cancer Lett 2006, 233:219–225.

41. Nebert DW, Dalton TP. The role of cytochrome P450 enzymes in endogenous signalling pathways and environmental carcinogenesis. Nat Rev Cancer 2006, 6:947–960.

42. Schuster VL. Prostaglandin transport. Prostaglandins Other Lipid Mediat 2002, 68–69:633–647.

43. Narumiya S, Sugimoto Y, Ushikubi F. Prostanoid receptors: structures, properties, and functions. Physiol Rev 1999, 79:1193–1226.

44. Kobayashi T, Narumiya S. Function of prostanoid receptors: studies on knockout mice. Prostaglandins Other Lipid Mediat 2002, 68–69:557–573.

45. Satoh T, Moroi R, Aritake K, et al. Prostaglandin D2 plays an essential role in chronic allergic inflammation of the skin via CRTH2 receptor. J Immunol 2006, 177:2621–2629.

46. Wang D, DuBois RN. Prostaglandins and cancer. Gut 2006, 55:115–122.

47. Erickson BA, Longo WE, Panesar N, et al. The effect of selective cyclooxygenase inhibitors on intestinal epithelial cell mitogenesis. J Surg Res 1999, 81:101–107.

48. Rizzo G, Fiorucci S. PPARs and other nuclear receptors in inflammation. Curr Opin Pharmacol 2006, 6:421–427.

49. Scher JU, Pillinger MH. 15d-PGJ2: the anti-inflammatory prostaglandin? Clin Immunol 2005, 114:100–109.

50. Osher E, Weisinger G, Limor R, et al. The 5 lipoxygenase system in the vasculature: emerging role in health and disease. Mol Cell Endocrinol 2006, 252:201–206.

51. Sun CW, Falck JR, Harder DR, et al. Role of tyrosine kinase and PKC in the vasoconstrictor response to 20-HETE in renal arterioles. Hypertension 1999, 33:414–418.

52. Dulin NO, Sorokin A, Douglas JG. Arachidonate-induced tyrosine phosphorylation of epidermal growth factor receptor and Shc-Grb2-Sos association. Hypertension 1998, 32:1089–1093.

53. Escalante B, Erlij D, Falck JR, et al. Cytochrome P-450 arachidonate metabolites affect ion fluxes in rabbit medullary thick ascending limb. Am J Physiol 1994, 266:C1775–C1782.

54. Yu M, Lopez B, Dos Santos EA, et al. Effects of 20-HETE on Na+ transport and Na+-K+-ATPase activity in the thick ascending loop of Henle. Am J Physiol 2007, 251(4005):799–802.

55. Egan K, Fitzgerald GA. Eicosanoids and the vascular endothelium. Handb Exp Pharmacol 2006, 189–211.

56. Baron JA, Sandler RS, Bresalier RS, et al. A randomized trial of rofecoxib for the chemoprevention of colorectal adenomas. Gastroenterology 2006, 131:1674–1682.

57. Lianos EA, Bresnahan BA, Pan C. Mesangial cell immune injury. Synthesis, origin, and role of eicosanoids. J Clin Invest 1991 88:623–631.

58. Miller SB. Prostaglandins in health and disease: an overview. Semin Arthritis Rheum 2006, 36:37–49.

59. DuBois RN, Eberhart CF, Williams CS. Introduction to eicosanoids and the gastroenteric tract. Gastroenterol Clin North Am 1996, 25:267–277.

60. Aggarwal BB, Shishodia S, Sandur SK, et al. Inflammation and cancer: how hot is the link? Biochem Pharmacol 2006, 72:1605–1621.

61. Calder PC. Polyunsaturated fatty acids and inflammation. Biochem Soc Trans 2005, 33:423–427.

62. Rajakariar R, Yaqoob MM, Gilroy DW. COX-2 in inflammation and resolution. Mol Interv 2006, 6:199–207.

63. Gualde N, Harizi H. Prostanoids and their receptors that modulate dendritic cell–mediated immunity. Immunol Cell Biol 2004, 82:353–360.

64. Narumiya S. Prostanoids in immunity: roles revealed by mice deficient in their receptors. Life Sci 2003, 74:391–395.

65. Angeli V, Staumont D, Charbonnier AS, et al. Activation of the D prostanoid receptor 1 regulates immune and skin allergic responses. J Immunol 2004, 172:3822–3829.

66. Eschwege P, de Ledinghen V, Camilli T, et al. [Arachidonic acid and prostaglandins, inflammation and oncology]. Presse Med 2001, 30:508–510.

67. Tsujii M, DuBois RN. Alterations in cellular adhesion and apoptosis in epithelial cells overexpressing prostaglandin endoperoxide synthase 2. Cell 1995, 83:493–501.

68. von Knethen A, Brune B. Attenuation of macrophage apoptosis by the cAMP-signaling system. Mol Cell Biochem 2000, 212:35–43.

69. Hao CM, Komhoff M, Guan Y, et al. Selective targeting of cyclooxygenase-2 reveals its role in renal medullary interstitial cell survival. Am J Physiol 1999, 277:F352–F359.

70. Liou JY, Ellent DP, Lee S, et al. Cyclooxygenase-2 derived PGE2 protects mouse embryonic stem cells from apoptosis. Stem Cells, 2007, 25:1096–1103.

71. Badawi AF. The role of prostaglandin synthesis in prostate cancer. BJU Int 200, 85:451–462.

72. Kamijo T, Sato T, Nagatomi Y, et al. Induction of apoptosis by cyclooxygenase-2 inhibitors in prostate cancer cell lines. Int J Urol 2001, 8:S35–S39.

73. Souza RF, Shewmake K, Beer DG, et al. Selective inhibition of cyclooxygenase-2 suppresses growth and induces apoptosis in human esophageal adenocarcinoma cells. Cancer Res 2000, 60:5767–5772.

74. Ding XZ, Tong WG, Adrian TE. Blockade of cyclooxygenase-2 inhibits proliferation and induces apoptosis in human pancreatic cancer cells. Anticancer Res 2000, 20:2625–2631.

75. McGinty A, Chang YW, Sorokin A, et al. Cyclooxygenase-2 expression inhibits trophic withdrawal apoptosis in nerve growth factor-differentiated PC12 cells. J Biol Chem 2000, 275:12095–12101.

76. Liu XH, Yao S, Kirschenbaum A, et al. NS398, a selective cyclooxygenase-2 inhibitor, induces apoptosis and down-regulates bcl-2 expression in LNCaP cells. Cancer Res 1998, 58:4245–4249.

77. Hsu AL, Ching TT, Wang DS, et al. The cyclooxygenase-2 inhibitor celecoxib induces apoptosis by blocking Akt activation in human prostate cancer cells independently of Bcl-2. J Biol Chem 2000, 275:11397–11403.

78. Lin MT, Lee RC, Yang PC, et al. Cyclooxygenase-2 inducing Mcl-1-dependent survival mechanism in human lung adenocarcinoma CL1.0 cells. Involvement of phosphatidylinositol 3-kinase/Akt pathway. J Biol Chem 2001, 276:48997–49002.

79. Ohd JF, Wikstrom K, Sjolander A. Leukotrienes induce cell-survival signaling in intestinal epithelial cells. Gastroenterology 2000, 119:1007–1018.

80. Paruchuri S, Mezhybovska M, Juhas M, et al. Endogenous production of leukotriene D4 mediates autocrine survival and proliferation via CysLT1 receptor signalling in intestinal epithelial cells. Oncogene 2006, 25:6660–6665.

81. Attiga FA, Fernandez PM, Weeraratna AT, et al. Inhibitors of prostaglandin synthesis inhibit human prostate tumor cell invasiveness and reduce the release of matrix metalloproteinases. Cancer Res 2000, 60:4629–4637.

82. Shureiqi I, Chen D, Lotan R, et al. 15-Lipoxygenase-1 mediates nonsteroidal anti-inflammatory drug-induced apoptosis independently of cyclooxygenase-2 in colon cancer cells. Cancer Res 2000, 60:6846–6850.

83. Goto HG, Nishizawa Y, Katayama H, et al. Induction of apoptosis in an estrogen-responsive mouse Leydig tumor cell by leukotriene. Oncol Rep 2007, 17:225–232.

84. Guo M, Roman RJ, Fenstermacher JD, et al. 9L Gliosarcoma cell proliferation and tumor growth in rats are suppressed by N-hydroxy-N'-(4-butyl-2-methyl-phenol) formamidine (HET0016), a selective inhibitor of CYP4A. J Pharmacol Exp Ther 2006, 317:97–108.

85. Hussain T, Gupta S, Mukhtar H. Cyclooxygenase-2 and prostate carcinogenesis. Cancer Lett 2003, 191:125–135.

86. Kirschenbaum A, Liu X, Yao S, et al. The role of cyclooxygenase-2 in prostate cancer. Urology 2001, 58:127–131.

87. Khan O, Hensby CN, Williams G. Prostacyclin in prostatic cancer: a better marker than bone scan or serum acid phosphatase? Br J Urol 1982, 54:26–31.

88. Liu XH, Kirschenbaum A, Yao S, et al. Inhibition of cyclooxygenase-2 suppresses angiogenesis and the growth of prostate cancer in vivo. J Urol 2000, 164:820–825.

89. Nithipatikom K, Laabs ND, Isbell MA, et al. Liquid chromatographic–mass spectrometric determination of cyclooxygenase metabolites of arachidonic acid in cultured cells. J Chromatogr B Analyt Technol Biomed Life Sci 2003, 785:135–145.

90. Tremblay C, Dore M, Bochsler PN, et al. Induction of prostaglandin G/H synthase-2 in a canine model of spontaneous prostatic adenocarcinoma. J Natl Cancer Inst 1999, 91:1398–1403.

91. Gupta S, Srivastava M, Ahmad N, et al. Over-expression of cyclooxygenase-2 in human prostate adenocarcinoma. Prostate 2000, 42:73–78.

92. Eisinger AL, Prescott SM, Jones DA, et al. The role of cyclooxygenase-2 and prostaglandins in colon cancer. Prostaglandins Other Lipid Mediat 2007, 82:147–154.

93. Wendum D, Masliah J, Trugnan G, et al. Cyclooxygenase-2 and its role in colorectal cancer development. Virchows Arch 2004, 445:327–333.

94. Eberhart CE, Coffey RJ, Radhika A, et al. Up-regulation of cyclooxygenase 2 gene expression in human colorectal adenomas and adenocarcinomas. Gastroenterology 1994, 107:1183–1188.

95. Oshima M, Dinchuk JE, Kargman SL, et al. Suppression of intestinal polyposis in Apc delta716 knockout mice by inhibition of cyclooxygenase 2 (COX-2). Cell 1996, 87:803–809.

96. Arber N, Eagle CJ, Spicak J, Iet al. 2006. Celecoxib for the prevention of colorectal adenomatous polyps. N Engl J Med 2006, 355:885–895.

97. Bertagnolli MM, Eagle CJ, Zauber AG, et al. Celecoxib for the prevention of sporadic colorectal adenomas. N Engl J Med 2006, 355:873–884.

98. Solomon SD, Pfeffer MA, McMurray JJ, et al. Effect of celecoxib on cardiovascular events and blood pressure in two trials for the prevention of colorectal adenomas. Circulation 2006, 114:1028–1035.

99. Bresalier RS, Sandler RS, Quan H, et al. Cardiovascular events associated with rofecoxib in a colorectal adenoma chemoprevention trial. N Engl J Med 2005, 352:1092–1102.

100. Solomon SD, McMurray JJ, Pfeffer MA, et al. Cardiovascular risk associated with celecoxib in a clinical trial for colorectal adenoma prevention. N Engl J Med 2005, 352:1071–1080.

101. Lanas A, Baron JA, Sandler RS, et al. Peptic ulcer and bleeding events associated with rofecoxib in a 3-year colorectal adenoma chemoprevention trial. Gastroenterology, 2006, 23(1):44–47.

102. Rodriguez-Nieto S, Zhivotovsky B. Role of alterations in the apoptotic machinery in sensitivity of cancer cells to treatment. Curr Pharm Des 2006, 12:4411–4425.

103. Oh WK, Kantoff PW. Management of hormone refractory prostate cancer: current standards and future prospects. J Urol 1998, 160:1220–1229.

104. Crawford ED, Eisenberger MA, McLeod DG, et al. A controlled trial of leuprolide with and without flutamide in prostatic carcinoma. N Engl J Med 1989, 321:419–424.

105. Raghavan D. Non-hormone chemotherapy for prostate cancer: principles of treatment and application to the testing of new drugs. Semin Oncol 1988, 15:371–389.

106. Cusack JC Jr. Overcoming antiapoptotic responses to promote chemosensitivity in metastatic colorectal cancer to the liver. Ann Surg Oncol 2003, 10:852–862.

107. Sheng H, Shao J, Washington MK, et al. Prostaglandin E2 increases growth and motility of colorectal carcinoma cells. J Biol Chem 2001 276:18075–18081.

108. Krysan K, Reckamp KL, Sharma S, et al. The potential and rationale for COX-2 inhibitors in lung cancer. Anticancer Agents Med Chem 2006, 6:209–220.

109. Sorokin A. Cyclooxygenase-2: potential role in regulation of drug efflux and multidrug resistance phenotype. Curr Pharmacol Des 2004, 10:647–657.

110. Gately S. The contributions of cyclooxygenase-2 to tumor angiogenesis. Cancer Metastasis Rev 2000, 19:19–27.

111. Dempke W, Rie C, Grothey A, et al. Cyclooxygenase-2: a novel target for cancer chemotherapy? J Cancer Res Clin Oncol 2001, 127:411–417.

112. Tyurina YY, Kapralov AA, Jiang J, et al. Oxidation and cytotoxicity of 6-OHDA are mediated by reactive intermediates of COX-2 overexpressed in PC12 cells. Brain Res 2006, 1093:71–82.

113. Grosch S, Tegeder I, Niederberger E, et al. COX22 independent induction of cell cycle arrest and apoptosis in colon cancer cells by the selective COX22 inhibitor celecoxib. FASEB J 2001, 15:2742–2744.

114. Fosslien E. Review: molecular pathology of cyclooxygenase-2 in cancer-induced angiogenesis. Ann Clin Lab Sci 2001, 31:325–348.

115. Tanaka S, Akaike T, Wu J, et al. Modulation of tumor-selective vascular blood flow and extravasation by the stable prostaglandin 12 analogue beraprost sodium. J Drug Target 2003, 11:45–52.

116. Bostanoglu S, Dincer S, Keskin A, et al. Beneficial effect of Iloprost on impaired colonic anastomotic healing induced by intraperitoneal 5-fluorouracil infusion. Dis Colon Rectum 1998, 41:642–648.

117. Lawson JA, Adams WJ, Morris DL. The effect of misoprostol on colon cancer. Aust N Z J Surg 1994, 64:197–201.

118. Di Popolo A, Memoli A, Apicella A, et al. IGF-II/IGF-I receptor pathway up-regulates COX-2 mRNA expression and PGE2 synthesis in Caco-2 human colon carcinoma cells. Oncogene 2000, 19:5517–5524.

119. Han C, Wu T. Cyclooxygenase-2-derived prostaglandin E2 promotes human cholangiocarcinoma cell growth and invasion through EP1 receptor-mediated activation of the epidermal growth factor

receptor and Akt. J Biol Chem 2005, 280:24053–24063.

120. Leahy KM, Ornberg RL, Wang Y, et al. Cyclooxygenase-2 inhibition by celecoxib reduces proliferation and induces apoptosis in angiogenic endothelial cells in vivo. Cancer Res 2002, 62:625–631.

121. Stenson WF. Prostaglandins and epithelial response to injury. Curr Opin Gastroenterol 2007, 23:107–110.

122. Cybulsky AV, Goodyer PR, Cyr MD, et al. Eicosanoids enhance epidermal growth factor receptor activation and proliferation in glomerular epithelial cells. Am J Physiol 1992, 262:F639–F646.

123. Jabbour HN, Boddy SC. Prostaglandin E2 induces proliferation of glandular epithelial cells of the human endometrium via extracellular regulated kinase 1/2-mediated pathway. J Clin Endocrinol Metab 2003, 88:4481–4487.

124. Floege J, Topley N, Resch K. Regulation of mesangial cell proliferation. Am J Kidney Dis 1991, 17:673–676.

125. Guo M, Roman RJ, Falck JR, et al. Human U251 glioma cell proliferation is suppressed by HET0016 [N-hydroxy-N'-(4-butyl-2-methylphenyl)formamidine], a selective inhibitor of CYP4A. J Pharmacol Exp Ther 2005, 315:526–533.

126. Xu L, Han C, Lim K, et al. - Cross-talk between peroxisome proliferator-activated receptor delta and cytosolic phospholipase A(2)alpha/cyclooxygenase-2/prostaglandin E(2) signaling pathways in human hepatocellular carcinoma cells. Cancer Res 2006, 66:11859–11868.

127. Nie D, Che M, Grignon D, et al. Role of eicosanoids in prostate cancer progression. Cancer Metastasis Rev 2001, 20:195–206.

128. Adam L, Mazumdar A, Sharma T, et al. A three-dimensional and temporo-spatial model to study invasiveness of cancer cells by heregulin and prostaglandin E2. Cancer Res 2001, 61:81–87.

129. Ono M, Kuwano M. Molecular mechanisms of epidermal growth factor receptor (EGFR) activation and response to gefitinib and other EGFR-targeting drugs. Clin Cancer Res 2006, 12:7242–7251.

130. Shchemelinin I, Sefc L, Necas E. Protein kinase inhibitors. Folia Biol (Praha) 2006, 52:137–148.

131. Sergina NV, Rausch M, Wang D, et al. Escape from HER-family tyrosine kinase inhibitor therapy by the kinase-inactive HER3. Nature 2007, 445:437–441.

132. Han C, Michalopoulos GK, Wu T. Prostaglandin E2 receptor EP1 transactivates EGFR/MET receptor tyrosine kinases and enhances invasiveness in human hepatocellular carcinoma cells. J Cell Physiol 2006, 207:261–270.

133. Ding YB, Shi RH, Tong JD, et al. PGE2 up-regulates vascular endothelial growth factor expression in MKN28 gastric cancer cells via epidermal growth factor receptor signaling system. Exp Oncol 2005, 27:108–113.

134. Pai R, Nakamura T, Moon WS, et al. Prostaglandins promote colon cancer cell invasion; signaling by cross-talk between two distinct growth factor receptors. FASEB J 2003, 17:1640–1647.

135. Pai R, Soreghan B, Szabo IL, et al. Prostaglandin E2 transactivates EGF receptor: a novel mechanism for promoting colon cancer growth and gastrointestinal hypertrophy. Nat Med 2002, 8:289–293.

136. Mendez M, LaPointe MC. PGE2-induced hypertrophy of cardiac myocytes involves EP4 receptor-dependent activation of p42/44 MAPK and EGFR transactivation. Am J Physiol Heart Circ Physiol 2005, 288:H2111–H2117.

137. Sales KJ, Maudsley S, Jabbour HN. Elevated prostaglandin EP2 receptor in endometrial adenocarcinoma cells promotes vascular endothelial growth factor expression via cyclic 3',5'-adenosine monophosphate-mediated transactivation of the epidermal growth factor receptor and extracellular signal-regulated kinase 1/2 signaling pathways. Mol Endocrinol 2004, 18:1533–1545.

138. Fischer OM, Hart S, Geschwind A, et al. EGFR signal transactivation in cancer cells. Biochem Soc Trans 2003, 31:1203–1208.

139. Reckamp KL, Krysan K, Morrow JD, et al. A phase I trial to determine the optimal biological dose of celecoxib when combined with erlotinib in advanced non-small cell lung cancer. Clin Cancer Res 2006, 12:3381–3388.

140. Buchanan FG, Wang D, Bargiacchi F, et al. Prostaglandin E2 regulates cell migration via the intracellular activation of the epidermal growth factor receptor. J Biol Chem 2003, 278:35451–35457.

141. Buchanan FG, Gorden DL, Matta P, et al. Role of beta-arrestin 1 in the metastatic progression of colorectal cancer. Proc Natl Acad Sci U S A 2006, 103:1492–1497.

142. Mann JR, Backlund MG, Buchanan FG, et al. Repression of prostaglandin dehydrogenase by epidermal growth factor and snail increases prostaglandin E2 and promotes cancer progression. Cancer Res 2006, 66:6649–6656.

143. Buchanan FG, DuBois RN. Connecting COX-2 and Wnt in cancer. Cancer Cell 2006, 9:6–8.

144. Borst P, Elferink RO. Mammalian ABC transporters in health and disease. Annu Rev Biochem 2002, 71:537–592.

145. Gottesman MM, Fojo T, Bates SE. Multidrug resistance in cancer: role of ATP-dependent transporters. Nat Rev Cancer 2002, 2:48–58.

146. Shabbits JA, Krishna R, Mayer LD. Molecular and pharmacological strategies to overcome multidrug

resistance. Expert Rev Anticancer Ther 2001, 1:585–594.

147. Johnstone RW, Ruefli AA, Smyth MJ. Multiple physiological functions for multidrug transporter P-glycoprotein? Trends Biochem Sci 2000, 25:1–6.

148. Sharom FJ. The P-glycoprotein efflux pump: how does it transport drugs? J Membr Biol 1997, 160:161–175.

149. Inui KI, Masuda S, Saito H. Cellular and molecular aspects of drug transport in the kidney. Kidney Int 2000, 58:944–958.

150. Pakunlu RI, Cook TJ, Minko T. Simultaneous modulation of multidrug resistance and antiapoptotic cellular defense by MDR1 and BCL-2 targeted antisense oligonucleotides enhances the anticancer efficacy of doxorubicin. Pharmacol Res 2003, 20:351–359.

151. Perez-Tomas R. Multidrug resistance: retrospect and prospects in anti-cancer drug treatment. Curr Med Chem 2006, 13:1859–1876.

152. Cnubben NH, Wortelboer HM, van Zanden JJ, et al. Metabolism of ATP-binding cassette drug transporter inhibitors: complicating factor for multidrug resistance. Expert Opin Drug Metab Toxicol 2005, 1:219–232.

153. van Brussel JP, Oomen MA, Vossebeld PJ, et al. Identification of multidrug resistance-associated protein 1 and glutathione as multidrug resistance mechanisms in human prostate cancer cells: chemosensitization with leukotriene D4 antagonists and buthionine sulfoximine. BJU Int 2004, 93:1333–1338.

154. Ziemann C, Schafer D, Rudell G, et al. The cyclooxygenase system participates in functional mdr1b overexpression in primary rat hepatocyte cultures. Hepatology 2002, 35:579–588.

155. Patel VA, Dunn MJ, Sorokin A. Regulation of MDR-1 (P-glycoprotein) by cyclooxygenase-2. J Biol Chem 2002, 277:38915–38920.

156. Ratnasinghe D, Daschner PJ, Anver MR, et al. Cyclooxygenase-2, P-glycoprotein-170 and drug resistance; is chemoprevention against multidrug resistance possible? Anticancer Res 2001, 21:2141–2147.

157. Surowiak P, Materna V, Denkert C, et al. Significance of cyclooxygenase 2 and MDR1/P-glycoprotein coexpression in ovarian cancers. Cancer Lett 2006, 235:272–280.

158. Surowiak P, Materna V, Matkowski R, et al. Relationship between the expression of cyclooxygenase 2 and MDR1/P-glycoprotein in invasive breast cancers and their prognostic significance. Breast Cancer Res 2005, 7:R862–R870.

159. Zatelli MC, Luchin A, Piccin D, et al. Cyclooxygenase-2 inhibitors reverse chemoresistance phenotype in medullary thyroid carcinoma by a permeability glycoprotein-mediated mechanism. J Clin Endocrinol Metab 2005, 90:5754–5760.

160. Kalyanaraman B, Joseph J, Kalivendi S, et al. Doxorubicin-induced apoptosis: implications in cardiotoxicity. Mol Cell Biochem 2002, 234–235:119–124.

161. Miller B, Patel VA, Sorokin A. Cyclooxygenase-2 rescues rat mesangial cells from apoptosis induced by Adriamycin via upregulation of multidrug resistance protein 1 (P-glycoprotein). J Am Soc Nephrol 2006, 17:977–985.

162. Gruol DJ, Bourgeois S. Expression of the mdr1 P-glycoprotein gene: a mechanism of escape from glucocorticoid-induced apoptosis. Biochem Cell Biol 1994, 72:561–571.

163. Johnstone RW, Ruefli AA, Tainton KM, et al. A role for P-glycoprotein in regulating cell death. Leuk Lymphoma 2000, 38:1–11.

164. Sakaeda T, Nakamura T, Hirai M, et al. MDR1 up-regulated by apoptotic stimuli suppresses apoptotic signaling. Pharmacol Res 2002, 19:1323–1329.

165. Thevenod F, Friedmann JM, Katsen AD, et al. Up-regulation of multidrug resistance P-glycoprotein via nuclear factor-kappaB activation protects kidney proximal tubule cells from cadmium- and reactive oxygen species-induced apoptosis. J Biol Chem 2000, 275:1887–1896.

166. Johnstone RW, Cretney E, Smyth MJ. P-glycoprotein protects leukemia cells against caspase-dependent, but not caspase-independent, cell death. Blood 1999, 93:1075–1085.

167. Cao Y, Prescott SM. Many actions of cyclooxygenase-2 in cellular dynamics and in cancer. J Cell Physiol 2002, 190:279–286.

168. Cao Y, Pearman AT, Zimmerman GA, et al. Intracellular unesterified arachidonic acid signals apoptosis. Proc Natl Acad Sci U S A 2000, 97:11280–11285.

169. Condorelli F, Canonico PL, Sortino MA. Distinct effects of ceramide-generating pathways in prostate adenocarcinoma cells. Br J Pharmacol 1999, 127:75–84.

170. Carson JP, Kulik G, Weber MJ. Antiapoptotic signaling in LNCaP prostate cancer cells: a survival signaling pathway independent of phosphatidylinositol 3′-kinase and Akt/protein kinase B. Cancer Res 1999, 59:1449–1453.

171. Vivanco I, Sawyers CL. The phosphatidylinositol 3-Kinase AKT pathway in human cancer. Nat Rev Cancer 2002, 2:489–501.

172. Hughes-Fulford M, Li CF, Boonyaratanakornkit J, et al. Arachidonic acid activates phosphatidylinositol 3-kinase signaling and induces gene expression in prostate cancer. Cancer Res 2006, 66:1427–1433.

173. Choi EM, Kwak SJ, Kim YM, et al. COX-2 inhibits anoikis by activation of the PI-3K/Akt pathway in human bladder cancer cells. Exp Mol Med 2005, 37:199–203.

174. Tong WG, Ding XZ, Talamonti MS, et al. LTB4 stimulates growth of human pancreatic cancer cells via MAPK and PI-3 kinase pathways. Biochem Biophys Res Commun 2005, 335:949–956.

175. Leng J, Han C, Demetris AJ, et al. Cyclooxygenase-2 promotes hepatocellular carcinoma cell growth through Akt activation: evidence for Akt inhibition in celecoxib-induced apoptosis. Hepatology 2003, 38:756–768.

176. Grosch S, Maier TJ, Schiffmann S, et al. Cyclooxygenase-2 (COX-2)-independent anticarcinogenic effects of selective COX-2 inhibitors. J Natl Cancer Inst 2006, 98:736–747.

177. Faddeeva MD, Beliaeva TN. [Sanguinarine and ellipticine cytotoxic alkaloids isolated from well-known antitumor plants. Intracellular targets of their action]. Tsitologiia 1997, 39:181–208.

178. Eun JP, Koh GY. Suppression of angiogenesis by the plant alkaloid, sanguinarine. Biochem Biophys Res Commun 2004, 317:618–624.

179. Lenfeld J, Kroutil M, Marsalek E, et al. Antiinflammatory activity of quaternary benzophenanthridine alkaloids from *Chelidonium majus*. Planta Med 1981, 43:161–165.

180. Ding Z, Tang SC, Weerasinghe P, et al. The alkaloid sanguinarine is effective against multidrug resistance in human cervical cells via bimodal cell death. Biochem Pharmacol 2002, 63:1415–1421.

181. Adhami VM, Aziz MH, Mukhtar H, et al. Activation of prodeath Bcl-2 family proteins and mitochondrial apoptosis pathway by sanguinarine in immortalized human HaCaT keratinocytes. Clin Cancer Res 2003, 9:3176–3182.

182. Ahmad N, Gupta S, Husain MM, et al. Differential antiproliferative and apoptotic response of sanguinarine for cancer cells versus normal cells. Clin Cancer Res 2000, 6:1524–1528.

183. Huh J, Liepins A, Zielonka J, et al. Cyclooxygenase 2 rescues LNCaP prostate cancer cells from sanguinarine-induced apoptosis by a mechanism involving inhibition of nitric oxide synthase activity. Cancer Res 2006, 66:3726–3736.

184. Chang YW, Jakobi R, McGinty A, et al. Cyclooxygenase 2 promotes cell survival by stimulation of dynein light chain expression and inhibition of neuronal nitric oxide synthase activity. Mol Cell Biol 2000, 20:8571–8579.

185. von Knethen A, Brune B. Cyclooxygenase-2: an essential regulator of NO-mediated apoptosis. FASEB J 1997, 11:887–895.

186. Miller B, Chang YW, Sorokin A. Cyclooxygenase 2 inhibits SAPK activation in neuronal apoptosis. Biochem Biophys Res Commun 2003, 300:884–888.

187. Ma XQ, Fu RF, Feng GQ, et al. Hypoxia-reoxygenation-induced apoptosis in cultured neonatal rat cardiomyocytes and the protective effect of prostaglandin E. Clin Exp Pharmacol Physiol 2005, 32:1124–1130.

188. Ikeguchi M, Maeta M, Kaibara N. Cisplatin combined with prostaglandin E1 chemotherapy in rat peritoneal carcinomatosis. Int J Cancer 2000, 88:474–478.

189. Kutuk O, Basaga H. Bcl-2 protein family: implications in vascular apoptosis and atherosclerosis. Apoptosis 2006, 11:1661–1675.

190. Danial NN, Korsmeyer SJ. Cell death: critical control points. Cell 2004, 116:205–219.

191. Shimizu D, Vallbohmer D, Kuramochi H, et al. Increasing cyclooxygenase-2 (cox-2) gene expression in the progression of Barrett's esophagus to adenocarcinoma correlates with that of Bcl-2. Int J Cancer 2006, 119:765–770.

192. Papa F, Scacco S, Vergari R, et al. Expression and subcellular distribution of Bcl-2 and BAX proteins in serum-starved human keratinocytes and mouth carcinoma epidermoid cultures. Life Sci 2003, 73:2865–2872.

193. Munkarah AR, Morris R, Baumann P, Get al. Effects of prostaglandin E(2) on proliferation and apoptosis of epithelial ovarian cancer cells. J Soc Gynecol Invest 2002, 9:168–173.

194. Godlewski MM, Motyl MA, Gajkowska B, et al. Subcellular redistribution of BAX during apoptosis induced by anticancer drugs. Anticancer Drugs 2001, 12:607–617.

195. Tessner TG, Muhale F, Riehl TE, et al. Prostaglandin E2 reduces radiation-induced epithelial apoptosis through a mechanism involving AKT activation and bax translocation. J Clin Invest 2004, 114:1676–1685.

196. Capra V, Ambrosio M, Riccioni G, et al. Cysteinyl-leukotriene receptor antagonists: present situation and future opportunities. Curr Med Chem 2006, 13:3213–3226.

197. Grossman EM, Longo WE, Panesar N, et al. The role of cyclooxygenase enzymes in the growth of human gall bladder cancer cells. Carcinogenesis 2000, 21:1403–1409.

198. Pruthi RS, Derksen E, Gaston K. Cyclooxygenase-2 as a potential target in the prevention and treatment

of genitourinary tumors: a review. J Urol 2003, 169:2352–2359.

199. Hayashi T, Nishiyama K, Shirahama T. Inhibition of 5-lipoxygenase pathway suppresses the growth of bladder cancer cells. Int J Urol 2006, 13:1086–1091.

200. Ikemoto S, Sugimura K, Kuratukuri K, et al. Antitumor effects of lipoxygenase inhibitors on murine bladder cancer cell line (MBT-2). Anticancer Res 2004, 24:733–736.

201. Mazhar D, Ang R, Waxman J. COX inhibitors and breast cancer. Br J Cancer 2006, 94:346–350.

202. Wang D, DuBois RN. Cyclooxygenase-2: a potential target in breast cancer. Semin Oncol 2004, 31:64–73.

203. Natarajan R, Nadler J. Role of lipoxygenases in breast cancer. Front Biosci 1998, 3:E81–E88.

204. Gupta RA, DuBois RN. Translational studies on Cox-2 inhibitors in the prevention and treatment of colon cancer. Ann N Y Acad Sci 2004, 910:196–204.

205. Sinicrope FA. Targeting cyclooxygenase-2 for prevention and therapy of colorectal cancer. Mol Carcinog 2006, 45:447–454.

206. Ikawa H, Kamitani H, Calvo BF, et al. Expression of 15-lipoxygenase-1 in human colorectal cancer. Cancer Res 1999, 59:360–366.

207. Kilic G, Gurates B, Garon J, et al. Expression of cyclooxygenase-2 in endometrial adenocarcinoma. Eur J Gynaecol Oncol 2005, 26:271–274.

208. Ohno S, Ohno Y, Suzuki N, et al. Multiple roles of cyclooxygenase-2 in endometrial cancer. Anticancer Res 2005, 25:3679–3687.

209. Mungan MU, Gurel D, Canda AE, et al. Expression of COX-2 in normal and pyelonephritic kidney, renal intraepithelial neoplasia, and renal cell carcinoma. Eur Urol 2006, 50:92–97.

210. Tuna B, Yorukoglu K, Gurel D, et al. Significance of COX-2 expression in human renal cell carcinoma. Urology 2004, 64:1116–1120.

211. Yoshimura R, Matsuyama M, Kawahito Y, et al. Study of cyclooxygenase-2 in renal cell carcinoma. Int J Mol Med 2004, 13:229–233.

212. Yoshimura R, Inoue K, Kawahito Y, et al. Expression of 12-lipoxygenase in human renal cell carcinoma and growth prevention by its inhibitor. Int J Mol Med 2004, 13:41–46.

213. Ryan EP, Pollock SJ, Kaur K, et al. Constitutive and activation-inducible cyclooxygenase-2 expression enhances survival of chronic lymphocytic leukemia B cells. Clin Immunol 2006, 120:76–90.

214. Vural F, Ozcan MA, Ozsan GH, et al. Cyclooxygenase 2 inhibitor, nabumetone, inhibits proliferation in chronic myeloid leukemia cell lines. Leuk Lymphoma 2005, 46:753–756.

215. Middleton MK, Zukas AM, Rubinstein T, et al. Identification of 12/15-lipoxygenase as a suppressor of myeloproliferative disease. J Exp Med 2006, 203:2529–2540.

216. Runarsson G, Liu A, Mahshid Y, et al. Leukotriene B4 plays a pivotal role in CD40-dependent activation of chronic B lymphocytic leukemia cells. Blood 2005, 105:1274–1279.

217. Snyder DS. Antiproliferative effects of lipoxygenase inhibitors on human leukemia cells. Adv Prostaglandin Thromboxane Leuk Res 1991, 21B:921–924.

218. Liao Z, Milas L. COX-2 and its inhibition as a molecular target in the prevention and treatment of lung cancer. Expert Rev Anticancer Ther 2004, 4:543–560.

219. Gonzalez AL, Roberts RL, Massion PP, et al. 15-Lipoxygenase-2 expression in benign and neoplastic lung: an immunohistochemical study and correlation with tumor grade and proliferation. Hum Pathol 2004, 35:840–849.

220. Denkert C, Kobel M, Berger S, et al. Expression of cyclooxygenase 2 in human malignant melanoma. Cancer Res 2001, 61:303–308.

221. Goulet AC, Einsphar JG, Alberts DS, et al. Analysis of cyclooxygenase 2 (COX-2) expression during malignant melanoma progression. Cancer Biol Ther 2003, 2:713–718.

222. Raso E, Dome B, Somlai B, et al. Molecular identification, localization and function of platelet-type 12-lipoxygenase in human melanoma progression, under experimental and clinical conditions. Melanoma Res 2004, 14:245–250.

223. Winer I, Normolle DP, Shureiqi I, et al. Expression of 12-lipoxygenase as a biomarker for melanoma carcinogenesis. Melanoma Res 2002, 12:429–434.

224. Hazar B, Ergin M, Seyrek E, et al. Cyclooxygenase-2 (Cox-2) expression in lymphomas. Leuk Lymphoma 2004, 45:1395–1399.

225. Wun T, McKnight H, Tuscano JM. Increased cyclooxygenase-2 (COX-2): a potential role in the pathogenesis of lymphoma. Leuk Res 2004, 28:179–190.

226. Ding XZ, Hennig R, Adrian TE. Lipoxygenase and cyclooxygenase metabolism: new insights in treatment and chemoprevention of pancreatic cancer. Mol Cancer 2003, 2:10.

227. Hennig R, Ding XZ, Tong WG, et al. 5-Lipoxygenase and leukotriene B(4) receptor are expressed in human pancreatic cancers but not in pancreatic ducts in normal tissue. Am J Pathol 2002, 161:421–428.

228. Kelavkar U, Glasgow W, Eling TE. The effect of 15-lipoxygenase-1 expression on cancer cells. Curr Urol Rep 2002, 3:207–214.

229. Myers CE, Ghosh J. Lipoxygenase inhibition in prostate cancer. Eur Urol 1999, 35:395–398.

230. Lee JL, Kim A, Kopelovich L, et al. Differential expression of E prostanoid receptors in murine and human non-melanoma skin cancer. J Invest Dermatol 2005, 125:818–825.

231. Nijsten T, Colpaert CG, Vermeulen PB, et al. Cyclooxygenase-2 expression and angiogenesis in squamous cell carcinoma of the skin and its precursors: a paired immunohistochemical study of 35 cases. Br J Dermatol 2004, 151:837–845.

232. Shappell SB, Keeney DS, Zhang J, et al. 15-Lipoxygenase-2 expression in benign and neoplastic sebaceous glands and other cutaneous adnexa. J Invest Dermatol 2001, 117:36–43.

233. Haynik DM, Prayson RA. Immunohistochemical expression of cyclooxygenase 2 in follicular carcinomas of the thyroid. Arch Pathol Lab Med 2005, 129:736–741.

234. Lo CY, Lam KY, Leung PP, et al. High prevalence of cyclooxygenase 2 expression in papillary thyroid carcinoma. Eur J Endocrinol 2005, 152: 545–550.

235. Nathoo N, Barnett GH, Golubic M. The eicosanoid cascade: possible role in gliomas and meningiomas. J Clin Pathol 2004, 57:6–13.

Part III
Sensitization via Transcription Factors

Chapter 10
The RKIP and STAT3 Axis in Cancer Chemotherapy: Opposites Attract

Devasis Chatterjee, Edmond Sabo, Murray B. Resnick,
Kam C. Yeung, and Y. Eugene Chin

1 Chemotherapy and the Regulation Apoptosis in Cancer

1.1 The Role of NF-κB

The acquisition of resistance to conventional therapies such as radiation and chemotherapeutic drugs remains a major obstacle in the successful treatment of cancer patients [1, 2]. In this regard, one of the major determinants of apoptosis sensitivity in cancer cells is the transcription factor nuclear factor kappa B (NFκB). Constitutive expression of NF-κB has been implicated in decreasing apoptosis in cancer cell lines [3]. NF-κB is an inducible transcription factor required for the upregulation of a large number of genes in response to inflammation, viral and bacterial infection, cell survival, cell adhesion, inflammation, differentiation, growth, and stress stimuli [4, 5]. Genes that are responsive to NF-κB activation include a variety of cytokines, cell adhesion molecules, acute phase response proteins, and apoptotic and anti-apoptotic proteins. It is believed that this reprogramming of gene expression is essential for cell survival during physiologic crisis situations. Active NF-κB is a dimer comprised of various members of the Rel family of proteins, and some form of NF-κB is expressed in most cell types. In unstimulated cells NF-κB is retained in the cytoplasm in an inactive form bound to a family of inhibitory proteins known as IκB (inhibitor of κB). Activation of NF-κB requires the phosphorylation and degradation of I-κB, which allows the NF-κB dimer to translocate to the nucleus [6, 7]. NF-κB can be activated by several signaling cascades and is subject to multiple levels of regulation. Considerable progress has been made in the identification of kinases that phosphorylate IκB and target it for subsequent degradation.

In addition to its critical role in re-programming gene expression in response to infection and other stresses, NF-κB also mediates cell survival signals, protecting cells from apoptosis [8–10]. For example, cells derived from NF-κB p65 knockout mice are significantly more sensitive to TNFα-induced cytotoxicity than normal cells. Specificity to NF-κB was demonstrated by transfecting p65 into the knockout cells that reversed the cytotoxicity [11, 12]. Similar effects were also observed when NF-κB activation was ablated with an IκB dominant-negative mutant. Cells with compromised NF-κB activation are also more vulnerable to other pro-apoptotic signals such as ionizing radiation and cancer chemotherapeutic agents [13]. NF-κB has also been implicated in downregulating apoptosis in cncer cell lines that constitutively express elevated NF-κB activity by the observation that ablation of NF-κB activity by a variety of means induced apoptosis [3]. Many of the target genes that are activated by NF-κB are critical to the

From: *Sensitization of Cancer Cells for Chemo/Immuno/Radio-therapy, 1st Edition.*
Edited by: Benjamin Bonavida © Humana Press, Totowa, NJ

establishment of early and late stages of aggressive cancers such as expression of cyclin D1, apoptosis suppressor proteins such as Bcl-2 and Bcl-XL and those required for metastasis and angiogenesis such as matrix metalloproteases (MMP) and vascular endothelial growth factor (VEGF).

It has been proposed that NF-κB suppresses apoptosis at least in part by inducing the expression of anti-apoptotic proteins, such as the inhibitor of apoptosis (IAP-1 and IAP-2) and TNF receptor associate protein (TRAF1 and TRAF2) [13]. IAP inhibits apoptosis by directly blocking the assembly and hence the activation of caspase-9 [14]. In addition to CD95L, the TNF death ligand is also able to transmit intracellular apoptotic signals via engagement of the TNFR and activation of caspase-8 apoptosis pathway. The TNFR-mediated activation of caspase-8 can be blocked by TRAF1 and 2 [15].

2   The Role of the JAK/STAT Pathway

A major target of NF-κB transcriptional activation in cancer cells is the pro-survival cytokine interleukin-6 (IL-6) [16, 17]. Aberrant and/or constitutive expression of IL-6 has been implicated in the progression of prostate cancer as well as acquired resistance to chemotherapy [18]. In addition, the unregulated production of this growth factor has been linked to cancer morbidity and mortality [19, 20]. By binding to its receptor, gp130, IL-6 activates the JAK (Janus kinases) and STAT (signal transducer and activators of transcription) signaling pathway for cell proliferation and survival [21]. The JAK family of receptor-associated tyrosine kinases consists of four members: JAK 1, JAK2, JAK3, and TYK2 (Tyrosine Kinase-2), which are activated by receptor dimerization or oligomerization [22]. Once activated, the JAK kinase protein causes tyrosine phosphorylation of the receptor cytoplasmic tails to provide docking sites for the recruitment of molecules recognizing phosphotyrosine via their phosphotyrosine binding domain (PTB) or SRC-homology-2 (SH2) domain [23]. Proteins bound to the receptor are phosphorylated, and thereby become activated. The STATs have an SH2 domain through which

they make contact with the receptor, dimerize, and move into the nucleus [24].

STATs proteins are activated also by growth factor receptors, such as epidermal growth factor (EGF), hepatocyte growth factor (HGF), platelet-derived growth factor (PDGF), and colony-stimulating factor-1 (CSF-1) receptors, which all possess an intrinsic tyrosine kinase activity [25]. These receptors may activate STAT proteins either indirectly, by means of JAK kinase proteins, or directly, as in the case of STAT1 activation by PDGF or EGF receptor [26], which may phosphorylate STAT1 in vitro [27, 28]. Constitutive activation of STAT proteins have been implicated in human cancers such as multiple myeloma, lymphomas, leukemias, and several solid tumors, which makes them logical targets for cancer therapy [29].

All seven STAT family members are highly conserved throughout evolution in their protein sequences, with each mediating a distinct process. Two novel genes in plants that carry STAT-type linker-SH2 domain indicates that the STAT linker-SH2 domain was formed prior to the divergence of animal and plant kingdoms and is apparently the oldest and fully developed SH2 domain [30].

2.1   STAT Activation and Inactivation Require Multiple Post-Translational Modifications

STATs in cells are tyrosine/serine phosphorylated upon cytokine or growth factor treatment. Phosphorylated-STAT proteins form homo- and/or heterodimers and are rapidly transported from the cytoplasm into the nucleus where they bind to DNA. Most STAT dimers recognize an inverted repeat DNA element with a consensus sequence of 5′-TTCxxxGAA-3′. Although C-terminal tyrosine residue phosphorylation is critical for STAT's function as a transcription factor, some evidence indicates that unphosphorylated STAT can also function as an adapter in mediating various cytokine receptors for signaling [31, 32].

Tyrosine kinases such as JAK, Src, and growth factor receptors are responsible for STAT tyrosine phosphorylation. STAT family members are also serine phosphorylated. Many cytokines and growth factors can trigger both STAT1 and STAT3 serine phosphorylation on Ser^{727} in the TA domain. In

STAT3, when a negative charge is introduced at position 727 by mutation of the Ser residue into Asp, the 65 C-terminal amino acids of STAT3 can function as an independent transcription activation domain. Such strong transcriptional activity of the C-terminal STAT3 Ser^{727} Asp is coupled to constitutive association with the co-activator p300 [33]. Both STAT1 and STAT3 phosphorylation on a single serine residue is required for maximal transcriptional activity [34]. Gene activation by STAT1 or STAT3, which requires tyrosine phosphorylation for activation, also depends on the activity of a presently unidentified serine kinase(s) for maximal activation.

Several PTPs have been reported to dephosphorylate STAT proteins. Using an antibody array, a number of PTP associated with STAT1 such as SHP-2, which can dephosphorylate STAT1 at both tyrosine and serine residues [35]. It is possible that SHP-1, which is homologous to SHP-2, can also dephosphorylate STAT [36]. Tyrosine phosphatase TC-45 purified from HeLa cell nuclear fractions can dephosphorylate both STAT1 and STAT3 [37]. Dephosphorylation of tyrosine-phosphorylated STAT1 is defective in TC-45-null mouse embryonic fibroblasts and primary thymocytes. Reconstitution of TC-45-null MEF with TC45 rescues the defect in STAT1 dephosphorylation [37]. These studies indicate that STAT activity can be monitored by PTK and PTP activities.

A recent publication demonstrated that p300 not only has HAT activity for protein acetylation but also carries ubiquitin E3 ligase activity for protein ubiquitination [38]. Polyubiquitinated STAT1 proteins were stabilized by a proteasome inhibitor in HeLa cells [39]. Thus, the ubiquitin-proteasome pathway may negatively regulate activated STAT proteins. The proteasome inhibitor MG132 stabilized IL-2–induced DNA-binding activity and tyrosine phosphorylation of STAT5, indicating that the kinase is targeted by the ubiquitin-proteasome pathway [40]. Protein kinase inhibitors can block this sustained STAT5 activation, which is consistent with the ability of the proteasome inhibitor to stabilize IL-2–induced tyrosine phosphorylation of JAK 1 and Jak3. However, no detectable ubiquitination of the STAT proteins is observed [40]. Another publication from indicates that by binding to SOCS proteins, protein tyrosine kinases such as EGFR and JAK are inhibited presumably due to PTK-ubiquitination and degradation [41]. However, it was noted that some SOCS factors interacted with STAT3 protein directly [41]. It is unclear whether SOCS-associated STAT3 proteins will be ubiquitinated and degraded by the proteasome.

STAT1 methylation at arginine residue (Arg^{31}) within the N-terminal region by the protein arginine methyl-transferase PRMT1 was reported as a novel requirement for IFN-induced transcription [42]. Methyl-thioadenosine, a methyl-transferase inhibitor that accumulates in transformed-cells, inhibits STAT1 responses triggered by interferon. This inhibition arises from impaired STAT1-DNA binding due to an increased association of protein inhibitor of activated STAT (PIAS1) with STAT1 dimers in the absence of methylation. Thus, Arg methylation is a post-translational modification event regulating STAT activity. Alteration of methylation is suspected to be responsible for the lack of interferon responsiveness in some malignancies. PIAS is a family of inhibitors that inactivate STAT proteins in nuclei by a presently unknown mechanism [43]. Recent findings demonstrated that PIAS family members bear SUMO (small ubiquitin–related modifier) E3 ligase activity [44], and thereby can target these proteins for ubiquitination and proteosome-dependent degradation. Thus, PIAS may modify STAT at lysine residues by adding SUMO. Two ΨKXE motifs have been identified in STAT proteins. STAT1 is sumoylated at one of these lysine residues [45]. STAT3 can recruit PIAS for sumoylation at two conserved lysine residues.

2.2 STAT Proteins Need Post-translational Modifications on Lysine Residues

It has been demonstrated that STAT1-Y^{701}F was nearly as effective as wild-type STAT1 in restoring caspase protein level in cells [46]. Hence, a modification event other than tyrosine phosphorylation is required for STAT activation. Acetylation/deacetylation can be such a post-translational modification event.

Approximately 20 different isoforms of HAT and 10 different isoforms of HDAC have been identified. HAT families include GCN5/PCAF family, TAFII250 family, p300/CBP family, SRC family, and

GNAT related proteins. HDACs are divided into three categories: type I RPD3-like (HDAC1, 2, 3, and 8); type II HDA1-like (HDAC4, 5, 6, 7, 9, and 10); and type III SIR2-like proteins. Transcription factors, such as Rb, E2F, p53, YY1, and NF-κB, are regulated by acetylation [47–53]. The Rel-A subunit of NF-κB is subject to inducible acetylation and acetylated Rel-A interacts weakly with IκB. Acetylated Rel-A is then deacetylated through a specific interaction with HDAC3 [50], p300/CBP and PCAF acetylate multiple Lys residues in p53 with varying functional consequences [49, 54]. HDAC1 has been found to deacetylate p53, thereby providing a mechanism for p53 downregulation [54]. Thus, the balance between p300/CBP and HDAC activities is implicated in the regulation of a number of onco- and anti-oncoproteins.

STAT proteins form complexes with p300/CBP in cells in a cytokine stimulation-dependent manner. The C-termini of STATs function as trans-activators capable of recruiting p300/CBP [55, 56]. Co-activators p300/CBP and NCoA-1 bind independently to specific regions within the STAT6 TA domain [57]. Thus, multiple contacts between NCoA-1, p300/CBP, and STAT6 are required for transcriptional activation. These findings provide mechanistic insights into how STAT6 can recruit co-activators required for IL-4–dependent trans-activation. Consistent with this model, STAT1 interacts directly with p300/CBP in cells, and microinjection of anti-CBP or anti-p300 antibody blocks transcriptional responses to IFN. Cells lacking STAT1 fail to inhibit AP-1 activity, and over-expression of CBP potentiates IFNα-dependent transcription and relieves AP-1/ets repression [58]. Thus, CBP and p300 integrate both positive and negative effects of IFN on gene expression by serving as essential coactivators of STAT1α, modulating gene-specific responses to simultaneous activation of two or more signal transduction pathways [58]. Ectopic expression of p300 enhances prolactin-induced STAT5-mediated transcription activation [59]. Adenovirus E1A protein, which binds to p300/CBP, suppresses STAT5- and STAT6-induced transcription [60]. This inhibition requires the STAT5 TA domain and the p300/CBP-binding site of E1A. Immuno-precipitation and mammalian two-hybrid assays reveal a direct interaction between the C-terminal TAD of STAT5 and p300/CBP. Direct interactions between STAT and these nuclear factors such as Nmi, estrogen receptor, and E1A are widely reported [58, 60].

Modification of lysine residue also includes ubiquitination and sumoylation. STAT1 polyubiquitination-dependent degradation has been reported; even though it may not play a major role in the recycling of STAT proteins [39]. Recently, STAT1 sumoylation on conserved lysine sites by PIAS family members have been shown to slightly inhibit STAT activity [45]. The ubiquitin system tags proteins for degradation by the proteosome but SUMO conjugation has a range of other functions, stabilizing some proteins and altering their subcellular localization. Sumoylation may also influence ubiquitination and protein stability indirectly.

2.3 STAT3 Is an Oncogene and Plays a Role In Both Cell Growth and Metastasis

Targeted disruption of the *stat3* gene leads to early embryonic lethality [61], whereas mice bearing STAT3 conditional knockout in skin, show impaired growth factor-dependent in vitro migration of keratinocytes despite normal proliferative responses [62]. However, a recent publication found that mice lacking STAT3 in their hematopoietic progenitors developed neutrophilia, and bone marrow cells were hyper-responsive to GM-CSF stimulation [63], suggesting that STAT3 may have a negative effect on cell proliferation. STAT3 is constitutively activated by overexpression of tyrosine kinases that carry threonine residue with the p+1 loop of the kinase domain [64] and further evidence indicates that STAT3 constitutive activation may play an important role in development of metastasis [64, 65].

Constitutive activation of STAT3 participates in tumor progression by stimulating cell proliferation, conferring resistance to apoptosis, or promoting angiogenesis and metastasis [66]. Decreased STAT3 expression by siRNA or by introduction a dominant negative STAT3 mutation can suppress tumor growth [29]. Downstream targets of STAT3 activation included proteins important for cell cycle regulation (i.e., cyclin D1, c-myc), apoptosis inhibition (Mcl1, Bcl-x$_L$) and angiogenesis and metastasis (MUC1, vascular endothelial growth factor (VEGF), matrix metalloproteinases (MMP), and CXCR4 [64, 67–69]. Thus, STAT3 protein represents a major mediator of tumorigenesis.

STAT3 is often constitutively phosphorylated in human cancers [70], although is not fully understood why STAT3 is constitutively active in tumor cells. Putative mechanisms that may account for the constitutive activity of STAT3 are dysregulation of signaling molecules or mutation or deletions in the protein that negatively regulates STAT3 (e.g., protein inhibitor of activated STAT3 or suppressor of cytokine signaling [SOCS]) [71, 72]. STAT3 requires not only tyrosine (Y^{705})/serine (S^{727}) phosphorylation but also lysine (K^{685}) acetylation for its maximum activation [73]. It has been shown that STAT3 acetylation on K^{685} and phosphorylation on Y^{705} are both required for STAT3 dimer formation leading to cell cycle and metastasis-related gene activation [73]. Although a detailed study of STAT acetylation/deacetylation process is just beginning, there are already some revealing results supporting the likely importance of acetylation modification for the metastatic phenotype.

2.4 Targeting STAT3 for Chemotherapy

Constitutively active STAT3 has been shown to lead to resistance to apoptosis induction in response to chemotherapeutic agents [74] possibly through the expression anti-apoptotic and cell cycle associated proteins [75, 76]. Similarly the role of STAT3 in tumorigenesis is mediated through the expression of various genes that suppress apoptosis, mediate proliferation, invasion, and angiogenesis [77–79]. The first validation that STAT3 is an appropriate target for chemotherapy came from a report showing that transfection of a dominant negative form of STAT3 induced cell death in mouse melanoma model [80]. Currently the availability of drugs that directly and specifically target STAT activity is very limited. Nonetheless, several chemotherapeutic agents exert their biological effects, in part, by the modulation of STAT activity. Drugs that affect STAT activity are chemically and structurally diverse and include compounds such as anthracyclines, rituximab, butyrate, sulindac, curcumin, cucurbitacin [81]. Another approach to inhibit STAT activity is to identify compounds that imitate the action of proteins inhibiting the JAK-STAT pathway (PTP, SOCS, PIAS), thus blocking STAT action, might open insights for anticancer therapy. The use of small peptides can disrupt STAT3 dimerization and signaling, and induce apoptosis

[82]. Similarly, the JAK tyrosine kinase inhibitor AG-490 can induce apoptosis while limiting STAT3 DNA binding [83]. In addition, several plant polyphenols have been identified that inhibit STAT3activation [71]. However, it is not known how these agents suppress STAT3 activation. Given the fact that STAT3 is persistently activated in numerous tumors, development of an approach to selectively block STAT3 signaling may enhance the efficacy of standard chemotherapeutic regimens. Thus, a targeted approach by the use of siRNA, specific kinase inhibitors of JAK/STAT activation or newly identified pharmacologic agents may enhance suppression of tumor growth with constitutive STAT activation.

3 RKIP, an Inhibitor of Cell Survival Pathways and Inducer of Chemotherapy-Triggered Apoptosis

3.1 RKIP: An Inhibitor of Raf and NF-κB Cell Survival Signaling

In addition to the JAK/STAT pathway, IL-6 can also regulate a Ras-dependent pathway, which includes intermediate steps involving Raf, MEK (Map kinase kinase), and MAPK. Studies have been performed to determine if Ras effectors contribute to NF-κB inhibition. Three of the most extensively studied Ras effectors are Raf kinase, phosphoinositol 3-kinase, and (PI3-kinase) and RalGDS [84, 85]. It has been demonstrated that NF-κB is activated by Ras [86]. However, the Ras effector, Raf, does not block NF-κB activation [87]. Beside its critical role in the control of cell proliferation and differentiation, the Raf signaling pathway also protects cell from apoptotic signals. The Raf pathway has been shown to promote cell survival in many cancer cells [88, 89]. Consistent with this observation in prostate tumors, the level of Raf-dependent activation of ERK correlates with increasing Gleason score and tumor stage [90]. RKIP was originally identified as a negative regulator of the Raf and NF-κB signaling pathways [91, 92]. RKIP is a highly species-conserved ubiquitously expressed protein involved in growth and differentiation signaling regulation. RKIP has been shown to disrupt the Raf-1–MEK1/2–ERK1/2 and NF-κB signaling pathways, via physical interaction with Raf-1–MEK1/2 and NIK or TAK1, respectively, thereby abrogating the survival and anti-apoptotic properties of these signaling pathways either by promoting or by inhibiting the formation of productive signaling complexes through protein–protein interactions [92, 93]. We have recently published a review on the properties, functions, and inhibition of the Ras–Raf-1–MEK1/2–ERK1/2 and NF-κB pathways by RKIP [94].

RKIP is a member of the family of phosphatidylethanolamine-binding protein (PEBP) expressed in brain tissue [95] with a binding affinity for phosphatidylethanolamine [96], nucleotides like GTP- and GDP- and small GTP-binding proteins and other hydrophobic ligands [97]. The PEBP family is a highly conserved group of proteins found in a variety of organisms from plants to *Drosophila* to mammals. There are 13 identified mammalian PEBP sequences with a highly conserved central region (residues 60–126) believed to be essential for PEBP function and binding to G proteins [98]. As such, RKIP is expressed in a wide variety of tissues including spleen, testis, prostate, breast, ovary, muscle, and stomach [99, 100].

RKIP is a much more abundant protein than Raf-1, and raises the possibility that it may have additional intracellular targets. RKIP has been recently reported to be a substrate for PKCα, βI, βII, γ, and ξ [101]. It was shown that phosphorylation of S153 on RKIP leads to its extension into the potential Raf-1 binding pocket, which suggests that phosphorylation at this site leads to the release of Raf-1 from RKIP and thus activation of the MAPK pathway [101]. RKIP has also been shown to be an inhibitor of G-protein–coupled receptor kinase 2 (GRK-2) [102]. After the stimulation of G-protein–coupled receptors, RKIP dissociates from Raf-1 to associate with GRK-2 to block its activity [43]. Consistent with the studies by Corbit et al. [101], this switch is triggered by PKC-dependent phosphorylation of S153 [101]. A recent report has shown that RKIP regulates the mitotic spindle assembly checkpoint by controlling Aurora B Kinase activity, and, consistent with previous findings, the mechanism involves Raf/MEK/ERK signaling [103]. Elevated MAP kinase signaling during the G1, S, or G2 phases of the cell cycle activates

checkpoints and induces arrest or senescence. The authors demonstrate that loss of RKIP during M phase leads to bypass of the spindle assembly checkpoint and the generation of chromosomal abnormalities [103]. These findings assign a role for RKIP and the MAP kinase cascade in ensuring the fidelity of chromosome segregation prior to cell division. These data indicate that the levels of MAP kinase activity are critical to ensure the integrity of the spindle assembly process and provide a mechanism for generating genomic instability in tumors [103]. Moreover, the RKIP status in tumors could influence the efficacy of treatments such as poisons that stimulate the Aurora B–dependent spindle assembly checkpoint [103].

3.2 RKIP Knockout Mouse Develops Olfaction Deficits

An RKIP knockout mouse has recently been generated [104]. These mice were derived from ES cells carrying a gene trap in intron 1 of RKIP-1. The mice homozygous for the mutated allele are viable and appear normal up to 10 months of age. It was reported that after this time they develop an olfaction deficit, a phenotype that correlates with the expression pattern of the gene in the brain [104]. This phenotype is thought to be evolutionarily reminiscent of the RKIP-1 function previously seen in *Drosophila* [100, 105]. In *Drosophila*, RKIP orthologues reside in antennae and olfactory hairs and play a role in odorant binding [105]. In plants, an orthologue of RKIP-1 was found to be involved in shoot growth and flowering [105]. The scent information generated from nasal mucosa is sensed by chemosensitive peripheries of olfactory neurons and then projected to the olfactory bulbs. These represent the brain centers affected in several age-related neurodegenerative disorders, including Alzheimer's disease (AD). RKIP-1 has been implicated in AD [106, 107], although it has not been linked to the loss of the sense of smell. RKIP-1 expression is reduced in AD-affected neurons [106]. Recently, an inverse correlation between the severity of AD and the RKIP-1 level in human and mouse brains has been reported [106, 108]. Thus far, targeted disruption of the *RKIP* gene has not been associated with embryonic lethality nor enhanced or ablated tumorigenic phenotype.

3.3 RKIP Inhibits Tumor Progression

Regulation of cell survival pathways by RKIP suggests that this protein may play a pivotal role in tumor progression. Studies by Fu et al. [109] have demonstrated that the low levels of RKIP mRNA and protein are correlated with the metastatic potential of human C4-2B prostate cancer cells when compared with parental non-metastatic LNCaP cells. Moreover, overexpression of RKIP in C4-2B cells decreased cell invasion in vitro and progression of lung metastases in vivo. In addition, increased levels of RKIP were associated with decreased vascular invasion in the primary tumor with no effect on primary tumor growth in mice [109, 110]. In malignant melanoma cells, loss of RKIP also correlates with enhanced invasion and progression of the disease [111]. Taken together, these results suggest that RKIP does not affect the tumorigenic properties of these prostate and melanoma cells but may represent a clinically significant suppressor of metastasis by decreasing vascular invasion [109].

3.4 RKIP Is a Metastasis Suppressor

In human breast cancer, RKIP is a metastasis suppressor protein whose expression must be downregulated for metastases to develop. Hagan et al. found that RKIP expression is reduced in lymph node metastases of 103 breast cancer patients [112]. Their results support the conclusion that RKIP is a metastasis suppressor protein, as its expression is consistently lost in lymph node metastases but not in primary tumors. Importantly, RKIP is independent of other established clinical markers, such as histologic type, tumor differentiation grade, size, or estrogen receptor status. The loss of RKIP expression in metastatic tissue suggests there is a lack of mutations in the RKIP coding sequence and could provide a novel therapeutic target for the treatment of metastatic breast cancer [91, 112].

In metastatic colorectal cancer (CRC), loss of cytoplasmic RKIP expression is associated with distant metastasis, vascular invasion, and poor prognosis [113]. In mismatch repair (MMR)–proficient and MMR deficient CRCs MMR loss of cytoplasmic RKIP was associated with distant metastasis and independently predicted worse survival [113]. Thus, the loss of RKIP expression is a marker of tumor progression and distant metastasis in MMR-proficient and -deficient CRCs.

Similar results have been observed in another study on metastatic CRC [114]. RKIP expression is downregulated in lymph node metastases. Once again, this result is consistent with recent data published on the role of RKIP and tumor progression in melanoma, prostate, and breast cancers and cancerous cell lines [109–116]. The results show a significant relationship among reduced RKIP expression, metastatic recurrence, and overall survival in colon cancer patients. This correlation is independent of Dukes' staging, in that Dukes' A and C patients with no or low RKIP expression had equivalent overall survival and vice versa [114].

Together, these findings support the recent study demonstrating that RKIP regulates the spindle checkpoints in cells [103]. Thus, it has been proposed that the loss of RKIP expression could lead to chromosomal instability, which in turn could influence tumor aggressiveness and its response to therapy [117]. However, the direct correlation between RKIP loss and chromosomal instability remains to be elaborated. Regardless of the mechanism, RKIP expression seems to be a suitable and easily determinable marker in the primary tumor that could predict the risk of early CRC to metastasize, and hence guide strategies for monitoring and therapy [117].

3.5 RKIP: A Regulator of Chemotherapy-Triggered Apoptosis

Numerous studies have provided evidence that regulation of apoptotic modulators within tumor cells correlates with sensitivity to traditional chemotherapies. Elucidation of cellular resistance mechanisms, where there is a lack of apoptosis induction, holds promise of leading to more effective treatment of cancer. Thus, because RKIP expression interferes with the survival pathways and the fact that chemotherapeutic agents also target the same survival pathways, the regulation of apoptosis by RKIP in response to chemotherapeutic agents has been examined. Indeed, while examining clinically relevant anticancer agents (i.e., camptothecin [CPT], rituximab, Taxol, cisplatin) and apoptosis induction it was discovered that the expression of RKIP was robustly induced in cancer cells [115, 116]. The direct correlation between RKIP expression and the extent of apoptosis suggests that the upregulation of RKIP is one of the mechanisms that sensitize cancer cells to apoptotic signals in response to chemotherapeutic agent–induced DNA damage. RKIP induction has been shown to sensitize human cancer cells, including prostate, breast, colon, melanoma, and non-Hodgkin's lymphoma cell lines to chemotherapy-triggered apoptosis [111, 115, 116, and unpublished observations]. Four lines of evidence support a crucial role for RKIP in apoptosis induction. First, the expression levels of RKIP correlate with the tumorigenic ability of cells when xenografted in nude mice, and the sensitivity of cancer cells to CPT-triggered apoptosis. Second, in human prostate cancer cells that are resistant to CPT-based analogues, RKIP expression levels do not change, there was a re-sensitization to undergo CPT-triggered apoptosis by overexpressing RKIP. Third, the ectopic expression of RKIP induced apoptosis in the absence of apoptotic signals. Fourth, downregulation of RKIP expression protected human prostate cancer cells from CPT-induced apoptosis.

4   RKIP and STAT3: Where's the Connection?

Although the molecular pathways have been partially delineated, little is known about the biological relevance of the inhibition of both the Raf-1–MEK1/2–ERK1/2 and the NF-κB pathways by RKIP. The inhibitory role of RKIP on both the NF-κB and ERK1/2 signaling pathways demonstrates that RKIP regulates cell proliferation and survival. We have reported that induction of RKIP is required for chemotherapy-triggered apoptosis in prostate cancer [115]. In prostate cancer, STAT3 is found to be constitutively activated [29]; thus, we hypothesized that in addition to inhibiting Raf and NF-κB cell survival pathways, RKIP may also affect JAK/STAT3 activation and signaling.

5   RKIP Inhibits IL-6–Mediated Cell Proliferation

Highly elevated expression of IL-6 has been observed in advanced prostate cancer tumors and cell lines. Whereas IL-6 promotes the growth of prostate and

other malignant cells primarily through the activation of the JAK/STAT pathway and phosphorylation of STAT3 [18, 20], treatment of DU145 human prostate cancer cells with the chemotherapeutic compound camptothecin (CPT) results in growth inhibition and apoptosis, which is dependent upon the induction of RKIP [115]. The growth of DU145 cells was measured by thymidine incorporation after IL-6 treatment. IL-6 exposure of DU145 cells resulted in a 12-fold increase in thymidine incorporation. The growth of DU145 cells was significantly inhibited by transient transfection of an RKIP expression vector (Fig. 10.1). Thus, RKIP can block IL-6–mediated proliferation in prostate cancer cells.

5.1   RKIP Inhibits IL-6–Mediated STAT3 Activation

IL-6 causes STAT3 transcriptional activation. We examined if RKIP of CPT could interfere with IL-6–mediated STAT3 transcriptional activity as measured by a STAT3-driven luciferase reporter protein assay. DU145 cells were transiently transfected with the *IRF1* luciferase reporter cDNA construct under the control of the STAT3-binding SIE fragment [73, 118]. Cells were treated with a CPT derivative or transiently transfected with an RKIP expression or CMV empty vector. Cells then were exposed to IL-6 in the presence or absence of CPT or HA-RKIP or CMV empty vector and luciferase activity was quantified. As shown in Fig. 10.2, IL-6 triggered a

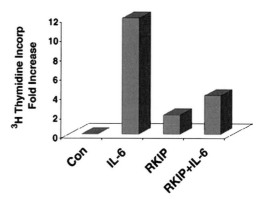

FIGURE 10.1. RKIP inhibits IL-6 mediated growth of prostate cells. ^{3}H thymidine incorporation was analyzed in DU145 cells treated with IL-6 or transfected with HA-RKIP then treated with IL-6. Data are reported as fold increase as compared with untreated control cells

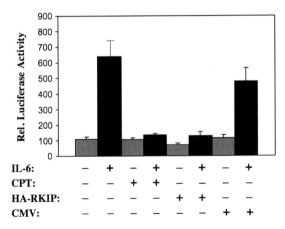

FIGURE 10.2. RKIP abrogates IL-6–triggered STAT3 activation. DU145 cells were co-transfected with Luc reporter construct containing 2× STAT3 binding SIE-fragment of the promoter region of mouse IRF1 gene and HA-RKIP or CMV empty vector. In addition, DU145 cells were treated with 50 nM of a CPT derivative in the presence and absence of IL-6. Forty-eight hours after transfection untreated and treated cell extracts were prepared and subjected to luciferase assay. The averages and standard deviations (SD, error bars) of the values of luciferase activities from triplicates of three representative experiments are shown. Luciferase activities in the cell extracts were measured using a luminometer to estimate transcriptional activity

700-fold increase in STAT3 transcriptional activation in IRF-1 or 600-fold in CMV transfected cells. In contrast, ectopic RKIP expression or 6-hour CPT treatment (where RKIP expression is induced) [115] reduced IL-6–induced luciferase activity by significantly (see Fig. 10.2). These results indicate that RKIP is a negative regulator of IL-6–mediated STAT3 activation.

5.2 RKIP Inhibits JAK 1, 2–Mediated STAT3 Activation

Having defined that RKIP was effective in inhibiting STAT3 activation as measured by the luciferase reporter assay, studies focused on that ability of RKIP and CPT to inhibit JAK kinase–mediated activation of STAT3 via luciferase reporter assay. DU145 cells were transfected with the IRF-1 plasmid along with cDNA encoding JAK 1, 2, 3, or TYK2 in the presence or absence of CPT or transiently expressed HA-RKIP or empty vector. Our data demonstrate that JAK 1, 2, 3, and

TYK2 were able to significantly stimulate STAT3 luciferase reporter activity (Fig. 10.3). Ectopic RKIP expression significantly inhibited STAT3 luciferase reporter activity in cells transfected with JAK 1 and 2 but not JAK3 or TYK2 (see Fig. 10.3). Treatment of IRF-1 and JAK 1, 2, 3, or TYK2 transfected cells with CPT from 0 to 6 hours did not affect STAT3 luciferase reporter activity. The induction of RKIP occurs between 3 and 6 hours of exposure to CPT [115], which indicates that inhibition of JAK activity is CPT-independent. The inhibition of STAT3 luciferase reporter assay by HA-RKIP was comparable with the JAK inhibitors tyrphostin and AG490 (data not shown).

5.3 Inverse Association Between RKIP and STAT3 with Clinical Outcome in Gastric Cancer Patients

Aberrant and/or constitutive expression of IL-6 and elevated serum levels has been implicated in the progression of gastric cancer as well as acquired resistance to chemotherapy [17, 119]. Serum IL-6 levels correlate with disease status of gastric cancer and has been proposed to represent a new tumor marker for monitoring treatment and response of gastric cancer patients [120]. Studies have shown that STAT3 is constitutively activated in gastric cancer cell lines and tumors [121–124]. The inhibition of constitutive STAT3 activation induces apoptosis in association with a reduction of survivin expression in gastric carcinoma cell lines [125]. Gong et al. showed that abnormally activated STAT3 expression represents a potential risk factor for poor prognosis and directly contributes to gastric cancer angiogenesis and patient survival (126). Since RKIP has been shown to be a novel and clinically relevant metastasis suppressor gene of prostate, breast, colon, and melanoma human tumor models (109–114) and STAT3 is frequently activated in gastric adenocarcinomas, we evaluated the association of RKIP and STAT3 expression in predicting the survival in gastric adenocarcinoma patients using a human gastric tumor tissue microarray. The influence of prognostic factors on tumor-related survival was assessed by Kaplan-Meier estimates, and subgroups were compared by the log-rank test for univariate analysis. Univariate analysis of survival indicated that in the gastric

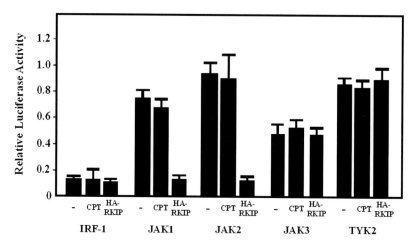

FIGURE 10.3. RKIP inhibits JAK 1, 2-triggered STAT3 activation. DU145 cells were transfected with Luc reporter construct containing 2× STAT3 binding SIE-fragment of the promoter region of mouse IRF1 gene. After 24 hours the cells were co-transfected with either JAK 1, 2, 3, or TYK2 in the presence or absence HA-RKIP. After 24 hours one set of cells transfected with JAK 1, 2, 3, or TYK2 were treated with 50 nM CPT for 6 hours. Those cells transfected with JAK 1, 2, 3, or TYK with HA-RKIP were treated for 6 hours with a non–drug containing vehicle. All the cells were harvested after the 6-hour treatment period, cell extracts were prepared and subjected to luciferase assay. The averages and standard deviations (SD, error bars) of the values of luciferase activities from triplicates of three representative experiments are shown. Luciferase activities in the cell extracts were measured using a luminometer to estimate transcriptional activity

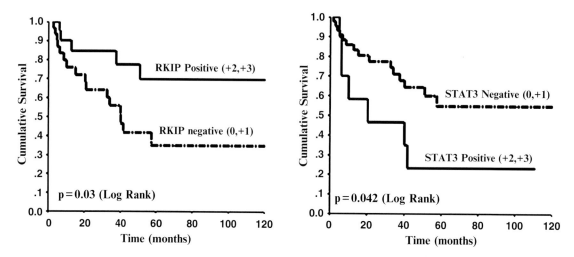

FIGURE 10.4. Kaplan-Meier univariate survival analysis of RKIP and STAT3 expression in gastric cancer. Positive, moderate/strong (+2, +3) and negative to weak (0, +1) staining for RKIP and STAT3 in intestinal type gastric adenocarcinoma. **A**, Univariate analysis of intestinal type cancer showing the relationship between RKIP positivity and cumulative patient survival. **B**, Direct positive relationship between STAT3 and survival in the intestinal tumor type. *p* Values represent the log-rank test

intestinal histological tumor type [127, 128] RKIP positively correlated with patient survival (Fig. 10.4). In the RKIP moderate to strongly positive (+2, +3) cases the 5-year survival rates were 71%

(median survival time not reached) as opposed to the RKIP negative or weakly positive (0, +1) cases displaying 5-year survival rates of 35% (median survival time: 40 months). The differences were

TABLE 10.1. Multivariate survival analysis (Cox proportional hazard model) for the intestinal tumor type.

Variable	Beta	Standard error	p Value	Odds ratio
Stage	1.72	0.53	0.001	5.59
RKIP	−1.50	0.57	0.009	0.22

statistically significant ($p = 0.029$, log-rank test). In this tumor type, STAT3 negatively correlated with survival: 5-year survival rates in the STAT3 positive (+2, +3) group was 23%, (median = 22 months) as opposed to the STAT3 negative group (0, +1) which showed 5-year survival rates of 55% (median survival not reached) ($p = 0.042$) (see Fig. 10.4). The multivariate Cox proportional hazard model was applied using a stepwise forward method to detect independent predictors of survival. Association between RKIP and STAT3 expression levels was tested using the Fisher's exact test. Two-tailed p values of 0.05 or less were considered to be statistically significant. Cox multivariate analysis was performed, using RKIP and STAT3 expression, tumor stage as well as other confounding variables such as age and patient gender. Significantly, in the intestinal tumor type, RKIP expression and stage were independently associated with patients' survival ($p = 0.009$, Odds Ratio = 0.22; Table 10.1). Notably, these results indicate the predictive and protective role of RKIP in gastric adenocarcinoma. Moreover, these results are the first to indicate an inverse association between RKIP and STAT3 expression and a positive correlation between RKIP and gastric patient survival.

6   Concluding Remarks

Dysregulation of apoptosis after treatment with different neoplastic agents remains variable and will allow metastatic cell survival in human cancer models. Understanding the role of endogenous intracellular factors involved in the regulation of genes and their products that are involved in apoptosis induction has the potential to lead to the development and selective application of novel mechanism–directed chemotherapeutic agents. Suppression of apoptosis induction may also occur through the constitutive activation of cell survival signal transduction pathways. STAT3 proteins, which are activated via post-translational modifications by the IL-6 family of cytokines, prevent apoptosis through the regulation of antiapoptotic proteins such as Bcl-2 and survivin. The growing list of cases in which suppression of STAT signaling leads to the demise of tumor cells establishes that these proteins contribute to the tumorigenic phenotype. Strict regulation of STAT3 activation is imperative for preventing tumorigenesis and maintaining normal cell growth. Targeting STAT3 for therapeutic intervention has been attempted by the use of oligonucleotides, kinase inhibitors, and small-molecule drugs, and have met with limited success. Our preliminary results indicate that another function of RKIP, in addition to Raf and NF-κB survival signaling inhibition and apoptosis induction, is the inhibition of STAT3 activation and signaling. Thus, targeting STAT3 by enhancing RKIP expression will assault cancer cells by simultaneously activating cell death and inhibiting STAT3-mediated cell survival. Moreover, these findings support the hypothesis that RKIP represents a critical element in determining the balance between cell death and survival of cancer cells. This is underscored by the fact that RKIP can reverse resistance to chemotherapy-triggered apoptosis in cancer cells. Establishing the mechanism by which RKIP inhibits STAT3 activation and metastasis may lead to the development of RKIP-directed therapeutic approach for the treatment of cancer. The manipulation of RKIP expression to target STAT3 activation represents a novel and important therapeutic approach and may provide insight to important clues for the progression and treatment of cancer.

Acknowledgments. The project described was supported by NIH COBRE grant P20 RR 17695 from the Institutional Development Award (IDeA) Program of the National Center for Research Resources (DC and ES) and NIH RO1 CA102128-02 (YEC).

References

1. Hanahan D, Weinberg RA. The hallmarks of cancer. Cell 2000, 100:57–70.
2. Reed JC, Apoptosis-targeted therapies for cancer. Cancer Cell 2003, 3:17–22.

3. Muenchen HJ, Lin DL, Walsh MA, et al. Tumor necrosis factor-alpha-induced apoptosis in prostate cancer cells through inhibition of nuclear factor-kappaB by an IkappaBalpha "super-repressor". Clin Cancer Res 2000, 6:1969–1977.

4. Courtois G, Gilmore TD. Mutations in the NF-kappaB signaling pathway: implications for human disease. Oncogene 2006, 51:6831–6843.

5. Aggarwal BB. Nuclear factor-B: the enemy within. Cancer Cell 2004, 6:203–208.

6. Kumar A, Takada Y, Boriek AM, et al. Nuclear factor-B: its role in health and disease. J Mol Med 2004, 82:434–448.

7. Ghosh S, May MJ, Kopp EB. NF-kB and Rel proteins: evolutionarily conserved mediators of immune responses. Annu Rev Immunol 1998, 16:225–260.

8. Baichwal VR, Baeuerle PA. Activate NF-kappa B or die? Curr Biol 1997, 2:94–96.

9. Karin M, CaoY, Greten FR, et al. NF-kB in cancer: from innocent bystander to major culprit. Nat Rev 2002, 2:301.

10. Foo SY, Nolan GP. NF-kappaB to the rescue: RELs, apoptosis and cellular transformation. Trends Genet 1999, 6:229–235.

11. Beg AA, Baltimore D. An essential role for NF-kappaB in preventing TNF-alpha-induced cell death. Science 1996, 5288:782–784.

12. Van Antwerp DJ, Martin SJ, Kafri TJ, et al. Suppression of TNF-alpha-induced apoptosis by NF-kappaB. Science 1996, 274:787–789.

13. Wang CY, Mayo WM, Baldwin AS Jr. TNF- and cancer therapy-induced apoptosis: potentiation by inhibition of NF-kappaB. Science 1996, 274:784–787.

14. Deveraux QL, Reed JC. IAP family proteins—suppressors of apoptosis. Genes Dev 1999, 13:239–252.

15. Wallach D, Varfolomeev EE, Malinin NL, et al. Tumor necrosis factor receptor and Fas signaling mechanisms. Annu Rev Immunol 1999, 17:331–367.

16. Xiao W, Hodge DR, Wang L, et al. Co-operative functions between nuclear factor NfkappaB and CCAT/enhancer-binding protein (C/EBP-beta) regulate the IL-6 promoter in autocrine human prostate cancer cells. Prostate 2004, 61:354–370.

17. Andela VB, Gordon AH, Zotalis G, et al. NFkappaB, a pivotal transcription factor in prostate cancer metastasis to bone. Clin Orthop 2003, 415:S75–85.

18. Siegall CB, Schwab G, Nordan RP, et al. Expression of interleukin 6 receptor and interleukin 6 in prostate cancer cell lines. Cancer Res 1990, 50:7786–7798.

19. Twillie DA, Eisenberger MA, Carducci MA. et al. Interleukin-6: a candidate mediator of human prostate cancer morbidity. Urology 1995, 45:542–549.

20. Michalaki V, Syrigos K, Charles P, et al. Serum levels of IL-6 and TNF-a correlate with clinicopathological features and patient survival in patients with prostate cancer. Br J Cancer 2004, 90:2312–2316.

21. Taga T, Hibi M, Hirata Y, et al. Interleukin-6 triggers the association of its receptor with a possible signal transducer, gp130. Cell 1989, 58:573–581.

22. Schindler C, Darnell JE Jr. Transcriptional responses to polypeptide ligands: the JAK–STAT pathway. Annu Rev Biochem 1995, 64:621–651.

23. Briscoe J, Kohlhuber F, Muller M. JAKs and STATs branch out. Trends Cell Biol 1996, 9:336–340.

24. Bromberg J, Darnell JE Jr. The role of STATs in transcriptional control and their impact on cellular function. Oncogene 2000, 21:2468–2473.

25. Catlett-Falcone R, Dalton WS, Jove R. STAT proteins as novel targets for cancer therapy. Signal transducer an activator of transcription. Curr Opin Oncol 1999, 6:490–496.

26. Fu XY, Zhang JJ. Transcription factor p91 interacts with the epidermal growth factor receptor and mediates activation of the c-fos gene promoter. Cell 1993, 61:135–145.

27. Quelle FW, Thierfelder W, Witthuhn BA, et al. Phosphorylation and activation of the DNA binding activity of purified Stat1 by the Janus protein-tyrosine kinases and the epidermal growth factor receptor. J Biol Chem 1995, 270:20775–20780.

28. Choudhury GG, Ghosh-Choudhury N, Abboud HE. Association and direct activation of signal transducer and activator of transcription1alpha by platelet-derived growth factor receptor. J Clin Invest 1998, 101:2751–2760.

29. Buettner R, Mora LB, Jove R. Activated signal transducers and activators of transcription signaling in human tumors provides novel molecular targets for therapeutic intervention. Clin Cancer Res 2002, 8:945–954.

30. Gao Q, Hua J, Kumuri R, et al. Identification of STAT's linker-SH2 domain as the SH2 domain origin using secondary structural analysis. Mol Cell Proteomics 2004, 3:704–714.

31. Pfeffer LM, Mullersman JE, Pfeffer SR, et al. STAT3 as an adapter to couple phosphatidylinositol 3-kinase to the IFNAR1 chain of the type I interferon receptor. Science 1997, 276:1418–1420.

32. Wang Y, Wu TR, Cai S, et al. Stat1 as a component of tumor necrosis factor alpha receptor 1-TRADD signaling complex to inhibit NF-κB activation. Mol Cell Biol 2000, 20:4505–4512.

33. Schepers H, Vellenga E, Kruijer W. Ser727-dependent transcriptional activation by association of p300 with STAT3 upon IL-6 stimulation. FEBS Lett 2001, 495:71–76.

34. Chang HM, Paulson M, Rice CM, et al. Induction of interferon-stimulated gene expression and antiviral

responses require protein deacetylase activity. Proc Natl Acad Sci U S A 2004, 101:9578–9583.

35. Wu TR, Hong YK, Wang XD, et al. SHP-2 is a dual-specificity phosphatase involved in Stat1 dephosphorylation at both tyrosine and serine residues in nuclei. J Biol Chem 2002, 277:47572–47580.

36. Aoki N, Matsuda T. A cytosolic protein-tyrosine phosphatase PTP1B specifically dephosphorylates and deactivates prolactin-activated STAT5a and STAT5b. J Biol Chem 2000, 275:39718–3926.

37. ten Hoeve J, de Jesus Ibarra-Sanchez M, Fu Y, et al. Identification of a nuclear Stat1 protein tyrosine phosphatase. Mol Cell Biol 2002, 22:5662–5668.

38. Grossman SR, Deato ME, Brignone C, et al. Polyubiquitination of p53 by a ubiquitin ligase activity of p300. Science 2003, 300:342–344.

39. Kim TK, Maniatis T. Regulation of interferon-gamma-activated STAT1 by the ubiquitin-proteasome pathway. Science 1996, 273:1717–1719.

40. Yu CL, Burakoff SJ. Involvement of proteasomes in regulating Jak-STAT pathways upon interleukin-2 stimulation. J Biol Chem 1997, 272:14017–14020.

41. Xia L, Wang L, Chung AS, et al. Identification of both positive and negative domains within the epidermal growth factor receptor COOH-terminal region for signal transducer and activator of transcription (STAT) activation. J Biol Chem 2002, 30716–30723.

42. Mowen KA, Tang J, Zhu W, et al. Arginine methylation of STAT1 modulates IFNαβ-induced transcription. Cell 2001, 104:731–741.

43. Chung CD, Liao J, Liu B, et al. Specific inhibition of Stat3 signal transduction by PIAS3. Science 1997, 278:1803–1805.

44. Johnson ES, Gupta AA. An E3-like factor that promotes SUMO conjugation to the yeast septins. Cell 2001, 106:735–744.

45. Rogers RS, Horvath CM, Matunis MJ. SUMO modification of STAT1 and its role in PIAS-mediated inhibition of gene activation. J Biol Chem 2003, 278:30091–30097.

46. Kumar A, Commane M, Flickinger TW, et al. Defective TNF-alpha-induced apoptosis in STAT1-null cells due to low constitutive levels of caspases. Science 1997, 278:1630–1632.

47. Gu W, Roeder RG. Activation of p53 sequence-specific DNA binding by acetylation of the p53 C-terminal domain. Cell 1997, 90:595–606.

48. Chen LF, Fischle W, Verdin E, et al. Duration of nuclear NF-kappaB action regulated by reversible acetylation. Science 2001, 293:1653–1657.

49. Yao YL, Yang WM, Seto E. Regulation of transcription factor YY1 by acetylation and deacetylation. Mol Cell Biol 2001, 21:5979–5991.

50. Morrison AJ, Sardet C, Herrera RE. Retinoblastoma protein transcriptional repression through histone deacetylation of a single nucleosome. Mol Cell Biol 2002, 22:856–865.

51. Martinez-Balbas MA, Bauer UM, Nielsen SJ, et al. Regulation of E2F1 activity by acetylation. EMBO J 2000, 19:662–671.

52. Chan HM, Krstic-Demonacos M, Smith L, et al. Acetylation control of the retinoblastoma tumour-suppressor protein. Nat Cell Biol 2001, 3:667–674.

53. Liu L, Scolnick DM, Trievel RC, et al. p53 sites acetylated in vitro by PCAF and p300 are acetylated in vivo in response to DNA damage. Mol Cell Biol 1999, 19:1202–1209.

54. Luo J, Su F, Chen D, et al. Deacetylation of p53 modulates its effect on cell growth and apoptosis. Nature 2000, 408:377–381.

55. Bhattacharya S, Eckner R, Grossman S, et al. Cooperation of Stat2 and p300/CBP in signalling induced by interferon-alpha. Nature 1996, 383:344–347.

56. Zhang JJ, Vinkemeier U, Gu W, et al. Two contact regions between Stat1 and CBP/p300 in interferon gamma signaling. Proc Natl Acad Sci U S A 1996, 93:15092–15096.

57. Litterst CM, Pfitzner E. Transcriptional activation by STAT6 requires the direct interaction with NCoA-1. J Biol Chem 2001, 276:45713–45721.

58. Horvai AE, Xu L, Korzus E, et al. Nuclear integration of JAK/STAT and Ras/AP-1 signaling by CBP and p300. Proc Natl Acad Sci U S A 1997, 94:1074–1079.

59. Zhu M, John S, Berg M, Leonard WJ. Functional association of Nmi with Stat5 and Stat1 in IL-2- and IFNgamma-mediated signaling. Cell 1999, 96:121–130.

60. Chakravarti D, Ogryzko V, Kao HY, et al. A viral mechanism for inhibition of p300 and PCAF acetyltransferase activity. Cell 1999, 96:393–403.

61. Takeda K, Noguchi K, Shi W, et al. Targeted disruption of the mouse Stat3 gene leads to early embryonic lethality. Proc Natl Acad Sci U S A 1997, 94:3801–3804.

62. Sano S, Itami S, Takeda K, et al. Keratinocyte-specific ablation of Stat3 exhibits impaired skin remodeling, but does not affect skin morphogenesis. EMBO J 1999, 18:4657–4668.

63. Lee CK, Raz R, Gimeno R, et al. STAT3 is a negative regulator of granulopoiesis but is not required for G-CSF-dependent differentiation. Immunity 2002, 17:63–72.

64. Yuan ZL, Guan YJ, Wang L, et al. Central role of threonine[P+1Loop] residue in RTK for constitutive STAT3 phosphorylation in metastatic cancer cells. Mol Cell Biol 2004, 27:9390–9400.

65. Yahata Y, Shirakata Y, Tokumaru S, et al. Nuclear translocation of phosphorylated STAT3 is essential

for vascular endothelial growth factor-induced human dermal microvascular endothelial cell migration and tube formation. J Biol Chem 2003, 278:40026–40031.

66. Bowman T, Broome MA, Sinibaldi D, et al. Stat3-mediated Myc expression is required for Src transformation and PDGF-induced mitogenesis. Proc Natl Acad Sci U S A 2001, 98:7319–7324.

67. Lin TS, Mahajan S, Frank DA. STAT signaling in the pathogenesis and treatment of leukemias. Oncogene 2000, 19:2496–504.

68. Hsieh FC, Cheng G, Lin J. Evaluation of potential Stat3-regulated genes in human breast cancer. Biochem Biophys Res Commun. 2005, 23:292–299.

69. Xie TX, Wei D, Liu M, et al. Stat3 activation regulates the expression of matrix metalloproteinase-2 and tumor invasion and metastasis. Oncogene 2004, 23:3550–3560.

70. Bromberg JF, Wrzeszczynska MH, Devgan G Jr, et al. Stat3 as an oncogene. Cell 1999, 98:295–303.

71. Aggarwal BB, Sethi G, Ahn KS, et al. Targeting signal-transducer-and-activator-of-transcription-3 for prevention and therapy of cancer: modern target but ancient solution. Ann N Y Acad Sci 2006, 1091:151–169.

72. Yoshikawa H, Matsubara K, Qian GS, et al. SOCS-1, a negative regulator of the JAK/STAT pathway, is silenced by methylation in human hepatocellular carcinoma and shows growth-suppression activity. Nat Genet 2001, 28:29–35.

73. Yuan Z, Guan Y, Chatterjee D, et al. Stat3 dimerization regulated by reversible acetylation of a single lysine residue. Science 2005, 307:269–273.

74. Catlette-Falcone, R, Landowski TH, Oshiro MM, et al. Constitutive activation of Stat3 signaling confers resistance to apoptosis in human U266 myeloma cells. Immunity 1999, 10:105–115.

75. Fujio Y, Kunisada K, Hirota H, et al. Signals through gp130 upregulate bcl-x gene expression via STAT1-binding cis-element in cardiac myocytes. J Clin Invest 1997, 99:2898–2905.

76. Masuda M, Suzui M, Yasumatu R, et al. Constitutive activation of signal transducers and activators of transcription 3 correlates with cyclin D1 overexpression and may provide a novel prognostic marker in head and neck squamous cell carcinoma. Cancer Res 2002, 62:3351–3355.

77. Turkson J, Bowman T, Garcia R, et al. Stat3 activation by Src induces specific gene regulation and is required for cell transformation. Mol Cell Biol 1998, 18:2545–2552.

78. Schlessinger K, Levy DE. Malignant transformation but not normal cell growth depends on signal transducer and activator of transcription 3. Cancer Res 2005, 65:5828–5834.

79. Niu, G., K.L. Wright, M. Huang, et al. Constitutive Stat3 activity up-regulates VEGF expression and tumor angiogenesis. Oncogene 2002, 21:2000–2008.

80. Niu G, Heller, R, Catlett-Falcone C, et al. Gene therapy with dominant negative STAT3 suppresses growth of the murine melanoma B16 tumor in vivo. Cancer Res 1999, 59:5059–5063.

81. Klampfer L. Signal transducers and activators of transcription 9STATs): novel targets of chemopreventive and chemotherapeutic drugs. Curr Cancer Drug Targets 2006, 6:107–121.

82. Primiano T, Baig M, Maliyyekkel A, et al. Identification of potential anticancer drugs through the selection of growth-inhibitory genetic elements. Cancer Cell 2003, 4:41–53.

83. Eplinmh-Burnette PK, Liu JH, Catlett-Falcone R, et al. Inhibition of stat3 signaling leads to apoptosis of leukemic large granular lymphocytes and decreased mcl-1 expression. J Clin Invest 2001, 107:351–362.

84. Campbell SL, Khosravi-Far R, Rossman KL, et al. Increasing complexity of Ras signaling. Oncogene 1998, 17:1395–1413.

85. Jonenson T, and Bar-Sagi DJ. Ras effectors and their role in mitogenesis and oncogenesis. Mol Med 1997, 75:587–593.

86. Finco TS, Westwick JK, Norris JL, et al. Oncogenic Ha-Ras-induced signaling activates NF-kappaB transcriptional activity, which is required for cellular transformation. J Biol Chem 1997, 272:24113–24116.

87. Hanson JL, Hawke NA, Kashatus D, et al. Oncoprotein suppression of tumor necropsies factor-induced NF-kB activation is independent of Raf-controlled pathways. J Biol Chem 2003, 278:34910–34917.

88. Bonni A, Brunet A, West AE, et al. Cell survival promoted by the Ras-MAPK signaling pathway by transcription-dependent and -independent mechanisms. Science 1999, 286:1358–1362.

89. Fang X, Yu S, Eder A, et al. Regulation of BAD phosphorylation at serine 112 by the Ras-mitogen-activated protein kinase pathway. Oncogene 1999, 48:6635–6640.

90. Gioeli D, Mandell JW, Petroni GR, et al. Activation of mitogen-activated protein kinase associated with prostate cancer progression. Cancer Res 1999, 59:279–284.

91. Yeung K, Seitz T, Li S, et al. Suppression of Raf-1 kinase activity and MAP kinase signalling by RKIP. Nature 1999, 401:173–177.

92. Yeung KC, Rose DW, Dhillon A, et al. Raf kinase inhibitor protein interacts with NF-kappaB-inducing kinase and TAK1 and inhibits NF-kappaB activation. Mol Cell Biol 2001; 21:7207–7217.

93. Yeung K, Janosch P, McFerran B, et al. Mechanism of suppression of the Raf/MEK/extracellular signal-regulated kinase pathway by the raf kinase inhibitor protein. Mol Cell Biol 2000, 20:3079–3085.

94. Odabaei G, Chatterjee D, Jazirehi AR, et al. Raf-1 kinase inhibitor protein (RKIP): structure, function, regulation of cell signaling and pivotal role in apoptosis. Adv Cancer Res 2004, 91:169–200.

95. Tohdoh N, Tojo S, Agui H, et al. Sequence homology of rat and human HCNP precursor proteins, bovine phosphatidylethanolamine-binding protein and rat 23-kDa protein associated with the opioid-binding protein. Brain Res Mol Brain Res 1995, 30:381–384.

96. Bernier I, Jolles P. Purification and characterization of a basic 23 kDa cytosolic protein from bovine brain. Biochim Biophys Acta 1984, 790:174–181.

97. Bucquoy S, Jolles P, Schoentgen F. Relationships between molecular interactions (nucleotides, lipids and proteins) and structural features of the bovine brain 21-kDa protein. Eur J Biochem 1994, 225:1203–1210.

98. Banfield MJ, Barker JJ, Perry AC, et al. Function from structure? The crystal structure of human phosphatidylethanolamine-binding protein suggests a role in membrane signal transduction. Structure 1998, 6:1245–1254.

99. Bollengier F, Mahler A. Localization of the novel neuropolypeptide h3 in subsets of tissues from different species. J Neurochem 1988, 50:1210–1214.

100. Frayne J, Ingram C, Love S, et al. Localisation of phosphoatidylethanolamine-binding protein in the brain and other tissues of the rat. Cell Tissue Res 1999, 298:415–423.

101. Corbit, KC, Trakul N, Eves EM, et al. Activation of Raf-1 signaling by protein kinase C through a mechanism involving Raf kinase inhibitory protein. J Biol Chem 2003, 278:13061–13068.

102. Lorenz K, Lohse MJ, Quittere U. Protein kinase C switches the Raf kinase inhibitor from Raf-1 to GRK-2. 2003, 426:574–579.

103. Eves EM, Shapiro P, Naik K, et al. Raf kinase inhibitory protein regulates aurora B kinase and the spindle checkpoint. Mol Cell 2006, 23:561–574.

104. Theroux S, Pereira M, Casten KS, et al. Raf kinase inhibitory protein knockout mice: expression in the brain and olfaction deficit. Brain Res Bull 2007, 71:559–567.

105. Pnueli L, Gutfinger T, Hareven D, et al. Tomato SP-interacting proteins define a conserved signaling system that regulates shoot architecture and flowering. Plant Cell 2001, 13:2687–2702

106. George AJ, Holsinger RM, McLean CA, et al. Decreased phosphatidylethanolamine binding protein expression correlates with Abeta accumulation in the Tg2576 mouse model of Alzheimer's disease. Neurobiol Aging 2006, 27:614–623.

107. Tsugu Y, Ojika K, Matsukawa N, et al. High levels of hippocampal cholinergic neurostimulating peptide (HCNP) in the CSF of some patients with Alzheimer's disease. Eur J Neurol 1998, 5:561–569.

108. Maki M, Matsukawa N, Yuasa H, et al. Decreased expression of hippocampal cholinergic neurostimulating peptide precursor protein mRNA in the hippocampus in Alzheimer disease. J Neuropathol Exp Neurol 2002, 61:176–185.

109. Fu Z, Smith PC, Zhang L, et al. Effects of Raf kinase inhibitor protein expression on suppression of prostate cancer metastasis. J Nat Cancer Inst 2003, 95:878–883.

110. Keller ET, Fu Z, Yeung K, et al. Raf kinase inhibitor protein: a prostate cancer metastasis suppressor gene. Cancer Lett 2004, 207:131–137.

111. Schuierer MM, Bataille F, Hagan S, et al. Reduction in Raf kinase inhibitor protein expression is associated with increased Ras-extracellular signal-regulated kinase signaling in melanoma cell lines. Cancer Res 2004, 64:5186–5192.

112. Hagan S, Al-Mulla F, Mallon E, et al. Reduction of Raf-1 kinase inhibitor protein expression correlates with breast cancer metastasis. Clin Cancer Res 2005, 11:7392–7397

113. Minoo P, Zlobec I, Baker K, et al. Loss of raf-1 kinase inhibitor protein expression is associated with tumor progression and metastasis in colorectal cancer. Am J Clin Pathol 2007, 127:820–827.

114. Al-Mulla F, Behbehani AI, Bitar MS, et al. Genetic profiling of stage I and II colorectal cancer may predict metastatic relapse. Mod Pathol 2006, 19:648–658.

115. Chatterjee D, Bai Y, Wang Z, et al. RKIP sensitizes prostate and breast cancer cells to drug-induced apoptosis. J Biol Chem 2004, 279:17515–17523.

116. Jazirehi AR, Chatterjee D, Odabaei G, et al. Rituximab triggers RKIP expression and inhibits MAPKs ERK1/2 and AP-1 activation in non-Hodgkin's lymphoma B-cells: Role in down-regulation of Bcl-x$_L$ expression. Cancer Res 2004, 64:7117–7126.

117. Galizia G, Lieto E, Ferraraccio F, et al. Determination of molecular marker expression can predict clinical outcome in colon carcinomas. Clin Cancer Res 2004, 10:3490–3499.

118. Andrews NC, Faller DV. A rapid micropreparation technique for extraction of DNA-binding proteins from limiting numbers of mammalian cells. Nucleic Acids Res 1991, 19:2499.

119. Xiao W, et al. Co-operative functions between nuclear factor NfkappaB and CCAT/enhnacer-binding

protein (C/EBP-beta) regulate the IL-6 promoter in autocrine human prostate cancer cells. Prostate 2004, 61:354–370.

120. De Vita F, Romano C, Orditura M, et al. Interleukin-6 serum level correlates with survival in advanced gastrointestinal cancer patients but is not an independent prognostic indicator. J Interferon Cytokine Res 2001, 21:45–52.

121. Wu CW, Wang SR, Chao MF, et al. Serum interleukin-6 levels reflect disease status of gastric cancer. Am J Gastroenterol 1996, 91:1417–1422.

122. Gong W, Wang L, Yao JC, et al. Expression of activated signal transducer and activator of transcription 3 predicts expression of vascular endothelial growth factor in and angiogenic phenotype of human gastric cancer. Clin Cancer Res 2005, 11:1386–1393.

123. Yu LF, Cheng Y, Qiao MM, et al. Activation of STAT3 signaling in human stomach adenocarcinoma drug-resistant cell line and its relationship with expression of vascular endothelial growth factor. World J Gastroenterol 2005, 11:875–879.

124. Jenkins BJ, Grail D, Nheu T, et al. Hyperactivation of Stat3 in gp130 mutant mice promotes gastric hyperproliferation and desensitizes TGF-beta signaling. Nat Med 2005, 11:845–852.

125. Kanda N, Seno H, Konda Y, et al. STAT3 is constitutively activated and supports cell survival in association with survivin expression in gastric cancer cells. Oncogene 2004, 23:4921–4929.

126. Gong W, Wang L, Yao JC, et al. Expression of activated signal transducer and activator of transcription 3 predicts expression of vascular endothelial growth factor in and angiogenic phenotype of human gastric cancer. Clin Cancer Res 2005, 11:1386–93.

127. Lardenoye JW, Kappetein AP, Lagaay MB, et al. Survival of proximal third gastric carcinoma. J Surg Oncol 1998, 68:183–186.

128. Lauren P. The two histological main types of gastric carcinoma: diffuse and so-called intestinal-type carcinoma. An attempt at a histoclassification. Acta Pathol Microbiol Scand 1965, 64:31–49.

Chapter 11

Targeting Transcription Factors with Decoy Oligonucleotides: Modulation of the Expression of Genes Involved in Chemotherapy Resistance of Tumor Cells

Roberto Gambari

1 Introduction

One of the major problems encountered in tumor chemotherapy is the development of tumor cell clones resistant toward the applied clinical treatment [1–9]. Therefore, a great effort has been dedicated in the last decade to the development of approaches aimed at finding novel drugs or drug combinations useful to inhibit in vitro and in vivo growth of tumor cells resistant to chemotherapy [10–13].

In this difficult task, the definition of the molecular mechanism(s) underlying both response and resistance of tumor cells to the anticancer therapy is very important [8, 9]. This will enable the identification of prognostic markers of great help in developing clinical protocols. In addition, the identification of resistance-associated markers is of great impact in the design and development of novel therapeutic agents able to reverse the resistance phenotype [13].

2 Apoptosis and Chemotherapy Resistance of Tumor Cells

It is well established that one of the major events associated with chemoresistance is a deregulation of apoptosis [14–21] as suggested by several

authors, including Boehrer et al. [14], Soengas et al. [17], and Cheng et al. [18]. Fully in agreement with the concept of a relationship between the lack to undergo apoptosis and chemotherapy resistance of tumor cells to anticancer agents, is the report by Devarajan et al. [22] showing a downregulation of caspase-3 in breast cancer–chemoresistant cells. These authors studied the expression of caspase-3 in primary breast tumor and normal breast parenchyma samples obtained from patients undergoing breast surgery, as well as in normal mammary epithelial cells and several established mammary cancer cell lines. The expression of caspase-3 was analyzed at the RNA level by reverse transcriptase–polymerase chain reaction and Northern blotting, and at the protein level by Western blot analysis. The results obtained demonstrated that approximately 75% of the tumor as well as morphologically normal peritumoral tissue samples lacked the caspase-3 transcript and caspase-3 protein expression. Interestingly, the sensitivity of caspase-3–deficient breast cancer (MCF-7) cells to undergo apoptosis in response to doxorubicin and other apoptotic stimuli was found augmented by reconstituting caspase-3 expression. These data are relevant, considering caspase-3 as a member of the cysteine protease family playing a crucial role in apoptotic pathways. In agreement with these

From: *Sensitization of Cancer Cells for Chemo/Immuno/Radio-therapy, 1st Edition.*
Edited by: Benjamin Bonavida © Humana Press, Totowa, NJ

data, the identification of molecules able to induce apoptosis or inhibit the expression of anti-apoptotic genes with different mechanism of action is a great value in defining approaches suitable to reverse tumor chemoresistance, both in vitro and in vivo.

3 Nuclear Factor kappa-B and Tumor Chemoresistance

Nuclear factor-kappaB (NF-kB) proteins represent a small class of dimeric transcription factors regulating the expression of several genes [23–32], several of which are involved in key biological processes, such as cell-adhesion molecules, angiogenesis modulators, growth factors, and antiapoptotic factors [33–35]. The most common p50/Rel-A (p65) dimer known "specifically" as NF-kB is relatively abundant as an inactive cytoplasmic complex bound to inhibitory proteins belonging to the so-called NF-kB inhibitor (IkB) family [36]. The inactive NF-kB-IkB complex is activated by a variety of stimuli, including proinflammatory cytokines, mitogens, growth factors, and stress-inducing agents [36]. The release of NF-kB from the NF-kB/IkB complex facilitates its translocation to the nucleus, where it promotes transcription of most of the target genes, despite the fact that in some cases NF-kB–mediated repression of gene expression has been reported [23–33, 36].

A large number of studies report that NF-kB transcription factors are involved in tumor chemoresistance, probably exerting a strong anti-apoptotic effect [37–44]. In fact, constitutive NF-kB activation has been described in a great number of solid tumors, and this activation appears to support cancer cell survival and reduce the sensitivity against chemotherapeutic drugs [37]. Additionally, some of these drugs induce this transcription factor and, through this mechanism, lower their antiproliferative and proapoptotic potential [37]. Chemoresistance linked to high NF-kB expression has been described in gastric [38], pancreatic [39], osteosarcoma [40], ovarian [41], and breast [42] cancer cells. Accordingly, inhibition of NF-kB by various means has been shown to enhance the sensitivity to antineoplastic- or radiation-induced apoptosis in vitro and in vivo. For instance, Bhardwaj et al. have demonstrated that resveratrol overcomes chemoresistance through down-regulation of the expression

of STAT3 and NF-kB–regulated antiapoptotic and cell survival genes in human multiple myeloma cells [46]; in addition, Singh et al. reported interesting evidences showing that oral silibinin forms a novel chemo-combination with doxorubicin targeting NF-kB–mediated inducible chemoresistance. A final example is the recent paper by Tamatani et al. [48], who examined the mechanisms underlying the enhancement of radiosensitivity and chemosensitivity to gamma-irradiation (IR) and 5-fluorouracil (5-FU) in human oral carcinoma cells (B88) in which NF-kB activity was constitutively suppressed. Three super-repressor forms of IkBα cDNA-transfected cell (B88mI) clones and 1 empty vector-transfected cell clone (B88neo) have been established [48]. They reported that the tumor-forming ability in nude mice of B88mI clones was significantly lower than that of B88 or B88neo. This suppressed ability in tumorigenicity was attributed to the downregulation of the expression of interleukin (IL)-1α, IL-6, IL-8, vascular endothelial growth factor (VEGF), and matrix metalloproteinase (MMP)-9. IR and 5-FU induced a much greater degree of apoptosis, as evidenced by flow cytometry analysis and annexin V staining, in B88mI cell clones than in B88 or B88neo [48]. After treatment of tumor-bearing nude with IR or 5-FU, tumor growth suppression was increased in B88mI cell clones as compared with that obtained employing B88 or B88neo cell clones [48]. ELISA analysis indicated that radiotherapy and chemotherapy-induced production of IL-6 and IL-8 was significantly suppressed in B88mI cell clones. These findings suggest that production of angiogenic factors and growth factors in response to radiotherapy and chemotherapy is a principal mechanism of inducible radioresistance and chemoresistance in human oral cancers, and establish that the inhibition of NF-kB might be a rational approach to improve conventional radiotherapy and chemotherapy outcomes.

4 Estrogen Receptors and Chemoresistance in Breast Cancer Cells

Very interesting findings published in the recent years are related to the cross-talk between estrogen receptor and NF-kB in breast cancer cells [49–51].

Most of the regulatory steps of NF-kB metabolism and regulation are indeed strongly affected by estrogen receptors (ERs), both ERα and ERβ [52–58] (Fig. 11.1). It has been indeed demonstrated that in tissues that express estrogen receptor (ER), 17β-estradiol inhibits NF-κB–driven transcription through multiple mechanisms that might include direct protein–protein interactions [59, 60], inhibition of NF-κB binding to DNA [61, 62], induction of IκB expression [63], or coactivator sharing [64, 65]. Therefore, both ERα and ERβ receptors can antagonize NF-κB transcriptional activity [66, 67]. This cross-talk between ERs and NF-kB has been recently reviewed by De Bosscher et al. [53] and by Kalaitzidis and Gilmore [56].

In conclusion, high levels of estrogen receptor interfere with NF-kB activity [54]. Interestingly, ER(–) breast cancer cells express at very high levels NF-kB and are resistant the estrogen chemotherapy. Accordingly, induction of ER expression in ER(–) breast cancer cells correlates with a more benign phenotype and reversion of chemoresistance [53].

5 Oligonucleotides and Chemoresistance of Tumor Cells

The identification of possible molecular targets for inducing a reversion of chemoresistance allows the proposal of several molecular approaches, based on very selective targets. In this respect, the use of DNA-based molecules appears of great interest. In general, inhibition of gene expression might be achieved by targeting mRNAs by an antisense approach using antisense oligonucleotides [68–70] or small-interfering RNAs [71–73] (siRNA), by targeting selected promoter sequences by the use of triple-helix forming oligonucleotides (TFOs) [74–76] or targeting transcription factors with

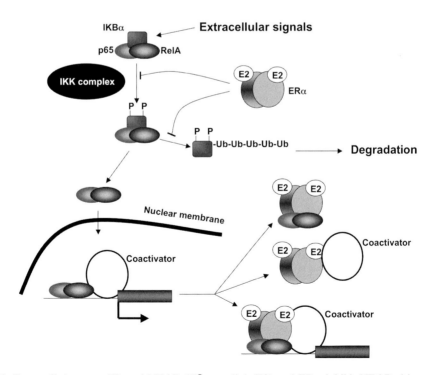

FIGURE 11.1. Cross-talk between ER and NF-kB 17β-estradiol (E2) and ERα inhibit NF-kB–driven transcription through multiple mechanisms that might include direct protein–protein interactions [59, 60], inhibition of NF-kB binding to DNA [61, 62], or coactivator sharing [64, 65]. Ub, ubiquitin

decoy oligonucleotides [77–79]. Several reports demonstrate that RNA- or DNA-based molecules can be employed to control cell growth of resistant tumor cells.

5.1   Antisense Oligonucleotides

Specific inhibition of gene expression through targeting of selected mRNA with antisense oligo-nucleotides (AS-ODNS) has been reported in several research papers and reviews. The mechanism of action of AS-ODN is the RNaseH-mediated degradation of the target mRNA, without other major effects on the overall gene expression. AS-ODN can be delivered and suitably modified to obtain therapeutic molecules clinically useful. Within the context of reversion of chemoresistance of tumor cells, several reports have been published based on AS-ODN. For instance, Bartholomeusz et al. [80] found that Bcl-2 antisense oligonucle-otides overcome resistance to E1A gene therapy in a low HER-2 expressing ovarian cancer tumor xenograft model. In this case, the clinical trial concerns E1A gene therapy for patients with ovarian cancer. The adenovirus type 5 E1A gene suppresses growth of ovarian cancer cells that overexpress HER-2/neu (HER2) and growth of some—but not all—that express low HER2. These authors hypothesized that Bcl-2 prolongs the viability of adenovirus host cells by inhibiting E1A-induced apoptosis. Accordingly, Bcl-2 is overexpressed in ovarian cancer and participates in chemoresistance, probably by inhibiting E1A-induced apoptosis. This hypothesis is sustained by the evidence that E1A suppressed colony forma-tion of ovarian cancer cells that express low levels of Bcl-2 and HER2, but enhanced colony formation in low HER2-, high Bcl-2–expressing ovarian cancer cells. However, when these resistant cells were treated with antisense oligonucleotide Bcl-2, E1A combined led to significant decreases in cell viability resulting from increased apoptosis rela-tive to cells treated with E1A alone. The increase in apoptosis was partly due to cytochrome c release and subsequently caspase-9 activation by Bcl-2-antisense oligonucleotides. Accordingly, in an ovarian cancer xenograft model, treatment with Bcl-2-AS-ODN did not prolong survival, but E1A plus Bcl-2-ASO did [80]. In conclusion, ovarian

tumors overexpressing Bcl-2 may not respond well to E1A gene therapy, but treatment with a combination of E1A and Bcl-2-AS-ODN may overcome this resistance. A second example is that reported by Hopkins-Donaldson et al. [81]. These authors investigated the role of the anti-apoptotic proteins Bcl-2 and Bcl-xL in the chemoresistance of cells derived from malignant pleural mesothe-lioma. First, they determined the basal expression levels of Bcl-2 and Bcl-xL in mesothelioma cells and examined the effect of its downregulation by antisense oligonucleotides. Bcl-xL mRNA and protein could be readily detected in mesothelioma cell lines, whereas only low levels of Bcl-2 mRNA and protein were found. Preferential downregula-tion of either Bcl-xL alone or of Bcl-xL and Bcl-2 simultaneously was achieved by treatment with antisense oligonucleotides, whereas the expres-sion of other apoptosis-relevant genes remained unaffected. Treatment with these oligonucleotides lowered the apoptosis threshold in mesothelioma cells, as indicated by an increase in cell death accompanied by increased caspase-3–like activity, a decrease of the mitochondrial transmembrane potential and the cleavage of procaspase-7 [80]. In addition to the direct induction of apoptosis, antisense treatment sensitized the cells to the cyto-static effect of cisplatin and gemcitabine. These and other similar experimental evidences firmly demonstrate that antisense therapy targeting apoptosis-related genes is able to reverse chemoresistance of tumor cells.

5.2   The RNA Interference Strategy

RNA interference is a recently discovered experi-mental approach of sequence-specific, post-transcriptional gene silencing that is initiated by double-stranded RNA molecules known as small interfering RNAs (siRNAs) [82]. The siRNA mol-ecules have an acceptable half-life in vitro and, in comparison with antisense oligonucleotides, and a very efficient transcript knockdown Bcl and thresh-old concentration. Following the paper published by Elbashir et al. [82], synthetic siRNAs have gained wide acceptance as a laboratory tool for target validation. Currently, there is considerable interest in the therapeutic use of siRNA, par-ticularly in areas of infectious disease and cancer.

In vitro and in vivo findings demonstrate the efficacy of siRNA knockdown of gene messages that are pivotal for tumor cell growth, metastasis, angiogenesis, and chemoresistance, leading to tumor growth suppression. However, siRNA-based cancer therapy faces similar pharmacokinetic limitations to antisense therapy with respect to the extent that siRNA accesses primary and metastatic target cells. In spite of these limitations, application of the siRNA approach to oncology has been proposed [82–85]. In line with this hypothesis, Leong et al. [83] presented data on silencing expression of UO-44 (CUZD1) using small interfering RNA, obtaining sensitization of human ovarian cancer cells to cisplatin in vitro.

As far as effects of siRNAs on the NF-kB system, Guo et al. demonstrated that increased sensitivity to the chemotherapeutic agent irinotecan in tumor cells in which downregulation of NF-kB was achieved. This is a very important point, since irinotecan belongs to the class of chemotherapy agents that activate NF-kB. This leads to increased expression of NF-kB regulated genes such as c-IAP1 and c-IAP2, which might be responsible for the inhibition of chemotherapy-induced apoptosis [84]. In their study, Guo et al. used a siRNA designed to downregulate the NF-kB p65 subunit in a colon cancer cell line. Reduction of endogenous p65 by siRNA treatment significantly impaired irinotecan-mediated NF-kB activation, enhanced apoptosis, and reduced colony formation in soft agar. Furthermore, the in vivo administration of p65 siRNA reduced tumor formation in xenograft models in the presence but not the absence of CPT-11 administration. These data indicate that the administration of siRNA directed against the p65 subunit of NF-kB can effectively enhance in vitro and in vivo sensitivity to chemotherapeutic agents [84, 85].

Among DNA-based non-viral gene therapy approaches, the delivery of double-stranded oligodeoxynucleotides (ODN) containing the consensus binding sequence of a transcription factor is a promising approach to treat several diseases by modulating transcription [74–76, 86, 87]. This transcription factor decoy (TFD) strategy results in the attenuation of the authentic cis–trans interaction, leading to the removal of transcription factors from the endogenous cis-element. TFD has been proved effective in vitro and in vivo, suggesting its use in therapy [88–90].

6 The Transcription Factor Decoy Approach for Alteration of Gene Expression

The transcription factor decoy (TFD) approach is based on the competition for trans-acting factors between endogenous cis-elements present within the regulatory regions of the target gene and exogenously added DNA decoys (or functionally active DNA analogues) mimicking the specific cis-elements [77–79, 84–90]. The objective of this molecular intervention is to cause an attenuation of the authentic interactions of trans-factors with their cis-elements, leading to a removal of the trans-factors from the endogenous cis-element inside the cell. The most important feature of potential decoy molecules is the ability to tightly bind to target transcriptions factors. In this case, the expression of genes directly regulated by the targeted transcription factors will be deeply altered. If the TFD molecules bind to a transcriptional activator, the gene transcription will be inhibited (Fig. 11.2, *left panel*); on the contrary, gene expression will be up-regulated if the TFD molecules is directed against a transcriptional repressor (see Fig. 11.2, *right panel*).

7 Inhibition of NF-kB Activity by TFD Molecules

The TFD strategy is a very promising approach to alter NF-kB regulated gene expression. Several studies reported TFD-mediated inhibition of the expression of specific genes. For instance, Shibuya et al. [91] described a double-strand decoy DNA oligomer for NF-kB able to inhibit TNFα-induced ICAM-1 expression in sinusoidal endothelial cells. Inhibition of TNFα induced expression was also reported by Tomita et al. [92].

These studies have practical implications, since it could be hypothesized an effect of NF-kB TFD molecules on cell cycle, as published by Gill et al. [93]. In this case, NF-kB decoy oligonucleotides were used with the aim to inhibit cell growth of the of glioblastoma (GBM) cell line growth. This approach was sustained by the finding that aberrant nuclear expression of NF-kB was found in a panel

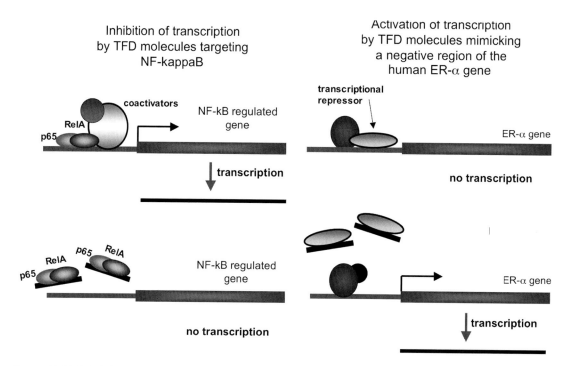

FIGURE 11.2. Modulation of gene expression using transcription factor decoy molecules. TFD molecules against NF-kB transcription factors usually causes inhibition of transcription of NF-kB regulated genes [86] *(left panel)*. TFD molecules mimicking a negative regulatory region of the human Era gene induce activation of transcription [87, 102]

of GBM cell lines, whereas untransformed glial cells did not display NF-kB activity. The reduction in cell number correlated with a decrease in cyclin D1 protein expression and a commensurate decrease in Cdk-4 activity.

In a recent paper [94] we used NF-kB decoy ODNs to alter the phenotype of human osteoclasts. In this case, mature osteoclasts and TRAP-positive precursors were treated with decoy molecules against NF-kB factors and then subjected to analysis of apoptosis. Morphologic analysis of NF-kB decoy-treated osteoclasts demonstrated cell retraction in comparison with unrelated ODN control–treated osteoclasts, indicative of apoptosis. Several nuclei showed morphologic changes consistent with nuclear damage. To confirm apoptosis, TUNEL staining of fragmented DNA was performed. The induction of apoptosis of primary osteoclasts treated with decoy ODN against NF-kB was associated with activation of caspase-3 and inhibition of IL-6.

In vivo applications of NF-kB decoy molecules have been reported by several groups, showing inhibition of vascular injury [95], cardioprotection [96], and regression of atopic dermatitis [97]. Alteration of specific gene expression in vivo was reported by Yoshimura et al. [98], and by Ueno et al. [99]. Yoshimura et al. described that vivo transfection of NFkB decoy oligodeoxynucleotides into balloon-injured rat carotid artery resulted in marked decrease of the expression of ICAM-1 and VCAM-1 genes in blood vessels transfected with NFkB decoy ODN compared with scrambled ODN [98]. Ueno et al. [99] demonstrated that introduction of NF-kB decoy ODN into rat brain neurons through the carotid artery during global brain ischemia inhibited the expression of TNFα-induced IL-1 beta and ICAM-1 messenger RNA 1 hour after global brain ischemia.

In conclusion, a growing number of reports demonstrate the possible use of NF-kB decoy molecules to treat several pathologies, especially related to inflammation and employing delivery systems effective.

As far as chemotherapy resistance, the TFD approach targeting NF-kB has been found useful.

For instance, Romano et al. [100] showed enhancement of cytosine arabinoside–induced apoptosis in human myeloblastic leukemia cells by NF-kB/Rel– specific decoy oligodeoxynucleotides. In this study, cell incubation with this ODN, but not its mutated (scrambled) form used as a control, resulted in abating the NF-kB/Rel nuclear levels in these cells, as verified by electrophoretic mobility shift assay (EMSA). Incubation of the leukemic cells with AraC in either the absence or presence of the decoy ODN demonstrated that the NF-kB decoy enhanced Ara-C–induced cellular apoptosis, suggesting that the inhibition of NF-kappa B/Rel factors can improve the effect of chemotherapy in AML. In addition, Uetsuka et al. [101] studied 5-FU induction of apoptosis of several cancer cell lines, investigating the relationship between activation of NF-kB and chemoresistance to 5-FU. They used two stomach cancer cell lines, NUGC3 (5-FU–sensitive) and NUGC3/5FU/L (5-FU–resistant), showing that inhibition of inducible NF-kB activation by using a NF-kB decoy could induce apoptosis and reduce chemoresistance against 5-FU. These results suggest that 5-FU chemoresistance can be overcome by inhibition of inducible NF-kB activation, and that the use of the NF-kB decoy combined with 5-FU treatment is a new molecular and gene therapeutic strategy aimed at treatment of human stomach cancers resistant to 5-FU.

8   Induction of ERα Gene Expression by TFD Molecules

The activation of ERα gene expression has been demonstrated very important on breast cancer cells resistant to hormone chemotherapy [49–51]. As far as modulation of ERα gene expression by the TFD approach, several reports have been published [102–108]. For instance, Lambertini et al. [104] address the hypothesis that transfection of oligonucleotide mimicking a negative regulatory sequence of promoter C of estrogen receptor alpha (ERα) gene was sufficient for its re-expression in ER-negative human cancer cell lines. They demonstrated that after this decoy treatment, the cells produced a functional ERα protein able to respond to 17-beta-estradiol and transactivate a transfected estrogen response element (ERE)–regulated reporter gene. The effects of reactivated ERα protein and its

estrogen dependence on endogenous target gene expression level, such as ERβ, have been also assessed. The proliferation of the cells transfected with low levels of decoy was significantly increased by estrogen and not by tamoxifen, suggesting that the levels of reactivated ERα in these decoy conditions confers a certain hormone sensitivity. On the contrary, high-level expression of ERα obtained at high doses of transfected decoy molecule produced a progressive decrease of cell proliferation. Decoy sequences mimicking the human ERα gene promoter were also demonstrated to induce apoptosis in human primary osteoclasts, but not in osteoblasts, in an estrogen-dependent manner, increasing also caspase 3 and Fas receptor levels [87]. These findings may be of relevance for a possible therapeutical approach for tumors, bone metastasis and osteopenic diseases.

In conclusion this research group has firmly demonstrated that the expression of estrogen receptor is under a negative transcriptional control, and that decoy ODNs against negative transcription factors could be employed for conversion of bresat cancer tumor cells from a ER(–) to an ER(+) phenotype, thereby reversing hormone chemoresistance.

9   Conclusions and Perspectives

The conclusion of this short chapter is that oligonucleotides and small interfering RNAs targeting pathways implicated in chemotherapy resistance might be employed in combined antitumor therapy. In the future, two issues deserve great efforts: (1) development of more stable and easily delivered molecules, and (2) identification of novel targets.

As for other ODN-based therapeutic strategies, the successful use of transcription factor decoys will almost always depend on stability of the employed molecules and efficient delivery to target tissues, cells and cellular compartments [109]. For instance, decoy ODNs might require nuclear localization if they are to prevent the transactivation of their target genes. Unfortunately, the endocytotic pathways translocate most of the employed ODN into lysosomal compartments, where they are efficiently degraded [110]. Therefore, the search for decoy biomolecules exhibiting on one hand efficient decoy activity, on the other hand resistance in serum and cellular extracts appears to be a major issue in pharmacological research. Examples are modified oligonucleotides, LNA (locked nucleic acids) [90] and peptide nucleic acids (PNA) based transcription factors decoy [109, 110].

For instance, peptide nucleic acids (PNAs) are recently described DNA mimics in which the sugar-phosphate backbone is replaced by N-(2-aminoethyl)glycine units [111]. These molecules efficiently hybridize with complementary DNA, forming Watson-Crick double helices. While the possible use of PNAs for targeting mRNA molecules and genomic sequences are well documented [112–114], the published reports on the possible use of PNA-based double stranded molecules to target transcription factors indicate that these molecules are not efficient [115]. By sharp contrast, PNA-DNA chimeras (Fig. 11.3) are of great interest for TFD pharmacotherapy. These molecules are composed of a part of PNA and a part of DNA. The PNA–DNA chimeras obey the Watson-Crick rules on binding to complementary DNA and RNA. We have designed and tested PNA-DNA-PNA (PDP) chimeras mimicking the NF-kB and Sp1 binding sites [116, 117]. In a first study, PDP chimeras carrying the NF-kB binding sites were analyzed by circular dichroism and gel shift assay [116]; the results obtained gave clear evidence for an efficient decoy activity by these molecules. Recent published observations of our research group show that PNA-DNA-PNA chimeras are more resistant than the DNA counterpart to endonucleases and when incubated in the presence of serum or cellular extracts [118, 119]. The results obtained by Borgatti et al. [119] indicate that only DNA/DNA decoy molecules are degraded in the presence of serum and cellular extracts isolated from human leukemic K562 cells, whereas DNA/PDP and PDP/PDP hybrids exhibit higher resistance levels. When PNA-DNA-PNA chimeras targeting NF-kB are used in cellular systems, we were able to demonstrate that PDP/DNA, DNA/PDP and PDP/PDP are all active in inducing apoptosis at a level similar to that obtained with DNA [21].

As far as identification of novel targets involved in chemotherapy resistance, proteomic approaches for studying chemoresistance in cancer are expected to be of great interest [120, 121]. For instance, Righetti et al. [120] recently reviewed proteomic papers reviewed supporting the concept that many metabolic pathways are affected during the chemoresistance process of tumor cells. Although the modulation of expression levels of such proteins is not clear proof of their role in drug resistance per se, at least some of the themes are very likely to be

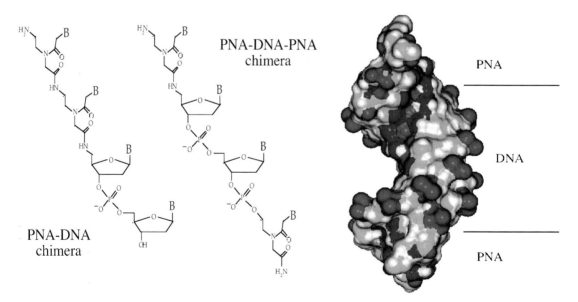

FIGURE 11.3. Structure of PNA-DNA-PNA chimeras targeting NF-kappaB–related proteins. The molecular structures and the sequence of the double-stranded PNA-DNA-PNA chimera mimicking NF-kappaB binding sites are modified from Borgatti et al. [116], Romanelli et al. [118], and Gambari et al. [109] (*See Color Plates*)

involved in the resistance phenotype, and thus may be potential targets for new drugs. Proteomic and transcriptomic analyzes are expected to identify in a near future all the molecular genes whose expression is causative of the resistance tumor cells to chemotherapy, allowing the identification of novel targets.

Acknowledgments. This work was supported by MIUR COFIN-2005, by AIRC, by Associazione Italiana per la Lotta alla Talassemia, by the Italian Cystic Fibrosis Research Foundation and by Fondazione CARIPARO.

References

1. Shah AN, Gallick GE. Src, chemoresistance and epithelial to mesenchymal transition: are they related? Anticancer Drugs 2007, 18:371–375.

2. Aggarwal BB, Sethi G, Ahn KS, et al. Targeting signal-transducer-and-activator-of-transcription-3 for prevention and therapy of cancer: modern target but ancient solution. Ann N Y Acad Sci U S A 2006, 1091:151–169.

3. Wilson TR, Longley DB, Johnston PG. Chemoresistance in solid tumours. Ann Oncol 2006, 17:315–324.

4. Teodoridis JM, Strathdee G, Plumb JA, et al. CpG-island methylation and epigenetic control of resistance to chemotherapy. Biochem Soc Trans 2004, 32:916–917.

5. La Porta CA. Drug resistance in melanoma: new perspectives. Curr Med Chem 2007, 14:387–391.

6. Efstathiou E, Logothetis CJ. Review of late complications of treatment and late relapse in testicular cancer. J Natl Compr Canc Netw 2006, 4:1059–1070.

7. Chou AJ, Gorlick R. Chemotherapy resistance in osteosarcoma: current challenges and future directions. Expert Rev Anticancer Ther 2006, 6:1075–1085.

8. Gatti L, Zunino F. Overview of tumor cell chemoresistance mechanisms. Methods Mol Med 2005, 111:127–148.

9. Luqmani YA. Mechanisms of drug resistance in cancer chemotherapy. Med Princ Pract 2005, 14:35–48.

10. Modok S, Mellor HR, Callaghan R. Modulation of multidrug resistance efflux pump activity to overcome chemoresistance in cancer. Curr Opin Pharmacol 2006, 6:350–354.

11. Takara K, Sakaeda T, Okumura K. An update on overcoming MDR1–mediated multidrug resistance in cancer chemotherapy. Curr Pharm Des 2006, 12:273–286.

12. Garg AK, Buchholz TA, Aggarwal BB. Chemosensitization and radiosensitization of tumors by plant polyphenols. Antiox Redox Signal 2005, 7:1630–1647.

13. Cheng JQ, Lindsley CW, Cheng GZ, et al. The Akt/PKB pathway: molecular target for cancer drug discovery. Oncogene 2005, 24:7482–7492.

14. Boehrer S, Nowak D, Hoelzer D, et al. Novel agents aiming at specific molecular targets increase chemosensitivity and overcome chemoresistance in hematopoietic malignancies. Curr Pharm Des 2006, 12:111–128.

15. Campioni M, Santini D, Tonini G, et al. Role of Apaf-1, a key regulator of apoptosis, in melanoma progression and chemoresistance. Exp Dermatol 2005, 14:811–818.

16. Fraser M, Leung B, Jahani-Asl A, et al. Chemoresistance in human ovarian cancer: the role of apoptotic regulators. Reprod Biol Endocrinol 2003, 1:66.

17. Soengas MS, Lowe SW. Apoptosis and melanoma chemoresistance. Oncogene 2003, 22:3138–3151.

18. Cheng JQ, Jiang X, Fraser M, et al. Role of X-linked inhibitor of apoptosis protein in chemoresistance in ovarian cancer: possible involvement of the phosphoinositide-3 kinase/Akt pathway. Drug Resist Update 2002, 5:131–146.

19. Schmitt CA, Lowe SW. Apoptosis and chemoresistance in transgenic cancer models. J Mol Med 2002, 80:137–146.

20. Deng X, Kornblau SM, Ruvolo PP, et al. Regulation of Bcl2 phosphorylation and potential significance for leukemic cell chemoresistance. J Natl Cancer Inst Monogr 2001, 28:30–37.

21. Krajewski S, Krajewska M, Turner BC, et al. Prognostic significance of apoptosis regulators in breast cancer. Endocr Relat Cancer 1999, 6:29–40.

22. Devarajan E, Sahin AA, Chen JS, et al. Down-regulation of caspase 3 in breast cancer: a possible mechanism for chemoresistance. Oncogene 2002, 21:8843–8851.

23. Bacher S, Schmitz ML. The NF-kappaB pathway as a potential target for autoimmune disease therapy. Curr Pharm Des 2004, 10:2827–2837.

24. Schmitz ML, Bacher S, Dienz O. NF-kappaB activation pathways induced by T cell costimulation. FASEB J 2003, 17:2187–2193.

25. Chen F, Demers LM, Shi X. Upstream signal transduction of NF-kappaB activation. Curr Drug Targets Inflamm Allergy 2002, 1:137–149.

26. Liou HC. Regulation of the immune system by NF-kappaB and IkappaB. J Biochem Mol Biol 2002, 35:537–546.

27. Tian B, Brasier AR. Identification of a nuclear factor kappa B-dependent gene network. Recent Prog Horm Res 2003, 58:95–130.

28. Storz P, Toker A. NF-kappaB signaling—an alternate pathway for oxidative stress responses. Cell Cycle 2003, 2:9–10.

29. Moscat J, Diaz-Meco MT, Rennert P. NF-kappaB activation by protein kinase C isoforms and B-cell function. EMBO Rep 2003, 4:31–36.

30. Hassa PO, Covic M, Hasan S, et al. The enzymatic and DNA binding activity of PARP-1 are not required for NF-kappa B coactivator function. J Biol Chem 2001, 276:45588–45597.

31. Hassa PO, Hottiger MO. The functional role of poly(ADP-ribose)polymerase 1 as novel coactivator of NF-kappaB in inflammatory disorders. Cell Mol Life Sci 2002, 59:1534–1553.

32. Wang T, Zhang X, Li JJ. The role of NF-kappaB in the regulation of cell stress responses. Int Immuno pharmacol 2002, 2:1509–1520.

33. Aggarwal BB, Takada Y, Shishodia S, et al. Nuclear transcription factor NF-kappa B: role in biology and medicine. Indian J Exp Biol 2004, 42:341–353.

34. Gaur U, Aggarwal BB. Regulation of proliferation, survival and apoptosis by members of the TNF superfamily. Biochem Pharmacol 2003, 66:1403–1408.

35. Li Q, Verma IM. NF-kappaB regulation in the immune system. Nat Rev Immunol 2002, 2:725–734.

36. Panwalkar A, Verstovsek S, Giles F. Nuclear factor-kappaB modulation as a therapeutic approach in hematologic malignancies. Cancer 2004, 100:1578–1589.

37. Arlt A, Schafer H. NFkappaB-dependent chemoresistance in solid tumors. Int J Clin Pharmacol Ther 2002, 40:336–347.

38. Camp ER, Li J, Minnich DJ, et al. Inducible nuclear factor-kappaB activation contributes to chemotherapy resistance in gastric cancer. J Am Coll Surg 2004, 199:249–258.

39. Muerkoster S, Arlt A, Sipos B, et al. Increased expression of the E3-ubiquitin ligase receptor subunit betaTRCP1 relates to constitutive nuclear factor-kappaB activation and chemoresistance in pancreatic carcinoma cells. Cancer Res 2005, 65:1316–1324.

40. Andela VB, Siddiqui F, Groman A, et al. An immunohistochemical analysis to evaluate an inverse correlation between Runx2/Cbfa1 and NF kappa B in human osteosarcoma. J Clin Pathol 2005, 58:328–330.

41. Salvatore C, Camarda G, Maggi CA, et al. NF-kappaB activation contributes to anthracycline resistance pathway in human ovarian carcinoma cell line A2780. Int J Oncol 2005, 27:799–806.

42. Montagut C, Tusquets I, Ferrer B. Activation of nuclear factor-kappa B is linked to resistance to neoadjuvant chemotherapy in breast cancer patients. Endocr Relat Cancer 2003, 13:607–616.

43. Grandage VL, Gale RE, Linch DC, et al. PI3-kinase/Akt is constitutively active in primary acute myeloid leukaemia cells and regulates survival and chemoresistance via NF-kappaB, Mapkinase and p53 pathways. Leukemia 2005, 19:586–594.

44. Bharti AC, Shishodia S, Reuben JM. Nuclear factor-kappaB and STAT3 are constitutively active in CD138+ cells derived from multiple myeloma patients, and suppression of these transcription factors leads to apoptosis. Blood 2004, 103:3175–3184.

45. Kim DS, Park SS, Nam BH, et al. Reversal of drug resistance in breast cancer cells by transglutaminase 2 inhibition and nuclear factor-kappaB inactivation. Cancer Res 2006, 66:10936–10943.

46. Bhardwaj A, Sethi G, Vadhan-Raj S, et al. Resveratrol inhibits proliferation, induces apoptosis, and overcomes chemoresistance through down-regulation of STAT3 and nuclear factor-kappaB-regulated antiapoptotic and cell survival gene products in human multiple myeloma cells. Blood 2007, 109:2293–2302.

47. Singh RP, Mallikarjuna GU, Sharma G. Oral silibinin inhibits lung tumor growth in athymic nude mice and forms a novel chemocombination with doxorubicin targeting nuclear factor kappaB-mediated inducible chemoresistance. Clin Cancer Res 2004, 10:8641–8647.

48. Tamatani T, Azuma M, Ashida Y. Enhanced radiosensitization and chemosensitization in NF-kappaB-suppressed human oral cancer cells via the inhibition of gamma-irradiation- and 5-FU-induced production of IL-6 and IL-8. Int J Cancer 2004, 108:912–921.

49. Holloway JN, Murthy S, El-Ashry D. A cytoplasmic substrate of mitogen-activated protein kinase is responsible for estrogen receptor-alpha down-regulation in breast cancer cells: the role of nuclear factor-kappaB. Mol Endocrinol 2004, 18:1396–1410.

50. Simstein R, Burow M, Parker A, et al. Apoptosis, chemoresistance, and breast cancer: insights from the MCF-7 cell model system. Exp Biol Med (Maywood) 2003, 228:995–1003.

51. Real PJ, Sierra A, De Juan A, et al. Resistance to chemotherapy via Stat3-dependent overexpression of Bcl-2 in metastatic breast cancer cells. Oncogene 2002, 21:7611–7618.

52. Zhou Y, Yau C, Gray JW, et al. Enhanced NF kappa B and AP-1 transcriptional activity associated with antiestrogen resistant breast cancer. BMC Cancer 2007, 7:59.

53. De Bosscher K, Vanden Berghe W, Haegeman G. Cross-talk between nuclear receptors and nuclear factor kappaB. Oncogene 2006, 25:6868–6886.

54. Keith JC Jr, Albert LM, Leathurby Y, et al. The utility of pathway selective estrogen receptor ligands that inhibit nuclear factor-kappa B transcriptional activity in models of rheumatoid arthritis. Arthritis Res Ther 2005, 7:427–438.

55. Ghisletti S, Meda C, Maggi A, et al. 17beta-estradiol inhibits inflammatory gene expression by controlling

NF-kappaB intracellular localization. Mol Cell Biol 2005, 25:2957–2968.

56. Kalaitzidis D, Gilmore TD. Transcription factor cross-talk: the estrogen receptor and NF-kappaB. Trends Endocrinol Metab 2005, 16:46–52.

57. Chadwick CC, Chippari S, Matelan E, et al. Identification of pathway-selective estrogen receptor ligands that inhibit NF-kappaB transcriptional activity. Proc Natl Acad Sci U S A 2005, 102:2543–2548.

58. Pratt MA, Bishop TE, White D, et al. Estrogen withdrawal-induced NF-kappaB activity and bcl-3 expression in breast cancer cells: roles in growth and hormone independence. Mol Cell Biol 2003, 23:6887–900.

59. Stein B, Yang MX. Repression of the interleukin–6 promoter by estrogen receptor is mediated by NF-kappa B and C/EBP beta. Mol Cell Biol 1995, 15:4971–4979.

60. Ray A, Prefontaine KE, Ray P. Down-modulation of interleukin-6 gene expression by 17 beta-estradiol in the absence of high affinity DNA binding by the estrogen receptor. J Biol Chem 1994, 269:12940–12946.

61. Ray P, Ghosh SK, Zhang DH, et al. Repression of interleukin-6 gene expression by 17 beta-estradiol: inhibition of the DNA-binding activity of the transcription factors NF-IL6 and NF-kappa B by the estrogen receptor. FEBS Lett 1997, 409:79–85

62. Deshpande R, Khalili H, Pergolizzi RG, et al. Estradiol down-regulates LPS-induced cytokine production and NFκB activation in murine macrophages. Am J Reprod Immunol 1997, 38:46–54.

63. Sun WH, Keller ET, Stebler BS, et al. Estrogen inhibits phorbol ester–induced I kappa B alpha transcription and protein degradation. Biochem Biophys Res Commun 1998, 244:691–695.

64. Harnish DC, Scicchitano MS, Adelman SJ, et al. The role of CBP in estrogen receptor cross-talk with nuclear factor-kappaB in HepG2 cells. Endocrinology 2000, 141:3403–3411.

65. Speir E, Yu ZX, Takeda K, et al. Competition for p300 regulates transcription by estrogen receptors and nuclear factor-kappa B in human coronary smooth muscle cells. Circ Res 2000, 87:1006–1011.

66. Tyree CM, Zou A, Allegretto EA. 17beta-Estradiol inhibits cytokine induction of the human E-selectin promoter. J Steroid Biochem Mol Biol 2002, 8:291–297.

67. Valentine JE, Kalkhoven E, White R, et al. Mutations in the estrogen receptor ligand binding domain discriminate between hormone-dependent transactivation and transrepression. J Biol Chem 2000, 275:25322–25329.

68. Prochownik EV. c-Myc as a therapeutic target in cancer. Expert Rev Anticancer Ther 2004, 4:289–302.

69. Biroccio A, Leonetti C, Zupi G. The future of antisense therapy: combination with anticancer treatments. Oncogene 2003, 22:6579–6588.

70. Cho-Chung YS. Antisense DNAs as targeted genetic medicine to treat cancer. Arch Pharm Res 2003, 26:183–191.

71. Milhavet O, Gary DS, Mattson MP. RNA interference in biology and medicine. Pharmacol Rev 2003, 55:629–648.

72. Chakraborty C. Potentiality of small interfering RNAs (siRNA) as recent therapeutic targets for gene-silencing. Curr Drug Targets 2007, 8:469–482.

73. Zaratiegui M, Irvine DV, Martienssen RA. Noncoding RNAs and gene silencing. Cell 2007, 128:763–776.

74. Rogers FA, Lloyd JA, Glazer PM. Triplex-forming oligonucleotides as potential tools for modulation of gene expression. Curr Med Chem Anticancer Agents 2005, 5:319–326.

75. Giovannangeli C, Helene C. Progress in developments of triplex-based strategies. Antisense Nucleic Acid Drug Dev 1997, 7:413–421.

76. Besch R, Giovannangeli C, Degitz K. Triplex-forming oligonucleotides—sequence-specific DNA ligands as tools for gene inhibition and for modulation of DNA-associated functions. Curr Drug Targets 2004, 5:691–703.

77. Morishita R, Sugimoto T, Aoki M, et al. In vivo transfection of cis element "decoy" against nuclear factor-kB binding site prevents myocardial infarction. Nat Med 1997, 3:894–899.

78. Mann MJ, Dzau VJ. Therapeutic application of transcriptional factor decoy oligonucleotides. J Clin Invest 2000, 106:1071–1075.

79. Piva R, Gambari R. Transcription factor decoy (TFD) in breast cancer research and treatment. Technol Cancer Res Treat 2002, 1:405–416.

80. Bartholomeusz C, Itamochi H, Yuan LX, et al. Bcl-2 antisense oligonucleotide overcomes resistance to E1A gene therapy in a low HER2--expressing ovarian cancer xenograft model. Cancer Res 2005, 65:8406–8413.

81. Hopkins-Donaldson S, Cathomas R, Simoes-Wust AP, et al. Induction of apoptosis and chemosensitization of mesothelioma cells by Bcl-2 and Bcl-xL antisense treatment. Int J Cancer 2003, 106:160–166.

82. Elbashir SM, Harborth J, Lendeckel W, et al. Duplexes of 21-nucleotide RNAs mediate RNA interference in cultured mammalian cells. Nature 2001, 411:494–498.

83. Leong CT, Ong CK, et al. Silencing expression of UO-44 (CUZD1) using small interfering RNA sensitizes human ovarian cancer cells to cisplatin in vitro. Oncogene 2007, 26:870–880.

84. Guo J, Verma UN, Gaynor RB, et al. Enhanced chemosensitivity to irinotecan by RNA interference-

mediated down-regulation of the nuclear factor-kappaB p65 subunit. Clin Cancer Res 2004, 10:3333–
3341.

85. Veiby OP, Read MA. Chemoresistance: impact of
nuclear factor (NF)-kappaB inhibition by small interfering RNA. Clin Cancer Res 2004, 10:3333–3341.

86. Piva R, Penolazzi L, Zennaro M, et al. Induction of
apoptosis of osteoclasts by targeting transcription
factors with decoy molecules. Ann NY Acad Sci
2006, 1091:509–516.

87. Piva R, Penolazzi L, Lambertini E, et al. Induction of
apoptosis of human primary osteoclasts treated with
a transcription factor decoy mimicking a promoter
region of estrogen receptor alpha. Apoptosis 2005,
10:1079–1094.

88. Borgatti M, Finotti A, Romanelli A, et al. Peptide
nucleic acids (PNA)-DNA chimeras targeting transcription factors as a tool to modify gene expression.
Curr Drug Targets 2004, 5:735–744.

89. Tomita N, Azuma H, Kaneda Y, et al. Application
of decoy oligodeoxynucleotides-based approach to
renal diseases. Curr Drug Targets 2004, 5:717–733.

90. Crinelli R, Bianchi M, Gentilini L, et al. Locked
nucleic acids (LNA): versatile tools for designing
oligonucleotide decoys with high stability and affinity. Curr Drug Targets 2004, 5:745–752.

91. Shibuya T, Takei Y, Hirose M, et al. A double-
strand decoy DNA oligomer for NF-kappaB inhibits
TNFalpha-induced ICAM-1 expression in sinusoidal
endothelial cells. Biochem Biophys Res Commun
2002, 298:10–16.

92. Tomita N, Morishita R, Yamamoto K, et al. Targeted
gene therapy for rat glomerulonephritis using HVJ-
immunoliposomes. J Gene Med 2002, 4:527–535.

93. Gill JS, Zhu X, Moore MJ, et al. Effects of NFkappaB
decoy oligonucleotides released from biodegradable
polymer microparticles on a glioblastoma cell line.
Biomaterials 2002, 23:2773–2781.

94. Penolazzi L, Lambertini E, Borgatti M, et al. Decoy
oligodeoxynucleotides targeting NF-kappaB transcription factors: induction of apoptosis in human
primary osteoclasts. Biochem Pharmacol 2003,
66:1189–1198.

95. Yamasaki K, Asai T, Shimizu M, et al. Inhibition of
NFkappaB activation using cis-element 'decoy' of
NFkappaB binding site reduces neointimal formation
in porcine balloon-injured coronary artery model.
Gene Ther 2003, 10:356–364.

96. Kupatt C, Wichels R, Deiss M, et al. Retroinfusion of
NFkappaB decoy oligonucleotide extends cardioprotection achieved by CD18 inhibition in a preclinical
study of myocardial ischemia and retroinfusion in
pigs. Gene Ther 2002, 9:518–526.

97. Nakamura H, Aoki M, Tamai K, et al. Prevention
and regression of atopic dermatitis by ointment
containing NF-kB decoy oligodeoxynucleotides
in NC/Nga atopic mouse model. Gene Ther 2002,
9:1221–1229.

98. Yoshimura S, Morishita R, Hayashi K, et al. Inhibition
of intimal hyperplasia after balloon injury in rat
carotid artery model using cis-element 'decoy' of
nuclear factor-kappaB binding site as a novel molecular strategy. Gene Ther 2001, 8(31):1635–1642.

99. Ueno T, Sawa Y, Kitagawa-Sakakida S, et al.
Nuclear factor-kappa B decoy attenuates neuronal
damage after global brain ischemia: a future strategy for brain protection during circulatory arrest.
J Thorac Cardiovasc Surg 2001, 122:720–727.

100. Romano MF, Lamberti A, Bisogni R, et al.
Enhancement of cytosine arabinoside–induced
apoptosis in human myeloblastic leukemia cells by
NF-kappa B/Rel-specific decoy oligodeoxynucleotides. Gene Ther 2000, 7:1234–1237.

101. Uetsuka H, Haisa M, Kimura M, et al. Inhibition of
inducible NF-kappaB activity reduces chemoresistance to 5-fluorouracil in human stomach cancer cell
line. Exp Cell Res 2003, 289:27–35.

102. Penolazzi L, Zennaro M, Lambertini E, et al. Induction
of estrogen receptor {alpha} expression with decoy
oligonucleotide targeted to NFATc1 binding sites in
osteoblasts. Mol Pharmacol 2007, 71:1457–1462.

103. Lambertini E, Lampronti I, Penolazzi L, et al.
Expression of estrogen receptor alpha gene in breast
cancer cells treated with transcription factor decoy
is modulated by Bangladeshi natural plant extracts.
Oncol Res 2005, 15:69–79.

104. Lambertini E, Penolazzi L, Magaldi S, et al.
Transcription factor decoy against promoter C of
estrogen receptor alpha gene induces a functional
ER alpha protein in breast cancer cells. Breast
Cancer Res Treat 2005, 92:125–132.

105. Lambertini E, Penolazzi L, Aguiari G, et al.
Osteoblastic differentiation induced by transcription
factor decoy against estrogen receptor alpha gene.
Biochem Biophys Res Commun 2002, 292:761–770.

106. Lambertini E, Penolazzi L, Sollazzo V, et al.
Modulation of gene expression in human osteoblasts by targeting a distal promoter region of human
estrogen receptor-alpha gene. J Endocrinol 2002,
172:683–693.

107. Piva R, del Senno L, Lambertini E, et al. Modulation
of estrogen receptor gene transcription in breast
cancer cells by liposome delivered decoy molecules.
J Steroid Biochem Mol Biol 2000, 75:121–128.

108. Penolazzi L, Lambertini E, Aguiari G, et al. Cis element 'decoy' against the upstream promoter of the

human estrogen receptor gene. Biochim Biophys Acta 200, 1492:560–567.

109. Gambari R. Biological activity and delivery of peptide nucleic acids (PNA)-DNA chimeras for transcription factor decoy (TFD) pharmacotherapy. Curr Med Chem 2004, 11:1253–1263.

110. Gambari R. New trends in the development of transcription factor decoy (TFD) pharmacotherapy. Curr Drug Targets 2004, 5:419–430.

111. Nielsen PE, Egholm M, Berg RH, Sequence-selective recognition of DNA by strand displacement with a thymine-substituted polyamide. Science 1991, 254:1497–1500.

112. Nielsen PE, Egholm M. An introduction to peptide nucleic acid. Curr Issues Mol Biol 1999, 1:89–104.

113. Egholm M, Buchardt O, Christensen L, et al. PNA hybridizes to complementary oligonucleotides obeying the Watson-Crick hydrogen-bonding rules. Nature 1993, 365:566–568.

114. Borgatti M, Boyd DD, Lampronti I, et al. Decoy molecules based on PNA–DNA chimeras and targeting Sp1 transcription factors inhibit the activity of urokinase-type plasminogen activator receptor (uPAR) promoter. Oncol Res 2005, 15:373–383.

115. Mischiati C, Borgatti M, Bianchi N, et al. Interaction of the human NF-kappaB p52 transcription factor with DNA-PNA hybrids mimicking the NF-kappaB binding

sites of the human immunodeficiency virus type 1 promoter. J Biol Chem 1999, 274:33114–33122.

116. Romanelli A, Pedone C, Saviano M, et al. Molecular interactions with nuclear factor kappaB (NF-kappaB) transcription factors of a PNA-DNA chimera mimicking NF-kappaB binding sites. Eur J Biochem 2001, 268:6066–6075.

117. Borgatti M, Lampronti I, Romanelli A, et al. Transcription factor decoy molecules based on a peptide nucleic acid (PNA)-DNA chimera mimicking Sp1 binding sites. J Biol Chem 2003, 278:7500–7509.

118. Borgatti M, Breda L, Cortesi R, et al. Cationic liposomes as delivery systems for double-stranded PNA-DNA chimeras exhibiting decoy activity against NF-kappaB transcription factors. Biochem Pharmacol 2002, 64:609–616.

119. Borgatti M, Romanelli A, Saviano M, et al. Resistance of decoy PNA-DNA chimeras to enzymatic degradation in cellular extracts and serum. Oncol Res 2003, 13:279–287.

120. Righetti PG, Castagna A, Antonioli P, et al. Proteomic approaches for studying chemoresistance in cancer. Expert Rev Proteomics 2005, 2:215–228.

121. Hutter G, Sinha P. Proteomics for studying cancer cells and the development of chemoresistance. Proteomics 2001, 1:1233–1248.

Chapter 12
p53 Inhibitors as Cancer Sensitizing Agents

Flavio Maina and Rosanna Dono

1 The p53 Player: A Stage Director for the Fate of Stressed Cells

The formation and the development of neoplasia is a Darwinian selection for cells that progressively lose negative regulators and simultaneously reinforce positive modulators of growth. For example, the regulatory mechanisms involved in checkpoint control and apoptosis are needed to guarantee genetic stability for healthy cells, but are detrimental for cancer cells. The possibility to pharmacologically tune such regulatory systems in cancer therapy represents a big challenge to selectively destroy cancer cells and protect healthy cells, thus favoring regeneration and tissue recovery.

One well-known key player of such regulatory mechanisms is p53. p53 is activated following different stresses, including altered signals from the environment, DNA damage by chemical and physical agents (genotoxic stress), altered assembly/disassembly of microtubules, oncogene activation and hypoxia [1-3]. All these cellular stresses result in cell-cycle arrest, apoptosis and/or DNA repair, depending on the cell type, their developmental time point, the character and the entity of the insult (Fig. 12.1) [4, 5]. However, the proper cellular stress-response is also orchestrated by p53 through the choice of its partners and its transcriptional readout [6–8].

Following cellular insults, there are three major decision points. The first one occurs after DNA damage and directs cells toward apoptosis or DNA repair. The second one occurs after DNA repair and establishes whether or not repair was successful and whether cells can resume proliferation. The third one happens when repair was unsuccessful. In this case cells must choose between irreversible growth arrest and mitotic catastrophe. In all three decision points, p53 is one major actor, and which function p53 ensures is also determined by the amount and duration of p53 activation. Indeed, strong and sustained p53 activation correlates with higher chances for apoptosis versus growth arrest [9]. The success of and the requirement for DNA damage repair after genotoxic stresses both largely depend on the proliferation rate of cells. For example, unrepaired chromosomal breaks can be lethal for dividing cells, but tolerated by quiescent cells that will never detect such damage being postmitotic. One of the best examples is the radiosensitivity of the brain, which is one of the most radioresistant organs in the adult, but the most sensitive one in embryos [10].

Thus, p53 has an important "social" function: ensuring DNA repair or elimination of irreversibly damaged and, consequently, dangerous cells, by inducing them to die for the benefit of the organism. All these functions unambiguously state the role of p53 as a key tumor suppressor gene. Mutation of the p53 gene is found in more than 50% of human

From: *Sensitization of Cancer Cells for Chemo/Immuno/Radio-therapy, 1st Edition.*
Edited by: Benjamin Bonavida © Humana Press, Totowa, NJ

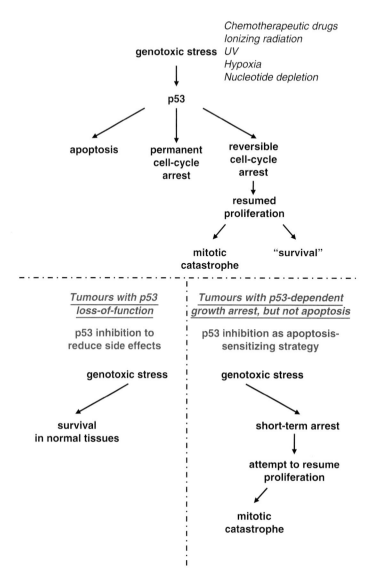

FIGURE 12.1. p53 is activated by several types of stress and mediates distinct biological responses, such as apoptosis, permanent or reversible cell cycle arrest. Such diversity on p53-mediated responses to genotoxic stress changes depending on the cellular context. In cancer therapies, inhibition of p53 prevents side effects caused by cell death in response to radiotherapy and chemotherapy. Such protective approach is applicable only when neoplastic cells have lost p53 function. Inhibition of p53 can also have a therapeutic potential in tumor cells where p53 induces cell cycle arrest to allow DNA repair caused by anticancer treatment. In such a context, p53 inhibition would sensitize neoplastic cells to radiotherapy and chemotherapy.

cancers and germ-line loss of function mutations in p53 account for the majority of inherited cancers, such as in the familiar Li-Fraumeni syndrome [11]. In addition to loss of wild-type p53 activity, a high percentage of human tumors are characterized by mutations that convert its tumor suppressor function into a dominant negative action or into an oncogenic signaling coordinator with the ability to induce gene expression distinct from the wild-type counterpart. These acquired novel functions enable mutant p53 to promote a large spectrum of cancer phenotypes (see reviews in *Oncogene* issue 26, 2007). By losing its full ability to induce

death genes and/or acquiring novel transcriptional properties that favor the growth of neoplastic cells, altered p53 function is a major event in cells doomed to neoplasia. Indeed, these newly acquired oncogenic properties can favor the insurgence, the maintenance, the spreading and the chemoresistance of malignant cells. Many recent efforts have been directed toward recovering p53 functions in tumors for therapeutic applications (when lost or mutated). The use of reversibly switchable p53 knock-in mice has recently highlighted the advantages and limitations of restoring p53 signaling [12–17].

It is currently believed that the threshold of p53 induction might lie at the balance between the aging process and the development of cancer. Indeed, mice with a superactive p53 locus never develop cancer, but show signs of premature aging. In contrast, mice in which the p53 locus has been inactivated or altered with mutations found in human tumors are extraordinarily cancer prone [18–21] (see also reviews in *Oncogene* issue 26, 2007). Such an increase in tumor incidence is thought to be caused by disorders in the mechanism that normally prevents accumulation of spontaneously mutating cells through irreversible cell-cycle arrest or apoptosis induction. Indeed, the population of p53-null cells is frequently characterized by chromosomal aberrations [22]. Thus, the optimal balance of p53 expression, presumably selected by profound evolutionary pressures, ensures that normal organisms have a prolonged fertile and active life, with cancer developing predominantly at an advanced age. For detailed information on molecular mechanisms and function of p53, refer to recent reviews [1–3, 5, 23–26].

2 Dominant or Discrete Action of p53 Depending on the Cellular Context

2.1 Lessons from Mouse Development

The crucial function of p53 to coordinate cellular responses is revealed by intriguing developmental studies on genetically engineered mice. Loss-of-function of p53 does not primarily impair mouse embryogenesis, since these mice do not show major developmental defects [19, 20]. Mild phenotypes, including defects in the embryonic eyes and teeth, exencephaly and anencephaly, have been observed in a small fraction of p53-null embryos and mice [27, 28]. Later studies demonstrated additional developmental roles of p53 during differentiation of B cells in the bone marrow [29], thymocytes [30], neurons and oligodendrocytes [31], and spermatogonia [32].

Lack of major defects in p53-null embryos suggests that during embryogenesis, p53 is also essentially a "guardian," dispensable if the signaling controlling cell behavior is correct, "exterminator"

in response to signaling alterations deleterious to the developing organism. This "guardian" role has so far limited studies on the temporal and spatial susceptibility of embryonic cells to p53 action. The first direct genetic proof of such a concept comes from the phenotypic analysis of mutant mice for the negative modulator of p53, the E3 ubiquitin ligase Mdm2. The early embryonic lethality of *mdm2* mutants was rescued by the absence of p53 [33, 34]. These genetic studies demonstrate that a precise amount of p53 is required to ensure correct signaling during embryogenesis. This has been further confirmed by recent studies using mutant mice with a hypomorphic version or a conditional allele of *mdm2* [35, 36]. For example, the use of embryos lacking *mdm2* in the central nervous system has recently revealed that the Mdm2/p53 interplay is required for neuronal survival [37]. Another example of such a role as "guardian" during development has been elegantly demonstrated by showing that genetic or pharmacologic inactivation of p53 rescues neural tube defects in Pax-3–deficient Splotch embryos [38].

By studying the signaling requirements ensuring survival of hepatocytes in developing livers, we have recently uncovered another crucial role of p53 during embryogenesis [39]. We found that the receptor tyrosine kinase Met, known as the master regulator of hepatocyte survival in vivo [40–46], acts on Mdm2 via the PI3K-Akt-mTOR pathway to prevent p53-dependent death in developing livers [39]. Since p53 is dispensable during normal liver development [47], our findings provide further evidence of the role that p53 plays as "guardian" in circumstances where normal survival signaling is perturbed. Moreover, our findings also contribute to further understanding of the great diversity of p53 activity depending on the cell type. Of note, p53 activity as a protective mechanism in rapidly proliferating cells such as embryonic hepatocytes prevents accumulation of genetically damaged cells in developing livers.

2.2 Human Diseases and p53: An Unavoidable Match

Several human diseases other than cancer, and injuries caused by various insults such as ischemia or excitotoxicity, have been found to rely on p53 activation. Neuronal death caused by p53 has been reported in several neurological disorders, such as stroke [48, 49], traumatic brain injury [50], Alzheimer's disease [51], and Parkinson's disease [52]. Studies on p53-deficient mice have genetically confirmed the essential role of p53 in neuronal apoptosis caused by ischemic and excitotoxic insults [53, 54]. In addition, activation of p53 plays a crucial role in regulating death of neurons in vitro following excitotoxic stimuli, including molecules such as glutamate or dopamine [53, 55–58]. Anticancer drugs, such as camptothecin and etoposide, irradiation, hypoxia, and oxidative stress can also kill neurons through the p53 signaling cascade [59, 60]. Moreover, signal by p53 has been also correlated to several other human pathologies, such as rheumatoid arthritis [61, 62], heart [63], pancreas [64], liver [65], and kidney diseases [66]. Therefore, inhibition of p53 is viewed as a promising therapeutic action for several pathologies associated with stress-mediated by p53 activation.

2.3   Side Effects of Cancer Therapy: Once Again p53 is Dictating the Rules

The vast majority of human cancers are currently treated with chemotherapy and radiation, but the extremely unpleasant side effects of such treatments often prevent effective therapy. Indeed, beside the desired killing effect on tumor cells, these treatments can also damage healthy tissues (see Fig. 12.1). During treatment with radiation and cytotoxic drugs, p53 is activated and plays its expected role: "guardian" of the genome in injured cells. Again, differences in sensitivity to genotoxic stresses of normal tissues appear to correlate with cell proliferation rates and once more, most resistant tissues consist of non-proliferating cells. In the brain, muscle, liver, lung, and kidney, proliferation of connective tissue and endothelial cells is limited to the time of tissue injury and regeneration. In such tissues, death of damaged differentiated cells following the genotoxic insult triggered by the treatment also affects proper regeneration by stem cells, known to be more resistant to these agents. In contrast, normal tissues known to be sensitive to radiotherapy and chemotherapy are those with high turnover rate, such as the hemopoietic and immune systems, epithelia of the small intestine, and the reproductive system. Consequently,

patients develop several problems such as anemia, infections, vomiting, and diarrhea.

The pivotal role of p53 in response to radiation or chemotherapeutic treatment with DNA damaging agents has been elegantly showed using mice expressing the LacZ reporter gene under the control of p53 promoter [67–69]. These treatments led to a pronounced activation of the p53 locus, measured by the LacZ readout. Such a paradigm allowed the correlation of the level of p53 activation to the level of induced damage, and revealed that the tissues previously known to show significant cell death in response to radiations/chemotherapy, showed a high level of p53-LacZ induction. In particular, activation of the transgene following radiation or anticancer drug injection was found in the spleen, thymus, hair follicles, oligodendroblasts of the spinal cord, and epithelia of the digestive track. Expression of the transgene (p53 activation) following cancer treatment was also reported in early embryos in which p53 activation does not occur in normal conditions (see Section 2.1). Moreover, the same correlation between sensitivity to radiation and p53-LacZ induction was also observed in embryonic cells, depending on their developmental time point. Early embryos are sensitive to irradiation and express the LacZ gene. However, as soon as organogenesis starts, tissue radiosensitivity changes, leading to a pattern similar to the one known in adults.

The different sensitivity of tissues to genotoxic stresses, reveled by the LacZ expression pattern, matches with the levels of stable p53 protein. Beside the fact that stabilized p53 protein is found in all tissues, the amount of p53 protein varies, with higher levels consistently found in radiosensitive tissues compared to those that are radioresistant [69–71]. Moreover, such levels of stable p53 protein appear to correlate with high levels of p53 mRNA expression. For example, p53 mRNA levels are much higher in spleen and thymus, which are radiosensitive, compared with those found in liver, brain and muscle, which are radioresistant [70, 72–76]. Such correlation has been also found in irradiated mouse embryos, with high levels of p53 mRNA in radiosensitive tissues [77, 78]. Finally, tissue sensitivity to genotoxic stresses is also determined by p53 responsive genes, with most of the p53 targets activated in a tissue-specific manner [79, 80].

The use of genetically modified mice allowed researchers to unequivocally show that p53 is responsible for many of the side effects of radio-therapy and chemotherapy during the first hours of treatments. Indeed, p53-null mice are resistant to radiation doses, which are lethal for control mice [81]. Therefore, conventional chemotherapy and radiation are often limited by the p53-dependent toxicity induced in normal tissues. The challenge facing researchers is then the following: How to deal with such a tumor suppressor, which compli-cates cancer treatment by triggering massive death of normal, but not always of neoplastic cells, dur-ing radiotherapy and chemotherapy? For therapy targeting p53-deficient cancer, temporal p53 sup-pression may avoid unwanted side effects, without necessarily leading to cancer predisposition as it has been shown in mice.

2.4 But p53 is Not Only a Killer

The p53 issue is even more complicated by the fact that p53 is found intact in those tumors that lack apoptotic pathway. A plausible explanation comes from the role of p53 in cells where apop-tosis is not observed in response to treatment. In such a context, when DNA damage is detected, p53 is activated to halt cells at different cell-cycle checkpoints (see Fig. 12.1). Simultaneously, DNA-damage sensors activate DNA-repair mechanisms to ensure repair before cell cycle progression occurs and avoid mitotic catastrophe. Therefore, p53 plays a protective role by preventing cells prematurely re-entering into mitosis and/or dying by mitotic catastrophe. Such a scenario elucidates how, in the absence of apoptosis, p53 might play a "survival" role by giving cells the time to repair damaged DNA before they re-enter mitosis. When this occurs in neoplastic cells, it is predictable that blockage of p53 activity prevents growth arrest, and consequently mutated cells progress toward mitotic catastrophe. Therefore, such a context would sensitize cancer cells to radiotherapeutic and chemotherapeutic intervention. One of the best examples of such a "survival" property of p53 is found in small intestinal epithelium, in which p53 is known to protect these cells from irradiation by allowing them to stay in growth arrest instead of induc-ing them to die, thus lowering the chance of mitotic

catastrophe [82]. Such an apparently odd situation has also been found in other types of neoplastic cells, in which p53 inactivation did not decrease cell survival in response to radiation [83]. Another example of the p53 "survival" property has been provided by the work of J. Folkman's and co-workers, who showed that effective chemotherapy of tumor vascular endothelium was higher in p53-null mice compared with wild-type mice [84]. Such a protective role of p53 in the absence of apoptosis appears to be true in different neoplastic cells. For example, human colon carcinoma and fibrosarcoma cells are more sensitive to DNA-damaging agents in the absence of p53, with a marked increase in the number of cells undergoing mitotic catastrophe [83, 85–87]. Moreover, an additional unexpected protective role of p53 has been reported in tumor endothelium when exposed to genotoxic stress, and might have profound clinical implication on how p53 can favor angiogenesis. However, p53 has also been proposed as an antiangiogenic factor since it activates two known suppressors of angiogenesis, thrombospondin 1 and 2 [88, 89]. Additional stud-ies will clarify how the p53 signal contributes to angiogenesis and whether p53 activation or inhibi-tion will enhance tumor angiogenesis.

Further support for the concept that p53 does not necessarily act as a killer, but can instead mediate "survival," comes from a study showing that tumor susceptibility to apoptosis depends on the tissue origin. By comparing the apoptotic response in mouse primary tumors derived from different tis-sues, it appears that radiation and chemotherapy treatments kill neoplastic cells only if they origi-nated from tissues susceptible to p53-dependent apoptosis. Therefore, susceptibility of neoplastic cells to the action of p53 in response to anticancer treatments need to be carefully evaluated to iden-tify those that are susceptible to p53-triggered cell death as opposed to those benefiting from p53-triggered "survival."

3 Is p53 Inhibition a Desirable Target for Human Cancer Therapy?

The role of p53-dependent apoptosis in tumor response to cancer therapy is quite modest con-sidering that most of the tumors, including those with wild-type p53, acquire resistance to apoptosis during their progression. Two major circumstances in which p53 inhibition can be beneficial are particularly attractive. The first one is when healthy cells (expressing p53) need to be protected against radiotherapy or chemotherapy, as mice do not show increased secondary tumor frequency. The second one corresponds to those cancers that express p53: inhibition of p53 function may have an apoptosis-sensitizing effect, for instance, when the cell cycle blockage mediated by p53 is abolished [90]. The next two paragraphs discusses these two circumstances to underline both beneficial and undesirable effects (see Fig. 12.1).

3.1 Inhibition of p53 to Reduce Side Effects of Radiotherapy and Chemotherapy

The pivotal role of p53 in the cellular response to genotoxic stress caused in normal tissues by radiotherapy and chemotherapy has led to the proposal that p53 inhibition could reduce side effects associated with cancer treatments. It is evident that this fascinating prospective can only be applicable to those tumors in which p53 is lost; hence, its suppression in the tumor should not have undesirable consequences. The number of tumors with this scenario is significant considering the high frequency of p53 alteration or loss in human cancer. Thus, p53 suppression should not affect the efficiency of treating these tumors. The approach is very attractive since inhibition of p53 can be used to mitigate the side effects of chemotherapy, such as hair loss, mucositis, myelosuppression, and intestinal damage. It is worth underlining that such beneficial effects can be achieved simply through a temporary reversible conversion of normal tissues to a p53-deficient state.

There are of course difficulties with such an approach that must be carefully evaluated. For example, it is considered that such a strategy is potentially pro-mutagenic. Healthy cells would be exposed to high doses of chemotherapeutic agents or radiations, and although the cells survive, they are more likely to be mutagenized. Therefore, temporal p53 inhibition can increase the risk of secondary tumor formation. With this respect, it is worth noting that survival of stem cells appears to be independent of p53. Therefore, temporary p53

inhibition should not increase the number of surviving stem cells with genetic alteration caused by genotoxic agents. In support of the temporary p53 inhibition to protect normal tissues, an additional consideration is that as tissue regeneration originates from stem cells through a long process, suppression of death in the population of differentiated cells should support function of normal tissues during the regeneration phase. The inhibition of p53 as an approach to reduce side effects of radiotherapy and chemotherapy would be particularly appropriate to those cells in permanent growth arrest, such as neurons. In these cells, there is no risk of cancer formation and they cannot be easily replaced.

It is worth taking into account that p53 is differentially implicated in the apoptosis of differentiated cells and stem cells. For example, acute p53-dependent apoptosis has been observed in differentiated epithelial cells of the intestine following irradiation, but only to a lesser extent in stem cells [91, 92]. These finding have led to the proposal that stem cells are less sensitive to p53-dependent apoptosis because in physiologic conditions they proliferate at low rates in contrast to differentiating cells. As a consequence, differentiated cells are more sensitive to damage of DNA and susceptible to p53 action.

It is clear that current cancer therapies are based on a careful balance of advantages and disadvantages. The rational of inhibiting p53 must be evaluated together with many other factors that altogether would ascertain the best combination for cancer therapy. For example, one could imagine that targeting p53 to reduce side effects in older patients would possibly be more beneficial than detrimental.

3.2 Inhibition of p53 as Apoptosis-Sensitizing Strategy

Inhibition of p53 is considered to have a therapeutic impact in neoplastic cells that retain p53-dependent growth arrest, but lack p53-dependent apoptosis. In these tumors, blockage of p53 activity would sensitize neoplastic cells to DNA-damaging agents by impairing growth arrest, thus leading to mitotic catastrophe. Such a context is likely to even increase the therapeutic effects of anti-cancer treatments. Consistently, p53 inhibition in human fibrosarcoma or colon carcinoma cells leads to mitotic catastrophe, thus sensitizing cells to genotoxic treatment [85, 93, 94]. The increased chemosensitivity of human cancer cells following p53 inhibition has been reported by several groups. For example, human glioma cells are more sensitive to chemotherapeutic agents such as cisplatin and temozolomide when p53 activity is impaired [95, 96]. Thus, p53 inhibition can have important therapeutic advantages in tumor cells, in which it predominantly causes growth arrest to allow tissue recovery. As previously pointed, careful evaluation of p53 function in tumor cells is mandatory to ensure beneficial therapeutic treatment of temporarily p53 inhibition associated to anticancer agents, reducing risk of inhibition in unwanted cells.

In conclusion, in view of its implication regarding side effects following radiation and chemotherapy, the biological properties of p53 must be well evaluated to correctly consider it, and therefore use it, for cancer therapy. In such scenario, the identification of novel p53 inhibitors able to selectively block either pro-apoptotic or growth arrest function would certainly give an impulse to their application for combined therapeutic treatment for cancer.

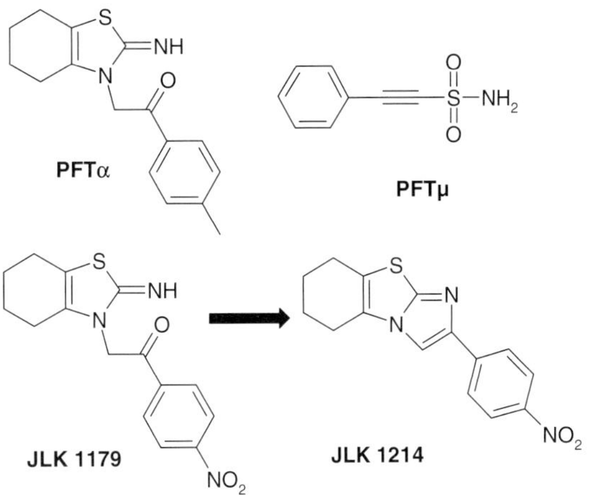

FIGURE 12.2. Schematic representation of the most potent p53 inhibitors identified and validated in vivo. The inactive prodrug JLK 1179 is converted in biological conditions into the active p53 inhibitor, the cyclic derivative JLK 1214

4 Chemistry and Biology Together for a Common Goal: Shoot p53

4.1 Pifithrin-α Opened the Way

The pioneer in such lines of research aiming at identifying potent p53 inhibitors has been the group of Gudkov, who identified a new inhibitor of p53 [97]. To do so, they screened for inhibitors of p53-induced transcription using a p53-reporter cell line in which a stably integrated reporter gene was activated in response to the damage of DNA induced by chemotherapeutic compounds, such as doxorubicin (Dox), etoposide, Taxol, AraC, and by irradiation. From a collection of 16,000 diverse small-chemical molecules, they identified a compound that showed these characteristics, which they called Pifithrin-a (PFTα, which stands for "p-fifty three inhibitor") (Fig. 12.2). Using cell-culture–based assays, they showed that temporary suppression of p53 inhibits apoptosis triggered

by DNA damage and thus enhances the proportion of cells surviving after stress. Consistently, PFTα impairs transcription of p53 target genes, such as Bax, p21, Mdm2, and cyclin G, known to modulate cell survival and death. The use of animal models allowed them to demonstrate that a single injection of PFTα protects 100% of mice from killing doses of gamma irradiation. Irradiated mice injected with PFTα lost less weight than those not pretreated with the drug. Beneficial effects of p53 inhibition were observed with both lethal and sublethal doses of whole-body gamma radiation. Remarkably, PFTα injection also prevented hair loss in mice exposed to irradiation. In addition to this, another striking result from this nice piece of work is that suppression of p53 activity by PFTα did not increase the risk of new cancer development and secondary tumor formation from surviving cells that were not eliminated by p53. Lack of protective action on p53-null mice confirmed that PFTα acts through a p53-dependent mechanism [97]. Although extensively studied in several cellular contexts, it is not clear yet how PFTα acts on p53. Studies have proposed that its mechanism of action is associated with the alteration in p53 nuclear transport (import, export, or both), leading to its depletion in the nucleus [97, 98], and possibly by modifying nuclear pores [99]. Moreover, PFTα also appears to inhibit p53 translocation to mitochondria and suppress mitochondrial dysfunction associated with reduced caspase activity [100, 101].

The impact of using a p53 inhibitor to prevent cell death following genotoxic stress has been broadly evaluated in several biological systems (Fig. 12.3). For example, PFTα prevented Dox-induced cell death in mouse hearts [102], opening the possibility for novel cardioprotective drug for treatment of cancer patients with pre-existing heart dysfunctions. The protective effects of PFTα have been evaluated following cisplatin treatment, which in addition to displaying a genotoxic response, is also known to cause neurotoxicity with consequent implication in hearing [76, 103]. Temporary inhibition of p53 by PFTα reduced the ototoxic, vestibulotoxic, and neurotoxic side effects of cisplatin [104]. The protective effects of PFTα on cisplatin exposure have also been reported in kidney cells [104], indicating that PFTα could be used to prevent the acute renal failure caused by cisplatin. Importantly, IP administered PFTα has been shown to protect neurons in vivo to genotoxic stress, demonstrating its ability to cross the blood–brain

barrier. Thus, the use of PFTα to temporarily suppress p53 appears to be unequivocally beneficial, without leading to cancer predisposition as in the context of p53 deficiency.

4.2 Derivatives of Pifithrin-α Can Do Even Better

The impressive list of protective activities of PFTα against cell death by genotoxic stresses both in vitro and in vivo has prompted researchers to identify derivatives, based on chemical modifications of its structure [105]. Moreover, taking into account the dual functions of p53 depending on the cellular context (death inducer of growth arrest-"survival" promoter), a big challenge is to identify new classes of p53 inhibitors capable of blocking selectively either its pro-apoptotic or growth arrest functions. Such a challenge can possibly be achieved because p53 induces apoptosis by acting also at the mitochondrial level, in addition to regulating gene expression. Recent work of Gudkov and colleagues showed that such an approach

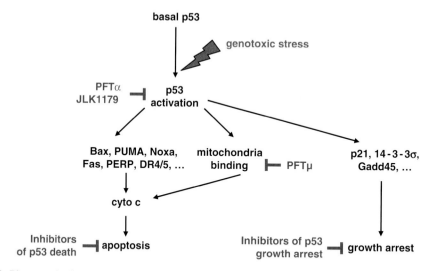

FIGURE 12.3. Pharmacologic suppression of p53 in cancer treatment. Cell cycle arrest by p53 occurs through enhanced expression of genes regulating cell cycle checkpoint control, such as p21, 14-3-3σ, and Gadd45. Cell death occurs through a transcriptional-dependent and -independent manner. p53 induces expression of pro-apoptotic genes, such as Bax, PUMA, Noxa, Fas, PERP, and DR4/5. Moreover, apoptosis is triggered by translocation of p53 into mitochondria. PFTα and JLK1179 inhibit both apoptosis and cell cycle arrest induced by genotoxic stress. In contrast, PFTμ impairs selectively p53 mitochondrial function. Chemical compounds able to block selectively p53-induced cell death or growth arrest have not been identified yet. The first type of compounds can be used to protect normal tissue from side effects of radiotherapy and chemotherapy. The second type of anticancer drugs may be used in combination with anticancer therapy to sensitize neoplastic cells to treatment

is feasible [106]. Indeed, they isolated a novel chemical compound, named PFTµ, which inhibits p53 binding to mitochondria [107] without acting on p53-dependent transactivation. This compound acts by reducing p53 affinity to anti-apoptotic proteins Bcl-xL and Bcl-2 (see Figs. 12.2 and 12.3). Consequently, PFTµ blocks p53-triggered death of cultured cells following radiation, and protects mice from doses of radiation that cause lethal hematopoietic syndrome. Thus, it is thought that blocking uniquely p53 binding to Bcl-xL and Bcl-2 only interferes with the capacity of p53 to trigger the apoptosis cascade, without interfering with other p53 functions [106].

Based on the imino-tetra-hydrobenzothiazole scaffold (the basic scaffold of the PFTα structure), novel possible p53 chemical inhibitors have been designed, synthesized, and biologically evaluated, leading to the identification of a novel inhibitor, namely JLK1179 (see Fig. 12.2) [108, 109]. Together with the group of J.L. Kraus, we have shown that JLK1179 is 2 logs more active than PFTα in preventing cell death in vivo [109]. Moreover, we have demonstrated that JLK1179 acts as a prodrug precursor to prevent p53-triggered cell death, without acting on other pro-apoptotic pathways. Studies on its mechanism of action show that the highly inhibitory property of JLK1179 results from a combination of stability and of its ability to be converted into an active form (Fig. 12.4). Using spectroscopy methods and a chemical model, we have demonstrated that the opened JLK1179 is stable in biological conditions. Moreover, we have shown that JLK1179 is irreversibly converted into its active cyclic form, JLK1214, through an intramolecular dehydration process. Thus, a balance between biological stability and intramolecular rearrangement underlies its efficiency to prevent cell death in vivo (see Fig. 12.4). JLK1179 can be considered as a new prodrug prototype that prevents p53-triggered cell death in several neuropathologies and possibly reduces cancer therapy side effects [108, 109]. Future studies will allow further understanding of how JLK1179 acts on p53 at mitochondrial level.

Additional work on pharmacologic modulations of p53 functions will allow a better evaluation of the protective effects of these and novel p53 inhibitors, a fuller understanding of their stability and mechanism of action, and testing whether they

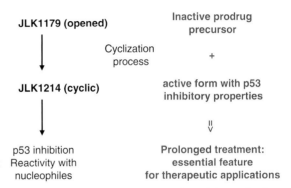

FIGURE 12.4. Schematic representation of the cyclization process leading to conversion of the inactive prodrug JLK1179 into the active p53 inhibitor, the cyclic derivative JLK1214. Degradation of JLK1214 most likely occurs by nucleophile attacks. A subtle balance between stability in biological conditions and intramolecular rearrangement underlies the inhibitory efficacy of JLK1179 in vivo

have additional activities besides blocking p53. This will lead to possible therapeutic applications for human diseases. It will be important to further complement such lines of research by searching for novel approaches able to specifically target p53 functions, gene regulators, and/or components of its pathway.

4.3 Proper Formulation of p53 Inhibitors Can Optimize Delivery and Efficacy, Reducing Risks

Whether or not p53 inhibitors will reach clinical applications certainly depends on their efficient formulation to optimize delivery and efficacy. Several parameters must be taken into account for optimal p53 inhibition as a strategy to reduce the damage of normal tissues by anticancer therapy. One is related to the timing of p53 inhibition. It is worth noting that after genotoxic stress, p53-dependent cell death occurs within the first hours. This first wave of death can be followed by a second wave that is p53 independent, depending on the tissue and irradiation dose [72, 76]. For example, p53-independent apoptosis in the small intestine is not observed after irradiation at low doses (1 Gy), but pronounced in mice exposed to high doses (8 Gy) [76]. Therefore, it is expected that "turning off" p53 activity within the first hours should be sufficient to: (1) kill tumor cells with p53 deficiency since their sensitivity to therapy should not by changed in the presence of p53 inhibitors; (2) reduce side effects of radiation and chemotherapy; and (3) avoid risk linked to prolonged inhibition.

Concerning the use of p53 inhibitors as apoptosis sensitizing strategies, it will be particularly important to ensure efficient and selective targeting in cancer cells, to sharpen their action and reduce side effects. In addition to developing novel anticancer agents, researchers are now focusing their efforts on directing drugs to neoplastic cells and finding the best formulation for delivery [110, 111]. Success in this direction will enhance the therapeutic impact of p53 inhibitors by combining its temporal and local inhibition.

As discussed, one of the main concerns about p53 inhibition is that survival of genetically altered cells, which would be normally eliminated, can constitute a dangerous subpopulation with tumorigenic potential. Experimental evaluation of such risk showed that radioprotection with temporary reversible inhibition of p53 can be relatively safe since increased tumor occurrence was not observed in mice [97]. Nevertheless, PFTα treatment is accompanied by an increased rate of chromosomal abnormalities in cultured cells [112]. Therefore, possible correlations between prolonged application of p53 inhibitors and tumor frequency must be carefully evaluated before clinical applications in cancer therapy. Moreover, the ratio between risks and benefits of using p53 inhibitors can vary significantly depending on the age of patients, type of disease and tumors.

5   Conclusion

The role of p53 in treating tumors and establishing susceptibility to treatment is not as simple as was originally thought. Its modulation in cancer therapy can be negative or positive, depending on its ability and requirement to promote death or "survival," according to the cellular context. Therefore, it is crucial to pose p53 diagnosis and prognosis in the context of type of tumors and with respect to other altered signals responsible for neoplasia. Are we ready to start using p53 inhibitors for cancer therapy? Scientists still need to carefully evaluate the benefits and risks before clinical applications can be envisaged. However,

we might not be too far from applying such a strategy in cancer therapy of specific tumors. Prediction of the responses of cancer cells to p53 inhibitors needs to establish whether p53 is intact or altered, and which functions it plays in these aberrant cells. The analysis of oncogenic molecular signatures to design the best combined cancer therapy and improve pharmacologic modulation of mechanisms governing neoplastic and normal cells should reduce this gap.

Acknowledgments. We thank member of F. Maina's team and O. Piccolo for helpful discussions, and L. Bardouillet and S. Richelme for technical supports to the lab. K. Dudley, F. Helmbacher, and F. Lamballe are particularly acknowledged for discussion and comments on this chapter. Work of our laboratory is supported by the Association Française contre les Myopathies (AFM), the Fondation pour la Recherche Médicale (FRM), the Association pour la Recherche sur le Cancer (ARC), the Fondation de France (FdF), the ACI, and the Institut National du Cancer (INCa) grant.

References

1. Fuster JJ, et al. Classic and novel roles of p53: prospects for anticancer therapy. Trends Mol Med 200, (68):777–782.
2. Kastan MB. Wild-type p53: tumors can't stand it. Cell 2007, 128(5):837–840.
3. Stiewe T. The p53 family in differentiation and tumorigenesis. Nat Rev Cancer 2007, 7(3):165–168.
4. Offer H, et al. The onset of p53-dependent DNA repair or apoptosis is determined by the level of accumulated damaged DNA. Carcinogenesis 2002, 23(6):1025–1032.
5. Vousden KH. p53: death star. Cell 2000, 103(5):691–694.
6. Wahl GM, Carr AM. The evolution of diverse biological responses to DNA damage: insights from yeast and p53. Nat Cell Biol 2001, 3(12):E277–286.
7. Samuels-Lev Y, et al. ASPP proteins specifically stimulate the apoptotic function of p53. Mol Cell 2001, 8(4):781–794.
8. Oda K, et al. p53AIP1, a potential mediator of p53-dependent apoptosis, and its regulation by Ser-46-phosphorylated p53. Cell 2000, 102(6):849–862.
9. Gudkov AV, Komarova EA. The role of p53 in determining sensitivity to radiotherapy. Nat Rev Cancer 2003, 3(2):117–129.
10. Schultheiss TE, et al. Radiation response of the central nervous system. Int J Radiat Oncol Biol Phys 1995, 31(5):1093–1112.

11. Iwakuma T, Lozano G, Flores ER. Li-Fraumeni syndrome: a p53 family affair. Cell Cycle 2005, 4(7):865–867.

12. Sharpless NE, DePinho RA. Cancer biology: gone but not forgotten. Nature 2007, 445(7128):606–607.

13. Christophorou MA, et al. The pathological response to DNA damage does not contribute to p53-mediated tumour suppression. Nature 2006, 443(7108):214–217.

14. Ringshausen I, et al. Mdm2 is critically and continuously required to suppress lethal p53 activity in vivo. Cancer Cell 2006, 10(6):501–514.

15. Martins CP, Brown-Swigart L, Evan GI. Modeling the therapeutic efficacy of p53 restoration in tumors. Cell 2006, 127(7):1323–1334.

16. Ventura A, et al. Restoration of p53 function leads to tumour regression in vivo. Nature 2007, 445(7128):661–665.

17. Xue W, et al. Senescence and tumour clearance is triggered by p53 restoration in murine liver carcinomas. Nature 2007, 445(7128):656–660.

18. Tyner SD, et al. p53 mutant mice that display early ageing-associated phenotypes. Nature 2002, 415(6867):45–53.

19. Jacks T, et al. Tumor spectrum analysis in p53-mutant mice. Curr Biol 1994, 4(1):1–7.

20. Donehower LA, et al. Mice deficient for p53 are developmentally normal but susceptible to spontaneous tumours. Nature 1992, 356(6366):215–221.

21. Lozano G. The oncogenic roles of p53 mutants in mouse models. Curr Opin Genet Dev 2007, 17(1):66–70.

22. Lee JM, et al. Susceptibility to radiation-carcinogenesis and accumulation of chromosomal breakage in p53 deficient mice. Oncogene 1994, 9(12):3731–3736.

23. Trotman LC, Pandolfi PP. PTEN and p53: who will get the upper hand? Cancer Cell 2003, 3(2):97–99.

24. Daujat S, Neel H, Piette J. MDM2: life without p53. Trends Genet 2001, 17(8):459–464.

25. Alarcon-Vargas D, Ronai Z. p53-Mdm2—the affair that never ends. Carcinogenesis 2002, 23(4):541–547.

26. Slee EA, O'Connor DJ, Lu X. To die or not to die: how does p53 decide? Oncogene 2004, 23(16):2809–2818.

27. Armstrong JF, et al. High-frequency developmental abnormalities in p53-deficient mice. Curr Biol 1995, 5(8):931–936.

28. Pan H, Griep AE. Temporally distinct patterns of p53-dependent and p53-independent apoptosis during mouse lens development. Genes Dev 1995, 9(17):2157–2169.

29. Shick L, et al. Decreased immunoglobulin deposition in tumors and increased immature B cells in p53-null mice. Cell Growth Diff 1997, 8(2):121–131.

30. Jiang D, Lenardo MJ, Zuniga-Pflucker JC. p53 prevents maturation to the CD4+CD8+ stage of thymocyte differentiation in the absence of T cell receptor rearrangement. J Exp Med 1996, 183(4):1923–1928.

31. Eizenberg O, et al. p53 plays a regulatory role in differentiation and apoptosis of central nervous system-associated cells. Mol Cell Biol 1996, 16(9):5178–5185.

32. Hendry JH, et al. P53 deficiency produces fewer regenerating spermatogenic tubules after irradiation. Int J Radiat Biol 1996, 70(6):677–682.

33. Jones SN, et al. Rescue of embryonic lethality in Mdm2-deficient mice by absence of p53. Nature 1995 378(6553):206–208.

34. Montes de Oca Luna R, Wagner DS, Lozano G. Rescue of early embryonic lethality in mdm2-deficient mice by deletion of p53. Nature 1995, 378(6553):203–206.

35. Mendrysa SM, et al. mdm2 Is critical for inhibition of p53 during lymphopoiesis and the response to ionizing irradiation. Mol Cell Biol 2003, 23(2):462–472.

36. Maetens M, et al. Distinct roles of Mdm2 and Mdm4 in red cell production. Blood 2007, 109(6):2630–2633.

37. Xiong S, et al. Synergistic roles of Mdm2 and Mdm4 for p53 inhibition in central nervous system development. Proc Natl Acad Sci U S A 2006, 103(9):3226–3231.

38. Pani L, Horal M, Loeken MR. Rescue of neural tube defects in Pax-3-deficient embryos by p53 loss of function: implications for Pax-3-dependent development and tumorigenesis. Genes Dev 2002, 16(6):676–680.

39. Moumen A, et al. Met acts on Mdm2 via mTOR to signal cell survival during development. Development, 2007, 134:1443–1451.

40. Schmidt C, et al. Scatter factor/hepatocyte growth factor is essential for liver development. Nature 1995, 373:699–702.

41. Bladt F, et al. Essential role for the c-met receptor in the migration of myogenic precursor cells into the limb bud. Nature 1995, 376:768–771.

42. Maina F, et al. Uncoupling of Grb2 from the Met receptor in vivo reveals complex roles in muscle development. Cell 1996, 87:531–542.

43. Huh CG, et al. Hepatocyte growth factor/c-met signaling pathway is required for efficient liver regeneration and repair. Proc Natl Acad Sci U S A 2004, 101(13):4477–4482.

44. Borowiak M, et al. Met provides essential signals for liver regeneration. Proc Natl Acad Sci U S A 2004, 101(29):10608–10613.

45. Maina F., et al. Coupling Met to specific pathways results in distinct developmental outcomes. Mol Cell 2001, 7(6):1293–1306.

46. Moumen A, et al. Met signals hepatocyte survival by preventing Fas-triggered FLIP degradation in a PI3k-Akt-dependent manner. Hepatology 2007, 45(5):1210–1217.

47. Lin P, et al. Tissue-specific regulation of Fas/APO-1/CD95 expression by p53. Int J Oncol 2002, 21(2):261–264.

48. Crumrine RC, Thomas AL, Morgan PF. Attenuation of p53 expression protects against focal ischemic damage in transgenic mice. J Cereb Blood Flow Metab 1994, 14(6):887–891.

49. Li Y, et al. p53-immunoreactive protein and p53 mRNA expression after transient middle cerebral artery occlusion in rats. Stroke 1994, 25(4):849–855; discussion 855–856.

50. Napieralski JA, Raghupathi R, McIntosh TK. The tumor-suppressor gene, p53, is induced in injured brain regions following experimental traumatic brain injury. Brain Res Mol Brain Res 1999, 71(1):78–86.

51. de la Monte SM, Sohn YK, Wands JR. Correlates of p53- and Fas (CD95)-mediated apoptosis in Alzheimer's disease. J Neurol Sci 1997, 152(1):73–83.

52. Jenner P, Olanow CW. Understanding cell death in Parkinson's disease. Ann Neurol 1998, 44(3 Suppl 1): S72–84.

53. Sakhi S, et al. Induction of tumor suppressor p53 and DNA fragmentation in organotypic hippocampal cultures following excitotoxin treatment. Exp Neurol 1997, 145(1):81–88.

54. Uberti D, et al. Induction of tumour-suppressor phosphoprotein p53 in the apoptosis of cultured rat cerebellar neurones triggered by excitatory amino acids. Eur J Neurosci 1998, 10(1):246–254.

55. Blum D, et al. p53 and Bax activation in 6-hydroxy-dopamine-induced apoptosis in PC12 cells. Brain Res 1997, 751(1):139–142.

56. Xiang H, et al. Bax involvement in p53-mediated neuronal cell death. J Neurosci 1998, 18(4):1363–1373.

57. Cregan SP, et al. Bax-dependent caspase-3 activation is a key determinant in p53-induced apoptosis in neurons. J Neurosci 1999, 19(18):7860–7869.

58. Daily D, et al. The involvement of p53 in dopamine-induced apoptosis of cerebellar granule neurons and leukemic cells overexpressing p53. Cell Mol Neurobiol 1999, 19(2):261–276.

59. Johnson MD, et al. Evidence for involvement of Bax and p53, but not caspases, in radiation-induced cell death of cultured postnatal hippocampal neurons. J Neurosci Res 1998, 54(6):721–733.

60. Jordan J, et al. p53 expression induces apoptosis in hippocampal pyramidal neuron cultures. J Neurosci 1997, 17(4):1397–1405.

61. Liu H, Pope RM. The role of apoptosis in rheumatoid arthritis. Curr Opin Pharmacol 2003, 3(3):317–322.

62. Firestein GS. Novel therapeutic strategies involving animals, arthritis, and apoptosis. Curr Opin Rheumatol 1998, 10(3):236–241.

63. Crow MT, et al. The mitochondrial death pathway and cardiac myocyte apoptosis. Circ Res 2004, 95(10):957–970.

64. Savkovic V, et al. The stress response of the exocrine pancreas. Dig Dis 2004, 22(3):239–246.

65. Schuchmann M, Galle PR. Apoptosis in liver disease. Eur J Gastroenterol Hepatol 2001, 13(7):785–790.

66. Famulski KS, Halloran PF. Molecular events in kidney ageing. Curr Opin Nephrol Hypertens 2005, 14(3):243–248.

67. Komarova EA, et al. Transgenic mice with p53-responsive lacZ: p53 activity varies dramatically during normal development and determines radiation and drug sensitivity in vivo. Embo J 1997, 16(6):1391–1400.

68. Gottlieb E, et al. Transgenic mouse model for studying the transcriptional activity of the p53 protein: age- and tissue-dependent changes in radiation-induced activation during embryogenesis. Embo J 1997, 16(6):1381–1390.

69. MacCallum DE, et al. The p53 response to ionising radiation in adult and developing murine tissues. Oncogene 1996, 13(12):2575–2587.

70. Chen X, et al. p53 levels, functional domains, and DNA damage determine the extent of the apoptotic response of tumor cells. Genes Dev 1996, 10(19):2438–2451.

71. Lassus P, et al. Anti-apoptotic activity of low levels of wild-type p53. Embo J 1996, 15(17):4566–4573.

72. Cui YF, et al. Apoptosis in bone marrow cells of mice with different p53 genotypes after gamma-rays irradiation in vitro. J Environ Pathol Toxicol Oncol 1995, 14(3–4):159–163.

73. Wang L, et al. Gamma-ray-induced cell killing and chromosome abnormalities in the bone marrow of p53-deficient mice. Radiat Res 1996, 146(3): 259–266.

74. Song S, Lambert PF. Different responses of epidermal and hair follicular cells to radiation correlate with distinct patterns of p53 and p21 induction. Am J Pathol 1999, 155(4):112–117.

75. Chow BM, Li YQ, Wong CS. Radiation-induced apoptosis in the adult central nervous system is p53-dependent. Cell Death Diff 2000, 7(8):712–720.

76. Merritt AJ, et al. The role of p53 in spontaneous and radiation-induced apoptosis in the gastrointestinal tract of normal and p53-deficient mice. Cancer Res 1994, 54(3):614–617.

77. Rogel A, et al. p53 cellular tumor antigen: analysis of mRNA levels in normal adult tissues, embryos, and tumors. Mol Cell Biol 1985, 5(10):2851–2855.

78. Schmid P, et al. Expression of p53 during mouse embryogenesis. Development 1991, 113(3):857–865.

79. Komarova EA, et al. Stress-induced secretion of growth inhibitors: a novel tumor suppressor function of p53. Oncogene 1998, 17(9):1089–1096.

80. Burns TF, Bernhard EJ, El-Deiry WS. Tissue specific expression of p53 target genes suggests a key role for KILLER/DR5 in p53-dependent apoptosis in vivo. Oncogene 2001, 20(34):4601–4612.

81. Westphal CH, et al. atm and p53 cooperate in apoptosis and suppression of tumorigenesis, but not in resistance to acute radiation toxicity. Nat Genet 1997, 16(4):397–401.

82. Komarova EA, et al. Dual effect of p53 on radiation sensitivity in vivo: p53 promotes hematopoietic injury, but protects from gastro-intestinal syndrome in mice. Oncogene 2004, 23(19):3265–3271.

83. Slichenmyer WJ, et al. Loss of a p53-associated G1 checkpoint does not decrease cell survival following DNA damage. Cancer Res 1993; 53(18):4164–4168.

84. Browder T, et al. Antiangiogenic scheduling of chemotherapy improves efficacy against experimental drug-resistant cancer. Cancer Res 2000, 60(7):1878–1886.

85. Bunz F, et al. Disruption of p53 in human cancer cells alters the responses to therapeutic agents. J Clin Invest 1999, 104(3):263–269.

86. Brachman DG, et al. p53 mutation does not correlate with radiosensitivity in 24 head and neck cancer cell lines. Cancer Res 1993, 53(16):3667–3669.

87. Roninson IB, Broude EV, Chang BD. If not apoptosis, then what? Treatment-induced senescence and mitotic catastrophe in tumor cells. Drug Resist Update 2001, 4(5):303–313.

88. Dameron KM, et al. Control of angiogenesis in fibroblasts by p53 regulation of thrombospondin-1. Science 1994, 265(5178):1582–1584.

89. Adolph KW, Liska DJ, Bornstein P. Analysis of the promoter and transcription start sites of the human thrombospondin 2 gene (THBS2). Gene 1997, 193(1):5–11.

90. Gudkov AV, Komarova EA. Prospective therapeutic applications of p53 inhibitors. Biochem Biophys Res Commun 2005, 331(3):726–736.

91. Hendry JH, et al. p53 deficiency sensitizes clonogenic cells to irradiation in the large but not the small intestine. Radiat Res 1997, 148(3):254–259.

92. Tron VA, et al. p53-regulated apoptosis is differentiation dependent in ultraviolet B-irradiated mouse keratinocytes. Am J Pathol 1998, 153(2):579–585.

93. Hawkins DS, Demers GW, Galloway DA. Inactivation of p53 enhances sensitivity to multiple chemotherapeutic agents. Cancer Res 1996, 56(4):892–898.

94. Palacios C, et al. The role of p53 in death of IL-3-dependent cells in response to cytotoxic drugs. Oncogene 2000, 19(31):3556–3559.

95. Datta K, et al. Sensitizing glioma cells to cisplatin by abrogating the p53 response with antisense oligonucleotides. Cancer Gene Ther 2004, 11(8):525–531.

96. Xu GW, Mymryk JS, Cairncross JG. Pharmaceutical-mediated inactivation of p53 sensitizes U87MG glioma cells to BCNU and temozolomide. Int J Cancer 2005, 116(2):187–192.

97. Komarov PG, et al. A chemical inhibitor of p53 that protects mice from the side effects of cancer therapy. Science 1999, 285(5434):1733–1737.

98. Lorenzo E, et al. Doxorubicin induces apoptosis and CD95 gene expression in human primary endothelial cells through a p53-dependent mechanism. J Biol Chem 2002, 277(13):10883–10892.

99. Feldherr CM, Akin D, Cohen RJ. Regulation of functional nuclear pore size in fibroblasts. J Cell Sci 2001, 114(Pt 24):4621–4627.

100. Dagher PC. Apoptosis in ischemic renal injury: roles of GTP depletion and p53. Kidney Int 2004, 66(2):506–509.

101. Kelly KJ, et al. P53 mediates the apoptotic response to GTP depletion after renal ischemia-reperfusion: protective role of a p53 inhibitor. J Am Soc Nephrol 2003, 14(1):128–138.

102. Kaji A, et al. Pifithrin-alpha promotes p53-mediated apoptosis in JB6 cells. Mol Carcinog 2003, 37(3):138–148.

103. Komarova EA, Gudkov AV. Could p53 be a target for therapeutic suppression? Semin Cancer Biol 1998, 8(5):389–400.

104. Zhang, M. et al. Pifithrin-alpha suppresses p53 and protects cochlear and vestibular hair cells from cisplatin-induced apoptosis. Neuroscience 2003, 120(1):191–205.

105. Zhu X, et al. Novel p53 inactivators with neuroprotective action: syntheses and pharmacological evaluation of 2-imino-2,3,4,5,6,7-hexahydrobenzothiazole and 2-imino-2,3,4,5,6,7-hexahydrobenzoxazole derivatives. J Med Chem 2002, 45(23):5090–5097.

106. Strom E, et al. Small-molecule inhibitor of p53 binding to mitochondria protects mice from gamma radiation. Nat Chem Biol 2006, 2(9):474479.

107. Mihara M, et al. p53 has a direct apoptogenic role at the mitochondria. Mol Cell 2003, 11(3):577–590.

108. Pietrancosta N, et al. Novel cyclized Pifithrin-alpha p53 inactivators: synthesis and biological studies. Bioorg Med Chem Lett 2005, 15(6):1561–1564.

109. Pietrancosta N, et al. Imino-tetrahydro-benzothiazole derivatives as p53 inhibitors: discovery of a highly potent in vivo inhibitor and its action mechanism. J Med Chem 2006, 49(12):3645–3652.

110. Minchinton AI, Tannock IF. Drug penetration in solid tumours. Nat Rev Cancer 2006, 6(8):583–592.

111. Torchilin VP. Recent advances with liposomes as pharmaceutical carriers. Nat Rev Drug Discov 2005, 4(2):145–160.

112. Bassi L, et al. Pifithrin-alpha, an inhibitor of p53, enhances the genetic instability induced by etoposide (VP16) in human lymphoblastoid cells treated in vitro. Mutat Res 2002, 499(2): 163–176.

Color Plates

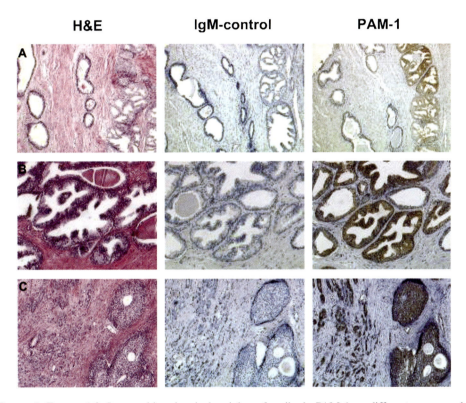

| | H&E | IgM-control | PAM-1 |

COLOR PLATE 1, FIGURE 1.2. Immunohistochemical staining of antibody PAM-1 on different precursor lesions of prostate carcinoma. Paraffin sections were stained with hematoxylin-eosin, unspecific human IgM as a negative control and antibody PAM-1: **A.** Normal prostate tissue and low grade prostate intraepithelial neoplasia (PIN). **B.** High-grade PIN. **D.** Prostate adenocarcinoma and high grade PIN (original magnification, ×100)

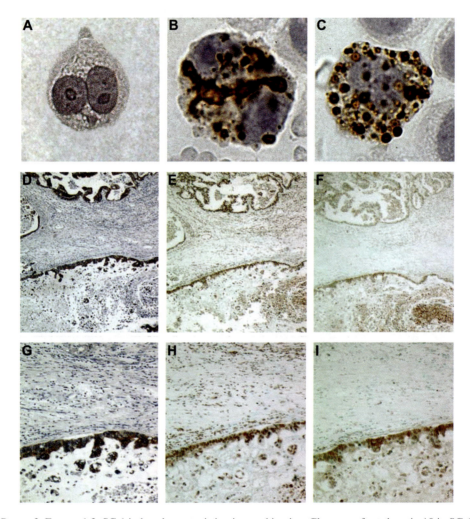

COLOR PLATE 2, FIGURE 1.3. SC-1 induced apoptosis in vitro and in vivo. Cleavage of cytokeratin 18 in SC-1-treated apoptotic stomach carcinoma cells in vitro. Immunohistochemical staining of cytospin preparations reveals that 24 hours after induction of apoptosis, cleavage of cytokeratin 18 starts (**B**), and after 48 hours, apoptotic bodies are released from the cells (**C**). In (**A**), a nonapoptotic cell is shown. (Original magnification × 400) DNA fragmentation in SC-1-treated apoptotic stomach carcinoma cells in vivo. Apoptotic stomach carcinoma cells in a metastasised tumour of a 50-year-old patient after treatment with the antibody SC-1. The patient received a single dose of antibody SC-1 and the tumour specimen was investigated for SC-1 induced apoptosis using the Klenow FragEL DNA fragmentation Kit (Oncogene, Boston). **D, G.** Control antibody CK8, tumour cells are stained. **E, H.** Positive control, all cell nuclei are stained. **F, I.** only the nuclei of apoptotic tumor cells are stained and normal not malignant tissue is not affected. (Original magnification, ×100 (*D, E, F*)/×200 (*G, H, I*))

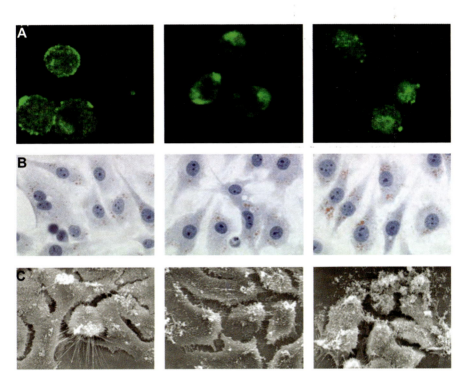

Color Plate 3, Figure 1.4. SAM-6 induced apoptosis: mode of action. Immunofluorescence of SAM-6 endocytosis. Pancreas carcinoma cells BXPC-3 were incubate with fluorochrome labeled SAM-6 antibody. After 30, 60, and 90 minutes cells were exposed on slides, fixed, and analyzed using confocal microscopy. 30 minutes, antibody binding; 60 minutes, "capping"; 90 minute antibody SAM-6 is completely internalized into the cell. Sudan III staining of neutral lipids in SAM-6 treated tumor cells. Pancreas carcinoma cells BXPC-3 were incubated with antibody SAM-6 antibody or for 2, 24, and 48 hours. An accumulation of red stained lipid droplets is visible in antibody SAM-6 treated tumor cells. Magnification ×200. Scanning electron microscopy of SAM-6 antibody-induced apoptosis. Stomach carcinoma cells 23132/87 were incubated with antibody SAM-6 for 2, 24 and 48 hours. Samples were proceeded for scanning electron microscopy and analyzed by ZEISS DSM 962. On the SAM-6 treated tumor cells apoptotic effects such as stress fibers, loss of cell-cell contacts, and clusters of apoptotic bodies are visible

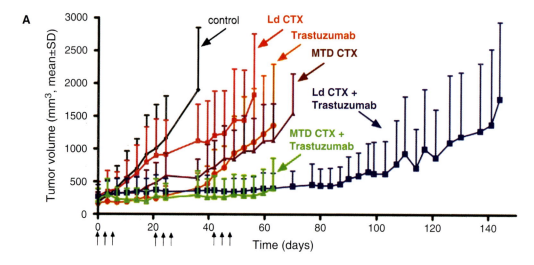

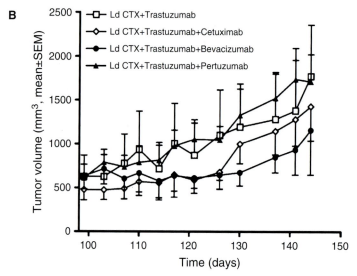

Color Plate 4, Figure 6.2. In vivo treatment of Her-2 positive MDA231-H2N human mammary tumors grown orthotopically in female SCID mice. **A.** Antitumor effects of low-dose metronomic cyclophosphamide (Ld CTX), maximum tolerated dose cyclophosphamide (MTD CTX), or trastuzumab alone, and combination regimens using low dose metronomic cyclophosphamide plus trastuzumab, or trastuzumab in combination with MTD cyclophosphamide. Arrows indicate time of MTD CTX dosing. **B.** Addition of second line therapies to tumors of MDA231-H2N that were starting to fail (as shown in top panel **[A]** after around 100 days) Ld CTX plus trastuzumab therapies. Second-line regimens were as indicated, with the addition of bevacizumab causing a further growth delay of approximately 4 weeks. (Adapted from du Manoir JM, Francia G, Man S, et al. Strategies for delaying or treating in vivo acquired resistance to trastuzumab in human breast cancer xenografts. Clin Cancer Res 2006, 12:904–916.)

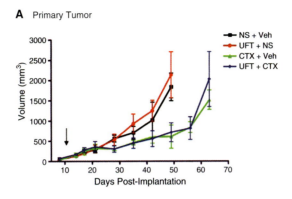

A Primary Tumor

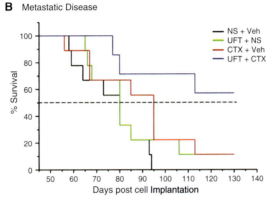

B Metastatic Disease

COLOR PLATE 5, FIGURE 6.3. An example of how chronic combination oral metronomic low-dose CTX and UFT prolongs survival of mice with advanced metastatic disease. (From Munoz R, Man S, Shaked Y, et al. Highly efficacious nontoxic preclinical treatment for advanced metastatic breast cancer using combination oral UFT-cyclophosphamide metronomic chemotherapy. Cancer Res 2006, 66:3386–3391.) **A.** 231/LM2-4 human breast metastatic variant cells were orthotopically injected into the MFPs of 6- to 8-week-old CB17 SCID mice. When tumors reached volumes of approximately $200\,mm^3$, treatment with either vehicle control, or $15\,mg/kg/day$ UFT by gavage, or $20\,mg/kg$ per day CTX through the drinking water, or a combination of CTX and UFT treatments was initiated. Tumors were measured weekly and tumor volume was plotted accordingly. Arrow indicates time of initiation of treatment. **B.** 6-week-old CB-17 SCID mice were recipients of 231/LM2-4 transplanted cells. When tumors reached $400\,mm^3$ (which took approximately 3 weeks) primary tumors were surgically removed. Treatment with vehicle control, $15\,mg/kg$ per day UFT by gavage, $20\,mg/kg$ per day CTX through the drinking water, or the daily combination of metronomic UFT and CTX, were initiated 3 weeks after surgery on a daily non-stop basis. For example, in the experiment shown in **(B),** the duration of the therapy was 140 days, and was initiated on day 43, 3 weeks after surgery, with termination at day 183. Mice were monitored frequently according to the institutional guidelines. A Kaplan-Meier survival curve was plotted accordingly for all treated groups, as indicated in the figure. **A, B.** n = 7–9/group. NS = normal saline, Veh = vehicle control (0.1% HPMC). Note that effects on primary tumor **(A)** were minor and in no way predictive of the survival benefits seen with UFT and CTX on metastatic disease (recorded as survival)

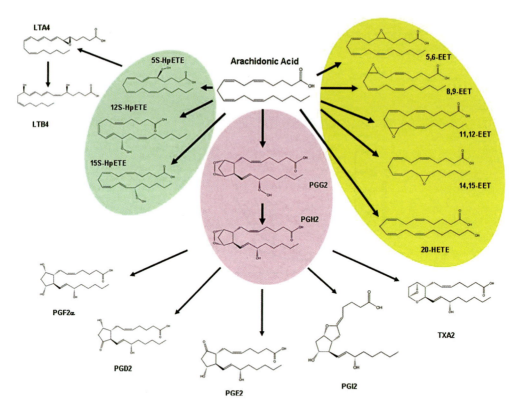

COLOR PLATE 6, FIGURE 9.1. Metabolism of arachidonic acid and synthesis of major eicosanoids. Reactions catalyzed by cyclooxygenases are shown in the pink field, reactions catalyzed by lipoxygenases are shown in the green field and reactions catalyzed by CYP monooxidases are shown in the yellow field. Also shown are five major prostanoids, which are synthesized by prostaglandin synthases from PGH_2, and LTA_4 converted from product of lipoxygenase reaction, which is further converted to LTB_4 by LTA_4 hydrolase

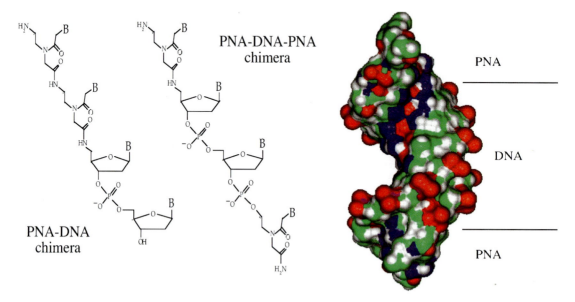

COLOR PLATE 7, FIGURE 11.3. Structure of PNA-DNA-PNA chimeras targeting NF-kappaB–related proteins. The molecular structures and the sequence of the double-stranded PNA-DNA-PNA chimera mimicking NF-kappaB binding sites are modified from Borgatti et al. [116], Romanelli et al. [118], and Gambari et al. [109]

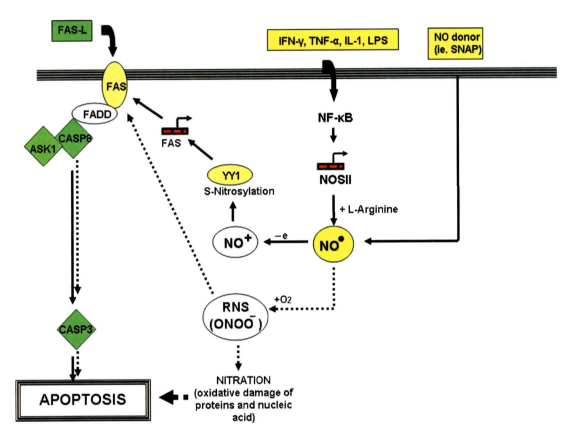

COLOR PLATE 8, FIGURE 13.1. Mechanism of tumor cell sensitization to Fas-L–induced apoptosis by IFN-γ: Pivotal role of NO. IFN-γ or other agents, such as TNF-a, IL-1, or LPS, upregulate NF-kB, which in turn regulates positively the transcription of NOSII. NOSII catalyses the biosynthesis of NO by L-arginine. NO can also be released in the cytosol by treatment of cells with an NO donor such as SNAP or DETANONOate. Free nitric oxide may react with O_2 (discontinuous line), resulting in the formation of reactive nitrogen species (RNS) such as ($ONOO^-$), which upregulate Fas and cause oxidative damage in protein and nucleic acids leading to apoptosis. Alternatively (continuous line), NO or NO^+ ion is capable of forming S-nitrosothiols resulting in S-nitrosylation of several proteins, including YY1, which acts as a repressor of Fas transcription. Thus, inducible levels of Fas by NO are able to overcome tumor resistance to Fas-L and sensitize them to Fas-L–mediated apoptosis

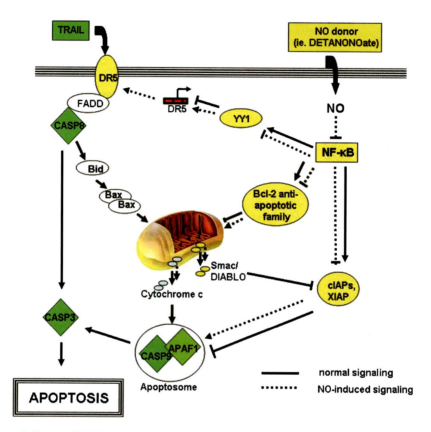

COLOR PLATE 9, FIGURE 13.2. Mechanism of tumor cell sensitization to TRAIL-induced apoptosis by NO. Treatment of several tumor cell lines with NO donors such as DETANONOate and TRAIL results in apoptosis and synergy is achieved. The synergy is the result of complementation in which each agent partially activates the apoptotic pathway and the combination results in apoptosis. The signal provided by NO partially inhibits NF-kB activity, and this leads to downregulation of antiapoptotic proteins of the Bcl-2 family such as Bcl-xL, and inhibition of cIAP family members (i.e., XIAP, cIAP-1, cIAP-2). In addition DETANONOate also partially activates the mitochondria and release of modest amounts of cytochrome C and Smac/DIABLO into the cytosol in the absence of caspase-9 activation. The NO-induced NF-kB suppression also inhibits the negative transcriptional regulator of DR5, YY1, resulting in DR5 upregulation. Thus, the combination treatment with TRAIL and DETANONOate results in significant activation of the mitochondria and release of high levels of cytochrome C and Smac/DIABLO, activation of caspases-9 and -3, promoting apoptosis. The role of Bcl-xL in the regulation of TRAIL apoptosis has been corroborated by the use of the chemical inhibitor 2MAM-A3 in several cell lines, which also sensitized the cells to apoptosis

can modify the apoptosis regulatory proteins to facilitate the apoptotic signaling cascade induced by the cytotoxic cells/ligands.

Host cytotoxic cells (NK, CTL, macrophages, etc.) mediate their cytotoxic killing by various mechanisms. These include the perforin/granzymes pathway and the death ligand family members (TNF-α, FasL, TRAIL), leading to necrosis/apoptosis. Thus, failure of tumor cells to respond to cytotoxic immunotherapy may be due to the development of resistance to death-induced stimuli by the cytotoxic cells/ligands. Hence, sensitizing agents that can modify the antiapoptotic regulatory mechanisms in tumor cells may be successfully used in combination with cytotoxic immunotherapy in the treatment of immune-resistant tumor cells.

2 Biological Activity of Nitric Oxide

Nitric oxide (NO) is a highly reactive free radical capable of mediating a multitude of reactions [2]. The free radical, NO, is an uncharged molecule containing an unpaired electron in its outermost orbital, allowing it to undergo several reactions functioning either as a weak oxidant (electron donor) or an antioxidant (electron acceptor). NO is able to react with other inorganic molecules (i.e., oxygen, superoxide, or transition metals), structures in DNA (pyrimidine bases), prosthetic groups (i.e., heme) or with proteins (leading to S-nitrosylation of thiol groups, nitration of tyrosine residues or disruption of metal-sulfide clusters such as zinc-finger domains or iron-sulfide complexes) [3]. In addition, NO can function as an antioxidant against reactive oxygen species (ROS) such as hydrogen peroxide (H_2O_2) and superoxide (O_2^-) by diffusing and concentrating into the hydrophobic core of low-density lipoprotein (LDL) [4]. It can react with several ROS, such as superoxide to form peroxynitrite ($ONOO^-$), a highly oxidizing and nitrating reactive nitrogen species (RNS) responsible for mediating protein oxidation reactions under physiologic conditions [5]. Another mechanism of NO-related reactivity is through the addition of an NO group to the thiol side-chain of cysteine residues within proteins and peptides, termed S-nitrosylation, which plays a significant role in the ubiquitous influence of NO on cellular signal

transduction [6]. NO or NO^+ ion is capable of forming S-nitrosothiols (RSNO; product of S nitrosylation), which function as potent platelet aggregation inhibitors and vasorelaxant compounds [7]. Other biological effects by NO have been recently reviewed [8].

3 Sensitization of Tumor Cells to Fas-L–Induced Apoptosis by IFN-γ: Pivotal Role of NO

Immunosensitization is the process by which cells are made sensitive to immune-mediated cytotoxicity (Fig. 13.1). Molecular mechanisms of immunosensitization such as transcriptional upregulation of proapoptotic proteins and downregulation of antiapoptotic proteins have been proposed to facilitate apoptosis by immunocytotoxic stimuli. Interestingly, NO has been found to be involved in the sensitization of tumor cells to various apoptotic stimuli, such as FasL (APO-1/CD95), TRAIL, and TNF-α. One mechanism responsible for the eradication of tumor cells by cytotoxic immune lymphocytes is Fas-mediated apoptosis, and Fukuo et al. [9] found that NO caused an increased expression of the Fas receptor in aortic vascular smooth muscle cells and increased sensitivity to FasL-mediated apoptosis. IFN-γ, together with many other proinflammatory cytokines (TNF-α, IL-1, LPS, etc.), can stimulate the induction of NOS II and the subsequent generation of NO. Through treatment with IFN-γ and the NO donor SNAP (alone or in combination), we have shown that human ovarian carcinoma cell lines (A2780 and AD10) were sensitized to FasL-mediated apoptosis by IFN-γ, partly due to NOS II induction and the consequent upregulation of Fas gene expression by RNS [10, 11]. These findings demonstrated that NO and RNS can regulate the sensitivity of tumor cells to FasL-mediated cytotoxic immune lymphocytes. A similar study by Park et al. [12], using ionizing radiation (IR) in combination with SNAP, showed sensitization to FasL-induced apoptotic cell death of HeLa human cervical cancer cells parallel to our findings with regard to the role of NO as an immunosensitizer. We have also previously reported, using a Fas promoter–driven luciferase reporter system, that the transcription factor Yin Yang 1

Chapter 13
Nitric Oxide–Induced Immunosensitization to Apoptosis by Fas-L and TRAIL

Benjamin Bonavida, Sara Huerta-Yepez, Mario I. Vega,
Demetrios A. Spandidos, and Stravoula Baritaki

1 Introduction

The incidence of cancer is only second to heart disease. Treatment of cancer remains one of the biggest challenges, particularly in view of increase in average life span, at least in developed countries; cancer is largely a disease of older individuals. Significant progress has been made in the development of novel therapeutics and the treatment of a large number of cancers. Hence, current therapeutics (chemotherapy, radiation, hormonal, and immunotherapy) have resulted in significant clinical responses and prolongation of life, albeit with little complete remission. One of the major problems in the eradication of cancer is the acquisition/development of resistance and refractoriness to conventional therapeutics. Cross-resistance develops since most cytotoxic therapeutics exerts their antitumor effect by inducing cell death by apoptosis and tumor cells develop mechanisms to resist apoptosis.

The failure to eradicate resistant tumors with current standard therapeutics calls for the use of alternative and less toxic novel therapies. For instance, a detailed understanding of the underlying molecular mechanisms of tumor drug resistance is critical for the development and design of new strategies to overcome the problem of resistance, thus improving the therapeutic outcome. The mechanisms of drug resistance are complex, and include among others poor vascular access and little drug penetration into the tumor mass, acquisition of multi-drug resistance phenotype in which the efflux of the drug is rapid, metabolic inactivation of the drugs, detoxification of accumulated toxic metabolites, enhanced DNA repair mechanism, and amplification of drug target genes [1]. The failure to cure chemoresistant tumors with conventional chemotherapeutic approaches has led to the introduction of immunotherapy. Immunotherapeutic strategies under investigation consider chemoresistant tumors to be sensitive to immunotherapy, as it has been assumed that cytotoxic immune cells attack tumor cells by different mechanisms of action and may not be subjected to the mechanisms of drug resistance. Despite the proposed advantages of immunotherapy over chemotherapy, immunotherapy today still fails to deliver a significant curative rate. It is unclear if drug-resistant tumors are actually sensitive to killing mediated by immune cytotoxic cells and whether cross-resistance is established. Tumor chemoresistance may actually reflect the general tumor resistance mechanism underlying a common cytotoxic pathway mediated by various cytotoxic stimuli, namely, programmed cell death or apoptosis. Such a resistance scheme to a central cytotoxic pathway may also lead to the cellular resistance to other cytotoxic mechanisms, including immunotherapy. With the premise that chemoresistant tumors develop general mechanisms of resistance to apoptosis-mediated stimuli, our hypothesis proposes ways to use immunosensitizing agents that

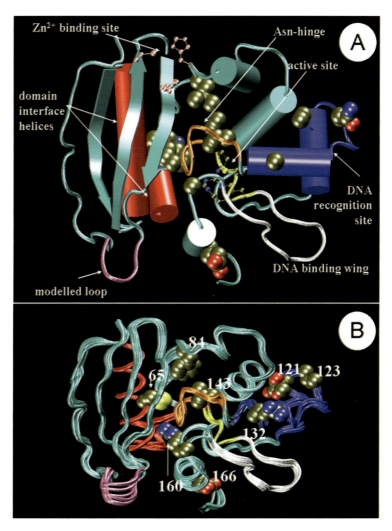

COLOR PLATE 11, FIGURE 21.4 Molecular modeling of SNP-based variants of MGMT. **A.** Wild-type structure of MGMT. Mutation sites are shown in van der Waals representation. Color codes of helices: red, N-terminal α-helices, blue, DNA recognition site with helix-turn-helix motif. The second DNA recognition site binds to the major groove of DNA, yellow, 3–10 helix with conserved Pro-Cys-His-Arg sequence. The active Cys145 is located here. Color code of loops: orange, Asn-hinge that joins the DNA recognition helix and the active site. It also provides 40% of the contact between N-terminal and C-terminal domains, white, DNA binding wing. O6-alkylguanine lies between the binding wing and the recognition helix. Color code of side chains shown in CPK representation: per atom, conserved active Pro-Cys-Arg sequence, pink, Zn²⁺ binding site (residues C5, C24, H29, H85). Cys5 is missing, since it is not resolved in the crystal structure 1QNT. **B.** Localization of amino acid change and overlap of structures of all mutants. Mutated side chains are shown in van der Waals representations. (Reprinted from Schwarzl SM, Smith JC, Kaina B, et al. Molecular modeling of O6-methylguanine-DNA methyltransferase mutant proteins encoded by single nucleotide polymorphisms. Int J Mol Med 2005, 16:553–557.)

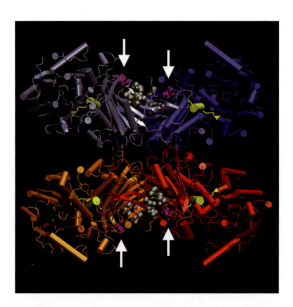

COLOR PLATE 10, FIGURE 21.1 Model of the G6PD Aachen tetramer. This G6PD variant has originally been described by Kahn et al. [253]. The mutation in the G6PD Aachen variant has been determined by us previously [254]. A mutation 1089C>G results in a predicted amino acid change 363Asn>Lys. The 1089C>G point mutation is unique, but produces the identical amino acid change found in another variant of G6PD deficiency, G6PD Loma Linda. The 363Asn>Lys exchange in G6PD Loma Linda is caused by a 1089C>A mutation [255]. Using the available three-dimensional structure of the human G6PD tetrameric protein complex [256], the location of the point mutation of amino acid 363 in G6PD Aachen is found at the surface of a monomer in close proximity to NADP$^+$ and more than 20Å away from the glucose-6-phosphate binding site. This residue is probably involved in NADP$^+$ binding that in turn is required for tetramer stability [256]. Thus, Arg363 may be required to indirectly maintain the structural integrity of the functional unit. Replacing it with a positively charged Lys residue would lead to charge–charge repulsion between Lys363 and NADP$^+$, thus affecting NADP$^+$ binding and tetramer formation. The two pairs of dimer-forming monomers are colored in ice blue/blue and in orange/red. The mutation site (Asn363) and the cofactor NADP$^+$ are shown in van der Waals representation in purple and grey, respectively. The conserved eight-residue peptide RIDHYLGK corresponding to the substrate binding site is colored in yellow. The figure was prepared from the crystal structure 1QKI using the program VMD [257]

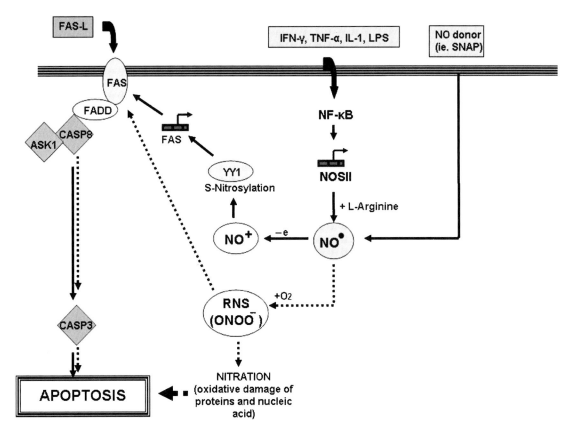

FIGURE 13.1. Mechanism of tumor cell sensitization to Fas-L–induced apoptosis by IFN-γ: Pivotal role of NO. IFN-γ or other agents, such as TNF-a, IL-1, or LPS, upregulate NF-kB, which in turn regulates positively the transcription of NOSII. NOSII catalyses the biosynthesis of NO by L-arginine. NO can also be released in the cytosol by treatment of cells with an NO donor such as SNAP or DETANONOate. Free nitric oxide may react with O_2 (discontinuous line), resulting in the formation of reactive nitrogen species (RNS) such as (ONOO⁻), which upregulate Fas and cause oxidative damage in protein and nucleic acids leading to apoptosis. Alternatively (continuous line), NO or NO⁺ ion is capable of forming S-nitrosothiols resulting in S-nitrosylation of several proteins, including YY1, which acts as a repressor of Fas transcription. Thus, inducible levels of Fas by NO are able to overcome tumor resistance to Fas-L and sensitize them to Fas-L–mediated apoptosis (*See Color Plates*)

(YY1) (which normally represses Fas expression by binding to a *cis*-element clustered at the silencer region of the Fas promoter) negatively regulates Fas expression through its interaction with the silencer region of the Fas promoter [13]. YY1 is a 414 amino acid Kruppel-related zinc transcription factor that binds to the CG (A/CC) CATNTT consensus DNA element located in promoters and enhancers of many cellular and virus genes [14]. YY1 physically interacts with and recruits histone-acetyl-transferase, histone-deacetylase and histone-methyl-transferase enzymes to the chromatin and may thus direct histone-acetylation, deacetylation and methylation at YY1 activated or repressed promoters [14]. NO-mediated inhibition of YY1 resulted in upregulation of Fas expression and sensitization of ovarian carcinoma cells to FasL-induced apoptosis [13]. Recently, we have found that the treatment of the B non-Hodgkin's lymphoma cell line (B-NHL), Ramos, with rituximab (chimeric anti-CD20 Ab) or with specific NF-κB inhibitors (e.g., Bay 11-7085 and DHMEQ) and/ or inhibition of YY1(through the use of the NO donor, DETA/NONOate), resulted in sensitization to FasL-induced apoptosis [15]. Noteworthy, the NO-mediated inhibition of YY1 activity (in the absence of rituximab) resulted in significant upregulation of surface Fas expression and sensitized Ramos cells to CH-11 (Fas agonist mAb)-induced apoptosis. Until now, the mechanism of YY1 inhibition by NO was unclear. However, Hongo et al. [16] demonstrated that treatment of prostate cancer (PC-3) cells with DETA/NONOate resulted in the S-nitrosylation of YY1, thereby upregulating Fas expression and sensitizing tumor cells to FasL-induced apoptosis through a direct NO-mediated mechanism.

4 Sensitization of Tumor Cells to TRAIL-Induced Apoptosis by NO: Roles of NF-kB and Bcl-xL

Tumor necrosis factor–related apoptosis-inducing ligand (TRAIL) is a cytotoxic molecule that has been shown to exert, selectively, antitumor cytotoxic effects both in vitro and in vivo with minimal toxicity to normal tissues [17, 18]. TRAIL has been considered a new therapeutic and preclinical studies demonstrate its antitumor activity alone or in combination with drugs [17, 19–21]. However, many tumor cells have been shown to be resistant to TRAIL [22–25]. We and others have reported that various sensitizing agents like chemotherapeutic drugs [1, 22, 26], cytokines [27], and inhibitors [28], are able to render TRAIL-resistant tumor cells sensitive to TRAIL apoptosis. Further, we [29] and others [30, 31] reported that (Z)-1-[2-(2-aminoethyl)-N-(2-ammonioethyl) amino] diazen-1-ium-1, 2-diolate (DETANONOate) can also sensitize tumor cells to TRAIL-mediated apoptosis.

The mechanism underlying the NO-mediated sensitization to TRAIL is not known. We hypothesized that NO-mediated sensitization of tumor cells to apoptosis may be due to NO-induced inhibition of constitutive NF-kB activity and this, in turn, results in the downregulation of the antiapoptotic resistant factor, Bcl-xL. Hence, downregulation of the antiapoptotic gene product Bcl-xL results in the sensitization of CaP cells to TRAIL-mediated apoptosis. The mechanism by which DETANONOate mediated the sensitization was examined. DETANONOate inhibited the constitutive NF-kB activity as assessed by EMSA. Also, p50 was S-nitrosylated by DETANONOate, resulting in inhibition of NF-kB. Inhibition of NF-kB activity by the chemical inhibitor Bay 11-7085, like DETANONOate, sensitized tumor cells to TRAIL-induced apoptosis. In addition, DETANONOate downregulated the expression of Bcl-2-related gene (Bcl-xL), which is under the transcriptional regulation of NF-kB. The regulation of NF-kB and Bcl-xL by DETANONOate was corroborated by the use of Bcl-xL and Bcl-xL kB reporter systems. DETANONOate inhibited luciferase activity in the wild-type and had no effect on the mutant cells. Inhibition of NF-kB resulted in downregulation of Bcl-xL expression and sensitized tumor cells to TRAIL-induced apoptosis. The role of Bcl-xL in the regulation of TRAIL apoptosis was corroborated by inhibiting Bcl-xL function by the chemical inhibitor 2-methoxyantimycin A3, and this resulted in sensitization of the cells to TRAIL apoptosis. Signaling by DETANONOate and TRAIL for apoptosis was also examined. DETANONOate altered the mitochondria by inducing membrane depolarization and releasing modest amounts of cytochrome c and Smac/DIABLO in the absence of downstream activation of caspases-9 and -3. However, the combination of DETANONOate and TRAIL resulted in activation of the mitochondrial pathway and activation of caspases-9 and -3, and induction of apoptosis [29]. These findings demonstrate that DETANONOate-mediated sensitization of tumor cells to TRAIL-induced apoptosis is via inhibition of constitutive NF-kB activity and Bcl-xL expression.

5 Sensitization of Tumor Cells to TRAIL-Induced Apoptosis by NO: Roles of YY1 and DR5

Several reports have revealed that treatment with certain sensitizing agents, such as chemotherapeutic drugs, upregulate DR4 and/or DR5 expression, and upregulation of these receptors correlated with sensitivity to TRAIL-induced apoptosis (Fig. 13.2) [32–34]. However, the molecular mechanisms by which these receptors are upregulated by sensitizing agents are not known. The transcriptional regulation of DR5 expression has been

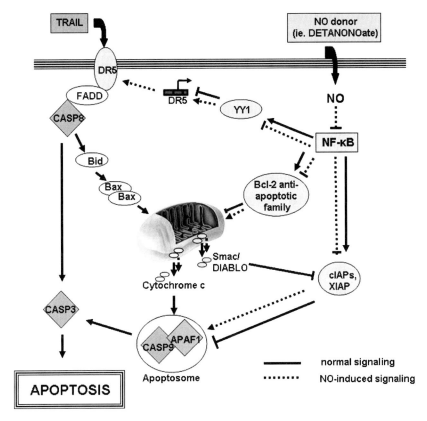

FIGURE 13.2. Mechanism of tumor cell sensitization to TRAIL-induced apoptosis by NO. Treatment of several tumor cell lines with NO donors such as DETANONOate and TRAIL results in apoptosis and synergy is achieved. The synergy is the result of complementation in which each agent partially activates the apoptotic pathway and the combination results in apoptosis. The signal provided by NO partially inhibits NF-kB activity, and this leads to downregulation of antiapoptotic proteins of the Bcl-2 family such as Bcl-xL, and inhibition of cIAP family members (i.e., XIAP, cIAP-1, cIAP-2). In addition DETANONOate also partially activates the mitochondria and release of modest amounts of cytochrome C and Smac/DIABLO into the cytosol in the absence of caspase-9 activation. The NO-induced NF-kB suppression also inhibits the negative transcriptional regulator of DR5, YY1, resulting in DR5 upregulation. Thus, the combination treatment with TRAIL and DETANONOate results in significant activation of the mitochondria and release of high levels of cytochrome C and Smac/DIABLO, activation of caspases-9 and -3, promoting apoptosis. The role of Bcl-xL in the regulation of TRAIL apoptosis has been corroborated by the use of the chemical inhibitor 2MAM-A3 in several cell lines, which also sensitized the cells to apoptosis (*See Color Plates*)

investigated by using a pDR5-reported system and demonstrated that the transcription factor Sp-1 is a major factor that regulates DR5 expression [35]. In a recent study, we have identified the transcription factor YY1 as a transcription repressor on the Fas promoter [10, 13]. Thus, we reasoned that upregulation of DR5 expression by sensitizing agents may be due to inactivation of a transcription repressor, such as YY1 on the DR5 promoter. Examination of the DR5 promoter revealed the presence of a putative DNA-binding site for YY1.

Preliminary findings demonstrated that the nitric oxide donor DETANONOate sensitized TRAIL-resistant tumor cells to TRAIL-induced apoptosis concomitant with DR5 upregulation. Thus, we hypothesized that the upregulation of DR5 expression by DETANONOate may be due to the inhibition of the YY1 repressor activity.

Treatment of TRAIL-resistant tumor cells with the nitric oxide donor DETANONOate sensitizes tumor cells to TRAIL-induced apoptosis concomitantly with DR5 upregulation. The mechanism of DR5 upregulation was examined based on the hypothesis that DETANONOate may inhibit a transcription repressor that negatively regulates DR5 transcription. Treatment of the prostate carcinoma cell line PC-3 cells with DETANONOate inhibited both NF-kappaB and YY1 DNA-binding activity concomitantly with upregulation of DR5 expression and TRAIL-induced apoptosis. The direct role of YY1 in the regulation of TRAIL resistance was demonstrated by transfection of PC-3 cells with YY1 siRNA. The cells exhibited upregulation of DR5 expression and were sensitized to TRAIL-induced apoptosis. The role of YY1 in the transcriptional regulation of DR5 was examined by a DR5 luciferase reporter system (pDR5) and two constructs, namely, the pDR5/-605 construct with a deletion of the putative YY1 DNA-binding region (-1224 to -605) and a construct pDR5-YY1 with a mutation of the YY1 DNA-binding site. Transfection of PC-3 cells with these two constructs resulted in comparable and significant (threefold) augmentation of luciferase activity over baseline transfection with pDR5. The present findings demonstrate that YY1 negatively regulates DR5 transcription and expression and these correlated with resistance to TRAIL-induced apoptosis (Huerta-Yepez et al., unpublished) [1]. Treatment with DETANONOate reverses resistance to TRAIL via inhibition of NF-kappaB and YY1. Inhibitors of YY1 may be used in combination with TRAIL in the treatment of TRAIL-resistant tumor cells.

The in vitro findings with DETANONOate on PC-3 cells, namely, inhibition of YY1 and upregulation of DR5, were examined for validation in an in vivo model of PC-3 tumor xenograft implanted SC into athymic nude mice. Mice were implanted with PC-3 cells and treated intratumorally with DETANONOate as described in Methods. The concentration of DETANONOate used for the in vivo administration was derived from reported studies in rats whereby DETANONOate was used in a noncancer model [36]. Following treatment, tumors were biopsied and analyzed by immunohistochemistry for the expression of YY1 and DR5. The findings demonstrate that treatment with DETANONOate inhibited YY1 expression, as compared with untreated controls, in which there was strong nuclear YY1 expression. The staining for YY1 was specific, as treatment with control IgG did not show any staining. In contrast, DR5 expression was augmented in treated mice as compared with untreated controls. The specificity for DR5 was demonstrated with the use of immunoglobulin control (Huerta-Yepez et al., unpublished). These findings demonstrate that the inhibition of YY1 and upregulation of DR5 following treatment of PC-3 cells with DETANONOate in vitro can also be reproduced in an in vivo model in mice bearing PC-3 tumor xenograft.

6 Concluding Remarks

NO plays several roles in cells and its effects vary depending on its concentration and selective modification of various gene products. Its ultimate manifestation results from a complex set of interactions depending on the type of cells studied. It is also clear from recent findings that NO can play a significant role as a chemopreventive agent in cancer development and cancer therapeutics. The application of NO donors as cancer therapeutics is a new venue that has not been appreciated in the past, as NO was primarily used for the treatment of blood vessel-related diseases and other non-cancer–related applications. The demonstration of NO-mediated cytotoxicity directly on cancer cells and/or indirectly in the tumor microenvironment through its antiproliferative and chemosensitizing roles, presents new challenges for its optimal use in cancer therapy. The data suggest that NO can be used as a chemosensitizing as well as an immunosensitizing agent; thus, one may consider its clinical application using combination treatment of NO donors and chemotherapy or immunotherapy resulting in synergistic activity in the treatment of cancer. It is also conceivable that one might use NO donors complexed with chemotherapeutic drugs or other cytotoxic agents. One may also consider using agents that can activate endogenous NO production via NOS II. Clearly, apart from the direct effects of NO on tumor cells, NO donors would also be functioning as vasodilators and thus have an even enhanced therapeutic potential. Possibly, novel NO donors may be administered orally and thus be more applicable to treatment. For certain tumors, it is also possible to administer NO donors intratumorally, thus reducing the systemic toxic effects that may arise from its route of administration. We expect that the application of NO donors in cancer therapy will be added to the armamentarium of cancer therapeutics in the near future.

Acknowledgments. This work was supported in part by the Department of Defense (DOD)/US Army DAMD 17-02-1-0023 and the Fogarty International Center Fellowship (D43 TW00013-14) (S.H.-Y., M.V., and U.C. MEXUS-Conacyt (S.H.-Y.), Research Supplement for Minorities from the PHS/NIH/NCI/CMBB ROI CA79976 to H.G., and a philanthropic contribution from the Ann C. Rosenfield Fund under the direction of David Leveton. We also acknowledge the assistance of Tania Golkar and Alina Katsman in the preparation of this manuscript.

References

1. Baritaki S, Huerta-Yepez S, Sakai T, et al. Chemotherapeutic drugs sensitize cancer cells to TRAIL-mediated apoptosis: up-regulation of DR5 and inhibition of Yin Yang 1. Mol Cancer Ther 2007, 6:1387–1399.
2. Blaise GA, Gauvin D, Gangal M, et al. Nitric oxide, cell signaling and cell death. Toxicology 2005, 208:177–192.
3. Bogdan C. Nitric oxide and the regulation of gene expression. Trends Cell Biol 2001, 11:66–75.
4. Benz S, Obermaier R, Wiessner R, et al. Effect of nitric oxide in ischemia/reperfusion of the pancreas. J Surg Res 2002, 106:46–53.
5. Tuteja N, Chandra M, Tuteja R, et al. Nitric oxide as a unique bioactive signaling messenger in physiology and pathophysiology. J Biomed Biotechnol 2004, 2004:227–237.
6. Hess DT, Matsumoto A, Kim SO, et al. Protein S-nitrosylation: purview and parameters. Nat Rev Mol Cell Biol 2005, 6:150–166.
7. Myers PR, Minor RL Jr, Guerra R Jr, et al. Vasorelaxant properties of the endothelium-derived relaxing factor more closely resemble S-nitroso-cysteine than nitric oxide. Nature 1990, 345:161–163.
8. Bonavida B, Khineche S, Huerta-Yepez S, et al. Therapeutic potential of nitric oxide in cancer. Drug Resist Updat 2006, 9:157–173.
9. Fukuo K, Hata S, Suhara T, et al. Nitric oxide induces upregulation of Fas and apoptosis in vascular smooth muscle. Hypertension 1996, 27:823–826.
10. Garban HJ, Bonavida B. Nitric oxide disrupts H_2O_2-dependent activation of nuclear factor kappa B. Role in sensitization of human tumor cells to tumor necrosis factor-alpha -induced cytotoxicity. J Biol Chem 2001, 276:8918–8923.
11. Garban HJ, Bonavida B. Nitric oxide sensitizes ovarian tumor cells to Fas-induced apoptosis. Gynecol Oncol 1999, 73:257–264.
12. Park IC, Woo SH, Park MJ, et al. Ionizing radiation and nitric oxide donor sensitize Fas-induced apoptosis via up-regulation of Fas in human cervical cancer cells. Oncol Rep 2003, 10:629–633.
13. Garban HJ, Bonavida B. Nitric oxide inhibits the transcription repressor Yin-Yang 1 binding activity at the silencer region of the Fas promoter: a pivotal role for nitric oxide in the up-regulation of Fas gene expression in human tumor cells. J Immunol 2001, 167:75–81.
14. Coull JJ, Romerio F, Sun JM, et al. The human factors YY1 and LSF repress the human immunodeficiency virus type 1 long terminal repeat via recruitment of histone deacetylase 1. J Virol 2000, 74:6790–6799.
15. Vega MI, Jazirehi AR, Huerta-Yepez S, et al. Rituximab-induced inhibition of YY1 and Bcl-xL expression in Ramos non-Hodgkin's lymphoma cell line via inhibition of NF-kappa B activity: role of YY1 and Bcl-xL in Fas resistance and chemoresistance, respectively. J Immunol 2005, 175:2174–2183.
16. Hongo F, Garban H, Huerta-Yepez S, et al. Inhibition of the transcription factor Yin Yang 1 activity by S-nitrosation. Biochem Biophys Res Commun 2005, 336:692–701.
17. Ashkenazi A, Dixit VM. Apoptosis control by death and decoy receptors. Curr Opin Cell Biol 1999, 11:255–260.
18. Ashkenazi A, Pai RC, Fong S, et al. Safety and anti-tumor activity of recombinant soluble Apo2 ligand. J Clin Invest 1999, 104:155–162.
19. de Jong S, Timmer T, Heijenbrok FJ, et al. Death receptor ligands, in particular TRAIL, to overcome drug resistance. Cancer Metastasis Rev 2001, 20:51–56.
20. Wajant H, Pfizenmaier K, Scheurich P. TNF-related apoptosis inducing ligand (TRAIL) and its receptors in tumor surveillance and cancer therapy. Apoptosis 2002, 7:449–459.
21. Chawla-Sarkar M, Bauer JA, Lupica JA, et al. Suppression of NF-kappa B survival signaling by nitrosylcobalamin sensitizes neoplasms to the anti-tumor effects of Apo2L/TRAIL. J Biol Chem 2003, 278:39461–39469.
22. Zisman A, Ng CP, Pantuck AJ, et al. Actinomycin D and gemcitabine synergistically sensitize androgen-independent prostate cancer cells to Apo2L/TRAIL-mediated apoptosis. J Immunother (1997) 2001, 24:459–471.

23. Ng CP, Zisman A, Bonavida B. Synergy is achieved by complementation with Apo2L/TRAIL and actinomycin D in Apo2L/TRAIL-mediated apoptosis of prostate cancer cells: role of XIAP in resistance. Prostate 2002, 53:286–299.

24. Bouralexis S, Findlay DM, Atkins GJ, et al A. Progressive resistance of BTK-143 osteosarcoma cells to Apo2L/TRAIL-induced apoptosis is mediated by acquisition of DcR2/TRAIL-R4 expression: resensitisation with chemotherapy. Br J Cancer 2003, 89:206–214.

25. Tillman DM, Izeradjene K, Szucs KS, et al. Rottlerin sensitizes colon carcinoma cells to tumor necrosis factor-related apoptosis-inducing ligand-induced apoptosis via uncoupling of the mitochondria independent of protein kinase C. Cancer Res 2003, 63:5118–5125.

26. Munshi A, McDonnell TJ, Meyn RE. Chemotherapeutic agents enhance TRAIL-induced apoptosis in prostate cancer cells. Cancer Chemother Pharmacol 2002, 50:46–52.

27. Park SY, Billiar TR, Seol DW. IFN-gamma inhibition of TRAIL-induced IAP-2 upregulation, a possible mechanism of IFN-gamma-enhanced TRAIL-induced apoptosis. Biochem Biophys Res Commun 2002, 291:233–236.

28. Nyormoi O, Mills L, Bar-Eli M. An MMP-2/MMP-9 inhibitor, 5a, enhances apoptosis induced by ligands of the TNF receptor superfamily in cancer cells. Cell Death Differ 2003, 10:558–569.

29. Huerta-Yepez S, Vega M, Jazirehi A, et al. Nitric oxide sensitizes prostate carcinoma cell lines to TRAIL-mediated apoptosis via inactivation of NF-kappa B and inhibition of Bcl-xl expression. Oncogene 2004, 23:4993–5003.

30. Lee YJ, Lee KH, Kim HR, et al. Sodium nitroprusside enhances TRAIL-induced apoptosis via a mitochondria-dependent pathway in human colorectal carcinoma CX-1 cells. Oncogene 2001, 20:1476–1485.

31. Secchiero P, Gonelli A, Celeghini C, et al. Activation of the nitric oxide synthase pathway represents a key component of tumor necrosis factor-related apoptosis-inducing ligand-mediated cytotoxicity on hematologic malignancies. Blood 2001, 98:2220–2228.

32. Shigeno M, Nakao K, Ichikawa T, et al. Interferon-alpha sensitizes human hepatoma cells to TRAIL-induced apoptosis through DR5 upregulation and NF-kappa B inactivation. Oncogene 2003, 22:1653–1662.

33. LaVallee TM, Zhan XH, Johnson MS, et al. 2-methoxyestradiol up-regulates death receptor 5 and induces apoptosis through activation of the extrinsic pathway. Cancer Res 2003, 63:468–475.

34. Johnston JB, Kabore AF, Strutinsky J, et al. Role of the TRAIL/APO2-L death receptors in chlorambucil- and fludarabine-induced apoptosis in chronic lymphocytic leukemia. Oncogene 2003, 22:8356–8369.

35. Yoshida T, Maeda A, Tani N, et al. Promoter structure and transcription initiation sites of the human death receptor 5/TRAIL-R2 gene. FEBS Lett 2001, 507:381–385.

36. Lu D, Mahmood A, Zhang R, et al. Upregulation of neurogenesis and reduction in functional deficits following administration of DEtA/NONOate, a nitric oxide donor, after traumatic brain injury in rats. J Neurosurg 2003, 99:351–361.

Chapter 14
Natural Agents That Can Sensitize Tumor Cells to Chemotherapy and Radiation Therapy

Ganesh Jagetia, Sunil Krishnan, and Bharat B. Aggarwal

1 Introduction

Cancer is a major public health problem in the United States and other developed countries. Currently, one in four deaths in the United States is due to cancer. Worldwide, cancer is the second leading cause of death after cardiovascular disease. The three primary modalities for treating cancer are surgery, chemotherapy, and radiotherapy. Yet better treatment and management strategies are needed. Unlike surgery, chemotherapy and radiotherapy are both highly toxic and must be modulated by reducing toxicity or improving therapeutic efficacy in order to enhance their beneficial effects. Such enhancement can be achieved by using other pharmacological agents to sensitize tumors to existing chemotherapeutic drugs (chemosensitization), ionizing radiation (radiosensitization), or both (chemoradiosensitization). One possibility is to use plant-based natural products to make tumors more amenable to chemotherapy or radiotherapy.

Natural plant products have been used throughout human history for various purposes. Having coevolved with life, these natural products are billions of years old. Tens of thousands of them are produced as secondary metabolites by the higher plants as a natural defense against disease and infection [1]. Medicines derived from plants have played a pivotal role in the health care of many cultures, both ancient and modern. The Indian system of holistic medicine known as ayurveda uses mainly plant-based drugs or formulations to treat ailments including cancer. Most of the small-molecule drugs introduced worldwide between 1981 and 2002 (i.e., 61% of 877 such drugs) can be traced back to their origins in natural products [2]. This is not surprising since plant-based drugs may be more suitable—at least in biochemical terms—for medicinal human use than the many exotic synthetic drugs produced through combinatorial chemistry. Nonetheless, modern medicine has neither held in very high esteem nor encouraged the medicinal use of natural products. Over the last two decades, however, successful attempts to better understand molecular mechanisms of action of some natural products have sparked interest in their therapeutic use in modern medical settings. Remarkably, most of the natural products experimentally evaluated so far have been found to be nontoxic or to have effective doses far below their toxic doses. The role of natural products in human health care cannot be underestimated. An estimated 80% of individuals in developing countries depend primarily on natural products to meet their healthcare needs [3]. Recent surveys suggest that one in three Americans uses medicinal natural products daily and that possibly one in two cancer patients (i.e., up to 50% of patients treated in cancer centers) use them [4, 5].

Among natural products used medicinally, polyphenols have been favored by many investigators because of their antioxidant properties [6, 7]. Produced as secondary metabolites by branched

From: *Sensitization of Cancer Cells for Chemo/Immuno/Radio-therapy, 1st Edition.*
Edited by: Benjamin Bonavida © Humana Press, Totowa, NJ

plants, polyphenols are essential for lignification, structure, pigmentation, pollination, allelopathy, pathogen and predator resistance, and growth [8]. More recently, polyphenols have been found to possess a wide array of other properties including chemopreventive, chemoprotective, chemosensitizing, radiosensitizing, and radioprotective activities. This chapter begins by discussing the molecular mechanisms responsible for sensitizing tumors to conventional chemotherapy and radiotherapy and then describes the sensitizing effects of various polyphenols known to exert such activity (See Figure 14.1; Tables 14.1 and 14.2).

2 Molecular Mechanisms of Chemoresistance and Radioresistance

2.1 Nuclear Factor-κB

Nuclear factor-κB (NF-κB) is a ubiquitous transcription factor important for its pleiotropic effects, inducible expression patterns, unique regulatory

mechanisms, and involvement in a large number of signaling and gene expression pathways [9–11]. The activation of NF-κB plays an important and crucial role in controlling a host of cellular functions, including immunity and inflammation, autoimmune disease, apoptosis, and cancer [12, 13]. There are two commonly heterodimeric forms of NF-κB: p50 (NF-κB1), which dimerizes with p65 (RelA), and p52 (NF-κB2), which dimerizes with RelB [14–16]. The p50/RelA complex is held by IκBα in the cellular cytoplasm in an inactive form until the signal for the nuclear translocation of NF-κB is received; that signal is regulated by upstream signaling events that converge for the most part on the IκB kinase (IKK) and its three related subunits (α, β, and g). The Iκκβ subunit phosphorylates Iκκα, which results in its ubiquitination, proteasomal degradation, and subsequent translocation of NF-κB into the nucleus and transcription of related genes [17]. NF-κB has been found to be constitutively expressed in several tumors, including cancers of the head and neck, breast, prostate, colon, and pancreas; leukemias; and lymphomas [18, 19]. This constitutive activation of NF-κB has

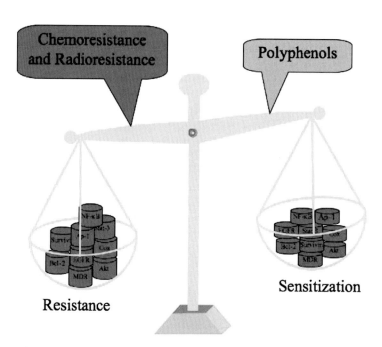

FIGURE 14.1. Balance between chemo/radiosensitization of tumors by polyphenols. Various mediators involved in resistance and sensitization of tumor cells are indicated

TABLE 14.1. Modulation of chemosensitivity by various polyphenols.

Curcumin

- Increases celecoxib-induced apoptosis in HT-29 IEC-18-K-ras and Caco-2 cells by blocking COX-2 [143].
- Increases cytotoxicity of doxorubicin in Hep2A cells by inhibiting glutathione-s-transferase and suppression of NF-κB activation [145, 146].
- Enhances sensitivity of multiple myeloma cells to melphalan and vincristine by suppressing NF-κB and *Bcl-2, Bcl-xL, cyclin D1*, and *IL-6* expression [147].
- Potentiates cytotoxic effects of doxorubicin, 5-fluouracil, and paclitaxel in prostate cancer cells by inhibiting NF-κB activation [148].
- Increases the sensitivity of paclitaxel-induced apoptosis, suppresses NFκB activation and expression of XIAP, IAP-1, IAP-2, Bcl-2, and Bcl-xL, COX-2, c-Myc, cyclin D1, VEGF, MMP-9, and AKT [149, 150].
- Enhances sensitivity of etoposide, mitoxantrone, topotecan, SN-38, and doxorubicin by suppressing MDR genes and drug efflux by P-glycoprotein in HEK cells [151, 152].
- Reduces camptothecin-, mechlorethamine-, and doxorubicin-induced apoptosis in MCF-7, MDA-MB-231, and BT-474 [153].

Resveratrol

- Increases sensitivity of gemcitabine, navelbine, cisplatinum (CDDP), paclitaxel, and TRAIL [166, 167].
- Enhances the effect of paclitaxel in A549, EBC-1, Lu65, NHL and MM cells [164, 165].
- Potentiates the effect of doxorubicin in MCF-7 and MDA-MB468 cells by G1 arrest [168].
- Increases sensitivity to temozolomide (TMZ) by increasing apoptosis and S-phase arrest in both the TMZ-sensitive M14 and TMZ-resistant SK-Mel-28 and PR-Mel cells [169].
- Elevates the therapeutic effects of cisplatin, doxorubicin, cycloheximide, busulfan, gemcitabine and paclitaxel [171–172].
- The low concentrations inhibit caspase activation, DNA fragmentation, and translocation of cytochrome *c* induced by hydrogen peroxide or C2, vincristine, and daunorubicin in human leukemia cells [170].

Genistein

- Potentiates the effect of cisplatin, docetaxel, doxorubicin, and gemcitabine in prostate, breast, pancreas, and lung cancer cells in vitro by suppression of NF−κB [183–186, 190].
- Sensitizes diffuse large cell lymphoma cells to cyclophosphamide, doxorubicin, vincristine, prednisone (CHOP) treatment [188, 194].
- Increases the effect of 5-fluorouracil in chemoresistant HT-29 colon cancer cells [189].
- Increases the effect of tamoxifen synergistically in dysplastic cells and reduces the side effects of tamoxifen through P450-mediated pathways [191, 192].
- Increases the effect of irinotecan by blocking CDK1 phosphorylation [193].
- Increases the effect of bleomycin by increasing the micronuclei frequency and DNA damage in HL-60 cells [195].

EGCG

- Increases the cytotoxic effect of doxorubicin in doxorubicin-resistant S180-dox and SW620-dox cells by inhibiting protein kinase C [223].
- Increases the effect of doxorubicin and reverses mutlidrug resistance by inhibitng the efflux of doxorubicin in vitro [224, 225, 228–231, 236].
- Sensitizes MDR human xenografted KB-A-1 tumor in athymic nude mice to doxorubicin by increasing the accumulation of doxorubicin [227, 228].
- Potentiates the effect of sulindac synergistically in Colon 26 cells [232, 236].
- Increases the cytotoxic effect of tamoxifen and 4-hydroxytamoxifen [232, 233, 236].
- Increases the effect of cisplatin in SKOV3, CAOV3, and C200 cells by increasing the oxidative stress [234].
- Increases the effect of low-dose As_2O_3 in human malignant B cells by inducing the higher production of ROS (O_2^- and H_2O_2) [235].

Silymarin

- Enhances the cytotoxicity of cisplatin and carboplatin [251–256].
- Increases the effect of doxorubicin [251, 255].

Quercetin

- Enhances cytotoxicity of topotecan in MCF-7 and MDA-MB-231 cells [264].
- Increases the cisplatin-induced cytotoxic effect in HeP2 cells by increasing the apoptosis [265].
- The genistein and quercetin induce synergistic effect in ovarian carcinoma cells [182].

(continued)

TABLE 14.1. (continued)

Emodin

- Sensitizes Her-2/neu-overexpressing breast cancer cells to paclitaxel [274, 284, 285].
- Increases the effect of cisplatin, doxorubicin, and etoposide in Her-2/neu-overexpressing non-small lung cancer cells by suppressing Her2/neu tyrosine kinase activity [286].
- Potentiates the effect of celecoxib synergistically by activation of caspases-9 and -3 and induction of apoptosis and inhibition of Akt activation [287].
- Increases the cytotoxic effect of cis-platinol (abiplastin), doxorubicin (adriablastin), and 5-fluorouracil, and STI 571 in Merkel cell carcinoma cells [288].
- Enhances the cytotoxic effect of arsenic trioxide in vitro and in vivo by increasing the generation of ROS and blockage of NF-kappa B activation [289–292].

Flavopiridol

- Increases the effect of paclitaxel and docetaxel in vitro and in vivo by activation of caspase 3, cleavage of poly (ADP-ribose) polymerase and inhibition of cyclin B1/cdc-2 kinase [310–312].
- Increases the effect of SN-38 and CPT-11 in vitro and in vivo by upregulating the activation of caspase-3 and greater cleavage of both p21 and XIAP [313].
- Increases the effect of irinotecan in cancer patients by modulating p21, Differentiation-related gene 1 (Drg1), and p53 genes [314].
- Enhances the effect of doxorubicin or cisplatin in MCF-7 breast cancer cells by suppressing survivin phosphorylation and increasing the tumor cell apoptosis [315].
- Increases the effect of gemcitabine [316].
- Increases the effect of trastuzumab in MDA453 (MDA-MB-453), BT474, and SKBR3 cells by elevating apoptosis [317].
- Increases the sensitivity of TRAIL in human myeloma and breast cancer cells by downregulating the transcriptional activation of FLIP(L) and Bak, Bax, and PUMA-alpha genes [318].

TABLE 14.2. Modulation of radiosensitivity by various polyphenols.

Curcumin

- Sensitizes PC-3 cells to irradiation by suppressing the radiation-induced NFκB activation and expression of Bcl2 protein and elevating radiation-induced activation of cytochrome c and caspase-9 and -3 enzymes and apoptosis [120].
- Increases the effect of irradiation in squamous cell carcinoma SSC2 cells [121].
- Increases radiation-induced chromosome aberrations in Chinese hamster ovary (CHO) cells [319].
- Increases the sensitivity of fibrosarcoma to radiation-induced cell killing by attenuating radiation-induced activation of NF-κB and MAPkinase [320].

Resveratrol

- Increases the radiosensitivity of HeLa and SiHa cells by inhibiting the radiation-induced COX-1 and arresting the cells in early S-phase [321].
- Enhances the cell killing effect of radiation in HeLa, K-562, and IM-9 cells by enhancing DNA degradation [322].
- Increases the effect of X-irradiation in AHH-1 cells by increasing the radiation-induced DNA damage, apoptosis, and blockage of cell cycle at early-S-phase [323].
- Increases the effect of radiation in NCI-H838 cells by suppressing NF-κB activation and arrest of cells in S-phase of the cell cycle [324].

Genistein

- Sensitizes TE-1 (p53, mutant) and TE-2 (p53, wild), against irradiation by abrogating the radiation-induced activation of p42/p44 extracellular signal-regulated kinase and AKT/PKB [325].
- Increases the effect of irradiation in DU145 cells by increasing G2/M arrest, reducing cyclin B1 expression and elevating p21/cip1 levels [326].
- Increases the effect of ionizing radiation in Bcr/abl-positive K562 leukemic cells by enhancing radiation-induced apoptosis and cell cycle arrest at G_2M phase of the cell cycle [327].
- Enhances the radiosensitivity of K562 by increasing the expression of thymidine kinase (TK1) [328].
- Increases the effect of irradiation in Me180 and CaSki cells by increasing the G2+M arrest and inhibiting the Mcl-1 and activating pAKT (Thr 308) [329].

(continued)

TABLE 14.2. (continued)

- Increases the radiosensitivity of prostate cancer in vivo and in vitro by G2/M arrest, elevating p21WAF1/Cip1 expression, reduced cyclin B1 expression and suppression of NF-κB [330, 331, 333].
- Increases the effect of radiation in vitro and in vivo in orthotropic model of human prostate cancer [332].

EGCG

- Increases the radiosensitivity of human embryonic endothelial cells by suppressing the type-1 matrix metalloproteinase (MT1-MMP) protein expression [334].
- Enhances the effect of X-irradiation in IM-9 and K-562 cells by increasing in the radiation-induced apoptosis [336].
- Increases X-irradiation-induced apoptotic and necrotic cell death in human leukaemic EOL-1 cells [336].
- Increases the radiation-induced death of endothelial cells by upregulated expression of cyclin dependent kinase inhibitors p21 (CIP/Waf1) and p27(Kip) and downregulation of RhoA expression. [337, 338].

Flavopiridol

- Increases the radiosensitivity of HCT-116 and MKN-74 cells by elevating radiation-induced apoptosis, caspase-3 activation, poly (ADP-ribose) polymerase cleavage, and cytochrome c release [339].
- Enhances the radiosensitivity in vitro and in vivo by blocking the repair of radiation-induced DNA damage, reducing Ku70 and Ku86 proteins, down-regulation of cyclin D1, cyclin E and cdk-9 and suppression of the phosphorylation of retinoblastoma protein [340, 341, 345].
- Increases the radiosensitivity of TE8, TE9 and KE4 cells by reducing the levels of cyclin D1 and Rb proteins and decreasing the Bcl-2 protein [342].
- Increases the radiosensitivity of DU145 and PC3 cells by increasing the γH2AX and inhibiting the DNA repair [343].
- Enhances the radiation effect in H460 cells and xenografted mice by increasing the G1 and G2/M arrest [344].
- Increases the radiosensitivity of GL261 glioma xenografted mouse [346].
- Increases the radiosensitivity of Zebrafish embryos by downregulation of cyclin D1 [347].

been linked to the resistance of tumors to conventional therapy [20, 21]. Theoretically then, natural products or phytochemicals that can block NF-κB activation may improve the therapeutic outcomes of various tumors by effectively combatting chemoresistance and radioresistance.

2.2 Activator Protein-1

The activator protein 1 (AP-1) is a dimeric transcription factor. Representatives of this class of transcription factor are proteins of the JUN, FOS, ATF, and MAF families. Fos and Jun are the most common proteins in mammalian cells [22–27]. The activation of AP-1 is generally induced by growth factors, cytokines, and oncoproteins [28, 29]. Once activated, AP-1 binds to the 12-O-tetradecanoylphorbol-13-acetate (TPA)-responsive element (TRE) and induces transcription of a number of genes involved in cellular proliferation, survival, differentiation, transformation, and apoptosis [30]. The development of chemoresistance and radioresistance can be traced to the activation of AP-1 since AP-1 is involved in transcription of the death antagonist

bcl-XL gene [31, 32]. This implies that an agent that could block the activation of AP-1 could also reduce chemoresistance and radioresistance.

2.3 Cyclooxygenase

Enzymes of the cyclooxygenase (COX) family are involved in the conversion of arachidonic acid into prostaglandins. There are two main isoforms: COX-1 and COX-2. COX-1 is usually produced constitutively by gastric mucosa, whereas COX-2 (also known as prostaglandin H2 synthase-2 [PGHS-2]), is usually induced by inflammation. Early investigations revealed that COX-2 could be induced by viral oncogenes [33] or by tumor promoters [34]. Subsequent investigations have found that COX-2 can also be induced by a variety of growth factors and mitogens [35, 36], thus making COX-2 particularly relevant to the processes of cell growth and cancer. Overexpression of COX-2 has been correlated with colorectal cancer [37] and with other cancers, including those of the lung, colon, and breast [38]. COX-2 expression has also been correlated with a more aggressive phenotype and resistance to therapy, especially in patients with cancers of the colon, breast, lung, pancreas, and head and neck [39]. Therefore, chemical agents that can inhibit COX-2 may facilitate improved outcomes by rendering tumors with chemoresistant or radioresistant phenotypes more amenable to therapy.

2.4   Bcl-2

First identified as a proto-oncogene in follicular B-cell lymphoma [40], the Bcl-2 protein confers a survival advantage on tumor cells in which it is transcriptionally upregulated and overexpressed. The overexpression of Bcl-2 has been observed in a number of solid tumors and hematologic malignancies, and it is associated with the development of chemoresistance and radioresistance and subsequent failure of conventional chemotherapy and radiotherapy [41–43]. Agents that can block the overexpression of Bcl-2 might help to facilitate tumor regression when combined with conventional chemotherapy or radiotherapy.

2.5   Survivin

Proteins that regulate apoptosis, such as survivin, are another factor that might be responsible for chemoresistance and radioresistance [43]. Survivin is a member of the IAP5 gene family [44]. It is overexpressed in most human tumors and correlates with reduced tumor cell apoptosis in vivo, shorter patient survival, accelerated recurrence rates, and increased chemoresistance and radioresistance. However, survivin is largely undetectable in normal differentiated tissues [45, 46]. In preclinical studies in vivo, molecular antagonism of survivin either alone or in combination with other anticancer therapies suppressed de novo tumor formation and inhibited the growth of established tumors [46, 47]. Indeed, use of molecular antagonists to the survivin pathway can effectively elicit the spontaneous death of tumor cells by apoptosis during chemotherapy and radiotherapy [46].

2.6   Epidermal Growth Factor Receptor

Epidermal growth factor receptor (EGFR) is a transmembrane glycoprotein belonging to a family of type I receptor tyrosine kinases that includes EGFR/ErbB1/HER, ErbB2/Neu/HER2, ErbB3/HER3, and ErbB4/HER4. EGFR plays an important role in signal transduction pathways that regulate key cellular functions such as survival and proliferation [48–50]. The known downstream effectors of EGFR include phosphatidylinositol 3-kinase (PI3K), RAS-RAF-MAPK p44/p42, and protein kinase C signaling pathways, among others [51–54]. EGFR or its ligands (and in some cases both) are overexpressed in several tumors including cancers of the head and neck, breast, and ovary as well as non-small cell lung cancer (NSCLC), and its upregulation has been linked with more aggressive disease, chemoresistance and radioresistance, and poor prognosis [51, 55–57]. Inhibitory targeting of EGFR has been found to improve the chemotherapeutic and radiotherapeutic outcomes of many tumors [56–65].

2.7   Akt

The PI3K/Akt pathway is a crucial regulator of cell survival, growth, proliferation, malignant transformation, vesicle trafficking, and cytoskeletal regulation [66, 67]. Akt has been identified as the crucial link between PI3K and the prevention of apoptosis [68, 69]. Akt prevents apoptosis by inhibiting the transcription factors Bad, caspase-9, and Forkhead, all of which are downstream targets of Akt and thus important components of the apoptotic machinery, and also by blocking the apoptotic effects of a wide range of apoptotic stimuli in many cell types. Studies using constitutively active and dominant-negative forms of Akt have conclusively demonstrated that Akt is essential for survival [70]. Upregulated signaling via the PI3K/Akt pathway has also been implicated in tumorigenesis [71] as well as in resistance of breast cancers to tamoxifen [72] and many other chemotherapeutic agents [73].

2.8   Signal Transducer and Activator of Transcription 3

The signal transducer and activator of transcription (STAT) protein family consists of seven highly conserved, latent transcription factors that appear to play an important role in cytokine stimulation-induced signal transduction [74]. The stimulation of STAT proteins results in their phosphorylation by the Janus kinases, which eventually leads to their dimerization and migration to the nucleus. In the nucleus, these proteins function as transcription factors that regulate the expression of a host of genes involved in cell proliferation and apoptosis [75]. One particular STAT protein, STAT3, has been shown to be an oncogene [74] that is constitutively activated in a wide variety of human neoplasms including non-Hodgkin's lymphomas [76–81]. The activation of STAT protein has been correlated with uncontrolled proliferation and apoptosis regulation. The DNA-binding activity of a STAT3-related protein was found to be enhanced in cells transformed by the v-Src oncoprotein [82, 83]. A number of experimental studies have provided evidence that STAT3 would be an effective mediator of chemoresistance and radioresistance. In studies using multiple myeloma, pancreatic cancer, and estrogen receptor-negative cells, the inactivation of STAT3 sensitized the cells to chemotherapy and radiotherapy and increased their rates of apoptosis [84, 85]. In another study using prostate cancer cells, inhibition of STAT3 correlated with increased radiation-induced apoptosis [86]. These studies indicate that targeting of STAT-3 may be a useful paradigm to reverse the chemoresistance or radioresistance of tumors.

2.9 Multidrug Resistance Protein

A major problem in cancer chemotherapy, one that limits the effectiveness of various anticancer drugs, is the development of multidrug resistance (MDR) [87, 88]. MDR and MDR-associated proteins represent two subfamilies of the ATP binding cassette (ABC) family of transporter proteins that control the pumping of drugs from cells at the expense of ATP hydrolysis [89, 90]. The overexpression of ABC transporters ABCB1 (P-glycoprotein) and ABCC1 (MRP1) confer chemoresistance by exporting chemotherapeutic drugs from cells in an ATP-dependent manner. ABCG2 (also called mitoxantrone resistance-associated protein, breast cancer resistance protein, and placental ABC transporter) is a half-transporter of the ABCG subfamily of ABC transporters whose overexpression causes malignant cells to become multidrug resistant [91–94]; as a result, ABCG2 confers resistance to various chemotherapeutic drugs, such as doxorubicin, mitoxantrone, topotecan,

and SN-38 [95–98]. Strategies that circumvent the development of MDR would indeed improve chemotherapeutic outcomes.

3 Chemosensitizing Polyphenols

3.1 Curcumin

Curcuminoids are the major active principles in the rhizome of turmeric *(Curcuma longa)* (also known as curry spice) and give turmeric its characteristic yellow color. One of the most important curcuminoids is the polyphenolic phytochemical curcumin (also known as diferuloylmethane); depending on the season of its harvest, turmeric usually contains 0.3–5.4% curcumin [99, 100]. Despite having no nutritive value, curcumin in the form of turmeric powder has been used by humans since at least Vedic times (approximately 1,500 BCE to 500 BCE) to give flavor and color to food. Curcumin's longstanding use and lack of any known toxicity in humans suggest that it is pharmacologically safe. Humans consume up to 100 mg of curcumin as a dietary spice each day [101], and recent phase I clinical trials indicate that a dose as high as 12 g/day can be tolerated with no toxic side effects [102, 103]. Curcumin reportedly possesses several pharmacologic properties including antiinflammatory, antimicrobial, antiviral, antifungal, antioxidant, chemosensitizing, radiosensitizing, and wound healing activity [104–123]. In experimental models, curcumin has been shown to suppress tumor initiation, promotion, and metastasis [124–134]. It has also been shown to act as an antiproliferative agent by interrupting the cell cycle, disrupting mitotic spindle structures, and inducing apoptosis and micronucleation [135–138].

In experiments in vitro, treatment of ovarian cancer cells (CAOV3 and SKOV3) with curcumin either simultaneously with or 24 hours before cisplatin treatment increased cisplatin-mediated cell killing by inhibiting the cellular production of IL-6 [139]. In another study, treatment of hepatic cancer HA22T/VGH cells with a combination of curcumin, cisplatin, and doxorubicin enhanced the effects of the chemotherapeutic drugs synergistically by downregulating the expression of c-myc, Bcl-XL, c-IAP-2, NAIP, and XIAP [140]. Similarly, curcumin treatment has been reported to increase the cytotoxic effects of cisplatin in SiHa cells but not in HeLa cells by increasing caspase activity and inhibiting the cisplatin-induced activation of NF-kB [141]. In a more recent study, curcumin treatment sensitized breast tumor cells to cisplatin by increasing apoptosis and inhibiting monoubiquitination of the FANCD2 protein [142].

In experiments investigating the effect of curcumin in the absence or presence of COX-2 expression, curcumin synergistically enhanced the cell-killing effect of celecoxib in both COX-2–expressing (HT-29, IEC-18-K-ras, Caco-2) and COX-2-deficient (SW-480) cell lines [143]. This enhanced effect was attributed to curcumin's ability to block COX-2 and other signaling pathways. Similarly, in subsequent experiments by the same group, combination of celecoxib and curcumin synergistically enhanced growth inhibition and apoptosis in HT-29 and SW480 cells by attenuating COX-2 activity [144]. In experiments by others, curcumin increased doxorubicin's cell-killing effects on doxorubicin-resistant Hep2A cells by inhibiting glutathione-*S*-transferase [145] and on Hep-3B cells by inhibiting doxorubicin-induced activation of NF-κB [146].

In our own laboratory, we have shown that curcumin treatment can elevate the sensitivity of multiple myeloma cells to melphalan and vincristine by abrogating NF-κB activity and reducing the expression of *Bcl-2, Bcl-xL, cyclin D1*, and *IL-6* [147]. Similarly, curcumin has been shown to enhance the cytotoxic effects of doxorubicin, 5-fluouracil, and paclitaxel in prostate cancer cells by inhibiting both constitutive and tumor necrosis factor (TNF)–induced activation of NF-κB [148], and increase the sensitivity of taxol-resistant breast cancer cells to paclitaxel by increasing apoptosis [149]. In the latter case, the increased apoptosis was mediated by curcumin's ability to block paclitaxel-induced activation of NF-κB and suppress paclitaxel-induced expression of antiapoptotic proteins (i.e., XIAP, IAP-1, IAP-2, Bcl-2, and Bcl-xL), cell cycle–related molecules (i.e., COX-2, c-Myc, and cyclin D1), vascular endothelial growth factor (VEGF), matrix metalloproteinase-9 (MMP9), and intercellular adhesion molecule-1 (ICAM1). In another study, combination of taxol and curcumin increased taxol's cytotoxic effect on HeLa cells via curcumin's inhibition of NF-κB activation and suppression of taxol-induced phosphorylation of the serine/threonine kinase Akt;

however, curcumin did not affect taxol-induced tubulin polymerization and cyclin-dependent kinase (cdk) Cdc2 activation [150].

In several studies by Chearwae and colleagues, curcumin was shown to attenuate the overexpression of MDR genes and subsequently suppresses P-glycoprotein–mediated drug efflux while increasing sensitivity to etoposide [151]. In other studies by the same investigators, curcumin, demethoxycurcumin, and bisdemethoxycurcumin were all shown to sensitize ABCG2 transporter-expressing HEK293 mutant cells and multidrug-resistant cells to treatment with mitoxantrone, topotecan, SN-38, and doxorubicin; of the three curcuminoids tested, however, curcumin had the greatest sensitizing effect [152].

In contrast, curcumin has also been shown to attenuate in a time- and concentration-dependent manner the apoptosis induced by camptothecin, mechlorethamine, and doxorubicin in MCF-7, MDA-MB-231, and BT-474 human breast cancer cells [153]. Similarly, curcumin-induced effect was observed in an in vivo model of cyclophosphamide-treated human breast cancer in which curcumin's action was attributed to its ability to reduce cyclophosphamide-induced apoptosis and JNK activation [153]. Together, these studies show curcumin's great potential for use as a therapeutic adjuvant to improve tumor control.

3.2   Resveratrol

Resveratrol (3,5,4′-trihydroxy-*trans*-stilbene), a phytoalexin present in grapes, peanuts, and pines, has antioxidant and anti-inflammatory activities [154]. It is the active ingredient in *Leguminosae* that inhibits cellular events associated with tumor initiation, promotion, and progression in a mouse skin cancer model [155]. In vitro, resveratrol inhibits the growth of various human tumor cells including those of oral squamous carcinoma [156], promyelocytic leukemia [157], human breast cancer [158], prostate cancer [159, 160] esophageal carcinoma [161], pancreatic cancer [162], and monocytic leukemia [163].

In a study in which resveratrol and paclitaxel were administered concurrently to A549, EBC-1, and Lu65 lung cancer cells, the addition of resveratrol did not increase the effect of paclitaxel [164]. However, when the cells were exposed neoadjuvantly to resveratrol for 3 days before treatment with paclitaxel, paclitaxel-induced cytotoxicity was markedly enhanced [164]. The same investigators also reported that the resveratrol pretreatment regimen almost quadrupled p21waf1 induction in EBC-1 cells. In a study in which drug-refractory non-Hodgkin's lymphoma (NHL) and multiple myeloma (MM) cell lines were treated with paclitaxel either alone or in combination with other drugs, the combination of resveratrol and paclitaxel effectively reduced tumor cell growth by arresting the cell cycle at the G2-M phase [165]. This effect was attributed to resveratrol's ability, even at a relatively low concentration, to sensitize the cells to paclitaxel-induced apoptosis by selectively down-regulating expression of the antiapoptotic proteins Bcl-xL and myeloid cell differentiation factor-1 (Mcl-1) and upregulating that of the proapoptotic proteins Bax and apoptosis protease-activating factor-1 (Apaf-1). Mechanistically, the combination of resveratrol and paclitaxel caused apoptosis by inducing the activation of tBid, the depolarization of mitochondrial membranes, the cytosolic release of cytochrome C and Smac/DIABLO, and the activation and cleavage of poly(ADP-ribose) polymerase. In contrast to its effect on NHL and MM cells, the combination of resveratrol with paclitaxel was very weakly cytotoxic to quiescent and mitogenically stimulated human peripheral blood mononuclear cells [165].

Resveratrol has also been found to synergistically sensitize the drug-refractory NHL Ramos cell line to various apoptosis-inducing agents including gemcitabine, navelbine, cisplatinum (CDDP), paclitaxel, and the TNF-related apoptosis inducing ligand (TRAIL) [166]. An earlier study had shown that resveratrol sensitizes tumor cells to TRAIL [167]. In another study, combination of grape seed extract (GSE) and doxorubicin enhanced doxorubicin's effect on receptor-positive MCF-7 and receptor-negative MDA-MB468 human breast carcinoma cells by arresting them in the G1 phase of the cell cycle [168]. Similarly, resveratrol treatment markedly increased the cytotoxic effect of temozolomide by increasing apoptosis and S-phase arrest in both the temozolomide-sensitive (M14) and TMZ-resistant (SK-Mel-28 and PR-Mel) cell lines [169].

In another study using human leukemia cells, low concentrations of resveratrol inhibited caspase activation, DNA fragmentation, and cytochrome *c* translocation induced by hydrogen peroxide or anticancer drugs such as C2, vincristine, and daunorubicin [170]. Similarly, resveratrol increased in an additive manner the effect of cisplatin and doxorubicin in human ovarian (OVCAR-3) and uterine (Ishikawa) cancer cells in vitro [171]. In yet another study, combination of resveratrol with doxorubicin, cycloheximide, busulfan, gemcitabine, and paclitaxel enhanced the drugs' cytotoxic effects on HL-60 cells and their multidrug-resistant (P-glycoprotein–positive) variant, HL60/VCR [172].

3.3 Genistein

Genistein is an isoflavone (i.e., a major metabolite of soy produced by intestinal bacteria) that is believed to be one of the anticancer agents found in soybeans [173, 174]. Genistein has a heterocyclic diphenolic structure similar to estrogen [175] and has antitumor and antiangiogenic activity [175, 176]. Epidemiologic studies have shown that Japanese and Chinese men who consume soy products have the lowest incidence of prostate carcinoma in the world [177].

In experimental studies using EGFR-expressing lung cancer cells, genistein combined with cisplatin, doxorubicin, or etoposide enhanced the drugs' antiproliferative effects and induced apoptosis [178]. An identical, synergistic effect was observed in a mouse model of lung cancer in which genistein was combined with cyclophosphamide [179]. In another study, genistein potentiated the effects of cisplatin and, to a lesser extent, those of vincristine in meduloblastoma cells [180]. In liver and colon cancer cell lines, combination of genistein with dexamethasone enhanced expression of the cyclin-dependent kinase 2 (Cdk2) inhibitor p21WAF1/CIP1, thereby halting cell-cycle progression [181]. The combination of genistein and quercetin also acted synergistically in ovarian carcinoma cells [182].

Genistein potentiated growth inhibition and apoptotic cell death induced by cisplatin, docetaxel, doxorubicin, and gemcitabine in prostate, breast, pancreas, and lung cancers in vitro [183–185]. This activity of genistein is mediated through genistein's inactivation of NF-κB [186]. Recently, one group of investigators reported that genistein significantly potentiates the antitumor, anti-invasive, and antimetastatic activities of docetaxel both in culture and in a severe combined immunodeficient (SCID) human model of experimental prostate cancer bone metastasis [187]. In particular, they found that genistein induced and docetaxel inhibited the expression of osteoprotegerin (OPG), whereas genistein significantly downregulated the expression and secretion of receptor activator of nuclear factor-kB (RANK) ligand (RANKL) and inhibited osteoclast formation. Moreover, they found that genistein downregulated the expression and activity of MMP-9 induced by docetaxel treatment and inhibited the invasive ability of PC-3 cells [187].

In another study in vivo, genistein was shown to increase the antitumor activities of other chemotherapeutic agents, including gemcitabine and docetaxel in pancreatic cancer in vivo by enhancing tumor cell killing and apoptotic cell death [184, 185]. In a study utilizing diffuse large cell lymphoma cells, genistein sensitized the cells to treatment with cyclophosphamide, doxorubicin, vincristine, and prednisone (CHOP) [188]. Likewise, the combination of genistein and 5-fluorouracil (5-FU) has been shown to increase the 5-FU-induced apoptosis synergistically in chemoresistant HT-29 colon cancer cells [189]. In another study, genistein treatment increased adriamycin-induced necrotic-like cell death in HER-2–overexpressing breast cancer cells [190]. In a study using dysplastic (MCF-10A1, MCF-ANeoT, MCF-T63B) and malignant epithelial breast cancer (MCF-7, MDA-231, MDA-435) cells, treatment with genistein at concentrations of 1–10 μg/ml inhibited the growth of both types of cells [191]. Furthermore, the addition of tamoxifen increased this effect synergistically in the dysplastic cells but additively in the malignant cells. However, none of these effects were modulated by estrogen receptor [191]. Meanwhile, genistein and its isoflavone analogs have been shown to reduce the side effects of tamoxifen through P450-mediated pathways [192].

In a very recent study, addition of genistein synergistically enhanced the cytotoxic effect of camptothecins by inhibiting camptothecin-induced G2/M arrest and increasing the rate of camptothecin-induced apoptosis [193]. Genistein also blocked CDK1 phosphorylation after irinotecan treatment [193]. Another study has shown that combination treatment with genistein and cisplatin enhances the sensitivity of human melanoma cells to cisplatin by significantly downregulating bcl-2 and bcl-xL protein expression and increasing Apaf-1 protein expression [194]. Similarly, genistein pretreatment has been reported to increase the cytotoxic effects of bleomycin by increasing micronuclei frequency and DNA damage in HL-60 cells [195].

3.4 Epigallocatechin Gallate

In Asia, green tea (Camellia sinensis) has a long history of medicinal use that stretches back more than 4,000 years. Green tea is rich in polyphenols, which make up 36% of its dry weight. The major green tea polyphenols are (–)-epigallocatechin gallate (EGCG), which constitutes more than 50% of green tea's polyphenolic content; (–)-epigallocatechin (EGC); (–)-epicatechin gallate (ECG); and (–)-epicatechin (EC) [196]. Green tea polyphenols reportedly exert antioxidant, anti-inflammatory, antibacterial, antiviral, antiaging, antiangiogenic, antimutagenic, and antidiabetic activity [197–206]. In addition, green tea polyphenols have been reported to have chemopreventive and antineoplastic properties [207–211].

EGCG causes G1 growth arrest, inhibits cyclin-dependent kinases (cdks), and induces the cdk inhibitors p21 and p27 in breast and prostate cancer cells [212, 213]. It also inhibits the signaling cascades of many growth factors, either directly by suppressing growth factor receptors or indirectly through downstream effects [214]. EGCG treatment can reportedly block transcription factor–mediated gene activation by blocking the activation of NF-κB and AP-1 [215]. Indeed, through their activity as scavengers of reactive oxygen species (ROS), green tea catechins inhibit NF-κB activation, which in turn inhibits the expression of proinflammatory and survival genes [216]. In addition, EGCG has been shown to directly inhibit proteasome activity [217], leading to accumulation of the NF-κB inhibitory protein IκB and other proapoptotic proteins such as Bax. EGCG also suppresses inducible nitric oxide synthase metalloproteinases as well as vascular endothelial growth factor receptor expression and signaling in tumor and endothelial cells [218–222]. Taken together, these properties suggest that the cytotoxic effects of chemotherapeutic agents may be enhanced by giving such drugs in combination with EGCG.

Recently, ECCG has drawn much attention because of its ability to prevent chemical carcinogenesis and induce cytotoxicity in neoplastic cells [207, 212, 213]. Therefore, its use in combination with other chemotherapeutic agents may be able to enhance cytotoxic effects on tumor cells while sparing normal cells. EGCG and EGC have been reported to increase the cytotoxic effect of doxorubicin in doxorubicin-resistant murine sarcoma (S180-dox) and human colon carcinoma (SW620-dox) cells by inhibiting protein kinase C [223]. In experiments using multidrug-resistant Ehrlich ascites carcinoma cells, EGCG reportedly enhanced the cytotoxic effect of doxorubicin and reversed the MDR phenotype by inhibiting the efflux of doxorubicin from the cells [224]. In another study, EGCG reportedly increased the influx of doxorubicin into drug-resistant KB-A-1 cells while simultaneously decreased its efflux out of them [225]. Identical results were observed in multidrug-resistant xenografted tumors in vivo. This reversal in MDR was apparently mediated by the downregulation MDR-1 mRNA and subsequently P-glycoprotein [225]. In more recent studies by the same investigators, EGCG reversed the doxorubicin resistance of multidrug-resistant KB-A-1 cells by more than twofold by increasing the intracellular concentration of ROS [226]. Likewise, in studies by other groups, EGCG was reportedly able to sensitize multidrug-resistant human KB-A-1 xenografts in athymic nude mice to doxorubicin, increasing the intratumoral accumulation of doxorubicin by 51% over that seen in xenografts treated with doxorubicin alone [227, 228]. In yet another study, EGCG reportedly enhanced doxorubicin's cytotoxic effect on a multidrug-resistant hepatocarcinoma cell line (BEL-7404/Adr) by downregulating MDR-1 gene expression [229]. Several reports suggest that ECGC's observed ability to reverse the MDR phenotype is mainly due to its ability to inhibit the P-glycoprotein [228, 230, 231].

In experiments using the mouse colon adenocarcinoma cell line Colon 26, combination treatment with sulindac and EGCG synergistically inhibited cell growth more strongly than did either agent alone. A similar effect on apoptosis was observed in PC-9 cells treated with a combination of tamoxifen and EGCG [232]. Similarly, in another study using MDA-MB-231 breast cancer cells, EGCG and EGC were found to increase synergistically the cytotoxic effects of 4-hydroxytamoxifen [233]. In a study using SKOV3, CAOV3, and C200 ovarian cancer cells, combination treatment with EGCG and cisplatin enhanced the cytotoxicity of cisplatin by increasing oxidative stress [234].

In one study, combination of low-dose As_2O_3 and EGCG significantly increased the cytotoxic effects on human malignant B-cell myeloma (RPMI8226 and IM9) and Burkitt's lymphoma (HS-sultan) by increasing apoptosis induction [235]. Compared with As_2O_3 or EGCG alone, their combination increased the production of ROS (i.e., O_2^- and H_2O_2) in HS-sultan and IM9 cells. The combination also completely blocked catalase activity and reduced GSH levels. In addition, combination of As_2O_3 and EGCG caused a decline in the levels of Mcl-1 and Bcl-2 in the myeloma cells [235]. However, in a report from another group, EGCG did not enhance the cytotoxic effects of tamoxifen, sulindac, or doxorubicin in HSC-2 and HSG head and neck cancer cells [236].

3.5 Silymarin

Milk thistle is one of the most popular herbs in the world, and it has been used for more than 2,000 years as a tonic, demulcent, antidepressant, and lactation enhancer [237]. It is currently the 10th top-selling herbal supplement in the United States [238]. In a relatively recent survey of cancer patients, 7% reported using milk thistle as an alternative to or in combination with cytotoxic chemotherapy [239], and it is the fourth most popular herb used by cancer patients.

The active component of milk thistle *(Silybum marianum)* is silymarin [240]. Silymarin is a flavonolignan complex mainly composed of silibinin and small amounts of its stereoisomers (i.e., isosilybin A, isosilybin B, silychristin, isosilychristin, and silydianin) [241]. Silymarin and silibinin have been found to be cytotoxic in cultured cancer cells and a number of rodent and human xenografted tumors in vivo. In addition, silymarin has been reported to exert antiangiogenic activities in a number of rodent and human cancers [242, 243] and to inhibit a variety of cytochrome P450 (CYP) isozymes including CYP isoform 3A4 (CYP3A4); UDP glucuronosyltransferases (UGT) including UGT isoform 1A1 (UGT1A1); and several efflux transporters including ABCB1 and P-glycoprotein [244–250].

In experimental cell-based studies, silybin enhanced the cytotoxicity of cisplatin in ovarian cancer cells and of doxorubicin in MCF-7 doxorubicin-resistant breast cancer cells [251]. Similarly, combination of silybin and cisplatin not only significantly enhanced cisplatin's cytotoxic effects in human ovarian cancer (A2780) cells and experimental mice, but also led to early recovery of body weight in treated mice [252]. A similar effect, attributed to cell-cycle arrest, was observed in human prostate cancer cells treated with silybin and cisplatin [253]. In another study in human prostate carcinoma DU145 cells, silibinin reportedly enhanced the cytotoxic effect of combination treatment with cisplatin and carboplatin by increasing the rates of growth inhibition and apoptosis [254].

In another laboratory study comparing its effects on estrogen-dependent and estrogen-independent human breast carcinoma cells (MCF-7 and MDA-MB468, respectively), silybin increased the therapeutic effects of doxorubicin, cisplatin, and carboplatin [255]. Each test agent alone inhibited the growth of both cell lines in a dose- and time-dependent manner, and combination of silybin and doxorubicin exerted the strongest synergistic effect in both cell lines by reducing cell growth. Similarly, combination of silybin and doxorubicin exerted a much greater apoptotic effect than either agent alone. Meanwhile, silybin did not enhance the apoptotic effect of cisplatin in either cell line and enhanced the apoptotic effect of carboplatin only in the estrogen-dependent cell (MCF-7) cell line. In contrast, in a clinical study in humans undergoing irinotecan therapy, milk thistle (200 mg three times a day for 14 days) had no enhancing effect [256].

3.6　Quercetin

Quercetin (3,3′,4′,5,7-pentahydroxyflavone) is a naturally occurring flavone that is found in most edible fruits and vegetables. The highest concentrations of quercetin are found in onions, apples, and red wine [257, 258]. Among quercetin's many biological activities are the ability to exert antitumor and antiproliferative effects on a wide range of human cancer cell lines and the ability to inhibit glycolysis, macromolecule synthesis, and enzymatic activity [259–263]. Combination of quercetin with

topotecan reportedly enhanced topotecan's cytotoxicity in human breast cancer cells (i.e., 1.4-fold in MCF-7 cells and 1.3-fold in MDA-MB-231 cells) by inducing significant elevations in ROS and nitrite levels [264]. In a study using human laryngeal HeP2 cells, quercetin enhanced cisplatin's cytotoxic effect by increasing the apoptosis rate via inhibition of Akt/PKB phosphorylation and consequently anti-apoptotic Bcl-2 and Bcl-XL activity, downregulation of Cu-Zn superoxide dismutase, and inhibition of heat shock proteins [265]. This activity of quercetin occurred independently of cisplatin's activity, which exerts its effects by increasing the activity, first of JNK and subsequently of endonucleases [265]. Moreover, as mentioned, genistein and quercetin have been reported to act synergistically in ovarian carcinoma cells [182].

3.7　Emodin

Emodin (1,3,8-trihydroxy-6-methylanthraquinone) is a naturally occurring anthraquinone present in the roots and bark of numerous plants of the genus *Rhamnus*. It is also present in the roots, bark, or dried leaves of buckthorn, senna, cascara, aloe, frangula, and rhubarb, plants that have been used as laxatives since ancient times and are now widely used in the preparation of herbal laxative preparations. Although the anthraquinone glycosides used in such preparations are poorly absorbed from the gastrointestinal tract, they are cleaved by gut bacteria to produce more readily absorbable and purgative aglycones (such as emodin) [266–268]. Emodin has been found to inhibit cell proliferation in various cancer cell lines [269]. This growth inhibitory effect of emodin has been linked to its ability to interfere with the progression of the cell cycle in a variety of cells, including human fibroblasts, smooth muscle cells, endothelial cells, and malignant cells [270–272]. Furthermore, emodin has been shown to inhibit HER-2/neu tyrosine kinase activity [273] and suppress the transformation of HER-2/neu-overexpressing breast cancer cells [274].

Emodin from aloe (aloe-emodin) has been shown to exert antineoplastic activity in neuroectodermal tumors, neuroblastomas, lung carcinoma, leukemia cells, promyeloleukemic HL-60 cells, hepatoma cells, and a Merkel carcinoma cell line [275–283]. Aloe-emodin acts by inducing a number of molecular processes including apoptosis [275, 277, 282, 283], the modulation of Bcl-2, Bax, and Fas proteins; the increased production of cytochrome c; the activation of caspase-3, -8, and -9; and the increased accumulation of p53 and p21 proteins [275, 277, 282].

In studies during the last decade, emodin has been shown to sensitize Her-2/neu-overexpressing breast cancer cells to paclitaxel therapy by suppressing cell growth and inducing cell differentiation, probably by inhibiting cellular Her2/neu tyrosine kinase activity [274, 284, 285]. Likewise, emodin has been reported to increase the cytotoxic effects of cisplatin, doxorubicin, and etoposide in Her-2/neu–overexpressing non-small lung cancer cells by suppressing Her2/neu tyrosine kinase activity [286]. In experimental rat studies comparing the anchorage-dependent and -independent growth of cholangiocarcinoma (C611B ChC) cells and neu-transformed WB-F344 rat-liver epithelial stem-like (WBneu) cells in culture, emodin and celecoxib acted synergistically to reduce cell growth in both lines beyond that induced by either agent alone [287]. This suppression of cell growth in combination-treated cells correlated directly with significant increases in the activation of caspases-9 and -3, induction of apoptosis, and inhibition of Akt activation [287]. In another study using Merkel cell carcinoma cells, the combination of aloe-emodin with cisplatinol (abiplastin), doxorubicin (adriablastin), 5-FU, and STI 571 increased the cytotoxic effect of these agents [288].

Emodin has also been reported to increase the cytotoxic effect of arsenic trioxide in HeLa cells in vitro and in tumor-bearing nude mice in vivo [289, 290] This sensitizing activity of emodin apparently correlated with its ability to increase the rate of arsenic-induced apoptosis in the HeLa cells by increasing the generation of ROS, which in turn caused a collapse in mitochondrial transmembrane potential and blocked activation of NF-kB. An almost identical effect of emodin has been observed in HeLa cells treated with arsenic trioxide [292]. However, other groups have observed no such effects on normal fibroblasts [291] or ovarian cancer cells [293].

3.8 Flavopiridol

Flavopiridol ((2)-cis-5,7-dihydroxy-2-(2-chlorphenyl)-8[4-(3-hydroxy-1-methyl) piperidinyl]-4H-benzopyran-4-on) is a semisynthetic flavone

and derivative of the alkaloid rohitukine found in the bark of the Indian tree *Dysoxylum binefacterium* [294]. Flavopiridol inhibits the proliferation of a wide variety of human tumors in vitro and in vivo [295, 296]. Flavopiridol is also a potent cdk inhibitor that inhibits several cdks (i.e., cdk1, -2, -4, -7, -8, and -9) with equal effectiveness [297, 298]. It acts as a competitive inhibitor of the ATP binding site formed by a cleft between the C- and N-terminal domains of the cdks [299]. It is cytostatic as well as cytotoxic and induces apoptosis equally in both resting and proliferating cells [295, 300]. Flavopiridol also downregulates Bcl-2 expression in several B-cell leukemia cell lines [301, 302]. The apoptosis induced by flavopiridol is independent of p53 and p16 status [303, 304]. Moreover, flavopiridol has also been reported to directly bind to DNA [305]. Recent studies have shown that flavopiridol can block the activation of NF-kB and thus the gene expression regulated by NF-kB [306].

Since flavopiridol exerts its antiproliferative effect via several mechanisms, its combination with other cytotoxic agents may be able to enhance treatment outcomes. In fact, several studies have reported that concurrent treatment with flavopiridol, 5FU, and cytarabine induces a significant synergistic effect [301, 307–309].

In studies using human gastric (MKN-74) and breast (MCF-7) cancer cells, neoadjuvant administration of flavopiridol followed by paclitaxel reportedly enhanced paclitaxel's apoptotic effect by inducing the activation of caspase-3 and the cleavage of poly(ADP-ribose) polymerase [310]. A similar augmentation in the effect of docetaxel has been reported in human gastric cells in vitro and in vivo, in which pretreatment with flavopiridol inhibited docetaxel-induced cyclin B1/cdc-2 kinase upregulation [311]. In one other study, docetaxel's cytotoxic effect on HCT116 colon cancer cells was enhanced by sequential combination of docetaxel for 1 hour, flavopiridol for 24 hours, and 5-FU for another 24 hours. This sequential combination regimen not only suppressed tumor cell growth more strongly than did other combinations, but also significantly decreased the colony-forming ability of the cancer cells in soft agar by enhancing apoptosis. Use of a similar regimen in a murine xenograft model delayed tumor growth and led to better survival [312].

In another study, again using HCT116 cells, sequential treatment first with the active metabolite of the DNA topoisomerase I inhibitor CPT-11 (i.e., SN-38) and then with flavopiridol increased SN-38's cytotoxic effect by upregulating the activation of caspase-3 and greater cleavage of both p21 and XIAP, an inhibitor of apoptosis. Similarly, in a human colon cancer xenograft model, sequential treatment with CPT-11 and flavopiridol enhanced the therapeutic index of CPT-11 when compared with other treatment regimens [313]. In a phase I clinical trial in cancer patients, flavopiridol enhanced the effect of irinotecan and by modulating the expression of p21, differentiation-related gene 1 (Drg1), and p53 genes [314].

In experiments using MCF-7 breast cancer cells, combination of flavopiridol with doxorubicin or cisplatin has been reported to increase the cytotoxic effect of either drug beyond that of either one alone [315]. This increased sensitivity to chemotherapy has been attributed to flavopiridol's ability to suppress survivin phosphorylation at Thr34 and enhancement of tumor cell apoptosis [315].

In another study, the sequential combination of gemcitabine and flavopiridol was shown to increase gemcitabine's cytotoxic effect in human pancreatic, gastric, and colon cancer cell lines by increasing apoptosis induction. This enhancement was due to attenuation of the expression of the RR-M2 protein responsible for gemcitabine resistance, which in turn correlated with the downregulation of E2F-1, a transcription factor that regulates RR-M2 transcription and hypophosphorylation of pRb [316]. In a study using erb2- or EGFR-overexpressing human breast carcinoma cell lines (MDA453 [MDA-MB-453], BT474, and SKBR3), combination treatment with trastuzumab and flavopiridol synergistically increased sensitivity to trastuzumab by elevating the apoptosis rate [317]. This enhancement was attributed to the complete suppression of EGFR in these cells that overexpress erb2 and EGFR [317]. Likewise, flavopiridol has been reported to increase the sensitivity of TRAIL in human myeloma and breast cancer cells by downregulating the transcriptional activation of FLIP(L) and Bak, Bax, and PUMA-α genes [318].

4 Radiosensitizing Polyphenols

The radiocurability of tumors is hindered by the development of radioresistant factors after repeated exposure to radiation. The molecular mechanisms underlying such radioresistance have become clearer in recent years thanks to advances in molecular radiobiology. It is now clear that radiotherapy principally kills cancer cells by damaging their genomes. Therefore, modulation of radiosensitivity is incumbent upon the molecules that are involved in DNA repair, cell-cycle checkpoint control, apoptosis, and signal transduction, as well as those factors (e.g., hypoxic and angiogenic) that help maintain the tumor microenvironment. Since radiation-induced cytotoxic pathways are very similar to those employed by chemotherapeutic agents, the molecular targets for radiation-induced cell killing are almost identical to those for chemotherapy; this has already been described in the preceding in detail.

4.1 Curcumin

The radiosensitizing potential of curcumin has been evaluated by several investigators. In one study, treatment of a p53 mutant prostate cancer cell line (PC-3) with 2 or 4 μM curcumin before irradiation to 5 Gy enhanced the radiation effect by inducing a decline in cell survival [120]. Treatment with curcumin before irradiation enhanced the apoptosis at 24 and 48 hours after irradiation. Irradiation of PC-3 cells upregulated TNF-α protein expression, leading to an increase in NF-κB activity and Bcl-2 protein levels. Conversely, pretreatment of PC-3 cells with curcumin suppressed TNF-α mediated NF-κB activation and the expression of Bcl-2 protein. However, the expression of Bax protein remained unaltered. In addition, curcumin enhanced radiation-induced activation of cytochrome c and caspase-9 and -3 enzymes and subsequently the apoptosis of PC-3 cells, thereby increasing the cell killing effect beyond that of irradiation alone [120].

In an in vitro study using the squamous cell carcinoma cell line SSC2, pretreatment with 3.5 μm curcumin for 48 hours enhanced radiation-induced declines in both cell survival and cell count [121]. In another study that examined the effect of turmeric and curcumin on the frequency of chromosome aberrations in Chinese hamster ovary (CHO) cells exposed to 2.5 Gy of gamma radiation [319], treatment with various concentrations of turmeric or curcumin before exposure to radiation increased the frequency of chromosome aberrations. Interestingly, while turmeric (500 mg/mL) and curcumin (10 mg/mL) both elevated the frequency of chromosome aberrations during G2/S phase, curcumin (10 mg/mL) also increased it during S phase. This clearly exacerbatory effect of the antioxidants turmeric and curcumin on radiation-induced clastogenicity suggests that they are also potentiating agents whose activity depends on the experimental conditions. Turmeric was not clastogenic by itself, whereas curcumin at 10 mg/mL enhanced the chromosomal damage frequency [319].

In a study using fibrosarcoma-bearing mice, curcumin treatment reportedly attenuated the radiation-induced activation of NF-κB and MAP kinase and enhanced the sensitivity of the fibrosarcomas to radiation-induced cell killing. Despite this, curcumin had no effect on the radiation-induced activation of PKC [320].

4.2   Resveratrol

Resveratrol has been reported to enhance the cytotoxic effect of ionizing radiation in HeLa and SiHa human cervical cancer cells by increasing the radiation-induced inhibition of cell growth and clonogenicity [321]. This effect was attributed to the inhibition of radiation-induced COX-1 inhibition and early S-phase arrest [321]. The effect of cotreatment with resveratrol has also been evaluated in HeLa, K-562, and IM-9 cells after X-irradiation [322]. In brief, resveratrol was found to increase the sensitivity of these cells to radiation-induced cell growth and apoptosis and to enhance polyploidy and DNA degradation in the irradiated cells as evidenced by a sub-G1 peak [322].

In a study using AHH-1 lymphoblastoid cells, treatment with resveratrol before and after exposure to X-rays increased the consequent radiation-induced DNA damage as assessed by Comet assay, apoptosis, and cell-cycle arrest at early S-phase [323]. In another study using human non-small lung cancer NCI-H838 cells, pretreatment with resveratrol increased the cytotoxic effect of radiation by significantly enhancing radiation-induced cell killing. This enhancement was mediated by resveratrol's suppression of NF-kB activation and arrest of cells in S-phase [324].

4.3   Genistein

In experiments using human esophageal squamous cell carcinoma cell lines expressing mutant p53 (TE-1) or wild-type p53 (TE-2), pretreatment with genistein increased the cytotoxic effect of radiation and enhanced radiation-induced apoptosis [325]. This enhancement was attributed to genistein's abrogation of the radiation-induced activation of p42/p44 extracellular signal–regulated kinase and AKT/PKB. Interestingly, the combination treatment apparently increased poly(ADP-ribose) polymerase cleavage (PARP) as well as Bax and Bcl-2 expression in TE-2, but not in TE-1, cells [325].

In a study using DU145 prostate cancer cells, genistein treatment before and after irradiation increased the cytotoxic effect of irradiation, as indicated by a reduction in colony-forming ability and an increase in apoptosis [326]. Furthermore, the combination therapy increased G2/M arrest, reduced cyclin B1 expression, and elevated p21cip1 levels [326]. Likewise, in another study using Bcr/abl-positive K562 leukemic cells, genistein reportedly increased the effect of ionizing radiation by enhancing radiation-induced apoptosis and cell-cycle arrest at G2/M phase [327]. In particular, the combination therapy caused the sustained and prolonged arrest of K562 cells at the G2 phase, a radiosensitizing effect that was attributed to increased expression of thymidine kinase (TK1) [328].

In experiments using the cervical tumor cell lines Me180 and CaSki, genistein had a radiosensitizing effect in both lines, but this effect was concentration dependent [329]. Genistein exerted its greatest radiosensitizing effects at concentrations of 20 and 40 mM in cells exposed to 200–800 cGy. This enhancement in radiosensitivity was linked to G2/M arrest, significant inhibition of Mcl-1, and activation of pAKT (Thr 308) [329].

In an earlier study in a murine xenograft model of human prostate cancer, genistein pretreatment was found to inhibit the DNA synthesis, cell growth, and clonogenicity of tumor cells exposed to low-dose photon or neutron irradiation [330]. In fact, genistein followed by radiation significantly increased the growth inhibition of primary PC-3 human prostrate xenografts beyond that induced by either genistein or radiation treatment alone. Genistein treatment also reduced the number of metastatic lymph nodes after irradiation. Interestingly, however, when genistein was administered by itself, an increase in the size of lymph nodes associated with heavy tumor infiltration was noted. Histologic examination of xenografted tumors subjected to irradiation alone revealed the typical effects of radiation treatment (i.e., areas of tumor destruction in which normal tissue had been replaced by fibrotic tissue and the presence of inflammatory cells and giant cells); conversely, histologic examination of genistein-treated, irradiated tumors revealed increased numbers of giant cells and inflammatory cells, enhanced apoptosis and fibrosis, and reduced tumor cell proliferation. In a subsequent study in the same model, genistein administered both before and after prostate tumor irradiation was shown to improve survival significantly [331].

In still other studies, genistein pretreatment reportedly increased the effect of radiation in vitro and in vivo in an orthotropic model of human prostate cancer in nude mice and RM9 prostate tumor in syngeneic C57BL/6 mice [332]. The combination treatment also inhibited spontaneous metastasis and increased the survival of mice [332].

In a recent study using PC-3 cells, combination of genistein with radiation increased the cytotoxic effect beyond that of genistein or radiation alone, as reflected in the enhanced inhibition of colony formation after combination treatment [333]. Moreover, pretreatment with genistein and continuous exposure to genistein after irradiation showed the greatest radiosensitizing effect. This increased radiation effect was attributed to G2/M arrest, elevated p21WAF1/Cip1 expression, and reduced cyclin B1 expression. Genistein pretreatment also suppressed the radiation-induced upregulation of NF-kB. Likewise, combination treatment significantly enhanced PARP cleavage, thus indicating that combination treatment increased the apoptosis of PC-3 cells [333].

4.4 EGCG

Few studies have investigated the modulation of radiosensitivity by EGCG. In experiments using human embryonic endothelial cells, EGCG suppressed the protein expression of type-1 matrix metalloproteinase (MT1-MMP), a crucial enzymatic promoter of endothelial cell migration and tube formation, and of caveolin-1, a protein that regulates tube formation after irradiation [334]. In a study using human multiple myeloma (IM-9) and chronic myelogenous leukemia (K-562) cells, combination therapy with X-irradiation and EGCG significantly enhanced the radiation-induced apoptosis and led to a corresponding decline in cell proliferation in both cell lines and to a lesser extent in HeLa cells

that received combined treatment [335]. A similar effect was observed in human leukemic EOL-1 cells, in which combination of EGCG with X-irradiation increased the rates of both apoptotic and necrotic cell death beyond those induced by either treatment alone [336].

In another study using endothelial cells, the combination of EGCG and ionizing radiation increased cell death by almost twofold beyond that induced by radiation treatment alone [337]. The combination treatment also upregulated expression of cdk inhibitors p21(CIP/Waf1) and p27(Kip). Interestingly, the synergistic reduction in cell survival did not correlate with increased activity of the pro-apoptotic caspase-3, -9, and cytochrome c proteins. Moreover, combination treatment augmented cell necrosis by fivefold but had no effect on late apoptosis [337]. The significance of this study lies in the inhibition of angiogenesis, since angiogenesis is also a major cause of radioresistance. Likewise, EGCG has been found to inhibit survivin-induced radioresistance in U-87 brain tumor cells [338]. Treatment of survivin-transfected cells with EGCG made them sensitive to the effect of ionizing radiation by causing the downregulation of RhoA expression [338]. Together, these studies demonstrate that EGCG has great potential as a radioresponse modulator.

4.5 Flavopiridol

The radiosensitizing effect of flavopiridol given before or after irradiation has been investigated by a number of groups. In a study using HCT-116 colon cancer cells and MKN-74 gastric cancer cells, following gamma irradiation, flavopiridol treatment enhanced the radiosensitivity of the cells [339]. This combination therapy significantly enhanced radiation-induced apoptosis, caspase-3 activation, PARP cleavage, and cytochrome c release in both cell lines [339]. In split-dose radiation experiments using murine ovarian cancer cell line OCA-I, flavopiridol increased radiosensitivity by enhancing radiation-induced inhibition of cell growth and clonogenic cell survival [340]. In short, flavopiridol blocked the repair of radiation-induced DNA damage by reducing the levels of Ku70 and Ku86 proteins in flavopiridol-treated irradiated cells. The increased radiosensitivity induced by flavopiridol correlated with the accumulation of cells in the G1 and G2 phases and a significant reduction in the S-phase fraction. Moreover, flavopiridol treatment downregulated the expression of cyclin D1, cyclin E, and cdk-9 and suppressed the phosphorylation of retinoblastoma protein [340]. The extension of this study in vivo into transplantable syngeneic mouse tumors such as mammary carcinoma (MCa-29), ovarian carcinoma (OCa-I), and a lymphoma (Ly-TH) has subsequently shown that flavopiridol's radiosensitizing effect also applies in vivo, where it is associated with antimetastatic effects and improved tumor cure rates [341].

In in vitro studies using TE8, TE9, and KE4 esophageal squamous cells, flavopiridol increased radiosensitivity and was directly linked to increased G2/M fractions, cellular nuclear fragmentation, and chromatin condensation [342]. Analysis of molecular pathways showed that flavopiridol reduced the levels of cyclin D1 and Rb protein in all three cell lines and decreased the levels of Bcl-2 protein in TE8 and KE4 cells [342]. In studies using DU145 and PC3 cells [343], continuous treatment with flavopiridol before and after irradiation increased the radiosensitivity, as evidenced by reduced clonogenic survival and enhanced apoptosis beyond those induced by either treatment alone. Interestingly, postirradiation exposure to flavopiridol did not significantly reduce cell proliferation, nor did it have any effect on radiation-induced apoptosis or activation of the G2 checkpoint. However, it did significantly increase the number of gH2AX foci in the irradiated cells, thus indicating that flavopiridol enhanced the radiosensitivity of human tumor cells by inhibiting DNA repair [343].

In a study using H460 lung carcinoma cells in vitro and H460 xenografts in mice in vivo, flavopiridol was found to enhance the effect of radiation on cancer cells [344]. However, as measured by clonogenic survival and apoptosis, flavopiridol was more effective when administered after irradiation rather than before. Further experiments by the same investigators found that docetaxel enhanced the effect of flavopiridol and radiation both in vitro and in vivo and that the greatest enhancement in radiosensitivity was observed when the sequence of treatment was docetaxel, radiation, and flavopiridol. Moreover, flavopiridol was shown to induce G1 and G2/M arrest in H460 cells [344].

In studies using human esophageal carcinoma SEG-1 cells, pretreatment with flavopiridol increased radiosensitivity and led to a reduction in survival beyond that induced by radiation treatment alone [345]. The increased radiosensitivity of the SEG-1 cells was attributed to their arrest in G1 phase by flavopiridol, which in turn correlated with reduced expression of cdk-1, cdk-2, cyclin D1, and Rb. Pretreatment with flavopiridol also augmented the levels of radiation-induced apoptosis and PARP cleavage while reducing the level of RNA polymerase II phosphorylation. Further studies by the same investigators in a murine xenograft model of esophageal adenocarcinoma revealed that flavopiridol enhanced the delay in tumor growth induced by irradiation, especially when flavopiridol was administered after, rather than before, irradiation [345]. In a study using a murine GL261 glioma xenograft model, combination of a single acute dose of radiation with flavopiridol increased the tumor cure and tumor-free survival rates of xenografted mice beyond those induced by either treatment alone [346]. The use of a conventional radiation-fractionation regimen with flavopiridol increased the radioresponse, as evaluated by tumor volume measurements; in fact, the greatest reduction in mean tumor volume occurred in the mice that received fractionated irradiation and flavopiridol [346]. In a study using *Zebrafish* embryos, treatment with flavopiridol after exposure to 40 Gy caused a twofold increase in embryonic mortality by 96 hours postfertilization [347]. The combination treatment resulted in a greater number of curled-up embryos (a distinct radiation-induced defect in midline development) than did radiation alone. This increased radiosensitivity was correlated with the downregulation of cyclin D1 by flavopiridol [347].

5   Conclusions

Polyphenols are a versatile group of phytochemicals found in higher plants that are consumed in one form or another by humans. After more than two decades of intensive research, their beneficial effects and the molecular mechanisms underlying them are slowly emerging. Some of these mechanisms exert a wide array of effects and are already known. The polyphenols covered in this chapter—curcumin, resveratrol, genistein, EGCG, silymarin, quercetin, emodin, and flavopiridol—all have the proven ability to sensitize tumor cells to most of today's frequently used chemotherapeutic agents. Evidence of their radiosensitizing activity, however, is scanty, even though there is ample evidence of their ability to increase the radioresponse and radiocurability of tumors. As a group, these polyphenolic sensitizing agents are known to exert their effects by suppressing the activities of several important transcription factors (e.g., NF-kB and AP-1) that in turn regulate several other genes. These agents are also known to use pathways related to molecules, such as COX, Bcl-2, survivin, EGFR, STAT3, and Akt, which have been implicated in the acquired resistance of tumors to chemotherapy and radiotherapy and in inflammation. Although it seems likely that these phytochemicals utilize other pathways as well, those pathways for now remain unknown.

Acknowledgments. The authors thank Jude Richard, ELS, for carefully editing the manuscript and providing valuable comments. Dr. Aggarwal is the Ransom Horne, Jr. Professor of Cancer Research. This work was supported by a grant from the Clayton Foundation for Research (to BBA), National Institutes of Health P01 grant CA91844 on lung chemoprevention (to BBA), and National Institutes of Health P50 Head and Neck SPORE grant P50CA97007 (to BBA).

References

1. Dixon RA. Natural products and plant disease resistance. Nature 2001, 411:843–844.
2. Newman DJ, Cragg GM, Snader KM. Natural products as sources of new drugs over the period 1981–2002. J Nat Prod 2003, 66:1022–1037.
3. Bannerman R, Burton J, Wein-Chieh C. Traditional medicine and healthcare coverage, a reader for health administrators and practioners. Geneva: World Health Organzation, 1983.
4. Cupp MJ. Herbal remedies: adverse effects and drug interactions. Am Family Phys 1999, 59(5):1239–1244.
5. Anonymus. Americans consume more dietary supplements than ever before. Bio|Analogics 2002, 2.
6. Duthie G, Crozier A. Plant-derived phenolic antioxidants. Curr Opin Clin Nutr Metab Care 2000, 3(6):447–451.
7. Moskaug JØ, Carlsen H, Myhrstad MCW, et al. Polyphenols and glutathione synthesis regulation. Am J Clin Nutr 2005, 81(Suppl):277S–283S.
8. Haslam E. Practical polyphenolics: from structure to molecular recognition and physiological action. Cambridge, UK: Cambridge University Press, 1998.
9. Hayden MS, Ghosh S. Signaling to NF-κB. Genes Dev 2004, 18:2195–2224.
10. Ravi R, Bedi A. NF-κB in cancer—a friend turned foe. Drug Resist Updat 2004, 7:53–67.
11. Verma IM. Nuclear factor (NF)-κB proteins: therapeutic targets. Ann Rheum Dis 2004, 63(Suppl 2): ii57–ii61.
12. Beg AA, Baltimore D. An essential role for NF-kappaB in preventing TNF-alpha-induced cell death. Science 1996, 274:782–784.
13. Barkett M, Gilmore TD. Control of apoptosis by Rel/NF-kappaB transcription factors. Oncogene 1999, 18:6910–6924.
14. Ghosh S, May MJ, Kopp EB. NF-kB and Rel proteins: evolutionarily conserved mediators of immune responses. Annu Rev Immunol 1998, 16:225–260.
15. Karin M, Ben-Neriah Y. Phosphorylation meets ubiquitination: the control of NF-[kappa]B activity. Annu Rev Immunol 2000, 18:621–663.
16. Ghosh S, Karin M. Missing pieces in the NF-kB puzzle. Cell 2002, 109:S81–S96.
17. Israel A. The IKK complex: an integrator of all signals that activate NF-kB. Trends Cell Biol 2000, 10:129–133.
18. Garg A, Aggarwal BB. Nuclear transcription factor kappaB as a target for cancer drug development. Leukemia 2002, 16:1053–1068.
19. Rayet B, Gelinas C. Aberrant rel/nfkb genes and activity in human cancer. Oncogene 1999, 18:6938–6947.
20. Wang CY, Mayo MW, Baldwin AS Jr. TNF- and cancer therapy-induced apoptosis: potentiation by inhibition of NF-kappaB. Science 1996, 274:784–787.
21. Baldwin AS Control of oncogenesis and cancer therapy resistance by the transcription factor NF-kB. J Clin Invest 2001, 107:241–246.
22. Angel P, Allegretto EA, Okino ST, et al. Oncogene jun encodes a sequence-specific trans-activator similar to AP-1. Nature 1988, 332:166–171.
23. Ryder K, Lanahan A, Perez-Albuerne E, et al. jun-D: a third member of the jun gene family. Proc Natl Acad Sci U S A 1989, 86:1500–1503.
24. Hirai SI, Ryseck RP, Mechta F, et al. Characterization of junD: a new member of thejun proto-oncogene family. EMBO J 1989, 8:1433–1439.
25. Curran T, Teich NM. Candidate product of the FBJ murine osteosarcoma virus oncogene: characterization of a 55,000-dalton phosphoprotein. J Virol 1982, 42:114–122.
26. Cohen DR, Curran T. fra-1: a serum-inducible, cellular immediate-early gene that encodes a fos-related antigen. Mol Cell Biol 1988, 8:2063–2069.

27. Nishina H, Sato H, Suzuki T, et al. Isolation and characterization of fra-2, an additional member of the fos gene family. Proc Natl Acad Sci U S A 1990, 87:3619–3623.

28. Shaw PE, Schroter H, Nordheim A. The ability of a ternary complex to form over the serum response element correlates with serum inducibility of the human c-fos promoter. Cell 1989, 56:563–572.

29. Gupta P, Prywes R. ATF1 phosphorylation by the ERK MAPK pathway is required for epidermal growth factor-induced c-jun expression. J Biol Chem 2002, 277:50550–50556.

30. Karin M, Liu Z, Zandi E. AP-1 function and regulation. Curr Opin Cell Biol 1999, 240–246.

31. Yang E, Korsmeyer SJ. Molecular thanatopsis: a discourse on the BCL2 family and cell death. Blood 1997, 88:386–401.

32. Takeuchi K, Motoda Y-I, Ito F. Role of transcription factor activator protein 1 (AP1) in epidermal growth factor-mediated protection against apoptosis induced by a DNA-damaging agent. FEBS J 2006, 273:3743–3755.

33. Xie W, Chipman JG, Robertson DL, et al. Expression of a mitogen-responsive gene encoding prostaglandin synthase is regulated by mRNA splicing. Proc. Natl Acad Sci U S A 1991, 88:2692–2696.

34. Kujubu DA, Fletcher BS, Varnum BC, et al. TIS10, a phorbol ester tumor promoter-inducible mRNA from Swiss 3T3 cells, encodes a novel prostaglandin synthase/cyclooxygenase homologue. J Biol Chem 1991, 266:12866–12872.

35. Bakhle YS, Botting RM. Cyclooxygenase-2 and its regulation in inflammation. Mediat Inflamm 1996, 5:305–323.

36. Smith WL. Dewitt DL, Garavito RM. Cyclooxygenases: structural, cellular and molecular biology. Annu Rev Biochem 2000, 69:145–182.

37. Kargman S, O'Neill G, Vickers P, et al. Expression of prostaglandin G/H synthase-1 and -2 protein in human colon cancer. Cancer Res 1995, 55:2556–2559.

38. Soslow RA, Dannenberg AJ, Rush D, et al. Cox-2 is expressed in human pulmonary, colonic and mammary tumors. Cancer 2000, 89:2637–2645.

39. Koki AT, Leahy KM, Masferrer JL. Potential utility of COX-2 inhibitors in chemoprevention and chemotherapy. Expert Opin Investig Drugs 1999, 8:1623–1638.

40. Tsujimoto Y, Cossman J, Jaffe E, et al. Involvement of the bcl-2 gene in human follicular lymphoma. Science 1985, 228:1440–1443.

41. Chao D, Korsmeyer S. Bcl-2 family: regulators of cell death. Annu Rev Immunol 1998, 16:395–419.

42. Adams J, Cory S. The Bcl-2 protein family: arbiters of cell survival. Science 1998, 281:1322–1326.

43. Reed JC. Mechanisms of apoptosis avoidance in cancer. Curr Opin Oncol 1999, 1:68–75.

44. Salvesen GS, Duckett CS. Apoptosis: IAP proteins: blocking the road to death's door. Nat Rev Mol Cell Biol 2002, 3:401–410.

45. Ambrosini G, Adida C, Altieri DC. A novel anti-apoptosis gene, survivin, expressed in cancer and lymphoma. Nat Med 1997, 3:917–921.

46. Altieri DC. Validating survivin as a cancer therapeutic target. Nat Rev Cancer 2003, 3: 46–54.

47. Andersen MH, Thor SP. Survivin: a universal tumor antigen. Histol Histopathol 2002, 17:669–675.

48. Riese DJ II, Stern DF. Specificity within the EGF family/ErbB receptor family signaling network. Bioessays 1998, 20:41–48.

49. Yarden Y, Sliwkowski MX. Untangling the ErbB signalling network. Nat Rev Mol Cell Biol 2001, 2:127–137.

50. Yarden Y. The EGFR family and its ligands in human cancer: signaling mechanisms and therapeutic opportunities. Eur J Cancer 2001, 37:3–8.

51. Salomon DS, Brandt R, Ciardiello F, et al. Epidermal growth factor-related peptides and their receptors in human malignancies. Crit Rev Oncol Hematol 1995, 19:183–232.

52. Hackel PO, Zwick E, Prenzel N, et al. Epidermal growth factor receptors: critical mediators of multiple receptor pathways. Curr Opin Cell Biol 1999, 11:184–189.

53. Sweeney C, Fambrough D, Huard C, et al. Growth factor-specific signaling pathway stimulation and gene expression mediated by ErbB receptors. J Biol Chem 2001, 276:22685–22698.

54. Wu CJ, Qian X, O'Rourke DM. Sustained mitogen-activated protein kinase activation is induced by transforming erbB receptor complexes. DNA Cell Biol 1999, 18(10):731–741.

55. Baselga J, Mendelsohn J. The epidermal growth factor receptor as a target for therapy in breast carcinoma. Breast Cancer Res Treat 1994, 29:127–138.

56. Bonner JA, Raisch K P, Trummell HQ. Enhanced apoptosis with combination C225/radiation treatment serves as the impetus for clinical investigation in head and neck cancers. J Clin Oncol 2000, 18:47s–53s.

57. Huang S-M, Harari PM. Modulation of radiation response after epidermal growth factor receptor blockade in squamous cell carcinomas: inhibition of damage repair, cell cycle kinetics, and tumor angiogenesis. Clin Cancer Res 2000, 6:2166–2174.

58. Baselga J, Pfister D, Cooper MR. Phase I studies of anti-epidermal growth factor receptor chimeric antibody C225 alone and in combination with cisplatin. J Clin Oncol 2000, 18:904–915.

59. Hidalgo M, Siu LL, Nemunaitis J, et al. Phase I and pharmacologic study of OSI-774, an epidermal growth factor receptor tyrosine kinase inhibitor, in patients with advanced solid malignancies. J Clin Oncol 2001, 19:3267–3279.

60. O'Rourke D, Kao GD, Singh N, et al. Conversion of a radio-resistant phenotype to a more sensitive one by disabling erbB receptor signaling in human cancer cells. Proc Natl Acad Sci U S A 1998, 95:10842–10847.

61. Milas L, Mason K, Hunter N, et al. In vivo enhancement of tumor radioresponse by C225 anti-epidermal growth factor receptor antibody. Clin Cancer Res 2000, 6:701–708.

62. Arteaga CL. The epidermal growth factor receptor: from mutant oncogene in non human cancers to therapeutic target in human neoplasia. J Clin Oncol 2001, 19:32–40.

63. Ciardiello F, Tortora G. A novel approach in the treatment of cancer: targeting the epidermal growth factor receptor. Clin Cancer Res 2001, 7:2958–2970.

64. Zwick E, Bange J, Ullrich A. Receptor tyrosine kinases as targets for anticancer drugs. Mol Med 2002, 8:17–23.

65. Mendelsohn J. Targeting the epidermal growth factor receptor for cancer therapy. J Clin Oncol 2002, 20:1–13.

66. Varticovski L, Druker B, Morrison D, et al. The colony stimulating factor-1 receptor associates with and activates phosphatidylinositol-3 kinase. Nature 1989, 342:699–702.

67. Cantley LC. The phosphoinositide 3-kinase pathway. Science 2002, 296:1655–1657.

68. Franke TF, Kaplan DR, Cantley LC, et al. Direct regulation of the Akt proto-oncogene product by phosphatidylinositol-3,4-bisphosphate. Science 275, 1997:665–668.

69. Dudek H, Datta SR, Franke TF, et al. Regulation of neuronal survival by the serinethreonine protein kinase Akt. Science 1997, 275:661–665.

70. Kim D, Chung J. Akt: versatile mediator of cell survival and beyond. Biochem Mol Biol 2002, 35:106–115.

71. Nicholson KM, Anderson NG. The protein kinase B/Akt signaling pathway in human malignancy. Cell Signal 2002, 14:381–395.

72. Campbell RA, Bhat-Nakshatri P, Patel NM, et al. Phosphatidylinositol 3-kinase/AKT-mediated activation of estrogen receptor alpha: a new model for anti-estrogen resistance. J Biol Chem 2001, 276:9817–9824.

73. Knuefermann C, Lu Y, Liu B, et al. HER2/PI-3K/Akt activation leads to a multidrug resistance in human breast adenocarcinoma cells. Oncogene 2003, 22:3205–3212.

74. Bromberg JF. Activation of STAT proteins and growth control. BioAssays 2001 23:161–169.

75. Darnell JE Jr, Kerr IM, Stark GR. Jak-STAT pathways and transcriptional activation in response to IFNs and other extracellular signaling proteins. Science 1994, 264:1415–1421.

76. Zong CR, Yan A, August JE, et al. Unique signal transduction of Eyk: constitutive stimulation of the JAK-STAT pathway by an oncogenic receptor-type tyrosine kinase. EMBO J 1996, 15:4515–4525.

77. Watson CJ, Miller WR. Elevated levels of members of the STAT family of transcription factors in breast carcinoma nuclear extracts. Br J Cancer 1995, 71:840–844.

78. Sartor CI, Dziubinski ML, Yu CL, et al. Role of epidermal growth factor receptor and STAT-3 activation in autonomous proliferation of SUM-102PT human breast cancer cells. Cancer Res 1997, 57:978–987.

79. Garcia R, Bowman TL, Niu G, et al. Constitutive activation of Stat3 by the Src and JAK tyrosine kinases participates in growth regulation of human breast carcinoma cells. Oncogene 2001, 20:2499–2513.

80. Zamo A, Chiarle R, Piva R, et al. Anaplastic lymphoma kinase (ALK) activates Stat3 and protects hematopoietic cells from cell death. Oncogene 2002, 21(7):1038–1047.

81. Eriksen KW, Kaltoft K, Mikkelsen G, et al. Constitutive STAT3-activation in Sezary syndrome: tyrphostin AG490 inhibits STAT3-activation, interleukin-2 receptor expression and growth of leukemic Sezary cells. Leukemia 2001, 15(5):787–793.

82. Yu CL, Meyer DJ, Campbell GS, et al. Enhanced DNA-binding activity of a Stat-3-related protein in cells transformed by the Src oncoprotein. Science 1995, 269(5220):81–83.

83. Cao X, Tay A, Guy GR, et al. Activation and association of Stat3 with Src in v-Src-transformed cell lines. Mol Cell Biol 1996, 16(4):1595–1603.

84. Bharti AC, Shishodia S, Reuben JM, et al. Nuclear factor-kappaB and STAT3 are constitutively active in CD138+ cells derived from multiple myeloma patients, and suppression of these transcription factors leads to apoptosis. Blood 2004, 103:3175–3184.

85. Greten FR, Weber CK, Greten TF, et al. Stat3 and NF-kappaB activation prevents apoptosis in pancreatic carcinogenesis. Gastroenterology 2002, 123:2052–2063.

86. Calvin DP, Nam S, Buettner R, et al. Inhibition of STAT3 activity with STAT3 antisense oligonucleotide (STAT3-ASO) enhances radiation-induced apoptosis in DU145 prostate cancer cells. Int J Radiat Oncol Biol Phys 2003, 57(Suppl):S297.

87. Ambudkar SV, Dey S, Hrycyna CA, et al. Biochemical, cellular, and pharmacological aspects

of the multidrug transporter. Annu Rev Pharmacol Toxicol 1999, 39:361–398.

88. Gottesman MM, Fojo T, Bates SE. Multidrug resistance in cancer: role of ATP-dependent transporters. Nat Rev Cancer 2002, 2:48–58.

89. Shabbits JA, Krishna R, Mayer LD. Molecular and pharmacological strategies to overcome multidrug resistance. Expert Rev Anticancer Ther 2001, 1:585–594.

90. Di Pietro A, Conseil G, Perez-Victoria JM, et al. Modulation by flavonoids of cell multidrug resistance mediated by P-glycoprotein and related ABC transporters. Cell Mol Life Sci 2002, 59:307–322.

91. Doyle LA, Yang W, Abruzzo LV, et al. A multidrug resistance transporter from human MCF-7 breast cancer cells. Proc Natl Acad Sci U S A 1998, 95: 15665–15670.

92. Maliepaard M, van Gastelen MA, de Jong LA, et al. Overexpression of the BCRP/MXR/ABCP gene in a topotecan-selected ovarian tumor cell line. Cancer Res 1999, 59:4559–4563.

93. Yang CH, Schneider E, Kuo ML, et al. BCRP/MXR/ABCP expression in topotecan-resistant human breast carcinoma cells. Biochem Pharmacol 2000, 60:731–783.

94. Haimeur A, Conseil G, Deeley RG, et al. The MRP-related and BCRP/ABCG2 multidrug resistance proteins: biology, substrate specificity and regulation. Curr Drug Metab 2004, 5:21–53.

95. Miyake K, Mickley L, Litman T, et al. Molecular cloning ofcDNAs which are highly overexpressed in mitoxantrone-resistant cells: demonstration of homology to ABC transport genes. Cancer Res 1999, 59:8–13.

96. Doyle LA, Ross DD. Multidrug resistance mediated by the breast cancer resistance protein BCRP (ABCG2). Oncogene 2003, 22:7340–7358.

97. Bates SE, Medina-Perez WY, Kohlhagen G, et al. ABCG2 mediates differential resistance to SN-38 (7-ethyl-10-hydroxycamptothecin) and homocamptothecins. J Pharmacol Exp Ther 2004, 310:836–842.

98. Yoshikawa M, Ikegami Y, Sano K, et al. Transport of SN-38 by the wild type of human ABC transporter ABCG2 and its inhibition by quercetin, a natural flavonoid. J Exp Ther Oncol 2004, 4:25–35.

99. Leung A. Encyclopedia of common natural ingredients used in food, drugs, and cosmetics. New York: John Wiley, 1980:313–314.

100. Lampe V, Milobedeska J. Ber Dtsch Chem Ges 1913, 46:2235.

101. Ammon HP, Wahl MA.Pharmacology of *Curcuma longa*. Planta Med 1991, 57(1):1–7.

102. Cheng AL, Hsu CH, Lin JK, et al. Phase I clinical trial of curcumin, a chemopreventive agent, in patients with high-risk or pre-malignant lesions. Anticancer Res 2001, 21(4B):2895–2900.

103. Lao CD, Ruffin MT 4th, Normolle D, et al. Dose escalation of a curcuminoid formulation. BMC Complement Altern Med 2006, 6:10.

104. Srimal RC, Dhawan BN. Pharmacology of diferuloyl methane (curcumin), a non-steroidal anti-inflammatory agent. J Pharm Pharmacol 1973, 25(6):447–452.

105. Arora R, Kapoor V, Basu N, et al. Anti-inflammatory studies on *Curcuma longa* (turmeric). Ind J Med Res 1971, 59:1289–1295.

106. Mukhopadhyay A, Basu N, Ghatak N. Anti-inflammatory and irritant activities of curcumin analogues rats. Agents Actions 1982, 12:508–515.

107. Srivastava R. Inhibition of neutrophil response by curcumin. Agents Actions 1989, 28:298–303.

108. Jobin C, Bradham CA, Russo MP, et al. Curcumin blocks cytokine-mediated NFkappaB activation and proinflammatory gene expression by inhibiting inhibitory factor I-κB kinase activity. J Immunol 1999, 163:3474–3483.

109. Negi PS, Jayaprakasha GK, Jagan Mohan Rao L, et al. Antibacterial activity of turmeric oil: a byproduct from curcumin manufacture. J Agric Food Chem 1999, 47:4297–4300.

110. Mazumder A, Raghavan K, Weinstein J, et al. Inhibition of human immunodeficiency virus type-1 integrase by curcumin. Biochem Pharmcol 1995, 49:1165–1170.

111. Bourne KZ, Bourne N, Reising SF, et al. Plant products as topical microbicide candidates: assessment of in vitro and in vivo activity against herpes simplex virus type 2. Antiviral Res 1999, 42:219–226.

112. Apisariyakul A, Vanittanakom N, Buddhasukh D. Antifungal activity of turmeric oil extracted from *Curcuma longa* (Zingiberaceae). J Ethnopharmacol 1995, 49:163–169.

113. Reddy AC, Lokesh BR.Effect of curcumin and eugenol on iron-induced hepatic toxicity in rats. Toxicology 1996, 107(1):39–45.

114. Ramsewak RS, DeWitt DL, Nair MG. Cytotoxicity, antioxidant and anti-inflammatory activities of curcumins I-III from *Curcuma longa*. Phytomedicine 2000, 7:303–308.

115. Ruby AJ, Kuttan G, Babu KD, et al. Anti-tumour and antioxidant activity of natural curcuminoids. Cancer Lett 1995, 94:79–83.

116. Balasubramanyam M, Koteswari AA, Kumar RS, et al. Curcumin-induced inhibition of cellular reactive oxygen species generation: novel therapeutic implications. J Biosci 2003, 28(6):715–721.

117. Garg AK, Buchholz TA, Aggarwal BB. Chemosensitization and radiosensitization of tumors by plant polyphenols. Antioxid Redox Signal 2005, 7(11–12): 1630–1647.

118. Abraham SK, Sarma L, Kesavan PC. Protective effects of chlorogenic acid, curcumin and beta-carotene against gamma-radiation-induced in vivo chromosomal damage. Mutat Res 1993, 303(3):109–112.

119. Rezvani M, Ross GA. Modification of radiation-induced acute oral mucositis in the rat. Int J Radiat Biol 2004, 80(2):177–182.

120. Chendil D, Ranga RS, Meigooni D, et al. Curcumin confers radiosensitizing effect in prostate cancer cell line PC-3. Oncogene 2004, 23:1599–1607.

121. Khafif A, Hurst R, Kyker K, et al. Curcumin: a new radiosensitizer of squamous cell carcinoma cells. Otolaryngol Head Neck Surg 2005, 132:317–321.

122. Sidhu GS, Singh AK, Thaloor D, et al. Enhancement of wound healing by curcumin in animals. Wound Repair Regen 1998, 6:167–177.

123. Sidhu GS, Mani H, Gaddipati JP, et al. Curcumin enhances wound healing in streptozotocin induced diabetic rats and genetically diabetic mice. Wound Repair Regen 1999, 7:362–374.

124. Jiang MC, Yang-Yen HF, Yen JJ, et al. Curcumin induces apoptosis in immortalized NIH 3T3 and malignant cancer cell lines. Nutri Cancer 1996, 26:111–120.

125. Huang M-T, Lou RY, Ma W, et al. Inhibitory effects of dietary curcumin on forestomach duodenal and colon carcinogenesis in mice, Cancer Res 1994, 54:5841–5847.

126. Huang M-T, Smart RC, Wong CQ, et al. Inhibitory effect of curcumin, chlorogenic acid, caffeic acid and ferulic acid on tumor promotion in mouse skin by 12-O-tetradecanoyl phorbol-13-acetate. Cancer Res 1988, 48:5941–5946.

127. Wang ZY, Georgiadis CA, Laskin JD, et al. Inhibitory effects of curcumin on tumor initiation by benzo(a)pyrene and 7,12-dimethylbenz(a)anthr acene. Carcinogenesis 1992, 54:5841–5847.

128. Rao CV, Rivenson A, Simi B, et al. Chemoprevention of colon carcinogenesis by dietary curcumin, a naturally occurring plant phenolic compound. Cancer Res 1995, 55:259–266.

129. Pereira MA, Grubbs CJ, Barnes LH, et al. Effects of the phytochemicals, curcumin and quercetin upon azoxymethane induced colon cancer and 7,12 dimethylbenz(a)anthracene induced mammary cancer in rats. Carcinogenesis 1996, 17:1305–1311.

130. Bhide SV, Azuine MA, Lahiri M, et al. Chemoprevention of mammary tumor virus-induced and chemical carcinogen-induced rodent mammary tumors by natural plant products. Breast Cancer Res Treat 1994, 30:233–242.

131. Singletary K, MacDonald C, Wallig M, et al. Inhibition of 7,12-dimethylbenz(a) anthracene (DMBA) induced mammary tumorigenesis and DMBA-DNA adduct formation by curcumin Cancer Lett 1996, 103: 137–141.

132. Mohan R, Sivak J, Ashton P, et al. Curcuminoids inhibit the angiogenic response stimulated by fibroblast growth factor-2, including expression of matrix metalloproteinase gelatinase B. J Biol Chem 2000, 275:10405–10412.

133. Perkins S, Verschoyle RD, Hill K, et al. Chemopreventive efficacy and pharmacokinetics of curcumin in the min/+ mouse, a model of familial adenomatous polyposis. Cancer Epidemiol Biomarkers Prev 2002, 11(6):535–540.

134. Siwak DR, Shishodia S, Aggarwal BB, et al. Curcumin-induced antiproliferative and proapoptotic effects in melanoma cells are associated with suppression of IkappaB kinase and nuclear factor kappaB activity and are independent of the B-Raf/mitogen-activated/extracellular signal-regulated protein kinase pathway and the Akt pathway. Cancer 2005, 104(4):879–890.

135. Holy JM. Curcumin disrupts mitotic spindle structure and induces micronucleation in MCF-7 breast cancer cells. Mutat Res 2002, 518:71–84.

136. Choudhuri T, Pal S, Agwarwal ML, et al. Curcumin induces apoptosis in human breast cancer cells through p53-dependent Bax induction. FEBS Lett 2002, 512:334–340.

137. Aggarwal BB, Kumar A, Bharti AC. Anticancer potential of curcumin: preclinical and clinical studies. Anticancer Res 2003, 23(1A):363–398.

138. LoTempio MM, Veena MS, Steele HL, et al. Curcumin suppresses growth of head and neck squamous cell carcinoma. Clin Cancer Res 2005, 11(19 Pt 1):6994–7002.

139. Chan MM, Fong D, Soprano KJ, et al. Inhibition of growth and sensitization to cisplatin-mediated killing of ovarian cancer cells by polyphenolic chemopreventive agents. J Cell Physiol 2002, 194:63–70.

140. Notarbartolo M, Poma P, Perri D, et al. Antitumor effects of curcumin, alone or in combination with cisplatin or doxorubicin, on human hepatic cancer cells. Analysis of their possible relationship to changes in NF-kB activation levels and in IAP gene expression. Cancer Lett 2005, 224:53–65.

141. Venkatraman M, Anto RJ, Nair A, et al. Biological and chemical inhibitors of NF-kB sensitize SiHa cells to cisplatin-induced apoptosis. Mole Carcinogen 2005, 44:51–59.

142. Chirnomas D, Taniguchi T, de la Vega M, et al. Chemosensitization to cisplatin by inhibitors of the Fanconi anemia/BRCA pathway. Mol Cancer Ther 2006, 5(4):952–961.

143. Lev-Ari S, Strier L, Kazanov D, et al. Celecoxib and curcumin synergistically inhibit the growth

of colorectal cancer cells. Clin Cancer Res 2005, 11(18):6738–6744.

144. Lev-Ari S, Strier L, Kazanov D, et al. Curcumin synergistically potentiates the growth-inhibitory and pro-apoptotic effects of celecoxib in osteoarthritis synovial adherent cells. Rheumatology 2006, 45:171–177.

145. Harbottle A, Daly AK, Atherton K, et al. Role of glutathione s-transferase p1, p-glycoprotein and multidrug resistance-associated protein 1 in acquired doxorubicin resistance. Int J Cancer 2001, 92:777–783.

146. Chuang SE, Yeh PY, Lu YS, et al. Basal levels and patterns of anticancer. Drug-induced activation of nuclear factor-kappaB (NFkappaB), and its attenuation by tamoxifen, dexamethasone, and curcumin in carcinoma cells. Biochem Pharmacol 2002, 63:1709–1716.

147. Bharti AC, Aggarwal BB. Chemopreventive agents induce suppression of nuclear factor-kappaB leading to chemosensitization. Ann N Y Acad Sci 2002, 973:392–395.

148. Hour TC, Chen J, Huang CY, et al. Curcumin enhances cytotoxicity of chemotherapeutic agents in prostate cancer cells by inducing p21(WAF1/CIP1) and C/EBPbeta expressions and suppressing NF-kappaB activation. Prostate 2002, 51:211–218.

149. Aggarwal BB, Shishodia S, Takada Y, et al. Curcumin suppresses the paclitaxel-induced nuclear factor-kB pathway in breast cancer cells and inhibits lung metastasis of human breast cancer in nude mice. Clin Cancer Res 2005, 11(20): 7490– 7498.

150. Bava SV, Puliappadamba VT, Deepti A, et al. Sensitization of Taxol-induced Apoptosis by curcumin involves down-regulation of nuclear factor-kB and the serine/threonine kinase Akt and is independent of tubulin polymerization. J Biol Chem 2005, 280:6301–6308.

151. Chearwae W, Wu C-P, Chu H-Y, et al. Curcuminoids purified from turmeric powder modulate the function of human multidrug resistance protein 1 (ABCC1). Cancer Chemother Pharmacol 2006, 57:376–388.

152. Chearwae W, Shukla S, Limtrakul P, et al. Modulation of the function of the multidrug resistance-linked ATP-binding cassette transporter ABCG2 by the cancer chemopreventive agent curcumin. Mol Cancer Ther 2006, 5(8):1995–2006.

153. Somasundaram S, Edmund NA, Moore DT, et al. Dietary curcumin inhibits chemotherapy-induced apoptosis in models of human breast cancer. Cancer Res. 2002, 62:3868–3875.

154. Jeandet P, Bessis R, Gautheron B. The production of resveratrol by grape berries in different developmental stages. Am J Enol Viticult 1991, 42:41.

155. Jang M, Cai L, Udeani G, et al. Cancer chemopreventive activity of resveratrol, a natural product derived from grapes. Science 1997, 275:218–220

156. Elattar T, Virji A.The effect of red wine and its components on growth and proliferation of human oral squamous carcinoma cells. Anticancer Res 1999, 19:5407–5414.

157. Surh Y, Hurh Y, Kang J, et al. Resveratrol, an antioxidant present in red wine, induces apoptosis in human promyelocytic leukemia (HL-60) cells. Cancer Lett 1999, 140:1–10.

158. Lu R, Serrero G. Resveratrol, a natural product derived from grape, exhibits antiestrogenic activity and inhibits the of human breast cancer cells. Cell Physiol 1999, 179:297–304.

159. Hsieh T, Wu J. Differential effects on growth, cell cycle arrest, and induction of apoptosis by resveratrol in human prostate cancer cell lines. Exp Cell Res 1999, 249:109–115.

160. Narayanan B, Narayanan N, Re G, et al. Differential expression of genes induced by resveratrol in LNCaP cells: P53-mediated molecular targets. Int J Cancer 2003, 104:204–212.

161. Zhou H, Yan Y, Sun Y, Zhu J. Resveratrol induces apoptosis in human esophageal carcinoma cells. World J Gastroenterol 2003, 9:408–411.

162. Ding X, Adrian T. Resveratrol inhibits proliferation and induces apoptosis in human pancreatic cancer cells. Pancreas 2002, 25:e71–e76.

163. Tsan M, White J, Maheshwari J, et al. Resveratrol induces Fas signalling-independent apoptosis in THP-1 human monocytic leukaemia cells. Br J Haematol 2000, 109:405–412.

164. Kubota T, Uemura Y, Kobayashi M, et al. Adsorption and desorption properties of trans-resveratrol on cellulose cotton. Ann Sci 2005, 21(2):183–186.

165. Jazirehi AR, Bonavida B. Resveratrol modifies the expression of apoptotic regulatory proteins and sensitizes non-Hodgkin's lymphoma and multiple myeloma cell lines to paclitaxel-induced apoptosis. Mol Cancer Ther 2004, 3(1):71–84.

166. Cal C, Garban H, Jazirehi A, et al. Resveratrol and cancer: chemoprevention, apoptosis, and chemoimmunosensitizing activities. Curr Med Chem Anti-Cancer Agents 2003, 3:77–93.

167. Fulda, S, Debatin KM. Natural agents can sensitize tumor cells to chemotherapy and radiation therapy. Proc Am Assoc Cancer Res 2002, 43:856.

168. Sharma G, Tyagi AK, Singh RP, et al. Synergistic anti-cancer effects of grape seed extract and conventional cytotoxic agent doxorubicin against human breast carcinoma cells. Breast Cancer Res Treat 2004, 85(1):1–12.

169. Fuggetta MP, Lanzilli G, Tricarico M, et al. In vitro antitumour activity of resveratrol in human melanoma cells sensitive or resistant to temozolomide. Melanoma Res 2004, 14(3):189–196.

170. Ahmad KA, Clement MV, Hanif IM, et al. Resveratrol inhibits drug-induced apoptosis in human leukemia cells by creating an intracellular milieu nonpermissive for death execution. Cancer Res 2004 64(4):1452–1459.

171. Rezk YA, Balulad SS, Keller RS, et al. Use of resveratrol to improve the effectiveness of cisplatin and doxorubicin: study in human gynecologic cancer cell lines and in rodent heart. Am J Obstet Gynecol 2006, 194(5):e23–e26.

172. Duraj J, Bodo J, Sulikova M, et al. Diverse resveratrol sensitization to apoptosis induced by anticancer drugs in sensitive and resistant leukemia cells. Neoplasma 2006, 53(5):384–392.

173. Mills R, Beeson W, Phillips R, et al. Cohort study of diet, lifestyle, and prostate cancer in Adventist men. Cancer 1989, 64:598–604.

174. Knight DC, Eden JA. A review of the clinical effects of phytoestrogens. Obstet Gynecol 1996, 87:897–904.

175. Adlercreutz CH, Goldin BR, Gorbach SL, et al. Soybean phytoestrogen intake and cancer risk. J Nutr 1995, 125:S757–770.

176. Tatsuta M, Iishi H, Baba M, et al. Attenuation by genistein of sodium-chloride-enhanced gastric carcinogenesis induced by N-methyl-NV-nitro-N-nitrosoguanidine in Wistar rats. Int J Cancer 1999, 80:396–399.

177. Giovannucci E. Epidemiologic characteristics of prostate cancer. Cancer 1995, 75:1766–1777.

178. Lei W, Mayotte JE, Levitt ML. Enhancement of chemosensitivity and programmed cell death by tyrosine kinase inhibitors correlates with EGFR expression in non-small cell lung cancer cells. Anticancer Res 1999, 19(1A):221–228.

179. Wietrzyk J, Boratynski J, Grynkiewicz G, et al. Antiangiogenic and antitumour effects in vivo of genistein applied alone or combined with cyclophosphamide. Anticancer Res 2001, 21:3893–3896.

180. Khoshyomn S, Manske GC, Lew SM, et al. Synergistic action of genistein and cisplatin on growth inhibition and cytotoxicity of human medulloblastoma cells. Pediatr Neurosurg 2000, 33:123–131.

181. Park JH, Oh EJ, Choi YH, et al. Synergistic effects of dexamethasone and genistein on the expression of Cdk inhibitor p21WAF1/CIP1 in human hepatocellular and colorectal carcinoma cells. Int J Oncol 2001, 18(5):997–1002.

182. Shen F, Weber G. Synergistic action of quercetin and genistein in human ovarian carcinoma cells. Oncol Res 1997, 9:597–602.

183. Li Y, Ellis KL, Ali S, et al. Apoptosis-inducing effect of chemotherapeutic agents is potentiated by soy isoflavone genistein, a natural inhibitor of NF-kappaB in BxPC-3 pancreatic cancer cell line. Pancreas 2004, 28:90–95.

184. Li Y, Ahmed F, Ali S, et al. Inactivation of nuclear factor kappaB by soy isoflavone genistein contributes to increased apoptosis induced by chemotherapeutic agents in human cancer cells. Cancer Res 2005, 65:6934–6942.

185. Banerjee S, Zhang Y, Ali S, et al. Molecular evidence for increased antitumor activity of gemcitabine by genistein in vitro and in vivo using an orthotopic model of pancreatic cancer. Cancer Res 2005, 65:9064–9072.

186. Li Y, Sarkar FH. Inhibition of nuclear factor kappaB activation in PC3 cells by genistein is mediated via Akt signaling pathway. Clin Cancer Res 2002, 8(7):2369–2377.

187. Li Y, Kucuk O, Hussain M, et al. Antitumor and antimetastatic activities of docetaxel are enhanced by genistein through regulation of osteoprotegerin/receptor activator of nuclear factor-kappaB (RANK)/RANK ligand/MMP-9 signaling in prostate cancer. Cancer Res 2006, 66(9):4816–4825.

188. Mohammad RM, Al-Katib A, Aboukameel A, et al. Genistein sensitizes diffuse large cell lymphoma to CHOP (cyclophosphamide, doxorubicin, vincristine, prednisone) chemotherapy. Mol Cancer Ther 2003, 2:1361–1368.

189. Hwang JT, Ha J, Park OJ. Combination of 5-fluorouraciland genistein induces apoptosis synergistically in chemo-resistant cancer cells through the modulation of AMPK and COX-2 signaling pathways. Biochem Biophys Res Commun 2005, 332:433–440.

190. Satoh H, Nishikawa K, Suzuki K, et al. Genistein, a soyisoflavone, enhances necrotic-like cell death in a breastcancer cell treated with a chemotherapeutic agent. Res Commun Mol Pathol Pharmacol 2003, 113–114:149–158.

191. Tanos V, Brzezinski A, Drize O, et al. Synergistic inhibitory effects of genistein and tamoxifen on human dysplastic and malignant epithelial breast cells in vitro. Eur J Obstet Gynecol Reprod Biol 2002, 102:188–194.

192. Chen J, Halls SC, Alfaro JF, et al. Potential beneficial metabolic interactions between tamoxifen and isoflavones via cytochrome P450-mediated pathways in female rat liver microsomes. Pharm Res 2004, 21:2095–2104.

193. Papazisisa KT, Kalemib TG, Zamboulia D, et al. Synergistic effects of protein tyrosine kinase inhibitor genistein with camptothecins against

three cell lines in vitro. Cancer Letts 2006, 233:255–264.

194. Tamura S, Bito T, Ichihashi M, et al. Genistein enhances the cisplatin-induced inhibition of cell growth and apoptosis in human malignant melanoma cells. Pigment Cell Res 2003, 16:470–476.

195. Lee R, Kim YJ, Lee YJ, et al. The selective effect of genistein on the toxicity of bleomycin in normal lymphocytes and HL-60 cells. Toxicology 2004, 195:87–95.

196. Graham HN. Green tea composition, consumption, and polyphenol chemistry. Preventive Medicine 1992, 21(3):334–350.

197. Yang CS. Tea and health. Nutrition 1999, 15(11–12): 946–949.

198. Dona M, Dell'Aica I, Calabrese F, et al. Neutrophil restraint by green tea: inhibition of inflammation, associated angiogenesis, and pulmonary fibrosis. J Immunol 2003, 170(8):4335–4341.

199. Nance CL, Shearer WT. Is green tea good for HIV-1 infection? J Allerg Clin Immunol 2003, 112(5):851–853.

200. Stapleton PD, Shah S, Anderson JC, et al. Modulation of beta-lactam resistance in Staphylococcus aureus by catechins and gallates. International J Antimicrob Agents 2004, 23(5):462–467.

201. Esposito E, Rotilio D, Di Matteo V, et al. A review of specific dietary antioxidants and the effects on biochemical mechanisms related to neurodegenerative processes. Neurobiol Aging 2002, 23(5):719–735.

202. Cao Y, Cao R. Angiogenesis inhibited by drinking tea. Nature 1999, 398(6726):381.

203. Pfeffer U, Ferrari N, Morini M, et al. Antiangiogenic activity of chemopreventive drugs. Int J Biol Mark 2003, 18(1):70–74.

204. Wang ZY, Cheng SJ, Zhou ZC, et al. Antimutagenic activity of green tea polyphenols. Mutat Res 1989, 223(3):273–285.

205. Han C. Screening of anticarcinogenic ingredients in tea polyphenols. Cancer Lett 1997, 114(1–2):153–158.

206. Wu LY, Juan CC, Hwang LS, et al. Green tea supplementation ameliorates insulin resistance and increases glucose transporter IV content in a fructose-fed rat model. Eur J Nutr 2004, 43(2):116–124.

207. Mukhtar H, Ahmad N. Tea polyphenols: prevention of cancer and optimizing health. Am J Clin Nutr 2000, 1(6 Suppl):1698S.

208. Yang G-yu, Liao J, Kim K, et al. Inhibition of growth and induction of apoptosis in human cancer cell lines by tea polyphenols. Carcinogenesis 1998, 19:611–616.

209. Yang CS. Maliakal P, Meng X. Inhibition of carcinogenesis by tea. Ann Rev Pharmacol Toxicol 2002, 42:25–54.

210. Stoner GD, Mukhtar H. Polyphenols as cancer chemopreventive agents. J Cell Biochem Suppl 1995, 22:169–180.

211. Lambert JD, Yang CS. Cancer chemopreventive activity and bioavailability of tea and tea polyphenols. Mutat Res 2003, 523–524:201–208.

212. Gupta S, Hussain T, Mukhtar H. Molecular pathway for (-)-epigallocatechin-3-gallate-induced cell cycle arrest and apoptosis of human prostate carcinoma cells. Arch Biochem Biophys 2004, 410(1):177–185.

213. Park AM, Dong Z. Signal transduction pathways: targets for green and black tea polyphenols. J Biochem Mol Biol 2003, 36(1):66–77.

214. Gouni-Berthold I, Sachinidis A. Molecular mechanisms explaining the preventive effects of catechins on the development of proliferative diseases. Curr Pharmaceut Des 2004, 10(11):1261–1271.

215. Ahmad N, Gupta S, Mukhtar H, Green tea polyphenol epigallocatechin-3-gallate differentially modulates nuclear factor kappaB in cancer cells versus normal cells. Arch Biochem Biophys 2000, 376(2):338–346.

216. Levites Y, Youdim MB, Maor G, et al. Attenuation of 6-hydroxydopamine (6-OHDA)-induced nuclear factor-kappaB (NF-kappaB) activation and cell death by tea extracts in neuronal cultures. Biochem Pharmacol 2002, 63(1):21–29.

217. Nam S, Smith DM, Dou QP. Ester bond-containing tea polyphenols potently inhibit proteasome activity in vitro and in vivo. J Biol Chem 2001, 276(16):13322–13330.

218. Singh R, Ahmed S, Islam N, et al. Epigallocatechin-3-gallate inhibits interleukin-1beta-induced expression of nitric oxide synthase and production of nitric oxide in human chondrocytes: suppression of nuclear factor kappaB activation by degradation of the inhibitor of nuclear factor kappaB. Arth Rheumatol 2002, 46(8):2079–2086.

219. Jung YD, Ellis LM.. Inhibition of tumour invasion and angiogenesis by epigallocatechin gallate (EGCG), a major component of green tea. Int J Exp Pathol 2001, 82(6):309–316.

220. Masuda M, Suzui M, Lim JT, et al. Epigallocatechin-3-gallate decreases VEGF production in head and neck and breast carcinoma cells by inhibiting EGFR-related pathways of signal transduction. J Exp Ther Oncol 2002, 2(6):350–359.

221. Kojima-Yuasa A, Hua JJ, Kennedy DO, et al. Green tea extract inhibits angiogenesis of human umbilical vein endothelial cells through reduction of expression of VEGF receptors. Life Sci 2003, 73(10):1299–1313.

222. Waleh N, Chao W-R, Bensari A, et al. Novel D-ring analog of epigallocatechin-3-gallate inhibits tumor

growth and VEGF expression in breast carcinoma cells. Anticancer Res 2005, 25(1A):397–402.

223. Stammler G, Volm M. Green tea catechins (EGCG and EGC) have modulating effects on the activity of doxorubicin in drug-resistant cell lines. Anticancer Drugs 1997, 8(3):265–268.

224. Sadzuka Y, Sugiyama T, Sonobe T. Efficacies of tea components on doxorubicin induced antitumor activity and reversal of multidrug resistance. Toxicol Lett 2000, 114:155–162.

225. Mei Y, Qian F, Wei D, Liu J. Reversal of cancer multidrug resistance by green tea polyphenols. J Parm Pharmacol 2004, 56:1307–1314.

226. Mei Y, Wei D, Liu J. Reversal of multidrug resistance in KB cells with tea polyphenol antioxidant capacity. Cancer Biol Ther 2005, 4(4):468–473.

227. Zhang Q, Wei D, Liu J. In vivo reversal of doxorubicin resistance by (-)-epigallocatechin gallate in a solid human carcinoma xenograft. Cancer Lett 2004, 208(2):179–186.

228. Qian F, Wei D, Zhang Q, et al. Modulation of P-glycoprotein function and reversal of multidrug resistance by (–)-epigallocatechin gallate in human cancer cells. Biomed Pharmacother 2005, 59:64–69.

229. Liang G, Zhang S, Huang ZM, et al. MDR-reversing effect of two components of catechin on human hepatocellular carcinoma BEL-7404/Adr in vitro. Ai Zheng 2004, 23(4):401–405.

230. Jodoin J, Demeule M, Beliveau R. Inhibition of the multidrug resistance P-glycoprotein activity by green tea polyphenols. Biochim Biophys Acta 2002, 1542:149–159.

231. Kitagawa S, Nabekura T, Kamiyama S. Inhibition of P-glycoprotein function by tea catechins in KB-C2 cells. JPP 2004, 56:1001–1005.

232. Suganuma M, Okabe S, Kai Y, et al. Synergistic effects of (–)-epigallocatechin gallate with (–)-epicatechin, sulindac, or tamoxifen on cancer-preventive activity in the human lung cancer cell line PC-9. Cancer Res 1999, 59(1):44–47.

233. Chisholm K, Bray BJ, Rosengren RJ. Tamoxifen and epigallocatechin gallate are synergistically cytotoxic to MDA-MB-231 human breast cancer cells. Anti-Cancer Drugs 2004, 15:889–897.

234. Chan MM, Soprano KJ, Weinstein K, et al. Epigallocatechin-3-gallate delivers hydrogen peroxide to induce death of ovarian cancer cells and enhances their cisplatin susceptibility. J Cell Physiol 2006, 207(2):389–396.

235. Nakazato T, Ito K, Ikeda Y, et al. Green tea component, catechin, induces apoptosis of human malignant B cells via production of reactive oxygen species. Clin Cancer Res 2005, 11:6040–6049.

236. Ishino A, Mita S, Watanabe S, et al. Effect of anti-cancer drugs, metals and antioxidants on cytotoxic activity of epigallocatechin gallate. Anticancer Res 1999, 19(5B):4343–4348.

237. Rambaldi A, Jacobs B, Iaquinto G, et al. Milk thistle for alcoholic and/or hepatitis B or C virus liver diseases. Cochrane Database Syst Rev 2005, CD003620.

238. Blumenthal M. Herb sales down 7.4 percent in mainstream market. HerbalGram 2005, 66:63.

239. Werneke U, Earl J, Seydel C, et al. Potential health risks of complementary alternative medicines in cancer patients. Br J Cancer 2004, 90:408–413.

240. Morazzoni P, Bombardelli E. *Silybum marianum (Carduus marianus)*. Fitoterapia 1995, 64:3–42.

241. Wagner H, Diesel P, Seitz M. Chemistry and analysis of silymarin from *Silybum marianum* (L.) Gaertn Arzneimit Forsch 1974, 24:466–471.

242. Tyagi A, Bhatia N, Condon M, et al. Antiproliferative and apoptotic effects of silibinin in rat prostate cancer cells. Prostate 2002, 53:211–217.

243. Singh RP, Dhanalakshmi S, Tyagi AK, et al. Dietary feeding of silibinin inhibits advance human prostate carcinoma growth in athymic nude mice and increases plasma insulin-like growth factor-binding protein-3 levels. Cancer Res 2002, 62:3063–3069.

244. Chrungoo VJ, Reen RK, Singh K, et al. Effects of silymarin on UDP-glucuronic acid and glucuronidation activity in the rat isolated hepatocytes and liver in relation to D-galactosamine toxicity. Indian J Exp Biol 1997, 35:256–263.

245. Venkataramanan R, Ramachandran V, Komoroski BJ, et al. Milk thistle, a herbal supplement, decreases the activity of CYP3A4 and uridine diphosphoglucuronosyl transferase in human hepatocyte cultures. Drug Metab Dispos 2000, 28:1270–1273.

246. Beckmann-Knopp S, Rietbrock S, Weyhenmeyer R, et al. Inhibitory effects of silibinin on cytochrome P- 450 enzymes in human liver microsomes. Pharmacol Toxicol 2000, 86:250–256.

247. Zuber R, Modriansky M, Dvorak Z, et al. Effect of silybin and its congeners on human liver microsomal cytochrome P450 activities. Phytother Res 2002, 16:632–638.

248. Zhang S, Morris ME. Effects of the flavonoids biochanin A, morin, phloretin, and silymarin on Pglycoprotein-mediated transport. J Pharmacol Exp Ther 2003, 304:1258–1267.

249. Nguyen H, Zhang S, Morris ME. Effect of flavonoids on MRP1-mediated transport in Panc-1cells. J Pharm Sci 2003, 92:250–257.

250. Sridar C, Goosen TC, Kent UM, et al. Silybin inactivates cytochromes P450 3A4 and 2C9 and inhib-

itsmajor hepatic glucuronosyltransferases. Drug Metab Dispos 2004, 32:587–594.

251. Scambia G, De Vincenzo R, Ranelletti FO, et al. Antiproliferative effect of silybin on gynecological malignancies: synergism with cisplatin and doxorubicin. Eur J Cancer 1996, 32A:877–882.

252. Giacomelli S, Gallo D, Apollonio P, et al. Silybin and its bioavailable complex (IdB 1016) potentiate *in vitro* and *in vivo* the activity of cisplatin. Life Sci 2002, 70:1447–1459.

253. Tyagi AK, Singh RP, Agarwal C, et al. Silibinin strongly synergizes human prostate carcinoma DU145 cells to doxorubicin-induced growth Inhibition, G2-M arrest, and apoptosis. *Clin Cancer Res* 2002, 8:3512–3519.

254. Dhanalakshmi S, Agarwal P, Glode LM, et al. Silibinin sensitizes human prostate carcinoma DU145 cells to cisplatin- and carboplatin-induced growth inhibition and apoptotic death. Int J Cancer 2003, 106:699–705.

255. Tyagi AK, Agarwal C, Chan DCF, et al. Synergistic anti-cancer effects of silibinin with conventional cytotoxic agents doxorubicin, cisplatin and carboplatin against human breast carcinoma MCF-7 and MDA-MB468 cells. Oncol Rep 2003, 11:493–499.

256. van Erp NPH, Baker SD, Zhao M, et al. Effect of milk thistle (*Silybum marianum*) on the pharmacokinetics of irinotecan. Clin Cancer Res 2005, 11(21):7800–7806.

257. Swain T. The flavonoids. London: Chapman & Hall, 1975.

258. Formica JV, Regelson W. Review of the biology of quercetin and related biofavonoids. Food ChemToxicol 1995, 33:1061–1080.

259. Castillo MH, Perkins E, Campbell JH, et al.The effects of the bioflavonoid quercetin on squamous cell carcinoma of head and neck origin. Am J Surg 1989, 158:351–335.

260. Scambia G, Ranelletti FO, Benedetti PP, et al. Quercetin inhibits the growth of a multidrug-resistant estrogen-receptor-negative MCF-7 human breast-cancer cell line expressing type II estrogen-binding sites. Cancer Chemother Pharmacol 1991, 28:255–258.

261. Ranelletti FO, Ricci R, Larocca LM, et al. Growth inhibitory effect of quercetin and presence of type-II estrogen-binding sites in human colon-cancer cell lines and primary colorectal tumors. Int J Cancer 1992, 50:486–492.

262. Elattar TM, Virji AS. The inhibitory effect of curcumin, genistein, quercetin and cisplatin on the growth of oral cancer cells in vitro. Anticancer Res 2000, 20:1733–1738.

263. Nair HK, Rao KV, Aalinkeel R, et al. Inhibition of prostate cancer cell colony formation by the flavonoid quercetin correlates with modulation of specific regulatory genes. Clin Diagn Lab Immunol 2004, 11:63–69.

264. Akbas SH, Timur M, Ozben T. The effect of quercetin on topotecan cytotoxicity in MCF-7 and MDA-MB 231 human breast cancer cells. J Surg Res 2005, 125:49–55.

265. Sharma H, Sen S, Singh N. The effect of quercetin on topotecan cytotoxicity in MCF-7 and MDA-MB 231 human breast cancer cells. Cancer Biol Ther 2005, 4(9):949–955.

266. Lee H, Tsai SJ. Effect of emodin on cooked-food mutagen activation. Food Chem Toxicol 1991, 29(11):765–770.

267. Sato M, Maulik G, Bagchi D, et al. Myocardial protection by protykin, a novel extract of transresveratrol and emodin. Free Radic Res 2000, 32(2):135–144.

268. Kuo YC, Tsai WJ, et al. Immune reponses in human mesangial cells regulated by emodin from Polygonum hypoleucum Ohwi. Life Sci 2001, 68(11):1271–1286.

269. Chang CJ, Ashendel CL, Geahlen RL, et al. Oncogene signal transduction inhibitors from medicinal plants. In Vivo 1996, 10:185–190.

270. Chan TC, Chang CJ, Koonchanok NM, et al. Selective inhibition of the growth of ras-transformed human bronchial epithelial cells by emodin, a protein-tyrosine kinase inhibitor. Biochem Biophys Res Commun 1993, 193(3):1152–1158.

271. Kamei H, Koide T, Kojima T, et al. Inhibition of cell growth in culture by quinones. Cancer Biother Radiopharm 1998, 13(3):185–188.

272. Qu Y, Yao P, Li TQ. Effects of emodin on lung fibroblast proliferation and cell cycle in vitro. Sichuan Da Xue Xue Bao Yi Xue Ban 2004, 35(1):74–76.

273. Jayasuriya H, Koonchanok NM, Geahlen RL, et al. Emodin, a protein tyrosine kinase inhibitor from *Polygonum cuspidatum*. J Nat Prod 1992, 55:696–698.

274. Zhang L, Chang CJ, Bacus SS, et al. Suppressed transformation and induced differentiation of HER-2/neu-overexpressing breast cancer cells by emodin. Cancer Res 1995, 55:3890–3896.

275. Pecere T, Gazzola MV, Mucignat C, et al. Aloe-emodin is a new type of anticancer agent with selective activity against neuroectodermal tumors. Cancer Res 2000, 60:2800– 2804.

276. Lee HZ, Hsu SL, Liu MC, et al. Effects and mechanisms of aloe-emodin on cell death in human lung squamous cell carcinoma. Eur J Pharmacol 2001, 431:287–295.

277. Lee H. Protein kinase C involvement in aloe-emodin and emodin-induced apoptosis in lung carcinoma cell. Br J Pharmacol 2001, 134:11–20.

278. Kupchan SM, Karim A. Tumor inhibitors 114 aloe emodin: antileukemic principle isolated from Rhamus frangula L. Lloydia 1976, 39:223–224.

279. Chung JG, Li YC, Lee YM, et al. Aloe-emodin inhibited N acetylation and DNA adduct of 2-aminofluorene and arylamine N-acetyltransferase gene expression in mouse leukemia L 1210 cells. Leuk Res 2003, 27:831–840.

280. Chen YC, Shen SC, Lee WR, et al. Emodin induces apoptosis in human promyeloleukemic HL-60 cells accompanied by activation of caspase 3 cascade but independent of reactive oxygen species production. Biochem Pharmacol 2002, 64:1713–1724.

281. Kuo P, Lin TC, Lin CC. The antiproliferative activity of aloe-emodin is through p53 dependent and p21-dependent apoptotic pathway in human hepatoma cell lines. Life Sci 2002, 71:1879–1892.

282. Wasserman L, Avigad S, Beery E, et al. The effect of aloe emodin on the proliferation of a new Merkel carcinoma cell line. Am J Dermatopathol 2002, 24:17–22.

283. Srinivas G, Anto RJ, Srinivas P, et al. Emodin induces apoptosis of human cervical cancer cells through poly(ADP-ribose) polymerase cleavage and activation of caspase-9. Eur J Pharmacol 2003, 473:117–125.

284. Zhang L, Lau YK, Xi L, et al. Tyrosine kinase inhibitors, emodin and its derivative repress HER-2/neu-induced cellular transformation and metastasisassociated properties. Oncogene 1998, 16:2855–2863.

285. Zhang L, Lau YK, Xia W, et al. Tyrosine kinase inhibitor emodin suppresses growth of HER-2/neu-overexpressing breast cancer cells in athymic mice and sensitizes these cells to the inhibitory effect of paclitaxel. Clin Cancer Res 1999, 5:343–353.

286. Zhang L, Hung MC. Sensitization of HER-2/neu-overexpressing non-small cell lung cancer cells to chemotherapeutic drugs by tyrosine kinase inhibitor emodin. Oncogene 1996, 12:571–576.

287. Lai G-H, Zhang Z, Sir AE. Celecoxib acts in a cyclooxygenase-2-independent manner and in synergy with emodin to suppress rat cholangiocarcinoma growth in vitro through a mechanism involving enhanced akt inactivation and increased activation of caspases-9 and -3. Mol Cancer Ther 2003, 2:265–271.

288. Fenig E, Nordenberg J, Beery E, et al. Combined effect of aloe-emodin and chemotherapeutic agents on the proliferation of an adherent variant cell line of Merkel cell carcinoma. Oncol Rep 2004, 11(1):213–217.

289. Yang J, Tang XM, Li H, et al. Emodin sensitizes HeLa cell to arsenic trioxide induced apoptosis via the reactive oxygen species-mediated signaling pathways. Shi Yan Sheng Wu Xue Bao 2003, 36(6):465–475.

290. Yang J, Li H, Chen YY, et al. Anthraquinones sensitize tumor cells to arsenic cytotoxicity in vitro and in vivo via reactive oxygen species mediated dual regulation of apoptosis. Free Radic Biol Med 2004, 37:2027–2041.

291. Jing YW, Yi J, Chen YY, et al. Dicumarol alters cellular redox state and inhibits nuclear factor kappa B to enhance arsenic trioxide-induced apoptosis. Acta Biochim Biophys Sin 2004, 36:235–242.

292. Yi J, Yang J, He R, et al. Emodin enhances arsenic trioxide-induced apoptosis via generation of reactive oxygen species and inhibition of survival signaling. Cancer Res 2004, 64:108–116.

293. Hengstler JG, Lange J, Kett A, et al. Contribution of c-erbB-2 and topoisomerase IIalpha to chemoresistance in ovarian cancer. Cancer Res 1999, 59:3206–3214.

294. Naik RG, Kattige SL, Bhat SV, et al. An anti-inflammatory cum immunomodulatory piperidinyl-benzopyranone from Dysoxvlum binectariferum: isolation, structure and total synthesis. Tetrahedron 1988, 44:2081.

295. Bible KC, Kaufmann SH. Flavopiridol: a cytotoxic flavone that induces cell death in noncycling A549 human lung carcinoma cells. Cancer Res 1996, 56:4856–4861.

296. Senderowicz A, Headlee D, Stinson S, et al. Phase I trial of continuous infusion flavopiridol, a novel cyclindependent kinase inhibitor, in patients with refractory neoplasms. J Clin Oncol 1998, 16:2986–2999.

297. Carlson BA, Dubay MM, Sausville EA, et al. De Azevedo Flavopiridol induces G1 arrest with inhibition of cyclin-dependent kinase (CDK) 2 and CDK4 in human breast carcinoma cells. Cancer Res 1996, 56:2973–2978.

298. Chao S, Fujinaga K, Marion JE, et al. Flavopiridol inhibits P-TEFb and blocks HIV-1 replication. J Biol Chem 2000, 275:28345–28348.

299. De Azevedo WFJ, Mueller DH, Schulze GU, et al. Structural basis for specificity and potency of a flavonoid inhibitor of human CDK2, a cell cycle kinase. Proc Natl Acad Sci U S A 1996, 93:2735–2740.

300. Kaur G, Stetle-Stevenson M, Sebers S, et al. Growth inhibition with reversible cell cycle arrest of carcinoma cells by flavone L86-8275. J Natl Cancer Inst 1992, 84:1736–1740.

301. Schwartz GK, Farsi K, Majhija S, et al. Sensitivity of tumor cells to the cyclin dependent kinase (CDK) inhibitor Flavopiridol (FLAVO) correlates

to loss of bcl-2 expression. Proc Am Assoc Cancer Res 1997, 38:472–473.

302. Kitada S, Tamm I, Andreeff M, et al. Protein kinase C inhibitor 7OH-staurosporine (UCN-01) and CDK-family kinase inhibitor flavopiridol down regulation expression of survival genes and induce apoptosis in B-CLL chronic lymphocyte leukemia (B-CLL). Blood 1998, 92(Suppl):102a.

303. Brusselbach S, Nettelbeck DM, Sedlacek HH, et al. Cell cycleindependent induction of apoptosis by the anti-tumor drug flavopiridol in endothelial cells. Int J Cancer 1998, 77:146–152.

304. Patel V, Senderowicz AM, Pinto D, et al. Flavopiridol, a novel cyclin dependent kinase inhibitor, suppresses the growth of head and neck squamous cell carcinomas by inducing apoptosis. J Clin Invest 1998, 102:1674–1681.

305. Bible KC, Bible RH Jr, Kottke TJ, et al. Flavopiridol binds to duplex DNA. Cancer Res 2000, 60: 2419–2428.

306. Lü X, Burgan WE, Cerra MA, et al. Transcriptional signature of flavopiridol-induced tumor cell death. Mol Cancer Ther 2004, 3(7):861–872.

307. Schwartz GK, Arkin H, Holland J, et al. Protein kinase C activity and multidrug resistance on MOLT-3 human lymphoblastic leukemia cells resistant to trimethotrexate. Cancer Res 1991, 51:55–61.

308. Schwartz GK, Farsi K, Daso D, et al. The protein kinase C (PKC) inhibitors UCN-01 and Flavopiridol (FLAVO) significantly enhance the cytotoxic effect of chemotherapy by promoting apoptosis in gastric and breast cancer cells. Proc Am Soc Oncol 1996, 15:501.

309. Bible KC, Kaufman SH. Cytotoxic synergy between Flavopiridol (NSC 649 890, L86-2875) and various antineoplastic agents: the importance of sequence of administration. Cancer Res 1997, 57:3375–3380.

310. Motwani M, Delohery TM, Schwartz GK. Sequential dependent enhancement of caspase activation and apoptosis by flavopiridol on paclitaxel-treated human gastric and breast cancer cells. Clin Cancer Res 1999, 5:1876–1883.

311. Motwani M, Rizzo C, Sirotnak FM, et al. Flavopiridol enhances the effect of docetaxel in vitro and in vivo in human gastric cancer cells. Mol Cancer Ther 2003, 2:549–555.

312. Guo Jun, Zhou An-wu, Fu Y-C, et al. Efficacy of sequential treatment of HCT116 colon cancer monolayers and xenografts with docetaxel, flavopiridol, and 5-fluorouracil. Acta Pharmacol Sin 2006, 27 (10):1375–1381.

313. Motwani M, Jung C, Sirotnak FM, et al. Augmentation of apoptosis and tumor regression by flavopiridol in the presence of CPT-11 in HCT-116 colon cancer monolayers and xenografts. Clin Cancer Res 2001, 7:4209–4219.

314. Shah MA, Kortmansky J, Motwani M, et al. A phase I clinical trial of the sequential combination of irinotecan followed by flavopiridol. Clin Cancer Res 2005, 11(10):3836–3845.

315. Wall N, O'Connor DS, Plescia J, et al. Suppression of survivin phosphorylation on Thr34 by flavopiridol enhances tumor cell apoptosis. Cancer Res 2003, 63:230–235.

316. Jung CP, Motwani MV, Schwartz GK. Flavopiridol increases sensitization to gemcitabine in human gastrointestinal cancer cell lines and correlates with down- regulation of ribonucleotide reductase M2 subunit. Clin Cancer Res 2001, 7:2527–2536.

317. Nahta R, Trent S, Yang C, et al. Epidermal growth factor receptor expression is a candidate target of the synergistic combination of trastuzumab and flavopiridol in breast cancer. Cancer Res 2003, 63:3626–3631.

318. Fandy TE, Ross DD, Gore SD, et al. Flavopiridol synergizes TRAIL cytotoxicity by downregulation of FLIP(L). Cancer Chemother Pharmacol 2006.

319. Araujo MCP, Dias FL, Takahashi CS. Potentiation by turmeric and curcumin of gamma - radiation-induced chromosome aberrations in Chinese Hamster ovary cells. Terat Carcinogen Mutag 1999, 19:9–18.

320. Kumar Mitra A, Krishna M. In vivo modulation of signaling factors involved in cell survival. J Radiat Res (Tokyo) 2004, 45(4):491–495.

321. Zoberi I, Bradbury CM, Curry HA, et al. Radiosensitizing and antiproliferative effects of resveratrol in two human cervical tumor cell lines. Cancer Lett 2002, 175:165–173.

322. Baatout S, Derradji H, Jacquet P, et al. Enhanced radiation-induced apoptosis of cancer cell lines after treatment with resveratrol. Int J Mol Med 2004, 13(6):895–902.

323. Fiore M, Festa F, Cornetta T, et al. Resveratrol affects X-ray induced apoptosis and cell cycle delay in human cells in vitro. Int J Mol Med 2005, 15:1005–1012.

324. Liao H-F, Kuo C-D, Yang Y-C, et al. Resveratrol enhances radiosensitivity of human non-small lung cancer NCI-H838 cells accompanied by inhibition of NF-kappaB activation. J Radiat Res 2005, 46:387–393.

325. Akimoto T, Nonaka T, Ishikawa H, et al. a tyrosine kinase inhibitor, enhanced radiosensitivity in human esophageal cancer cell lines in vitro: possible involvement of inhibition of survival signal transduction pathways. Int J Radiat Oncol Biol Phys 2001, 50:195–201.

326. Yan S-X, Ejima Y, Sasaki R, et al. Combination of genistein with ionizing radiation on androgen-inde-

pendent prostate cancer cells. Asian J Androl 2004, 6:285–290.

327. Papazisis KT, Zambouli D, Kimoundri OT, et al. Protein tyrosine kinase inhibitor, genistein, enhances apoptosis and cell cycle arrest in K562 cells treated with gamma-irradiation. Cancer Lett 2000, 160:107–113.

328. Jeong M H, Jin YH, Kang EY, et al. The modulation of radiation-induced cell death by genistein in K562 cells: Activation of thymidine kinase 1. Cell Res 2004, 14(4):295–302.

329. Yashar CM, Spanos WJ, Taylor DD, et al. Potentiation of the radiation effect with genistein in cervical cancer cells. Gynecol Oncol 2005, 99:199–205.

330. Hillman GG, Forman JD, Kucuk O, et al. Genistein potentiates the radiation effect on prostate carcinoma cells. Clin Cancer Res 2001, 7:382–390.

331. Hillman GG Wang Y, Kucuk O, et al. Genistein potentiates inhibition of tumor growth by radiation in a prostate cancer orthotopic model. Mol Cancer Ther 2004, 3(10):1271–1279.

332. Wang Y, Raffoul JJ, Che M, et al. Prostate cancer treatment is enhanced by genistein *in vitro* and *in vivo* in a syngeneic orthotopic tumor model. Radiat Res 2006, 166:73–80.

333. Raffoul J, Wang Y, Kucuk O, et al. Genistein inhibits radiation-induced activation of NF-κB in prostate cancer cells promoting apoptosis and G2/M cell cycle arrest. BMC Cancer 2006, 6:107.

334. Annabi B, Lee YT, Martel C, et al. Radiation induced-tubulogenesis in endothelialcells is antagonized by the antiangiogenic properties of green tea polyphenol (-) epigallocatechin-3-gallate. Cancer Biol Ther 2003 2:642–649.

335. Baatout S, Jacquet P, Derradji H, et al. Study of the combined effect of X-irradiation and epigallocatechin-gallate (a tea component) on the growth inhibition and induction of apoptosis in human cancer cell lines. Oncol Rep 2004, 12(1):1591–1567.

336. Baatout S, Derradji H, Jacquet P, et al. Increased radiation sensitivity of an eosinophilic cell line following treatment with epigallocatechin-gallate, resveratrol and curcuma. Int J Mol Med 2005, 15:337–352.

337. McLaughlin N, Annabi B, Lachambre M-P, et al. Combined low dose ionizing radiation and green tea-derived epigallocatechin-3-gallate treatment induces human brain endothelial cells death. J Neurooncol 2006, 80:111–121.

338. McLaughlin N, Annabi B, Bouzeghrane M, et al. The Survivin-mediated radioresistant phenotype of glioblastomas is regulated by RhoA and inhibited by the green tea polyphenol (–)-epigallocatechin-3-gallate. Brain Res 2006, 1071:1–9.

339. Jung C, Motwani M, Kortmansky J, et al. The cyclin-dependent kinase inhibitor flavopiridol potentiates g-irradiation-induced apoptosis in colon and gastric cancer cells. Clin Cancer Res 2003, 9:6052–6061.

340. Raju U, Nakata E, Mason KA, et al. Flavopiridol, a cyclin-dependent kinase inhibitor, enhances radiosensitivity of ovarian carcinoma cells. Cancer Res 2003, 63:3263–3267.

341. Mason KA, Hunter NR. Raju U, et al. Flavopiridol increases therapeutic ratio of radiotherapy by preferentially enhancing tumor radioresponse. Int J Radiat Oncol Biol Phys 2004, 59(4):1181–1189.

342. Sato S, Kajiyama Y, Sugano M, et al. Flavopiridol as a radio-sensitizer for esophageal cancer cell lines. Dis Esophag 2004, 17:338–344.

343. Camphausen K, Brady KJ, Burgan WE, et al. Flavopiridol enhances human tumor cell radiosensitivity and prolongs expression of gH2AX foci. Mol Cancer Ther 2004, 3(4):409–416.

344. Kim J-C, Saha D, Cao Q, et al. Enhancement of radiation effects by combined docetaxel and flavopiridol treatment in lung cancer cells. Radiother Oncol 2004, 71:213–221.

345. Raju U, Ariga Hi, Koto M, et al. Improvement of esophageal adenocarcinoma cell and xenograft responses to radiation by targeting cyclin-dependent kinases. Radiother Oncol 2006, 80:185–191.

346. Newcomb EW, Lymberis SC, Lukyanov Y, et al. Radiation sensitivity of gl261 murine glioma model and enhanced radiation response by flavopiridol. Cell Cycle 2006, 5:93–99.

347. MCAleer MF, Duffy KT, Davidson WR, et al. Antisense inhibition of cyclin d1 expression is equivalent to Flavopiridol for radiosensitization of Zebrafish embryos. Int J Radiat Oncol Biol Phys 2006, 66:546–551.

Part IV
Sensitization via Targeting Apoptotic Pathways

Chapter 15
Inhibitors of the Bcl-2 Protein Family as Sensitizers to Anticancer Agents

Daniel E. Johnson

1 Introduction

Chemotherapy drugs and radiation induce cellular apoptosis by activating the intrinsic, mitochondrial-mediated pathway of apoptosis. Activation of this pathway results in activation of the caspase protease cascade, leading to the internal destruction of the cell. The intrinsic apoptosis pathway is tightly regulated by members of the Bcl-2 protein family, which consists of both pro-apoptotic and anti-apoptotic members. Pro-apoptotic members of this family stimulate the intrinsic apoptosis pathway, while anti-apoptotic members, including Bcl-2, Bcl-X$_L$, Mcl-1, Bcl-w, and A1/Bfl-1, prevent activation of the intrinsic pathway. Evidence accumulated over the past 15 years has shown that anti-apoptotic Bcl-2 family members are commonly overexpressed in both hematopoietic and solid tumor malignancies, giving rise to enhanced chemoresistance and radiation resistance in the tumor cells. Thus, considerable attention has focused on developing methodologies and reagents to inhibit the expression or function of anti-apoptotic Bcl-2 family members as a means of restoring cancer cell sensitivity to cytotoxic agents. This chapter reviews the exciting approaches currently being used to inhibit anti-apoptotic Bcl-2 family members, and the biologically active small molecules that have been discovered or designed to specifically target this important family of proteins. The abilities of these novel compounds to act as chemosensitizers are discussed.

2 Pathways of Apoptosis

Apoptosis is a genetically defined and active form of cell death [1]. During eukaryotic development, apoptosis plays an important role in sculpting and molding developing tissues and removal of excess cells [2]. For example, roughly half of human central nervous system neurons are removed via apoptosis during the first few years of development, preventing overcrowding and inappropriate target innervation. In adults, apoptosis is important for maintaining homeostasis of cell numbers, providing a counterbalance to ongoing proliferation in tissues. Additionally, apoptosis is used by the immune system to remove transformed or infected cells, as well as self-reactive lymphocytes. A regulatory feedback pathway serves to induce apoptosis in activated cytotoxic T lymphocytes and natural killer cells, which allows rapid downmodulation of immune responses. Apoptosis is also important for elimination of cells that have become damaged by insults such as chemical agents (e.g., chemotherapy drugs) or radiation. Cells that have been damaged by these types of insults can either repair the damage, or activate a pathway of apoptosis leading to their own destruction. The removal of damaged cells spares the host organism the potential for expanding a population of defective, and potentially neoplastic, cells.

Two molecular signaling pathways of apoptotic cell death have been defined: the extrinsic pathway and the intrinsic, or mitochondrial-mediated, pathway (Fig. 15.1). The extrinsic pathway of

From: *Sensitization of Cancer Cells for Chemo/Immuno/Radio-therapy, 1st Edition.*
Edited by: Benjamin Bonavida © Humana Press, Totowa, NJ

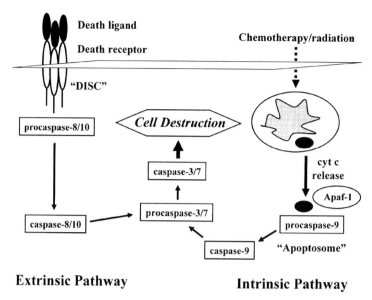

FIGURE 15.1. The extrinsic and intrinsic pathways of apoptosis

apoptosis is utilized primarily by the immune system, and may also be important for some aspects of apoptosis during development. The extrinsic pathway is activated by proteins called death ligands [1, 3]. Death ligands can be either soluble secreted proteins, or cell surface integral membrane proteins. Examples of death ligands include tumor necrosis factor (TNF), Fas ligand, and TNF-related apoptosis-inducing ligand (TRAIL). Death ligands bind to cognate receptors (death receptors) present on the surface of target cells. Following binding, the death receptors undergo trimerization, and begin recruiting a number of proteins to their cytoplasmic regions, forming a complex referred to as the death-inducing signaling complex (DISC) [1]. Among the proteins recruited to the DISC are the zymogen forms of the initiator caspases, procaspase-8 or procaspase-10 [4–6]. Recruitment of these initiator procaspases to the DISC results in their processing and activation, and once activated they cleave and activate downstream executioner procaspases, including procaspase-3 and -7 [7]. The activated executioner caspases then cleave a variety of cellular substrate proteins, leading to fragmentation of genomic DNA and internal destruction of the cell.

The intrinsic pathway of apoptosis is activated and important for death induced by a variety of insults, including chemotherapy drugs, radiation, oxidative stress, and growth factor withdrawal. The intrinsic pathway is also commonly referred to as the mitochondrial-mediated pathway since the mitochondria play a key role in propagating the death signal [1, 8]. When a cell has become damaged by chemotherapy drugs or radiation, or is responding to oxidative stress or deprivation of an essential cytokine or neurotrophic factor, the cell may choose to undergo apoptosis, a decision that likely hinges on the extent of the damage or the presence of other environmental cues. If apoptosis is chosen, then signals (typically activated proapoptotic proteins, see the following) are relayed to the mitochondria, provoking the mitochondria to release into the cytosol several apoptogenic proteins, including cytochrome c, apoptosis-inducing factor (AIF), and second mitochondria-derived activator or SMAC (9–13). Once present in the cytosol, cytochrome c associates with apoptotic protease activating factor-1 (Apaf-1) and the zymogen form of the initiator caspase, procaspase-9 (14–16). The cytochrome c/Apaf-1/procaspase-9 complex is referred to as the "apoptosome," and formation of the apoptosome leads to processing and activation of procaspase-9. Activated caspase-9 then cleaves and activates the downstream executioners, procaspase-3 and -7, leading to amplification of the caspase cascade and destruction of the cell. Activation of the executioner caspases represents the point of convergence for the extrinsic and intrinsic apoptosis pathways. However, in some cells an earlier point of cross-talk occurs. Active caspase-8 has the ability to cleave a cellular protein called Bid, and the Bid cleavage product can migrate to the mitochondria and provoke cytochrome c release [17, 18].

3 The Bcl-2 Protein Family

Proteins belonging to the Bcl-2 protein family act as key regulators of the intrinsic, mitochondrial-mediated apoptosis pathway [19]. Structurally, members of this family share homology via conserved domains referred to as Bcl-2 Homology domains, or BH domains (Fig. 15.2). Some Bcl-2 family members contain up to four different BH domains (BH1, BH2, BH3, BH4), while others contain only

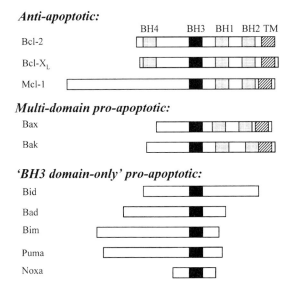

Anti-apoptotic:

Bcl-2

Bcl-X$_L$

Mcl-1

Multi-domain pro-apoptotic:

Bax

Bak

'BH3 domain-only' pro-apoptotic:

Bid

Bad

Bim

Puma

Noxa

FIGURE 15.2. The Bcl-2 protein family. A partial listing of antiapoptotic and proapoptotic members of the Bcl-2 protein family is shown. The figure depicts the BH1, BH2, BH3, BH4, and transmembrane (TM) domains

a single BH domain (typically a BH3 domain). Functionally, Bcl-2 family members possess either antiapoptotic properties or proapoptotic properties. Antiapoptotic members of this family contain multiple BH domains, and include Bcl-2 [20, 21], Bcl-X$_L$ [22], Mcl-1 [23], A1/Bfl-1 [24, 25], and Bcl-w [26]. These antiapoptotic proteins also typically contain a carboxyl-terminal membrane anchoring domain that allows for anchoring in the outer mitochondrial membrane, and to a lesser degree, membranes of the nucleus and endoplasmic reticulum. Antiapoptotic Bcl-2 family members inhibit apoptosis primarily by inhibiting the release of cytochrome c into the cytosol, thereby preventing formation of the apoptosome and activation of the caspase protease cascade [10, 11].

Proapoptotic members of the Bcl-2 protein family can be subdivided into two categories, multidomain proteins and "BH3 domain-only" proteins. The multidomain proteins, exemplified by Bax [27] and Bak [28], contain BH1, BH2, and BH3 domains. While Bak is integrally associated with the mitochondrial outer membrane, Bax is normally located in the cytoplasm. However, when cells are treated with an apoptotic stimulus Bax translocates to the mitochondria. Similarly, "BH3 domain-only" proteins are normally localized in the mitochondria, but translocate to the mitochondria in response to cellular damage or stress. Collectively, proapoptotic Bcl-2 family members promote apoptosis by promoting cytochrome c release from mitochondria.

The abilities of Bcl-2 family members to regulate cytochrome c release and the intrinsic apoptosis pathway are modulated by protein–protein interactions that occur among the different family members. Bax and Bak have been shown to homo-oligomerize, forming pores in the outer mitochondrial membrane that facilitate efflux of cytochrome c into the cytosol [29–33]. Genetic inactivation studies have shown that cells lacking expression of both Bax and Bak fail to undergo apoptosis in response to a variety of stimuli that activate the intrinsic pathway [34, 35]. Thus, Bax and Bak play critically important roles in mediating apoptosis induced by chemotherapy drugs and radiation. Antiapoptotic Bcl-2 family members, such as Bcl-2 and Bcl-X$_L$, form tight associations with both multi-domain and "BH3 domain-only" proapoptotic proteins [19]. By binding to proapoptotic Bcl-2 family members, Bcl-2 and Bcl-X$_L$ are thought to constrain and neutralize the proapoptotic activities of these proteins. Indeed, numerous studies have shown that enforced over-expression of Bcl-2 or Bcl-X$_L$ confers resistance to both chemotherapy drugs and radiation. The precise mechanism of action of the "BH3 domain-only" proteins remains a subject of intensive investigation, although several models have been proposed. In one model, certain "BH3 domain-only" proteins, namely Bid, Bim, and PUMA, act as "activators" by directly binding and inducing the oligomerization of Bax and/or Bak [36–39]. Additionally, other "BH3 domain-only" proteins, such as Bad and Noxa, act as "sensitizers," binding to antiapoptotic proteins and displacing bound "activators." In another model, the "BH3 domain-only" proteins activate Bax and Bak indirectly by binding antiapoptotic proteins and freeing up sequestered Bax and Bak [40–42].

In view of the central role that antiapoptotic Bcl-2 family members play in regulating the intrinsic apoptosis pathway, it is perhaps not surprising that their expression is commonly found to be dys-regulated in human cancers. Bcl-2, the prototypic member of this family, was initially discovered to be overexpressed as a result of a specific chromo-somal translocation (t14;18) found in patients with follicular B cell lymphoma [20, 21]. Greater than 85% of patients with this disease exhibit Bcl-2 overexpression [43, 44]. Subsequently, overex-pression of Bcl-2 has been observed in a variety of other hematopoietic and solid tumor malignan-cies, including AML, B-CLL, ALCL, multiple myeloma, melanoma SCLC, neuroblastoma, and cancers of the breast, prostate, colon, bladder, and ovaries [45]. Similarly, overexpression of Bcl-X$_L$ has been observed in AML, multiple myeloma, melanoma, squamous cell carcinoma of the head and neck, and cancers of the bladder, breast, and pancreas [45]. In many cases, overexpression of Bcl-2 and/or Bcl-X$_L$ correlates with resistance to chemotherapy or radiation and poor clinical prog-nosis. Because of this, considerable effort has been invested in developing methodologies and reagents that can be used to inhibit the expression or func-tion of these overexpressed proteins. The remain-der of this chapter describes current successes in achieving these goals.

4 Antisense Strategies to Inhibit Expression of Bcl-2 Family Members

Antisense approaches have been used extensively to downregulate the expression of antiapoptotic Bcl-2 family members in cancer cells. These stud-ies have focused largely on downregulating expres-sion of Bcl-2, although several studies have also investigated the impact of Bcl-X$_L$ or Mcl-1 down-regulation. Numerous investigations have shown that antisense-mediated downregulation of Bcl-2 serves to sensitize cancer cells to chemotherapeutic agents. Similarly, xenograft studies in immuno-deficient mice have revealed promising antitumor activities of antisense Bcl-2 molecules when used in combination with chemotherapy. However, clini-cal trials of Bcl-2 antisense have primarily demon-strated only modest anticancer activity.

The impact of Bcl-2 antisense molecules was first demonstrated in 1990 by Reed et al. [46], who found that a 20-nucleotide molecule directed against the translation start site of Bcl-2 mRNA promoted apoptosis in leukemia cell lines, an effect that was enhanced upon phosphorothioate modification of the 20-mer. Treatment of primary acute myelogenous leukemia (AML) cells with this antisense molecule resulted in enhanced sensitivity to cytosine arabinoside (Ara-C) [47–49]. A dif-ferent Bcl-2 antisense molecule, targeting codons 141–147, also enhanced the chemosensitivities of B-cell chronic lymphocytic (CLL) cell lines and primary specimens, and promoted apoptosis of SW2 lung cancer cells [50–52].

To date, most preclinical studies, and all of the clinical trials incorporating Bcl-2 antisense have utilized G3139 (also called Genasense or oblim-ersen sodium), an 18-nucleotide molecule target-ing codons 1-6 of Bcl-2 mRNA. First reported by Kitada et al. [53], G3139 efficiently downregulated Bcl-2 expression in cells harboring the t(14;18) translocation, and enhanced Ara-C- and methotrex-ate-induced killing of these cells. In severe combined immunodeficient (SCID) mice, G3139 has shown activity against Epstein-Barr virus–associated lym-phoproliferative disease [54]. Similarly, the combi-nation of G3139 and low-dose cyclophosphamide caused elimination of non-Hodgkin's lymphoma in a SCID mouse model [55]. G3139 has also shown impressive activity in xenograft models of solid tumors. In combination with doxorubicin, or a P-glycoprotein inhibitor, G3139 inhibited the growth of human breast cancer xenografts [56, 57]. In combination with cisplatin, G3139 exhibited antitumor activity against gastric cancer xenografts, and enhanced the chemosensitivity of human melanoma xenografts [58, 59]. Acting as a monotherapeutic, or in combination with paclitaxel, G3139 caused tumor shrinkage in mice bearing androgen-dependent prostate cancer [60–62].

Initial clinical trials of G3139 demonstrated moderate monotherapeutic activity against relapsing NHL [63, 64] and chemosensitizing activity against relapsed AML and ALL [65]. A phase III trial of G3139 in combination with fludarabine and cyclophosphamide resulted in increased complete response and partial response rates in patients with relapsed or refractory CLL [66]. A randomized trial of 771 patients with advanced melanoma showed that addition of G3139 to dacarbazine treatment modestly improved clinical response rates [67]. Similarly, a phase II trial of G3139 in combination with docetaxel yielded encouraging response rates in patients with androgen-independent prostate cancer [68]. Preliminary evaluation of G3139 in combination with chemotherapy has been reported in clinical trials of several other malignancies, including relapsed multiple myeloma [69], advanced breast cancer [70], metastatic colorectal cancer [71], and small cell lung cancer [72]. It should be noted that the ability of G3139 to downregulate Bcl-2 expression in tumors has been variably detected in the reported clinical trials. This has raised some suspicion that the antitumor effects of G3139 may be due, in part, to effects of this molecule that are unrelated to antisense-mediated downregulation of Bcl-2 [73]. Of particular concern is the possibility that CpG sequences in G3139 may be stimulating an immune response. Two lines of preclinical evidence suggest that induction of an immune response cannot fully explain the antitumor activity of G3139. First, a modified G3139 molecule containing methylated cytosine residues, and thereby ineffective at stimulating an immune response, retained antitumor activity against melanoma xenografts [74]. Second, unmodified G3139 has shown antitumor

activity in multiple models using immune-deficient strains of mice. Thus, the biological activity of G3139 in clinical trials may reflect some combination of specific downregulation of Bcl-2 and non-antisense-mediated effects.

The impact of antisense molecules targeting Bcl-X_L mRNA on a variety of cancer cell lines has also been reported. Bcl-X_L antisense molecules enhance the chemosensitivities of leukemia, breast, NSCLC, prostate, head and neck, colorectal, and glioblastoma cancer cell lines [75–82]. A particularly promising anticancer strategy may involve simultaneous downregulation of both Bcl-2 and Bcl-X_L, since many tumors overexpress both of these anti-apoptotic proteins. In this regard, a bispecific antisense molecule targeting both Bcl-2 and Bcl-X_L mRNA has shown substantial pro-apoptotic and chemosensitizing activity in preclinical models of lung cancer, colon cancer, breast cancer, and melanoma [83–88].

5   Structural Studies of Bcl-2 Family Members

Efforts to identify and design small molecule inhibitors of antiapoptotic Bcl-2 family members have benefited from structure–function studies and structural analyses of these proteins. Particularly useful have been those studies that have elucidated the domains and tertiary structures important for binding and neutralizing pro-apoptotic Bax, Bak, and the "BH3 domain-only" proteins. Early studies utilized deletion and site-directed mutagenesis, and demonstrated that heterodimerization between Bcl-2 and either Bax or Bak required all four BH domains present in the Bcl-2 protein [89–92]. On the other hand, only the BH3 domains of Bax or Bak were needed to form Bcl-2/Bax or Bcl-2/Bak complexes [93–97]. These studies also revealed a critical requirement for the BH3 domains of Bax or Bak for the proapoptotic activities of these proteins [93, 94, 97]. In the same way, the BH3 domains of "BH3 domain-only" proteins are both necessary and sufficient for binding to Bcl-2 and Bcl-X_L, and are absolutely required for the proapoptotic activities of these proteins [19, 98–101].

Bcl-X_L was the first anti-apoptotic Bcl-2 family member whose tertiary structure was determined.

Muchmore et al. [102] elucidated both the X-ray and NMR structure of Bcl-X$_L$, and found that the protein was comprised of two core hydrophobic alpha helices, surrounded by five shorter alpha helices. Subsequently, the NMR structures of Bcl-2 [103], Mcl-1 [104], and Bcl-w [105] have been reported, and have revealed a similar overall structure to that of Bcl-X$_L$. Importantly, in each of these three prosurvival proteins, the BH1, BH2, and BH3 domains reside in close proximity and form an elongated hydrophobic cleft. Structural determination of a complex between antiapoptotic Bcl-X$_L$ and a peptide derived from the BH3 domain of proapoptotic Bak revealed that this hydrophobic cleft serves as the docking site for the alpha-helical BH3 domains of proapoptotic proteins [106]. Hence, the cleft formed by the BH1, BH2, and BH3 domains of antiapoptotic proteins has come to be known as the BH3 binding pocket. Mutational analyses of BH3 binding pockets in Bcl-2, Bcl-X$_L$, and Mcl-1 have confirmed the importance of this region for determining binding affinities and specificities for different proapoptotic proteins.

Since antiapoptotic members of the Bcl-2 protein family suppress apoptosis, in large part, by binding and neutralizing proapoptotic proteins, attempts to develop inhibitors of the antiapoptotic proteins have focused on identifying agents that can physically disrupt the interactions between anti-apoptotic and proapoptotic proteins. The elucidation of the BH3 binding pocket as a key component of heterodimerization interactions has provided a landscape for designing and screening for compounds with the potential to disrupt these interactions. Three general strategies have evolved for identifying such compounds. In the first strategy, peptides (and derivatives thereof) derived from the BH3 domains of proapoptotic proteins have been evaluated. The second strategy has involved screening of chemical libraries to identify small organic molecules that disrupt interactions between anti-apoptotic and proapoptotic proteins in an optimized screening assay. Lastly, a third strategy has involved virtual screening algorithms to identify compounds with the potential to bind in the BH3 binding pocket of antiapoptotic Bcl-2 family members. In each of these three strategies, once a compound has been identified with the ability to disrupt heterodimerization between anti-apoptotic and proapoptotic proteins, it is usually evaluated in a cell-based assay for the ability to promote

apoptosis or the ability to sensitize cells to other apoptotic stimuli, including cytotoxic agents. If promising results are obtained, the compounds are typically investigated for antitumor activity in a xenograft model. A few of the currently identified small molecule inhibitors have now entered clinical trials. The remainder of this chapter describes the most promising small molecule inhibitors of anti-apoptotic Bcl-2 family members that have been reported, along with their antitumor, chemosensitizing and radiation-sensitizing activities.

6 BH3 Peptides

The discovery of the BH3 binding pocket in antiapoptotic Bcl-2 family members, and the realization that this pocket forms the docking site for the BH3 domains of proapoptotic proteins, led investigators to propose that peptides derived from the BH3 domains of proapoptotic proteins may be able to disrupt key heterodimerization interactions. In so doing, it was hypothesized that BH3 peptides may liberate sequestered proapoptotic proteins and sensitize cells to apoptotic stimuli. Experimental evidence has strongly supported this hypothesis.

Early studies using cell-free binding assays demonstrated that a 21-amino acid Bax BH3 peptide and a 16-amino acid Bak BH3 peptide were capable of disrupting Bax/Bcl-2 and Bax/Bcl-X$_L$ interactions [96, 98]. Related experiments showed that Bad BH3 was more effective at disrupting Bax/Bcl-X$_L$ than Bax/Bcl-2 [107]. More recent studies have clarified in detail the binding specificities of different BH3 peptides for different anti-apoptotic proteins [38, 40, 42]. These investigations have shown that peptides derived from the BH3 domains of Bid, Bim, and PUMA bind to all five of the tested antiapoptotic proteins (Bcl-2, Bcl-X$_L$, Bcl-w, Mcl-1, A1/Bfl-1), while peptides derived from Bad and Noxa exhibit more selectivity in binding [38].

To evaluate the impact of BH3 peptides on intact cells, a variety of approaches have been used to enable entry into the cell: (1) conjugation of the peptide to decanoic acid 108]; (2) fusion to peptide transduction domains such as *Drosophila* Antennapedia (Ant) or polyarginine sequences [36, 109–111]; (3) transduction using cationic lipids [107]; (4) transduction using polyethylene glycol [112]; and (5) electroporation [113]. In an early report, cell-permeable Ant-Bak BH3 peptide was found to induce caspase activation and apoptosis in HeLa cells engineered to overexpress Bcl-X$_L$ [109]. Moreover, low doses of Ant-Bak BH3 were found to sensitize cells to Fas ligand-induced apoptosis [109]. Subsequent reports confirmed the abilities of BH3 peptides to override the antiapoptotic effects resulting from Bcl-2 or Bcl-X$_L$ overexpression. Bad BH3 peptide fused to decanoic acid promoted apoptosis in HL-60 cells overexpressing Bcl-2, and inhibited the growth of leukemic cells in a SCID mouse model [108]. Cell-permeable Bax BH3 and Bak BH3 peptides induced caspase-mediated apoptosis in prostate cancer cells through disruption of Bcl-2/Bak complexes [113], while Bax BH3 and Bad BH3 peptides were found to induce apoptosis in Jurkat T leukemic cells engineered for overexpression of Bcl-2 and Bcl-X$_L$ [107]. Moreover, a BH3 peptide sensitized A2780 ovarian carcinoma cells to camptothecin [112].

The application of BH3 peptides as potential therapeutic agents is limited by several factors, including issues of solubility, size, and stability. The inherent susceptibility of peptides to cellular proteases and their poor solubility can potentially be solved through the generation of peptidomimetic or other types of derivatives. Recent evidence indicates that modifications which stabilize the alpha-helical structure of the BH3 peptide, including hydrocarbon stapling and generation of hydrogen-bond surrogates, may dramatically enhance the protease resistance and biological activities of BH3 peptides [114–116]. Of particular note, Walensky et al. [114] observed that intravenous injection of a 23-amino acid, hydrocarbon-stapled, Bid BH3 peptide potently inhibited the growth of leukemia xenografts in SCID mice. However, it remains to be determined whether considerably shorter BH3 peptides, or peptidomimetic derivatives, will prove efficacious in either xenograft models or patients.

7 Nonpeptidic, Small Molecule Inhibitors of Bcl-2 Family Members

In view of the difficulties associated with developing peptide therapeutics, there has been considerable interest in identifying and developing nonpeptidic, small organic molecules that can act as inhibitors of antiapoptotic Bcl-2 family members. Both naturally occurring and rationally designed inhibitors have been identified in the past 6 years. For the most part, these compounds exhibit inhibitory activity by binding to the BH3 binding pocket in antiapoptotic proteins.

7.1 ABT-737

Among the more promising small molecule inhibitors is ABT-737 (Fig. 15.3), developed by Abbott Laboratories [117]. ABT-737 was generated by an NMR-based strategy for identifying and optimizing compounds that bind to the BH3 binding pocket of Bcl-X$_L$. Binding assays showed that ABT-737 binds with high affinity to Bcl-X$_L$, Bcl-2, and Bcl-w, but has very weak affinity for Mcl-1 and A1/Bfl-1 [117]. ABT-737 was able to compete the binding of fluorescently labeled Bad BH3 peptide to Bcl-X$_L$ and Bcl-2 with IC$_{50}$ values of 35 and 103 nM, respectively [117]. Additionally, 0.1 µM ABT-737 was sufficient to cause nearly 50% disruption of a Gal4-Bcl-X$_L$/VP-16-Bcl-X$_S$ complex in a mammalian two-hybrid system [117]. In cell culture, ABT-737 potently induces apoptosis as a monotherapy in leukemia, lymphoma, and SCLC cell lines, and in primary cells from CLL, follicular lymphoma, multiple myeloma, and AML [117–122]. ABT-737 typically exhibits IC$_{50}$ values <1 µM, and often in the low nanomolar range, with these sensitive cell types. On the other hand, with the exception of SCLC cells, most solid tumor cell lines have shown higher resistance to ABT-737, although IC$_{50}$ values in the approximate range of 1–10 µM are not uncommon. Several studies have indicated that the sensitivity of cells to ABT-737 may depend on the levels of expression of antiapoptotic Mcl-1. Downregulation of Mcl-1 with siRNA or roscovitine treatment significantly enhances ABT-737 killing activity, while enforced overexpression of Mcl-1 dramatically inhibits ABT-737-induced apoptosis [118, 119, 123–125]. Collectively, these studies suggest that dual targeting with ABT-737 and a small molecule inhibitor of Mcl-1 may have considerable impact.

In addition to monotherapeutic activity against some cell types, ABT-737 also synergizes with cytotoxic agents in the killing of cancer cells. ABT-737 has shown synergy with: (1) paclitaxel in the killing of A549 NSCLC cells 117]; (2) Ara-C and doxorubicin in AML cell lines [119]; (3) carboplatin/etoposide in SCLC cells [125]; and (4)

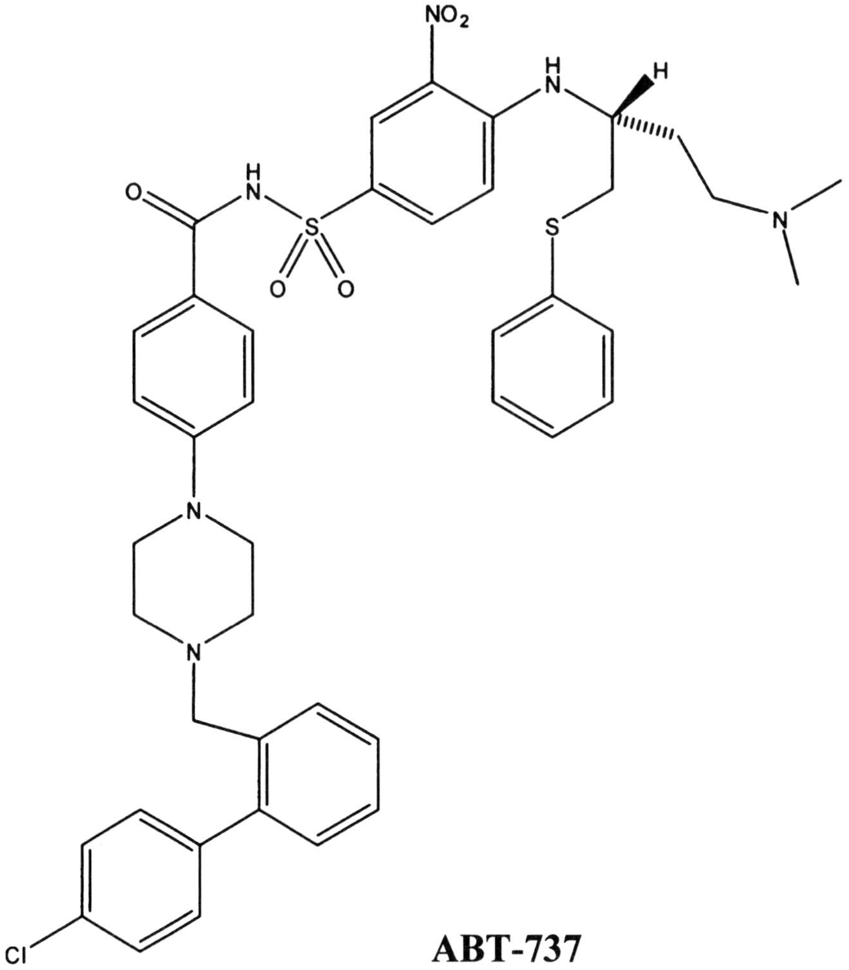

ABT-737

FIGURE 15.3. Chemical structure of ABT-737

dexamethasone and melphalan in the killing of multiple myeloma cell lines [126]. The therapeutic potential of ABT-737 as a monotherapy, or in combination with chemotherapy, has been heightened by recent experiments with mouse models. ABT-737 induced regression of SCLC xenografts in SCID mice, without causing significant side effects [117]. Similarly, ABT-737 inhibited the growth of leukemia xenografts, but did not cause significant changes in blood counts and chemistries, and did not alter the histologies of normal liver, spleen, and kidney [119]. However, both in vivo studies revealed reduced platelet counts in ABT-737–treated mice. None the less, these promising investigations underscore the importance of further testing of this compound.

7.2   Gossypol, Apogossypol, and TW-37

Recent studies have shown that gossypol (Fig. 15.4), a natural product found in the seeds and roots of cotton plants, is an inhibitor of Bcl-2 and Bcl-X_L. Using a combination of virtual screening, NMR-based binding assays, and fluorescence polarization assays, two groups determined that gossypol binds to the BH3 binding pocket of antiapoptotic proteins [127, 128]. Gossypol was found to exhibit IC_{50} values of roughly 0.5 µM for displacing fluorescent BH3 peptides from purified Bcl-2 or Bcl-X_L [127, 128]. Docking studies revealed that (−)-gossypol (see Fig. 15.4), the negative enantiomer, likely binds with much higher affinity than (+)-gossypol [127]. This is consistent with previous

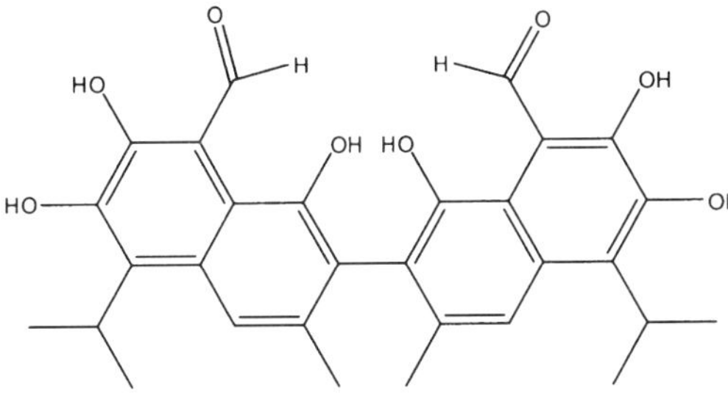

(−)–Gossypol

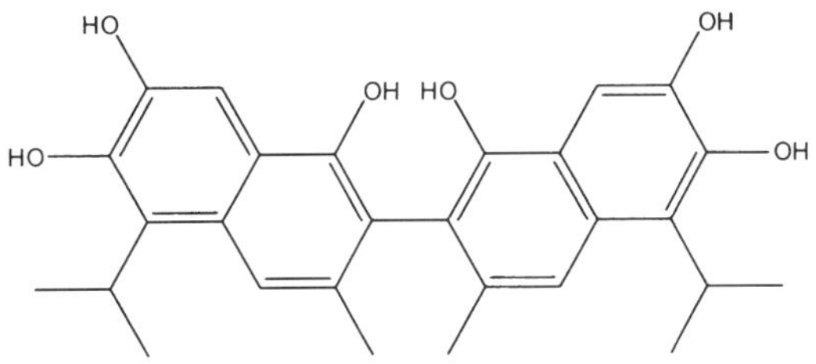

Apogossypol

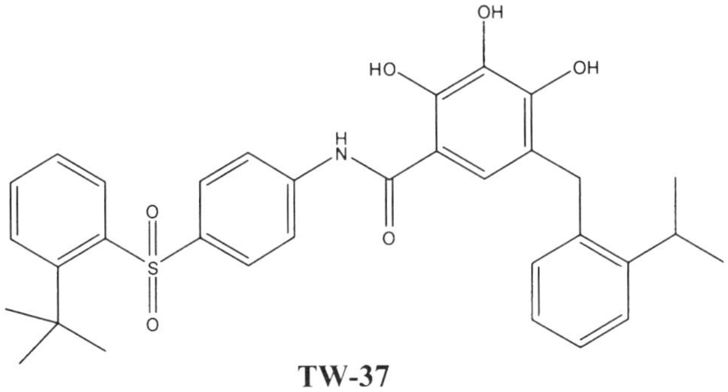

TW-37

FIGURE 15.4. Chemical structures of (−)-gossypol, apogossypol, and TW-37

observations showing that (−)-gossypol is more potently pro-apoptotic than (+)-gossypol [129].

The apoptosis inducing activity of both racemic gossypol and (−)-gossypol has been observed in vitro and in vivo using a variety of cancer cell lines representing breast, colon, prostate, pancreatic, and head and neck cancers, as well as leukemia and lymphoma [130–140]. In combination studies, gossypol has been shown to enhance the killing of solid tumor cell lines by cisplatin, genistein, radiation, and TRAIL [134, 135, 141, 142]. In vivo studies indicate that gossypol can be administered orally, and is generally well tolerated. Early clinical trials of gossypol were conducted prior to eluci-dation of the mechanism of the compound, and indicated moderate clinical benefit in recurrent gliomas [143] and metastatic adrenal cancer [144], but negligible activity against anthracycline- and taxane-resistant metastatic breast cancer [145]. These trials supported conclusions from animal studies that oral gossypol is well tolerated.

With the elucidation of the Bcl-2- and Bcl-X$_L$-inhibitory activity of gossypol, there is renewed interest in evaluating this compound and in gen-erating derivatives with more potent anti-cancer

activities and drug-like properties. Becattini et al. (146) have reported the synthesis of apogossypol (see Fig. 15.4), wherein the reactive aldehyde groups of gossypol have been removed. The apog-ossypol derivative exhibits similar docking energies (to the BH3 binding pocket) compared with gos-sypol, and similar activity against a leukemia cell line harboring the t(14;18) translocation [146]. In addition, apogossypol enhanced the activity of F-ara-A against CLL cells. A different derivative of gossypol, compound TW-37 (see Fig. 15.4), has been reported by Wang et al. [128]. TW-37 exhibits similar binding to Bcl-2, Bcl-X$_L$, and Mcl-1 compared with gossypol, and induces apoptosis in PC-3 prostate cancer cells with an IC$_{50}$ between 1 and 5 μM [128]. TW-37 also potently inhibits the growth of chemoresistant lymphoma cells in vitro and in vivo, and synergizes with MEK inhibitors in the killing of melanoma cells [147, 148].

7.3   Antimycin A$_3$

The screening of mitochondrial electron transport inhibitors for compounds capable of inducing apoptosis in Bcl-X$_L$–overexpressing cells led to the discovery of antimycin A_3 (Fig. 15.5) as an inhibitor of Bcl-X_L and Bcl-2 [149]. Antimycin A_3 is an antibiotic isolated from *Streptomyces*, and represents a subtype of the antimycin A compounds. Molecular docking experiments have determined that antimycin A_3 interacts with the BH3 binding pocket of antiapoptotic proteins, and fluorescence polarization experiments have shown that antimycin A_3 competes with Bak BH3 peptide for binding to Bcl-2 [149]. Antimycin A induces apoptosis in hepatocytes overexpressing Bcl-X_L [149], As 4.1 juxtaglomerular cells [150], and primary CLL cells [151]. The growth of HeLa cells is also inhibited by antimycin A, albeit with IC_{50} values in the range of 50 µM [152]. In mesothelioma cell lines antimycin A_3 has been reported to synergize with cisplatin [153].

7.4 HA14-1

HA14-1 (see Fig. 15.5) was identified by virtual screening of a chemical library for compounds with the potential to bind in the BH3 binding pocket of Bcl-2 [154]. Subsequent assays showed that HA14-1 competes with Bak BH3 peptide for binding to Bcl-2, with an IC_{50} of approximately 9 µM [154]. The proapoptotic activity of HA14-1 against both hematopoietic and solid

tumor cancer cells has been reported in multiple systems. HA14-1 promotes apoptosis in AML, follicular lymphoma, multiple myeloma, colon carcinoma, glioma, and prostate cancer cell lines [154–161]. In addition, HA14-1-induced apoptosis has been reported in patient-derived primary cells representing AML and CLL [151, 155]. HA14-1 has also been shown to sensitize cancer cells to a variety of agents, including MEK inhibitors, Ara-C, bortezomib, TRAIL, flavopiridol, doxorubicin, dexamethasone, sulindac, and radiation [155–159, 161–164]. However, relatively high concentrations of HA14-1 are generally required to induce apoptosis (typically 10 µM or higher), and the specificity of this compound for Bcl-2 and Bcl-X_L remains unknown.

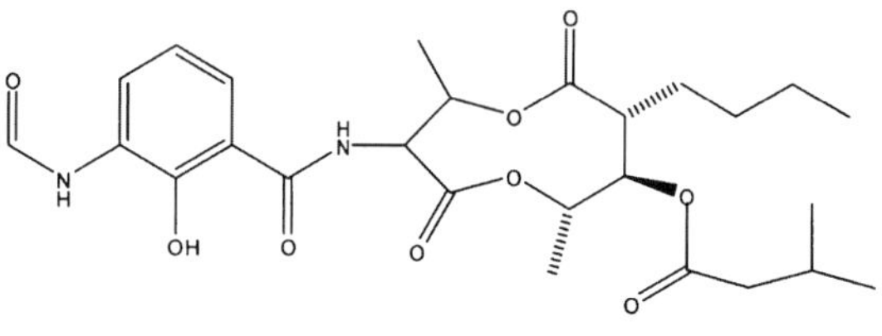

7.5 BH3Is

Degterev et al. [165] used fluorescent polarization assays to screen a library of 16,320 compounds to identify compounds capable of displacing Bak BH3 peptide from Bcl-X_L. Seven compounds were identified and called BH3 inhibitors or BH3Is. The seven BH3Is fell into two categories, with three resembling BH3I-1 and four resembling BH3I-2. The BH3I-1 and BH3I-2 lead compounds were found to disrupt Bax/Bcl-X_L interactions in intact cells, and their abilities to induce apoptosis in leukemic cells correlated with their potencies for displacing Bak BH3 peptide from Bcl-X_L [165]. Follow-up evaluations of BH3Is have been rather limited, but have demonstrated the abilities of these compounds to sensitize leukemia and prostate cancer cell lines to TRAIL, and to enhance the sensitivities of NSCLC cell lines to radiation [164, 166, 167].

Antimycin A_3

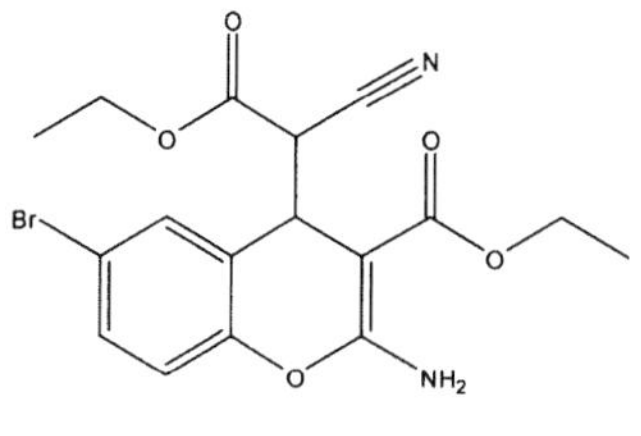

HA14-1

FIGURE 15.5. Chemical structures of antimycin A_3 and HA14-1

7.6 Tea Polyphenols

A number of studies have indicated antiproliferative and proapoptotic effects of polyphenol compounds derived from green or black tea. Particular attention has focused on the catechin compound (−)-epigallocatechin-3-gallate (EGCG). EGCG induces apoptosis in cell lines derived from melanoma, hepatocellular carcinoma, prostate cancer, and bladder cancer, and enhances the sensitivities of head and neck squamous cell carcinoma cells to

5-fluorouracil [168–172]. Leone et al. [173] have employed fluorescence polarization assays (BH3 peptide displacement) and computational docking to evaluate the binding of a large number of tea polyphenol compounds to the BH3 binding pockets of Bcl-X_L and Bcl-2. These studies have revealed that theaflavinin, theaflavin-3′ gallate, theaflavin, ECG, CG, and EGCG all exhibit binding to Bcl-X_L and Bcl-2, with Ki values for displacing BH3 peptide ranging from 120 nM to 1.23 μM. Collectively, these polyphenol compounds hold significant promise as both chemopreventive agents and as therapeutics for established cancers.

7.7 Chelerythrine and Sanguinarine

Chelerythrine and sanguinarine are naturally occurring and structurally related benzophenanthridine alkaloids. A high-throughput screen of 107,423 extracts from natural products identified chelerythrine as a Bcl-X_L binding compound [174]. Chelerythrine inhibited Bak BH3 binding to Bcl-X_L with an IC_{50} value of 1.5 μM [174]. Subsequent docking studies and NMR experiments verified the binding of both chelerythrine and sanguinarine to Bcl-X_L [175]. Interestingly, both compounds appear to bind to a site distinct from the BH3-binding pocket. Chelerythrine has demonstrated proapoptotic activity against HL-60, SH-SY5Y neuroblastoma, HCT116 colon carcinoma, and MCF7 breast cancer cell lines [174, 176]. However, these results should be interpreted with some degree of caution, since chelerythrine is a known inhibitor of protein kinase C [177]. Sanguinarine has been shown to induce apoptosis in pancreatic carcinoma cell lines and immortalized keratinocytes, and to sensitize keratinocytes to radiation [178–180].

7.8 Other Inhibitors

Various other small molecule inhibitors of antiapoptotic Bcl-2 family members have been reported, but have not been extensively characterized in biological assays. YC137 was identified by virtual screening for compounds that bind to the BH3 binding pocket of Bcl-2 [181]. In competition binding assays YC137 displaced Bid BH3 peptide from Bcl-2 with an IC_{50} of 1.3 μM [181]. YC137 induced apoptosis in FL5.12 hematopoietic

cells and breast cancer cell lines expressing high levels of Bcl-2 [181]. A-385358 was identified on the basis of NMR and fluorescence polarization assays [182]. In the absence of serum, A-385358 displaced Bad BH3 peptide from Bcl-X_L and Bcl-2 with Kis of 0.8 and 67 nM, respectively. A-385358 enhanced the chemosensitivities of a large variety of solid tumor cell lines to paclitaxel, doxorubicin, etoposide, and cisplatin [182]. Similarly, A-385358 enhanced the antitumor efficacy of paclitaxel in SCID mice harboring A549 NSCLC xenografts [182]. Lastly, GX015-070 is a novel small molecule that binds to the BH3-binding pocket. GX015-070 has been reported to induce apoptosis in 15 out of 16 myeloma cell lines, with a mean IC_{50} value of 246 nM [183]. Activity against primary cells from myeloma patients was also observed. Notably, GX015-070 induced apoptosis in myeloma cell lines that were resistant to melphalan and dexamethasone, and also enhanced the proapoptotic activities of these compounds, as well as the antimyeloma activity of the proteasome inhibitor bortezomib.

8 Conclusions

Evidence is rapidly accumulating from cell line models, animal models, and even clinical trials showing the value of inhibiting the expression or function of antiapoptotic Bcl-2 family members. Numerous studies have demonstrated monotherapeutic, or chemosensitizing and radiation-sensitizing activities of Bcl-2 inhibitors. The use of Bcl-2 antisense (G3139) in clinical trials has shown only modest positive benefit, but the development of strategies utilizing siRNA molecules may provide an improvement on this general approach. Similarly, although BH3 peptides are likely to be highly specific, the clinical application of these peptides is currently hindered by their large sizes and relative insolubilities. Further development of the BH3 peptides through the generation of helix-stabilized compounds and peptidomimetic derivatives will likely be necessary for successful translation to the clinic.

The most promising avenue for the development of inhibitors of antiapoptotic proteins is the identification and optimization of nonpeptidic, small organic molecules. Several promising lead compounds have been identified, including rationally designed molecules and molecules obtained from natural sources. Early comparative studies have indicated that the different small molecule inhibitors exhibit considerable differences in selectivity of binding to different antiapoptotic proteins and different potencies in cellular and animal models [184]. Further characterization of the potential toxicities, pharmacodynamic, and pharmacokinetic properties of these novel compounds is needed. Additionally, combined treatment with different inhibitors may be necessary to achieve potent inhibition of all the antiapoptotic Bcl-2 family members present in the cell. To date, relatively few combination studies of different Bcl-2 family inhibitors have been performed. None the less, based on encouraging preclinical studies, it is likely that the next few years will see the implementation of several important clinical trials involving these inhibitors. Given the success that has already been achieved in cell line and animal models, there is considerable enthusiasm that molecular targeting of antiapoptotic Bcl-2 family members will provide a groundbreaking, rational and feasible approach for combating chemotherapy and radiation resistance in cancer patients.

Acknowledgments. This work was supported by NIH grants CA108904 and P50-CA097190.

References

1. Danial NN, Korsmeyer SJ. Cell death: critical control points. Cell 2004, 116(2):205–219.
2. Jacobson MD, Weil M, Raff MC. Programmed cell death in animal development. Cell 1997, 88(3): 347–354.
3. Wajant H. The Fas signaling pathway: more than a paradigm. Science 2002, 296(5573):1635–1636.
4. Muzio M, Stockwell BR, Stennicke HR, et al. An induced proximity model for caspase–8 activation. The Journal of biological chemistry 1998, 273(5):2926–30.
5. Martin DA, Siegel RM, Zheng L, et al. Membrane oligomerization and cleavage activates the caspase-8 (FLICE/MACHalpha1) death signal. J Biol Chem 1998, 273(8):4345–4349.
6. Yang X, Chang HY, Baltimore D. Autoproteolytic activation of pro-caspases by oligomerization. Mol Cell 1998, 1(2):319–325.
7. Stennicke HR, Jurgensmeier JM, Shin H, et al. Pro-caspase-3 is a major physiologic target of caspase-8. J Biol Chem 1998, 273(42):27084–27090.
8. Green DR. Apoptotic pathways: ten minutes to dead. Cell 2005, 121(5):671–674.
9. Liu X, Kim CN, Yang J, et al. Induction of apoptotic program in cell-free extracts: requirement for dATP and cytochrome c. Cell 1996, 86(1):147–157.
10. Yang J, Liu X, Bhalla K, et al. Prevention of apoptosis by Bcl-2: release of cytochrome c from mitochondria blocked. Science 1997, 275(5303):1129–1132.
11. Kluck RM, Bossy-Wetzel E, Green DR, et al. The release of cytochrome c from mitochondria: a primary site for Bcl-2 regulation of apoptosis. Science 1997, 275(5303):1132–1136.
12. Susin SA, Lorenzo HK, Zamzami N, et al. Molecular characterization of mitochondrial apoptosis-inducing factor. Nature 1999, 397(6718):441–446.
13. Du C, Fang M, Li Y, et al. Smac, a mitochondrial protein that promotes cytochrome c–dependent caspase activation by eliminating IAP inhibition. Cell 2000, 102(1):33–42.
14. Zou H, Henzel WJ, Liu X, et al. Apaf-1, a human protein homologous to *C. elegans* CED-4, participates in cytochrome c-dependent activation of caspase-3. Cell 1997, 90(3):405–413.
15. Li P, Nijhawan D, Budihardjo I, et al. Cytochrome c and dATP-dependent formation of Apaf-1/caspase-9 complex initiates an apoptotic protease cascade. Cell 1997, 91(4):479–489.
16. Yoshida H, Kong YY, Yoshida R, et al. Apaf1 is required for mitochondrial pathways of apoptosis and brain development. Cell 1998, 94(6):739–750.
17. Li H, Zhu H, Xu CJ, et al. Cleavage of BID by caspase 8 mediates the mitochondrial damage in the Fas pathway of apoptosis. Cell 1998, 94(4): 491–501.
18. Luo X, Budihardjo I, Zou H, et al. Bid, a Bcl2 interacting protein, mediates cytochrome c release from mitochondria in response to activation of cell surface death receptors. Cell 1998, 94(4):481–490.
19. Cory S, Huang DC, Adams JM. The Bcl-2 family: roles in cell survival and oncogenesis. Oncogene 2003, 22(53):8590–8607.
20. Tsujimoto Y, Croce CM. Analysis of the structure, transcripts, and protein products of bcl-2, the gene involved in human follicular lymphoma. Proc Natl Acad Sci U S A 1986, 83(14):5214–5218.
21. Tsujimoto Y, Finger LR, Yunis J, et al. Cloning of the chromosome breakpoint of neoplastic B cells with the t(14, 18) chromosome translocation. Science 1984, 226(4678):1097–1099.
22. Boise LH, Gonzalez-Garcia M, Postema CE, et al. bcl-x, a bcl-2-related gene that functions as a

dominant regulator of apoptotic cell death. Cell 1993, 74(4):597–608.

23. Kozopas KM, Yang T, Buchan HL, et al. MCL1, a gene expressed in programmed myeloid cell differentiation, has sequence similarity to BCL2. Proc Natl Acad Sci U S A 1993, 90(8):3516–3520.

24. Lin EY, Orlofsky A, Berger MS, et al. Characterization of A1, a novel hemopoietic-specific early-response gene with sequence similarity to bcl-2. J Immunol 1993, 151(4):1979–1988.

25. Choi SS, Park IC, Yun JW, et al. A novel Bcl-2 related gene, Bfl-1, is overexpressed in stomach cancer and preferentially expressed in bone marrow. Oncogene 1995, 11(9):1693–1698.

26. Gibson L, Holmgreen SP, Huang DC, et al. bcl-w, a novel member of the bcl-2 family, promotes cell survival. Oncogene 1996, 13(4):665–675.

27. Oltvai ZN, Milliman CL, Korsmeyer SJ. Bcl-2 heterodimerizes in vivo with a conserved homolog, Bax, that accelerates programmed cell death. Cell 1993, 74(4):609–619.

28. Chittenden T, Harrington EA, O'Connor R, et al. Induction of apoptosis by the Bcl-2 homologue Bak. Nature 1995, 374(6524):733–736.

29. Annis MG, Soucie EL, Dlugosz PJ, et al. Bax forms multispanning monomers that oligomerize to permeabilize membranes during apoptosis. EMBO J 2005, 24(12):2096–2103.

30. Korsmeyer SJ, Wei MC, Saito M, et al. Pro-apoptotic cascade activates BID, which oligomerizes BAK or BAX into pores that result in the release of cytochrome c. Cell death and differentiation 2000, 7(12):1166–1173.

31. Wei MC, Lindsten T, Mootha VK, et al. tBID, a membrane-targeted death ligand, oligomerizes BAK to release cytochrome c. Genes Dev 2000, 14(16):2060–2071.

32. Sharpe JC, Arnoult D, Youle RJ. Control of mitochondrial permeability by Bcl-2 family members. Biochim Biophys Acta 2004, 1644(2–3):107–113.

33. Kuwana T, Mackey MR, Perkins G, et al. Bid, Bax, and lipids cooperate to form supramolecular openings in the outer mitochondrial membrane. Cell 2002, 111(3):331–342.

34. Wei MC, Zong WX, Cheng EH, et al. Pro-apoptotic BAX and BAK: a requisite gateway to mitochondrial dysfunction and death. Science 2001, 292(5517):727–730.

35. Zong WX, Lindsten T, Ross AJ, et al. BH3-only proteins that bind pro-survival Bcl-2 family members fail to induce apoptosis in the absence of Bax and Bak. Gene Dev 2001, 15(12):1481–1486.

36. Letai A, Bassik MC, Walensky LD, et al. Distinct BH3 domains either sensitize or activate mitochondrial

apoptosis, serving as prototype cancer therapeutics. Cancer Cell 2002, 2(3):183–192.

37. Kuwana T, Bouchier-Hayes L, Chipuk JE, et al. BH3 domains of BH3-only proteins differentially regulate Bax-mediated mitochondrial membrane permeabilization both directly and indirectly. Mol Cell 2005, 17(4):525–535.

38. Certo M, Del Gaizo Moore V, Nishino M, et al. Mitochondria primed by death signals determine cellular addiction to anti-apoptotic BCL-2 family members. Cancer Cell 2006, 9(5):351–365.

39. Cartron PF, Gallenne T, Bougras G, et al. The first alpha helix of Bax plays a necessary role in its ligand-induced activation by the BH3-only proteins Bid and PUMA. Mol Cell 2004, 16(5):807–818.

40. Chen L, Willis SN, Wei A, et al. Differential targeting of prosurvival Bcl-2 proteins by their BH3-only ligands allows complementary apoptotic function. Mol Cell 2005, 17(3):393–403.

41. Willis SN, Chen L, Dewson G, et al. Pro-apoptotic Bak is sequestered by Mcl-1 and Bcl-xL, but not Bcl-2, until displaced by BH3-only proteins. Gene Dev 2005, 19(11):1294–1305.

42. Willis SN, Fletcher JI, Kaufmann T, et al. Apoptosis initiated when BH3 ligands engage multiple Bcl-2 homologs, not Bax or Bak. Science 2007, 315(5813):856–859.

43. Yunis JJ, Frizzera G, Oken MM, et al. Multiple recurrent genomic defects in follicular lymphoma. A possible model for cancer. N Engl J Med 1987, 316(2):79–84.

44. Crisan D. BCL-2 gene rearrangements in lymphoid malignancies. Clin Lab Med 1996, 16(1):23–47.

45. Shangary S, Johnson DE. Recent advances in the development of anticancer agents targeting cell death inhibitors in the Bcl-2 protein family. Leukemia 2003, 17(8):1470–1481.

46. Reed JC, Stein C, Subasinghe C, et al. Antisense-mediated inhibition of BCL2 protooncogene expression and leukemic cell growth and survival: comparisons of phosphodiester and phosphorothioate oligodeoxynucleotides. Cancer Res 1990, 50(20):6565–6570.

47. Campos L, Sabido O, Rouault JP, et al. Effects of BCL-2 antisense oligodeoxynucleotides on in vitro proliferation and survival of normal marrow progenitors and leukemic cells. Blood 1994, 84(2):595–600.

48. Keith FJ, Bradbury DA, Zhu YM, et al. Inhibition of bcl-2 with antisense oligonucleotides induces apoptosis and increases the sensitivity of AML blasts to Ara-C. Leukemia 1995, 9(1):131–138.

49. Konopleva M, Tari AM, Estrov Z, et al. Liposomal Bcl-2 antisense oligonucleotides enhance proliferation,

sensitize acute myeloid leukemia to cytosine-arabinoside, and induce apoptosis independent of other anti-apoptotic proteins. Blood 2000, 95(12):3929–3938.

50. Ziegler A, Luedke GH, Fabbro D, et al. Induction of apoptosis in small-cell lung cancer cells by an antisense oligodeoxynucleotide targeting the Bcl-2 coding sequence. J Natl Cancer Inst 1997, 89(14):1027–1036.

51. Pepper C, Thomas A, Hoy T, et al. Antisense-mediated suppression of Bcl-2 highlights its pivotal role in failed apoptosis in B-cell chronic lymphocytic leukaemia. Br J Haematol 1999, 107(3):611–615.

52. Vu UE, Pavletic ZS, Wang X, et al. Increased cytotoxicity against B-chronic lymphocytic leukemia by cellular manipulations: potentials for therapeutic use. Leukemia Lymph 2000, 39(5–6):573–582.

53. Kitada S, Takayama S, De Riel K, et al. Reversal of chemoresistance of lymphoma cells by antisense-mediated reduction of bcl-2 gene expression. Antisense Res Dev 1994, 4(2):71–79.

54. Guinness ME, Kenney JL, Reiss M, et al. Bcl-2 antisense oligodeoxynucleotide therapy of Epstein-Barr virus-associated lymphoproliferative disease in severe combined immunodeficient mice. Cancer Res 2000, 60(19):5354–5358.

55. Klasa RJ, Bally MB, Ng R, et al. Eradication of human non-Hodgkin's lymphoma in SCID mice by BCL-2 antisense oligonucleotides combined with low-dose cyclophosphamide. Clin Cancer Res 2000, 6(6):2492–2500.

56. Lopes de Menezes DE, Hudon N, McIntosh N, et al. Molecular and pharmacokinetic properties associated with the therapeutics of bcl-2 antisense oligonucleotide G3139 combined with free and liposomal doxorubicin. Clin Cancer Res 2000, 6(7):2891–2902.

57. Lopes de Menezes DE, Hu Y, Mayer LD. Combined treatment of Bcl-2 antisense oligodeoxynucleotides (G3139), p-glycoprotein inhibitor (PSC833), and sterically stabilized liposomal doxorubicin suppresses growth of drug-resistant growth of drug-resistant breast cancer in severely combined immunodeficient mice. J Exp Ther Oncol 2003, 3(2):72–82.

58. Wacheck V, Heere-Ress E, Halaschek-Wiener J, et al. Bcl-2 antisense oligonucleotides chemosensitize human gastric cancer in a SCID mouse xenotransplantation model. J Mol Med (Berlin) 2001, 79(10):587–593.

59. Jansen B, Schlagbauer-Wadl H, Brown BD, et al. bcl-2 antisense therapy chemosensitizes human melanoma in SCID mice. Nat Med 1998, 4(2):232–234.

60. Gleave M, Tolcher A, Miyake H, et al. Progression to androgen independence is delayed by adjuvant treatment with antisense Bcl-2 oligodeoxynucleotides after castration in the LNCaP prostate tumor model. Clin Cancer Res 1999, 5(10):2891–2898.

61. Miyake H, Tolcher A, Gleave ME. Antisense Bcl-2 oligodeoxynucleotides inhibit progression to androgen–independence after castration in the Shionogi tumor model. Cancer Res 1999, 59(16):4030–4034.

62. Leung S, Miyake H, Zellweger T, et al. Synergistic chemosensitization and inhibition of progression to androgen independence by antisense Bcl-2 oligodeoxynucleotide and paclitaxel in the LNCaP prostate tumor model. Int J Cancer 2001, 91(6):846–850.

63. Webb A, Cunningham D, Cotter F, et al. BCL-2 antisense therapy in patients with non-Hodgkin lymphoma. Lancet 1997, 349(9059):1137–1141.

64. Waters JS, Webb A, Cunningham D, et al. Phase I clinical and pharmacokinetic study of bcl-2 antisense oligonucleotide therapy in patients with non-Hodgkin's lymphoma. J Clin Oncol 2000, 18(9):1812–1823.

65. Marcucci G, Byrd JC, Dai G, et al. Phase 1 and pharmacodynamic studies of G3139, a Bcl–2 antisense oligonucleotide, in combination with chemotherapy in refractory or relapsed acute leukemia. Blood 2003, 101(2):425–432.

66. O'Brien S, Moore JO, Boyd TE, et al. Randomized phase III trial of fludarabine plus cyclophosphamide with or without oblimersen sodium (Bcl-2 antisense) in patients with relapsed or refractory chronic lymphocytic leukemia. J Clin Oncol 2007, 25(9):1114–1120.

67. Bedikian AY, Millward M, Pehamberger H, et al. Bcl-2 antisense (oblimersen sodium) plus dacarbazine in patients with advanced melanoma: the Oblimersen Melanoma Study Group. J Clin Oncol 2006, 24(29):4738–4745.

68. Tolcher AW, Chi K, Kuhn J, et al. A phase II, pharmacokinetic, and biological correlative study of oblimersen sodium and docetaxel in patients with hormone-refractory prostate cancer. Clin Cancer Res 2005, 11(10):3854–3861.

69. Badros AZ, Goloubeva O, Rapoport AP, et al. Phase II study of G3139, a Bcl-2 antisense oligonucleotide, in combination with dexamethasone and thalidomide in relapsed multiple myeloma patients. J Clin Oncol 2005, 23(18):4089–4099.

70. Marshall J, Chen H, Yang D, et al. A phase I trial of a Bcl-2 antisense (G3139) and weekly docetaxel in patients with advanced breast cancer and other solid tumors. Ann Oncol 2004, 15(8):1274–1283.

71. Mita MM, Ochoa L, Rowinsky EK, et al. A phase I, pharmacokinetic and biologic correlative study of oblimersen sodium (Genasense, G3139) and irinotecan in patients with metastatic colorectal cancer. Ann Oncol 2006, 17(2):313–321.

72. Rudin CM, Kozloff M, Hoffman PC, et al. Phase I study of G3139, a bcl-2 antisense oligonucleotide, combined with carboplatin and etoposide in patients with small-cell lung cancer. J Clin Oncol 2004, 22(6):1110–1117.

73. Kim R, Emi M, Matsuura K, et al. Antisense and nonantisense effects of antisense Bcl-2 on multiple roles of Bcl-2 as a chemosensitizer in cancer therapy. Cancer Gene Ther 2007, 14(1):1–11.

74. Wacheck V, Krepler C, Strommer S, et al. Antitumor effect of G3139 Bcl-2 antisense oligonucleotide is independent of its immune stimulation by CpG motifs in SCID mice. Antisense Nucl Acid Drug Dev 2002, 12(6):359–367.

75. Amarante-Mendes GP, McGahon AJ, Nishioka WK, et al. Bcl-2-independent Bcr-Abl–mediated resistance to apoptosis: protection is correlated with up regulation of Bcl-xL. Oncogene 1998, 16(11):1383–1390.

76. Fennell DA, Corbo MV, Dean NM, et al. In vivo suppression of Bcl-XL expression facilitates chemotherapy-induced leukaemia cell death in a SCID/NOD-Hu model. Br J Haematol 2001, 112(3):706–713.

77. Simoes-Wust AP, Olie RA, Gautschi O, et al. Bcl-xl antisense treatment induces apoptosis in breast carcinoma cells. Int J Cancer 2000, 87(4):582–590.

78. Leech SH, Olie RA, Gautschi O, et al. Induction of apoptosis in lung-cancer cells following bcl-xL anti-sense treatment. Int J Cancer 2000, 86(4):570–576.

79. Lebedeva I, Rando R, Ojwang J, et al. Bcl–xL in prostate cancer cells: effects of overexpression and down-regulation on chemosensitivity. Cancer Res 2000, 60(21):6052–6060.

80. Sharma H, Sen S, Lo Muzio L, et al N. Antisense-mediated downregulation of anti–apoptotic proteins induces apoptosis and sensitizes head and neck squamous cell carcinoma cells to chemotherapy. Cancer Biol Ther 2005, 4(7):720–727.

81. Hayward RL, Macpherson JS, Cummings J, et al. Antisense Bcl-xl down-regulation switches the response to topoisomerase I inhibition from senescence to apoptosis in colorectal cancer cells, enhancing global cytotoxicity. Clin Cancer Res 2003, 9(7):2856–2865.

82. Guensberg P, Wacheck V, Lucas T, et al. Bcl-xL antisense oligonucleotides chemosensitize human glioblastoma cells. Chemotherapy 2002, 48(4):189–195.

83. Zangemeister-Wittke U, Leech SH, Olie RA, et al. A novel bispecific antisense oligonucleotide inhibiting both bcl-2 and bcl-xL expression efficiently induces apoptosis in tumor cells. Clin Cancer Res 2000, 6(6):2547–2555.

84. Gautschi O, Tschopp S, Olie RA, et al. Activity of a novel bcl-2/bcl-xL-bispecific antisense oligonucleotide against tumors of diverse histologic origins. J Natl Cancer Inst 2001, 93(6):463–471.

85. Simoes-Wust AP, Schurpf T, Hall J, et al. Bcl-2/bcl-xL bispecific antisense treatment sensitizes breast carcinoma cells to doxorubicin, paclitaxel and cyclophosphamide. Breast Cancer Res Treat 2002, 76(2):157–166.

86. Olie RA, Hafner C, Kuttel R, et al. Bcl-2 and bcl-xL antisense oligonucleotides induce apoptosis in melanoma cells of different clinical stages. J Invest Dermatol 2002, 118(3):505–512.

87. Strasberg Rieber M, Zangemeister-Wittke U, Rieber M. p53-Independent induction of apoptosis in human melanoma cells by a bcl-2/bcl-xL bispecific antisense oligonucleotide. Clin Cancer Res 2001, 7(5):1446–1451.

88. Del Bufalo D, Trisciuoglio D, Scarsella M, et al. Treatment of melanoma cells with a bcl-2/bcl-xL antisense oligonucleotide induces antiangiogenic activity. Oncogene 2003, 22(52):8441–8447.

89. Yin XM, Oltvai ZN, Korsmeyer SJ. BH1 and BH2 domains of Bcl-2 are required for inhibition of apoptosis and heterodimerization with Bax. Nature 1994, 369(6478):321–323.

90. Hanada M, Aime-Sempe C, Sato T, et al. Structure-function analysis of Bcl-2 protein. Identification of conserved domains important for homodimerization with Bcl-2 and heterodimerization with Bax. J Biol Chem 1995, 270(20):11962–11969.

91. Hunter JJ, Bond BL, Parslow TG. Functional dissection of the human Bcl2 protein: sequence requirements for inhibition of apoptosis. Mol Cell Biol 1996, 16(3):877–883.

92. Hirotani M, Zhang Y, Fujita N, et al. NH2-terminal BH4 domain of Bcl-2 is functional for heterodimerization with Bax and inhibition of apoptosis. J Biol Chem 1999, 274(29):20415–20420.

93. Chittenden T, Flemington C, Houghton AB, et al. A conserved domain in Bak, distinct from BH1 and BH2, mediates cell death and protein binding functions. EMBO J 1995, 14(22):5589–5596.

94. Hunter JJ, Parslow TG. A peptide sequence from Bax that converts Bcl-2 into an activator of apoptosis. J Biol Chem 1996, 271(15):8521–8524.

95. Zha H, Aime-Sempe C, Sato T, et al. Pro-apoptotic protein Bax heterodimerizes with Bcl-2 and homodimerizes with Bax via a novel domain (BH3) distinct from BH1 and BH2. J Biol Chem 1996, 271(13):7440–7444.

96. Diaz JL, Oltersdorf T, Horne W, et al. A common binding site mediates heterodimerization and homodimerization of Bcl-2 family members. J Biol Chem 1997, 272(17):11350–11355.

97. Simonen M, Keller H, Heim J. The BH3 domain of Bax is sufficient for interaction of Bax with itself and with other family members and it is required for induction of apoptosis. Eur J Biochem FEBS 1997, 249(1):85–91.

98. Ottilie S, Diaz JL, Horne W, et al. Dimerization properties of human BAD. Identification of a BH-3 domain and analysis of its binding to mutant BCL-2 and BCL-XL proteins. J Biol Chem 1997, 272(49):30866–30872.

99. Zha J, Harada H, Osipov K, et al. BH3 domain of BAD is required for heterodimerization with BCL-XL and pro-apoptotic activity. J Biol Chem 1997, 272(39):24101–24104.

100. O'Connor L, Strasser A, O'Reilly LA, et al. Bim: a novel member of the Bcl-2 family that promotes apoptosis. EMBO J 1998, 17(2):384–395.

101. Yu J, Zhang L, Hwang PM, et al. PUMA induces the rapid apoptosis of colorectal cancer cells. Mol Cell 2001, 7(3):673–682.

102. Muchmore SW, Sattler M, Liang H, et al. X-ray and NMR structure of human Bcl-xL, an inhibitor of programmed cell death. Nature 1996, 381(6580): 335–341.

103. Petros AM, Medek A, Nettesheim DG, et al. Solution structure of the anti-apoptotic protein bcl-2. Proc Natl Acad Sci U S A 2001, 98(6):3012–3017.

104. Day CL, Chen L, Richardson SJ, et al. Solution structure of prosurvival Mcl-1 and characterization of its binding by pro-apoptotic BH3-only ligands. J Biol Chem 2005, 280(6):4738–4744.

105. Denisov AY, Madiraju MS, Chen G, et al. Solution structure of human BCL-w: modulation of ligand binding by the C-terminal helix. J Biol Chem 2003, 278(23):21124–21128.

106. Sattler M, Liang H, Nettesheim D, et al. Structure of Bcl-xL-Bak peptide complex: recognition between regulators of apoptosis. Science 1997, 275(5302):983–986.

107. Shangary S, Johnson DE. Peptides derived from BH3 domains of Bcl-2 family members: a comparative analysis of inhibition of Bcl-2, Bcl-x(L) and Bax oligomerization, induction of cytochrome c release, and activation of cell death. Biochemistry 2002, 41(30):9485–9495.

108. Wang JL, Zhang ZJ, Choksi S, et al. Cell permeable Bcl-2 binding peptides: a chemical approach to apoptosis induction in tumor cells. Cancer Res 2000, 60(6):1498–1502.

109. Holinger EP, Chittenden T, Lutz RJ. Bak BH3 peptides antagonize Bcl-xL function and induce apoptosis through cytochrome c-independent activation of caspases. J Biol Chem 1999, 274(19):13298–13304.

110. Vieira HL, Boya P, Cohen I, et al. Cell permeable BH3-peptides overcome the cytoprotective effect of Bcl-2 and Bcl-X(L). Oncogene 2002, 21(13):1963–1977.

111. Shangary S, Oliver CL, Tillman TS, et al. Sequence and helicity requirements for the pro-apoptotic activity of Bax BH3 peptides. Mol Cancer Ther 2004, 3(11):1343–1354.

112. Dharap SS, Chandna P, Wang Y, et al. Molecular targeting of BCL2 and BCLXL proteins by synthetic BCL2 homology 3 domain peptide enhances the efficacy of chemotherapy. J Pharmacol Exp Ther 2006, 316(3):992–998.

113. Finnegan NM, Curtin JF, Prevost G, et al. Induction of apoptosis in prostate carcinoma cells by BH3 peptides which inhibit Bak/Bcl-2 interactions. Br J Cancer 2001, 85(1):115–121.

114. Walensky LD, Kung AL, Escher I, et al. Activation of apoptosis in vivo by a hydrocarbon-stapled BH3 helix. Science 2004, 305(5689):1466–1470.

115. Wang D, Liao W, Arora PS. Enhanced metabolic stability and protein-binding properties of artificial alpha helices derived from a hydrogen-bond surrogate: application to Bcl–xL. Angewandte Chemie 2005, 44(40):6525–6529.

116. Wang D, Chen K, Kulp Iii JL, et al. Evaluation of biologically relevant short alpha-helices stabilized by a main-chain hydrogen-bond surrogate. J Amer Chem Soc 2006, 128(28):9248–9256.

117. Oltersdorf T, Elmore SW, Shoemaker AR, et al. An inhibitor of Bcl-2 family proteins induces regression of solid tumours. Nature 2005, 435(7042):677–681.

118. van Delft MF, Wei AH, Mason KD, et al. The BH3 mimetic ABT-737 targets selective Bcl-2 proteins and efficiently induces apoptosis via Bak/Bax if Mcl-1 is neutralized. Cancer Cell 2006, 10(5): 389–399.

119. Konopleva M, Contractor R, Tsao T, et al. Mechanisms of apoptosis sensitivity and resistance to the BH3 mimetic ABT-737 in acute myeloid leukemia. Cancer Cell 2006, 10(5):375–388.

120. Kuroda J, Puthalakath H, Cragg MS, et al. Bim and Bad mediate imatinib–induced killing of Bcr/Abl+ leukemic cells, and resistance due to their loss is overcome by a BH3 mimetic. Proc Natl Acad Sci U S A 2006, 103(40):14907–14912.

121. Chauhan D, Velankar M, Brahmandam M, et al. A novel Bcl-2/Bcl-X(L)/Bcl-w inhibitor ABT-737 as therapy in multiple myeloma. Oncogene 2007, 26(16):2374–2380.

122. Del Gaizo Moore V, Brown JR, Certo M, et al. Chronic lymphocytic leukemia requires BCL2 to sequester prodeath BIM, explaining sensitivity to BCL2 antagonist ABT–737. J Clin Invest 2007, 117(1):112–121.

123. Lin X, Morgan-Lappe S, Huang X, et al. 'Seed' analysis of off-target siRNAs reveals an essential role of Mcl-1 in resistance to the small-molecule Bcl-2/Bcl-X(L) inhibitor ABT-737. Oncogene 2006.

124. Chen S, Dai Y, Harada H, et al. Mcl-1 down-regulation potentiates ABT-737 lethality by cooperatively inducing Bak activation and Bax translocation. Cancer Res 2007, 67(2):782–791.

125. Tahir SK, Yang X, Anderson MG, et al. Influence of Bcl-2 family members on the cellular response of small-cell lung cancer cell lines to ABT-737. Cancer Res 2007, 67(3):1176–1183.

126. Trudel S, Stewart AK, Li Z, et al. The Bcl-2 family protein inhibitor, ABT-737, has substantial antimyeloma activity and shows synergistic effect with dexamethasone and melphalan. Clin Cancer Res 2007, 13(2 Pt 1):621–629.

127. Kitada S, Leone M, Sareth S, et al. Discovery, characterization, and structure-activity relationships studies of pro-apoptotic polyphenols targeting B-cell lymphocyte/leukemia-2 proteins. J Med Chem 2003, 46(20):4259–4264.

128. Wang G, Nikolovska-Coleska Z, Yang CY, et al. Structure-based design of potent small–molecule inhibitors of anti-apoptotic Bcl-2 proteins. J Med Chem 2006, 49(21):6139–6142.

129. Qiu J, Levin LR, Buck J, et al. Different pathways of cell killing by gossypol enantiomers. Exp Biol Med 2002, 227(6):398–401.

130. Gilbert NE, O'Reilly JE, Chang CJ, et al. Antiproliferative activity of gossypol and gossypolone on human breast cancer cells. Life Sci 1995, 57(1):61–67.

131. Thomas M, von Hagen V, Moustafa Y, et al. Effects of gossypol on the cell cycle phases in T-47D human breast cancer cells. AntiCancer Res 1991, 11(4):1469–1475.

132. Wang X, Wang J, Wong SC, et al. Cytotoxic effect of gossypol on colon carcinoma cells. Life Sci 2000, 67(22):2663–2671.

133. Zhang M, Liu H, Guo R, et al. Molecular mechanism of gossypol-induced cell growth inhibition and cell death of HT-29 human colon carcinoma cells. Biochem Pharmacol 2003, 66(1):93–103.

134. Xu L, Yang D, Wang S, et al. (−)-Gossypol enhances response to radiation therapy and results in tumor regression of human prostate cancer. Mol Cancer Ther 2005, 4(2):197–205.

135. Mohammad RM, Wang S, Banerjee S, et al. Nonpeptidic small-molecule inhibitor of Bcl-2 and Bcl-XL, (−)-Gossypol, enhances biological effect of genistein against BxPC-3 human pancreatic cancer cell line. Pancreas 2005, 31(4):317–324.

136. Oliver CL, Bauer JA, Wolter KG, et al. In vitro effects of the BH3 mimetic, (−)-gossypol, on head and neck squamous cell carcinoma cells. Clin Cancer Res 2004, 10(22):7757–7763.

137. Wolter KG, Wang SJ, Henson BS, et al. (−)-Gossypol inhibits growth and promotes apoptosis of human head and neck squamous cell carcinoma in vivo. Neoplasia 2006, 8(3):163–172.

138. Le Blanc M, Russo J, Kudelka AP, et al. An in vitro study of inhibitory activity of gossypol, a cottonseed extract, in human carcinoma cell lines. Pharmacol Res 2002, 46(6):551–555.

139. Mohammad RM, Wang S, Aboukameel A, et al. Preclinical studies of a nonpeptidic small-molecule inhibitor of Bcl-2 and Bcl-X(L) [(−)-gossypol] against diffuse large cell lymphoma. Mol Cancer Ther 2005, 4(1):13–21.

140. Oliver CL, Miranda MB, Shangary S, et al. (−)-Gossypol acts directly on the mitochondria to overcome Bcl-2- and Bcl-X(L)-mediated apoptosis resistance. Mol Cancer Ther 2005, 4(1):23–31.

141. Bauer JA, Trask DK, Kumar B, et al. Reversal of cisplatin resistance with a BH3 mimetic, (−)-gossypol, in head and neck cancer cells: role of wild-type p53 and Bcl-xL. Mol Cancer Ther 2005, 4(7):1096–1104.

142. Yeow WS, Baras A, Chua A, et al. Gossypol, a phytochemical with BH3-mimetic property, sensitizes cultured thoracic cancer cells to Apo2 ligand/tumor necrosis factor-related apoptosis-inducing ligand. J Thorac Cardiovasc Surg 2006, 132(6):1356–1362.

143. Bushunow P, Reidenberg MM, Wasenko J, et al. Gossypol treatment of recurrent adult malignant gliomas. J Neuro-oncol 1999, 43(1):79–86.

144. Flack MR, Pyle RG, Mullen NM, et al. Oral gossypol in the treatment of metastatic adrenal cancer. J Clin Endocrinol Metab 1993, 76(4):1019–1024.

145. Van Poznak C, Seidman AD, Reidenberg MM, et al. Oral gossypol in the treatment of patients with refractory metastatic breast cancer: a phase I/II clinical trial. Breast Cancer Res Treat 2001, 66(3):239–248.

146. Becattini B, Kitada S, Leone M, et al. Rational design and real time, in-cell detection of the pro-apoptotic activity of a novel compound targeting Bcl-X(L). Chem Biol 2004, 11(3):389–395.

147. Mohammad RM, Goustin AS, Aboukameel A, et al. Preclinical studies of TW-37, a new nonpeptidic small-molecule inhibitor of Bcl-2, in diffuse large cell lymphoma xenograft model reveal drug action on both Bcl-2 and Mcl-1. Clin Cancer Res 2007, 13(7):2226–2235.

148. Verhaegen M, Bauer JA, Martin de la Vega C, et al. A novel BH3 mimetic reveals a mitogen-activated

protein kinase-dependent mechanism of melanoma cell death controlled by p53 and reactive oxygen species. Cancer Res 2006, 66(23):11348–113 59.

149. Tzung SP, Kim KM, Basanez G, et al. Antimycin A mimics a cell-death-inducing Bcl-2 homology domain 3. Nat Cell Biol 2001, 3(2):183–191.

150. Park WH, Han YW, Kim SW, et al. Antimycin A induces apoptosis in As4.1 juxtaglomerular cells. Cancer Lett 2006.

151. Campas C, Cosialls AM, Barragan M, et al. Bcl-2 inhibitors induce apoptosis in chronic lymphocytic leukemia cells. Exper Hematol 2006, 34(12): 1663–1669.

152. Park WH, Han YW, Kim SH, et al. An ROS generator, antimycin A, inhibits the growth of HeLa cells via apoptosis. J Cell Biochem 2007.

153. Cao X, Rodarte C, Zhang L, et al. Bcl2/bcl-x(L) inhibitor engenders apoptosis and increases chemosensitivity in mesothelioma. Cancer Biol Ther 2007, 6(2):246–252.

154. Wang JL, Liu D, Zhang ZJ, et al. Structure–based discovery of an organic compound that binds Bcl-2 protein and induces apoptosis of tumor cells. Proc Natl Acad Sci U S A 2000, 97(13):7124–7129.

155. Milella M, Estrov Z, Kornblau SM, et al. Synergistic induction of apoptosis by simultaneous disruption of the Bcl-2 and MEK/MAPK pathways in acute myelogenous leukemia. Blood 2002, 99(9): 3461–3464.

156. Skommer J, Wlodkowic D, Matto M, et al. HA14-1, a small molecule Bcl-2 antagonist, induces apoptosis and modulates action of selected anticancer drugs in follicular lymphoma B cells. Leukemia Res 2006, 30(3):322–331.

157. Pei XY, Dai Y, Grant S. The small-molecule Bcl-2 inhibitor HA14-1 interacts synergistically with flavopiridol to induce mitochondrial injury and apoptosis in human myeloma cells through a free radical-dependent and Jun NH2-terminal kinasedependent mechanism. Mol Cancer Ther 2004, 3(12):1513–1524.

158. Sinicrope FA, Penington RC, Tang XM. Tumor necrosis factor-related apoptosis-inducing ligand-induced apoptosis is inhibited by Bcl-2 but restored by the small molecule Bcl-2 inhibitor, HA 14-1, in human colon cancer cells. Clin Cancer Res 2004, 10(24):8284–8292.

159. Sinicrope FA, Penington RC. Sulindac sulfide-induced apoptosis is enhanced by a small-molecule Bcl-2 inhibitor and by TRAIL in human colon cancer cells overexpressing Bcl-2. Mol Cancer Ther 2005, 4(10):1475–1483.

160. Manero F, Gautier F, Gallenne T, et al. The small organic compound HA14-1 prevents Bcl-2 interaction with Bax to sensitize malignant glioma

cells to induction of cell death. Cancer Res 2006, 66(5):2757–2764.

161. An J, Chervin AS, Nie A, et al. Overcoming the radioresistance of prostate cancer cells with a novel Bcl-2 inhibitor. Oncogene 2007, 26(5):652–661.

162. Lickliter JD, Wood NJ, Johnson L, et al. HA14-1 selectively induces apoptosis in Bcl-2-overexpressing leukemia/lymphoma cells, and enhances cytarabine-induced cell death. Leukemia 2003, 17(11):2074–2080.

163. Pei XY, Dai Y, Grant S. The proteasome inhibitor bortezomib promotes mitochondrial injury and apoptosis induced by the small molecule Bcl-2 inhibitor HA14-1 in multiple myeloma cells. Leukemia 2003, 17(10):2036–2045.

164. Hao JH, Yu M, Liu FT, et al. Bcl-2 inhibitors sensitize tumor necrosis factor-related apoptosis-inducing ligand-induced apoptosis by uncoupling of mitochondrial respiration in human leukemic CEM cells. Cancer Res 2004, 64(10):3607–3616.

165. Degterev A, Lugovskoy A, Cardone M, et al. Identification of small-molecule inhibitors of interaction between the BH3 domain and Bcl-xL. Nat Cell Biol 2001, 3(2):173–182.

166. Ray S, Bucur O, Almasan A. Sensitization of prostate carcinoma cells to Apo2L/TRAIL by a Bcl-2 family protein inhibitor. Apoptosis 2005, 10(6):1411–1418.

167. Roa W, Chen H, Alexander A, et al. Enhancement of radiation sensitivity with BH3I-1 in non-small cell lung cancer. Clin Invest Med 2005, 28(2): 55–63.

168. Nihal M, Ahmad N, Mukhtar H, Wood GS. Antiproliferative and pro-apoptotic effects of (−)-epigallocatechin-3-gallate on human melanoma: possible implications for the chemoprevention of melanoma. Int J Cancer 2005, 114(4):513–521.

169. Nishikawa T, Nakajima T, Moriguchi M, et al. A green tea polyphenol, epigalocatechin-3-gallate, induces apoptosis of human hepatocellular carcinoma, possibly through inhibition of Bcl-2 family proteins. J Hepatol 2006, 44(6):1074–1082.

170. Chung LY, Cheung TC, Kong SK, et al. Induction of apoptosis by green tea catechins in human prostate cancer DU145 cells. Life Sci 2001, 68(10): 1207–1214.

171. Qin J, Xie LP, Zheng XY, et al. A component of green tea, (−)-epigallocatechin-3-gallate, promotes apoptosis in T24 human bladder cancer cells via modulation of the PI3K/Akt pathway and Bcl-2 family proteins. Biochem Biopphys Res Commun 2007, 354(4):852–857.

172. Masuda M, Suzui M, Weinstein IB. Effects of epigallocatechin-3-gallate on growth, epidermal growth factor receptor signaling pathways, gene

expression, and chemosensitivity in human head and neck squamous cell carcinoma cell lines. Clin Cancer Res 2001, 7(12):4220–4229.

173. Leone M, Zhai D, Sareth S, et al. Cancer prevention by tea polyphenols is linked to their direct inhibition of anti-apoptotic Bcl-2-family proteins. Cancer Res 2003, 63(23):8118–8121.

174. Chan SL, Lee MC, Tan KO, et al. Identification of chelerythrine as an inhibitor of BclXL function. J Biol Chem 2003, 278(23):20453–20456.

175. Zhang YH, Bhunia A, Wan KF, et al. Chelerythrine and sanguinarine dock at distinct sites on BclXL that are not the classic BH3 binding cleft. J Mol Biol 2006, 364(3):536–549.

176. Jarvis WD, Turner AJ, Povirk LF, et al. Induction of apoptotic DNA fragmentation and cell death in HL-60 human promyelocytic leukemia cells by pharmacological inhibitors of protein kinase C. Cancer Res 1994, 54(7):1707–1714.

177. Herbert JM, Augereau JM, Gleye J, et al. Chelerythrine is a potent and specific inhibitor of protein kinase C. Biochem Biophys Res Commun 1990, 172(3):993–999.

178. Ahsan H, Reagan-Shaw S, Breur J, et al. Sanguinarine induces apoptosis of human pancreatic carcinoma AsPC-1 and BxPC-3 cells via

modulations in Bcl-2 family proteins. Cancer Lett 2007, 249(2):198–208.

179. Adhami VM, Aziz MH, Mukhtar H, et al. Activation of prodeath Bcl-2 family proteins and mitochondrial apoptosis pathway by sanguinarine in immortalized human HaCaT keratinocytes. Clin Cancer Res 2003, 9(8):3176–3182.

180. Reagan-Shaw S, Breur J, Ahmad N. Enhancement of UVB radiation-mediated apoptosis by sanguinarine in HaCaT human immortalized keratinocytes. Mol Cancer Ther 2006, 5(2):418–429.

181. Real PJ, Cao Y, Wang R, et al. Breast cancer cells can evade apoptosis-mediated selective killing by a novel small molecule inhibitor of Bcl-2. Cancer Res 2004, 64(21):7947–7953.

182. Shoemaker AR, Oleksijew A, Bauch J, et al. A small-molecule inhibitor of Bcl–XL potentiates the activity of cytotoxic drugs in vitro and in vivo. Cancer Res 2006, 66(17):8731–8739.

183. Trudel S, Li ZH, Rauw J, et al. Pre-clinical studies of the pan-Bcl inhibitor obatoclax (GX015-070) in multiple myeloma. Blood 2007.

184. Zhai D, Jin C, Satterthwait AC, et al. Comparison of chemical inhibitors of anti-apoptotic Bcl-2-family proteins. Cell Death Diff 2006, 13(8): 1419–1421.

Chapter 16
Therapeutic Targeting of Apoptosis in Cancer

Timothy R. Wilson, Daniel B. Longley, and Patrick G. Johnston

1   Introduction

Evasion of apoptosis is one the six critical alterations in cell physiology that result in the progression of a "normal" cell into a metastatic tumor [1]. Furthermore, defects in the apoptotic pathway may limit the effectiveness of chemotherapies used in the treatment of cancer. When a tumor fails to respond to chemotherapy, it becomes drug resistant. Tumors may be inherently resistant to chemotherapies (intrinsic), or may become resistant over the course of a treatment regimen (acquired). Unfortunately, tumors that have acquired resistance to one chemotherapy are often resistant to other chemotherapies with different mechanisms of action. Clearly, if drug resistance could be overcome the survival benefits for patients would be immense.

Tumors may be resistant to chemotherapy for many reasons. The amount of active drug reaching the tumor may be limited, for example due to increased drug efflux by multi-drug resistant proteins, which can actively transport cytotoxic drugs out of cells. Drug inactivation can also occur: 80% of 5-fluorouracil is degraded by dihydropyrimidine dehydrogenase in the liver before it reaches the tumor. Drug resistance can also be caused by defects in response to the damage caused by the anticancer agents, such as mutations in mismatch repair genes *hMLH1* and *hMSH2* that result in increased tolerance to certain DNA-damaging drugs. Many factors enable a tumor cell to evade apoptosis, such as increased expression of anti-apoptotic proteins,

such as Bcl-2, cellular FLICE-inhibitory protein (c-FLIP) and inhibitors of apoptosis proteins (IAPs), or decreased expression of pro-apoptotic proteins, such as Fas and Bax. This chapter focuses on apoptotic pathways and discusses how the apoptotic machinery may be targeted to enhance the effectiveness of commonly used chemotherapies.

1.1   Apoptosis

Apoptosis, or programmed cell death, is a genetically regulated and active process, in which an individual cell responds to an internal or external signal by committing suicide [2]. Apoptosis is a vital part of life in multicellular organisms, playing key roles in embryonic development and normal tissue homeostasis [3]. In adults, approximately 10 billion cells die daily to maintain the balance with the production of new cells arising from the body's stem cell population [4]. Apoptosis is distinguished from necrotic cell death by the absence of an associated inflammatory response, and is characterized by chromatin condensation, DNA fragmentation, plasma membrane blebbing, and cell shrinkage [5]. Eventually, the cell breaks into small membrane-bound fragments (apoptotic bodies) that are cleared by a process known as phagocytosis. As mentioned, dysregulation of apoptosis is associated with cancer development, and has also been shown to be associated with AIDS, neurodegenerative diseases, and ischemic stroke [6]. Apoptosis can be divided into a series of events.

The first event being the initiation of apoptosis by stimuli such as DNA damage, hypoxia, UV radiation, growth factor withdrawal, heat shock, and death receptor activation [7]. Second, the initial stimuli must provoke an intracellular signal. This is mediated by the activation of cysteine proteases (caspases), which are able to amplify the signal, causing cleavage of key proteins, resulting in their inactivation (anti-apoptotic proteins) or activation (pro-apoptotic proteins) [8]. Next, intracellular calcium stores are released, ATP is depleted, and the morphologic characteristics of apoptosis are seen. A crucial step appears to be the loss of cell membrane asymmetry leading to extracellular exposure of anionic phospholipids and phosphatidylserine [9]. Eventually, the plasma membrane of the cell begins to bud off and encapsulate the contents of the cell to form apoptotic bodies, which must be recognized and engulfed by the phagocytic cells to prevent an inflammatory response. There are two distinct pathways that can initiate the caspase cascade, resulting in apoptosis: the intrinsic (or mitochondrial) pathway and the extrinsic (or death receptor) pathway (Fig. 16.1).

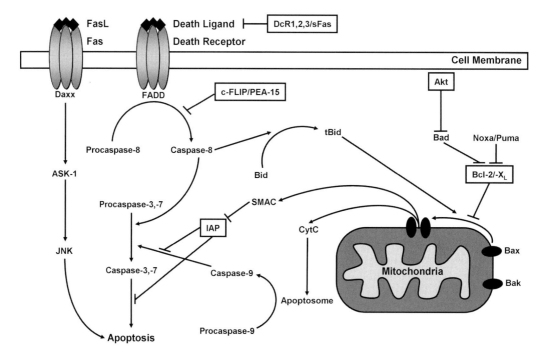

FIGURE 16.1. The intrinsic and extrinsic apoptotic pathway. When death receptors, such as Fas, DR4, and DR5, are bound by their ligand, they recruit the adapter molecule FADD. FADD can in turn recruit caspase-8 molecules that become cleaved and activated to initiate apoptosis. Activated caspase-8 can directly cleave executioner caspases-3 and -7, or cleave its substrate Bid. Alternatively, Fas ligation can result in the recruitment of Daxx, and subsequent activation of ASK-1. Activated ASK-1 in turn phosphorylates JNK, resulting in apoptosis. Initiation of apoptosis by the mitochondria is controlled by the homodimerization and heterodimerization of the Bcl-2 family members Bak and Bax, which form a pore large enough for the release of cytochrome c and SMAC. Cytochrome c, together with APAF-1 and procaspase-9 form the apoptosome, which results in the activation of caspase-9. Activated caspase-9 can subsequently activate the executioner caspases-3 and -7. SMAC can bind the IAPs to prevent their inhibitory effect and promote apoptosis. Regulation of the intrinsic pathway is tightly controlled. Bcl-2 and Bcl-X_L bind to and inhibit dimerization of Bax and Bak. Upon an apoptotic stimulus, Bad, Noxa, and Puma translocate to the mitochondria and bind Bcl-2 and Bcl-X_L to prevent their anti-apoptotic function. Cleaved Bid can initiate Bax and Bak dimerization, thereby providing cross-talk between the extrinsic and intrinsic pathways. Various apoptosis inhibitors exist: The decoy receptors prevent death receptor ligation; c-FLIP and PEA-15 bind to FADD at the DISC and prevent caspase-8 activation; the IAPs bind to caspases-9, -3, and -7 to prevent their full activation; activated Akt can phosphorylate Bad to prevent its inhibition on Bcl–X_L

1.2   Intrinsic Pathway

Mitochondria are affected early in the apoptotic process and are now thought to act as central coordinators of cell death [10]. Several factors can induce mitochondrial-mediated apoptosis, including chemotherapy, UV, DNA damage, reactive oxygen species, and growth factor withdrawal. Mitochondria are organelles that consist of a matrix surrounded by the inner membrane (IM), the inner membrane space and the outer membrane (OM). The IM contains various molecules involved in respiration, with the OM containing a voltage-dependent ion channel. The inner membrane space contains several molecules such as cytochrome c, second mitochondria–derived activator of caspases (SMAC)/direct IAP binding protein with low Pi (DIABLO), and apoptosis inducing factor (AIF). Mitochondrial OM permeabilization is a critical event for initiating the release of these molecules into the cytoplasm. The release of cytochrome c promotes the formation of the apoptosome and activation of caspase-9, the initiator caspase of mitochondrial-mediated apoptosis [11]. The components of the apoptosome include cytochrome c, ATP, an adapter molecule termed APAF-1 (apoptotic protease-activating factor) and procaspase-9. In the apoptosome, procaspase-9 is cleaved and activated, with only the caspase-9 molecules bound to the apoptosome able to cleave and activate downstream executioner caspases such as caspase-3 and -7 [12].

The Bcl-2 family of proteins regulates the efflux of pro-apoptotic proteins from the mitochondria and is comprised of over a dozen proteins, which have been classified into three functional groups. Members of the first group, such as Bcl-2 and Bcl-X$_L$, are characterized by four short conserved Bcl-2 homology (BH) domains (BH1–BH4) and all possess anti-apoptotic activity. They also possess a C-terminal hydrophobic tail that localizes to the outer surface of the mitochondria, or occasionally the endoplasmic reticulum, with the protein facing out toward the cytoplasm [13]. The second group is pro-apoptotic and has a similar structure to group I proteins, containing BH1–BH3 domains and the hydrophobic tail, but not the BH4 domain at the N-terminus [14]. Members of this group include Bax and Bak. Group III is a large and diverse family that contains an amphipathic helix BH3 domain, which interacts with other Bcl-2 family members [15]. These "BH3-only" molecules often reside in the cytoplasm, where they can translocate to the mitochondria in response to an apoptotic stimulus. Members of this subgroup include Bik, Bad, Bid, Bim, Noxa, and Puma [16].

The pro- and anti-apoptotic Bcl-2 family proteins can physically bind to each other, forming a complex network of homodimers and heterodimers. It is the relative expression of these proteins to one another that determines cellular fate. Bcl-2 and Bcl-X$_L$ are anchored to the mitochondrial OM and suppress apoptosis by forming a heterodimeric complex with Bax, preventing the formation of Bax homodimers [17]. Bax, which is present as a monomer in the cytoplasm, translocates to the mitochondria, where it can form a dimer or higher-order oligomer that promotes cytochrome c release. Bak can also associate with the OM in the presence of an apoptotic stimulus [18]. Bim, present in microtubules, also translocates to the OM during apoptosis, where it can bind and activate Bax or Bak. Phosphorylated Bad is sequestered in the cytoplasm and, at the onset of apoptosis, becomes dephosphorylated and translocates to the OM, where it binds and heterodimerizes with anti-apoptotic family member, Bcl-X$_L$, thereby inhibiting the anti-apoptotic effect of Bcl-X$_L$ on Bax [19]. Bid, which can be truncated by caspase-8 (providing cross-talk between the extrinsic and intrinsic pathways), translocates to the mitochondria, where it binds to and activates Bak or Bax. The BH3-only proteins are suggested to act as critical death-inducing ligands via interactions with Bak or Bax at the OM [20]. Activation of the pro-apoptotic molecules Bax or Bak, results in a conformational change that triggers their oligomerization and increased membrane insertion [21, 22]. Oligomerized Bax or Bak subsequently forms a pore that allows the release of pro-apoptotic molecules from the mitochondria.

1.3   Extrinsic Pathway

Death receptors are type I trans-membrane proteins and members of the tumor necrosis factor (TNF) receptor superfamily. Members of this family include: Fas (CD95/Apo-1) and the TNF-related apoptosis inducing ligand (TRAIL) receptor-1 (TRAIL-R1/death receptor 4 (DR4)), and -2 (TRAIL-R2/ DR5) [23]. Death receptors contain one to five cysteine-rich repeats in their extracellular domain and a cytoplasmic domain of approximately 80 amino acids termed the death domain (DD) [24]. The corresponding ligands for the death receptors are members of the TNF superfamily and include Fas ligand (FasL) and TRAIL [25]. The ligands are type II membrane proteins, which can be cleaved by metalloproteases to generate soluble molecules that are also capable of binding the receptors. When the death receptors are ligated, they form a death inducing signalling complex (DISC) [26]. A vital component of the DISC is the Fas-associated death domain (FADD) adapter protein. FADD contains a C-terminal death domain (DD) that interacts homotypically with the DD in the cytoplasmic region of the death receptors [27]. FADD also contains a second protein–protein interaction domain at its N-terminus, called the death effector domain (DED). This domain is required for the recruitment of molecules containing a DED to the DISC, including the DED containing caspases procaspases-8 and -10 [28]. At the DISC, procaspase-8 is cleaved and activated to initiate apoptosis via direct cleavage of procaspase-3 and -7, or by cleavage of its substrate Bid, resulting in amplification of the apoptotic signal through the mitochondria [29]. In addition to activating caspase-8, Fas ligation can regulate a second apoptotic pathway. Death-associated protein (Daxx) can bind Fas through death domain interactions [30]. Binding of Daxx to Fas, allows recruitment and activation of apoptosis signal-regulating kinase 1 (ASK-1) to initiate a phosphorylation cascade that ultimately results in the phosphorylation and subsequent activation of c-Jun N-terminal kinase (JNK), resulting in apoptosis [31, 32].

Death receptor–mediated apoptosis may be inhibited at various levels. Initially, the death ligand needs to bind to the receptor to initiate the caspase cascade. Competition for death ligand binding occurs for both FasL and TRAIL. Decoy Receptor 3 (DcR3) and soluble Fas (sFas, a splice variant of Fas) can bind to FasL to prevent initiation of the Fas death receptor pathway [33, 34]. Similarly, the non-apoptotic TRAIL receptors DcR1, DcR2 and osteoprotegerin, inhibit ligation of DR4 and DR5 by TRAIL [35]. sFas and osteoprotegerin are soluble receptors that are secreted from the cell, whereas DcR1, 2, and 3 are membrane-bound receptors that lack an active cytoplasmic domain. Downstream, a key regulator of death receptor-mediated

apoptosis is the caspase-8/-10 inhibitor, termed cellular FLICE-inhibitory protein (c-FLIP). c-FLIP exists as two main isoforms, a long (c-FLIP$_L$) and a short (c-FLIP$_S$) form, with both variants containing DEDs [36]. Both isoforms of c-FLIP may be recruited to both the Fas and TRAIL DISCs and inhibit full activation of caspase-8, preventing apoptosis. A second downstream inhibitor of caspase-8 activation is phosphoprotein enriched in astrocytes 15 (PEA-15), which was first identified in brain astrocytes [37]. PEA-15 contains an N-terminus DED, which can bind FADD to prevent TRAIL-induced apoptosis in glioma cells [38] and TNF-induced apoptosis in astrocytes [39]. Regulation of death receptor–mediated apoptosis by PEA-15 depends on its phosphorylation status. PEA-15 can be phosphorylated at Serine-104 by protein kinase C, which allows PEA-15 to bind and regulate the ERK/MAPK pathway [40]; or can be phosphorylated at Serine-116 by calmodulin-dependent protein kinase II (CaMKII) and Akt, which enables PEA-15 to translocate to the DISC and inhibit caspase-8 activation [38, 39, 41].

1.4   Inhibitors of Apoptosis Proteins

Once the initiator caspases have been activated, further downstream inhibitors of apoptosis may prevent activation of executioner caspases. Such molecules include the inhibitors of apoptosis proteins (IAPs), which contain at least one copy of a BIR (baculovirus IAP repeat) domain and a zinc-binding fold [42]. Members of this family include cIAP1, cIAP2, XIAP and survivin. XIAP, cIAP1 and cIAP2 can bind to and inhibit activated caspases-3, -7 and -9, preventing apoptosis [43, 44]. Suppression of the effector caspases (-3 and -7) is mediated through binding of the caspases to the first two BIR domains of the IAP and results in masking of the active site in the caspase and subsequent steric occlusion of normal caspase substrates [45]. Importantly, IAPs are selective caspase inhibitors and lack activity against many members of the caspase family [46]. For example XIAP, cIAP1 and cIAP2 appear not to inhibit caspases-1, -6, -8, or -10 [47]. The role of survivin in inhibiting apoptosis through caspase inhibition is controversial. Several reports have indicated that survivin can inhibit caspase-9, whereas other reports have indicated caspase-3 [48, 49]. As mentioned, triggering of the intrinsic pathway and subsequent mitochondrial depolarization releases SMAC/DIABLO, which is a negative regulator of the IAPs. SMAC/DIABLO can bind to the IAPs and prevent their inhibition of activated caspases, thereby promoting apoptosis [50].

1.5 Survival Pathways

The balance between apoptotic and survival signaling is critical for a "successful" metastatic tumor. Protein tyrosine kinases (PTKs) have an important role in regulating anti-apoptotic pathways (51). Some of the most well characterized PTKs include the epidermal growth factor receptor (EGFR) family, which include EGFR (Her1), Her2, Her3, and Her4 (52). Binding of growth factors, such as transforming growth factor-alpha (TGF-α), epidermal growth factor (EGF), or the heregulins results in homodimerization and heterodimerization of the EGFR family members. Receptor dimerization results in phosphorylation of tyrosine residues in the cytoplasmic domain and subsequent activation of downstream anti-apoptotic signals. Such signaling cascades include the phosphatidylinositol 3-kinase (PI3K)/Akt (protein kinase B (PKB)) and the signal transducers and activators of transcription (STAT) pathways [53]. Activation of the PI3K/Akt pathway inhibits apoptosis in numerous ways. First, Akt may phosphorylate the BH3 family member Bad, which prevents its inhibition of Bcl-X$_L$. Other anti-apoptotic roles of Akt include the inhibition of caspase-9, activation of the transcription factor nuclear factor-κB (NFκB), and inactivation of the Forkhead transcription factor [54]. NFκB can positively regulate the expression of numerous anti-apoptotic genes such as those encoding the proteins for the IAPs, c-FLIP, Bcl-2, and Bcl-X$_L$ [55], while the Forkhead transcription factors, upregulate pro-apoptotic genes such as *FasL* and *Bim* [56]. Akt can also regulate p53 activity, as it promotes phosphorylation of human double minute-2 (HDM-2), a negative regulator of p53 [57]. Once HDM-2 is phosphorylated, it translocates to the nucleus and targets p53 for ubiquitin-mediated degradation via the proteasome, thus keeping levels of p53 low in resting cells [58]. Activation of STAT-3 and STAT-5 results in their translocation to the nucleus, where they upregulate anti-apoptotic genes such as *Bcl-X$_L$*, *Mcl-1* and *survivin* [59–61].

2 Correlation of Apoptotic Proteins with Disease Response

2.1 Death Receptor Pathway

Expression of the death receptor Fas has been found to be diminished in 50% of colonic tumors compared to normal epithelia [62]. However, Backus and colleagues demonstrated that 5-fluorouracil treatment upregulated Fas expression in colorectal tumors [63], suggesting that Fas might be an important determinant of response to 5-fluorouracil. Elevated FasL expression has been found in the more advanced stages of several cancers, including colorectal and pancreatic [64, 65]. Downregulation of Fas and upregulation of FasL have been shown to correlate with disease progression in several cancers [66–69]. It has been hypothesized that the downregulation of Fas and upregulation of FasL result in "immune escape" from Fas-sensitive T cells [70]. As activated T cells express Fas and are sensitive to Fas-mediated apoptosis [71], the tumor cell upregulation of FasL may enable the tumor cells to kill tumor-infiltrating lymphocytes, resulting in tumor immunity. Indeed, it has recently been shown that downregulation of FasL in colorectal cancer cells impeded tumor growth in immune competent mice [72], providing more evidence for the "Fas counter-attack" model. A recent study indicated that low Fas expression, but high FasL expression in breast tumors correlated with significantly shorter disease-free survival and overall survival following adjuvant chemotherapy treatment [73]; however, a number of studies have failed to find any correlation [74–76]. The Fas decoy receptors sFas has been shown to be upregulated in numerous cancers, including colon, breast, and bladder cancers [77–79]. Furthermore, an increase in sFas expression was reported in the serum of cancer patients that was directly related to tumor stage and burden, suggesting a role for sFas in the malignancy of the disease [80].

High expression of the TRAIL receptors (DR4 and DR5) has been observed in numerous carcinomas, including colorectal, breast, and acute myeloid leukemia (AML) [81–83]. In a phase III colorectal cancer study, van Geelen and colleagues demonstrated that high DR4 expression was associated with a shorter overall survival and time to recurrence in patients following adjuvant 5-fluorouracil–based treatment [81]. Whereas TRAIL expression was observed in a high number of tumors in this study, expression of TRAIL and DR5 had no prognostic value. In a similar study by Strater et al., expression of TRAIL and DR5 in colonic carcinomas showed no significant correlation with disease-free survival. However, low expression of DR4 was associated with a poor prognosis in this study [84]. In breast cancer, DR4 expression had no association with survival; however, high DR5 expression correlated with decreased survival and lymph node metastasis [83]. In AML patients treated with cytosine arabinoside and idarubicin, expression of DR4 and DR5 showed no correlation with any clinical parameters, such as response [82]. Clearly, the relationship between the TRAIL receptors and prognosis has yet to be proved, however, a more accurate predictive model may be achieved by assessing the ratios among DR4, DR5, and the decoy receptors DcR1 and DcR2.

The caspase-8 inhibitor c-FLIP has been demonstrated to be overexpressed in certain carcinomas, including colorectal, gastric, and melanoma [85–88]. In Burkitt's lymphomas, patients whose tumors overexpressed c-FLIP$_L$ had a significantly lower 2-year survival rate following chemotherapy treatment than in patients with no c-FLIP$_L$ expression [89]. Further studies are awaited in other diseases.

2.2   Bcl-2 Family

Disruption of the intrinsic apoptotic pathway has been extensively studied in relation to chemotherapy response. In vitro studies have demonstrated that Bcl-2 overexpression renders cancer cells less sensitive to a range of chemotherapies [90, 91]. In mouse models, Bcl-2 and Bcl-X$_L$ overexpression has been demonstrated to facilitate tumorigenesis [92]. Similar to overexpression of the anti-apoptotic Bcl-2 family members, decreased expression of pro-apoptotic Bcl-2 family members can have the same effect. For example, loss of Bax expression can also decrease chemotherapy response in vitro [93] and enhance tumor progression in vivo [92]. Clinically, the relationship between Bcl-2 family expression and prognosis is controversial. In two independent breast cancer studies, Bcl-2 expression correlated with estrogen receptor expression [94,

95]. Unexpectedly, Bcl-2 expression has been correlated positively with patient survival [95, 96]. However, this may be due to the favorable clinical parameters that coincide with estrogen receptor positivity. In bladder carcinoma, the role of Bcl-2 is also ambiguous. In a study by Hussain et al., overexpression of Bcl-2 was a poor prognostic marker for patients receiving chemoradiotherapy [97], whereas Maluf et al. found no correlation between Bcl-2 expression and survival following neoadjuvant chemotherapy treatment [98]. Numerous studies have shown that Bax expression does not correlate with prognosis in breast [99], head and neck [100], and ovarian carcinomas [101]. However, a study by Skirnisdottir et al. demonstrated that Bax expression was a significant prognostic factor in patients with ovarian carcinoma who received adjuvant radiotherapy [102]. Whereas the Bcl-2 family members clearly play an important role in regulating chemotherapy response/tumorigenesis, analyzing one family member may be insufficient. Relative ratios of pro-apoptotic to anti-apoptotic members may provide a better prognostic marker of response/survival.

2.3   Inhibitors of Apoptosis Proteins

Overexpression of survivin has been demonstrated to decrease sensitivity to chemotherapy in vitro [103]. In complementary experiments, downregulation of survivin was shown to enhance sensitivity to chemotherapy [104, 105]. Similarly, XIAP expression has been demonstrated to modulate chemotherapy sensitivity in lung cancer cell lines [106, 107]. In vivo, decreasing the expression of either survivin or XIAP has been shown to enhance sensitivity to chemotherapy [107] and radiotherapy [108] in lung cancer xenografts. Clinically, survivin is more highly expressed in numerous human cancers, but not in normal cells [109], and low levels of survivin have been correlated with a better prognosis and improved response to chemotherapy [104, 110, 111]. In AML, high expression of XIAP has been shown to be a poor prognostic factor [112, 113]. However, Ferreira and colleagues found no correlation between cIAP-1, cIAP-2, or XIAP expression and overall survival in NSCLC patients [114].

3 Targeting Apoptosis

Despite advances in chemotherapies, a plateau of effectiveness has been reached in the treatment of advanced human cancers. Numerous trials have demonstrated that substituting one type of chemotherapy for another in a regimen does not necessarily improve response rates or overall survival. Thus, new therapeutic strategies are urgently needed to increase the effectiveness of anticancer therapies. One strategy is to target apoptosis (Fig. 16.2).

3.1 Extrinsic Pathway

Targeting of the death receptors to induce apoptosis has received a lot of attention. However, targeting of the Fas receptor in vivo has failed to prove a promising therapeutic strategy. The observation that 5-fluorouracil treatment upregulated Fas expression in colorectal tumors has also been observed in vitro [115, 116]. Moreover, combination treatment with chemotherapy and Fas agonistic antibodies results in synergistic interactions. However, this observation has yet to be proved in vivo, as systemic injections of Fas-targeted agents have been shown to result in fatal liver damage in mice [117–119]. As such, most preclinical models are now focused on local administration of recombinant FasL, or the use of FasL-expressing vectors. Despite this, one group has successfully generated a nonhepatotoxic agonistic Fas antibody, suggesting that antibody targeting of the Fas receptor may be possible [120]. With the problems of targeting Fas, much interest has arisen concerning TRAIL

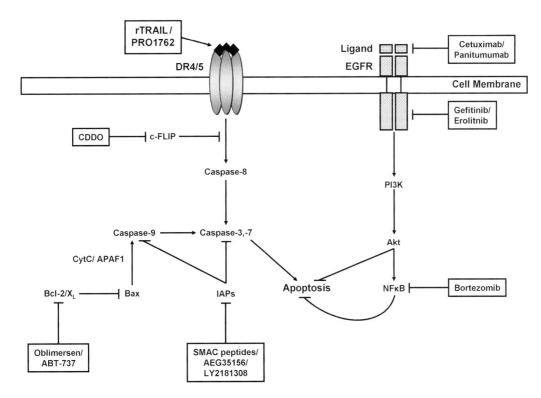

FIGURE 16.2. Targeting the apoptotic pathway. Activating the extrinsic pathway may occur via ligand binding by rTRAIL and PRO1762, or by decreasing the expression of c-FLIP via CDDO compounds. Regulation of the intrinsic pathway can be achieved by antisense targeting of Bcl-2 (oblimersen), or BH-3 peptides (ABT-737), which bind Bcl-2/X$_L$ to inhibit their anti-apoptotic function. SMAC peptides and IAP antisense strategies (AEG35156 and LY2181308) prevent IAP-mediated inhibition of caspase-9, -3, and -7. The proteosome inhibitor bortezomib prevents NFκB-mediated gene transcription. Finally, inhibiting growth and proliferation signals from receptor tyrosine kinases, such as EGFR can be achieved by disrupting ligand binding with specific antibodies (cetuximab and panitumumab), or by inhibiting receptor tyrosine kinase activity (gefitinib and erolitnib)

as a novel therapy. Recombinant human TRAIL has been shown to induce apoptosis in numerous cancer cells, but not normal cells [35, 121], and generated no toxicity when systemically injected into rodents or nonhuman primates [121–123]. Furthermore, administration of TRAIL in tumor-bearing mice was shown to induce tumor regression, especially when used in conjunction with chemotherapy [121, 124–127]. As such, phase I and II studies are ongoing for the targeting of the TRAIL receptors. Genentech has begun a phase I trial using soluble TRAIL (PRO1762). Human Genome Sciences has generated a DR5-specific agonistic antibody HGS-ETR2 and is starting a phase I trial in both the United States and the United Kingdom. After preclinical success with a specific DR4 monoclonal antibody HGS-ETR1, Human Genome Sciences has enrolled patients in phase II trials for the treatment of NSCLC, colorectal cancer, and non-Hodgkin's lymphoma. So far, no hepatic or hematologic toxicity has been observed in these patients [128]. The death ligand tumor necrosis factor (TNF), a type II membrane protein, binds its receptors TNF-receptor 1 (TNF-R1) and TNF-R2. Ligation of TNF-R1 results in the recruitment of TNF-related associated death domain (TRADD), which can either result in cell survival or apoptosis. For example, TRADD can subsequently recruit TNF-receptor associating factor 2 (TRAF2), which activates NFκB or the mitogen-activated protein kinase (MAPK) pathway [129]. Alternatively, FADD may be recruited to TRADD, to result in caspase-8 activation. Similar to TNF-R1, ligation of TNF-R2 results in direct association with TRAF2 and activation of the NFκB and MAPK pathways. Preclinical models using TNF-coupled to a endothelial cell peptide ligand (CD13) resulted in complete regression of established xenograft lymphomas [130], indicating that specific targeting could enhance its therapeutic potential. Specific targeting is crucial, as in a phase II study, systemic injection of TNF caused an inflammatory response resembling septic shock in humans [131]. Despite this, Lejeune and colleagues developed a regimen of high-dose TNF in combination with chemotherapy, applied by isolated limb perfusion, to treat locally advanced melanomas and sarcomas [132, 133]. TNF in combination with melphalan is awaiting approval for treatment of localized melanomas following a phase III trial in the United States.

In vitro, we and others have shown that decreasing c-FLIP expression can enhance sensitivity to chemotherapy [134, 135], and c-FLIP$_L$ overexpression has been demonstrated to decrease sensitivity to chemotherapy [134, 136, 137]. In addition, our laboratory has demonstrated that growth inhibition of colorectal cancer xenografts in response to 5-fluorouracil and oxaliplatin is overcome by c-FLIP$_L$ overexpression [181]. We have further demonstrated that *c-FLIP* gene silencing induces spontaneous apoptosis in a panel of colorectal cancer, NSCLC and prostate cancer cell lines in vitro, and retarded colorectal cancer xenograft growth in vivo. The synthetic triterpenoid CDDO has been demonstrated to sensitize TRAIL-resistant cancer cells to TRAIL-mediated apoptosis in vitro, and interacted synergistically with TRAIL in vivo to reduce breast cancer xenograft burden [138]. The authors demonstrated that treatment with CDDO reduced c-FLIP levels and upregulated expression of DR4 and DR5. The same group subsequently screened a 50,000-compound library for c-FLIP inhibitors [139]. Eight compounds were identified, two of which decreased c-FLIP expression, and also sensitized to TRAIL-induced apoptosis. Targeting c-FLIP appears to be an interesting strategy for sensitizing cancer cells to chemotherapy and death ligand therapy; however, further preclinical studies are needed to develop this approach.

3.2 Intrinsic Pathway

Control of the intrinsic apoptotic pathway is controlled by the relative ratio of pro-apoptotic to anti-apoptotic Bcl-2 family members, which are therefore potential therapeutic targets for cancer therapy. One approach that has been tested is reducing the levels of anti-apoptotic Bcl-2 using the antisense oligonucleotide oblimersen. Preclinical studies have demonstrated that oblimersen interacts synergistically with many cytotoxic drugs in a wide range of solid tumors and hematologic malignancies [140]. However, so far clinical trials have not been very successful. Oblimersen failed to be approved by the FDA in the United States as no survival difference was observed following a phase III trial between melanoma patients treated with dacarbazine alone or in combination with oblimersen [141]. Furthermore, no clinical benefit was observed in chemorefractory multiple myeloma patients who received oblimersen in combination with dexamethasone versus dexamethasone alone in a phase III trial [142]. Despite these failures, improved response rates have been observed in patients with chronic lymphocytic leukemia who received oblimersen in combination with fludarabine and cyclophosphamide compared with chemotherapy alone [143] and in patients with hormone refractory prostate cancer treated with oblimersen and docetaxel [144]. Preclinical studies are underway using an antisense oligonucleotide targeted against both Bcl-2 and Bcl-X$_L$ [145]. A second approach to target the intrinsic pathway involves mimicking the pro-apoptotic BH3-only peptides. Synthetic Bid-based BH3 peptides have been designed that bind to Bcl-2 and Bcl-X$_L$ and have been shown to induce apoptosis in cells and inhibit growth of human leukemia cells in vivo, prolonging the survival of mice [146]. Recently, ABT-737, a potent Bcl-2, Bcl-X$_L$, and Bcl-W inhibitor, has been shown to synergistically interact with cytotoxic drugs and radiation in various cell lines [147]. In mouse models, ABT-737 produced complete regression of non-Hodgkin's lymphoma when administered in combination with a range of cytotoxic drugs and complete regression of small cell lung carcinoma xenografts when given as a monotherapy [147]. As the Bcl-2 group of proteins contains numerous members, this strategy of targeting multiple members appears the most promising.

3.3 IAPs

Inhibition of active caspases by IAPs is an integral checkpoint for cell survival. Two of the most studied IAPs for potential therapeutic targets are XIAP and survivin. XIAP contains three BIR domains which inhibit caspase-3, -7, and -9 [148]. The BIR-3 domain of XIAP binds and inhibits caspase-9, whereas the BIR-2 region (together with the linker region) binds and inhibits caspases-3 and -7. Downregulation of XIAP by siRNA or antisense oligonucleotides has been demonstrated to restore chemosensitivity in a range of cancer cell lines [149–151]. In mouse models, XIAP antisense therapy in combination with vinorelbine produced promising results in lung cancer xenografts models [107]. Aegera pharmaceuticals has developed a second-generation XIAP antisense oligonucleotide AEG35156 that is currently in a phase I clinical trial in the United Kingdom [152]. In vitro pre-clinical

models have shown that AEG35156 can sensitize pancreatic cancer cells to TRAIL-induced apoptosis [153]. AEG35156 had only modest antitumoral activity in vivo as a monotherapy, but produced complete tumor regression when combined with taxanes [153]. "SMAC-peptides" have also been developed that bind to XIAP and inhibit its anti-apoptotic function. SMAC-peptides may not induce apoptosis as a single agent, but have been shown to sensitize cancer cells lines and xenografts to TRAIL and cytotoxic drugs [128]. Survivin is the smallest IAP family member, containing only one BIR domain. The anti-apoptotic role of survivin is still in question [154]. Similar to XIAP, strategies to downregulate survivin using antisense oligonucleotides have been developed. In vitro, survivin-targeted antisense oligonucleotides have been demonstrated to induce apoptosis in lung and mesothelioma cell lines [155]. In vivo, downregulation of survivin using an antisense approach has been shown to reduce xenograft growth in survivin expressing tumors [156]. In addition, treatment of lung cancer xenografts with survivin antisense oligonucleotides has been shown to synergize with cisplatin [157]. ISIS pharmaceuticals/Eli Lilly has developed a second-generation survivin antisense oligonucleotide LY2181308, which after promising activity in early studies is intended to start a phase II trial at the end of 2006 or early 2007 (www.isispharm.com). However, if clinical outcome depends on inhibition of numerous IAPs, perhaps the generation of pan-IAP inhibitors might prove to be more therapeutically valuable.

3.4 Survival Pathways

Chemotherapy-induced apoptosis can be enhanced by inhibiting pro-survival signalling such as those mediated by receptor and non-receptor tyrosine kinases, such as EGFR, Her2, VEGFR, and Src. These tyrosine kinases can activate anti-apoptotic signalling pathways, such as those mediated by PI3K/Akt, STAT-3/-5, and NFκB. Many tumors overexpress EGFR and/or have activating EGFR mutations including non-small cell lung carcinoma (NSCLC), prostate cancer, breast cancer, and colorectal cancer [158]. Whereas EGFR plays an important role in maintaining normal cellular function, dysregulation of the EGFR pathway has been shown to contribute to tumor malignancies

[159]. Therefore targeting of the EGFR receptor is an attractive proposition for the treatment of cancer. Two EGFR-targeted approaches have been explored: first, monoclonal antibodies such as cetuximab and panitumumab, which target the EGFR extracellular domain and inhibit ligand binding; and second, small molecule tyrosine kinase inhibitors, such as gefitinib and erlotinib, which bind intracellularly to the kinase domain to prevent downstream signaling. A large number of preclinical xenograft studies have demonstrated that combining EGFR inhibitors with radiotherapy and/or chemotherapy enhances tumor regression and retarded tumor growth compared with radiotherapy or chemotherapy alone [160–164]. In colorectal cancer, impressive results have been obtained for cetuximab as a monotherapy [165], and in combination with irinotecan [166]. However, no significant difference was observed in several NSCLC trials assessing the use of EGFR inhibitors [167–172]. NFκB is an important anti-apoptotic transcription factor that regulates the expression of genes mediating tumorigenesis. It has been demonstrated that the proteasome inhibitor bortezomib, through the stabilization of the NFκB inhibitory molecule IκB, results in NFκB inhibition [173]. After demonstrating antitumor effects in both cell line and xenografts models [174], bortezomib is the first proteasome inhibitor to be investigated in clinical trials. In several phase I and II trials, bortezomib has been well tolerated [175, 176], and is currently being assessed in combination with chemotherapy and radiotherapy in advanced solid tumors. The PI3K/Akt pathway regulates many proteins that control both the extrinsic and intrinsic apoptotic pathways, and is deregulated in many cancers [177, 178]. Whereas inhibitors of the PI3K/Akt pathway (e.g., LY294002) have shown antitumoral activity in cell line and xenografts models, these compounds lack specificity and are more often used as a tool to study the function of this pathway [178]. Specific Akt inhibitors are awaited.

4   Conclusions and Future Perspectives

In the post-genomic era, individualization of cancer treatment will become a reality. In this regard, an in-depth knowledge of cellular mechanisms involved in regulating the apoptotic response to chemotherapy will play a key role in the individualization of cancer therapies. Targeting apoptosis has produced some very promising preclinical results, most often when used in conjunction with conventional chemotherapy. Tumor-specific delivery is a major problem for therapeutic targeting of apoptosis, particularly when the targeting agent is an siRNA, antisense oligonucleotide, or expression vector. Studies are underway to design tumor-specific vectors or liposomes. One such strategy involves targeting the transferrin receptor, which is often overexpressed in tumors, making it an attractive tumor-specific target [179]. Cationic liposomes complexed to an antitransferrin single-chain antibody fragment have been developed that selectively deliver siRNA into tumor-bearing mice (both primary and metastatic) [180]. Use of these types of technologies will be fundamental for the delivery of molecules that target the apoptotic machinery. Profiling patients (e.g., using cDNA microarray) may identify the best combination of chemotherapy and apoptosis-targeting strategy to use and ultimately result in a better prognosis. In conclusion, the combination of conventional chemotherapy with agents targeting apoptosis will play an increasingly important role in future anticancer strategies.

References

1. Hanahan D, Weinberg RA. The hallmarks of cancer. Cell 2000, 100:57–70.
2. White E. Life, death, and the pursuit of apoptosis. Genes Dev 1996, 10:1–15.
3. Jacobson MD, Weil M, Raff MC. Programmed cell death in animal development. Cell 1997, 88:347–354.
4. Renehan AG, Booth C, Potten CS. What is apoptosis, and why is it important? BMJ 2001, 322:1536–1538.
5. Kerr JF, Wyllie AH, Currie AR. Apoptosis: a basic biological phenomenon with wide-ranging implications in tissue kinetics. Br J Cancer 1972, 26:239–257.
6. Thompson CB. Apoptosis in the pathogenesis and treatment of disease. Science 1995, 267:1456–1462.
7. Rowan S, Fisher DE. Mechanisms of apoptotic cell death. Leukemia 1997, 11:457–465.
8. Hofmann K, Bucher P, Tschopp J. The CARD domain: a new apoptotic signalling motif. Trends Biochem Sci 1997, 22:155–156.
9. Savill J. Apoptosis in resolution of inflammation. J Leukoc Biol 1997, 61:375–380.
10. Green DR, Reed JC. Mitochondria and apoptosis. Science 1998, 281:1309–1312.

11. Zou H, Henzel WJ, Liu X, et al. Apaf-1, a human protein homologous to *C. elegans* CED-4, participates in cytochrome c-dependent activation of caspase-3. Cell 1997, 90:405–413.

12. Rodriguez J, Lazebnik Y. Caspase-9 and APAF-1 form an active holoenzyme. Genes Dev 1999, 13:3179–3184.

13. Adams JM, Cory S. The Bcl-2 protein family: arbiters of cell survival. Science 1998, 281:1322–1326.

14. Antonsson B, Martinou JC. The Bcl-2 protein family. Exp Cell Res 2000, 256:50–57.

15. Huang DC, Strasser A. BH3-Only proteins—essential initiators of apoptotic cell death. Cell 2000, 103:839–842.

16. Puthalakath H, Strasser A. Keeping killers on a tight leash: transcriptional and post-translational control of the pro-apoptotic activity of BH3-only proteins. Cell Death Differ 2002, 9:505–512.

17. Hanada M, Aime-Sempe C, Sato T, et al. Structure-function analysis of Bcl-2 protein. Identification of conserved domains important for homodimerization with Bcl-2 and heterodimerization with Bax. J Biol Chem 1995, 270:11962–11969.

18. Nechushtan A, Smith CL, Hsu YT, et al. Conformation of the Bax C-terminus regulates subcellular location and cell death. EMBO J 1999, 18:2330–2341.

19. Franke TF, Cantley LC. Apoptosis. A Bad kinase makes good. Nature 1997, 390:116–117.

20. Degli Esposti M, Dive C. Mitochondrial membrane permeabilisation by Bax/Bak. Biochem Biophys Res Commun 2003, 304:455–461.

21. Kelekar A, Thompson CB. Bcl-2-family proteins: the role of the BH3 domain in apoptosis. Trends Cell Biol 1998, 8:324–330.

22. Korsmeyer SJ, Wei MC, Saito M, et al. Pro-apoptotic cascade activates BID, which oligomerizes BAK or BAX into pores that result in the release of cytochrome c. Cell Death Differ 2000, 7:1166–1173.

23. Ashkenazi A, Dixit VM. Death receptors: signaling and modulation. Science 1998, 281:1305–1308.

24. Gupta S. Molecular signaling in death receptor and mitochondrial pathways of apoptosis (Review). Int J Oncol 2003, 22:15–20.

25. Beutler B, van Huffel C. Unraveling function in the TNF ligand and receptor families. Science 1994, 264:667–668.

26. Kischkel FC, Hellbardt S, Behrmann I, et al. Cytotoxicity-dependent APO-1 (Fas/CD95)-associated proteins form a death-inducing signaling complex (DISC) with the receptor. EMBO J 1995, 14:5579–5588.

27. Chinnaiyan AM, Tepper CG, Seldin MF, et al. FADD/MORT1 is a common mediator of CD95 (Fas/APO-1) and tumor necrosis factor receptor-induced apoptosis. J Biol Chem 1996, 271:4961–4965.

28. Sprick MR, Rieser E, Stahl H, et al. Caspase-10 is recruited to and activated at the native TRAIL and CD95 death-inducing signalling complexes in a FADD-dependent manner but can not functionally substitute caspase-8. EMBO J 2002, 21:4520–4530.

29. Li H, Zhu H, Xu CJ, et al. Cleavage of BID by caspase 8 mediates the mitochondrial damage in the Fas pathway of apoptosis. Cell 1998, 94:491–501.

30. Yang X, Khosravi-Far R, Chang HY, et al. Daxx, a novel Fas-binding protein that activates JNK and apoptosis. Cell 1997, 89:1067–1076.

31. Chang HY, Nishitoh H, Yang X, et al. Activation of apoptosis signal-regulating kinase 1 (ASK1) by the adapter protein Daxx. Science 1998, 281:1860–1863.

32. Curtin JF, Cotter TG. Live and let die: regulatory mechanisms in Fas-mediated apoptosis. Cell Signal 2003, 15:983–992.

33. Cascino II, Papoff G, Eramo A, et al. Soluble Fas/Apo-1 splicing variants and apoptosis. Front Biosci 1996, 1:12–18.

34. Pitti RM, Marsters SA, Lawrence DA, et al. Genomic amplification of a decoy receptor for Fas ligand in lung and colon cancer. Nature 1998, 396:699–703.

35. Walczak H, Krammer PH. The CD95 (APO-1/Fas) and the TRAIL (APO-2L) apoptosis systems. Exp Cell Res 2000, 256:58–66.

36. Krueger A, Baumann S, Krammer PH, et al. FLICE-inhibitory proteins: regulators of death receptor-mediated apoptosis. Mol Cell Biol 2001, 21:8247–8254.

37. Araujo H, Danziger N, Cordier J, et al. Characterization of PEA–15, a major substrate for protein kinase C in astrocytes. J Biol Chem 1993, 268:5911–5920.

38. Xiao C, Yang BF, Asadi N, et al. Tumor necrosis factor-related apoptosis-inducing ligand-induced death-inducing signaling complex and its modulation by c-FLIP and PED/PEA-15 in glioma cells. J Biol Chem 2002, 277:25020–25025.

39. Kitsberg D, Formstecher E, Fauquet M, et al. Knock-out of the neural death effector domain protein PEA-15 demonstrates that its expression protects astrocytes from TNFalpha-induced apoptosis. J Neurosci 1999, 19:8244–8251.

40. Formstecher E, Ramos JW, Fauquet M, et al. PEA-15 mediates cytoplasmic sequestration of ERK MAP kinase. Dev Cell 2001, 1:239–250.

41. Condorelli G, Vigliotta G, Cafieri A, et al. PED/PEA-15: an anti-apoptotic molecule that regulates FAS/TNFR1-induced apoptosis. Oncogene 1999, 18:4409–4415.

42. Hinds MG, Norton RS, Vaux DL, et al. Solution structure of a baculoviral inhibitor of apoptosis (IAP) repeat. Nat Struct Biol 1999, 6:648–651.

43. Deveraux QL, Leo E, Stennicke HR, et al. Cleavage of human inhibitor of apoptosis protein XIAP results in fragments with distinct specificities for caspases. EMBO J 1999, 18:5242–5251.

44. Roy N, Deveraux QL, Takahashi R, et al. The c-IAP-1 and c-IAP-2 proteins are direct inhibitors of specific caspases. EMBO J 1997, 16:6914–6925.

45. Huang Y, Park YC, Rich RL, et al. Structural basis of caspase inhibition by XIAP: differential roles of the linker versus the BIR domain. Cell 2001, 104:781–790.

46. Reed JC. Mechanisms of apoptosis. Am J Pathol 2000, 157:1415–1430.

47. Hay BA. Understanding IAP function and regulation: a view from *Drosophila*. Cell Death Differ 2000, 7:1045–1056.

48. Shi Y. Survivin structure: crystal unclear. Nat Struct Biol 2000, 7:620–623.

49. Shin S, Sung BJ, Cho YS, et al. An anti-apoptotic protein human survivin is a direct inhibitor of caspase-3 and -7. Biochemistry 2001, 40:1117–1123.

50. Verhagen AM, Ekert PG, Pakusch M, et al. Identification of DIABLO, a mammalian protein that promotes apoptosis by binding to and antagonizing IAP proteins. Cell 2000, 102:43–53.

51. Blume-Jensen P, Hunter T. Oncogenic kinase signalling. Nature 2001, 411:355–365.

52. Jorissen RN, Walker F, Pouliot N, et al. Epidermal growth factor receptor: mechanisms of activation and signalling. Exp Cell Res 2003, 284:31–53.

53. Olayioye MA, Neve RM, Lane HA, et al. The ErbB signaling network: receptor heterodimerization in development and cancer. EMBO J 2000, 19:3159–3167.

54. Fresno Vara JA, Casado E, de Castro J, et al. PI3K/Akt signalling pathway and cancer. Cancer Treat Rev 2004, 30:193–204.

55. Lin A, Karin M. NF-kappaB in cancer: a marked target. Semin Cancer Biol 2003, 13:107–114.

56. Burgering BM, Medema RH. Decisions on life and death: FOXO Forkhead transcription factors are in command when PKB/Akt is off duty. J Leukoc Biol 2003, 73:689–701.

57. Feng J, Tamaskovic R, Yang Z, et al. Stabilization of Mdm2 via decreased ubiquitination is mediated by protein kinase B/Akt-dependent phosphorylation. J Biol Chem 2004, 279:35510–35517.

58. Slee EA, O'Connor DJ, Lu X. To die or not to die: how does p53 decide? Oncogene 2004, 23:2809–2818.

59. Grandis JR, Drenning SD, Zeng Q, et al. Constitutive activation of Stat3 signaling abrogates apoptosis in squamous cell carcinogenesis in vivo. Proc Natl Acad Sci U S A 2000, 97:4227–4232.

60. Aoki Y, Feldman GM, Tosato G. Inhibition of STAT3 signaling induces apoptosis and decreases survivin expression in primary effusion lymphoma. Blood 2003, 101:1535–1542.

61. Yu H, Jove R. The STATs of cancer—new molecular targets come of age. Nat Rev Cancer 2004, 4:97–105.

62. Moller P, Koretz K, Leithauser F, et al. Expression of APO-1 (CD95), a member of the NGF/TNF receptor superfamily, in normal and neoplastic colon epithelium. Int J Cancer 1994, 57:371–377.

63. Backus HH, Dukers DF, van Groeningen CJ, et al. 5-Fluorouracil induced Fas upregulation associated with apoptosis in liver metastases of colorectal cancer patients. Ann Oncol 2001, 12:209–216.

64. Pernick NL, Sarkar FH, Tabaczka P, et al. Fas and Fas ligand expression in pancreatic adenocarcinoma. Pancreas 2002, 25:e36–41.

65. Bennett MW, O'Connell J, Houston A, et al. Fas ligand upregulation is an early event in colonic carcinogenesis. J Clin Pathol 2001, 54:598–604.

66. von Reyher U, Strater J, Kittstein W, et al. Colon carcinoma cells use different mechanisms to escape CD95-mediated apoptosis. Cancer Res 1998, 58:526–534.

67. Gratas C, Tohma Y, Barnas C, et al. Up-regulation of Fas (APO-1/CD95) ligand and down-regulation of Fas expression in human esophageal cancer. Cancer Res 1998, 58:2057–2062.

68. Mottolese M, Buglioni S, Bracalenti C, et al. Prognostic relevance of altered Fas (CD95)-system in human breast cancer. Int J Cancer 2000, 89:127–132.

69. Niehans GA, Brunner T, Frizelle SP, et al. Human lung carcinomas express Fas ligand. Cancer Res 1997, 57:1007–1012.

70. O'Connell J, Bennett MW, O'Sullivan GC, et al. The Fas counterattack: a molecular mechanism of tumor immune privilege. Mol Med 1997, 3:294–300.

71. Krammer PH. CD95's deadly mission in the immune system. Nature 2000, 407:789–795.

72. Ryan AE, Shanahan F, O'Connell J, et al. Fas ligand promotes tumor immune evasion of colon cancer in vivo. Cell Cycle 2006, 5:246–249.

73. Botti C, Buglioni S, Benevolo M, et al. Altered expression of FAS system is related to adverse clinical outcome in stage I-II breast cancer patients treated with adjuvant anthracycline–based chemotherapy. Clin Cancer Res 2004, 10:1360–1365.

74. Pernick NL, Biernat L, Du W, et al. Clinicopathologic analysis of fas, fas ligand, and other biomarkers in locally advanced breast carcinoma. Breast J 2000, 6:233–241.

75. Bezulier K, Fina F, Roussel M, et al. Fas/FasL expression in tumor biopsies: a prognostic response factor to fluoropyrimidines? J Clin Pharm Ther 2003, 28:403–408.

76. Sjostrom J, Blomqvist C, von Boguslawski K, et al. The predictive value of bcl-2, bax, bcl-xL, bag-1, fas, and fasL for chemotherapy response in advanced breast cancer. Clin Cancer Res 2002, 8:811–816.

77. Kushlinskii NE, Britvin TA, Abbasova SG, et al. Soluble Fas antigen in the serum of patients with colon cancer. Bull Exp Biol Med 2001, 131:361–363.

78. Ueno T, Toi M, Tominaga T. Circulating soluble Fas concentration in breast cancer patients. Clin Cancer Res 1999, 5:3529–3533.

79. Mizutani Y, Yoshida O, Bonavida B. Prognostic significance of soluble Fas in the serum of patients with bladder cancer. J Urol 1998, 160:571–576.

80. Midis GP, Shen Y, Owen-Schaub LB. Elevated soluble Fas (sFas) levels in nonhematopoietic human malignancy. Cancer Res 1996, 56:3870–3874.

81. van Geelen CM, Westra JL, de Vries EG, et al. Prognostic significance of tumor necrosis factor-related apoptosis-inducing ligand and its receptors in adjuvantly treated stage III colon cancer patients. J Clin Oncol 2006, 24:4998–5004.

82. Min YJ, Lee JH, Choi SJ, et al. Prognostic significance of Fas (CD95) and TRAIL receptors (DR4/DR5) expression in acute myelogenous leukemia. Leuk Res 2004, 28:359–365.

83. McCarthy MM, Sznol M, DiVito KA, et al. Evaluating the expression and prognostic value of TRAIL-R1 and TRAIL-R2 in breast cancer. Clin Cancer Res 2005, 11:5188–5194.

84. Strater J, Hinz U, Walczak H, et al. Expression of TRAIL and TRAIL receptors in colon carcinoma: TRAIL-R1 is an independent prognostic parameter. Clin Cancer Res 2002, 8:3734–3740.

85. Bullani RR, Huard B, Viard-Leveugle I, et al. Selective expression of FLIP in malignant melanocytic skin lesions. J Invest Dermatol 2001, 117:360–364.

86. Ryu BK, Lee MG, Chi SG, et al. Increased expression of cFLIP(L) in colonic adenocarcinoma. J Pathol 2001, 194:15–19.

87. Zhou XD, Yu JP, Chen HX, et al. Expression of cellular FLICE–inhibitory protein and its association with p53 mutation in colon cancer. World J Gastroenterol 2005, 11:2482–2485.

88. Zhou XD, Yu JP, Liu J, et al. Overexpression of cellular FLICE-inhibitory protein (FLIP) in gastric adenocarcinoma. Clin Sci (Lond) 2004, 106:397–405.

89. Valnet-Rabier MB, Challier B, Thiebault S, et al. c-Flip protein expression in Burkitt's lymphomas is associated with a poor clinical outcome. Br J Haematol 2005, 128:767–773.

90. Teixeira C, Reed JC, Pratt MA. Estrogen promotes chemotherapeutic drug resistance by a mechanism involving Bcl-2 proto-oncogene expression in human breast cancer cells. Cancer Res 1995, 55:3902–3907.

91. Miyashita T, Reed JC. bcl-2 gene transfer increases relative resistance of S49.1 and WEHI7.2 lymphoid cells to cell death and DNA fragmentation induced by glucocorticoids and multiple chemotherapeutic drugs. Cancer Res 1992, 52:5407–5411.

92. Heiser D, Labi V, Erlacher M, et al. The Bcl-2 protein family and its role in the development of neoplastic disease. Exp Gerontol 2004, 39:1125–1135.

93. Zhang L, Yu J, Park BH, et al. Role of BAX in the apoptotic response to anticancer agents. Science 2000, 290:989–992.

94. Linjawi A, Kontogiannea M, Halwani F, et al. Prognostic significance of p53, bcl-2, and Bax expression in early breast cancer. J Am Coll Surg 2004, 198:83–90.

95. Yang Q, Sakurai T, Yoshimura G, et al. Prognostic value of Bcl-2 in invasive breast cancer receiving chemotherapy and endocrine therapy. Oncol Rep 2003, 10:121–125.

96. Bukholm IR, Bukholm G, Nesland JM. Reduced expression of both Bax and Bcl-2 is independently associated with lymph node metastasis in human breast carcinomas. Apmis 2002, 110:214–220.

97. Hussain SA, Ganesan R, Hiller L, et al. BCL2 expression predicts survival in patients receiving synchronous chemoradiotherapy in advanced transitional cell carcinoma of the bladder. Oncol Rep 2003, 10:571–576.

98. Maluf FC, Cordon-Cardo C, Verbel DA, et al. Assessing interactions between mdm-2, p53, and bcl-2 as prognostic variables in muscle-invasive bladder cancer treated with neo-adjuvant chemotherapy followed by locoregional surgical treatment. Ann Oncol 2006, 17:1677–1686.

99. Veronese S, Mauri FA, Caffo O, et al. Bax immunohistochemical expression in breast carcinoma: a study with long term follow-up. Int J Cancer 1998, 79:13–18.

100. Casado S, Forteza J, Dominguez S, et al. Predictive value of P53, BCL-2, and BAX in advanced head and neck carcinoma. Am J Clin Oncol 2002, 25:588–590.

101. Skirnisdottir I, Seidal T, Gerdin E, et al. The prognostic importance of p53, bcl-2, and bax in early stage epithelial ovarian carcinoma treated with adjuvant chemotherapy. Int J Gynecol Cancer 2002, 12:265–276.

102. Skirnisdottir I, Sorbe B, Seidal T. P53, bcl-2, and bax: their relationship and effect on prognosis in early stage epithelial ovarian carcinoma. Int J Gynecol Cancer 2001, 11:147–158.

103. Tamm I, Wang Y, Sausville E, et al. IAP-family protein survivin inhibits caspase activity and apoptosis induced by Fas (CD95), Bax, caspases, and anticancer drugs. Cancer Res 1998, 58:5315–5320.

104. Nakamura M, Tsuji N, Asanuma K, et al. Survivin as a predictor of cis-diamminedichloroplatinum sensitivity in gastric cancer patients. Cancer Sci 2004, 95:44–51.

105. Zangemeister-Wittke U. Antisense to apoptosis inhibitors facilitates chemotherapy and TRAIL-induced death signaling. Ann N Y Acad Sci 2003, 1002:90–94.

106. Olie RA, Simoes-Wust AP, Baumann B, et al. A novel antisense oligonucleotide targeting survivin expression induces apoptosis and sensitizes lung cancer cells to chemotherapy. Cancer Res 2000, 60:2805–2809.

107. Hu Y, Cherton-Horvat G, Dragowska V, et al. Antisense oligonucleotides targeting XIAP induce apoptosis and enhance chemotherapeutic activity against human lung cancer cells in vitro and in vivo. Clin Cancer Res 2003, 9:2826–2836.

108. Cao C, Mu Y, Hallahan DE, et al. XIAP and survivin as therapeutic targets for radiation sensitization in preclinical models of lung cancer. Oncogene 2004, 23:7047–7052.

109. Ambrosini G, Adida C, Altieri DC. A novel anti-apoptosis gene, survivin, expressed in cancer and lymphoma. Nat Med 1997, 3:917–921.

110. Kato J, Kuwabara Y, Mitani M, et al. Expression of survivin in esophageal cancer: correlation with the prognosis and response to chemotherapy. Int J Cancer 2001, 95:92–95.

111. Zaffaroni N, Pennati M, Colella G, et al. Expression of the anti-apoptotic gene survivin correlates with taxol resistance in human ovarian cancer. Cell Mol Life Sci 2002, 59:1406–1412.

112. Tamm I, Richter S, Oltersdorf D, et al. High expression levels of x-linked inhibitor of apoptosis protein and survivin correlate with poor overall survival in childhood de novo acute myeloid leukemia. Clin Cancer Res 2004, 10:3737–3744.

113. Tamm I, Richter S, Scholz F, et al. XIAP expression correlates with monocytic differentiation in adult de novo AML: impact on prognosis. Hematol J 2004, 5:489–495.

114. Ferreira CG, van der Valk P, Span SW, et al. Assessment of IAP (inhibitor of apoptosis) proteins as predictors of response to chemotherapy in advanced non-small-cell lung cancer patients. Ann Oncol 2001, 12:799–805.

115. Longley DB, Allen WL, McDermott U, et al. The roles of thymidylate synthase and p53 in regulating Fas-mediated apoptosis in response to antimetabolites. Clin Cancer Res 2004, 10: 3562–3571.

116. McDermott U, Longley DB, Galligan L, et al. Effect of p53 status and STAT1 on chemotherapy-induced, Fas-mediated apoptosis in colorectal cancer. Cancer Res 2005, 65:8951–8960.

117. Rensing-Ehl A, Frei K, Flury R, et al. Local Fas/APO-1 (CD95) ligand-mediated tumor cell killing in vivo. Eur J Immunol 1995, 25:2253–2258.

118. Ogasawara J, Watanabe-Fukunaga R, Adachi M, et al. Lethal effect of the anti-Fas antibody in mice. Nature 1993, 364:806–809.

119. Timmer T, de Vries EG, de Jong S. Fas receptor-mediated apoptosis: a clinical application? J Pathol 2002, 196:125–134.

120. Ichikawa K, Yoshida-Kato H, Ohtsuki M, et al. A novel murine anti-human Fas mAb which mitigates lymphadenopathy without hepatotoxicity. Int Immunol 2000, 12:555–562.

121. Ashkenazi A, Pai RC, Fong S, et al. Safety and antitumor activity of recombinant soluble Apo2 ligand. J Clin Invest 1999, 104:155–162.

122. Walczak H, Miller RE, Ariail K, et al. Tumoricidal activity of tumor necrosis factor-related apoptosis-inducing ligand in vivo. Nat Med 1999, 5:157–163.

123. Kelley SK, Harris LA, Xie D, et al. Preclinical studies to predict the disposition of Apo2L/tumor necrosis factor-related apoptosis-inducing ligand in humans: characterization of in vivo efficacy, pharmacokinetics, and safety. J Pharmacol Exp Ther 2001, 299:31–38.

124. Wajant H, Pfizenmaier K, Scheurich P. TNF-related apoptosis inducing ligand (TRAIL) and its receptors in tumor surveillance and cancer therapy. Apoptosis 2002, 7:449–459.

125. Bonavida B, Ng CP, Jazirehi A, et al. Selectivity of TRAIL-mediated apoptosis of cancer cells and synergy with drugs: the trail to non-toxic cancer therapeutics (review). Int J Oncol 1999, 15: 793–802.

126. Shankar S, Srivastava RK. Enhancement of therapeutic potential of TRAIL by cancer chemotherapy and irradiation: mechanisms and clinical implications. Drug Resist Updat 2004, 7:139–156.

127. Gliniak B, Le T. Tumor necrosis factor-related apoptosis-inducing ligand's antitumor activity in vivo is enhanced by the chemotherapeutic agent CPT-11. Cancer Res 1999, 59:6153–6158.

128. Fischer U, Schulze-Osthoff K. New approaches and therapeutics targeting apoptosis in disease. Pharmacol Rev 2005, 57:187–215.

129. MacEwan DJ. TNF ligands and receptors—a matter of life and death. Br J Pharmacol 2002, 135:855–875.

130. Curnis F, Sacchi A, Borgna L, et al. Enhancement of tumor necrosis factor alpha antitumor immunotherapeutic properties by targeted delivery to aminopeptidase N (CD13). Nat Biotechnol 2000, 18:1185–1190.

131. Hersh EM, Metch BS, Muggia FM, et al. Phase II studies of recombinant human tumor necrosis factor alpha in patients with malignant disease: a summary of the Southwest Oncology Group experience. J Immunother 1991, 10:426–431.

132. Lienard D, Ewalenko P, Delmotte JJ, et al. High-dose recombinant tumor necrosis factor alpha in combination with interferon gamma and melphalan

in isolation perfusion of the limbs for melanoma and sarcoma. J Clin Oncol 1992, 10:52–60.

133. Renard N, Lienard D, Lespagnard L, et al. Early endothelium activation and polymorphonuclear cell invasion precede specific necrosis of human melanoma and sarcoma treated by intravascular high-dose tumor necrosis factor alpha (rTNF alpha). Int J Cancer 1994, 57:656–663.

134. Longley DB, Wilson TR, McEwan M, et al. c-FLIP inhibits chemotherapy-induced colorectal cancer cell death. Oncogene 2006, 25:838–848.

135. Kamarajan P, Sun NK, Chao CC. Up-regulation of FLIP in cisplatin-selected HeLa cells causes cross-resistance to CD95/Fas death signalling. Biochem J 2003, 376:253–260.

136. Conticello C, Pedini F, Zeuner A, et al. IL-4 protects tumor cells from anti-CD95 and chemotherapeutic agents via up-regulation of antiapoptotic proteins. J Immunol 2004, 172:5467–5477.

137. Matta H, Eby MT, Gazdar AF, et al. Role of MRIT/cFLIP in protection against chemotherapy-induced apoptosis. Cancer Biol Ther 2002, 1:652–660.

138. Hyer ML, Croxton R, Krajewska M, et al. Synthetic triterpenoids cooperate with tumor necrosis factor-related apoptosis-inducing ligand to induce apoptosis of breast cancer cells. Cancer Res 2005, 65:4799–4808.

139. Schimmer AD, Thomas MP, Hurren R, et al. Identification of small molecules that sensitize resistant tumor cells to tumor necrosis factor-family death receptors. Cancer Res 2006, 66:2367–2375.

140. Klasa RJ, Gillum AM, Klem RE, et al. Oblimersen Bcl-2 antisense: facilitating apoptosis in anticancer treatment. Antisense Nucleic Acid Drug Dev 2002, 12:193–213.

141. Frantz S. Lessons learnt from Genasense's failure. Nat Rev Drug Discov 2004, 3:542–543.

142. Gleave ME, Monia BP. Antisense therapy for cancer. Nat Rev Cancer 2005, 5:468–479.

143. Fesik SW. Promoting apoptosis as a strategy for cancer drug discovery. Nat Rev Cancer 2005, 5:876–885.

144. Tolcher AW, Chi K, Kuhn J, et al. A phase II, pharmacokinetic, and biological correlative study of oblimersen sodium and docetaxel in patients with hormone–refractory prostate cancer. Clin Cancer Res 2005, 11:3854–3861.

145. Del Bufalo D, Trisciuoglio D, Scarsella M, et al. Treatment of melanoma cells with a bcl-2/bcl-xL antisense oligonucleotide induces antiangiogenic activity. Oncogene 2003, 22:8441–8447.

146. Walensky LD, Kung AL, Escher I, et al. Activation of apoptosis in vivo by a hydrocarbon-stapled BH3 helix. Science 2004, 305:1466–1470.

147. Oltersdorf T, Elmore SW, Shoemaker AR, et al. An inhibitor of Bcl-2 family proteins induces regression of solid tumours. Nature 2005, 435:677–681.

148. Schimmer AD, Dalili S, Batey RA, et al. Targeting XIAP for the treatment of malignancy. Cell Death Differ 2006, 13:179–188.

149. Sasaki H, Sheng Y, Kotsuji F, et al. Down-regulation of X-linked inhibitor of apoptosis protein induces apoptosis in chemoresistant human ovarian cancer cells. Cancer Res 2000, 60:5659–5666.

150. McManus DC, Lefebvre CA, Cherton-Horvat G, et al. Loss of XIAP protein expression by RNAi and antisense approaches sensitizes cancer cells to functionally diverse chemotherapeutics. Oncogene 2004, 23:8105–8117.

151. Holcik M, Yeh C, Korneluk RG, et al. Translational upregulation of X-linked inhibitor of apoptosis (XIAP) increases resistance to radiation induced cell death. Oncogene 2000, 19:4174–4177.

152. Cummings J, Ranson M, Lacasse E, et al. Method validation and preliminary qualification of pharmacodynamic biomarkers employed to evaluate the clinical efficacy of an antisense compound (AEG35156) targeted to the X-linked inhibitor of apoptosis protein XIAP. Br J Cancer 2006, 95:42–48.

153. LaCasse EC, Cherton-Horvat GG, Hewitt KE, et al. Preclinical characterization of AEG35156/GEM 640, a second-generation antisense oligonucleotide targeting X-linked inhibitor of apoptosis. Clin Cancer Res 2006, 12:5231–5241.

154. Reed JC, Bischoff JR. BIRinging chromosomes through cell division—and survivin' the experience. Cell 2000, 102:545–548.

155. Schimmer AD. Inhibitor of apoptosis proteins: translating basic knowledge into clinical practice. Cancer Res 2004, 64:7183–7190.

156. Sun Y, Lin R, Dai J, et al. Suppression of tumor growth using antisense oligonucleotide against survivin in an orthotopic transplant model of human hepatocellular carcinoma in nude mice. Oligonucleotides 2006, 16:365–374.

157. Zhang MC, Hu CP, Chen Q. [Effect of down-regulation of survivin gene on apoptosis and cisplatin resistance in cisplatin resistant human lung adenocarcinoma A549/CDDP cells]. Zhonghua Zhong Liu Za Zhi 2006, 28:408–412.

158. Salomon DS, Brandt R, Ciardiello F, et al. Epidermal growth factor-related peptides and their receptors in human malignancies. Crit Rev Oncol Hematol 1995, 19:183–232.

159. Herbst RS, Shin DM. Monoclonal antibodies to target epidermal growth factor receptor-positive tumors: a new paradigm for cancer therapy. Cancer 2002, 94:1593–1611.

160. Bruns CJ, Harbison MT, Davis DW, et al. Epidermal growth factor receptor blockade with C225 plus gemcitabine results in regression of human pancreatic carcinoma growing orthotopically in nude mice by antiangiogenic mechanisms. Clin Cancer Res 2000, 6:1936–1948.

161. Milas L, Mason K, Hunter N, et al. In vivo enhancement of tumor radioresponse by C225 antiepidermal growth factor receptor antibody. Clin Cancer Res 2000, 6:701–708.

162. Huang SM, Li J, Armstrong EA, et al. Modulation of radiation response and tumor-induced angiogenesis after epidermal growth factor receptor inhibition by ZD1839 (Iressa). Cancer Res 2002, 62:4300–4306.

163. Williams KJ, Telfer BA, Stratford IJ, et al. ZD1839 ('Iressa'), a specific oral epidermal growth factor receptor-tyrosine kinase inhibitor, potentiates radiotherapy in a human colorectal cancer xenograft model. Br J Cancer 2002, 86:1157–1161.

164. Ciardiello F, Caputo R, Bianco R, et al. Antitumor effect and potentiation of cytotoxic drugs activity in human cancer cells by ZD-1839 (Iressa), an epidermal growth factor receptor-selective tyrosine kinase inhibitor. Clin Cancer Res 2000, 6:2053–2063.

165. Saltz LB, Meropol NJ, Loehrer PJ Sr, et al. Phase II trial of cetuximab in patients with refractory colorectal cancer that expresses the epidermal growth factor receptor. J Clin Oncol 2004, 22:1201–1208.

166. Cunningham D, Humblet Y, Siena S, et al. Cetuximab monotherapy and cetuximab plus irinotecan in irinotecan-refractory metastatic colorectal cancer. N Engl J Med 2004, 351:337–345.

167. Fukuoka M, Yano S, Giaccone G, et al. Multi-institutional randomized phase II trial of gefitinib for previously treated patients with advanced non-small-cell lung cancer (The IDEAL 1 Trial) [corrected]. J Clin Oncol 2003, 21:2237–2246.

168. Kris MG, Natale RB, Herbst RS, et al. Efficacy of gefitinib, an inhibitor of the epidermal growth factor receptor tyrosine kinase, in symptomatic patients with non–small cell lung cancer: a randomized trial. Jama 2003, 290:2149–2158.

169. Giaccone G, Herbst RS, Manegold C, et al. Gefitinib in combination with gemcitabine and cisplatin in advanced non-small-cell lung cancer: a phase III trial—INTACT 1. J Clin Oncol 2004, 22:777–784.

170. Herbst RS, Giaccone G, Schiller JH, et al. Gefitinib in combination with paclitaxel and carboplatin in advanced non-small-cell lung cancer: a phase III trial—INTACT 2. J Clin Oncol 2004, 22:785–794.

171. Herbst RS, Prager D, Hermann R, et al. TRIBUTE: a phase III trial of erlotinib hydrochloride (OSI-774) combined with carboplatin and paclitaxel chemotherapy in advanced non-small-cell lung cancer. J Clin Oncol 2005, 23:5892–5899.

172. Gatzemeier U, Pluzanska A, Tabata M. Results of a phase III trial of erlotinib (OSI-774) combined with cisplatin and gemcitabine (GC) chemotherapy in advanced non-small cell lung cancer (NSCLC) [abstract]. J Clin Oncol 2004, 22(14S):

173. Berenson JR, Ma HM, Vescio R. The role of nuclear factor-kappaB in the biology and treatment of multiple myeloma. Semin Oncol 2001, 28:626–633.

174. LeBlanc R, Catley LP, Hideshima T, et al. Proteasome inhibitor PS-341 inhibits human myeloma cell growth in vivo and prolongs survival in a murine model. Cancer Res 2002, 62:4996–5000.

175. Adams J. Proteasome inhibition in cancer: development of PS-341. Semin Oncol 2001, 28:613–619.

176. L'Allemain G. [Update on … the proteasome inhibitor PS341]. Bull Cancer 2002, 89:29–30.

177. Lu Y, Wang H, Mills GB. Targeting PI3K-AKT pathway for cancer therapy. Rev Clin Exp Hematol 2003, 7:205–228.

178. Mitsiades CS, Mitsiades N, Koutsilieris M. The Akt pathway: molecular targets for anti-cancer drug development. Curr Cancer Drug Targets 2004, 4:235–256.

179. Ponka P, Lok CN. The transferrin receptor: role in health and disease. Int J Biochem Cell Biol 1999, 31:1111–1137.

180. Pirollo KF, Zon G, Rait A, et al. Tumor-targeting nanoimmunoliposome complex for short interfering RNA delivery. Hum Gene Ther 2006, 17:117–124.

181. Wilson TR, McLaughlin KM, McEwan M, et al. C-FLIP: a key regulator of colorectal cancer cell death. Cancer Res 2007, 67:5754–5762.

Chapter 17
Peptides and Peptidomimetics as Cancer Therapy Sensitizing Agents

Shantanu Banerji, Sudharsana Rao Ande, Subbareddy Maddika, Versha Banerji, Iran Rashedi, Neil W. Owens, Anne Zuse, Frank Schweizer, and Marek Los

1   Introduction

Apoptosis, or programmed cell death, is the body's normal method of disposing of unwanted or damaged cells, and represents a crucial step in the life cycle of a cell [1]. Disruption of this process is believed to be associated with tumor growth and resistance to chemotherapy, and so has recently become a major focus of research [2]. While the process is still not completely understood, apoptosis is mediated through a cascade of events induced from extracellular or intracellular signals [3] and converge to induce the activity of a group of enzymes, specifically the caspases, which dismantle the cell once activated.

Treatment of malignancy has traditionally involved the use of cytotoxic chemotherapy drugs directed against several key cell growth functions such as DNA synthesis and cell division. Cancer cells are believed to take advantage of an impaired genetic machinery and cell cycle control, which normally allow any interference in these critical mechanisms to lead to cell death, primarily through the activation of apoptosis [4]. Using cultured murine leukemia L1210 cells with a growth fraction of 100%, Skipper et al. first demonstrated that a given concentration of cytotoxic drug killed a fixed proportion of cancer cells in vitro following logarithmic cell-kill kinetics [5]. However, years of clinical experience, particularly in patients with metastatic disease, has revealed that cytotoxic chemotherapy itself is incapable of eradicating the disease in the majority of cancer subtypes

in vivo. Proposed reasons for this failure of therapy include innate drug resistance mechanisms such as escape from active cellular growth and expression of surface transporters leading to efflux of drug from the cell [6]. Also, impaired delivery of drug in vivo further limits the concentration of drug in target tissue, thereby limiting its effectiveness. Finally, the relatively nonspecific nature of these agents leads to injury of noncancer tissues. The resulting toxicity significantly restricts the therapeutic index of each individual drug [7]. Therefore, there is an urgent need for the development of new anticancer agents that would preferentially target malignant cells. New developments are frequently inspired by rapidly increasing knowledge of the pathomechanism of uncontrolled cancerous growth, malfunction of apoptotic processes, and deregulation of angiogenesis at the tumor site.

2   Antiangiogenic Peptides and Peptidomimetics as Cancer Sensitizing Agents

The neovascularization of tumors can occur via three different processes: angiogenesis, vasculogenesis, and intussusception [8, 9]. Angiogenesis is the proliferation of endothelial cells to form new blood vessels from pre-existing ones [10]. Under physiologically conditions a balance between inducers of angiogenesis and inhibiting factors are

From: *Sensitization of Cancer Cells for Chemo/Immuno/Radio-therapy, 1st Edition.*
Edited by: Benjamin Bonavida © Humana Press, Totowa, NJ

maintained. This occurs normally during various processes in the human body such as wound healing, embryogenesis, female ovulatory cycle, and development of collateral blood circulation. Only up to 1–2 mm of the tumor is able to absorb nutrients and oxygen by diffusion, but the tumor cannot expand to a significant size without the sprouting of a novel vascular network [11]. Also, the removal of metabolic products through the newly formed vessels is necessary for the tumor growth and metastasis. The switch to the angiogenic stage is dependent on angiogenic factors such as vascular endothelial growth factor (VEGF), which stimulates capillary formation in vivo. It has also been shown to direct mitogenic activities that are restricted to endothelial cells [12]. The key receptor which triggers angiogenesis and vascular permeability is VEGFR2 [13]. Cutting off the supply of tumor cells by inhibiting the formation of vascular network represent a well-recognized approach in cancer therapy [9, 14, 15]. The efficacy of conventional chemotherapy can be increased by the combination of classical anti-cancer treatment with anti-angiogenic peptides [17]. Also, properly designed peptides represent ideal carriers of therapeutics because they have an excellent tumor penetration. Advantages of using angiogenesis inhibitors are that these agents are less toxic than the conventional therapeutic approach and have a lower risk to develop drug resistance [16]. The formation of new blood vessels is a multiple stage process [12]. The specific receptors on the surface of the endothelial cells are activated through the binding of angiogenic growth factors leading to signaling from cell surface to the nucleus. This produces several molecules such as enzymes necessary for development of the vascular network. After dissolving tiny holes in the basement membranes, the endothelial cells can now begin to proliferate and migrate toward the tumor tissue. The sprouting of new blood vessels is supported by adhesion molecules or integrins while holes are formed in the tissue with the help of matrix metalloprotein kinases (MMP). Through rounding up of the sprouting cells the new blood vessel tube is formed and stabilized by smooth muscle cells, or pericytes. Anti-angiogenic therapy can target different events in the angiogenic cascade including blockade of the soluble angiogenic factors such as VEGF, e.g., the monoclonal antibodies Bevacizumab [14] or VEGF Trap [12], interruption

of downstream signaling by tyrosine kinase inhibitors [15], mimetic antiangiogenesis inhibitors such as endostatin [17], or preventing the binding of ligands to endothelial receptors with anti-VEGFR antibody [9].

Several endogenous angiogenesis inhibitors in plasma and in extracellular matrix exist such as arrestin, interferon-β, pigment epithelial derived factor (PEDF), platelet factor 4 and tumstatin [17–19]. Angiostatin was one of the first direct inhibitors described to specifically inhibit the growth of the endothelial cell in vitro and angiogenesis in vivo. Angiostatin is a collective name for different internal fragments of plasminogen. The cleavage by proteases results in the formation of 38- and 45-kDa peptides, which contain homologous triple-disulfide bridged kringle domains [17]. Four potential targets for angiostatin including ATP synthase, $\alpha_v\beta_3$ integrin, angiomotin, and NG2 chondroitin sulfate proteoglycan (CSPG) have been described [20]. Cell adhesion to several extracellular matrix proteins is mediated by $\alpha_v\beta_3$ integrins that are upregulated on the surface of proliferating endothelial cells in angiogenic microvessels. The non-matrix derived inhibitor binds specifically to $\alpha_v\beta_3$ integrin, thereby reducing the cells attachment/adhesion and inhibits plasmin-induced cell migration [20]. Although it contains the Arg-Gly-Asp (RGD) motif, angiostatin leads to RGD-independent induction of focal adhesion kinase (FAK) activity. The molecule binds to ATP synthase on the surface of endothelial cells and seems to be responsible for the reduced pH in the cell that can trigger apoptosis [17]. Through inhibiting the synthase activity, the source of cellular energy and the proangiogenic process is blocked. Angiostatin also binds to the 72-kDa molecule angiomotin which is a cell-surface–associated protein that is co-localized with FAK [20]. Another mechanism of action for angiostatin could be through its antagonism to angiomotin thereby inhibiting the interaction with promigratory proteins. A further endogenous inhibitor is the 20-kDa endostatin that represents a carboxyl-terminal proteolytic cleavage fragment of collagen type XVIII [20, 21]. The agent blocks angiogenesis, and suppresses tumor growth and metastasis in experimental animal models [22]. Not all receptor mediated biological effects of endostatin are known. For example, TNF-induced activation of c-Jun NH_2-terminal kinase and c-Jun

NH_2-terminal kinase-dependent pro-oncogenic gene expression is inhibited by endostatin [17].

Pex, a 210 amino acid, noncatalytic COOH-terminal hemopexin-like fragment of MMP-2 belongs to the family of endogenous angiogenesis inhibitors [8]. Endothelial cell invasion is promoted by the interaction of MMP-2 and $\alpha_v\beta_3$ integrin. Proteolytic activity on the cell surface can be blocked by preventing the interaction of the two molecules [23] Angiogenesis, cell invasion, formation of capillary-like structures, and tumor growth in nude mice are inhibited by Pex [17, 20]. Furthermore, platelet factor-4 (Pf-4) represents an endogenous protein that is released from platelet α-granules during platelet aggregation and has anti-angiogenic properties through blocking the dimerization of FGF-2 and its binding to endothelial cells [17, 26]. The carboxyl-terminal fragment of Pf-4 shows greater inhibitory activity than the parent molecule [24]. Further mechanisms include the binding to anti-angiogenic growth factors such as VEGF [20], competitive inhibition of bFGF, binding to HSPG, or binding to chemokine receptor CXCR3-B which results in increased levels of cAMP and p21[CIP1] [24]. Recombinant human platelet factor-4 (rhPF4) inhibits blood vessel proliferation in the chicken chorioallantoic membrane in a dose-dependent manner. The inhibition of growth factor–stimulated endothelial cell proliferation seems to be responsible for the angiostaxic effect of rhPF4, while the inhibitory activities are associated with the carboxyl-terminal, heparin-binding region of the molecule. One more endogenous molecule is PEDF, a 50-kDa glycoprotein, which is a noninhibitory member of the serpin (serine protease inhibitors) family. The exact action of the protein is not known but studies have shown that collagen- and heparin-binding regions mediate interactions with different proteins and are important for the activity of the molecule [20].

Furthermore, angiogenesis may be modulated by the regulation of endothelial cell apoptosis via c-FLIP controlled by NFAT (nuclear factor of activated T cells) and its upstream regulator JNK. Increase in the activity of the Fas death receptor pathway in endothelial cells has anti-angiogenic effects [17, 20]. Under pro-angiogenic conditions endothelial cells would be protected because NFAT would upregulate the expression of c-FLIP, a competitive inhibitor of caspase-8 [20]. PEDF, however, induces upregulation of FasL on endothelial cells, activates

c-Jun N-terminal protein kinase, and thereby blocks the signaling of transcription factor NFAT [25] As a consequence, the lack of c-FLIP, which has a very short half-life, caspase-8 is no longer inhibited and apoptosis of endothelial cells may occur.

Finally, thrombospondin-1 and -2 (TSP-1/-2) belong to the larger family of TSP proteins, containing three type 1 repeat domains and represent members of the endogenous angiogenic inhibitors [20]. TSP-1 is a 90-kD polypeptide and was named when it was noted that TSP is released from α-granules in platelets following exposure to thrombin [26]. It was the first discovered natural protein with inhibiting effect on angiogenesis. The protein also has some effects on other biological processes such as cell adhesion, cell proliferation, and survival. Transcriptional activation of TSP-1 occurs through the tumor suppressor protein p53 [26]. Inhibition of cell migration and induction of apoptosis is mediated by binding to CD36 on the surface of endothelial cells. TSP1 also inhibits angiogenesis by upregulation of FasL expression [27]. TSP2 inhibits growth of human microvascular endothelial cells (HMVECs) causing cell cycle arrest at the G_0/G_1 phase through inhibition of cyclin D1, in a caspase-independent manner [28]. Since TSP is a large protein, its use as therapeutic agent has been limited. This lead to the development of a derivative from TSP-1, the nonapeptide GVITRIR (ABT-510), whose structure is based on the NH_2-terminal portion of the molecule that is responsible for the anti-angiogenic activity [26].

Recent advances in understanding functions of cell adhesion molecules (CAM) has led to the development of novel anticancer agents that target the extracellular matrix (ECM). There are four classes of CAMs: cadherins, integrins, selectins, and the immunoglobulin superfamily [29]. A further approach to inhibit the formation of new blood vessels is to attack cadherins, which represent a family of transmembrane glycoproteins mediating calcium-dependent homophilic cell–cell interactions regulating cell motility, proliferation, and apoptosis. The cadherin superfamily is composed of various subfamilies [30]. Cadherins are able to bind catenins and further anchor cytoskeletal and signaling molecules through their intracellular domain [31]. On the homophilic binding site of classical cadherins type I is found the His-Ala-Val

(HAV) sequence [32]. Peptides that contain this motif are able to inhibit actions of cadherins such as cell adhesion. The strength of the cell adhesion is regulated by the tyrosine phosphorylation status of cadherin, β-catenin, and p120^{ctn} [33]. One example of active inhibitors is the cyclic peptide N-Ac-CHAVC-NH$_2$ that interrupts the endothelial cell actions derived from cadherin [31, 34]. It induces cell-density–dependent apoptosis and affects distribution of various junctional components such as β-catenin in different ways. It was shown that in endothelial cells the activation of FGF receptor might be involved in cadherin-derived signaling because peptide-induced apoptosis is blocked by bFGF. Further suppression of tyrosine phosphorylation of FRS2, a target of FGFR has been observed [31]. It was found that a linear peptide mimetic of a short sequence in ECD1 (first extracellular domain) of N-cadherin (INPISGQ) represents a highly specific N-cadherin antagonist. It has also been shown that a short cyclic peptide containing the INP motif of N-cadherin also shows a strong antagonistic effect for the molecule [35]. Several short cyclic HAV peptides which inhibit N-cadherin function are identified. It has been demonstrated that flanking of the HAV motif with different amino acids has influence on cadherin specificity [34].

Further molecules for targeting cell adhesion and inhibiting angiogenesis are integrins which belong to a family of cell surface glycoproteins. These molecules are heterodimeric surface proteins that mediate the interactions between the cell surface and surrounding ECM. A survival signal from the ECM is transduced by $\alpha_v\beta_3$ integrin to the proliferating vascular cells [36]. Endothelial cell migration is mediated by integrin through interactions with the tripeptide motif Arg-Gly-Asp (RGD). Antagonizing the function of $\alpha_v\beta_3$ integrin can be achieved either by RGD-containing peptides or by monoclonal antibodies against integrin [37, 38]. The synthetic cyclic RGD-containing peptide (cyclo(-RGDf=V-)) inhibits angiogenesis through binding of $\alpha_v\beta_3$ integrin. Furthermore, cyclo(-RGDfV-) (fV) and cyclo(-RGDf-MeV-)(fMeV) inhibit adhesion and growth of HUVEC cells in a dose-dependent manner [36]. Cyclic RGD-pentapeptides are expected to be more potent inhibitors of integrin because a higher affinity of cyclic compounds compared with linear peptides are presumed [39]. Various RGD peptides have been developed includ-

ing the cyclic peptides cRGDfK, cRGDyK, and RGDC4 that effectively inhibit angiogenesis in solid tumors [29]. Selective targeting of the solid tumor is possible because endothelial cells in these tumors express α_v integrins that are not detectable in normal blood vessels [29]. The five amino acid peptide ATN-161 (Ac-PHSCN-NH$_2$) represents a β-integrin antagonist which was recently tested in phase I trials in patients with solid tumors [40]. The peptide is derived from the synergy region of fibronectin and the integrin-suppressing effect might result from forming disulfide bonds with the integrin target through its free cysteine thiol [40]. Through screening of a conformational constrained bicyclic lactam RGD-containing pseudopeptides library, a novel molecule, ST1646, was identified as a highly specific antagonist of $\alpha_v\beta_3/\alpha_v\beta_5$ integrin [41]. Further integrin interacting agents were found in animals. These include salmosin, which represents a novel disintegrin, isolated from Korean snake (Agkistrodon halys brevicaudus) venom, which inhibits platelet aggregation, endothelial cell proliferation, and tumor progression through antagonism of $\alpha_v\beta_3$ integrin–derived cell interactions. Disintegrins are a family of cysteine-rich, low-molecular-weight RGD-containing peptides found in different snake venoms. They are able to inhibit fibrinogen binding to glycoprotein IIb/IIIa complex as well as binding of other ligands to RGD-dependent integrins [42]. Salmosin comprises 73 amino acids, including 12 cysteines and an RGD sequence. The molecule significantly inhibits solid tumor growth by blocking angiogenesis and proliferation of tumor cells while no effect on the normal cell growth is observed [43]. When tested in a murine melanoma model, subcutaneous administration of salmosin in cationic liposomes resulted in systemic expression of the gene product and concomitant inhibition of the growth of B16BL6 melanoma cells. Suppression of pulmonary metastases, verified by experimental and spontaneous metastasis models in mice, also resulted from salmosin gene treatment [44]. The transfection of the disintegrin gene into an animal model showed that the expressed product was able to inhibit angiogenesis in vivo.

Another disintegrin found in snake venom is Saxatilin, a 7.7-kDa member of homologous proteins with six disulfide bonds, only three of them seem to be important for biological activity [45]. Saxatilin strongly inhibit endothelial cell proliferation and smooth muscle cell migration and has only little effect on normal cell growth [46]. Interaction of human umbilical vein cell with immobilized vitronectin is also inhibited by binding of saxatilin to $\alpha_v\beta_3$ integrin [47]. In addition, proliferation of cancer cell line MDAH 2774 induced by tumor necrosis factor-alpha (TNF) is significantly suppressed by treatment with saxatilin [48].

MMPs are a family of zinc-dependent neutral endopeptidases which are involved in the remodeling of ECM and in neo-vascularization [32]. Batimastat (BB-94) represents an effective broad-spectrum inhibitor of MMPs, which is a substituted analogue peptide derived from one side of type I collagen [49]. Treatment with BB-94 results in resolution of ascites, tumor growth delay in different animal models, and dose-dependent increase in survivals of nude mice in xenograft models [49]. While the relatively nonselective inhibitor, Marimastat, showed high musculoskeletal toxicity in some studies, other more specific agents are under development [9].

Anginex represents a 33-residue designed peptide with anti-angiogenic, apoptotic and anti-tumoral activities with structural relationship to various anti-angiogenic cytokines [50, 51]. Further inhibition of adhesion and migration in endothelial cells is also described. The beta-sheet–forming peptide is composed of numerous cationic and hydrophobic amino acids [51]. It is known that amphiphilic antimicrobial peptides bind to membranes and destroy lipid bilayers by micellization and pore formation [52]. Based on its structural similarities to antimicrobial peptides, its ability to act with cell membranes was determined. Data have shown that Angionex binds to lipid bilayers with preference to anionic and zwitterionic phospholipids causing membrane depolarization and permeabilization [51]. It synergistically improves chemotherapy with carboplatin and its combination with angiostatin results in higher tumor growth arrest [53].

Angiogenesis plays a key role in many pathological conditions. The discovery that tumor growth and metastasis are angiogenesis-dependent presents a novel therapeutic approach to block tumor growth and prevent metastasis by cutting off the tumors supply. The combination of potent inhibitors of angiogenesis with conventional chemotherapy, radiation, or gene therapy has been demonstrated to be successful [53].

3 Peptides Targeting Bcl-2 and Bcl-xL Proteins

B-cell lymphoma 2 (Bcl-2) is a proto-oncogene, identified at the chromosomal break point (16; 18) (q32; q21) in B-cell lymphomas [54]. It is one of the most extensively studied members among the Bcl-2 family of proteins. There are about 20 proteins known so far that belong to the Bcl-2 family. The family functions as key regulators of apoptosis. Based on their role in apoptotic cell demise, they are categorized into pro-apoptotic proteins and anti-apoptotic proteins. Pro-apoptotic proteins initiate the apoptotic process and include Bax, Bad, Bim, Bak, Bok, Bid, and anti-apoptotic members include Bcl-2 and Bcl-xL. Anti-apoptotic family of proteins, inhibit the apoptotic process by preventing the release of cytochrome c from mitochondria [55] and also form heterodimers with the pro-apoptotic Bcl-2 proteins so that they no longer induce the cells to undergo apoptosis [56]. Important characteristic feature of the Bcl-2 family is that they carry at least one of the B-cell homology (BH) domains, BH1-BH4. Anti-apoptotic proteins carry all four domains, whereas pro-apoptotic proteins are divided into the ones that carry only the BH3 domain, which include Bad, Bid, Bim, etc. and others that carry all the BH domains except BH4. The latter include, Bax, Bak, and Bok.

The BH3 domain of pro-apoptotic Bcl-2 family proteins interacts with the surface pocket of Bcl-2 and Bcl-xL proteins. This interaction leads to the elimination of the anti-apoptotic function of Bcl-2 and Bcl-xL proteins [57]. In most cancers, these Bcl-2 and Bcl-xL proteins are overexpressed hence preventing the cells from undergoing apoptosis [58]. Thus, targeting of Bcl-2 and Bcl-xL proteins using various strategies that downregulate their anti-apoptotic function is a valid approach for the development of anticancer therapies. Among several strategies currently under development, peptide based inhibitors of anti-apoptotic Bcl-2 family members gained significant attention within both industry and academia.

Short and minimal BH3 domain peptides of Bak and Bax are sufficient to facilitate apoptosis in a cell-free system [59]. For example, BH3 domain peptides of pro-apoptotic proteins Bax and Bak activated caspases by releasing cytochrome c from mitochondria. However, the BH3 domain peptide of Bcl-2 protein fails to induce apoptosis. BH3 domain peptides of pro-apoptotic proteins Bax and Bak efficiently bind to Bcl-2 family cell death suppressors thereby disrupting their anti-apoptotic function, followed by induction of apoptosis [59].

A synthetic peptide corresponding to the BH3 domain of pro-apoptotic protein Bak (Bak-BH3) is the potential inducer of apoptosis in HeLa cells. Microinjection of Bcl-xL, a key cell death suppressor, into HeLa cells protected the cells from undergoing death receptor induced apoptosis [60]. However, co-injection of Bcl-xL and Bak-BH3 peptide did not protect the cells from receptor-mediated apoptosis, indicating the Bak-BH3 peptide's role in mediating the apoptotic process. Co-injection of mutant peptide, which does not have the ability to bind Bcl-xL along with Bcl-xL protein retained the protective effect against receptor-mediated apoptosis, indicating that Bak-BH3 peptide promotes apoptosis by binding to Bcl-xL protein. In addition, fusion of Bak-BH3 peptide with the internalizing sequence of antennapedia protein (Ant) delivered the Ant-Bak-BH3 peptide into intact HeLa cells. Internalization of this peptide greatly reduced the viability of the cells [60]. However, treatment of cells with internalizing sequence (Ant) alone or Bak-BH3 peptide alone had no effect on the viability of the cells, indicating that internalization is essential for the cells to be vulnerable to cell demise. In this experiment, overexpression of Bcl-xL, prevented the cells from Ant-Bak-BH3 induced apoptosis. Yet, Ant-Bak-BH3 could re-sensitize the Bcl-xL–overexpressing cells to receptor-mediated apoptosis. However, internalization of mutant Ant-Bak-BH3 protein, which does not have the ability to bind Bcl-xL, was ineffective in inducing the cell death. It can be concluded from these experiments that Bak-BH3 peptides interact with Bcl-xL death suppressors, thereby possibly inducing conformational changes so that they no longer function as death suppressors, hence causing the cancer cells to undergo the safest mode of cell death by apoptosis.

The BH3 domain of pro-apoptotic protein Bad has been fused to a decanoic acid, which can effectively deliver the peptide into the cell [61]. This cell permeable peptide (cpm-1285) induced all hallmarks of apoptosis like activation of caspase-3, apoptotic DNA-fragmentation, and cleavage of PARP in the human myeloid HL-60 cell line. The induction of apoptosis was mediated by binding of the Bad-derived, modified BH3-peptide to Bcl-2, and inhibiting it. Interestingly, cpm-1285 had a little effect in

non-tumor cells. In vivo experiments revealed that, cpm-1285 could decrease the tumor growth in the mice bearing human myeloid leukemia [61].

Targeting of peptides by fusing them to transport-proteins, that can direct the peptide into the cell is another interesting strategy. For example, VP22, a structural protein of herpes simplex virus, form a complex particle of protein and oligonucleotides when incubated with oligonucleotides. These particles are cell permeable and stable inside the cells [62]. When exposed to light, these particles dissociate by themselves. Using this phenomenon, BH3-peptide was fused with VP22-protein generating BH3-VP22 fusion protein. When incubated with oligonucleotides, the particles composed of BH3-VP22 and oligonucleotides were formed and masked the BH3-domain. These particles stably entered the cells, whereupon activation by light unmasked the BH3-peptide allowing it to induce apoptosis in adenocarcinoma cell lines [63]. In vivo experiments revealed that the tumor growth was reduced in immunodeficient mice, when injected with oligonucleotide-coated BH3-VP22 particles with subsequent exposure to the light [63].

Another interesting and effective result of targeting Bcl-2 anti-apoptotic family proteins using peptides comes from the report of Walensky and colleagues (2004). The basic requirements for a peptide to perform its targeting function are protease resistance and cell permeability. In order to serve these requirements, they stapled the BH3 domain of Bid with a hydrocarbon chain [64]. This stapled peptide induced apoptosis in leukemia cells. In Bcl-2–overexpressing leukemia cells, lower concentrations of the stapled peptide could not stimulate the cells to apoptotic cell death; however at higher concentrations, it efficiently induced apoptosis by downregulating the anti-apoptotic function of Bcl-2. In addition, hydrocarbon-stapled Bid-BH3 peptide suppressed tumor growth in the immunodeficient mice xenografted with lymphoid leukemia [64].

4 Peptide-Based Targeting of Protein Kinases in Cancer Therapy

The protein kinases accounting for nearly 2% of the human genome contain a large superfamily of enzymes that catalyze the transfer of the terminal phosphoryl group of ATP to their specific protein substrates [65, 66]. They play a critical role in various signal transduction pathways and regulate different cellular functions, such as metabolism, cell division and survival, transcription, differentiation, and cell death [67]. Protein kinases are now the second largest group of drug targets, approximating almost 25% of current drug development efforts by various groups [68]. To date, most protein kinase inhibitors in clinical development such as Imatinib target the ATP-binding site of the kinase. However, the disadvantage of this strategy is that these ATP-binding site inhibitors must compete with high intracellular ATP concentrations. Furthermore, if these inhibitors are to be specific, they must discriminate between the ATP-binding sites of other protein kinases as well as more than 500 other human cellular proteins that also utilize ATP, thus making these compounds liable for nonspecific inhibition of signaling pathways. Thus, peptides and the peptidomimetics that can interfere with the protein–protein interactions and can target sites other than the ATP-binding sites may provide a powerful means of inhibiting various signaling events that can be utilized in successful cancer therapy.

So far, though in early stages, several peptide inhibitors have been developed and characterized for different protein kinases that are involved in the process of early tumorigenesis, cancer progression, cancer cell survival, and metastasis (Table 17.1). ErbB2 and the epidermal growth factor receptor (EGFR), the EGF receptor tyrokine kinase family members are overexpressed in several tumors such as breast, lung and colon cancers and correlate with poor prognosis (69). EC-1, a peptide inhibitor identified by phage display random peptide library, targets the extracellular domain of Erb2 and was shown to selectively inhibit the proliferation of ErbB2-overexpressing breast cancer cells by effectively downregulating the tyrosine phosphorylation of ErbB2 receptor at Y1248 and Y877 residues [70]. Recently, a 12-mer peptide (Tkip) mimetic of SOCS-1 (suppressor of cytokine signaling) was described to suppress cytokine signaling via binding to Janus kinase-2 (JAK2) and inhibiting its autophosphorylation event that has been implicated in the progression of several human cancers [71]. The Tkip 12-mer peptide is also shown to inhibit the autophosphorylation of EGFR and thus shown to inhibit the growth of prostate cancer cells. Subsequently, the N-terminal palmitoylated Tkip peptide (to enhance the membrane permeability) has been shown to inhibit both the constitutive and IL-6 induced activation of STAT-3, and to inhibit the proliferation of

prostate cancer cell lines DU145 and LNCap through suppression of the G1/S progression of the cell cycle [72]. In addition, using a random peptide library, two independent peptide inhibitors KDI-1 and AII-7 for the intracellular domains of EGFR and ErbB2 receptors, respectively have been recently identified [73, 74]. The aptameric fusion of these peptides with bacterial thioredoxin and their intracellular delivery efficiently blocked the tyrosine phosphorylation of EGFR and ErbB2 and subsequently led to growth inhibition of tumor cells via the downregulation of STAT activation and Akt signaling.

Akt/PKB is a serine/threonine kinase, downstream molecule of PI3-kinase signaling, which is overexpressed and implicated to have a role in various cancers. An optimal peptide substrate sequence (ARKRERTYSFGHHA) identified in an oriented peptide library screening approach has been shown to bind and inhibit Akt1 with a K_i of $12\,\mu M$ [75]. Later, several small peptide-based inhibitors of Akt have been developed including the most effective inhibitors that showed the K_i values in the range of $0.1\,\mu M$. The potent Akt peptide inhibitor fused with a polyarginine membrane permeable sequence readily inhibits the growth of Hela and other cancer cells via persistent Akt inhibition. Among other studies, Niv and colleagues using a "KinAce" strategy (discovery of potent peptide inhibitors based on the substrate binding region of the kinase), have identified nearly 30 potential peptide inhibitors for both the tyrosine kinases c-Kit and Lyn and the serine/threonine kinases PDK1 and Akt. KRX-702-H105 was found to be the most potent inhibitor of PDK1, the kinase that phosphorylates the activatory threonine 308 residue of Akt. This peptide inhibited the Akt phosphorylation in insulin-like growth factor (IGF) stimulated DU145 and PC-3 prostate cancer cells. KRX-014-H151 was the most effective peptide Akt inhibitor in the KinAce screen, which resulted in 90% inhibition of MDA-MB-231 breast cancer cell line proliferation. Among the tyrosine kinase peptide inhibitors in the KinAce screen, KRX-147-D103 peptide, an inhibitor of c-Kit and the KRX-055-G106 peptide, an inhibitor of Lyn tyrosine kinase selectively inhibited their respective kinase activities and subsequently inhibited the proliferation of various cancer cells. Although the use of inhibitory proteins has been the starting point in designing the inhibitory peptides for protein kinases during effective cancer therapy strategies, other protein

TABLE 17.1. Peptide-based inhibitors of protein kinases.

Peptide name	Target protein kinase	Peptide sequence
EC-1	ErbB2	WTGWCLNPEESTWGFCTGSF [70]
Tkip	EGF-R, JAK2	WLVFFVIFYFFR [72]
KDI-1	EGF-R	Trx-VFGVSWVVGFWCQMHRRLVC-Trx [73]
Peptide 2	Akt	VELDPEFEPRARERTYAFGH [161]
Peptide 4	Akt	VELDPEFEPRARERAYAFGH [161]
Akt-in	Akt	AVTDHPDRLWAWEKF [162]
KRX-014.H151	Akt	Myr-GGYNQNHQKLFQL-amide [163]
KRX-702.H105	PDK1	Myr-GGRAGNQYL-amide [163]
KRX 123.101	Lyn	Myr-GIVTYGKI-amide [163]
KRX-147.D103	c-Kit	Myr-GNLLNFLRRK-amide [163])
n.d.	c-Src	Y-c[Pen-YGSFC]KK-amide [78]
n.d.	c-Src	Y-c[D-Pen (3-Iodo-Y) GSFC]KR-amide
n.d.	c-Src	Y-c[D-Pen (3, 5-diIodo-Y) -GSFC]KR-amide
n.d.	c-Src	Myr-EFL-amide and Myr-EFLYGVFF [164]
n.d.	PKC	Myr-RKRCLRRL [165]
Myr-ΨPKC	PKCα/β/γ	Myr-FARKGALRQ [83]
Myr-ΨPKCζ	PKCζ	Myr-RRGARRWRK [82]
n.d.	PKCα/β/γ	RFARKGALRQKNV [84]
MLC (11-19)	MLCK	KKRAARATS [87]
n.d.	MLCK	AKKLSKDRMKKYMARRKWQKTG [166]
n.d.	MLCK	KRRFKK [166]
TI-JIP	JNK	RPKRPTTLNLF [90]
CAMKII (290–302)	CAMKII	LKKFNARRKLKGA [92]
C4	CDK2	TYTKKQVLRMAHLVLKVLTFDL [167]
P1	Casein kinase 2	CRNCTVIQFSC [93]
n.d.	ERK2	MPKKKPTPIQLNPAPDG [168]

partners may have the potential to inhibit activity and/or activation of a specific kinase. The proto-oncogene TCL1 has been identified as a co-activator of Akt [76]; however, a 15-mer Akt-in peptide derived from TCL-1 inhibits Akt activity although requiring high micromolar (400 µM) peptide concentrations. Biochemical analyses indicated that Akt-in peptide binds to the pleckstrin homology (PH) domain of Akt and suppresses Akt interaction with the phosphatidylinositol and the plasma membrane translocation required for its activation. The Akt-in peptide fused with a cell permeable TAT (HIV-derived protein transduction domain) peptide inhibited the cell growth and cell viability and showed an enhancement in mitochondrial depolarization in various cancer cells. The inhibitory effects of Akt-in peptide have also been tested in vivo following injection of Akt-in peptide directly into a xenograft tumor every second day for 2 weeks, which dramatically reduced tumor growth and enhanced the number of apoptotic cells.

Several peptide-based inhibitors for the intracellular nonreceptor tyrosine kinases have also been identified and characterized, which have a great potential in turning into effective anticancer molecules. The peptides inhibitors for Lyn and c-Kit identified based on the KinAce screen have been described before. In addition to these peptides, Lam and colleagues screened a 7-mer peptide library in vitro for inhibitors of c-Src and identified a potential peptide inhibitor with an IC_{50} of approximately 3 µM [77]. Subsequently, two peptidomimetic derivatives of the potential c-Src inhibitor have been developed with IC_{50} values of 0.13 and 0.54 µM [78]. c-Src is a nonreceptor tyrosine kinase involved in cell proliferation and mitotic progression and its deregulation plays a critical role in the development of various tumors. Thus, the preliminary in vitro studies with the c-Src peptide inhibitors have revealed the suppression of cell proliferation and enhanced mitotic arrest in different cancer cells. These peptidomimetic derivatives showed higher affinity for c-Src when compared with the original parent peptide, and a non-selective inhibition of Lck, another family member of nonreceptor tyrosine kinases. In another study, a 7-mer peptide has also been identified based on the enzyme-substrate interaction site of c-Src, which neither inhibited Lck nor Lyn [79]. To further improve the cell permeability of this 7-mer c-Src peptide inhibitor, a 4-mer hydrophobic peptide has been developed, which retained the c-Src inhibitory properties, yet the effectiveness of the 4-mer peptide needs to be evaluated in further studies.

Protein kinase C (PKC) is another family of serine/threonine kinases that is involved in the transduction of signals for cell proliferation and differentiation [80]. The important role of PKC in processes relevant to neoplastic transformation, carcinogenesis and tumor cell invasion renders it a potentially suitable target for anticancer therapy. Furthermore, there is accumulating evidence that selective inhibition of PKC may improve the therapeutic efficacy of established neoplastic agents and sensitize cells to ionizing radiation. Several peptide inhibitors for different PKC-isoforms (PKCα, PKCβ, PKCγ, PKCδ, and PKCζ) with different K_i values have been identified (Table 17.1) [81–84]. The N-terminal myristylated PKC peptide inhibitors have been shown to enter the cells readily in both the drug-sensitive and multi-drug resistant cancer cell lines and induce apoptosis. Interestingly, in addition to cancer cell growth inhibition, PKC inhibitory peptides have also been shown to influence other physiologic functions such as NF-κB activation, insulin stimulated glucose transport, angiotensin II actions to increase expression of c-Fos, taurocholate transport in hepatocytes and the late phase of long-term potentiation in synaptic transmission, implicating their usefulness in a wide variety of cellular processes [85].

Myosin light chain kinase (MLCK), a crucial serine/threonine kinase in smooth muscle contraction, has also been implicated to have a role in cancer cell invasion and metastasis [86]. The inhibition of MLCK using either specific small molecule inhibitors or siRNA leads to cancer cell apoptosis and reduced cell migration in invasion assays. A substrate-based peptide inhibitor of MLCK has been described based on the myosin light chain sequence corresponding to residues 11–19 with an estimated K_i of 10 µM [87]. Subsequently, other higher potency inhibitory peptides for MLCK based on the pseudosubstrate sequences have been developed, but many of these inhibitors have not been characterized for their pro-apoptotic and antimetastatic activities and thus await further studies.

c-Jun N-terminal kinase (JNK) plays a critical role in various cellular processes including apoptosis, cell survival, and tumor development [88]. JNK-interacting protein (JIP), a scaffold in the JNK pathway, was originally described as a cytosolic inhibitor of JNK action [89]. Based on this knowledge, a 11-mer ATP non-competitive inhibitor peptide (TI-JIP), with a K_i value of 0.39 µM was derived from JIP and was subsequently shown to selectively inhibit the JNK activity [90]. The TAT-JIP fusion peptide [91] was subsequently studied both in vitro in cultured cells and in vivo in animal models of the disease and was shown to be a successful JNK peptide inhibitor. Among other kinases, potential peptide-based inhibitors have been identified for calmodulin-dependent protein kinase [92], cyclin dependent kinase 2, casein kinase 2 [93], and ERK2. All these kinases have been assigned a role in either cancer cell survival or cell cycle and thus their inhibition leads to growth inhibition and/or apoptosis. Although some of these peptides have been tested in vitro for their antitumor activity, further studies need to be performed to evaluate the full potential of these kinase peptide inhibitors as an effective cancer therapy.

5 Inducing Tumor Apoptosis Using Smac Peptidomimetics

Caspases are the "proteolytic engines" of apoptotic process that is triggered both by natural stimuli and cancer treatment. Once activated, caspases are controlled by a group of inhibitors of apoptosis proteins (IAPs), which contain multiple baculovirus inhibitor repeat (BIR) domains that drape across the active sites of the caspases [94]. It has been found that IAP proteins are overexpressed in many tumor cells, suggesting they may be prolonging the survival of the tumor cells [95–101].

It has been found recently that IAP proteins are regulated by a small protein called the second mitochondria-derived activator of caspase (Smac), or direct IAP binding protein with low PI (DIABLO). Further investigation revealed that Smac binds to the BIR domain of IAP proteins through a short four-amino acid segment (Ala-Val-Pro-Ile) exposed at the N-terminus (Fig. 17.1) [102, 103].

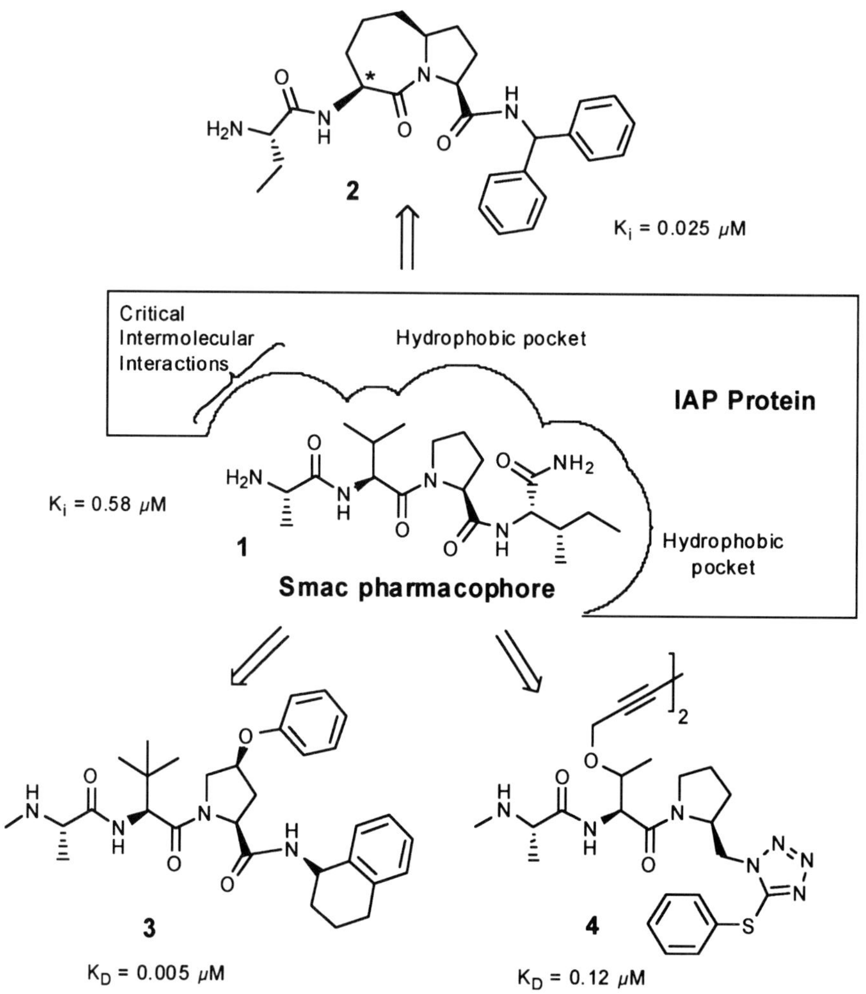

FIGURE 17.1. Smac peptidomimetics with enhanced bioactivity. The number 1 is the backbone molecule. Peptidomimetic 2 is a conformationally locked peptidomimetic in which the side chains of valine and proline are presented on a bicyclic lactam scaffold. Peptidomimetic 3 incorporates a hydropically enhanced proline residue (4-phenoxy-proline). Mimetic 4 is a dimerized Smac peptidomimetic

This observation evoked the idea of developing small molecule inhibitors of IAP, which can mimic the action of Smac.

Experimental studies showed that Smac-based peptides, tethered to carrier proteins for transporting into the cell, could inhibit IAP in tumor cells where IAP was overexpressed, and consequently improved the effectiveness of conventional chemotherapy both in vitro and in vivo [104–107]. Interestingly, these compounds showed little toxicity to normal cells/tissues.

There are, however, several serious limitations for the use of peptide-based drugs. These include rapid degradation by proteases and other enzymes, low bioavailability and membrane permeability as well as relatively high production costs. One solution to these problems has been the development of peptidomimetics: synthetic compounds that mimic protein or peptide interactions, but exhibit improved resistance to biodegradation, cell permeability, and potency [108, 109].

Over the years, various strategies for the design of peptidomimetics have been developed [110]. For instance, replacement of natural by unnatural side chains, substitution of L- for D-amino acids, isosteric or isoelectric exchange of units in the peptide backbone, and introduction of additional fragments have been widely used for designing peptidomimetics. Also, the reduction of the conformational flexibility of peptides is frequently used to enhance the "drug-like" properties of peptides [111]. This is typically achieved by installation of conformation-stabilizing elements such as rings and bridges that serve as conformational locks and fix the position of side chains in the molecule [112, 113]. Additionally, scaffold peptidomimetics have appeared in which a completely unnatural framework is used to support the side-chains of the original peptide [114, 115]. In particular, *gluco*-configured sugar scaffolds have attracted significant interest due to their rigidity and polyfunctional character. Sugar scaffolds are highly functionalized scaffolds that can be easily derivatized to present peptide side-chains or other pharmacophores [116]. Finally, nonpeptide mimetics have been discovered in course of random screening of combinatorial libraries. Nonpeptide mimetics belong to various classes of organic compounds and have no apparent structural relationship with the natural peptide ligand [117, 118].

One of the major goals in the design of peptidomimetics is to prevent the lysis of the amide bond by proteases. Since proteases have a specific binding pocket to recognize the side-chain of natural amino acids, they usually do not bind unnatural amino acids or nonpeptide bonds. This means that peptidomimetics have a lower tendency to be metabolized as normal peptides, conferring improved stability. Also, since peptidomimetics only replicate the pharmacophore of a protein, they are inherently shorter than the native protein, and thus show improved cell permeability.

Perhaps one of the most prominent approaches that follow the design of peptidomimetics for the treatment of cancer is the development of Smac peptidomimetics to alleviate the inhibition of caspase family of pro-apoptotic proteases in tumor cells by IAPs. Such projects gained momentum when the 3D structure of the Smac/IAP complex was established [103], which deciphered specific interactions between the two proteins leading to the engineering of more potent analogues. Further structure activity relationship (SAR) studies revealed several critical binding interactions [119], including the hydrophobic interaction between the Ala methyl group and W310 of the BIR3 domain, an intermolecular hydrogen bond involving the amide NH of valine,

and the contact between the 4 (gamma) and 5 (delta) methylene groups of proline with Trp-323. The isoleucine residue is minimally involved in binding [120, 121]. Subsequently, an iterative approach using molecular modeling, organic synthesis, and focused combinatorial chemistry has produced very potent inhibitors of IAP. These compounds have proved to be effective at killing various cancer cell lines both in vitro and in vivo.

Using the Smac tetrapeptide pharmacophore Ala-Val-Pro-Ile [122] as an example, we discuss several recent representative examples of peptidomimetics developed for the treatment of cancer, that are based on our knowledge about protein-protein interactions. In 2004, Roller and Wang et al. developed a series of Smac peptidomimetics [121]. Through analysis of the Smac/IAP complex, they found that isoleucine was bound in a hydrophobic pocket, but the carbonyl group was not involved in any binding interactions. Therefore, they replaced isoleucine with a more hydrophobic aromatic amine and saw a twofold increase the binding affinity of their analogue to the IAP protein (see Fig. 17.1). They then effectively fused the valine and proline side-chains in six- and seven-membered lactams to reduce the conformational freedom of the peptide (see Fig. 17.1). The six-membered lactam had lower binding affinity than the native peptide, but the seven-membered lactam, together with another modification of alanine, had a K_i of 25 nM, or a 23-fold increase in the potency over the native peptide (see Fig. 17.1). Interestingly, changing the stereochemistry at one chiral center (marked with * in Fig. 17.1) showed a marked decrease in the binding affinity (680× less), indicating that the binding of Smac is highly stereospecific. In vitro tests showed that the analogue specifically targets IAP and NMR studies showed the analogue binds in the same BIR region as Smac.

In 2004, Fesik and his colleagues used a similar strategy of incorporating unnatural amino acids to improve the binding affinity of Smac analogues, and reduce their peptide character [119]. They used a combinatorial approach to build libraries of Smac analogues. Amino acids were varied to determine what characteristics at each position in the pharmacophore contributed most to binding affinity. Most notably, from the 3D Smac/IAP complex, they noticed a hydrophobic groove in the binding site near the proline residue not exploited by the native peptide. Therefore, they replaced proline with an unnatural proline analogue in which a hydrophobic appendage was attached (see Fig. 17.1) resulting in a threefold improvement in the binding affinity over their most active analogue. They tested their compounds on 59 human cancer cells lines in vitro, and found remarkable cytotoxic activity in several cell types. Further in vivo testing showed that Smac analogues inhibited the growth of tumors in an MDA-MB-231 breast cancer murine xenograft model.

Finally, in 2004, after building a 180-member library of Smac peptidomimetics in a combinatorial fashion, Li et al. found optimal binding affinity by replacing isoleucine with a tetrazole thioether. Unexpectedly, one analogue dimerized through an unnatural alkyne-modified amino acid, but produced highly potent IAP inhibitors (see Fig. 17.1) [120]. It is assumed that the analogue interacts with multiple BIR domains simultaneously, a trait which is also implicated for the Smac protein [122]. Together with TNF-related apoptosis-inducing ligand (TRAIL) (50 ng/mL), the peptidomimetic was active in vivo against T98G human glioblastoma cells resistant to DNA damage-induced apoptosis, at concentrations as low as 100 pM (1×10^{-10} M). Treatment with TRAIL alone caused no apoptosis, or caspase activity. The compounds also had no detectable effect on human skin fibroblasts (10 μM) alone or with TRAIL. Further tests confirmed that the analogue bound to the BIR domains of IAP proteins.

Although, the development of Smac-mimetics as novel cancer therapeutics is still at an early stage, significant progress has been made since the establishment of the 3D structure of the Smac/IAP complex in 2000. Together, careful consideration of the 3D structure of Smac/IAP complex and combinatorial screening of various molecular motifs, revealed to be the basic requirements for optimal binding. Following synthesis, the in vitro and in vivo efficacy of these peptidomimetics has been confirmed through biological testing. While in most cases peptide-based drugs seem to be impractical, peptidomimetics have many advantages including increased cell permeability, prolonged bioavailability, and increased potency and specificity. Hopefully, Smac-based peptidomimetics will effectively contribute toward developing new treatments of cancer in the future.

6 Clinical Utility of Peptides

The limited effectiveness of cytotoxic drugs along with our growing understanding of cellular processes such as growth, differentiation, and survival have led to a better search for more refined cancer treatment and focusing on more rational cellular targets. The first generation of successful "targeted therapy" in the clinical setting has consisted of monoclonal antibodies (mAB) directed against specific cell surface markers or receptors and synthetic molecules directed against the tyrosine kinase domain of specific growth factor receptors [123, 124]. The anti-CD20 rituximab and tositumomab are regularly used for the treatment of non-Hodgkin's lymphoma and chronic lymphocytic leukemia resulting in an increase in tumor response and more durable remissions [125–128]. Anti-CD52 alemtuzumab and anti-CD33 gemtuzumab ozogamicin has shown to produce some clinical remissions in the treatment of refractory chronic lymphoproliferative disorders and acute myeloid leukemia respectively [129, 130]. Bevacizumab, directed against VEGF in randomized studies combined with cytotoxic chemotherapy has been shown to improve survival in patients with either metastatic colorectal or non-squamous non-small cell carcinoma of the lung [131, 132]. Cetuximab, the mAB against the EGFR, is actively being studied in treatment of refractory metastatic colorectal carcinoma as well as head and neck carcinoma [133, 134]. Finally, the humanized ErbB2 antibody, trastuzumab, resulted in decreased event rates and improved survival when used for adjuvant therapy or treatment of metastatic breast cancer, respectively [135, 136]. Although mABs are very effective at delivering targeted therapy, they have a number of clinical drawbacks. Their high molecular weight (160,000 kDa) leads to poor tissue penetration, particularly at the center of large tumors where the blood supply is inadequate [7]. Moreover, the big molecular size limits their use to cell surface targets rather than specific intracellular molecules. In addition, despite being either humanized or chimeric, mABs can still produce dangerous hypersensitivity reactions (immune response against such antibodies). Also, they are taken up by the reticular endothelial system (RES) such as the liver, spleen, and bone marrow, resulting in dose-limiting toxicity to these organs [137].

7 Roles for Peptide-Based Targeting in Cancer Therapy

Similar to mABs, tissue-targeting peptides can be designed to recognize and bind specific cell surface targets unique to a particular cancer cell. However, unlike mABs, peptides generally have much lower molecular weights and therefore have the capacity to be internalized by the cell [7]. This makes these short peptides ideal for targeted delivery of a number of therapeutic modalities directly into the cancer cells. However, there are a number of limitations to the development and use of these peptides. Firstly, large combinatorial peptide libraries need to be laboriously examined to identify suitable peptides, which should further be examined to determine the specific cellular ligand. In further steps, an effective system by which the peptide is delivered in vivo needs to be designed with particular focus on limiting natural protein degradation in the cell.

7.1 Tumor-Specific Targeting

Currently, gene expression profiling is readily being used to determine key differences between genes activated or downregulated in cancer as opposed to its normal counterpart [138]. The products of these genetic aberrations (intracellular or extracellular) unique to the cancer cell are potential targets for peptide-based therapy. Also, as discussed earlier in this chapter, peptides targeted against active endothelial cells or cell adhesion molecules may restrict the process of angiogenesis/growth or tumor spread, respectively [29, 139].

7.2 Drug Delivery

Recent advances have enabled the use of small peptides in targeted delivery of many different therapeutic agents directly to cancer cells with minimal toxicity to surrounding normal cells [7]:

- Cytotoxic agents. Drug or pro-drug directly conjugated to peptide or drug containing liposome coated with peptide [140–142]
- Oligonucleotides. Interfere with gene expression and could be used to suppress individual oncogenes activated in malignant cells [143, 144]
- Pro-apoptotic agents. Combination of peptide with amphipathic peptides designed to disrupt prokaryotic membranes such as the mitochondrial

membrane [145, 146]. Other possibilities include specifically binding and disrupting the IAP family of proteins, which bind caspase proteins and prevent apoptosis [100, 147]
- Gene therapy. Improve the tropism of viral vectors to target cells by modifying the viral coat proteins [148]

8 Experience with Peptide-Targeted Therapy

Experience with peptide-targeted therapy has been limited to date. Significant work has been done over the past 20 years particularly using murine cancer models of human cell lines. Since research into this area started prior to the modern genomic era, the focus has mainly consisted of the conjugation of cytotoxic drugs to several known peptide hormones and their analogues. Specific hormone receptors are overexpressed in certain malignancies and therefore may offer a means by which to deliver cytotoxic therapy directly to the cancer cells with minimal toxicity to surrounding tissue [149].

One such factor that has been identified is somatostatin. Currently somatostatin analogues have found widespread use in the treatment of symptoms related to a variety of neuroendocrine tumors. Five different somatostatin receptors are known (SST 1–5) and the level of expression of each subtype differs based on tumor type. Currently three major analogues are available: octreotide, vapreotide, and lanreotide. Each shares a high affinity for the SST2 receptor with vapreotide also having high affinity for SST3 and SST5. Interestingly, it has been observed that treatment with somatostatin analogues alone has partial antiproliferative effects on cancer cells. The mechanism is suspected to involve either a direct receptor-mediated antiproliferative effect or indirectly via the inhibition of growth hormone production, resulting in lower levels of IGF-1, which in turn leads to increased levels of apoptosis as a result of IGF-1 inhibition. In addition to its antiproliferative effects, somatostatin also inhibits angiogenesis by reducing baseline levels of VEGF [150]. Tumors currently under study include neuroendocrine tumors as well as colon, breast, prostate, brain, lung, pancreatic, and gastric tumors [149]. Cytotoxic drugs such as doxorubicin have been conjugated to the somatostatin analogues and have been shown to inhibit growth in murine- and human-derived breast and prostate cancer cell line. The use of these analogues is not limited to the delivery of cytotoxic drugs. When combined with a radionucleotide, octreotide can be used to identify somatostatin receptor–expressing tumors by scintigraphy [149].

Similar to the somatostatin analogues, the LHRH agonists have been used to target sex hormone sensitive tumors such as ovarian, endometrial, breast, and prostate cancers [149]. These LHRH agonists also appear to have a direct effect on cancer cell growth by inhibition of the hormone receptor. The LHRH peptide has been modified to carry a cytotoxic drug such as melphalan, methotrexate, nitrogen mustard, or doxorubicin by modifying the amino acid at position 6. Promising results have included significant inhibition of growth using ovarian, breast, and prostate cell lines. Interestingly, if an equimolar concentration of conjugated drug was delivered in an unconjugated fashion, ineffective growth inhibition was seen. This observation further supports the principle of peptide-targeted cytotoxic drug delivery [149]. Finally, the analogues of bombesin/gastrin-releasing peptide (GRP) have also been modified in a similar fashion to the LHRH agonists. Bombesin/GRP receptors are expressed in a wide variety of malignancies, including small cell lung carcinoma, neuroblastoma, malignant melanoma, colorectal, gastric, pancreatic, prostate, ovarian, breast, thyroid, and renal cell cancer [149, 151]. Initial promising results have been seen in nude mice xenografted with a small cell lung cancer cell line.

9   Limitations to the Use of Peptide Targeted Therapy

The use of peptides in vivo has been hampered by innate properties of proteins. This often culminates in poor bioavailability, poor stability, immunogenicity, poor penetration across biologic membranes, and short half-life in serum [137, 152]. Proper production and storage of peptides must ensure that minimal alteration of the protein structure takes place. Unfolding of peptides during storage may lead to dramatic changes to the specificity of the peptide [153]. Oral bioavailability is

hampered by poor transport across the gastrointestinal tract. Once in circulation, these peptides are also subject to common mechanisms of clearance either through renal excretion or liver metabolism. Delivery may be improved with either the use of controlled intravenous infusions or implantable devices at the tumor site. Bioavailability may be improved somehow with the advent of "designer" peptides conjugated with polymers such as water-soluble polymers, poly (ethylene glycol), or poly (styrene co-maleic acid anhydride). Also, the peptides are subject to damaging shear forces in circulation and proteolytic degradation. The latter can be reduced by masking these peptides from proteases by means of blocking both the N- and C-terminus, cyclization of the peptide molecule, or synthesizing the peptide with D-amino acids or other unnatural amino acids [137].

Once bound to a target cell surface receptor, peptides need to survive the process of cellular internalization and deliver their therapeutic payload in an unadulterated form. This may be quite complex given the presence of low intracellular pH and need to escape from lysosomal or endosomal compartments. This escape potentially can be accomplished via conjugation with fusogenic negatively charged peptides or amphiphilic photosensitizers [153]. Peptides with intracellular targets must overcome the poor permeability of cell membranes [154]. This can be accomplished either via encapsulation of peptides within liposomes or smaller micelles. Alternatively, further screening of peptide libraries may be required to identify peptides more specific for intracellular targets. The continuous search for better peptides and drug delivery systems both prolongs the process of development and escalates the cost associated with identifying peptides candidates for targeted therapy [155].

10   The Use of Peptides in Vaccination Against Cancer

As described, there are numerous limitations to the use of peptides for cancer therapy in vivo. This has led to considering peptides for the development of anticancer vaccines. The practice of widespread vaccination has changed the overall perspective of human disease. Numerous life-threatening infectious

TABLE 17.2. Peptide-based anticancer approaches under evaluation in clinical trials.

Peptide	Delivery	Study	Population
NGR-TNF	Antiangiogenic	Phase I	Advanced solid tumors
NGR-hTNF	Antiangiogenic	Phase I	Advanced or metastatic solid tumor previously treated with a noncumulative dose of doxorubicin
GRP/bombesin	Gastrin-releasing antagonist	Phase I	Advanced solid tumor
MART-1 peptide	Immuno-therapy	Phase I/II	Stage IV or relapsed malignant melanoma
In-labeled peptide	Radiolabeling	Phase I	CEA-producing tumor
Peptide 946, tetanus peptide melanoma vaccine	Vaccine	Phase I	High-risk melanoma
CEA peptide	Vaccine	Phase II	Stage I–III adenocarcinoma of pancreas
HLA-A1, -A2, and -B35 restricted survivin peptides	Vaccine	Phase I/II	Advanced melanoma, pancreatic, colon, and cervical
Ovarian cancer synthetic peptides, tetanus toxoid helper peptide	Vaccine	Phase I	Previously treated ovarian epithelial or primary peritoneal cancer
ESO-1: 157–165 (165V) peptide	Vaccine	Phase II	Metastatic melanoma
NY-ESO-1b peptide	Vaccine	Phase I	Cancer expressing NY-ESO-1 or LAGE-1
Melanoma-derived helper peptides, class I MHC-restricted peptide, multi-epitope melanoma peptides, tetanus toxoid helper peptide	Vaccine	Phase I	Resected melanoma
Carcinoembryonic antigen peptide 1-6D, CMV pp65 peptide	Vaccine	Phase I	Refractory stage IV CEA-expressing malignancies
Melanoma peptides	Vaccine	Phase II	Metastatic melanoma
gp100:ES209-217	Vaccine	Phase	Metastatic melanoma
gp100 Antigen, MAGE-3, and tyrosinase peptide	Vaccine	Phase II	Resected stage IIC, III, or IV melanoma
MART-1, GP100, tyrosinase	Vaccine	Phase I/II	Metastatic melanoma
MAGE-3.A1 peptide	Vaccine	Phase I/II	Metastatic melanoma
gp100 Peptide	Vaccine	Phase I	Metastatic melanoma
Epitopes of the MART-1 or gp100 melanoma	Vaccine	Phase II	High risk for recurrence of melanoma
MAGE-1.A2, MAGE-3.A2, MAGE-4.A2, MAGE-10.A2, MAGE-C2.A2, NA17.A2, NY-ESO-1.A2, Tyrosinase.A2	Vaccine	Phase I/II	Metastatic cutaneous melanoma
Wild-type p53 peptides, T-helper peptide epitope	Vaccine	Phase I	SCC head and neck
Tyrosinase, gp100: 209–217, OVA BiP Peptide	Vaccine	Phase I	Stage III or IV melanoma
Immunodominant tyrosinase, gp100 peptides	Vaccine	Phase II	Recurrent or refractory metastatic melanoma
Heat shock protein peptide complex-96	Vaccine	Phase I/II	Recurrent high-grade glioma
Five synthetic ovarian cancer–associated peptides	Vaccine	Phase I	Advanced ovarian, primary peritoneal or fallopian tube cancer
Multi-epitope melanoma peptides, tetanus toxoid helper peptide	Vaccine	Phase I	Resected stage II–IV melanoma
Antigenic peptides	Vaccine	Phase I	Skin melanoma
Melan-A, Mage-10, NY-ESO	Vaccine	Phase I	Stage III–IV malignant melanoma
Lipidated human papillomavirus 16 E7 peptide	Vaccine	Phase I	Recurrent or persistent cervical cancer
EP2101 Peptide	Vaccine	Phase I	Stage III colon cancer
gp100-Tyrosinase	Vaccine	Phase II	Advanced melanoma
Melanoma peptide vaccine, Melan-A ELA, NY-ESO-1a, NY-ESO-1b MAGE-10.A2	Vaccine	Phase I	Stage IIb or IIIa non-small cell lung cancer (NSCLC)
HPV16 E6 or E7 peptide	Vaccine	Phase II	Advanced or recurrent carcinoma of the cervix or other tumors carrying HPV16
Ras peptide	Vaccine	Phase I	Advanced pancreatic or colorectal adenocarcinoma
Tyrosinase/gp100/MART-1 peptides	Vaccine	Phase I/II	Resected stage III or IV melanoma

(continued)

TABLE 17.2. (continued)

Peptide	Delivery	Study	Population
Synthetic peptide corresponding to the tumor's p53 or ras mutation	Vaccine	Phase II	Tumors expressing mutant p53 or Ras
gp100: 209–217, HPV-16 E7 (12–20) Peptide	Booster Vaccine	Phase I	Melanoma
Peptide pulsed dendritic cell vaccine	Vaccine	Phase I	Stage IV melanoma
Tyrosinase peptide, gp100 antigen	Vaccine	Phase II	Resectable stage IIA or IIB melanoma
Telomerase: 540–548 peptide vaccine	Vaccine	Phase II	HLA-A*0201-expressing patients with metastatic cancer
MART-1/gp100/tyrosinase/NY-ESO-1	Vaccine	Phase II	Chemotherapy-naïve metastatic melanoma
WT-1 peptides	Vaccine	Phase I	Thoracic and myeloid neoplasms
gp100 antigen, tyrosinase, and tetanus peptides	Vaccine	Phase II	Advanced melanoma
Melan-A/MART-1 peptide	Vaccine	Phase I	Advanced melanoma
Tyrosinase: 368–376, gp100: 209–217 (210M) Antigen, and MART-1: 27–35 Peptide	Vaccine	Phase III	Locally advanced or metastatic melanoma
Prostate specific antigen peptide 3A	Vaccine	Phase II	Recurrent prostate cancer
NY-ESO-1 peptide vaccine	Vaccine	Phase I	Transitional cell carcinoma of the bladder expressing NY-ESO-1 or LAGE-1 antigen
Tumor-specific mutated Ras peptides	Vaccine	Phase II	Metastatic solid tumors
MART-1/gp100/tyrosinase peptide	Vaccine	Phase II	Metastatic melanoma
gp100 peptides	Vaccine	Phase II	Previously treated HLA-A* 0201 positive subjects with stage IV melanoma
PR1 leukemia peptide	Vaccine	Phase I/II	Chronic myeloid leukemia, acute myeloid leukemia, or myelodysplastic syndromes
Oncopeptide pulsed DC, p53, survivin and telomerase peptides	Vaccine	Phase I/II	Metastatic breast cancer
MART-1: 27–35, gp100: 209–217 (210M), and tyrosinase: 368–376 (370D) peptides	Vaccine	Phase II	HLA-A2 positive patients with metastatic melanoma
Tyrosinase peptide, gp100 antigen, and MART-1 antigen	Vaccine	Phase II	Resected stage IIB, IIC, III, or IV melanoma
Heat shock protein–peptide complex	Vaccine	Phase III	Stage IV melanoma
EBV-LMP-2	Vaccine	Phase I/II	Patients at high risk for recurrence of anaplastic nasopharyngeal cancer
Nine synthetic breast cancer peptides	Vaccine	Phase I/II	Advanced breast cancer
Melanoma peptide vaccines	Vaccine	Phase II	Melanoma
(LMP)-2: 340–349 peptide or LMP-2: 419–427 peptide	Vaccine	Phase I/II	Anaplastic nasopharyngeal carcinoma at high risk for recurrence
HER-2-neu and carcinoembryonic antigen synthetic peptides	Vaccine	Phase I/II	Stage IIB, III, or IV colorectal cancer
Tyrosinase, gp100, and MART-1 peptides	Vaccine	Phase II	Resected stage IIB, IIC, III, or IV melanoma
WT1 and PR1 peptide	Vaccine	Phase I	Myeloid malignancies
gp100 Peptides	Vaccine	Phase II	Stage IV melanoma
Ras peptide cancer vaccine	Vaccine	Phase I	Myelodysplastic syndromes
Peptide-pulsing	Vaccine	Phase I/II	Melanoma
Telomerase peptide	Vaccine	Phase I	Advanced breast cancer
Ras peptide cancer vaccine	Vaccine	Phase I	Stage IB–IV non-small cell lung cancer
Telomerase: 540–548 peptide	Vaccine	Phase I	Sarcoma or brain tumor
Ras peptide cancer vaccine	Vaccine	Phase I/II	Locally advanced or metastatic colorectal cancer
p53 peptide vaccine	Vaccine	Phase I/II	Stage IV, recurrent, or progressive breast or ovarian cancer
Heat shock protein peptide complex	Vaccine	Phase I	Resected pancreatic adenocarcinoma
bcr/abl Breakpoint peptide vaccine	Vaccine	Phase II	Chronic myelogenous leukemia

(continued)

TABLE 17.2. (continued)

Peptide	Delivery	Study	Population
Heat shock protein peptide vaccine	Vaccine	Phase II	Recurrent soft tissue sarcoma
Heat shock protein 70-peptide complexes	Vaccine	Phase I	High risk breast cancer
Mutant p53 peptide	Vaccine	Phase II	Locally advanced non-small cell lung cancer
HLA-A2 peptides, MAGE-1.A2,MAGE-3.A2, MAGE-4.A2, MAGE-10.A2, MAGE-C2.A2, NA17.A2, tyrosinase.A2 and NY-ESO-1.A2 peptides	Vaccine	Phase I/II	Disease-free melanoma
HER-2/Neu peptide	Vaccine	Phase I	Stage III or IV HER-2/Neu expressing cancers
Tyrosinase: 368–376 peptide, gp100: 209–217 antigen, MART-1: 26–35	Vaccine	Phase I	Completely resected stage III or IV melanoma
HER-2/Neu (HER2) intracellular domain (ICD) peptide	Vaccine	Phase II	HER2 positive breast cancer patients receiving trastuzumab monotherapy
Multi-epitope peptide	Vaccine	Phase I/II	Stage IIB, IIC, III, or IV Melanoma
Carcinoembryonic antigen (CEA) peptide	Vaccine	Phase II	HLA-A2+ patients with CEA-producing adenocarcinomas of gastrointestinal tract origin
Carcinoembryonic antigen (CEA) peptide	Vaccine	Phase I/II	Colorectal cancer patients with liver metastases
Tyrosinase, MART-1, gp100	Vaccine	Phase I	Stage IV malignant melanoma
Tyrosinase, gp100	Vaccine	Phase II	Resected stage III or IV melanoma
Mutant VHL peptide, VHL peptide	Vaccine	Phase II	Renal cell carcinoma
Synthetic tumor-specific breakpoint peptide	Vaccine	Phase II	Chronic myeloid leukemia (CML) and minimal residual disease
Tyrosinase, MART-1, gp100	Vaccine	Phase II	Resected stage III or stage IV melanoma
HER-2/Neu peptides	Vaccine	Phase I	Advanced stage HER-2/Neu expressing cancers
G250 peptide	Vaccine	Phase II	Following surgical resection of locally advanced or metastatic renal cell carcinoma
Melanoma peptides	Vaccine	Phase I	Melanoma
Tyrosinase: 368–376, gp100: 209–217, MART-1: 26–35	Vaccine	Phase I	Resected stage III or IV melanoma
NY-ESO-1 peptide	Vaccine	Phase I	Stage II, III, or IV soft tissue sarcoma expressing NY-ESO-1 or LAGE antigen
MAGE-A3/HPV 16 peptide	Vaccine	Phase I	Squamous cell carcinoma of the head and neck
TA-HPV	Vaccine	Phase II	Early cervical cancer
Melanoma peptide	Vaccine	Phase III	Previously treated unresectable stage III or IV melanoma
HLA-A2 and HLA-A3-binding peptides	Vaccine	Phase I	Renal cancer
Synthetic melanoma peptide	Vaccine	Phase II	Stage IIB–IV melanoma
Modified CEA (mCEA) peptide	Vaccine	Phase II	Locally advanced or surgically resected adenocarcinoma of the pancreas
EGFRvIII peptide	Vaccine	Phase I	EGFRvIII-expressing cancer
ImMucin	Vaccine	Phase I	MUC1-expressing tumor malignancies
gp100: 209–217 (210M), gp100: 17–25 Antigen, Tyrosinase: 368–376 (370D), Tyrosinase: 240–251 (244S), Tyrosinase: 206–214, Tyrosinase-related protein-1 (ORF3):	Vaccine	Phase I	Melanoma
Multiple synthetic melanoma peptides	Vaccine	Phase II	Unresectable stage III or IV melanoma
ESO-1 peptide	Vaccine	Phase II	Refractory metastatic melanoma expressing ESO-1

(continued)

Table 17.2. (continued)

Peptide	Delivery	Study	Population
Multiple synthetic melanoma peptides	Vaccine	Phase I/II	Stage IIIB, IIIC, or IV melanoma
gp100: 209–217, MART-1: 27–35	Vaccine	Phase II	Metastatic melanoma
NA17.A2 Antigen, melanoma differentiation peptides	Vaccine	Phase III	Metastatic melanoma
MAGE-3/Melan-A/gp 100/NA17 Peptide	Vaccine	Phase II	Metastatic melanoma
Recombinant HER2/neu peptides	Vaccine	Phase II	DCIS
Modified gp100 melanoma peptide	Vaccine	Phase I	Metastatic melanoma
HPV16 E7 peptide	Vaccine	Phase I	Recurrent cervical cancers
gp100 antigen, MART-1	Vaccine	Phase I	Metastatic melanoma
gp100: 209–217 (210M)	Vaccine	Phase I	High-risk melanoma
Multiepitope peptide vaccination	Vaccine	Phase II	Metastatic melanoma
Autologous human tumor–derived HSPPC-96	Vaccine	Phase II	Indolent lymphoma
gp100, MART-1	Vaccine	Phase II	Metastatic melanoma
gp100 Peptides	Vaccine	Phase II	Stage IV melanoma
Recombinant fowlpox-CEA-TRICOM, CMV pp65 peptide	Vaccine	Phase I	Advanced or metastatic malignancies expressing CEA
gp100 and Tyrosinase peptides	Vaccine	Phase I/II	Metastatic melanoma
Human alpha fetoprotein (hAFP) peptide	Vaccine	Phase I/II	HLA-A*0201 positive patients with hepatocellular carcinoma
WT1 126–134 peptide	Vaccine	Phase II	Acute myeloid leukemia
Alpha-fetoprotein peptide	Vaccine	Phase I/II	HLA-A*0201-positive patients with hepatocellular carcinoma
gp100: 209–217(210M)	Vaccine	Phase II	Advanced melanoma
gp100 Peptide	Vaccine	Phase II	Melanoma
HLA A*0201-restricted melanoma antigens	Vaccine	Phase II	Melanoma
MAGE-3 or Melan-A	Vaccine	Phase I/II	Metastatic melanoma
MART-1, gp100, tyrosinase, MAGE-3	Vaccine	Phase I	Metastatic melanoma
MART-1	Vaccine	Phase I	Metastatic melanoma
p53 Vaccine	Vaccine	Phase I/II	Adenocarcinoma of the ovary who have no evidence of disease or marker disease only
HER-2	Vaccine	Phase I	Breast Cancer
GP100: 209–217 (210M)	Vaccine	Phase III	Metastatic melanoma
MART-1	Vaccine	Phase I	Advanced melanoma
gp100 Peptide	Vaccine	Phase II	Recurrent or refractory metastatic melanoma
gp100: 209–217 (210M), human papilloma virus (HPV)-16 E7(12–20)	Vaccine	Phase II	HLA-A2.1-positive patients with at least 1 mm melanoma on initial biopsy
gp100: 209–217 (210M)	Vaccine	Phase II	HLA-A2.1-positive patients with metastatic melanoma
PR1: 169–177, WT1: 126–134	Vaccine	Phase I/II	Myelodysplastic syndrome
Tumor peptide-loaded	Vaccine	Phase I/II	HLA-A1 and/or -A2+ stage III or IV melanoma
Tumor specific mutated Ras peptides	Vaccine	Phase I	Patient's tumor-specific mutated Ras peptides
MAGE-3, Melan-A, gp100, NA17-A	Vaccine	Phase II	Metastatic melanoma
Recombinant gp100, gp100: 209–217 (210M)	Vaccine	Phase II	Metastatic melanoma
Tyrosinase-related protein-2: 180–188 peptide	Vaccine	Phase II	HLA-A0201-positive Refractory metastatic melanoma
gp100 peptide	Vaccine	Phase II	Metastatic melanoma
Peptide-pulsed dendritic cells	Vaccine	Phase I	Melanoma
gp209-2M	Vaccine	Phase I	Metastatic melanoma who have failed prior vaccine therapy
MART-1, gp100	Vaccine	Phase II	Metastatic Melanoma
Tumor and influenza antigen peptides	Vaccine	Phase I/II	HLA-A1 and/or HLA-A2.1 positive stage III or IV melanoma

(continued)

TABLE 17.2. (continued)

Peptide	Delivery	Study	Population
gp100: 44–59, gp100: 209–217 (210M) MART-1	Vaccine	Phase II	Metastatic melanoma
MAGE-A3, NY-ESO-1	Vaccine	Phase I/II	Multiple myeloma
HER2 peptide	Vaccine	Phase I/II	Her 2 positive cancers
EP-2101	Vaccine	Phase II	Stage IIIB or IV or recurrent non-small cell lung cancer
Oncopeptide pulsed DC, survivin and telomerase peptides	Vaccine	Phase I/II	Renal cell carcinoma
Melanoma epitopes, pan-DR epitope (PADRE)	Vaccine	Phase I	Metastatic melanoma
PADRE, oncopeptide pulsed DC, p53, survivin, telomerase	Vaccine	Phase I/II	Malignant melanoma
HSPPC-70	Vaccine	Phase II	Chronic myelogenous leukemia (CML) in chronic phase who are cytogenetically positive after treatment with Gleevec™
Peptide vaccine (MUC-1)	Vaccine	Phase I	MUC1-expressing tumor malignancies

diseases have been nearly eradicated, thereby improving the overall public health of the population. The use of vaccination in cancer has been postulated since the early 1980s, but the first landmark vaccine was only approved in 2006 specifically for cervical cancer prevention.

Therapeutic vaccination for cancer is still in its early stages of development. The key concept underlying cancer vaccines is that the host is capable of initiating an immune response against specific tumor antigen(s) of choice. Thus, many cellular markers, including specific peptides, could be used to generate antibodies. The ultimate objective would be to obtain a graft versus tumor–like effect without any graft versus host toxicity [156]. Key limitations to this process include the rapid profiling of gene expression in both tumor and normal tissues to identify suitable peptides. The ideal peptide needs to be overexpressed in the malignant clone and absent or expressed at very low levels in normal tissue. However, gene products often undergo significant post-translational modification; thus, the resulting identified peptide may not adequately represent the ideal antigen [156].

Currently, there are over 150 trials using peptides in cancer clinical trials (Table 17.2); all except two are vaccine related. The most common approaches respect the application of tumor-specific peptides as vaccines and peptides used as markers or drug targets (e.g., Her2/Neu). Melanoma, which is felt to have a significant immune pathogenesis, seems to be the target for most of these trials, whereas clinical trials of vaccines in hematologic malignancies

are few mainly since a functional immune system, which is a basic necessity in vaccination is normally compromised in most leukemias and lymphomas.

11 Epilogue

Rapid development of large scale screening methods, such as DNA/RNA arrays, proteomics, protein arrays, and more recently metabolomics not only deepens our knowledge about cancer as a disease, but also allows us to appreciate the complexity and variety of these malignancies and offers new hints and molecular targets that could be used to fight it [157, 158]. Future cancer therapy will most likely be much more patient-tailored, with an array of molecular markers tested prior to the first injection of any given anticancer agent, followed closely by monitoring of cell death markers in blood and individual dose adjustments [158–160]. It is hard to predict how long "classical" cytotoxic chemotherapy will be used, but certainly different variants of anticancer drugs will arise. Some of them kill the tumor, others will prevent its growth and spread, others will reverse the resistance of tumor cells against certain therapies, and still others will just prevent from expansion between the cycles of more aggressive chemotherapy. In the future, although more patients will truly be cured of cancer, others will have their disease converted from a rapidly progressing state to a more stationary, chronic diseases that could for decades be maintained by oral nontoxic chemotherapy.

Acknowledgments. ML thankfully acknowledges the support by the CFI-Canada Research Chair program, PCRFC-, CCMF-, MHRC, and CIHR-foundation–financed programs. SM thankfully acknowledges the support by the MHRC-, CCMF-, and University of Manitoba–funded fellowships. SP, EW, and KW thankfully acknowledge the support by CIHR-training fellowship. AZ, SP, and ME thankfully acknowledge generous CCMF-funded fellowships.

References

1. Los M, Wesselborg S, Schulze-Osthoff K. The role of caspases in development, immunity, and apoptotic signal transduction: lessons from knockout mice. Immunity 1999, 10:629–639.

2. Philchenkov A, Zavelevich M, Kroczak TJ, et al. Caspases and cancer: mechanisms of inactivation and new treatment modalities. Exp Oncol 2004, 26(2):82–97.

3. Wang X. The expanding role of mitochondria in apoptosis. Genes Dev 2001, 15(22):2922–2933.

4. Chu E, DeVita VT. Principles of medical oncology. In: DeVita VT, Hellman S, Rosenberg SA (eds.). Cancer: principles and practice of oncology, 7th ed. Philadelphia: Lippincott Williams & Wilkins, 2005.

5. Skipper HE. Kinetics of mammary tumor cell growth and implications for therapy. Cancer 1971, 28(6):1479–1499.

6. Ling V. Multidrug resistance: molecular mechanisms and clinical relevance. Cancer Chemother Pharmacol 1997, 40(Suppl):S3–8.

7. Shadidi M, Sioud M. Selective targeting of cancer cells using synthetic peptides. Drug Resist Updat 2003, 6(6):363–371.

8. Cao Y. Endogenous angiogenesis inhibitors and their therapeutic implications. Int J Biochem Cell Biol 2001, 33(4):357–369.

9. Rosenblatt MI, Azar DT. Anti-angiogenic therapy: prospects for treatment of ocular tumors. Semin Ophthalmol 2006, 21(3):151–160.

10. Gibaldi M. Regulating angiogenesis: a new therapeutic strategy. J Clin Pharmacol 1998, 38(10):898–903.

11. Folkman J. Tumor angiogenesis: therapeutic implications. N Engl J Med 1971, 285(21):1182–1186.

12. Pandya NM, Dhalla NS, Santani DD. Angiogenesis—a new target for future therapy. Vascul Pharmacol 2006, 44(5):265–274.

13. Bikfalvi A, Bicknell R. Recent advances in angiogenesis, anti-angiogenesis and vascular targeting. Trends Pharmacol Sci 2002, 23(12):576–582.

14. Gille J. Antiangiogenic cancer therapies get their act together: current developments and future prospects of growth factor- and growth factor receptor-targeted approaches. Exp Dermatol 2006, 15(3):175–186.

15. Sivakumar B, Harry LE, Paleolog EM. Modulating angiogenesis: more vs less. JAMA 2004, 292(8):972–977.

16. O'Reilly MS. The combination of antiangiogenic therapy with other modalities. Cancer J 2002, 8(Suppl 1):S89–99.

17. Nyberg P, Xie L, Kalluri R. Endogenous inhibitors of angiogenesis. Cancer Res 2005, 65(10):3967–3979.

18. Folkman J. Endogenous angiogenesis inhibitors. Apmis 2004, 112(7–8):496–507.

19. Folkman J. Antiangiogenesis in cancer therapy—endostatin and its mechanisms of action. Exp Cell Res 2006, 312(5):594–607.

20. Rege TA, Fears CY, Gladson CL. Endogenous inhibitors of angiogenesis in malignant gliomas: nature's antiangiogenic therapy. Neuro-oncol 2005, 7(2):106–121.

21. Hajitou A, Grignet C, Devy L, et al. The antitumoral effect of endostatin and angiostatin is associated with a down-regulation of vascular endothelial growth factor expression in tumor cells. FASEB J 2002, 16(13):1802–1804.

22. Folkman J. Antiangiogenesis in cancer therapy—endostatin and its mechanisms of action. Exp Cell Res 2006, 312(5):594–607.

23. Brooks PC, Stromblad S, Sanders LC, et al. Localization of matrix metalloproteinase MMP-2 to the surface of invasive cells by interaction with integrin alpha v beta 3. Cell 1996, 85(5):683–693.

24. Bikfalvi A. Recent developments in the inhibition of angiogenesis: examples from studies on platelet factor-4 and the VEGF/VEGFR system. Biochem Pharmacol 2004, 68(6):1017–1021.

25. Zaichuk TA, Shroff EH, Emmanuel R, et al. Nuclear factor of activated T cells balances angiogenesis activation and inhibition. J Exp Med 2004, 199(11):1513–1522.

26. Staton CA, Lewis CE. Angiogenesis inhibitors found within the haemostasis pathway. J Cell Mol Med 2005, 9(2):286–302.

27. Volpert OV, Zaichuk T, Zhou W, et al. Inducer-stimulated Fas targets activated endothelium for destruction by anti-angiogenic thrombospondin-1 and pigment epithelium-derived factor. Nat Med 2002, 8(4):349–357.

28. Armstrong LC, Bjorkblom B, Hankenson KD, et al. Thrombospondin 2 inhibits microvascular endothelial cell proliferation by a caspase-independent mechanism. Mol Biol Cell 2002, 13(6):1893–1905.

29. Dunehoo AL, Anderson M, Majumdar S, et al. Cell adhesion molecules for targeted drug delivery. J Pharm Sci 2006, 95(9):1856–1872.

30. Blaschuk OW, Rowlands TM. Plasma membrane components of adherens junctions (Review). Mol Membr Biol 2002, 19(2):75–80.

31. Erez N, Zamir E, Gour BJ, et al. Induction of apoptosis in cultured endothelial cells by a cadherin antagonist peptide: involvement of fibroblast growth factor receptor-mediated signalling. Exp Cell Res 2004, 294(2):366–378.

32. Mendoza FJ, Espino PS, Cann KL, et al. Anti-tumor chemotherapy utilizing peptide-based approaches-apoptotic pathways, kinases, and proteasome as targets. Arch Immunol Ther Exp 2005, 53(1):47–60.

33. Blaschuk OW, Rowlands TM. Cadherins as modulators of angiogenesis and the structural integrity of blood vessels. Cancer Metastasis Rev 2000, 19(1–2):1–5.

34. Williams E, Williams G, Gour BJ, et al. A novel family of cyclic peptide antagonists suggests that N-cadherin specificity is determined by amino acids that flank the HAV motif. J Biol Chem 2000, 275(6):4007–4012.

35. Williams EJ, Williams G, Gour B, et al. INP, a novel N-cadherin antagonist targeted to the amino acids that flank the HAV motif. Mol Cell Neurosci 2000, 15(5):456–464.

36. Kawaguchi M, Hosotani R, Ohishi S, et al. A novel synthetic Arg-Gly-Asp-containing peptide cyclo(-RGDfV-) is the potent inhibitor of angiogenesis. Biochem Biophys Res Commun 2001, 288(3):711–717.

37. Friedlander M, Brooks PC, Shaffer RW, et al. Definition of two angiogenic pathways by distinct alpha v integrins. Science 1995, 270(5241):1500–1502.

38. Nicosia RF, Bonanno E. Inhibition of angiogenesis in vitro by Arg-Gly-Asp-containing synthetic peptide. Am J Pathol 1991, 138(4):829–833.

39. Aumailley M, Gurrath M, Muller G, et al. Arg-Gly-Asp constrained within cyclic pentapeptides. Strong and selective inhibitors of cell adhesion to vitronectin and laminin fragment P1. FEBS Lett 1991, 291(1):50–54.

40. Cianfrocca ME, Kimmel KA, Gallo J, et al. Phase 1 trial of the antiangiogenic peptide ATN-161 (Ac-PHSCN-NH(2)), a beta integrin antagonist, in patients with solid tumours. Br J Cancer 2006, 94(11):1621–1626.

41. Belvisi L, Riccioni T, Marcellini M, et al. Biological and molecular properties of a new alpha(v)beta3/alpha(v)beta5 integrin antagonist. Mol Cancer Ther 2005, 4(11):1670–1680.

42. Niewiarowski S, McLane MA, Kloczewiak M, et al. Disintegrins and other naturally occurring antagonists of platelet fibrinogen receptors. Semin Hematol 1994, 31(4):289–300.

43. Kang IC, Lee YD, Kim DS. A novel disintegrin salmosin inhibits tumor angiogenesis. Cancer Res 1999, 59(15):3754–3760.

44. Kim SI, Kim KS, Kim HS, et al. Inhibitory effect of the salmosin gene transferred by cationic liposomes on the progression of B16BL6 tumors. Cancer Res 2003, 63(19):6458–6462.

45. Hong SY, Sohn YD, Chung KH, et al. Structural and functional significance of disulfide bonds in saxatilin, a 7.7 kDa disintegrin. Biochem Biophys Res Commun 2002, 293(1):530–536.

46. Kim KS, Kim DS, Chung KH, et al. Inhibition of angiogenesis and tumor progression by hydrodynamic cotransfection of angiostatin K1-3, endostatin, and saxatilin genes. Cancer Gene Ther 2006, 13(6):563–571.

47. Hong SY, Koh YS, Chung KH, et al. Snake venom disintegrin, saxatilin, inhibits platelet aggregation, human umbilical vein endothelial cell proliferation, and smooth muscle cell migration. Thromb Res 2002, 105(1):79–86.

48. Kim DS, Jang YJ, Jeon OH, et al. Saxatilin inhibits TNF-alpha-induced proliferation by suppressing AP-1-dependent IL-8 expression in the ovarian cancer cell line MDAH 2774. Mol Immunol 2006.

49. Macaulay VM, O'Byrne KJ, Saunders MP, et al. Phase I study of intrapleural batimastat (BB-94), a matrix metalloproteinase inhibitor, in the treatment of malignant pleural effusions. Clin Cancer Res 1999, 5(3):513–520.

50. Griffioen AW, van der Schaft DW, Barendsz-Janson AF, et al. Anginex, a designed peptide that inhibits angiogenesis. Biochem J 2001, 354(Pt 2):233–242.

51. Pilch J, Franzin CM, Knowles LM, et al. The anti-angiogenic peptide anginex disrupts the cell membrane. J Mol Biol 2006, 356(4):876–885.

52. Shai Y, Oren Z. From "carpet" mechanism to de-novo designed diastereomeric cell-selective antimicrobial peptides. Peptides 2001, 22(10):1629–1641.

53. Dings RP, Yokoyama Y, Ramakrishnan S, et al. The designed angiostatic peptide anginex synergistically improves chemotherapy and antiangiogenesis therapy with angiostatin. Cancer Res 2003, 63(2):382–385.

54. Tsujimoto Y, Cossman J, Jaffe E, et al. Involvement of the bcl-2 gene in human follicular lymphoma. Science 1985, 228(4706):1440–1443.

55. Yang J, Liu X, Bhalla K, et al. Prevention of apoptosis by Bcl-2: release of cytochrome c from mitochondria blocked. Science 1997, 275(5303):1129–1132.

56. Oltvai ZN, Milliman CL, Korsmeyer SJ. Bcl-2 heterodimerizes in vivo with a conserved homolog, Bax, that accelerates programmed cell death. Cell 1993, 74(4):609–619.

57. Kelekar A, Thompson CB. Bcl-2-family proteins: the role of the BH3 domain in apoptosis. Trends Cell Biol 1998, 8(8):324–330.

58. Manion MK, Hockenbery DM. Targeting BCL-2-related proteins in cancer therapy. Cancer Biol Ther 2003, 2(4 Suppl 1):S105–114.

59. Cosulich SC, Worrall V, Hedge PJ, et al. Regulation of apoptosis by BH3 domains in a cell-free system. Curr Biol 1997, 7(12):913–920.

60. Holinger EP, Chittenden T, Lutz RJ. Bak BH3 peptides antagonize Bcl-xL function and induce apoptosis through cytochrome c-independent activation of caspases. J Biol Chem 1999, 274(19):13298–13304.

61. Wang JL, Zhang ZJ, Choksi S, et al. Cell permeable Bcl-2 binding peptides: a chemical approach to apoptosis induction in tumor cells. Cancer Res 2000, 60(6):1498–1502.

62. Zavaglia D, Normand N, Brewis N, et al. VP22-mediated and light-activated delivery of an anti-c-raf1 antisense oligonucleotide improves its activity after intratumoral injection in nude mice. Mol Ther 2003, 8(5):840–845.

63. Brewis ND, Phelan A, Normand N, et al. Particle assembly incorporating a VP22-BH3 fusion protein, facilitating intracellular delivery, regulated release, and apoptosis. Mol Ther 2003, 7(2):262–270.

64. Walensky LD, Kung AL, Escher I, et al. Activation of apoptosis in vivo by a hydrocarbon-stapled BH3 helix. Science 2004, 305(5689):1466–1470.

65. Hanks SK, Quinn AM, Hunter T. The protein kinase family: conserved features and deduced phylogeny of the catalytic domains. Science 1988, 241(4861):42–52.

66. Manning G, Whyte DB, Martinez R, et al. The protein kinase complement of the human genome. Science 2002, 298(5600):1912–1934.

67. Maddika S, Ande SR, Panigrahi S, et al. Cell survival, cell death and cell cycle pathways are interconnected: Implications for cancer therapy. Drug Resist Updat 2007, 10: in press.

68. Cohen P. Protein kinases: the major drug targets of the twenty-first century? Nat Rev Drug Discov 2002, 1(4):309–315.

69. Paez JG, Janne PA, Lee JC, et al. EGFR mutations in lung cancer: correlation with clinical response to gefitinib therapy. Science 2004, 304(5676):1497–1500.

70. Pero SC, Shukla GS, Armstrong AL, et al. Identification of a small peptide that inhibits the phosphorylation of ErbB2 and proliferation of ErbB2 overexpressing breast cancer cells. Int J Cancer 2004, 111(6):951–960.

71. Flowers LO, Johnson HM, Mujtaba MG, et al. Characterization of a peptide inhibitor of Janus kinase 2 that mimics suppressor of cytokine signaling 1 function. J Immunol 2004, 172(12):7510–7518.

72. Flowers LO, Subramaniam PS, Johnson HM. A SOCS-1 peptide mimetic inhibits both constitutive and IL-6 induced activation of STAT3 in prostate cancer cells. Oncogene 2005, 24(12):2114–2120.

73. Buerger C, Nagel-Wolfrum K, Kunz C, et al. Sequence-specific peptide aptamers, interacting with the intracellular domain of the epidermal growth factor receptor, interfere with Stat3 activation and inhibit the growth of tumor cells. J Biol Chem 2003, 278(39):37610–37621.

74. Kunz C, Borghouts C, Buerger C, et al. Peptide aptamers with binding specificity for the intracellular domain of the ErbB2 receptor interfere with AKT signaling and sensitize breast cancer cells to Taxol. Mol Cancer Res 2006, 4(12):983–998.

75. Obata T, Yaffe MB, Leparc GG, et al. Peptide and protein library screening defines optimal substrate motifs for AKT/PKB. J Biol Chem 2000, 275(46):36108–36115.

76. Laine J, Kunstle G, Obata T, et al. The protooncogene TCL1 is an Akt kinase coactivator. Mol Cell 2000, 6(2):395–407.

77. Lam KS, Wu J, Lou Q. Identification and characterization of a novel synthetic peptide substrate specific for Src-family protein tyrosine kinases. Int J Pept Protein Res 1995, 45(6):587–592.

78. Alfaro-Lopez J, Yuan W, Phan BC, et al. Discovery of a novel series of potent and selective substrate-based inhibitors of p60c-src protein tyrosine kinase: conformational and topographical constraints in peptide design. J Med Chem 1998, 41(13):2252–2260.

79. Kamath JR, Liu R, Enstrom AM, et al. Development and characterization of potent and specific peptide inhibitors of p60c-src protein tyrosine kinase using pseudosubstrate-based inhibitor design approach. J Pept Res 2003, 62(6):260–268.

80. Coussens L, Parker PJ, Rhee L, et al. Multiple, distinct forms of bovine and human protein kinase C suggest diversity in cellular signaling pathways. Science 1986, 233(4766):859–866.

81. Chen L, Hahn H, Wu G, et al. Opposing cardioprotective actions and parallel hypertrophic effects of delta PKC and epsilon PKC. Proc Natl Acad Sci USA 2001, 98(20):11114–11119.

82. Dominguez I, Diaz-Meco MT, Municio MM, et al. Evidence for a role of protein kinase C zeta subspecies in maturation of Xenopus laevis oocytes. Mol Cell Biol 1992, 12(9):3776–3783.

83. Eichholtz T, de Bont DB, de Widt J, et al. A myristoylated pseudosubstrate peptide, a novel protein kinase C inhibitor. J Biol Chem 1993, 268(3):1982–1986.

84. House C, Kemp BE. Protein kinase C contains a pseudosubstrate prototope in its regulatory domain. Science 1987, 238(4834):1726–1728.

85. Bogoyevitch MA, Barr RK, Ketterman AJ. Peptide inhibitors of protein kinases-discovery, characterisation and use. Biochim Biophys Acta 2005, 1754(1–2): 79–99.

86. Tohtong R, Phattarasakul K, Jiraviriyakul A, et al. Dependence of metastatic cancer cell invasion on

MLCK-catalyzed phosphorylation of myosin regulatory light chain. Prostate Cancer Prostatic Dis 2003, 6(3):212–216.

87. Pearson RB, Misconi LY, Kemp BE. Smooth muscle myosin kinase requires residues on the COOH-terminal side of the phosphorylation site. Peptide inhibitors. J Biol Chem 1986, 261(1):25–27.

88. Barr RK, Bogoyevitch MA. The c-Jun N-terminal protein kinase family of mitogen-activated protein kinases (JNK MAPKs). Int J Biochem Cell Biol 2001, 33(11):1047–1063.

89. Dickens M, Rogers JS, Cavanagh J, et al. A cytoplasmic inhibitor of the JNK signal transduction pathway. Science 1997, 277(5326):693–696.

90. Barr RK, Kendrick TS, Bogoyevitch MA. Identification of the critical features of a small peptide inhibitor of JNK activity. J Biol Chem 2002, 277(13):10987–10997.

91. Bonny C, Oberson A, Negri S, et al. Cell-permeable peptide inhibitors of JNK: novel blockers of beta-cell death. Diabetes 2001, 50(1):77–82.

92. Payne ME, Fong YL, Ono T, et al. Calcium/calmodulin-dependent protein kinase II. Characterization of distinct calmodulin binding and inhibitory domains. J Biol Chem 1988, 263(15):7190–7195.

93. Perea SE, Reyes O, Puchades Y, et al. Antitumor effect of a novel proapoptotic peptide that impairs the phosphorylation by the protein kinase 2 (casein kinase 2). Cancer Res 2004, 64(19):7127–7129.

94. Cassens U, Lewinski G, Samraj AK, et al. Viral modulation of cell death by inhibition of caspases. Arch Immunol Ther Exp 2003, 51(1):19–27.

95. Cellier F, Conejero G, Breitler JC, et al. Molecular and physiological responses to water deficit in drought-tolerant and drought-sensitive lines of sunflower. Accumulation of dehydrin transcripts correlates with tolerance. Plant Physiol 1998, 116(1):319–328.

96. Dai Z, Zhu WG, Morrison CD, et al. A comprehensive search for DNA amplification in lung cancer identifies inhibitors of apoptosis cIAP1 and cIAP2 as candidate oncogenes. Hum Mol Genet 2003, 12(7):791–801.

97. Hasegawa T, Suzuki K, Sakamoto C, et al. Expression of the inhibitor of apoptosis (IAP) family members in human neutrophils: up-regulation of cIAP2 by granulocyte colony-stimulating factor and overexpression of cIAP2 in chronic neutrophilic leukemia. Blood 2003, 101(3):1164–1171.

98. Imoto I, Yang ZQ, Pimkhaokham A, et al. Identification of cIAP1 as a candidate target gene within an amplicon at 11q22 in esophageal squamous cell carcinomas. Cancer Res 2001, 61(18):6629–6634.

99. Krajewska M, Krajewski S, Banares S, et al. Elevated expression of inhibitor of apoptosis proteins in prostate cancer. Clin Cancer Res 2003, 9(13):4914–4925.

100. Salvesen GS, Duckett CS. IAP proteins: blocking the road to death's door. Nat Rev Mol Cell Biol 2002, 3(6):401–410.

101. Tamm I, Kornblau SM, Segall H, et al. Expression and prognostic significance of IAP-family genes in human cancers and myeloid leukemias. Clin Cancer Res 2000, 6(5):1796–1803.

102. Liu Z, Sun C, Olejniczak ET, et al. Structural basis for binding of Smac/DIABLO to the XIAP BIR3 domain. Nature 2000, 408(6815):1004–1008.

103. Wu G, Chai J, Suber TL, et al. Structural basis of IAP recognition by Smac/DIABLO. Nature 2000, 408(6815):1008–1012.

104. Arnt CR, Chiorean MV, Heldebrant MP, et al. Synthetic Smac/DIABLO peptides enhance the effects of chemotherapeutic agents by binding XIAP and cIAP1 in situ. J Biol Chem 2002, 277(46):44236–44243.

105. Fulda S, Wick W, Weller M, et al. Smac agonists sensitize for Apo2L/TRAIL- or anticancer drug-induced apoptosis and induce regression of malignant glioma in vivo. Nat Med 2002, 8(8):808–815.

106. Guo F, Nimmanapalli R, Paranawithana S, et al. Ectopic overexpression of second mitochondria-derived activator of caspases (Smac/DIABLO) or cotreatment with N-terminus of Smac/DIABLO peptide potentiates epothilone B derivative-(BMS 247550) and Apo-2L/TRAIL-induced apoptosis. Blood 2002, 99(9):3419–3426.

107. Yang L, Mashima T, Sato S, et al. Predominant suppression of apoptosome by inhibitor of apoptosis protein in non-small cell lung cancer H460 cells: therapeutic effect of a novel polyarginine-conjugated Smac peptide. Cancer Res 2003, 63(4):831–837.

108. Denicourt C, Dowdy SF. Medicine. Targeting apoptotic pathways in cancer cells. Science 2004, 305(5689):1411–1413.

109. Patch JA, Barron AE. Mimicry of bioactive peptides via non-natural, sequence-specific peptidomimetic oligomers. Curr Opin Chem Biol 2002, 6(6):872–877.

110. Gante J. Peptidomimetics: tailored enzyme inhibitors. Angewandte Chemie [International Edition in English] 1994, 33(17):1699–1720.

111. Arnt CR, Kaufmann SH. The saintly side of Smac/DIABLO: giving anticancer drug-induced apoptosis a boost. Cell Death Diff 2003, 10(10):1118–1120.

112. Freidinger RM, Veber DF, Perlow DS, et al. Bioactive conformation of luteinizing hormone-releasing hormone: evidence from a conformationally constrained analogue. Science 1980, 210(4470):656–658.

113. Sukumaran DK. A molecular constraint that generates a cis peptide bond. J Amer Chem Soc 1991, 113(2):706–707.

114. Alig L, Edenhofer A, Hadvary P, et al. Low molecular weight, non-peptide fibrinogen receptor antagonists. J Med Chem 1992, 35(23):4393–4407.

115. Marshall GR. A hierarchical approach to peptidomimetic design. Tetrahedron 1993, 49:3547–3558.

116. Hirschmann R. De novo design and synthesis of somatostatin non-peptide peptidomimetics utilizing beta-D-glucose as a novel scaffolding. J Am Chem Soc 1993, 115(26):12550–12568.

117. De B, Plattner JJ, Bush EN, et al. LH-RH antagonists: design and synthesis of a novel series of peptidomimetics. J Med Chem 1989, 32(9):2036–2038.

118. Weinstock J, Keenan RM, Samanen J, et al. 1-(carboxybenzyl)imidazole-5-acrylic acids: potent and selective angiotensin II receptor antagonists. J Med Chem 1991, 34(4):1514–1517.

119. Oost TK, Sun C, Armstrong RC, et al. Discovery of potent antagonists of the antiapoptotic protein XIAP for the treatment of cancer. J Med Chem 2004, 47(18):4417–4426.

120. Li L, Thomas RM, Suzuki H, et al. A small molecule Smac mimic potentiates TRAIL- and TNFalpha-mediated cell death. Science 2004, 305(5689):1471–1474.

121. Sun H, Nikolovska-Coleska Z, Yang CY, et al. Structure-based design of potent, conformationally constrained Smac mimetics. J Am Chem Soc 2004, 126(51):16686–16687.

122. Huang Y, Rich RL, Myszka DG, et al. Requirement of both the second and third BIR domains for the relief of X-linked inhibitor of apoptosis protein (XIAP)-mediated caspase inhibition by Smac. J Biol Chem 2003, 278(49):49517–49522.

123. Booy EP, Johar D, Maddika S, et al. Monoclonal and bispecific antibodies as novel therapeutics. Arch Immunol Ther Exp 2006, 54:1–17.

124. Johnston JB, Navaratnam S, Pitz MW, et al. Targeting the EGFR pathway for cancer therapy. Curr Med Chem 2006, 13:3483–3492.

125. Byrd JC, Rai K, Peterson BL, et al. Addition of rituximab to fludarabine may prolong progression-free survival and overall survival in patients with previously untreated chronic lymphocytic leukemia: an updated retrospective comparative analysis of CALGB 9712 and CALGB 9011. Blood 2005, 105(1):49–53.

126. Hiddemann W, Kneba M, Dreyling M, et al. Frontline therapy with rituximab added to the combination of cyclophosphamide, doxorubicin, vincristine, and prednisone (CHOP) significantly improves the outcome for patients with advanced-stage follicular lymphoma compared with therapy with CHOP alone: results of a prospective randomized study of the German Low-Grade Lymphoma Study Group. Blood 2005, 106(12):3725–3732.

127. Kaminski MS, Tuck M, Estes J, et al. 131I-tositumomab therapy as initial treatment for follicular lymphoma. N Engl J Med 2005, 352(5):441–449.

128. Marcus R, Imrie K, Belch A, et al. CVP chemotherapy plus rituximab compared with CVP as first-line treatment for advanced follicular lymphoma. Blood 2005, 105(4):1417–1423.

129. Larson RA, Sievers EL, Stadtmauer EA, et al. Final report of the efficacy and safety of gemtuzumab ozogamicin (Mylotarg) in patients with CD33-positive acute myeloid leukemia in first recurrence. Cancer 2005, 104(7):1442–1452.

130. O'Brien SM, Kantarjian HM, Thomas DA, et al. Alemtuzumab as treatment for residual disease after chemotherapy in patients with chronic lymphocytic leukemia. Cancer 2003, 98(12):2657–2663.

131. Hurwitz H, Fehrenbacher L, Novotny W, et al. Bevacizumab plus irinotecan, fluorouracil, and leucovorin for metastatic colorectal cancer. N Engl J Med 2004, 350(23):2335–2342.

132. Sandler A, Gray R, Perry MC, et al. Paclitaxel-carboplatin alone or with bevacizumab for non-small-cell lung cancer. N Engl J Med 2006, 355(24):2542–2550.

133. Bonner JA, Harari PM, Giralt J, et al. Radiotherapy plus cetuximab for squamous-cell carcinoma of the head and neck. N Engl J Med 2006, 354(6):567–578.

134. Cunningham D, Allum WH, Stenning SP, et al. Perioperative chemotherapy versus surgery alone for resectable gastroesophageal cancer. N Engl J Med 2006, 355(1):11–20.

135. Piccart-Gebhart MJ, Procter M, Leyland-Jones B, et al. Trastuzumab after adjuvant chemotherapy in HER2-positive breast cancer. N Engl J Med 2005, 353(16):1659–1672.

136. Slamon DJ, Leyland-Jones B, Shak S, et al. Use of chemotherapy plus a monoclonal antibody against HER2 for metastatic breast cancer that overexpresses HER2. N Engl J Med 2001, 344(11):783–792.

137. Aina OH, Sroka TC, Chen ML, et al. Therapeutic cancer targeting peptides. Biopolymers 2002, 66(3):184–199.

138. Beckman RA, Loeb LA. Genetic instability in cancer: theory and experiment. Semin Cancer Biol 2005, 15(6):423–435.

139. D'Andrea LD, Del Gatto A, Pedone C, et al. Peptide-based molecules in angiogenesis. Chem Biol Drug Des 2006, 67(2):115–126.

140. Arap W, Pasqualini R, Ruoslahti E. Cancer treatment by targeted drug delivery to tumor vasculature in a mouse model. Science 1998, 279(5349):377–380.

141. Janssen AP, Schiffelers RM, ten Hagen TL, et al. Peptide-targeted PEG-liposomes in anti-angiogenic therapy. Int J Pharm 2003, 254(1):55–58.

142. Schiffelers SL, Akkermans JA, Saris WH, et al. Lipolytic and nutritive blood flow response to beta-adrenoceptor stimulation in situ in subcutaneous abdominal adipose tissue in obese men. Int J Obes Relat Metab Disord 2003, 27(2):227–231.

143. Opalinska JB, Gewirtz AM. Nucleic-acid therapeutics: basic principles and recent applications. Nat Rev Drug Discov 2002, 1(7):503–514.

144. Sorensen DR, Leirdal M, Sioud M. Gene silencing by systemic delivery of synthetic siRNAs in adult mice. J Mol Biol 2003, 327(4):761–766.

145. Ellerby HM, Arap W, Ellerby LM, et al. Anti-cancer activity of targeted pro-apoptotic peptides. Nat Med 1999, 5(9):1032–1038.

146. Javadpour MM, Juban MM, Lo WC, et al. De novo antimicrobial peptides with low mammalian cell toxicity. J Med Chem 1996, 39(16):3107–3113.

147. Fulda S, Debatin KM. Apoptosis pathways: turned on their heads? Drug Resist Updat 2003, 6(1):1–3.

148. Grifman M, Trepel M, Speece P, et al. Incorporation of tumor-targeting peptides into recombinant adeno-associated virus capsids. Mol Ther 2001, 3(6):964–975.

149. Schally AV, Nagy A. Cancer chemotherapy based on targeting of cytotoxic peptide conjugates to their receptors on tumors. Eur J Endocrinol 1999, 141(1):1–14.

150. Kvols LK, Woltering EA. Role of somatostatin analogues in the clinical management of non-neuroendocrine solid tumors. Anticancer Drugs 2006, 17(6):601–608.

151. Yegen BC. Bombesin-like peptides: candidates as diagnostic and therapeutic tools. Curr Pharm Des 2003, 9(12):1013–1022.

152. Yamamoto Y, Tsutsumi Y, Mayumi T. Molecular design of bioconjugated cell adhesion peptide with a water-soluble polymeric modifier for enhancement of antimetastatic effect. Curr Drug Targets 2002, 3(2):123–130.

153. Lu Y, Yang J, Sega E. Issues related to targeted delivery of proteins and peptides. Aaps J 2006, 8(3):E466–478.

154. Torchilin VP, Lukyanov AN. Peptide and protein drug delivery to and into tumors: challenges and solutions. Drug Discov Today 2003, 8(6):259–266.

155. Dass CR, Choong PF. Carrier-mediated delivery of peptidic drugs for cancer therapy. Peptides 2006, 27(11):3020–3028.

156. Rammensee HG. Some considerations on the use of peptides and mRNA for therapeutic vaccination against cancer. Immunol Cell Biol 2006, 84(3):290–294.

157. Anderson JE, Hansen LL, Mooren FC, et al. Methods and biomarkers for the diagnosis and prognosis of cancer and other diseases: Towards personalized medicine. Drug Resist Updat 2006, 9(4–5):198–210.

158. Kroczak TJ, Baran J, Pryjma JS, et al. The emerging importance of DNA mapping and other comprehensive screening techniques as tools to identify new drug targets and as a mean of (cancer) therapy personalization. Expert Opin Ther Targets 2006, 10:289–302.

159. Barczyk K, Kreuter M, Pryjma J, et al. Serum cytochrome c indicates in vivo apoptosis and can serve as a prognostic marker during cancer therapy. Int J Cancer 2005, 116(2):167–173.

160. Ghavami S, Hashemi M, Kadkhoda K, et al. Apoptosis in liver diseases - detection and therapeutic applications. Med Sci Monit 2005, 11(11): RA337–3345.

161. Luo Y, Smith RA, Guan R, et al. Pseudosubstrate peptides inhibit Akt and induce cell growth inhibition. Biochemistry 2004, 43(5):1254–1263.

162. Hiromura M, Okada F, Obata T, et al. Inhibition of Akt kinase activity by a peptide spanning the betaA strand of the proto-oncogene TCL1. J Biol Chem 2004, 279(51):53407–53418.

163. Niv MY, Rubin H, Cohen J, et al. Sequence-based design of kinase inhibitors applicable for therapeutics and target identification. J Biol Chem 2004, 279(2):1242–1255.

164. Ramdas L, Obeyesekere NU, Sun G, et al. N-myristoylation of a peptide substrate for Src converts it into an apparent slow-binding bisubstrate-type inhibitor. J Pept Res 1999, 53(5):569–577.

165. Ward NE, Gravitt KR, O'Brian CA. Irreversible inactivation of protein kinase C by a peptide-substrate analogue. J Biol Chem 1995, 270(14):8056–8060.

166. Kemp BE, Pearson RB, Guerriero V Jr, et al. The calmodulin binding domain of chicken smooth muscle myosin light chain kinase contains a pseudosubstrate sequence. J Biol Chem 1987, 262(6):2542–2548.

167. Gondeau C, Gerbal-Chaloin S, Bello P, et al. Design of a novel class of peptide inhibitors of cyclin-dependent kinase/cyclin activation. J Biol Chem 2005, 280(14):13793–13800.

168. Bardwell AJ, Flatauer LJ, Matsukuma K, et al. A conserved docking site in MEKs mediates high-affinity binding to MAP kinases and cooperates with a scaffold protein to enhance signal transmission. J Biol Chem 2001, 276(13):10374–10386.

Chapter 18
Non-Peptidic Mimetics as Cancer-Sensitizing Agents

Ruud P.M. Dings, Mark Klein, and Kevin H. Mayo

1 Introduction

Successful therapeutic treatment of solid tumors has been limited during the past decade, and available anticancer strategies aimed directly at killing tumor cells leave much room for improvement. The killing of tumor cells is effective only when the antitumor drug can reach tumor cells from the circulation. Due to a solid tumor's architecture and high interstitial pressure, however, these requirements are not always met, resulting in limited success of chemotherapeutic, immunotoxin strategies, and radiation due to the lack of radiosensitizing oxygen. Moreover, tumor cells are genetically unstable and often develop resistance to chemical agents targeted against them. Because angiogenesis is a prerequisite for the development of metastases and the outgrowth of tumors, the use of anti-angiogenic therapy has come to the forefront, first exemplified in the clinic with Avastin® [1–4].

Regardless of the therapeutic approach or drug of choice, monotherapy usually has limited effectiveness in treating cancer. Moreover, standard chemotherapy and radiation therapy always raise concerns related to systemic effects and toxicity. In this regard, use of some targeted therapy in combination with radiation and chemotherapy is likely to hold the key to future successes in the clinic. This chapter focuses on the development of non-peptidic mimetics to sensitize tumors in order to reduce toxicity from chemotherapy and radiation therapy and improve outcome. There are many potential advantages that would attend such structurally simplified, smaller therapeutic agents, the most important being increased in vivo exposure and reduced immunogenicity, as well as ease and cost of production.

2 Designing a Non-Peptidic Mimetic

The first step to design a peptidomimetic or non-peptidic mimetic is usually the analysis of the protein structure of interest, either that of the target or of the ligand, or both if one is fortunate enough to have them. Normally structure analysis of the ligand yields more useful information, as one would like to mimic the activity of the ligand, usually to design an antagonist that tends to be easier than to design an agonist. This phase of the design often leads to structural hypotheses that are tested normally by producing various amino acid substituted variants of the native sequence. In the case of a native protein, site-directed mutagenesis is usually employed by expressing the variant protein using an appropriate expression system. In the case of a peptide, variants may be synthesized chemically, and amino acid residues can be substituted, most often by using alanine scanning in which native residues are substituted with alanines one at a time, and a series of such peptides is produced that basically "scan" the native sequence. In either instance, variants of the native sequence are then assessed

From: *Sensitization of Cancer Cells for Chemo/Immuno/Radio-therapy, 1st Edition*.
Edited by: Benjamin Bonavida © Humana Press, Totowa, NJ

for structural maintenance, and in a functional and/or binding assay to assess the effect on activity. This later point results in an understanding of the structure–activity relationships (SAR) of the protein/peptide of interest.

Overall, the key is to delineate those features of the biomolecular interface between ligand and its target (the pharmacophore site), and then to mimic that interfacial surface with the peptidomimetic. The following sections briefly outline these various aspects to the design process.

2.1 Structure Analysis

The primary techniques used to elucidate molecular structure at high resolution are X-ray crystallography [5] and nuclear magnetic resonance (NMR) spectroscopy [6–8]. Many high resolution structures are presently known and are available to the public in, e.g., the Protein Data Base (PDB). Although it does not provide high-resolution structures, circular dichroism (CD) spectropolarimetry is often used to assess the presence of some conformational fold, most useful to drug design when working with peptides. The end result from the structural analysis is to have a good working model of the structure of interest.

2.2 Structure–Activity Relationships

Once one has a good working model of the structure, one needs to assess structure–activity relationships (SAR) of the protein or peptide to identify those residues most responsible for activity. There are two basic approaches used to assess activity: a "receptor" binding assay and a functional assay. The receptor binding assay assesses how well one molecule interacts or binds another, and there are a number of physical techniques that can be used to assess this, e.g., isothermal calorimetry, BIAcore analysis, or through a titration using some spectroscopic technique such as NMR. At best, this analysis provides a thermodynamic binding constant that can be compared among compounds of interest. A functional assay is any assay (usually cell-based or enzyme-based) that provides a readout of some biological activity. One or both of these approaches may be used.

With proteins, the usual approach is to use site-directed mutagenesis to assess which region(s)

of the protein structure or amino acid sequence holds functionally important residues. This may be proceeded by some amino acid sequence analysis of homologous sequences to provide some initial insight and reduce workload. In general, site-directed mutagenesis is used to replace one or a few amino acid residues at a time with alanine (called alanine scanning) or glycine, or some other residue that has different chemical properties. After the variant protein is checked for native folding using some biophysical technique for structural analysis, the activity readout identifies which residues are most important to function. With peptides, one normally uses the solid phase peptide synthesis approach to make variant peptides using alanine or glycine scanning through the amino acid sequence.

2.3 The Pharmacophore Site

Interactions among biomolecules dictate biological events. For optimal activity, these interactions should be complementary in terms of their surface topology and chemical composition. With proteins, the interaction surface is defined by the spatial arrangement of amino acid side chains (i.e., chemical groups) that are scaffolded in place by the peptide backbone, which by itself is usually irrelevant (with the noted exception of enzyme active sites) to the interaction surface. By using SAR information, the investigator can define the pharmacophore site [9], which for all intents and purposes is that surface or site on a protein or peptide that imparts all, most, or at least some of the biological activity. Once identified, the pharmacophore site is invaluable to design or identify smaller entities to mimic that site, either fully or partly. Most often, peptide or non-peptidic mimetics function as antagonists of the protein–"receptor" interaction in order to abrogate the normal biological response.

2.4 Non-Peptidic Mimetic Design Strategies

The usual approach to develop non-peptidic mimetics is first to identify the functionally important amino acid residues and/or pharmacophore site in a native protein of interest, as outlined. After that, approaches vary widely, but usually are developed about a few common themes. One of these is to reduce the native protein to a smaller peptide and constrain it in the bioactive conformation, either by cyclizing the peptide or using a scaffold to present minimal sequences. Another theme is to work with the pharmacophore site and directly design the mimetic, either de novo or via database screening. In either of these instances, the initial design phase would normally go through some iterative process of lead optimization.

A number of studies have focused on the former approach. For example, Gentilucci et al. took a solid-phase peptide synthesis approach to construct a library of fibronectin-derived RGD tripeptide mimetics that target integrins [10].

Fasan et al. used a β-hairpin motif to mimic an α-helix in targeting the p53-HDM2 protein–protein interaction [11]. Andrews et al. created cyclic peptides to target the cyclin-binding groove (CBG) of CDK2 [12]. Walensky et al. had the novel idea to "staple" peptides by introducing non-natural amino acids to chemically "tether" a peptide conformation. This approach is exemplified by a hydrocarbon-stapled (by reduction of α,α-disubstituted olefinic side-chains) peptide mimetic derived from the anti-apoptotic protein BID [13].

Other investigators have used various organic scaffolds to tether peptides. This is demonstrated nicely by calixarene scaffolded mimetics that target the growth factors VEGF and PDGF [14–16]. Mayo et al. used dibenzofuran as a β-turn mimetic to link active peptides derived from the designed antiangiogenic peptide anginex [17]. In the apoptosis arena, Yin et al. used terphenyl to make scaffolded analogs to mimic the BH3 domain of Bak and the HDM2-binding domain of p53 [18, 19]. One should take note that cyclic and scaffolded peptides have some advantages over linear peptides in that their conformation is not only constrained, hopefully appropriately, but the structure is also more resistant to protease degradation, which can improve in vivo exposure.

The latter approach has many more variants. For example, Moerke et al. used a multifaceted approach (high throughput screening and structure-based design) to find a small molecule inhibitor of the protein–protein interaction of translational initiation factors eIF4E and eIF4G [20]. Tatsuta et al. also took a structure-based in silico screening approach to find lead molecules that could inhibit the Syk C-terminal SH2 domain [21].

A new application of high throughput methodology via NMR was utilized to search for inhibitors of the BH3-binding domain of Bcl-X_L [22]. These authors used an approach to find fragments that could bind separate but proximal sites on Bcl-x_L, then chemically linked them to create a molecule with nanomolar (nM) affinity. Orsini et al. used a pharmacophore-based HTS approach to identify antagonists and agonists of metastin [23]. There is a tremendous advantage to include a structure-based approach prior to HTS screening, as it was shown to improve the HTS hit rate from 0.6% to 3.8% of compounds screened.

The following section provides a more thorough overview of these and other examples to design non-peptidic mimetics.

3   Non-Peptidic Mimetics in Action

Even though selectivity might be compromised, minimizing peptides to non-peptidic mimetics, the premise of their implementation [considering, e.g., clearance, degradation, metabolism, but also practical arguments such as costs, scale-up production], overall, seem to favor clinical testing and development of nonpeptidic mimetics.

3.1   p53-HDM2 Mimetic

HDM2 binds to p53 to facilitate ubiquitination, thereby removing it from its role in promoting cell cycle arrest or apoptosis [24, 25]. Disrupting this protein–protein interaction could allow p53 a longer half-life in the cell, thereby possibly inhibiting cell processes that promote tumor growth.

3.1.1   Cyclized Peptide Mimetic

Fasan et al. designed a cyclic peptide in the form of a β-hairpin to mimic an α-helix in targeting the p53-HDM2 protein–protein interaction [11]. The rationale was based on the fact that in a crystal structure of a complex of HDM2 (residues 17–125) and a peptide derived from p53 (residues 15–29), the peptide assumes an α-helical conformation and inserts residues F19, W23, and L26 into a surface pocket on HDM2. The distance between the Cα atoms of the $i, i + 4$

residues F19 and W23 approximate the distance between the $i, i + 2$ residues of a beta-hairpin. A series of ten-residue cyclic peptides incorporating a D-Pro-L-Pro stabilizing template and the corresponding F1, W3, and L4 in the appropriate positions were designed, synthesized, and evaluated via BIAcore. The series of analogs showed that the preceding three residues made important contributions in addition to a K at the 6 position. The tightest binding peptide had an IC_{50} of 0.15 μM and incorporated a 6-chloro-tryptophan in place of W3. Two-dimensional NMR HSQC (heteronuclear single quantum correlation) spectroscopy chemical shift mapping of this peptide in solution with HDM2 demonstrated involvement of several HDM2 residues corresponding to the p53 binding site in addition to some more remote from the site. NMR spectroscopy of a related analog confirmed the cyclic nature of the peptide series.

3.1.2 Terphenyl Scaffolded Mimetic

Terphenyl derivatives have been designed to mimic the BH3 domain of Bak and the HDM2-binding domain of p53 (Fig. 18.1) [18, 19]. A similar approach was used to mimic the p53 residues F19, W23, and L26 that are key to the interaction of p53 with HDM2 [27]. It is interesting that the tightest binding compound from the terphenyl scaffold library only differed from that above in that the 2-naphthalene-substituted compound exhibited the tightest binding (0.18 μM) in a fluorescence polarization assay [18, 19].

3.2 CDK2 Mimetic

The cell cycle involves progression through the various phases, G1, S, G2, and M, and is critical for cell growth and division. Cyclin-dependent kinases and cyclins are two families of proteins that interact to regulate this process by acting as so-called "checkpoint" regulators. Abnormalities in this regulation can lead to unchecked cancer growth; therefore, there has been interest in developing peptidomimetics as therapeutics that target this process.

In an approach to inhibit the cyclin–CDK2 interaction, Andrews et al. designed cyclic peptides to target the cyclin-binding groove (CBG) of CDK2 (Fig. 18.2) [12]. Proteins that bind the CBG have

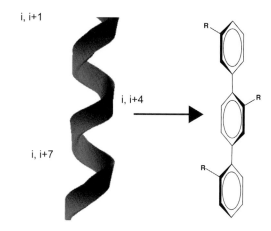

FIGURE 18.1. Terphenyl scaffolded mimetic

a cyclin recognition motif (CRM) that consists of R-Xaa-L-Yaa-Zaa, where X may be any residue while at least one of Yaa or Zaa must be hydrophobic [28]. In this study, a series of peptides were designed based on the p21^WAF1 CRM. Xaa-L-F-G was cyclized and Xaa (Lys, Orn peptide, A_2bu, or A_2pr) was also linked to amino acid sequences of lengths between 2 and 9 residues. In a competitive cyclin A binding and a CDK2-pRb kinase assay the peptide with the most activity was Ac-A-A-Abu-R-K-L-F-G with an IC_{50} of 0.63 and 0.53 μM, respectively. Two-dimensional NMR spectroscopy confirmed the cyclic nature of the peptides, and a crystal structure of CDK2/Ac-R-K-L-F-G demonstrated the peptide bound to the CBG with the L and F making important contacts with the CDK2 CBG.

3.3 Mimetics of Proteins Involved in Apoptosis

Apoptosis is a critical part of normal cellular life cycles. Many malignancies have abnormalities in the myriad of proteins that interact to progress a

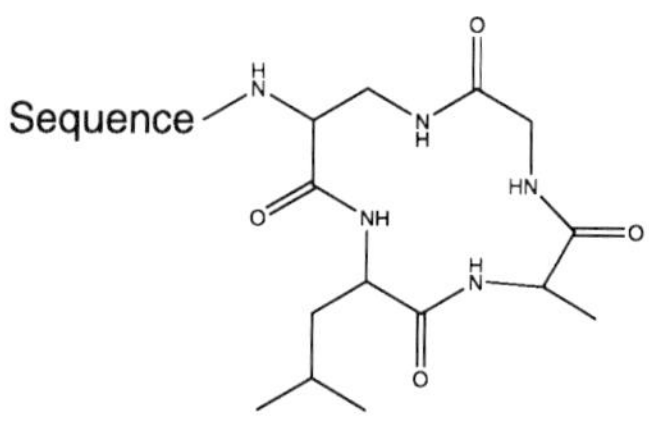

FIGURE 18.2. CDK2 inhibitor binds the CBG

cell through apoptosis. However, many of the regulatory events involve protein–protein interactions that involve large, shallow, hydrophobic surfaces, which are difficult to disrupt via small molecules. Hence, there has been a need to develop larger peptidomimetic inhibitors that cannot only disrupt these interactions but also be able to pass the pharmacokinetic standards of existing drugs.

3.3.1 Pro-apoptotic Protein BID

Aside from using cyclization to stabilize peptide conformation, introduction of non-natural amino acids can be used to "tether" a peptide conformation. An example is illustrated in Fig. 18.3. A notable example is that of a stabilized peptide mimetic of the anti-apoptotic protein BID [13]. Here, a 23-amino acid peptide–derived from the BH3 domain of the BID protein was synthesized with α,α-disubstituted amino acids containing olefin groups. A ruthenium catalyzed metathesis reaction was used to link the amino acids and form the tether. The authors demonstrated via circular dichroism that the tethered peptide had a higher helical content (35–87%) compared with the untethered peptide. Two-dimensional HSQC NMR spectroscopy demonstrated similar spectra for the native protein and the tethered peptide. BCL-2 affinity for the tethered peptide was greater than six times that of the untethered peptide (as determined via fluorescence polarization). It was also demonstrated that the (fluorescein isothiocyanate)-labeled tethered peptide could penetrate Jurkat T leukemia cells and was likely due to pinocytosis instead of endocytosis. The peptide also demonstrated pro-apoptotic activity in MTT (3-(4,5-dimethylthiazol-2-yl)-2,5-diphenyltetrazolium bromide) proliferation assays of Jurkat T, B cell, and mixed lineage leukemia cells. Mice with leukemia xenografts treated with the tethered peptide lived 11 days compared to 5 days for controls [13].

3.3.2 Mimetics of the BH3 Domain of Bak

Terphenyl derivatives have been designed to mimic the BH3 domain of Bak and the HDM2-binding domain of p53 [27]. A library of terphylene scaffolds was 3,2′,2″-substituted to correspond to the $i, i + 4, i + 7$ side-chains of an α-helix. The library was designed to explore the best combination to mimic the V74, L78, and I81 residues of Bak that are critical in binding

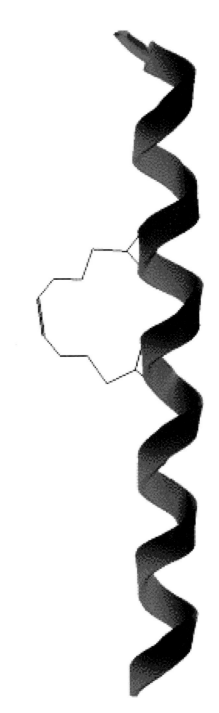

FIGURE 18.3. Pro-apoptotic "stapled" peptide mimetic of BID

Bcl-x_L [29]. The tightest binding compound demonstrated a K_i 0.114 μM via fluorescence polarization and inhibition of Bcl-x_L/BAK association in HEK293 transfected with HA-Bcl-x_L and flag-Bax.

3.3.3 BH3 Mimetics

A new application of high-throughput methodology via NMR was utilized to search for inhibitors of the BH3-binding domain of Bcl-X$_L$ [22]. The authors used an approach to find fragments that could bind separate but proximal sites on Bcl-x$_L$, and then link them to create a molecule with nanomolar (nM) affinity. The compound ABT-737 (Fig. 18.4) was created this way and while the fragments have K_d = 0.3 and 4.3 mM, respectively, ABT-737 has a K_i = < 1 nM for Bcl-X$_L$, Bcl-2, and Bcl-w. Of note, ABT-737 was shown to mimic the effect of a Bad-derived BH3 in blocking Bid-mediated cytochrome c release. It also displayed activity against cell lines derived from small cell lung cancer (SCLC) and lymphoid malignancies (EC$_{50}$ all in sub-µM range) in addition to displaying synergism with paclitaxel, doxorubicin, cisplatin, and etoposide in vitro in a variety of cell lines. The most exciting results came from SCLC xenograft experiments in which ABT-737 caused complete regression that lasted >1 month in a high percentage of mice [22].

3.3.4 Smac Mimetics

Smac (or DIABLO) is a pro-apoptotic protein that interacts with XIAP to inhibit its anti-apoptotic

function [30–37]. Sun et al. have designed a cell permeable mimetic of Smac that has a K_i = 61 nM [38]. This was a second-generation compound in that it was based on a primary amine that likely formed a hydrogen bond network with D309 and E314 in XIAP; however, this compound had poor cell permeability. The new compound, SM-131 had a secondary amine instead. In addition to its high affinity for XIAP, it also demonstrated activity in a caspase-9 functional assay and against MDA-MB-231 in vitro (IC$_{50}$ = 100 nM) [38].

3.4 Tumor Suppressor Mimetic

Metastin is a tumor suppressor peptide 54 amino acids in length that has decreased expression in a number of small tumors [39–42]. The minimal sequence length that mimics the activity of the 54-mer is 10 amino acids at the C-terminus [12]. Orsini et al. were able to use structure–activity relationships to discover three potential pharmacophores from the C-terminus of truncated Mestatin to search the Johnson and Johnson, Inc. chemical database for structural mimetics of Mestatin [23]. Twenty-one compounds with a K_i for the receptor <1 µM were discovered as potential lead compounds. An example is shown in Fig. 18.5.

3.5 Mimetics of Translational Initiation Factors

Translation is a critical cell process and is a bottleneck in cell activity. Targeting this critical cell pathway may lead to potential cancer therapeutics. Moerke et al. used a multidimensional approach to

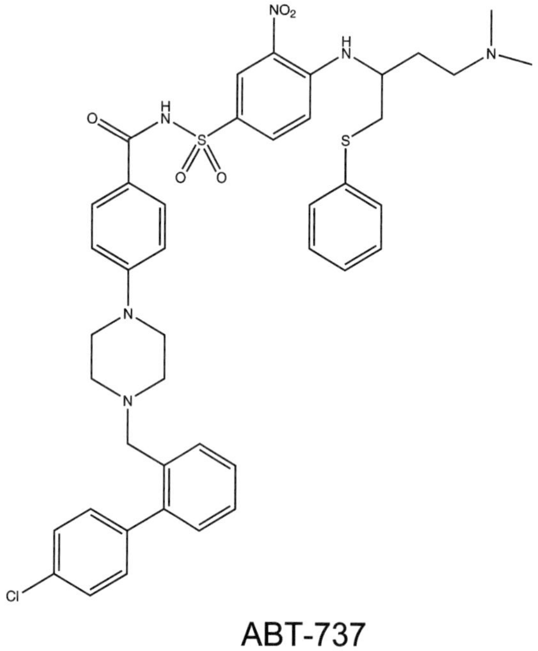

ABT-737

FIGURE 18.4. ABT-737 is active at two sites on Bcl-x$_L$

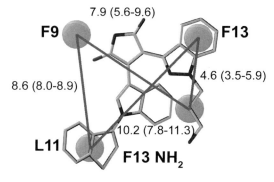

FIGURE 18.5. A representative example of a pharmacophore of the active site of Mestatin

finding a small molecule inhibitor of the protein-protein interaction of translational initiation factors eIF4E and eIF4G [20]. To find a lead compound, the authors employed a fluorescence polarization assay to screen the 16,000 Chembridge Diver-Set E library for compounds that would disrupt the interaction between eIF4E and a fluoresceinated peptide (KYTYDELFQLK) that represents the $Y(X)_4L\Phi$ motif of eIF4G. A compound, denoted 4EGI-1 was identified that has a K_D for eIF4E of $25\,\mu M$, displaces eIF4G from eIF4E in reticulocyte lysate, inhibits in vitro cap-dependent translation, and inhibits Jurkat cell growth ($IC_{50} = 6\,\mu M$). It was also demonstrated to interact with eIF4E via NMR chemical shift mapping [20].

3.6 Mimetic of the Syk C-Terminal SH2 Domain

Many of the previous examples have been directly designed in silico from knowledge of the crystal structure. Tatsuta et al. took a structure-based in silico screening approach to find lead molecules that could inhibit the Syk C-terminal SH2 domain [21]. Syk contains two SH2 domains that bind immunoreceptor tyrosine-based activation motifs to mediate downstream effects [43]. In this study the three-dimensional NMR-derived structure of Syk CSH2 was determined to consist of a central anti-parallel β-sheet flanked by two α-helices with three major binding pockets, pY, pY + 1, pY + 3. Initial in silico screening did not yield significant hits, which was thought to be due to partial occupation of the binding sites as determined by computational docking. Therefore, separate in silico screening experiments were conducted for each pocket in an effort to identify fragments.

3.7 Mimetics of Proteins Involved in Angiogenesis

Inhibition of angiogenesis can be accomplished in several ways [44]. The first, and as yet most successful, is the search for agents that specifically block EC proliferation. Progress was made with the demonstration that inhibition of angiogenesis prevents tumor growth and leads to regression of tumors. This was initially exemplified by the endogenous inhibitor endostatin, a fragment of collagen XVIII, in which treatment of mice with tumors of different

origins resulted in regression [45, 46]. A number of other antiangiogenic agents are being tested in the lab and/or in human trials as potential antitumor drugs. Vascular endothelial growth factor (VEGF) is generally accepted as the dominant growth factor in tumor angiogenesis [47–49], and antiangiogenic strategies that target VEGF signaling have shown some recent clinical success [1–4]. Unfortunately, tumors may be able to develop resistance or partially escape VEGF blockade by employing other growth factors, including fibroblast growth factor (FGF) and platelet-derived growth factor (PDGF) [50, 51]. Other compounds have been developed to inhibit angiogenesis by blocking multiple growth factor pathways. Unfortunately, the toxicity of several of these compounds may be detrimental in implementing their use in the clinic [52].

Another strategy for inhibiting tumor growth is to interfere with EC adhesion and migration. EC adhesion to extracellular matrix components can be prevented by blocking endothelial adhesion molecules (EAMs), which has been demonstrated by the use of anti-$\alpha_v\beta_3$ antibodies [53, 54] and RGD mimetics [55]. Migration of EC can also be prevented by the use of specific inhibitors of proteases that dissolve the matrix to facilitate vessel formation [56, 57]. Due to the plethora of antiangiogenic molecules currently under preclinical and clinical investigation, this section focuses on exemplary molecules that have been used as a model to eventually, in some cases already successfully, create non-peptidic mimetics.

3.7.1 Mimetics of Growth Factors

Many critical cellular growth processes involve the binding of an extracellular growth factor to a receptor on a cell. Examples of growth factor–dependent cancers include breast and prostate cancer. Also, angiogenesis is a growth factor–dependent process as well. Efforts to mimic this interaction in a selective way can not only lead to receptor inhibition, but also possibly a decrease in side effects.

3.7.1.1 Growth Factor Mimetics

Calix[4]arenes are scaffolds consisting of four phenyl rings linked via methylene groups in a circular fashion. The scaffold itself can approximate the diameter of an α-helix or a β-sheet. Different functional groups can be added to the phenyl moieties to mimic amino acid side chains. Much of the work done thus far has been geared to angiogenesis. Previously, work done by the Mayo lab in targeting galactin-1 on endothelial cells was described. VEGF and PDGF have also been targeted via calix[4]arenes [14–16]. The Hamilton group has designed calixarenes capable of selective binding of PDGF or VEGF (Fig. 18.6) [14, 15]. A series of calixarenes with alkyl ethers ortho- to the methylene linkers and four-member cyclized peptides meta- to the methylene linkers were designed and synthesized. A unique aspect of these molecules is the large surface area (approximately 500 Å^2) that can cover large, shallow protein–protein interfaces. It was demonstrated via growth factor–dependent receptor tyrosine phosphorylation assays that the calixarene (GFB-111) containing the cyclized peptide moiety GDGY had selective binding for PDGF (IC$_{50}$ 0.25 μM) and the calixarene (GFA-116) containing the moiety GKGK was selective for VEGF (IC$_{50}$ 0.50 μM). Both compounds were shown to inhibit angiogenesis and tumor growth in vivo in nude mice xenografts. Because PDGF and VEGF affect different aspects of angiogenesis, Sun et al. synthesized calixarenes that could bind both PDGF and VEGF [15]. The difference between these calixarenes and those previously described is the

substitution of a 3-aminomethylbenzoate for the peptide moieties. The library contained substitutions at both para- positions to the amino linker. GFB-204 showed the highest binding affinity for both PDGF and VEGF (IC$_{50}$ 0.19 μM for PDGF and 0.48 μM for VEGF). The authors demonstrated that this was not a general growth factor phenomenon in that *Erk1, Erk2, Akt,* and *STAT3* activation was inhibited via VEGF and PDGF sequestration but not via IGF-1, EGF, or bFGF. Angiogenesis and tumor growth was decreased in mouse xenograft models utilizing lung cancer A-549 cells [15].

3.7.1.2 Tyrosine Kinase Inhibitors

Among the several angiogenic growth factors know to date, VEGF, FGF, and PDGF, are by far the most studied. These growth factors induce EC proliferation by binding to high affinity receptor tyrosine kinases (RTKs) on EC. In the case of VEGF, it can bind to VEGFR-1 (Flt-1), and VEGFR-2 (Flk-1/KDR) to stimulate EC proliferation trough a paracrine and autocrine matter [58, 59]. Hence designing non-peptidic molecules to interfere with the ligand receptor binding was a very promising avenue [60]. Most of the design, development, and screening has been executed by applying molecular

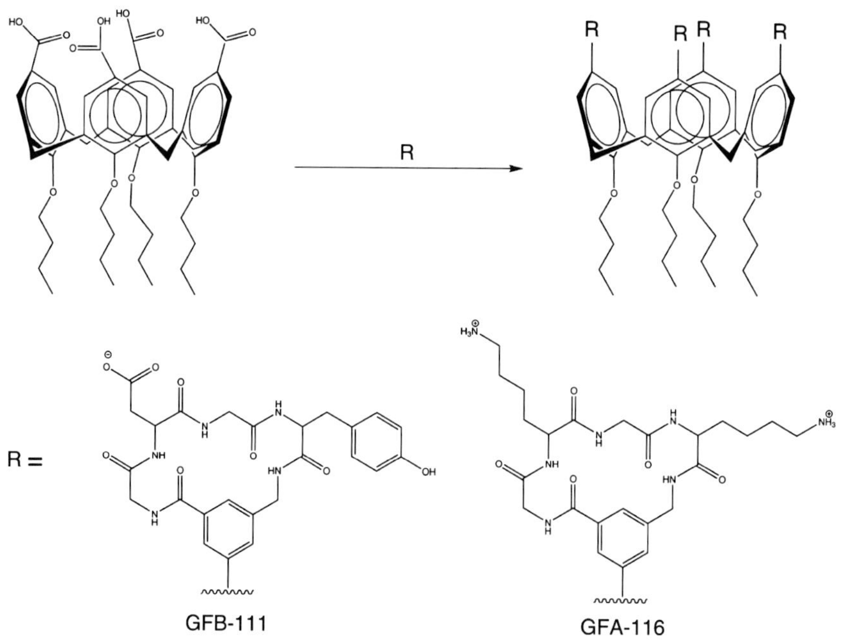

FIGURE 18.6. Calix[4]arene-derivatives that can inhibit PDGF (GFB-111) or VEGF (GFA-116)

modeling, NMR, crystallography, and medicinal chemistry [60]. Indeed, blocking this interaction has demonstrated to be a useful strategy to block angiogenesis in vivo [61]. While some tyrosine kinase inhibitors are broad spectrum and therefore are able to block multiple growth factor pathways, such as small molecule SU6666 (blocking VEGF, FGF, and PDGF RTKs) [62], SU11248 (blocking VEGF and PDGF RTKs) [63], SU14813, and Pazopanib (blocking VEGF, PDGF, KIT, and FLT3 RTKs) [64], PTK787/ZK 222584 (blocking VEGF, and other class III RTKs at lower concentrations) [65], ZD6474 (blocking VEGF and EGF) [66], SU11657 (blocking class III/V RTKs) [67], others are more specific to inhibit one RTK, such as SU5416 (blocking VEGFR-2) [61]. The list of RTKs mentioned here are by no means exhaustive, since pharmaceutical industries have put a lot of effort in generating compounds with interfere with growth factor signaling after the success and FDA approval of VEGF-inhibitor bevacizumab [44].

3.7.2 TNP-470

Non-peptidic small molecule TNP-470 (also referred to as AGM-1470) was derived from the pesticide drug fumagillin [68]. Due to toxicity noted in animals the parental drug was not applicable for further development, whereas TNP-470 showed no toxic side effects at efficacious doses. This non-peptidic mimetic inhibits EC proliferation by inhibiting methionylaminopeptidase-2 [69], and was able to block tumor progression and metastases in multiple tumor models and species [69].

3.7.3 Thalidomide

Initially thalidomide was used as a sedative hypnotic but due to its teratogenic and non-hematologic side effects stopped being prescribed in the clinic. Yet it was the first effective drug to become available for multiple myeloma in over two decades due to its anti-angiogenic and immunomodulatory activity [70]. Thalidomide undergoes bio-transformation by non-enzymatic hydrolysis and enzyme-mediated hydroxylation to form a multitude of bioactive metabolites. Hence, several analogues were developed by a drug discovery program with the intent of enhancing the anti-angiogenic and immunomodulatory effect while minimizing the teratogenic risk (referred to as immunomodulatory drugs (IMiDs))

[70, 71]. Lenalidomide (CC-5013) and Actimid (CC-4047) were the first such analogues to undergo clinical testing. Lenalidomide has shown impressive activity in relapsed refractory myeloma as well as newly diagnosed disease [70]. The precise mechanism of anti-MM activity of thalidomide and the IMiDs is not clear, but studies suggest that several other mechanisms besides anti-angiogenic effects may play a role [70, 71].

3.7.4 Squalamine

Squalamine, a water-soluble aminosterol, is derived from the cartilage and liver of the dogfish shark, *Squalus acanthus*. Squalamine is a cationic steroid characterized by a condensation of an anionic bile salt intermediate with spermidine [72]. It inhibits angiogenesis by inhibiting the EC membrane Na(+)/H(+) exchanger and may further function as a calmodulin chaperone. In addition it is orally available and nontoxic in animal models [73].

3.7.5 Galectin-1 Antagonist

Galectins are a family of carbohydrate-binding proteins that share a conserved carbohydrate recognition domain of approximately 130 amino acids [74, 75]. Galectins can be secreted and, depending on the cell type or state of differentiation, they have been found in the nucleus, cytoplasm, or extracellular matrix (ECM). It has been proposed that galectin-1 (gal-1) acts as a mediator of cell adhesion and migration [76] and is involved in processes like proliferation [77], apoptosis [78], and even mRNA splicing [79]. Previous studies have demonstrated that gal-1 expression facilitates interactions between tumor cells and endothelial cells (EC) in vitro, a phenomenon that could underlie metastasis formation in vivo [80, 81]. In addition, gal-1 can protect the tumor against infiltrating cytotoxic leukocytes [78, 82]. Elevated stromal expression of gal-1 has been reported in several cancers, including cancer of the breast [83], ovaries [84], colon [81], and prostate [85]. Since gal-1 is crucial in several processes required for tumor growth, compounds that target gal-1 may sensitize tumors on multiple levels.

We have rationally designed anginex, a peptide 33mer, which targets gal-1 and is potently antiangiogenic and inhibits tumor growth by about 80% in mice at a dose of 10 mg/kg of body weight/day. Subsequently, we designed a next generation

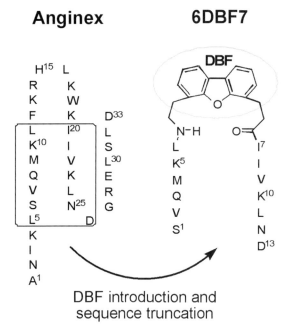

Anginex   6DBF7

H15	L
R	K
K	W
F	K	D33

L |20   L
K10   I   S
M   V   L30
Q   K   E
V   L   R
S   N25   G
L5   D
K
I
N
A1

DBF

N–H   O
L |7
K5   I
M   V
Q   K10
V   L
S1   N
               D13

**DBF introduction and
sequence truncation**

FIGURE 18.7. The amino acid sequences of anginex and 6DF7

analog series to minimize the molecular size while containing the key amino acids by introducing a non-peptidic molecular scaffold, dibenzofuran (DBF) [17]. Key residues reside within the boxed portion of anginex as shown in Fig. 18.7 [17]. Dibenzofuran was developed by Kelly et al. [86, 87] and showed that when DBF is inserted between a pair of lipophilic residues, a "hydrophobic cluster" is created wherein the side-chains of those residues are nested within the hydrophobic pocket created by the aromatic rings of the canted DBF subunit [86, 87]. Inserting the DBF moiety in place of eight amino acid residues of anginex served to significantly reduce the molecular weight of the analogs and the burden of their synthesis [17].

Of the DBF-based anginex analogs, 6DBF7 showed good in vitro and in vivo activity, superior to anginex, inhibiting both tumor growth and tumor angiogenesis about 2-fold better than anginex [17].

3.7.6   RGD Mimetic

Many integrins are important in cellular–extracellular matrix interactions and recognize ligand targets via an RGD tripeptide motifs [10]. Antagonists of this interaction could have potential in the treatment of tumors [55]. An example of this application was demonstrated by Gentilucci et al., who took a solid-phase synthesis approach to construct a library of RGD mimetics consisting of a heterochiral sequence Xaa-D-Pro-Yaa, in which Xaa is R or an R-mimetic and Yaa is a derivative of D. Proline was chosen to replace G due to its ability to induce beta- or gamma-turn-like structures [55]. A library of resulting compounds was tested for the ability to inhibit the interaction between fibronectin and a human malignant melanoma cell line (SK-MEL-24) expressing $\alpha_v\beta_3$ integrin. A fluorometric assay yielded only one active peptide (IC$_{50}$ 1.5 10^{-7} M). A second library based on this active peptide was constructed that eliminated a previously introduced N-tosyl moiety to part of the library and increased the distance between the C-terminal carboxylic acid and the N-terminal guanidinium group. Screening of this second library using the same fluorometric assay yielded two compounds with improved activity (IC$_{50}$ = 2.6 × 10^{-8} M and 3.5 × 10^{-8} M, respectively).

3.7.7   Others

Many antiangiogenic proteins consist primarily of β-sheet structures with significant hydrophobic and cationic character, usually in the context of an amphipathic structure [88]. These characteristics were used in the design of non-peptidic, calixarene-based compounds to mimic this overall structure of small units of ß-sheet structure with antiangiogenic activity [89]. The design approximated the molecular dimensions and surface topology of spatially related, key amino acid residues in antiangiogenic peptides/proteins. From a library of 23 calixarene-based compounds, two compounds (0118 and 1097) showed to be potent anticancer agents. For example, in the B16F10 melanoma tumor mouse model, 0118 and 1097 inhibited tumor growth on average by up to about 80%, without displaying any signs of toxicity [89]. While topomimetics 0118 and 1097 potently inhibit angiogenesis tumor growth, presently their target(s) are unknown. It appears that 0118 and 1097 do target adhesion/migration receptors on the surface of EC, taken in account results from an EC migration assay [89]. Interestingly, 1097 is also cytotoxic toward various tumor cell lines whereas 0118 is only cytotoxic against EC [89]. Therefore, these protein surface calixarene–based topomimetics may be used to antagonize several protein–biomolecular interactions. For example, the composition and chemical character of either or both surfaces (hydrophilic/hydrophobic) can be readily modified to be designed to target a specific receptor.

3.7.7.1 Almost There

There are several groups working to reduce the effective size of larger proteins involved in angiogenesis as antiangiogenic agents. At this point, these are peptides derived from the proteins, but even at this stage, they seem quite successful.

The notations of compounds such as thrombospondin, prolactin/growth hormone family molecules and endostatin, which were minimized from proteins to small peptides and are promising candidates for modeling and generating non-peptidic mimetics, were added for their laudable sensitizing activity when combined with conventional treatments (Section 4).

3.7.7.2 Thrombospondin-1

Thrombospondin-1 (TSP-1) is a relative large modular endogenous extracellular protein containing 3 identical disulfide-linked 180-kDa chains [90, 91]. Initially CD36 was recognized as the cellular receptor of TSP-1, however recently it was found that TSP-1 is also able to inhibit VEGF and FGF-2 mediated angiogenesis independent of CD36 [92]. Subsequently ß1-integrins were recognized as additional receptors [93] and since TSP-1 contains a RGD motif, $\alpha_v\beta_3$ integrin also binds [94]. Although highly potent (activities at the sub-nanomolar both in vitro as in vivo) the use of TSP-1 as a therapeutic is not practical due to its size nor desirable due to its diverse and multifunctional activities [95]. Internal fragments derived from TSP-1 rendered solutions to both problems. After first identifying, a central 50-kDa proteolytic fragment containing the procollagen homology region and properdin type 1 repeat was found to harbor anti-angiogenic activity [91], peptide sequences 424–442 (Mal I) and 481–499 (Mal-II) were localized to contain angiostatic activity. Further deduction identified the three

peptides sequences, WSHWSPW, VTCG, and GVITRIR to be responsible for the anti-angiogenic activity [91, 96]. Although the activity dropped to micromolar concentrations of these peptides, this loss in activity could be recovered by substituting an L-amino acid by its D-isoleucyl enantiomer [95]. Additional modification, such as introduction of the non-natural amino acid norvaline and ethylamide-capping resulted in a potency increase similar to parental TSP-1. This semi non-peptidic molecule was able to inhibit tumor growth in vivo [97]. Based on this GVITRIR sequence Abbott Laboratories has now a nonamer thrombospondin-mimetic, ABT-510, in clinical trials for the treatment of soft tissue sarcomas and melanomas [98, 99].

3.7.7.3 Prolactin/Growth Hormone Family Members

Although parental molecules prolactin (PRL), growth hormone (GH), and placental lactogen promote angiogenesis, proteolytic cleavage peptide fragments show distinct anti-angiogenic properties [100–102]. Recently it was demonstrated that a 14 amino acid–sequence within these 16-kDa fragments seems to be responsible for this angiostatic activity [103]. These short helical peptides were characterized as tilted peptides since they share the peculiar distribution of hydrophobic residues: a net hydrophobicity increase from one end to the other end of the peptide and an amphipathic topology. These tilted peptides of both PRL and GH are able to destabilize endothelial cell membranes and lipid cores due to their asymmetric distribution of hydrophobic residues along their helical axis [103]. Even tilted peptide derived from non-angiogenesis related proteins (i.e., Alzheimer's ß-amyloid and simian immunodeficiency virus gp32) showed anti-angiogenic effects in vitro and in the chicken chorioallantoic membrane assay [103]. Since no molecular target has been identified for these peptides, specificity might be a concern, although the authors favor the thought a specific protein–protein interaction over a generic protein–membrane interaction [103]. Nonetheless, these structural and compositional characteristics of tilted peptides appear to be a very promising model for the design and further development of derived non-peptidic mimetics.

3.7.7.4 Endostatin

Endostatin is a 183 amino acid (20-kDa) C-terminal proteolytic fragment of the non-collagenous region 1 domain of collagen XVIII [45], which inhibits EC proliferation, migration, and tube formation leading to antitumor effect in vivo without apparent toxic side effects [104]. The precise mechanism of action has not been elucidated, yet many surface proteins have been identified to interact with endostatin, such as glypicans, albeit with low affinity [105], integrin $\alpha_5\beta_1$ [106], and to some extend $\alpha_v\beta_3$ and $\alpha_v\beta_5$ [107]. On a signaling pathway level endostatin downregulates proliferation stimulating genes such as MAPK1, MAPK2 or c-myc [108], it causes G1 arrest through inhibition of cyclin D1 [109], induces reorganization of the actin cytoskeleton via downregulation of RhoA activity [110], blocks VEGF signaling [111], and the activation and catalytic activity of matrix metalloproteinase [112], and inhibits the wnt-signaling pathway [113]. In China endostatin is currently (in a slightly modified form and under the name Endostar) approved for the treatment of lung cancer in the clinic [44]. In order to minimize the compound and to capture the activity mediated through integrin binding, Wickström et al. synthesized five N-terminal 11–13-mers, which were postulated to interact with EC surface ß1-integrins and/or heparan sulfate proteoglycans. More specifically these peptides, termed ES-1 through ES-5 were selected to represent sequences containing areas of surface exposed basic amino acids, predominantly, arginines (R), and lysines (K). ES-2, IVRRADRAAVP proved to be the most potent anti-migratory peptide. A drop in activity pointed out the importance of the arginine residues within the peptide, after substituting the arginines residues with alanines residues. Its mechanism of action was mediated through ß1-integrin and heparin-dependent mechanisms [114]. The knowledge of the importance of the basic and surface exposed residues will aid in the development of non-peptidic mimetics of endostatin.

3.7.8 MMP Inhibitors

Matrix metalloproteinases (MMPs) are zinc-containing enzymes involved in the turnover and remodeling of the extracellular matrix and capillary basement membrane, an important step during angiogenesis [69]. MMPs are initially secreted as pro-enzymes, requiring modification or removal of a 10-kDa amino terminal domain for expression of enzymatic activity. Once MMPs are activated, general protease inhibitors (i.e., tissue inhibitors of metalloproteinases, TIMPs) may inhibit them. During tumor growth and metastasis, there is an excessive MMPs expression and activation. This allows local expansion of the tumor mass through the disruption of normal tissue structure and boundaries, facilitating the invasion of blood vessels and lymphatics, creating an aggressive and metastatic phenotype. By blocking these activities, an MMP inhibitor (MMPI) is expected to constrict and contain tumor growth. Batimastat (BB-94) is the first synthetic nonpeptic MMPI mimetic tested in cancer patients [115]. Batimastat is a broad-spectrum inhibitor with potent inhibitory activity against multiple MMPs, such as angiotensin converting enzyme [69]. In addition, at least 24 MMP inhibitors have been pursued as clinical candidates in the treatment of cancer since the late 1970s when the first drug discovery program targeting this enzyme family began [116]. Unfortunately, most of them have been discontinued due to the inadequate assessment of the therapeutic index (the ratio of dose required for efficacy versus that for toxicology). The musculoskeletal syndrome has been the most pronounced dose-limiting side effect that has hampered MMP inhibitor development, such was the case for, e.g., marimastat [116]. Marimastat is closely related to batimastat with similar properties but was the first oral availability MMP inhibitor in the clinic [115].

4 Combination Therapy

Sensitizing tumors cells to make them more receptive for conventional treatments has been of particular interest in the angiogenesis field. In animal models, combining angiogenesis inhibitors with chemotherapy [117–119], gene therapy [120], or radiation therapy [121, 122] has been shown to be potentially beneficial and an improvement over monotherapy alone. In fact, the results of the first successful phase III clinical trial using antiangiogenic agent bevacizumab (Avastin) were acquired in combination with chemotherapy [1–3]. This trial first proved that angiogenesis inhibitors, used in conjunction with conventional chemotherapy, are very powerful tools to combat cancer in patients and subsequently lengthen the life span.

Even though the preclinical and clinical testing of sensitizing anti-angiogenic agents shows promise, the need for better angiogenesis inhibitors is still persistent and the optimization of their implementation is of great importance [123]. This progress is likely decelerated by the scarcity of biomarkers to monitor possible effects, such as local sensitizing of antiangiogenic therapy [44, 47]. For years now, clinical evaluation of antiangiogenic therapy has been limited to the methods developed to evaluate traditional chemotherapeutics, predominantly dictated by toxicities. Unfortunately, most of these means do not apply since many antiangiogenic therapies might not be most efficacious at their maximum tolerated dose since they most of the time do not show any toxicities, in contrast to conventional chemotherapeutics [104, 124, 125]. Recently new appropriate methods and biomarkers have been developed, that should aid in determining and monitoring antiangiogenic treatment efficacy [124, 126].

Thus far, several mechanisms (molecular, cellular, and physiologic) have been suggested by which antiangiogenic agents can sensitize tumors. For example, they can directly sensitization of endothelium, change pO_2 in tumors, decrease the interstitial pressure, change in permeability, inhibit repair of drug or radiation-induced damage in endothelium, and "normalize" the vasculature [127–130]. Likely none of these mechanisms is mutual exclusive and the effects seen might be a result of a combination of them. Moreover, time of assessment might be crucial to identify a particular mechanism since it is not unlikely that these mechanisms coincide and are temporal in nature.

4.1 Radiation Therapy

Several clinical and preclinical studies have shown that combining various types of sensitizing antiangiogenic therapy to single dose or fractionated radiotherapy can synergistically improve the radiation tumor response (in both human and murine tumors) [2, 121, 122, 129, 131, 132].

Independent of the tumor oxygenation state, apoptosis of endothelial cells has been accepted as a critical component of the radiation response [133–135]. Therefore, treating with anti-angiogenic

agents to target tumor endothelium in conjunction with radiation therapy to cumulative increase tumor endothelial cell and subsequently tumor cell apoptosis is a rational and clinically relevant strategy [129]. The list of antiangiogenic agents demonstrated to make tumors more susceptible to radiation therapy includes but is not limited to thrombospondin-1 [136], endostatin [131], angiostatin [121, 132], various receptor tyrosine kinase inhibitors [122, 137], and anti-VEGF and VEGFR antibodies [135, 138]. Not as many non-peptidic mimetics have been tested in combination with radiation as compared chemotherapy. For instance, we were only able to find three reports on radiation combination therapy with TNP-470 against Lewis lung and C3H mammary carcinoma as well as U87 human glioblastoma [139–141]. Thalidomide was just recently for the first time tested against a fibrosarcoma combined with three doses of 20 Gy [142]. Ansiaux et al. found that thalidomide was able to radiosensitize tumors by modifying the tumor microenvironment: modulation of tumor perfusion and interstitial pressure led to an increase in tumor oxygenation, without radiosensitizing the tumor cells directly [142]. The RTK inhibitors have been studied more extensively, especially the importance of scheduling of RTK inhibitors and radiation [137]. Some studies showed that initiating radiation after RTK administration, in this case SU11657, gained more benefit than starting the SU11657 administration one day post a single dose of radiation [143]. Whereas another study using fractionated radiation and ZD6474 treatment over a 2-week period showed the combination effect was additive regardless ZD6474 was given prior, during, or after the radiation treatment [144].

However, due to the multiple variables that can contribute to the sensitivity of tumors to radiation as a result of non-peptidic mimetics and or anti-angiogenic treatment (e.g., changes in O_2, pH, bloodflow, perfusion) it has been difficult to develop a consensus on how to combine and or schedule these therapies [123, 145, 146]. Some studies showed that sensitization and killing of endothelial cells just before exposure to radiation may be the most effective way to improve radiation response [122, 128–130, 142, 147, 148], while others revealed that blocking survival signaling in endothelial cells after irradiation is also highly effective in increasing the radiation response [121, 138].

In addition, several examples can be found in the literature where the angiogenesis inhibitor was given in a combination of before, during, and/or after radiation [136, 144, 149–151]. However, the concern is that in certain particular scenarios of combination treatments, the effect of one therapy (e.g., induction of hypoxia via blood vessel damage) may be detrimental to another (e.g., hypoxic radioprotection or reduced access of chemotherapy to the tumor). Therefore, a logical and proven rationale for optimal combination or even multimodality therapy is a necessity for efficient translation of successful preclinical strategies to human clinical applications (145, 146, 152).

Since, our initial investigations demonstrated that the antiangiogenic peptide anginex effectively radiosensitized human ovarian (MA148), murine breast (SCK), and melanoma (B16F10) tumors in mice [130], the non-peptidic topomimetic 0118 modeled after anginex was tested in similar combination studies.

When B16F10 tumors reached an average size of 100 mm³, animals were given IP injections of 0118 (10 mg/kg) in combination with a single dose of radiation (5 Gy) on day 2 after the start of 0118 administration. This resulted in tumor growth inhibition significantly more than either monotherapy alone (Fig. 18.8). Mechanistically, it appears that anginex and 0118 both function as endothelial cell-specific radiosensitizers, since anginex and 0118 showed no effect on in vitro radiosensitivity of B16F10, SCK, or MA148 tumor cells, whereas both molecules significantly enhanced the in vitro radiosensitivity of endothelial cell types [129, 130].

4.2 Chemotherapy

Being one of the first described synthetic non-peptidic compound TNP-470, the fumagillin derivative, was studied extensively. Over the years TNP-470 has been combined with chemotherapeutics such as cisplatin, docetaxel, gemcitabine, melphalan, cyclophosphamide, Adriamycin, mitomycin C, 5-FU, etoposide, and prednisolone, against a myriad of tumor types, such as murine Lewis lung carcinoma, B16 melanoma, EMT6 mammary carcinoma, ISOS-1 angiosarcoma, rat 9L glioblastoma, and human bladder and pancreatic carcinoma [139, 153–157]. Here, TNP-470 was predominantly administered in doses ranging from daily 2.5 mg/kg to only two doses of 75 mk/kg SC with regimes of prior, during, and after a particular chemotherapeutic was given [139, 153–157].

Thalidomide has been used as in combination with chemotherapeutics such as cyclophosphamide, paclitaxel, and dacarbazine against colon, colorectal, melanoma, and liver tumors of both human and murine origin [158–161]. Interestingly, Seegers et al. showed that the most pronounced anti-tumor effect of cyclophosphamide coincided with the maximal tumor oxygenation increase induced by thalidomide [161].

Endostatin and thrombospondin have been tested in conjunction with carboplatin, cyclophosphamide, doxorubicin, irinotecan, and gemcitabine [162–165] against human testicular and colorectal tumors, but also pancreatic, ovarian, and lung carcinomas. Interestingly endostatin was given as low as 50 µg per day to 10 mg/kg/d SC, whereas thrombospondin was given 3–20 mg/kg/d IP.

The tyrosine kinase inhibitors SU6668, SU5416, and SU11657 have been combined with paclitaxel, gemcitabine, and pemetrexed against human epithelial, pancreatic, and ovarian carcinomas in doses up to 200 mg/kg IP as well as PO (143, 166, 167).

Since anginex in combination with a suboptimal dose of carboplatin led to an improved outcome with a synergistic effect compared to either treatment alone [168], we subsequently tested if non-peptidic topomimetic of anginex, 0118, also was capable of regressing tumors in combination with chemotherapy. We performed several studies

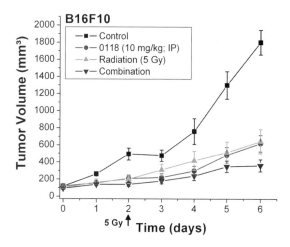

FIGURE 18.8. Tumor growth inhibition by combination treatment of 0118 and radiation (5 Gy)

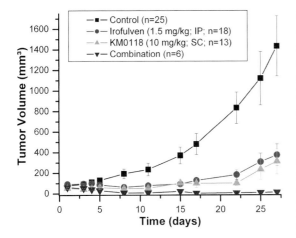

FIGURE 18.9. Tumor growth inhibition by combination treatment of 0118 and irofulven

any number of these therapeutic agents to clinical use. Nevertheless, it is likely that their use as monotherapies will not be as successful as when they are used to sensitize the growing tumor to chemotherapy and/or radiation therapy. This "double punch" approach will likely be the future for greater success against cancer.

Acknowledgments. This work was supported by research grants to KHM from the National Institutes of Health: NCI grant ROI CA-096090, and by the NIH/NIAID Regional Center of Excellence for Bio-defense and Emerging Infectious Diseases Research (RCE) Program, Region V 'Great Lakes' RCE (NIH award 1-U54-AI-057153).

to determine whether our non-peptidic mimetic would work synergistically in combination with standard chemotherapeutics at suboptimal doses to reduce their highly toxic effects while still promoting tumor regression.

For this, we used two chemotherapeutic agents: carboplatin and irofulven (169). Both function by forming adducts with DNA, thereby killing cells. Results using irofulven and 0118 are presented in Fig. 18.9, and similar results occur using carboplatin (not shown). Human ovarian MA148 tumors were allowed to grow to an average size of 80 mm³ prior to initiation of treatment. Irofulven or carboplatin was administered IP on days 1, 4, 7, and 10 at a suboptimal dose of 1.5 mg/kg, whereas 0118 (10 mg/kg) was administered SC using osmotic minipumps (Alzet) implanted on day 0. While irofulven and 0118 at their respective doses show similar results in tumor growth inhibition, their combination clearly shows increased benefit. By day 8, all tumors became impalpable (see Fig. 18.9).

5   Future Directions

Hopefully this chapter has provided a reasonably good overview of a number of interesting non-peptidic mimetic compounds that may be useful to combat cancer. Some of the compounds discussed have not yet evolved into full non-peptidic mimetics, but they do hold considerable promise. In this regard, the prognosis is very good to bring

References

1. Yang JC, Haworth L, Sherry RM, et al. A randomized trial of bevacizumab, an anti-vascular endothelial growth factor antibody, for metastatic renal cancer. N Engl J Med 2003, 349:427–434.

2. Willett CG, Boucher Y, di Tomaso E, et al. Direct evidence that the VEGF-specific antibody bevacizumab has antivascular effects in human rectal cancer. Nat Med 2004, 10:145–147.

3. Willett CG, Boucher Y, Duda DG, et al. Surrogate markers for antiangiogenic therapy and dose-limiting toxicities for bevacizumab with radiation and chemotherapy: continued experience of a phase I trial in rectal cancer patients. J Clin Oncol 2005, 23:8136–8139.

4. Bicknell R. The realisation of targeted antitumour therapy. Br J Cancer 2005, 92(Suppl 1):S2–5.

5. Rossman MG, Arnold E. Volume F: crystallography of biological molecules. international tables for crystallography. Norwell, MA: Kluwer Academic Publishers, 2001.

6. Wuthrich K. Protein structure determination in solution by nuclear magnetic resonance spectroscopy. Science 1989, 243:45–50.

7. Wuthrich K. NMR of proteins and nucleic acids. Hoboken, NJ: John Wiley & Sons, Inc., 1986.

8. Cavanaugh J, Fairbrother WJ, Palmer III AG, et al. Protein NMR spectroscopy: principles and practice. San Diego: Academic Press, 1996.

9. Guner O. Pharmacophore perception, development, and use in drug design. La Jolla: International University Line, 2000.

10. Giancotti FG, Ruoslahti E. Integrin signaling. Science 1999, 285:1028–1032.

11. Fasan R, Dias RL, Moehle K, et al. Using a beta-hairpin to mimic an alpha-helix: cyclic peptidomimetic

inhibitors of the p53-HDM2 protein-protein interaction. Angew Chem Int Ed Engl 2004, 43:2109–2112.

12. Andrews MJ, McInnes C, Kontopidis G, et al. Design, synthesis, biological activity and structural analysis of cyclic peptide inhibitors targeting the substrate recruitment site of cyclin-dependent kinase complexes. Org Biomol Chem 2004, 2:2735–2741.

13. Walensky LD, Kung AL, Escher I, et al. Activation of apoptosis in vivo by a hydrocarbon-stapled BH3 helix. Science 2004, 305:1466–1470.

14. Blaskovich MA, Lin Q, Delarue FL, et al. Design of GFB-111, a platelet-derived growth factor binding molecule with antiangiogenic and anticancer activity against human tumors in mice. Nat Biotechnol 2000, 18:1065–1070.

15. Sun J, Blaskovich MA, Jain RK, et al. Blocking angiogenesis and tumorigenesis with GFA-116, a synthetic molecule that inhibits binding of vascular endothelial growth factor to its receptor. Cancer Res 2004, 64:3586–3592.

16. Sun J, Wang DA, Jain RK, et al. Inhibiting angiogenesis and tumorigenesis by a synthetic molecule that blocks binding of both VEGF and PDGF to their receptors. Oncogene 2005, 24:4701–4709.

17. Mayo KH, Dings RP, Flader C, et al. Design of a partial peptide mimetic of anginex with antiangiogenic and anticancer activity. J Biol Chem 2003, 278:45746–45752.

18. Yin H, Lee GI, Sedey KA, et al. Terphenyl-based Bak BH3 alpha-helical proteomimetics as low-molecular-weight antagonists of Bcl-xL. J Am Chem Soc 2005, 127:10191–10196.

19. Yin H, Lee GI, Sedey KA, et al. Terephthalamide derivatives as mimetics of helical peptides: disruption of the Bcl-x(L)/Bak interaction. J Am Chem Soc 2005, 127:5463–5468.

20. Moerke NJ, Aktas H, Chen H, et al. Small-molecule inhibition of the interaction between the translation initiation factors eIF4E and eIF4G. Cell 2007, 128:257–267.

21. Niimi T, Orita M, Okazawa-Igarashi M, et al. Design and synthesis of non-peptidic inhibitors for the Syk C-terminal SH2 domain based on structure-based in-silico screening. J Med Chem 2001, 44:4737–4740.

22. Oltersdorf T, Elmore SW, Shoemaker AR, et al. An inhibitor of Bcl-2 family proteins induces regression of solid tumours. Nature 2005, 435:677–681.

23. Orsini MJ, Klein MA, Beavers MP, et al. Metastin (KiSS-1) mimetics identified from peptide structure-activity relationship-derived pharmacophores and directed small molecule database screening. J Med Chem 2007, 50:462–471.

24. Zhang Y, Xiong Y. Control of p53 ubiquitination and nuclear export by MDM2 and ARF. Cell Growth Differ 2001, 12:175–186.

25. Chen L, Yin H, Farooqi B, et al. p53 alpha-Helix mimetics antagonize p53/MDM2 interaction and activate p53. Mol Cancer Ther 2005, 4:1019–1025.

26. Kussie PH, Gorina S, Marechal V, et al. Structure of the MDM2 oncoprotein bound to the p53 tumor suppressor transactivation domain. Science 1996, 274:948–953.

27. Yin H, Lee GI, Park HS, et al. Terphenyl-based helical mimetics that disrupt the p53/HDM2 interaction. Angew Chem Int Ed Engl 2005, 44:2704–2707.

28. McInnes C, Andrews MJ, Zheleva DI, et al. Peptidomimetic design of CDK inhibitors targeting the recruitment site of the cyclin subunit. Curr Med Chem Anticancer Agents 2003, 3:57–69.

29. Muchmore SW, Sattler M, Liang H, et al. X-ray and NMR structure of human Bcl-xL, an inhibitor of programmed cell death. Nature 1996, 381:335–341.

30. Liu Z, Sun C, Olejniczak ET, et al. Structural basis for binding of Smac/DIABLO to the XIAP BIR3 domain. Nature 2000, 408:1004–1008.

31. Deveraux QL, Reed JC. IAP family proteins—suppressors of apoptosis. Genes Dev 1999, 13:239–252.

32. Salvesen GS, Duckett CS. IAP proteins: blocking the road to death's door. Nat Rev Mol Cell Biol 2002, 3:401–410.

33. Huang Y, Park YC, Rich RL, et al. Structural basis of caspase inhibition by XIAP: differential roles of the linker versus the BIR domain. Cell 2001, 104:781–790.

34. Du C, Fang M, Li Y, et al. Smac, a mitochondrial protein that promotes cytochrome c-dependent caspase activation by eliminating IAP inhibition. Cell 2000, 102:33–42.

35. Verhagen AM, Ekert PG, Pakusch M, et al. Identification of DIABLO, a mammalian protein that promotes apoptosis by binding to and antagonizing IAP proteins. Cell 2000, 102:43–53.

36. Ekert PG, Silke J, Hawkins CJ, et al. DIABLO promotes apoptosis by removing MIHA/XIAP from processed caspase 9. J Cell Biol 2001, 152:483–490.

37. Wu G, Chai J, Suber TL, et al. Structural basis of IAP recognition by Smac/DIABLO. Nature 2000, 408:1008–1012.

38. Sun H, Nikolovska-Coleska Z, Lu J, et al. Design, synthesis, and evaluation of a potent, cell-permeable, conformationally constrained second mitochondria derived activator of caspase (Smac) mimetic. J Med Chem 2006, 49:7916–7920.

39. Ringel MD, Hardy E, Bernet VJ, et al. Metastin receptor is overexpressed in papillary thyroid cancer and activates MAP kinase in thyroid cancer cells. J Clin Endocrinol Metab 2002, 87:2399.

40. Sanchez-Carbayo M, Capodieci P, Cordon-Cardo C. Tumor suppressor role of KiSS-1 in bladder cancer: loss of KiSS-1 expression is associated with bladder cancer progression and clinical outcome. Am J Pathol 2003, 162:609–617.

41. Ikeguchi M, Yamaguchi K, Kaibara N. Clinical significance of the loss of KiSS-1 and orphan G-protein-coupled receptor (hOT7T175) gene expression in esophageal squamous cell carcinoma. Clin Cancer Res 2004, 10:1379–1383.

42. Dhar DK, Naora H, Kubota H, et al. Downregulation of KiSS-1 expression is responsible for tumor invasion and worse prognosis in gastric carcinoma. Int J Cancer 2004, 111:868–872.

43. Bu JY, Shaw AS, Chan AC. Analysis of the interaction of ZAP-70 and syk protein-tyrosine kinases with the T-cell antigen receptor by plasmon resonance. Proc Natl Acad Sci U S A 1995, 92:5106–5110.

44. Griffin RJ, Molema G, Dings RP. Angiogenesis treatment, new concepts on the horizon. Angiogenesis 2006, 9:67–72.

45. O'Reilly MS, Boehm T, Shing Y, et al. Endostatin: an endogenous inhibitor of angiogenesis and tumor growth. Cell 1997, 88:277–285.

46. Boehm T, Folkman J, Browder T, et al. Antiangiogenic therapy of experimental cancer does not induce acquired drug resistance. Nature 1997, 390:404–407.

47. Kerbel RS. Tumor angiogenesis: past, present and the near future. Carcinogenesis 2000, 21:505–515.

48. Ferrara N, Davis-Smyth T. The biology of vascular endothelial growth factor. Endocr Rev 1997, 18:4–25.

49. Benjamin LE, Keshet E. Conditional switching of vascular endothelial growth factor (VEGF) expression in tumors: induction of endothelial cell shedding and regression of hemangioblastoma-like vessels by VEGF withdrawal. Proc Natl Acad Sci U S A 1997, 94:8761–8766.

50. Viloria-Petit A, Crombet T, Jothy S, et al. Acquired resistance to the antitumor effect of epidermal growth factor receptor-blocking antibodies in vivo: a role for altered tumor angiogenesis. Cancer Res 2001, 61:5090–5101.

51. Kerbel RS. Clinical trials of antiangiogenic drugs: opportunities, problems, and assessment of initial results. J Clin Oncol 2001, 19:45S–51S.

52. Liekens S, De Clercq E, Neyts J. Angiogenesis: regulators and clinical applications. Biochem Pharmacol 2001, 61:253–270.

53. Friedlander M, Brooks PC, Shaffer RW, et al. Definition of two angiogenic pathways by distinct alpha v integrins. Science 1995, 270:1500–1502.

54. Brooks PC, Montgomery AM, Rosenfeld M, et al. Integrin alpha v beta 3 antagonists promote tumor regression by inducing apoptosis of angiogenic blood vessels. Cell 1994, 79:1157–1164.

55. Gentilucci L, Cardillo G, Squassabia F, et al. Inhibition of cancer cell adhesion by heterochiral Pro-containing RGD mimetics. Bioorg Med Chem Lett 2007.

56. Rasmussen HS, McCann PP. Matrix metalloproteinase inhibition as a novel anticancer strategy: a review with special focus on batimastat and marimastat. Pharmacol Ther 1997, 75:69–75.

57. Brooks PC, Silletti S, von Schalscha TL, et al. Disruption of angiogenesis by PEX, a noncatalytic metalloproteinase fragment with integrin binding activity. Cell 1998, 92:391–400.

58. Fong GH, Rossant J, Gertsenstein M, et al. Role of the Flt-1 receptor tyrosine kinase in regulating the assembly of vascular endothelium. Nature 1995, 376:66–70.

59. McMahon G. VEGF receptor signaling in tumor angiogenesis. Oncologist 2000, 5 Suppl 1:3–10.

60. Sun L, Tran N, Liang C, et al. Design, synthesis, and evaluations of substituted 3-[(3- or 4-carbox-yethylpyrrol-2-yl)methylidenyl]indolin-2-ones as inhibitors of VEGF, FGF, and PDGF receptor tyrosine kinases. J Med Chem 1999, 42:5120–5130.

61. Fong TA, Shawver LK, Sun L, et al. SU5416 is a potent and selective inhibitor of the vascular endothelial growth factor receptor (Flk-1/KDR) that inhibits tyrosine kinase catalysis, tumor vascularization, and growth of multiple tumor types. Cancer Res 1999, 59:99–106.

62. Laird AD, Vajkoczy P, Shawver LK, et al. SU6668 is a potent antiangiogenic and antitumor agent that induces regression of established tumors. Cancer Res 2000, 60:4152–4160.

63. Mendel DB, Laird AD, Xin X, et al. In vivo antitumor activity of SU11248, a novel tyrosine kinase inhibitor targeting vascular endothelial growth factor and platelet-derived growth factor receptors: determination of a pharmacokinetic/pharmacodynamic relationship. Clin Cancer Res 2003, 9:327–337.

64. Patyna S, Laird AD, Mendel DB, et al. SU14813: a novel multiple receptor tyrosine kinase inhibitor with potent antiangiogenic and antitumor activity. Mol Cancer Ther 2006, 5:1774–1782.

65. Wood JM, Bold G, Buchdunger E, et al. PTK787/ZK 222584, a novel and potent inhibitor of vascular endothelial growth factor receptor tyrosine kinases, impairs vascular endothelial growth factor-induced responses and tumor growth after oral administration. Cancer Res 2000, 60:2178–2189.

66. McCarty MF, Wey J, Stoeltzing O, et al. ZD6474, a vascular endothelial growth factor receptor tyrosine kinase inhibitor with additional activity against epidermal growth factor receptor tyrosine kinase, inhibits orthotopic growth and angiogenesis of gastric cancer. Mol Cancer Ther 2004, 3:1041–1048.

67. Cain JA, Grisolano JL, Laird AD, et al. Complete remission of TEL-PDGFRB-induced myeloproliferative

disease in mice by receptor tyrosine kinase inhibitor SU11657. Blood 2004, 104:561–564.

68. Kusaka M, Sudo K, Fujita T, et al. Potent anti-angiogenic action of AGM-1470: comparison to the fumagillin parent. Biochem Biophys Res Commun 1991, 174:1070–1076.

69. Bergers G, Javaherian K, Lo KM, et al. Effects of angiogenesis inhibitors on multistage carcinogenesis in mice. Science 1999, 284:808–812.

70. Kumar S, Rajkumar SV. Thalidomide and lenalidomide in the treatment of multiple myeloma. Eur J Cancer 2006, 42:1612–1622.

71. Knight R. IMiDs: a novel class of immunomodulators. Semin Oncol 2005, 32:S24–30.

72. Moore KS, Wehrli S, Roder H, et al. Squalamine: an aminosterol antibiotic from the shark. Proc Natl Acad Sci U S A 1993, 90:1354–1358.

73. Pietras RJ, Weinberg OK. Antiangiogenic steroids in human cancer therapy. Evid Based Complement Alternat Med 2005, 2:49–57.

74. Barondes SH, Castronovo V, Cooper DN, et al. Galectins: a family of animal beta-galactoside-binding lectins. Cell 1994, 76:597–598.

75. Drickamer K. Two distinct classes of carbohydrate-recognition domains in animal lectins. J Biol Chem 1988, 263:9557–9560.

76. Hughes RC. Galectins as modulators of cell adhesion. Biochimie 2001, 83:667–676.

77. Scott K, Weinberg C. Galectin-1: a bifunctional regulator of cellular proliferation. Glycoconj J 2004, 19:467–77.

78. Perillo NL, Pace KE, Seilhamer JJ, et al. Apoptosis of T cells mediated by galectin-1. Nature 1995, 378:736–739.

79. Park JW, Voss PG, Grabski S, et al. Association of galectin-1 and galectin-3 with Gemin4 in complexes containing the SMN protein. Nucleic Acids Res 2001, 29:3595–3602.

80. Clausse N, van den Brule F, Waltregny D, et al. Galectin-1 expression in prostate tumor-associated capillary endothelial cells is increased by prostate carcinoma cells and modulates heterotypic cell-cell adhesion. Angiogenesis 1999, 3:317–325.

81. Lotan R, Matsushita Y, Ohannesian D, et al. Lactose-binding lectin expression in human colorectal carcinomas. Relation to tumor progression. Carbohydr Res 1991, 213:47–57.

82. Rubinstein N, Alvarez M, Zwirner NW, et al. Targeted inhibition of galectin-1 gene expression in tumor cells results in heightened T cell-mediated rejection. A potential mechanism of tumor-immune privilege. Cancer Cell 2004, 5:241–251.

83. Gabius HJ, Brehler R, Schauer A, et al. Localization of endogenous lectins in normal human breast, benign breast lesions and mammary carcinomas. Virchows Arch B Cell Pathol Incl Mol Pathol 1986, 52:107–115.

84. Allen HJ, Sucato D, Woynarowska B, et al. Role of galaptin in ovarian carcinoma adhesion to extracellular matrix in vitro. J Cell Biochem 1990, 43:43–57.

85. van den Brule FA, Waltregny D, Castronovo V. Increased expression of galectin-1 in carcinoma-associated stroma predicts poor outcome in prostate carcinoma patients. J Pathol 2001, 193:80–87.

86. Diaz H, Tsang KY, et al. Design, synthesis, and partial characterization of water-soluble ß-sheets stabilized by a dibenzofuran-based amino acid. J. Am. Chem. Soc. 1993, 115:3790–3791.

87. Tsang KY, Diaz H, Graciani N, et al. Hydrophobic cluster formation is necessary for dibezofuran-based amino acids to function as ß-sheet nucleators. J Am Chem Soc 1994, 116:3988–4005.

88. Dings RP, Nesmelova I, Griffioen AW, et al. Discovery and development of anti-angiogenic peptides: A structural link. Angiogenesis 2003, 6:83–91.

89. Dings RP, Chen X, Hellebrekers DM, et al. Design of non-peptidic topomimetics of antiangiogenic proteins with antitumor activities. J Natl Cancer Inst 2006, 98:932–936.

90. Good DJ, Polverini PJ, Rastinejad F, et al. A tumor suppressor-dependent inhibitor of angiogenesis is immunologically and functionally indistinguishable from a fragment of thrombospondin. Proc Natl Acad Sci U S A 1990, 87:6624–6628.

91. Tolsma SS, Volpert OV, Good DJ, et al. Peptides derived from two separate domains of the matrix protein thrombospondin-1 have anti-angiogenic activity. J Cell Biol 1993, 122:497–511.

92. Ren B, Yee KO, Lawler J, et al. Regulation of tumor angiogenesis by thrombospondin-1. Biochim Biophys Acta 2006, 1765:178–188.

93. Short SM, Derrien A, Narsimhan RP, et al. Inhibition of endothelial cell migration by thrombospondin-1 type-1 repeats is mediated by beta1 integrins. J Cell Biol 2005, 168:643–653.

94. Lawler J, Weinstein R, Hynes RO. Cell attachment to thrombospondin: the role of ARG-GLY-ASP, calcium, and integrin receptors. J Cell Biol 1988, 107:2351–2361.

95. Dawson DW, Volpert OV, Pearce SF, et al. Three distinct D-amino acid substitutions confer potent antiangiogenic activity on an inactive peptide derived from a thrombospondin-1 type 1 repeat. Mol Pharmacol 1999, 55:332–338.

96. Dawson DW, Pearce SF, Zhong R, et al. CD36 mediates the In vitro inhibitory effects of thrombospondin-1 on endothelial cells. J Cell Biol 1997, 138:707–717.

97. Reiher FK, Volpert OV, Jimenez B, et al. Inhibition of tumor growth by systemic treatment with

thrombospondin-1 peptide mimetics. Int J Cancer 2002, 98:682–689.

98. Baker LH, Demetri GD, Mendelson DS, et al. A randomized phase II study of the thrombospondin-mimetic peptide ABT-510 in patients with advanced soft tissue sarcoma (STS). J Clin Oncol 2005, (23):9013.

99. Markovic S, Suman V, Rao R, et al. A phase II study of ABT-510 for the treatment of metastatic melanoma. J Clin Oncol 2006, (24):8041.

100. Ferrara N, Clapp C, Weiner R. The 16K fragment of prolactin specifically inhibits basal or fibroblast growth factor stimulated growth of capillary endothelial cells. Endocrinology 1991, 129:896–900.

101. Corbacho AM, Martinez De La Escalera G, et al. Roles of prolactin and related members of the prolactin/growth hormone/placental lactogen family in angiogenesis. J Endocrinol 2002, 173:219–238.

102. Struman I, Bentzien F, Lee H, et al. Opposing actions of intact and N-terminal fragments of the human prolactin/growth hormone family members on angiogenesis: an efficient mechanism for the regulation of angiogenesis. Proc Natl Acad Sci U S A 1999, 96:1246–1251.

103. Nguyen NQ, Tabruyn SP, Lins L, et al. Prolactin/growth hormone-derived antiangiogenic peptides highlight a potential role of tilted peptides in angiogenesis. Proc Natl Acad Sci U S A 2006, 103:14319–14324.

104. Tjin Tham Sjin RM, Naspinski J, Birsner AE, et al. Endostatin therapy reveals a U-shaped curve for anti-tumor activity. Cancer Gene Ther 2006, 13:619–627.

105. Karumanchi SA, Jha V, Ramchandran R, et al. Cell surface glypicans are low-affinity endostatin receptors. Mol Cell 2001, 7:811–822.

106. Wickstrom SA, Alitalo K, Keski-Oja J. Endostatin associates with integrin alpha5beta1 and caveolin-1, and activates Src via a tyrosyl phosphatase-dependent pathway in human endothelial cells. Cancer Res 2002, 62:5580–5589.

107. Rehn M, Veikkola T, Kukk-Valdre E, et al. Interaction of endostatin with integrins implicated in angiogenesis. Proc Natl Acad Sci U S A 2001, 98:1024–1029.

108. Shichiri M, Hirata Y. Antiangiogenesis signals by endostatin. Faseb J 2001, 15:1044–1053.

109. Hanai J, Dhanabal M, Karumanchi SA, et al. Endostatin causes G1 arrest of endothelial cells through inhibition of cyclin D1. J Biol Chem 2002, 277:16464–16469.

110. Wickstrom SA, Alitalo K, Keski-Oja J. Endostatin associates with lipid rafts and induces reorganization of the actin cytoskeleton via down-regulation of RhoA activity. J Biol Chem 2003, 278:37895–37901.

111. Kim YM, Hwang S, Pyun BJ, et al. Endostatin blocks vascular endothelial growth factor-mediated signaling via direct interaction with KDR/Flk-1. J Biol Chem 2002, 277:27872–27879.

112. Kim YM, Jang JW, Lee OH, et al. Endostatin inhibits endothelial and tumor cellular invasion by blocking the activation and catalytic activity of matrix metalloproteinase. Cancer Res 2000, 60:5410–5413.

113. Hanai J, Gloy J, Karumanchi SA, et al. Endostatin is a potential inhibitor of Wnt signaling. J Cell Biol 2002, 158:529–539.

114. Wickstrom SA, Alitalo K, Keski-Oja J. An endostatin-derived peptide interacts with integrins and regulates actin cytoskeleton and migration of endothelial cells. J Biol Chem 2004, 279:20178–20185.

115. Wojtowicz-Praga S. Clinical potential of matrix metalloprotease inhibitors. Drugs R D 1999, 1:117–129.

116. Peterson JT. The importance of estimating the therapeutic index in the development of matrix metalloproteinase inhibitors. Cardiovasc Res 2006, 69:677–687.

117. Teicher BA, Sotomayor EA, Huang ZD. Antiangiogenic agents potentiate cytotoxic cancer therapies against primary and metastatic disease. Cancer Res 1992, 52:6702–6704.

118. Teicher BA, Emi Y, Kakeji Y, et al. TNP-470/minocycline/cytotoxic therapy: a systems approach to cancer therapy. Eur J Cancer 1996, 32A:2461–2466.

119. Herbst RS, Takeuchi H, Teicher BA. Paclitaxel/carboplatin administration along with antiangiogenic therapy in non-small-cell lung and breast carcinoma models. Cancer Chemother Pharmacol 1998, 41:497–504.

120. Wilczynska U, Kucharska A, Szary J, et al. Combined delivery of an antiangiogenic protein (angiostatin) and an immunomodulatory gene (interleukin-12) in the treatment of murine cancer. Acta Biochim Pol 2001, 48:1077–1084.

121. Mauceri HJ, Hanna NN, Beckett MA, et al. Combined effects of angiostatin and ionizing radiation in antitumour therapy. Nature 1998, 394:287–291.

122. Griffin RJ, Williams BW, Wild R, et al. Simultaneous inhibition of the receptor kinase activity of vascular endothelial, fibroblast, and platelet-derived growth factors suppresses tumor growth and enhances tumor radiation response. Cancer Res 2002, 62:1702–1706.

123. Fogarty M. Learning from angiogenesis trial failures. Scientist 2002, 16:33–35.

124. Shaked Y, Bocci G, Munoz R, et al. Cellular and molecular surrogate markers to monitor targeted and non-targeted antiangiogenic drug activity and determine optimal biologic dose. Curr Cancer Drug Targets 2005, 5:551–559.

125. Shaked Y, Bertolini F, Man S, et al. Genetic heterogeneity of the vasculogenic phenotype parallels

angiogenesis, Implications for cellular surrogate marker analysis of antiangiogenesis. Cancer Cell 2005, 7:101–111.

126. O'Connor JP, Jackson A, Parker GJ, et al. DCE-MRI biomarkers in the clinical evaluation of antiangiogenic and vascular disrupting agents. Br J Cancer 2007, 96:189–195.

127. Teicher BA, Holden SA, Ara G, et al. Response of the FSaII fibrosarcoma to antiangiogenic modulators plus cytotoxic agents. Anticancer Res 1993, 13:2101–2106.

128. Jain RK. Normalizing tumor vasculature with antiangiogenic therapy: a new paradigm for combination therapy. Nat Med 2001, 7:987–989.

129. Dings RP, Williams BW, Song CW, et al. Anginex synergizes with radiation therapy to inhibit tumor growth by radiosensitizing endothelial cells. Int J Cancer 2005, 115:312–319.

130. Dings RPM, Loren M, Heun H, et al. Scheduling of radiation with angiogenesis inhibitors Anginex and Avastin improves therapeutic outcome via vessel normalization. Clin Cancer Res 2007, in press.

131. Hanna NN, Seetharam S, Mauceri HJ, et al. Antitumor interaction of short-course endostatin and ionizing radiation. Cancer J 2000, 6:287–293.

132. Gorski DH, Mauceri HJ, Salloum RM, et al. Potentiation of the antitumor effect of ionizing radiation by brief concomitant exposures to angiostatin. Cancer Res 1998, 58:5686–5689.

133. Garcia-Barros M, Paris F, Cordon-Cardo C, et al. Tumor response to radiotherapy regulated by endothelial cell apoptosis. Science 2003, 300:1155–1159.

134. Kisker O, Becker CM, Prox D, et al. Continuous administration of endostatin by intraperitoneally implanted osmotic pump improves the efficacy and potency of therapy in a mouse xenograft tumor model. Cancer Res 2001, 61:7669–7674.

135. Lee CG, Heijn M, di Tomaso E, et al. Anti-vascular endothelial growth factor treatment augments tumor radiation response under normoxic or hypoxic conditions. Cancer Res 2000, 60:5565–5570.

136. Rofstad EK, Henriksen K, Galappathi KBM. Antiangiogenic treatment with thrombospondin-1 enhances primary tumor radiation response and prevents growth of dormant pulmonary micrometastases after curative radiation therapy in human melanoma xenografts. Cancer Res 2003, 63:4055–4061.

137. Ning S, Laird D, Cherrington JM, et al. The antiangiogenic agents SU5416 and SU6668 increase the antitumor effects of fractionated irradiation. Radiat Res 2002, 157:45–51.

138. Gorski DH, Beckett MA, Jaskowiak NT, et al. Blockage of the vascular endothelial growth factor stress response increases the antitumor effects of ionizing radiation. Cancer Res 1999, 59:3374–3378.

139. Teicher BA, Holden SA, Ara G, et al. Potentiation of cytotoxic cancer therapies by TNP-470 alone and with other anti-angiogenic agents. Int J Cancer 1994, 57:920–925.

140. Murata R, Nishimura Y, Hiraoka M. An antiangiogenic agent (TNP-470) inhibited reoxygenation during fractionated radiotherapy of murine mammary carcinoma. Int J Radiat Oncol Biol Phys 1997, 37:1107–1113.

141. Lund EL, Bastholm L, Kristjansen PE. Therapeutic synergy of TNP-470 and ionizing radiation: effects on tumor growth, vessel morphology, and angiogenesis in human glioblastoma multiforme xenografts. Clin Cancer Res 2000, 6:971–978.

142. Ansiaux R, Baudelet C, Jordan BF, et al. Thalidomide radiosensitizes tumors through early changes in the tumor microenvironment. Clin Cancer Res 2005, 11:743–750.

143. Huber PE, Bischof M, Jenne J, et al. Trimodal cancer treatment: beneficial effects of combined antiangiogenesis, radiation, and chemotherapy. Cancer Res 2005, 65:3643–3655.

144. Brazelle WD, Shi W, Siemann DW. VEGF-associated tyrosine kinase inhibition increases the tumor response to single and fractionated dose radiotherapy. Int J Radiat Oncol Biol Phys 2006, 65:836–841.

145. Citrin D, Menard C, Camphausen K. Combining radiotherapy and angiogenesis inhibitors: clinical trial design. Int J Radiat Oncol Biol Phys 2006, 64:15–25.

146. Carter SK. Clinical strategy for the development of angiogenesis inhibitors. Oncologist 2000, 5 Suppl 1:51–54.

147. Abdollahi A, Lipson KE, Sckell A, et al. Combined therapy with direct and indirect angiogenesis inhibition results in enhanced antiangiogenic and antitumor effects. Cancer Res 2003, 63:8890–8898.

148. Hess C, Vuong V, Hegyi I, et al. Effect of VEGF receptor inhibitor PTK787/ZK222584 [correction of ZK222548] combined with ionizing radiation on endothelial cells and tumour growth. Br J Cancer 2001, 85:2010–2016.

149. Schuuring J, Bussink J, Bernsen HJ, et al. Irradiation combined with SU5416: microvascular changes and growth delay in a human xenograft glioblastoma tumor line. Int J Radiat Oncol Biol Phys 2005, 61:529–534.

150. Kozin SV, Boucher Y, Hicklin DJ, et al. Vascular endothelial growth factor receptor-2-blocking antibody potentiates radiation-induced long-term control of human tumor xenografts. Cancer Res 2001, 61:39–44.

151. Gorski DH, Mauceri HJ, Salloum RM, et al. Prolonged treatment with angiostatin reduces metastatic burden during radiation therapy. Cancer Res 2003, 63:308–311.

152. Horsman MR, Siemann DW. Pathophysiologic effects of vascular-targeting agents and the implications for combination with conventional therapies. Cancer Res 2006, 66:11520–11539.

153. Teicher BA, Holden SA, Dupuis NP, et al. Potentiation of cytotoxic therapies by TNP-470 and minocycline in mice bearing EMT-6 mammary carcinoma. Breast Cancer Res Treat 1995, 36:227–236.

154. Kato T, Sato K, Kakinuma H, et al. Enhanced suppression of tumor growth by combination of angiogenesis inhibitor O-(chloroacetyl-carbamoyl)fumagillol (TNP-470) and cytotoxic agents in mice. Cancer Res 1994, 54:5143–5147.

155. Shishido T, Yasoshima T, Denno R, et al. Inhibition of liver metastasis of human pancreatic carcinoma by angiogenesis inhibitor TNP-470 in combination with cisplatin. Jpn J Cancer Res 1998, 89:963–969.

156. Teicher BA, Holden SA, Ara G, et al. Influence of an anti-angiogenic treatment on 9L gliosarcoma: oxygenation and response to cytotoxic therapy. Int J Cancer 1995, 61:732–737.

157. Ma G, Masuzawa M, Hamada Y, et al. Treatment of murine angiosarcoma with etoposide, TNP-470 and prednisolone. J Dermatol Sci 2000, 24:126–133.

158. Ding Q, Kestell P, Baguley BC, et al. Potentiation of the antitumour effect of cyclophosphamide in mice by thalidomide. Cancer Chemother Pharmacol 2002, 50:186–192.

159. Fujii T, Tachibana M, Dhar DK, et al. Combination therapy with paclitaxel and thalidomide inhibits angiogenesis and growth of human colon cancer xenograft in mice. Anticancer Res 2003, 23:2405–2411.

160. Heere-Ress E, Boehm J, Thallinger C, et al. Thalidomide enhances the anti-tumor activity of standard chemotherapy in a human melanoma xenotransplantation model. J Invest Dermatol 2005, 125:201–206.

161. Segers J, Fazio VD, Ansiaux R, et al. Potentiation of cyclophosphamide chemotherapy using the anti-angiogenic drug thalidomide: importance of optimal scheduling to exploit the 'normalization' window of the tumor vasculature. Cancer Lett 2006, 244:129–135.

162. Abraham D, Abri S, Hofmann M, et al. Low dose carboplatin combined with angiostatic agents prevents metastasis in human testicular germ cell tumor xenografts. J Urol 2003, 170:1388–1393.

163. Bertolini F, Fusetti L, Mancuso P, et al. Endostatin, an antiangiogenic drug, induces tumor stabilization after chemotherapy or anti-CD20 therapy in a NOD/SCID mouse model of human high-grade non-Hodgkin lymphoma. Blood 2000, 96:282–287.

164. Zhang X, Galardi E, Duquette M, et al. Antiangiogenic treatment with three thrombospondin-1 type 1 repeats versus gemcitabine in an orthotopic human pancreatic cancer model. Clin Cancer Res 2005, 11:5622–5630.

165. Allegrini G, Goulette FA, Darnowski JW, et al. Thrombospondin-1 plus irinotecan: a novel antiangiogenic-chemotherapeutic combination that inhibits the growth of advanced human colon tumor xenografts in mice. Cancer Chemother Pharmacol 2004, 53:261–266.

166. Bocci G, Danesi R, Marangoni G, et al. Antiangiogenic versus cytotoxic therapeutic approaches to human pancreas cancer: an experimental study with a vascular endothelial growth factor receptor-2 tyrosine kinase inhibitor and gemcitabine. Eur J Pharmacol 2004, 498:9–18.

167. Garofalo A, Naumova E, Manenti L, et al. The combination of the tyrosine kinase receptor inhibitor SU6668 with paclitaxel affects ascites formation and tumor spread in ovarian carcinoma xenografts growing orthotopically. Clin Cancer Res 2003, 9:3476–3485.

168. Dings RP, Yokoyama Y, Ramakrishnan S, et al. The designed angiostatic peptide anginex synergistically improves chemotherapy and antiangiogenesis therapy with angiostatin. Cancer Res 2003, 63:382–385.

169. Dings RP, van Laar ES, Webber J, Waters SJ, MacDonald JR, and Mayo KH. Ovarian tumor growth regression using a combination of vascular targeting agents anginex or 0118 and the chemotherapeutic irofulven. Cancer Letters 2008 Mar 29 [Epub ahead of print].

Chapter 19
Sensitization of Cancer Cells to Cancer Therapies by Isoflavone and Its Synthetic Derivatives

Fazlul H. Sarkar and Yiwei Li

1   Introduction

Despite the huge efforts made on the development of cancer therapies over the past several decades, cancer still remains the second leading cause of death in the United States [1]. Conventional cancer therapies primarily consist of surgery, chemotherapy, and radiation therapy. Physicians and scientists have vigorously investigated ways to improve the efficacy of these therapies. These include modifying surgical techniques, developing new chemotherapy agents, refining therapeutic strategies, etc. To achieve higher efficacy, combination treatments using combined therapies or combined agents with distinct targets are considered more promising, resulting in better survival of patients diagnosed with cancers. However, the combination treatment is also associated with a certain degree of dose-related toxicity. Therefore, the development of mechanism-based and targeted therapeutic strategies to improve therapeutic efficacy and reduce side effects is very important. In targeted combination therapy, combined agents target specific cell signaling pathways that play important roles in the development and progression of cancer. By attacking or blocking these important targets that are responsible for promoting cancer cell survival and growth, the targeted cancer therapy will more effectively inhibit the proliferation of cancer cells and show strong potency against cancers.

In recent years, more dietary compounds, i.e., isoflavone genistein, 3,3′-diindolylmethane (DIM), indole-3-carbinol (I3C), curcumin, (–)-epigallo-catechin-3-gallate (EGCG), resveratrol, etc, have been recognized as cancer chemopreventative agents because of their anticarcinogenic activity [2]. These compounds exert antitumor activities through regulation of different cell signaling transduction pathways that are involved in the development and progression of cancer. By regulating cell signaling transduction, these dietary compounds can sensitize cancer cells to apoptotic death. Therefore, conventional cancer therapies combined with these dietary compounds may exert enhanced antitumor activity through synergic action or compensation of inverse properties [3]. The combination treatment may also decrease the systemic toxicity caused by chemotherapy or radiotherapy because lower doses of therapeutic agents could be used and no systemic toxicity has been found from dietary compounds. Emerging in vitro and in vivo data have shown that isoflavone and its synthetic derivatives enhance the efficacy of chemotherapy and radiotherapy in various cancers through regulation of Akt, NF-κB, Cox-2, OPG/RANKL/MMP-9, and apoptotic pathways, suggesting a novel targeted therapeutic strategy against cancers [4–9].

From: *Sensitization of Cancer Cells for Chemo/Immuno/Radio-therapy, 1st Edition*.
Edited by: Benjamin Bonavida © Humana Press, Totowa, NJ

2 Antitumor Activity of Conventional Cancer Therapies Are Enhanced by Isoflavone and Its Derivatives

Growing evidence has demonstrated that isoflavone and its derivatives can potentiate antitumor activities of chemotherapeutics or other conventional therapies, suggesting that isoflavone and its derivatives could be very useful in the treatment of cancer.

2.1 Potentiation of Chemotherapeutic Effects by Isoflavone and Its Derivatives

2.1.1 Isoflavone Genistein

The published studies have shown that isoflavone genistein could potentiate the antitumor effects of chemotherapeutic agents in various cancers, in vitro and in vivo. We have reported that isoflavone genistein in vitro enhanced growth inhibition and apoptotic cell death caused by chemotherapeutic agents, including cisplatin, docetaxel, doxorubicin, and gemcitabine in prostate, breast, pancreas, and lung cancers [4–7, 10, 11]. The study showed that pretreatment of cancer cells with isoflavone genistein prior to treatment with lower doses of chemotherapeutic agents caused a significantly greater degree of growth inhibition and apoptosis, suggesting that increased anticancer activities of chemotherapeutic agents with lower toxicity to normal cells could be achieved by combination treatment with isoflavone genistein and conventional chemotherapeutic agents. To investigate if same effect of isoflavone genistein could be observed in vivo, we conducted animal studies and found that dietary isoflavone genistein could also enhance the antitumor activities of gemcitabine and docetaxel in an animal tumor model, leading to greater apoptotic cell death and tumor growth inhibition [4, 6]. In addition to solid tumor, we also found that isoflavone genistein could sensitize diffuse large cell lymphoma to CHOP (cyclophosphamide, doxorubicin, vincristine, and prednisone) chemotherapy, resulting in greater inhibition of lymphoma cell growth [12]. More importantly, we also found that antitumor, anti-invasion, and antimetastatic activities of docetaxel could be enhanced by isoflavone

genistein through the inhibition of osteoclastic bone resorption and prostate cancer bone metastasis [7], suggesting that isoflavone genistein could be used in combination with chemotherapeutics for the treatment of metastatic prostate cancer.

Other investigators have reported similar results showing that the antitumor effects of chemotherapeutics could be enhanced by isoflavone genistein. Hwang et al. showed that the combination of genistein and 5-fluorouracil synergistically induced apoptotic cell death in chemoresistant HT-29 colon cancer cells [8]. Isoflavone genistein also enhanced necrotic-like cell death in HER-2 overexpressing breast cancer cells treated with Adriamycin [13]. Genistein also enhanced the antitumor effect of bleomycin in HL-60 cells, but not in normal lymphocytes in an in vitro study [14]. Tanos et al. reported that isoflavone genistein inhibited the growth of dysplastic and malignant epithelial breast cancer cells in vitro and that the addition of tamoxifen had a synergistic inhibitory effect on breast cancer growth [15]. The synergistic action of genistein and cisplatin or BCNU on growth inhibition of glioblastoma and medulloblastoma cells has also been observed [16, 17]. These reports support our findings and suggest that isoflavone genistein is not just a chemopreventive agent but could also be used as a potent therapeutic agent in combination with other chemotherapeutics for cancer treatment.

Several phase I clinical trials have been conducted to investigate the toxicity and effects of isoflavones in patients with prostate cancer [18–21]. Busby et al. tested the safety of purified unconjugated genistein, daidzein, and glycitein, and defined pharmacokinetic parameters for their absorption and metabolism using a single dose. They observed that dietary supplements of purified unconjugated isoflavones administered to humans in single doses exceeding normal dietary intake by many fold resulted in minimal clinical toxicity [19]. Fischer et al. also showed similar results by multiple-dose administration to men with prostate cancer [20]. Takimoto et al. reported that oral administration of soy isoflavones gave plasma concentrations of genistein that have been associated with antimetastatic activity in vitro [18]. The observed maximum plasma genistein concentration ranged from 2.7 to 27.4 μM [19]. These results suggest that isoflavone genistein can be administrated safely with minimal side effects and that the plasma genistein concentration achieved is comparable to the concentrations used in most in vitro experiments. Kumar et al. showed that supplementing early-stage prostate cancer patients with soy isoflavones altered surrogate markers of proliferation such as serum PSA and free testosterone in a larger number of subjects in the isoflavone supplemented group than the group receiving placebo, suggesting the beneficial effects of isoflavone on early stage prostate cancer [21]. So far, no phase II clinical trial has been reported for assessing the effects of isoflavone genistein and likewise no clinical trial has been conducted to investigate the effects of isoflavone genistein in combination with conventional therapy. However, the encouraging clinical evidence from phase I trials urges consideration of phase II and III clinical trials investigating the efficacy of soy isoflavones alone and/or in combination with other therapeutics.

2.1.2 Isoflavone Derivatives

To enhance the antitumor activity of isoflavone, several isoflavone derivatives have been synthesized and used in experimental systems in vitro and in vivo, and in clinical trials. These compounds have shown low IC_{50} in the inhibition of cancer cell growth in vitro. They have been found to target specific cell signaling transduction pathways. Moreover, these compounds at low concentration were able to enhance the antitumor activity of clinically available conventional chemotherapeutic agents, suggesting their potent effects as therapeutic agents for combination treatment.

2.1.2.1 Phenoxodiol

Phenoxodiol is an analog of isoflavone genistein and has shown a broad-spectrum anticancer effect. It is developed as a therapeutic agent for early- and late-stage prostate cancer, including hormone-refractory prostate cancer, early stage cervical and vaginal cancer, chemoresistant ovarian cancer, and renal cancer [22, 23]. In an animal study, phenoxodiol inhibited dimethylbenzene[a]anthracene (DMBA)–induced mammary carcinogenesis in female Sprague-Dawley rats, suggesting that phenoxodiol is an effective chemopreventive agent against DMBA–induced mammary carcinogenesis [24]. In experimental studies and clinical trials,

phenoxodiol has been used both as a monotherapy and in combination with standard chemotherapeutics. In some cancers, phenoxodiol appears to be strong enough to work on its own as a monotherapy. However, one of the major benefits of phenoxodiol is its ability to sensitize cancer cells to the antitumor effects of standard chemotherapeutics [25]. It has been found that in cancer cells that are susceptible to the effects of standard chemotherapeutics, phenoxodiol increases their sensitivity to those agents. In cancer cells that have become resistant to the effects of standard chemotherapeutics, phenoxodiol restores chemosensitivity [9, 26]. By exposing chemoresistant cancer cells to phenoxodiol first, long-standing drug resistance is removed, making cancer cells susceptible once again to standard chemotherapeutics, such as cisplatin, carboplatin, taxanes, and gemcitabine. Another benefit of phenoxodiol is its high specificity for cancer cells. This is because the primary target of phenoxodiol is tNOX (NADH oxidase), which is present only on cancer cells [27]. So far, phenoxodiol has been tested and shown effective in breast, prostate, ovarian, lung, colorectal, and head and neck carcinomas, mesotheliomas, rhabdomyosarcoma, and neuroglioma. However, it was not suitable for colorectal cancer and leukemia. A derivative drug of phenoxodiol, NV5063, is undergoing development for colorectal cancer and leukemia. Phenoxodiol is currently undergoing clinical studies in the United States and Australia. So far, phase I and II clinical trials have shown some disease stabilization without severe toxicity [23].

It has been found that the primary target of phenoxodiol is a protein receptor known as tNOX [27]. This receptor plays important roles in helping maintain high levels of anti-apoptotic proteins in the cancer cell. By binding to the tNOX receptor and blocking its function, phenoxodiol eventually inhibits the anti-apoptotic proteins XIAP and FLICE inhibitory protein (FLIP) through the Akt signal transduction pathway [9]. This, in turn, leads to activation of caspases, resulting in the induction of apoptotic cell death. Experiments have shown that cancer cells display signs of distress after several hours of exposure to phenoxodiol. The cancer cells stop mitosis and then show apoptotic death after 24–48 hours of treatment. In addition to the caspase-dependent apoptotic pathway, another mechanism is also involved in the cell death caused by phenoxodiol. It has been found that phenoxodiol induces G1 arrest by specific loss in cyclin-dependent kinase 2 activity through p53-independent induction of p21[WAF1/CIP1] [28]. In addition, phenoxodiol also inhibits the catalytic activity of topo II in a dose-dependent manner and stabilizes the topo II–mediated cleavable complex, demonstrating that this agent is an inhibitor of topo II [29]. The mechanism of phenoxodiol action needs to be further investigated.

2.1.2.2 FPA-120 to FPA-127

In order to achieve higher antitumor activity, we have also synthesized several derivatives of isoflavone genistein, named FPA-120 to FPA-127 [30, 31]. We designed an assembly process of an evolving soy isoflavone genistein motif and its possible variations, conjugating them with copper and providing the molecules capable of killing tumor cells efficiently in breast, prostate, and pancreatic cancers. Specifically, the Schiff base ligands, FPA-120 to FPA-123, were synthesized and recrystallized. The ligands further interacted with copper chloride in the stoichiometric ratio to precipitate out the corresponding copper conjugates, FPA-124 to FPA-127, which were purified by the chromatographic work-ups. The amines we used for synthesis of FPA-120 to FPA-123 are effective pharmacophores found in many therapeutic compounds currently used in the clinic, and they serve as spacers in the present design, keeping the cytotoxic metal conjugates away from the genistein moiety. Such a strategy is found to be useful for retaining pharmacologic properties of both the carrier and cytotoxic moieties.

We have tested the effects of synthetic derivatives of genistein, FPA-120 to FPA-127, on the growth of various cancers, including BT20 breast, PC-3 prostate, Colo357, and BxPC-3 pancreatic cancer cells [30, 31]. We found that the synthetic derivatives of genistein, FPA-120 to FPA-127, inhibited cell proliferation in all cancer cell lines tested. The inhibition was dose dependent in BT20 and PC-3 cells, showing 50% growth inhibition at $10\,\mu M$. This IC_{50} value is much lower than that of parent genistein (at $30–50\,\mu M$), suggesting that these derivatives of isoflavone genistein are more potent for the inhibition of cancer cell growth than the parent molecule genistein. More importantly,

we observed a higher degree of antiproliferative activities of copper-conjugated derivatives of genistein, FPA-124 to FPA-127, compared with their corresponding parent ligands. Significantly lower IC_{50} values were obtained using copper-conjugated derivatives of isoflavone genistein, indicating therapeutically achievable efficacy of these conjugates. By ELISA apoptosis assay we found that these synthetic derivatives of isoflavone genistein significantly induced apoptotic cell death in BT20 breast and PC-3 prostate cancer cells. Moreover, we compared apoptosis-inducing activities between synthetic derivatives of isoflavone genistein and parent genistein. We found that synthetic derivatives of isoflavone at lower doses induced more apoptosis compared to parent genistein, suggesting more potent activity of synthetic derivatives of isoflavone genistein, which could be useful as a better chemopreventive or therapeutic agent against cancer.

To investigate the molecular mechanism of action of synthetic isoflavone derivatives, we tested the effects of FPA-124 to FPA-127 on Akt and NF-κB pathways, which are known to play important roles in cell survival and apoptosis [30, 31]. We found that all synthetic derivatives of isoflavone exhibited 50% inhibition of Akt kinase activity at a dose $<15\,\mu M$, whereas $70\,\mu M$ of genistein was needed to show similar activity against Akt. The compound FPA-124 showed the lowest IC_{50} value ($0.1\,\mu M$) compared with other copper conjugates in the inhibition of Akt kinase activity. The effects of FPA-124 on tumor growth and NF-κB activity in an orthotropic pancreatic tumor model using Colo 357 pancreatic cancer cells were also tested. The results showed that FPA-124 had no apparent animal toxicity, but did cause a decrease in tumor weight, suggesting the inhibitory effect of FPA-124 on tumor growth in vivo. Moreover, FPA-124 also significantly inhibited NF-κB DNA binding activity in animal tumor tissues. These results suggest that the growth inhibitory and apoptosis-inducing effects of FPA-124 to FPA-127 are partly mediated by the inactivation of Akt and NF-κB pathways. Because these synthetic derivatives of isoflavone show potent antitumor activity through inhibition of Akt and NF-κB signaling, they could be used in combination of chemotherapeutics for cancer therapy, suggesting that they could be excellent new anticancer drugs without any adverse effects.

2.1.2.3   Other Isoflavone Derivatives

Phytoestrogen biochanin A is another isoflavone derivative with anticarcinogenic properties. It has been found that treating the cells with biochanin A alone caused the accumulation of CYP1A1 mRNA and an increase in CYP1A1-specific 7-ethoxyresorufin O-deethylase (EROD) activity in a dose dependent manner. A concomitant treatment with 7,12-dimethylbenz[a]anthracene (DMBA) and biochanin A markedly reduced the DMBA-inducible EROD activity and CYP1A1 mRNA level, suggesting that it could be a useful anticarcinogenic agent that could lead to sensitization of cancer cells to death [32]. Biochanin A also inhibited N-nitroso-N-methyl urea-induced rat mammary carcinogenesis [33]. Recently, Vasselin et al. reported a new series of synthetic fluoro- and amino-substituted isoflavones as potential antitumor agents based on structural similarities to known isoflavones. These isoflavone derivatives have shown significant potentiation of growth inhibitory activity with very low IC_{50} (<1 μM) in breast cancer cells, demonstrating its potent antitumor activity [34]. Certain oxime- and methyloxime-containing isoflavone derivatives were also synthesized and evaluated for their antiproliferative activity against cervical, hepatocellular, and oral epithelial cell carcinoma. They showed strong antiproliferative activities with GI_{50} values of <1 μM [35]. The growth inhibition achieved by these compounds was due to G2/M phase arrest and induction of apoptosis. Whatmore et al. also synthesized several isoflavone derivatives and showed that the derivatives, SU1433, SU 5416, and SU6668 are more potent inhibitors of VEGF-induced angiogenesis than naturally occurring isoflavones. Cytotoxicity studies indicated that none of these compounds examined exhibited cytotoxicity at antiangiogenic concentrations, suggesting the potent antitumor and antiangiogenesis effects of these compounds [36]. It has been reported that the synthetic isoflavone derivative ipriflavone is able to reduce bone loss in various types of animal models of experimental osteoporosis, suggesting that it could be used in combination with other chemotherapeutics to reduce bone loss in the case of cancer osteoclastic bone metastasis [37]. In addition, to enhance the intrinsic activity of genistein, its glycosidic derivatives were synthesized. On the basis of GI_{50} and LC_{50} values, two of the most active derivative glycosides were found to be several times more potent in their cytostatic and cytotoxic effect than genistein [38], suggesting their potent effect on the inhibition of cancer cell growth. These reports demonstrate better antitumor activity of synthetic isoflavone derivatives than parent genistein and suggest that these synthetic isoflavone derivatives could be used in combination with chemotherapeutics for better cancer treatment.

2.2   Potentiation of Radiotherapy Effects by Isoflavone

The experimental studies also showed that the efficacy of radiotherapy could be enhanced by isoflavones. Hillman's laboratory has investigated the effect of the combination of isoflavone genistein and radiation on prostate cancer cells. They found that the combination of isoflavone genistein and radiation showed enhanced inhibitory effects on DNA synthesis and cell growth [39]. Furthermore, they found that isoflavone genistein combined with radiation led to a greater control of the growth of the primary tumor and metastasis to lymph nodes than either isoflavone genistein or radiation alone, suggesting that isoflavone genistein enhanced the radiosensitivity of PC-3 prostate cancer cells [40]. Genistein combined with radiation also caused greater inhibition in PC-3 colony formation with more cancer cell death compared with either genistein or radiation alone. Mechanistic studies showed that increased cell death by genistein and radiation occurred via inhibition of NF-κB, leading to altered expression of regulatory cell cycle proteins, cyclin B and p21$^{WAF1/Cip1}$, thus promoting G_2/M arrest and increased radiosensitivity [41]. Soy isoflavone also enhanced prostate cancer radiotherapy through downregulation of apurinic/apyrimidinic endonuclease 1/redox factor-1 expression [42]. They also found that pure genistein caused adverse effects on inducing lymph node metastasis [43], whereas such adverse effects were not observed with isoflavone mixture [44]. These findings support the important and novel strategy of combining isoflavone genistein with radiation for the treatment of prostate cancer.

A similar report by other investigators showed that isoflavone genistein enhanced the radiosensitivity of cervical cancer cells through increased apoptosis, prolonged cell cycle arrest and impaired repair of DNA damage [45]. Isoflavone genistein was also shown to enhance radiosensitivity in human esophageal cancer cells in vitro [46], suggesting that the enhancement of radiosensitivity by isoflavone genistein is not cell-type dependent. These reports demonstrate that radiotherapy combined with isoflavone can cause more growth inhibition and apoptotic cell death of various cancers compared with monotreatment.

3   Molecular Mechanisms of Cancer Cell Sensitization to Conventional Cancer Therapy by Isoflavone and Its Derivatives

The molecular mechanisms by which isoflavone and its derivatives enhance the antitumor efficacy of conventional cancer therapies have not yet been fully elucidated. It has been known that chemotherapy and radiotherapy can induce drug resistance in cancer cells, leading to treatment failure. Emerging evidence has demonstrated that NF-κB, Akt, and some molecules in apoptotic pathway are involved in the development of drug resistance. Experimental studies have shown that the enhanced antitumor effects by isoflavone and its derivatives could be, in part, through the regulation of NF-κB, Akt, COX-2, and OPG/RANKL/MMP-9 signaling, which are known to play important roles in cell survival and tumor metastasis. Isoflavone and its derivatives could also sensitize cancer cells to apoptosis by regulating several important molecules, i.e., Bcl-2, Bcl-X$_L$, survivin, XIAP, FLIO, caspases, p21^{WAF1}, etc., in the apoptotic pathway.

3.1   Regulation of the Akt Pathway

The Akt pathway is an important cell signaling pathway that controls cell survival and drug resistance. It has been reported that isoflavone genistein enhanced necrotic-like cell death concomitant with significant inhibition of Akt activity in breast cancer cells treated with isoflavone genistein and Adriamycin, suggesting that the enhanced growth inhibition of combination is due to the inactivation of the Akt pathway [13]. We and other investigators have also shown that activated Akt was inhibited

by isoflavone genistein combined with gemcitabine or radiation in pancreatic, cervical, and esophageal cancer cells, suggesting that enhancement of chemotherapeutic or radiation effects by isoflavone genistein may be partially mediated by the inhibition of Akt signaling [4, 45, 46]. Kamsteeg et al. reported that phenoxodiol, one of the synthetic derivatives of genistein, could inhibit Akt signaling transduction and subsequently activate caspase system, inhibiting XIAP, an inhibitor of apoptosis, disrupting FLICE inhibitory protein (FLIP) expression, and leading to increased chemosensitization [9].

3.2   Regulation of NF-κB Pathway

It has been well known that many chemotherapeutic agents induce activity of NF-κB, which causes drug resistance in cancer cells [47]. From in vitro and in vivo experimental studies, we observed that NF-κB activity was significantly increased by cisplatin, docetaxel, gemcitabine, and radiation treatment and that the NF-κB–inducing activity of these agents was completely abrogated by isoflavone genistein treatment in prostate, breast, lung, and pancreatic cancer cells, suggesting that isoflavone genistein pretreatment inactivates NF-κB and may contribute to increased growth inhibition and apoptosis induced by these agents [4–6, 39, 41]. We also found that isoflavone genistein enhanced the antitumor activity of CHOP by inhibition of NF-κB in lymphoma cells [12], suggesting that the inhibition of NF-κB is important for chemosensitization.

3.3   Regulation of Apoptosis Pathways

It has been reported that the genistein derivative phenoxodiol can bind to the tNOX receptor, block its function, and subsequently inhibit the anti-apoptotic proteins XIAP and FLICE inhibitory protein (FLIP), eventually inducing apoptotic cell death [9]. We and others have also found that isoflavone genistein combined with docetaxel or gemcitabine significantly inhibited Bcl-2, Bcl-X$_L$, and survivin, and induced p21^{WAF1}, suggesting that the enhanced antitumor effect in combination treatment is through the regulation of these important molecules in the apoptotic pathway [4, 6].

3.4 Regulation of Other Pathways

We have reported that the antitumor and antimetastatic activities of docetaxel are enhanced by isoflavone genistein through regulation of OPG/RANK/RANKL/MMP-9 signaling in prostate cancer, suggesting that isoflavone genistein could be a promising nontoxic agent to improve the treatment outcome of metastatic prostate cancer with docetaxel [7]. It has been found that the combination of 5-FU and isoflavone genistein enhanced therapeutic effects in colon cancers through the COX-2 pathway [8]. Apart from the COX-2 pathway, isoflavone genistein and its isoflavone analogs showed the potential to decrease side effects of tamoxifen through metabolic interactions that inhibit the formation of α-hydroxytamoxifen via inhibition of CYP1A2 [48], resulting in the beneficial effects of isoflavone genistein in combination with tamoxifen.

4 Conclusion and Perspective

Growing evidences from in vitro and in vivo studies published so far all suggest that isoflavone and its derivatives may serve as potent agents for enhancing the therapeutic efficacy of chemotherapy, radiotherapy, or other conventional therapies for the treatment of human cancers. However, further in-depth mechanistic studies, in vivo animal experiments, and clinical trials are needed to test the value of isoflavone and its derivatives in combination therapy of human cancers. More potent synthetic derivatives of isoflavone with low IC_{50} are also needed to improve the efficacy of mechanism-based and targeted therapeutic strategies against human cancers.

Acknowledgments. The authors' work cited in this chapter was funded by grants from the National Cancer Institute, NIH (5R01CA083695, 5R01CA101870, and 5R01CA108535 awarded to FHS), a subcontract award to FHS from the University of Texas MD Anderson Cancer Center through a SPORE grant (5P20-CA101936) on pancreatic cancer awarded to James Abbruzzese, and a grant from the Department of Defense (DOD Prostate Cancer Research Program DAMD17-03-1-0042) awarded to FHS.

References

1. American Cancer Society. Cancer facts & figures 2007. Atlanta: American Cancer Society, 2007.
2. Surh YJ. Cancer chemoprevention with dietary phytochemicals. Nat Rev Cancer 2003, 3:768–780.
3. Sarkar FH, Li Y. Using chemopreventive agents to enhance the efficacy of cancer therapy. Cancer Res 2006, 66:3347–3350.
4. Banerjee S, Zhang Y, Ali S, et al. Molecular evidence for increased antitumor activity of gemcitabine by genistein in vitro and in vivo using an orthotopic model of pancreatic cancer. Cancer Res 2005, 65:9064–9072.
5. Li Y, Ellis KL, Ali S, et al. Apoptosis-inducing effect of chemotherapeutic agents is potentiated by soy isoflavone genistein, a natural inhibitor of NF-kappaB in BxPC-3 pancreatic cancer cell line. Pancreas 2004, 28:e90–e95.
6. Li Y, Ahmed F, Ali S, et al. Inactivation of nuclear factor kappaB by soy isoflavone genistein contributes to increased apoptosis induced by chemotherapeutic agents in human cancer cells. Cancer Res 2005, 65:6934–6942.
7. Li Y, Kucuk O, Hussain M, et al. Antitumor and antimetastatic activities of docetaxel are enhanced by genistein through regulation of osteoprotegerin/receptor activator of nuclear factor-kappaB (RANK)/RANK ligand/MMP-9 signaling in prostate cancer. Cancer Res 2006, 66:4816–4825.
8. Hwang JT, Ha J, Park OJ. Combination of 5-fluorouracil and genistein induces apoptosis synergistically in chemo-resistant cancer cells through the modulation of AMPK and COX-2 signaling pathways. Biochem Biophys Res Commun 2005, 332:433–440.
9. Kamsteeg M, Rutherford T, Sapi E, et al. Phenoxodiol—an isoflavone analog—induces apoptosis in chemoresistant ovarian cancer cells. Oncogene 2003, 22:2611–2620.
10. Banerjee S, Zhang Y, Wang Z, et al. In vitro and in vivo molecular evidence of genistein action in augmenting the efficacy of cisplatin in pancreatic cancer. Int J Cancer 2007, 120:906–917.
11. El–Rayes BF, Ali S, Ali IF, et al. Potentiation of the effect of erlotinib by genistein in pancreatic cancer: the role of Akt and nuclear factor-kappaB. Cancer Res 2006, 66:10553–10559.
12. Mohammad RM, Al-Katib A, Aboukameel A, et al. Genistein sensitizes diffuse large cell lymphoma to CHOP (cyclophosphamide, doxorubicin, vincristine, prednisone) chemotherapy. Mol Cancer Ther 2003, 2:1361–1368.
13. Satoh H, Nishikawa K, Suzuki K, et al. Genistein, a soy isoflavone, enhances necrotic–like cell death in

a breast cancer cell treated with a chemotherapeutic agent. Res Commun Mol Pathol Pharmacol 2003, 113–114:149–158.

14. Lee R, Kim YJ, Lee YJ, et al. The selective effect of genistein on the toxicity of bleomycin in normal lymphocytes and HL-60 cells. Toxicology 2004, 195:87–95.

15. Tanos V, Brzezinski A, Drize O, et al. Synergistic inhibitory effects of genistein and tamoxifen on human dysplastic and malignant epithelial breast cells in vitro. Eur J Obstet Gynecol Reprod Biol 2002, 102:188–194.

16. Khoshyomn S, Manske GC, Lew SM, et al. Synergistic action of genistein and cisplatin on growth inhibition and cytotoxicity of human medulloblastoma cells. Pediatr Neurosurg 2000, 33:123–131.

17. Khoshyomn S, Nathan D, Manske GC, et al. Synergistic effect of genistein and BCNU on growth inhibition and cytotoxicity of glioblastoma cells. J Neurooncol 2002, 57:193–200.

18. Takimoto CH, Glover K, Huang X, et al. Phase I pharmacokinetic and pharmacodynamic analysis of unconjugated soy isoflavones administered to individuals with cancer. Cancer Epidemiol Biomarkers Prev 2003, 12:1213–1221.

19. Busby MG, Jeffcoat AR, Bloedon LT, et al. Clinical characteristics and pharmacokinetics of purified soy isoflavones: single-dose administration to healthy men. Am J Clin Nutr 2002, 75:126–136.

20. Fischer L, Mahoney C, Jeffcoat AR, et al. Clinical characteristics and pharmacokinetics of purified soy isoflavones: multiple-dose administration to men with prostate neoplasia. Nutr Cancer 2004, 48:160–170.

21. Kumar NB, Cantor A, Allen K, et al. The specific role of isoflavones in reducing prostate cancer risk. Prostate 2004, 59:141–147.

22. Mor G, Fu HH, Alvero AB. Phenoxodiol, a novel approach for the treatment of ovarian cancer. Curr Opin Investig Drugs 2006, 7:542–548.

23. Choueiri TK, Wesolowski R, Mekhail TM. Phenoxodiol: isoflavone analog with antineoplastic activity. Curr Oncol Rep 2006, 8:104–107.

24. Constantinou AI, Mehta R, Husband A. Phenoxodiol, a novel isoflavone derivative, inhibits dimethylbenz[a]anthracene (DMBA)-induced mammary carcinogenesis in female Sprague-Dawley rats. Eur J Cancer 2003, 39:1012–1018.

25. Alvero AB, O'Malley D, Brown D, et al. Molecular mechanism of phenoxodiol-induced apoptosis in ovarian carcinoma cells. Cancer 2006, 106:599–608.

26. Sapi E, Alvero AB, Chen W, et al. Resistance of ovarian carcinoma cells to docetaxel is XIAP dependent and reversible by phenoxodiol. Oncol Res 2004, 14:567–578.

27. Davies SL, Bozzo J. Spotlight on tNOX: a tumor-selective target for cancer therapies. Drug News Perspect 2006, 19:223–225.

28. Aguero MF, Facchinetti MM, Sheleg Z, et al. Phenoxodiol, a novel isoflavone, induces G1 arrest by specific loss in cyclin-dependent kinase 2 activity by p53-independent induction of p21WAF1/CIP1. Cancer Res 2005, 65:3364–3373.

29. Constantinou AI, Husband A. Phenoxodiol (2H-1-benzopyran-7-0,1,3-(4-hydroxyphenyl)), a novel isoflavone derivative, inhibits DNA topoisomerase II by stabilizing the cleavable complex. Anticancer Res 2002, 22:2581–2585.

30. Barve V, Ahmed F, Adsule S, et al. Synthesis, molecular characterization, and biological activity of novel synthetic derivatives of chromen-4-one in human cancer cells. J Med Chem 2006, 49:3800–3808.

31. Sarkar FH, Adsule S, Padhye S, et al. The role of genistein and synthetic derivatives of isoflavone in cancer prevention and therapy. Mini Rev Med Chem 2006, 6:401–407.

32. Han EH, Kim JY, Jeong HG. Effect of biochanin A on the aryl hydrocarbon receptor and cytochrome P450 1A1 in MCF-7 human breast carcinoma cells. Arch Pharm Res 2006, 29:570–576.

33. Gotoh T, Yamada K, Yin H, et al. Chemoprevention of N-nitroso-N-methylurea-induced rat mammary carcinogenesis by soy foods or biochanin A. Jpn J Cancer Res 1998, 89:137–142.

34. Vasselin DA, Westwell AD, Matthews CS, et al. Structural studies on bioactive compounds. 40. (1) Synthesis and biological properties of fluoro-, methoxyl-, and amino-substituted 3-phenyl-4H-1-benzopyran-4-ones and a comparison of their antitumor activities with the activities of related 2-phenylbenzothiazoles. J Med Chem 2006, 49:3973–3981.

35. Wang TC, Chen IL, Lu PJ, et al. Synthesis, antiproliferative, and antiplatelet activities of oxime- and methyloxime-containing flavone and isoflavone derivatives. Bioorg Med Chem 2005, 13:6045–6053.

36. Whatmore JL, Swann E, Barraja P, et al. Comparative study of isoflavone, quinoxaline and oxindole families of anti-angiogenic agents. Angiogenesis 2002, 5:45–51.

37. Brandi ML. Natural and synthetic isoflavones in the prevention and treatment of chronic diseases. Calcif Tissue Int 1997, 61 Suppl 1:S5–S8.

38. Polkowski K, Popiolkiewicz J, Krzeczynski P, et al. Cytostatic and cytotoxic activity of synthetic genistein glycosides against human cancer cell lines. Cancer Lett 2004, 203:59–69.

39. Hillman GG, Forman JD, Kucuk O, et al. Genistein potentiates the radiation effect on prostate carcinoma cells. Clin Cancer Res 2001, 7:382–390.

40. Hillman GG, Wang Y, Kucuk O, et al. Genistein potentiates inhibition of tumor growth by radiation in a prostate cancer orthotopic model. Mol Cancer Ther 2004, 3:1271–1279.

41. Raffoul JJ, Wang Y, Kucuk O, et al. Genistein inhibits radiation-induced activation of NF–kappaB in prostate cancer cells promoting apoptosis and G2/M cell cycle arrest. BMC Cancer 2006, 6:107.

42. Raffoul JJ, Banerjee S, Singh-Gupta V, et al. Down-regulation of apurinic/apyrimidinic endonuclease 1/redox factor-1 expression by soy isoflavones enhances prostate cancer radiotherapy in vitro and in vivo. Cancer Res 2007, in press.

43. Wang Y, Raffoul JJ, Che M, et al. Prostate cancer treatment is enhanced by genistein in vitro and in vivo in a syngeneic orthotopic tumor model. Radiat Res 2006, 166:73–80.

44. Raffoul JJ, Banerjee S, Che M, et al. Soy isoflavones enhance radiotherapy in a metastatic prostate cancer model. Int J Cancer 2007, in press.

45. Yashar CM, Spanos WJ, Taylor DD, et al. Potentiation of the radiation effect with genistein in cervical cancer cells. Gynecol Oncol 2005, 99:199–205.

46. Akimoto T, Nonaka T, Ishikawa H, et al. Genistein, a tyrosine kinase inhibitor, enhanced radiosensitivity in human esophageal cancer cell lines in vitro: possible involvement of inhibition of survival signal transduction pathways. Int J Radiat Oncol Biol Phys 2001, 50:195–201.

47. Chuang SE, Yeh PY, Lu YS, et al. Basal levels and patterns of anticancer drug-induced activation of nuclear factor-kappaB (NF-kappaB), and its attenuation by tamoxifen, dexamethasone, and curcumin in carcinoma cells. Biochem Pharmacol 2002, 63:1709–1716.

48. Chen J, Halls SC, Alfaro JF, et al. Potential beneficial metabolic interactions between tamoxifen and isoflavones via cytochrome P450-mediated pathways in female rat liver microsomes. Pharm Res 2004, 21:2095–2104.

Chapter 20
Antisense Oligonucleotides and siRNA as Specific Inhibitors of Gene Expression: Mechanisms of Action and Therapeutic Potential

Yvonne Förster and Bernd Schwenzer

1 Introduction

Several diseases, including cancer, are associated with abnormalities in gene expression. Numerous approaches inhibiting the aberrantly expressed genes based on the application of specific nucleic acids have been developed over the years. Today, antisense technology is broadly used as a powerful tool not only for therapeutic purposes, but also for functional genomics and target validation. Three types of sequence-specific inhibition of gene expression can be distinguished: (1) single-stranded antisense oligonucleotides (AS-ODN); (2) RNA cleavage by catalytically active nucleic acids called ribozyme or DNA enzyme; and (3) RNA interference induced by small interfering RNA molecules (siRNA).

The significant advantages of such DNA/RNA–based drugs compared with conventional low-molecular-weight pharmaceuticals are their selective recognition of molecular targets and pathways imparting tremendous specificity of action and consequentially a minimized potential for negative side effects. Most of the conventional drugs bind to protein and modulate their function. In contrast, AS-ODN or siRNA prevent translation through interacting with the target mRNA. However, problems remain, such as stability in vivo, effective cellular uptake, and toxicity; and

despite the theory, nonspecific effects hamper the use of antisense agents in many cases and need to be solved. This chapter discusses current and new developments in the field of AS-ODN and siRNA and examines their therapeutic potential.

2 Mechanisms of Action

2.1 Antisense Oligonucleotides

Typically, AS-ODN consist of 15–20 nucleotides, which are complementary to their target mRNA. The mechanisms of interaction between AS-ODN and target mRNA are complex and numerous [1] (Table 20.1). RNase H cleavage and blocking of translation are two major mechanisms that contribute to antisense activity. RNase H is a ubiquitous enzyme that has been identified in organisms as viruses and human cells [2]. It cleaves the mRNA strand of a DNA/RNA duplex [3, 4]. The activity of RNase H was found in the nucleus as well as cytoplasm, but seems to be greater in the nucleus [5]. The precise recognition elements for RNase H are not known. The enzyme does not require a fully complementary hybrid between the target mRNA and the AS-ODN. It has been shown that short duplexes as small as six nucleotides can be recognized by the enzyme [6]. Virtually only

From: *Sensitization of Cancer Cells for Chemo/Immuno/Radio-therapy, 1st Edition*.
Edited by: Benjamin Bonavida © Humana Press, Totowa, NJ

TABLE 20.1. Molecular mechanisms of action.

Occupancy-only mediated mechanism	Binding of AS-ODNs to specific target RNA/DNA sequences may inhibit the interaction of this nucleic acid with proteins or other factors required for metabolism or transfer of genetic information.
Transcriptional arrest	AS-ODNs may bind to DNA and prevent effective binding of factors required for initiation or elongation steps in transcription.
Inhibition of splicing	AS-ODNs bind to sequences of pre-mRNA, which are necessary to excision of introns.
Translational arrest	Binding position of AS-ODNs is within the area of the initiation codon of mRNA.
Disruption of essential RNA structures	Binding of AS-ODNs cause destruction of intramolecular mRNA structures with crucial roles for their stability, distribution, transport, metabolism, and recognition.
Destabilization of mRNA by hindrance of 5′ cap function or polyadenylation	AS-ODNs bind near the cap site or in the 3′ terminal region of pre-mRNA.
Induction of RNase H activity	RNase H degrades RNA in the DNA/RNA duplex

From: Crooke ST. Antisense strategies. Curr Mol Med 2004, 4:465–487.

phosphodiester and phosphorothioate oligonucleotides can activate RNase H. Alterations in sugar or backbone influence the ability of recognition. Uncharged AS-ODNs especially are not recognized by the enzyme. In order to overcome these problems, chimeric AS-ODNs were designed with greater affinity to the target mRNA and enhanced specificity [7, 8]. AS-ODN, which cannot induce RNase H cleavage, can be used to inhibit translation by steric blockade of the ribosome [9, 10]. Generally, a very high affinity of the AS-ODNs for their target mRNA is required to mediate a successful translational arrest.

2.2 siRNA

RNA interference (RNAi) is described as a biological protective reaction found in fungi, plants, and animals against transposons and repetitive elements, antiviral response, and developmental regulation [11]. RNAi is triggered by siRNA, small 21–23 nucleotides double-stranded RNA molecules with an overhang of two nucleotides at the 3′ end, which can be generated in three ways:

1. Long double-stranded RNA molecules are processed into siRNA by the enzyme Dicer.
2. Chemically or in vitro transcribed siRNA duplexes are transfected into cells.
3. siRNA molecules can be generated in vivo from plasmids or retroviral vectors [12].

As mentioned, the generation of siRNA is achieved by the action of Dicer, a RNase III enzyme [13, 14]. In mammalian systems, the introduction of long double-stranded RNA (>30 bp) results in

the activation of the interferon response, leading to global shutoff protein synthesis and nonspecific mRNA degradation [14, 15]. This obstacle could be overcome by the exogenous delivery of siRNA or expression from vectors. The siRNA is then incorporated into a multiprotein complex called RNA-induced silencing complex (RISC). The RISC-associated helicase activity unwinds the duplexes and enables the antisense strand within RISC to recognize the target mRNA through base pairing [16–18] (Fig. 20.1). Cleavage of the target mRNA occurs in the middle of the complementary region, 10 nucleotides upstream of the nucleotide paired with the 5′ end of the guide siRNA [19]. The cleavage product is released and degraded because of either lack of stabilization cap or poly(A) tail [14]. The RISC complex can now further reduce the available pool of target mRNA [20].

3 Chemical Modification and Pharmacologic Properties

3.1 Antisense Oligonucleotides

The nucleases of unmodified AS-ODNs were rapidly attacked in biological fluids. Many chemical modifications were developed to enhance nuclease resistance, increase affinity and potency, prolong tissue half-life, and reduce non-sequence–specific toxicity. In general, three types of modifications can be distinguished:

1. Altered phosphate backbones
2. Analogs with unnatural bases
3. Modified sugars [21].

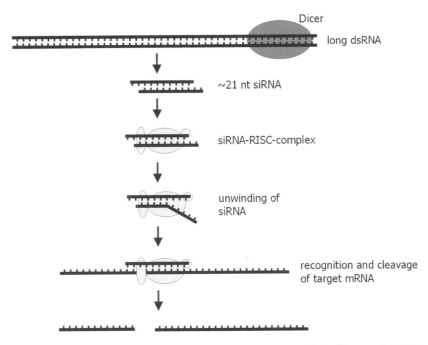

FIGURE 20.1. Mechanism of siRNA. Long dsRNA is cleaved by Dicer to small double-stranded RNA, incorporated into RISC. After unwinding of the siRNA, recognition and cleavage of the target take place. The cleavage products are released and loaded RISC complex can cleave a further target mRNA molecule. (Modified from Zhou D, He QS, Wang C, et al. RNA interference and potential applications. Curr Top Med Chem 2006, 6:901–911.)

3.1.1   First-Generation AS-ODNs

Phosphorothioate-modified AS-ODN (PS-ODN) is the major representative of first-generation AS-ODNs. In this class a non-bridging oxygen in the phosphodiester bond is replaced by sulfur (Fig. 20.2) [22]. Phosphorothioate modification was primarily intended to enhance nuclease resistance, leading to higher bioavailability of oligonucleotides. The half-life of PS-ODN in serum is 9–10 hours, compared with 1 hour for unmodified oligonucleotides [23–25]. In addition, PS-ODNs promote RNase H-mediated cleavage of target mRNA. However, the modification may slightly reduce the affinity to the target mRNA because the melting temperature of the PS-ODN/mRNA duplex is decreased by approximately 0.5°C per modified nucleotide [26]. At least it has been shown that PS-ODN has an affinity for nonspecific binding to proteins such as vascular endothelial growth factor (VEGF) and serum albumin [27–29].

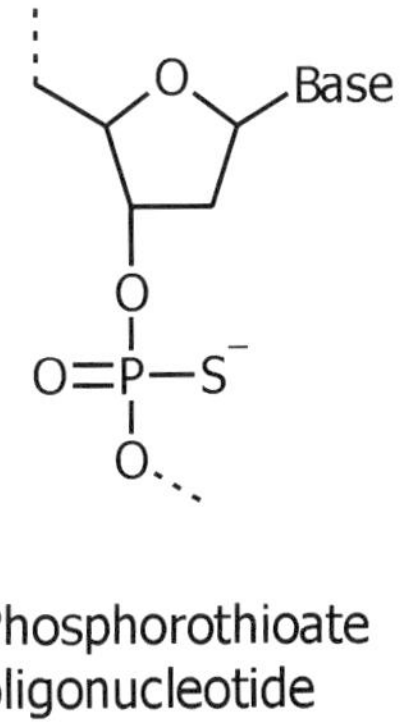

Although it is likely that the nonspecific binding of nucleotides leads to toxicity, it is this binding that probably leads to reduced oligonucleotide

Phosphorothioate oligonucleotide

FIGURE 20.2. First-generation AS-ODN

clearance through the kidney and its relatively prolonged plasma half-life [30]. Major sites of accumulation are the liver, kidney, spleen, and lung, whereas only a minor distribution is found in brain [31–33]. PS-ODNs bind serum proteins in a sequence-independent manner. Following intravenous administration rapid clearance from plasma with plasma half-life of 30–60 minutes is observed [34]. Clearance of PS-ODNs within the tissue is due primarily to exonuclease metabolism, and metabolites are eliminated through the urine [31, 33].

3.1.2 Second-Generation AS-ODNs

Second-generation AS-ODNs were developed to enhance nuclease resistance and increase binding affinity for target mRNA. AS-ODNs of this class contain nucleotides with alkyl modifications at the 2′ position of the ribose, including 2′-O-methyl and 2′-O-methoxyethyl RNA (Fig. 20.3). They are less toxic than PS-ODN and have a slightly enhanced affinity to their target mRNA [25, 35]. However, 2′-O-alkyl RNA cannot induce RNase H cleavage of the target, so the antisense effect can only be due to steric blockade. Mechanistic studies of the RNase H reaction revealed that availability of the 2′-OH group is required for RNase H cleavage [36]. To circumvent this shortcoming chimeric strategies or gapmer technology has been developed. Chimeric AS-ODN analogs bring together the beneficial properties of two types of chemistry. In general, they consist of two segments: a central "gap" region consisting of PS-ODN to activate RNase H cleavage, flanked by a 2′-O-alkyl–modified RNA-enhancing affinity to target mRNA [37].

Gapmers are substantially more stable than PS-ODNs. In mice, rats, and monkeys, the elimination half-life is nearly 30 days in plasma and several tissues [2]. Comparison of a gapmer consisting of PS-ODN that was modified with 2′-MOE at the 3′- and 5′-end and a phosphodiester fully modified with 2′-MOE showed that plasma protein binding was lower for the phosphodiester nucleotide, resulting in higher urinary elimination [38]. In summary, less data are known about pharmacology in comparison to PS-ODNs, because there is only a 5- to 15-fold increase in potency both in vitro and in vivo [2].

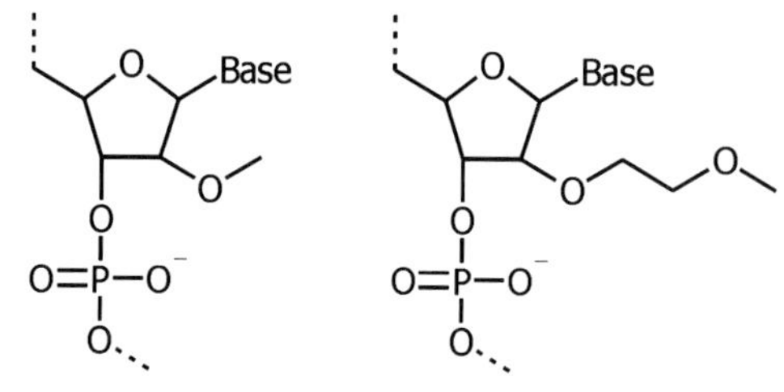

2′-O-methyl RNA   2′-O-methoxyethyl RNA

FIGURE 20.3. Second-generation AS-ODN

3.1.3 Third-Generation AS-ODNs

Most third-generation AS-ODNs contain deoxyribose ring modification. They have enhanced target affinity, nuclease resistance, bioavailability, and pharmacokinetics. Peptide nucleic acid (PNA), locked nucleic acid (LNA), and phosphoramidite morpholino oligomers (PMO) are the three most studied members of this class [2, 39]. Further AS-ODNs are tricyclo-DNA (tr-DNA), cyclohexene nucleic acid (CeNA), N3′-P5′-phosphoroamidates (NPs), and 2′-deoxy-2′-fluoro-β-D-arabino nucleic acid (FANA) (Fig. 20.4) [12]. PNA is a synthetic DNA mimic in which the deoxyribose phosphate backbone is completely replaced with a polyamide linkage. The nucleobases are attached to the backbone via methylene carbonyl linkage [40, 41]. Peptide nucleic acids are non-charged molecules with higher affinity and specificity than unmodified DNA/RNA or DNA/DNA duplexes, but do not elicit target mRNA by RNase H cleavage. Locked nucleic acid is a ribonucleotide containing a methylene bridge that connects the 2′-oxygen of the ribose with the 4′-carbon [42–44]. The affinity toward the target mRNA is greatly enhanced with an increase in the thermal stability of the duplex [45]. However, it is not a substrate for RNase H. Last but not least, PMO are non-charged DNA mimics, in which ribose is replaced by a morpholino moiety and the phosphodiester bond is replaced by phosphoroamidite linkage [46]. They do not support RNase H activity. PMO possess excellent resistance to nucleases and proteases in biological fluids [47–50]. The pattern of distribution of PMOs is independent of sequence and length of PMO, as well as the route of administration. After a single dose injection, accumulation occurs in kidney and liver followed by spleen, lung, and heart. PMO has been shown to be eliminated mainly in the urine [51]. They demonstrate a sufficient biological stability independent in route of administration [51, 52].

3.2 siRNA

As described, a number of modifications are known to stabilize oligonucleotides against degradation and improve pharmacokinetics. Therefore it may also be possible to enhance stability of siRNA through chemical modifications. Extensive studies were made to examine the effects of 2′-modifications

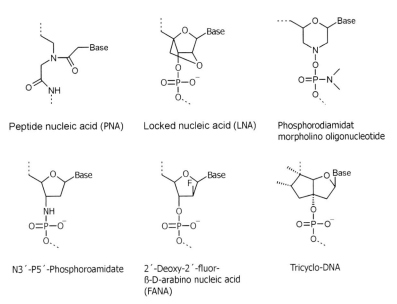

FIGURE 20.4. Third-generation AS-ODN

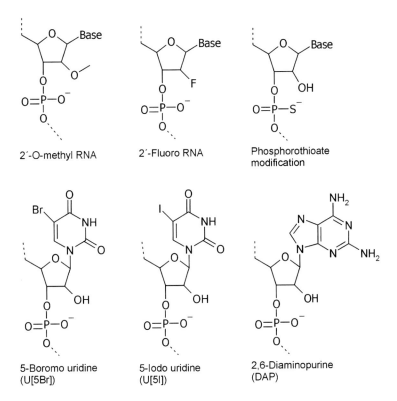

FIGURE 20.5. Modifications of siRNA

[53–59]. Figure 20.5 summarizes structures and nomenclature of modifications. 2′-Fluoro substitution has been demonstrated to maintain siRNA activity regardless of position of modification in the antisense strand [55]. For 2′-OMe modifications differences have to be made. An extensive modification with 2′-OMe abolishes the siRNA activity, while alternate or partial modification of the 3′-terminus increases the duration of siRNA efficiency [54, 59, 60]. Using phosphorothioate modification increases half-life and prolongs efficacy in vivo. However, extensive substitution results in cellular toxicity and nonspecific effects [53, 54, 56]. Studies of the effect of modifications confirmed that 2′-OH of siRNA is not required for RISC activity as long as they do not distort the A-form of the helix. They further demonstrated that the integrity of the 5′-half of the duplex, as defined by the antisense strand, is functionally more important than that of the 3′-half [53, 61].

4   AS-ODN and siRNA Selection

For antisense inhibition, specific Watson-Crick base pairing between an oligonucleotide and the target mRNA is essential. Unfortunately, the successful use of AS-ODNs is somewhat limited since only a small portion of all possible AS-ODNs against a target mRNA show efficient gene knockdown [62–64]. Currently, it seems to be accepted that efficiency of AS-ODN is among other things dependent on local structure of target mRNA and chemical properties of AS-ODN. In order to increase the "hit rate" Chan et al. (2006) considered four parameters [65]:

1. Prediction of the secondary structure of the target mRNA
2. Identification of accessible target sites
3. Motif searching and calculation of GC content
4. Prediction of binding energy ($\Delta G^0(37°C)$)

Secondary structure can be determined by computational analyses. A major disadvantage of such theoretical approaches is the lack of reliability of RNA structure prediction [66]. A widely used program is mfold, which predicts optimal and suboptimal structures of the target mRNA on the basis of minimal free energy [67]. Accessible sites for AS-ODNs are usually located at the terminal ends, internal loops, hairpins, and bulges of 10 or more single-stranded nucleotides [68]. However, results from the literature indicate that optimal local structure is not fully understood. Lima et al. and Thierry et al. found that single-stranded hairpin loops in mRNA were the best target sites, whereas Laptev et al. (1994) suggested that sequences predicted to form clustered double-stranded structures in mRNA had great potential [69–71]. Additionally, a shift of the sequence of a successful AS-ODN by

one nucleotide along the target mRNA can fully abolish the effect [68]. In contrast to Forster et al. (2004), Patzel et al. (1999) favor the idea that hybridization of the 3′-end of an AS-ODN in the single-stranded region of the target mRNA is more effective than the 5′-end [68, 72]. Furthermore, AS-ODNs should be preferred, which contain sequence motifs such as CCAC, TCCC, ACTC, GCCA, and CTCT. These motifs showed a positive correlation between presence and knockdown of target mRNA. On the other site, sequence motifs such as GGGG, ACTG, AAA, and TAA should be avoid as analyzed from more than 1,000 experiments [73]. Moreover, efficient inhibition of gene expression by AS-ODN is dependent on GC content of the AS-ODN. High GC content results in efficient gene knockdown [74, 75]. To calculate duplex stability between AS-ODN and the target mRNA the program "Oligo walk" can be used [76]. AS-ODNs that form stable duplexes with target mRNA and have small self-interactions are suggested to be more effective than AS-ODNs, which form less stable duplexes or have more self-interactions [75]. Figure 20.6 shows a flow chart that summarizes the steps of theoretical design of effective AS-ODNs that may be improved or executed by computational analysis. Nevertheless, accessible sites in the target mRNA can be found by other methods. "Walk the gene" or "sequence walking" use several AS-ODNs (up to 100) targeted to various regions in the target mRNA in order to find the most potent sequence [64, 77]. However, this method is expensive and time consuming and does not show any accessible target sites. With this approach only 2–5% of the randomly chosen AS-ODNs showed efficient gene knockdown [78]. RNase H mapping also promises selection of accessible target sites. A random library of AS-ODNs is allowed to hybridize with target mRNA and subsequently cleaved by RNase H. The resulting fragments are analyzed by sequencing to identify potential target sites [79–81]. Oligonucleotide scanning arrays are used for screening of accessible sites in a rapid way. The results show a good correlation between binding strength and efficacy of AS-ODN in different systems [82, 83].

Similarly, siRNAs show a spectrum of efficiency. Therefore small shifts of the sequence of a successful siRNA were sufficient to alter gene knockdown [58, 84, 85], and different strategies to identify accessible sites for siRNA have been developed. On one hand, several databases containing

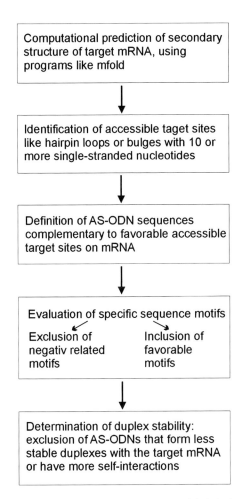

FIGURE 20.6. Flow chart of a computer-aided design of AS-ODNs. (Modified from Zhou D, He QS, Wang C, et al. RNA interference and potential applications. Curr Top Med Chem 2006, 6:901–911.)

experimentally tested siRNAs were known from the literature [86, 87]. Furthermore, prevalidated siRNAs are provided by different vendors, but the user still has to examine their efficiency [88]. On the other hand, experimental strategies to select efficient siRNAs have been designed. Several researchers have developed algorithms that rely on stability features and intrinsic sequence of the siRNA [85, 89–94]. As analyzed in different studies, most effective siRNAs have a GC content of 30–52% [85, 93]. A GC content that is too low may destabilize the siRNA duplex, whereas a high GC content may hinder RISC loading. Low internal stability is preferred at positions 9–14 of the antisense strand. It promotes cleavage of the target mRNA mediated by the loaded RISC complex [93, 95]. Low stability of the 5′ end of the antisense strand promotes its incorporation into RISC [96]. Additionally, several analyses of siRNAs revealed base preferences at specific positions in the sense strand (Fig. 20.7) [85, 89, 91, 92, 94, 97]. Besides, secondary structure prediction and accessibility of the target mRNA are thought to be important in the development of efficient siRNA [98–102]. The efficiency of siRNA was about tenfold higher for the group of siRNA, where target sites were defined as accessible by computational approach [102]. Schubert et al. analyzed the necessity of accessibility in a systematically study. The target sites of siRNAs were included in well-defined structural elements rendering them either highly accessible or completely involved in base-pairing. They were able to show that the accessibility of the target site is important for siRNA efficiency [101].

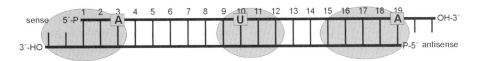

Biophysical, thermodynamic and structural considerations

- low to medium GC content (32-50%)
- low internal stability at 5′ antisense strand
- high internal stability at 5′ sense strand
- no internal repeats and palindroms
- A-form helix between siRNA and target mRNA

Base preferences at specific positions in the sense strand

- presence of an A at positions 3 and 19 in the sense strand
- no G or C at positions 19 in the sense strand
- presence of a U at positions 10 in the sense strand
- no G at position 13 in the sense strand

FIGURE 20.7. Design of effective siRNA. The small double-stranded RNA contains a 19 nt duplexed region and 2 nt 3′ overhangs. The position of the 19nt duplexed region of the sense strand are shown. Furthermore, mechanism-based rules are shown that when applied to siRNA design are expected to show maximum suppression of target mRNA. (Modified from Mittal V. Improving the efficiency of RNA interference in mammals. Nat Rev Genet 2004, 5:355–365.)

5   Toxicity of AS-ODN and siRNA

Toxicity from AS-ODNs can be induced by antisense effects (resulting from hybridization of the AS-ODN with its target mRNA) or non-antisense effects (effects independent on hybridization). Antisense effects may result when reduction of gene expression by AS-ODNs results in undesirable biological effects. Theoretically, hybridization of AS-ODN with other mRNAs than the indented target mRNA might also induce toxicity. However, this type should be extremely rare, because of the specificity of AS-ODN and data base alignment [31].

Non-antisense effects of AS-ODNs include among other things interactions with proteins. These toxicities would be generally less dependent on sequence but more dependent on the chemical class of the AS-ODN. The most common acute toxicities associated with AS-ODN treatment are activation of the complement cascade and inhibition of the clotting cascade [103]. Both toxic effects are thought to be related to the negative charge of the backbone and are largely due to the nonspecific interactions with proteins at high plasma concentrations [31, 104]. Another sequence-dependent toxic effect of AS-ODNs is induction of the immunologic response. The CpG motif flanked by two purines at the 5′ and two pyrimidines at the 3′ has shown to induce B-cell proliferation and the release of cytokines [105–107].

Although RNAi strategies are reliant on a high degree of specificity, interferon genes can be activated, particularly at high concentrations of siRNA [108–111]. Binding to Toll-like receptors could result in toxicity from the ensuring inflammatory process, which activates interferons and cytokines [108, 109, 111]. Chemical modifications of the sugar backbone may solve this problem without decreasing efficiency [112].

6   Therapeutic Potential of AS-ODNs and siRNAs

6.1   Brain Cancer

The prevalence of central nervous system neoplasms is 1.3 per 10,000 in Western countries. Approximately 60% of them are gliomas. Half are glioblastomas, which is the most invasive and aggressive form of these tumors [113–115]. The median survival is about 12 months [116]. Glioblastoma diffusely infiltrate brain tissue and cannot be cured by surgical resection [117]. Normal tissue can only tolerate 60 Gy of radiation, but this is below the threshold required to kill malignant tumor cells. Additionally, blood–brain barrier hinders the transport of chemotherapeutics into the central nervous system [118]. Therefore, new therapies for treatment of glioblastoma and other cancers of the brain are needed.

6.1.1   Protein Kinase c-Alpha

Protein kinase c-alpha (PKC-α) is a phospholipid-dependent cytoplasmic serine threonine kinase that triggers many cellular responses, including gene expression, cellular differentiation, and proliferation. In in vitro experiments with glia cells it plays a critical role in malignant transformation, proliferation, and invasiveness (119–121). Aprinocarsen (Isis 3521, LY900003) is a PS-ODN targeted against the 3′ untranslated region of PKC-α mRNA. In a phase II study no complete or partial responses were seen in any patients; but the observed toxicities associated with aprinocarsen were mild, reversible, and manageable [122, 123]. However, aprinocarsen is applied not only for glioblastoma, but also for other malignancies, for example, ovarian cancer and non-small cell lung cancer [124, 125].

6.1.2   Epidermal Growth Factor Receptor

Epidermal growth factor receptor (EGFR) plays an angiogenic role in 90% of primary brain cancer and 70% of solid cancer outside the brain [126, 127]. The gene is the most overexpressed oncogene in glioblastoma. It has been shown that the interruption of interaction between EGFR and its ligands may induce cell death [128]. For this reason EGFR is an excellent target for antisense or siRNA therapy. Inhibition of EGFR in U87 glioblastoma cells with AS-ODNs results in reduced tumor growth [129]. In an U251 subcutaneous mice model, tumor volumes were significantly smaller when treated with EGFR antisense or siRNA constructs [130]. Treatment of mice with plasmid containing EGFR in antisense orientation resulted in an increased life span [131]. Intravenous injection of a plasmid encoding a short hairpin RNA directed against mRNA led to a prolonged survival time of mice with advanced intracranial brain cancer [132].

6.1.3 Transforming Growth Factor-β

Transforming growth factor-betas (TGF-β) are polypeptides which play various roles in cell function including cell growth and differentiation, proliferation, migration and they are key regulators of the immune system (133). TGF-βs were expressed in three mammalian isoforms: TGF-β1, TGF-β2, and TGF-β3 [134, 135]. Using cDNA of TGF-β in antisense orientation in an intracranial rat C6 glioma model resulted in increased survival of the rats [136]. AP 12009 is a 18mer PS-ODN targeted to TGF-β2 mRNA. It has been developed for the treatment of malignant glioma, pancreatic cancer, and other TGF-β2 overexpressing tumors. It is being tested in a clinical phase IIb study for local treatment of high-grade glioma. In April 2005 another international, open-label, randomized parallel-group phase IIb study started with 145 patients [137]. At the time of preparing this manuscript patients were either on treatment or follow-up.

6.2 Prostate Cancer

Prostate cancer (PCa) is one of the most frequently occurring tumors in men over 60 years of age in most industrial countries. Currently, metastatic PCa is incurable and ultimately claims the life of patients [138–141]. The development of a progressive hormone refractory disease occurs within a few years, in most cases worsening survival; therefore, therapy strategies against advanced PCa have to be developed [142, 143].

6.2.1 Bcl-2

Bcl-2 belongs to a family of genes that are important in the regulation of apoptosis. They function as either death agonists like bax, or cell death antagonists like bcl-2 [144]. Bcl-2 is overexpressed in a variety of human cancers [145–147]. In PCa bcl-2 has been found to be associated with tumor aggressiveness [148, 149]. Moreover, the increased expression of bcl-2 is associated with the development of androgen independence, treatment resistance, and poor prognosis [150–152]. Oblimersen

(Genasense, G3139, Genta) is an 18mer PS-ODN that hybridizes to the first six codons of the bcl-2 mRNA. In a clinical phase I study with oblimersen alone, no major antitumor response was observed [153]. Clinical trials oblimersen in combination with chemotherapy in patients with PCa have shown interesting response rates. In a clinical study phase II antitumor activity in 52% of the patients was detected after treatment with oblimersen and docetaxel [154]. Inhibition of bcl-2 enhanced paclitaxel cytotoxicity in an LNCaP tumor model [155]. Further, an in vitro study showed that decreasing of bcl-2 expression with oblimersen sensitizes several PCa cells to radiation [156].

6.2.2 Clusterin

In cancer, clusterin has been defined as an antiapoptotic protein that is activated after therapeutic stress [157–159]. It functions as a cytoprotective chaperone and stabilizes conformations of proteins in time of cell stress. In human PCa increased expression of clusterin is associated with high Gleason scores [160]. In a PCa xenograft model, AS-ODN against clusterin mRNA enhanced tumor regression in combination with chemotherapeutic agents [161, 162]. Furthermore, administration of clusterin AS-ODNs combined with radiation inhibited tumor growth in vivo synergistically [159]. OGX-011 is a MOE-gapmer complementary to the clusterin mRNA translation initiation site. It strongly inhibits the expression of clusterin in vitro and in vivo [163]. A clinical phase I pharmacokinetic and pharmacodynamic study showed that OGX-011 is well tolerated and reduces clusterin expression in primary prostate tumors [164]. Additionally, clinical phase II studies of OGX-011 in combination with chemotherapy and/or hormonal therapy are planned to start in patients with prostate, breast, and lung cancer [165].

6.2.3 Urokinase Plasminogen Activator Receptor

Urokinase plasminogen activator (uPA) and its receptor uPAR are suggested to play a key role in tissue degradation, cell migration, angiogenesis, cancer invasion, and metastasis [166]. uPA-uPAR interaction activates a number of pathways. In many cells uPA-uPAR signaling activates the extracellular-signal related kinase 1/2 pathway, which controls proliferation and cancer invasion [167, 168]. Activation of Jak/Stat signaling cascade is involved in proliferation, survival, and apoptosis [169, 170]. Inhibition of uPAR expression by AS-ODNs blocked proliferation in PC3 cells in vitro. In vivo downregulation of uPAR produced a prophylactic effect against prostate cancer bone metastasis [171]. Treatment of PC3 cells with shRNA complementary to uPAR mRNA results in decreased mRNA and protein. Moreover, reduced cell invasion was shown after uPA-uPAR inhibition in a Matrigel invasion assay [172].

6.2.4 PKC-α

Increased expression of PKC-α in prostate cancer results in enhanced tumor growth and invasiveness [173]. It is suggested to be a marker of malignant transformation, because PKC-α can be found in early prostate cancer but not in adjacent benign prostate epithelium [174]. Affinitak (Isis 3521) was tested in a clinical phase II study to determine antitumor activity in 15 patients after continuous i.v. infusion for 21 days. Two patients show stable disease and three had no PSA increases over 25% for more than 120 days. There was no objective response. However, PKC-α was not assessed before or at the end of therapy [175].

6.2.5 c-raf or raf-1

c-raf or raf-1 is a member of the raf family that is a key signal transducer of several extracellular stimuli resulting in activation of MAPK signaling pathways. Raf kinase play a critical role in proliferation, differentiation, and apoptosis [176]. Isis 5132 is a 20mer PS-ODN complementary to 3′ untranslated region of c-raf mRNA [177]. In hormone refractory prostate cancer 16 patients were treated with Isis 5132. There was no objective response or PSA response, but three patients showed prolonged stable disease up to 7 months [175].

6.2.6 Heat Shock Protein 27

Hsp27 is an effective protein to protect cells against various cytotoxic conditions, including oxidative stress and cytotoxic chemotherapy [178]. In prostate cancer it is thought to play a critical role in progression to hormone refractory disease [179, 180]. Furthermore, hsp27 is among the most consistently overexpressed genes in hormone refractory prostate cancer xenografts [181]. Rocchi et al. showed that inhibition of hsp27 by AS-ODNs or siRNAs decreased cell growth and enhanced apoptosis in PC3 cells [179]. Both AS-ODNs and siRNAs also enhance paclitaxel chemosensitivity in vitro. In mice AS-ODN administration results in decreased PC3 tumor progression. OGX-427 is a 2′-methoxyethyl modified AS-ODN that inhibited hsp27 expression and enhanced drug efficacy in xenograft model [182]. A clinical phase study I study will start in 2007. Further, siRNA induced gene knockdown significantly induces apoptosis in prostate cancer cell lines even in hormone refractory prostate cancer. In vivo evaluation will show the therapeutic potential [178].

6.3 Bladder Cancer

Bladder cancer is the second most common malignancy in urology, with a high recurrence rate [183]. After transurethral resection (TUR) with or without intravesical chemotherapy, 50–70% of superficial bladder cancer will recur [184]. Moreover, 50% of patients with carcinoma in situ develop invasive bladder cancer within 5 years [185]. However, bladder cancer is a good model for testing AS-ODNs or siRNAs because administration can made locally and diagnosis can occur through urine cytology, biopsy, and cytoscopy.

6.3.1 Clusterin

Clusterin is not only expressed in prostate cancer but also in bladder cancer mainly after TUR and cystectomy [186]. In bladder cancer overexpression of clusterin is associated with recurrence and progression [187]. Different groups could show that inhibition of clusterin expression by AS-ODNs chemosensitizes different bladder cancer cell lines in vitro [186, 188, 189]. In an orthotopic bladder cancer model treatment with AS-ODN reduce IC50 of gemcitabine by 90% compared with chemotherapeutic agent alone [190]. These data indicate that clusterin is a suitable target to enhance chemosensitivity in bladder cancer.

6.3.2 Hsp27

Expression of hsp27 is associated with multidrug resistance and direct inhibition of apoptosis [191–193]. In bladder cancer low hsp27 expression is suggested to be a marker that the tumor is sensitive for radiotherapy [194]. So et al. showed that AS-ODNs were able to reduce protein and mRNA in different bladder cancer cell lines. In vivo studies using xenograft models showed chemosensitizing effects resulting from inhibition of hsp27 [186]. Bladder cancer was part of the clinical phase I study using OGX-427 in 2007.

6.3.3 Vascular Endothelial Growth Factor

Among many triggers of tumor angiogenesis, vascular endothelial growth factor (VEGF) is thought to be one of the major angiogenic factors [195–197]. VEGF-mediated enhanced tumor growth was not only observed in bladder cancer, but also in breast and pancreatic carcinoma [198–200]. Our in vitro data showed that treatment of EJ28 bladder cancer cell line by PS-ODN reduced protein by more than 80% [62]. Pre-treatment of EJ28 with AS-ODNs further enhances cytotoxicity of mitomycin C, gemcitabine, and cisplatin [201].

6.4 Lung Cancer

Lung cancer causes about 30% of all cancer-related deaths. Non-small cell lung cancer (NSCLC) amounts to 85% of all new cases of lung cancer. Patients, having an advanced stage diagnosis, have a poor prognosis. They were treated with platinum-based chemotherapy with limited efficacy [202]. So, antisense technology is thought to be a promising strategy in treatment not only in lung cancer.

6.4.1 PKC-α

PKC-α is not only overexpressed in glioma but also in lung cancer. Inhibition of PKC-α by aprinocarsen (Affinitak, Isis 3521), a 20mer AS-ODN, results in sequence-specific reduction of protein and mRNA in NSCLC cell line A549 [203]. A clinical phase I study demonstrated stable disease as the best response in NSCLC [125]. However, clinical phase I/II study with aprinocarsen, carboplatin, and paclitaxel showed a 42% response rate,

suggesting potentiation of chemotherapy activity [204]. Therefore, clinical phase III study was made to determine whether aprinocarsen improves survival when added to chemotherapy in NSCLC. In this study aprinocarsen did not enhance the overall survival compared with chemotherapy alone [202].

6.4.2 Bcl-2

Overexpression of bcl-2 is seen in small cell lung cancer as well as in NSCLC. However, it is not suitable as prognostic marker, because the impact of bcl-2 expression on outcome has not been shown as significant as that of p53 [205]. In an ectopic and orthotopic murine xenograft model, inhibition of tumor growth and prolongation of survival by oblimersen was demonstrated. In the ectopic model bcl-2, suppression was measured in isolated tumor tissue. Additionally, when oblimersen was combined with the chemotherapeutic vinorelbine 33% of mice survived >90 days (206). A clinical phase study I showed that oblimersen, carboplatin, and etoposide are well tolerated. Treatment of patients with small cell lung cancer results in an encouraging response rate and time to progression [207].

6.5 Breast Cancer

Breast cancer is the most common malignancy in women in western countries. Almost all breast cancers were found to accompany gene mutations and/or overexpression. Countless studies demonstrated that a relationship exists between overexpression and pathogenesis, development, and prognosis of breast cancer. Chemotherapy has been used in clinical practice with limited success because of the development of chemoresistance; therefore, new therapeutics strategies have to be developed.

6.5.1 Her-2 (c-erbB-2, Neu)

Her-2 is a member of a transmembrane receptor tyrosine kinase family, in which her-2 was found to be the most relevant in cancer [208]. It is overexpressed in about 30% of cases resulting in worse prognosis [209, 210]. An increased expression of her-2 is suggested to be associated with increased chemoresistance [211]. In early 1990, AS-ODNs directed to translation initiation site and the 3' untranslated region of her-2 mRNA were positively tested in breast cancer cell lines [212]. In cell lines with high her-2 expression, inhibition of her-2 with AS-ODNs chemosensitizes the cells. In contrast, in breast cancer cell lines with low her-2 expression no synergistic effect could be detected. The same report also demonstrated a significant inhibition of tumor growth in nude mice xenografts in combination with chemotherapy [213]. Systemically delivered AS-ODN that was complementary to translation initiation site, inhibited in combination with docetaxel tumor growth in an aggressive breast cancer model that does not overexpress her-2 [214]. The successful inhibition of her-2 expression was also demonstrated in mice using a modified siRNA [215].

6.5.2 Bcl-2

Bcl-2 expression is significantly higher in primary breast cancer. It might play a pivotal role in tumor initiation, progression, and resistance to chemotherapy and radiotherapy [216, 217]. The effect of AS-ODN targeted to bcl-2 mRNA was evaluated in four human breast cancer cell lines in vitro and in vivo. It was shown that bcl-2 inhibition enhanced the chemosensitivity to taxanes and doxorubicin. Additionally, bcl-2 AS-ODNs augment anticancer drug–induced signal transduction pathways leading to apoptosis [218]. In another study chemosensitivity to tamoxifen of several breast cancer cell lines was enhanced, when treated with bcl-2 PS-ODN [219].

6.5.3 Integrins

Integrins are a group of cell adhesion proteins, which consist of two noncovalently bound transmembrane subunits (alpha and beta) [220]. They are not only implicated in cell–cell and cell–extracellular matrix interaction, but have been shown to play a critical role in cell signaling, migration, differentiation, proliferation, and survival [221]. Townsend et al. tested a PS-ODN complementary to alpha$_v$ integrin. The antisense treated cells showed evidence of programmed cell death and reduced alpha$_v$ protein and mRNA [222]. A first study using siRNA demonstrated that siRNA silencing of integrin alpha$_v$ enhanced the radiosensitivity of breast cancer cells [223].

6.5.4 Mouse Double Minute 2

Mouse double minute 2 (MDM2) is induced by p53 and binds to p53 with high affinity, resulting in inhibition of the ability of p53 to act as transcription factor. It indicates that MDM2 functions as a negative regulator of p53 [224–226]. The MDM2 gene is overexpressed in several tumors and has been identified as a prognostic factor in cancer patients. Moreover, MDM2 overexpression is associated with invasive tumors, metastasis, recurrence, and high-grade/late stage tumors [227–230]. Wang et al. selected mixed-backbone AS-ODN complementary to MDM2 mRNA. The 20mer was 2'-OMe modified at the 3' and 5' end, whereas the other nucleotides were phosphorothioate modified. In MCF-7 and MDA-MB-468 cells MDM2 was inhibited, resulting in p53 elevation. In nude mice bearing MCF-7 and MDA-MB-468 xenografts, MDM2 antisense dose-dependently decreased tumor growth. When AS-ODN treatment was combined with chemotherapeutics such as 5-fluorouracil or paclitaxel additive or synergistic effects were observed [231]. The same AS-ODN was used to evaluate the effect on radiation therapy. In several breast cancer cell lines radiation-induced antiproliferation effect were significantly increased after MDM2 AS-ODN treatment. Moreover, in SCID or nude mice xenograft radiation–induced effects on tumor growth were also increased [232].

6.5.5 CXCR4

Usually breast cancer cells metastasize to the regional lymph nodes, bone marrow, lungs, and liver in an organ-selective process. The chemokine receptor, CXCR4, is one of the critical factors for breast cancer metastasis through interaction with its ligand [233]. Inhibition of CXCR4 mRNA with siRNA impaired the invasion of MDA-MB-231 cells in a Matrigel invasion assay. Furthermore, lung metastases of MDA-MB-231 cells were blocked in vivo [234].

6.5.6 Type I Insulin-Like Growth Factor

Type I insulin-like growth factor (IGF-IR) is thought to play a key role in breast cancer development. Additionally, it interacts with steroid hormones that regulate breast tumor growth. It is overexpressed in breast tumors and breast cancer cell lines [235]. Decreasing IGF-IR mRNA with PS-ODNs inhibited tumor growth in C4HD tumors in vivo [236]. Because of IFG-IR mRNA inhibition activation of several signaling pathways was blocked in vivo.

6.6 Pancreatic Cancer

Pancreatic cancer is the fourth commonest cause of cancer-related death in men and women [237]. At the time of diagnosis most patients have locally advanced or metastatic disease leading to a poor overall 5-year survival rate of <5% [238]. However, pancreatic cancer results from accumulated genetic alterations and the number of mutations identified, is higher than in other common tumor types [239].

6.6.1   K-ras

K-ras is a member of ras gene family. In >90% of pancreatic cancers the protein is activated because of mutations [240]. The most common mutation is located in codon 12. Treatment of pancreatic cancer cell lines with mutation-specific AS-ODN cell proliferation was reduced and invasiveness was significantly inhibited [241]. Adenovirus-mediated expression of antisense K-ras mRNA induced apoptosis in pancreatic cancer cells in the peritoneal cavity of immune-deficient mice was suppressed after intraperitoneal injection [242]. Targeting a specific K-ras mutation with siRNA induced cell death or reversed the neoplastic cellular phenotype [243]. An adenovirus-delivered siRNA termed internavec against K-ras mutation reduced the growth of subcutaneous H79 pancreatic cancer xenografts in mice. The activity of internavec contributed to cell cycle blockage and increased apoptosis [244].

6.6.2   Hypoxia-Inducible Factor 1α

Hypoxia-inducible factor 1α (hif-1α) is a transcription factor that plays an important role in solid tumor formation in vivo by promoting angiogenesis and anaerobic metabolism [245]. The expression of hif-1α was demonstrated in several tumors, including breast, bladder, colon, glia, renal, prostate, and pancreatic tumors [246]. Hif-1α expression can be induced by low oxygen [247]. Transfection of BxPc-3 cells with antisense hif-1α plasmid reduced hif-1α, surviving and beta1-integrin. In a xenograft mice model, control group tumors grew more rapidly, and volume and weight were obviously higher than in the hif-1α antisense–treated group, but there was no statistical significance [248]. Transient transfection of siRNA expression vectors targeting hif-1α mRNA reduced both mRNA and protein in pancreatic cancer cell lines. In vivo, tumors were barely formed when the siRNA vector was administered [249].

6.6.3   Survivin

Survivin is a member of the inhibitor of apoptosis (IAP) family. It is involved in control of cell division and inhibition of apoptosis [250]. Survivin is expressed during fetal development as well as in the majority of human cancers, but is undetectable in most adult tissues [251]. The expression of survivin is associated with shorter survival and was reported to be a prognostic marker for pancreatic cancer patients [252]. A surviving siRNA containing plasmid was used to inhibit survivin expression in AsPC-1 cells. There was a significant decrease in survivin mRNA and an increase in caspase-3 activity. Furthermore, cells were more sensitive to radiation compared with non-silencing control siRNA [253]. In another study, a siRNA plasmid expression vector was transfected in PC-2 cells. Both, survivin mRNA and protein were reduced significantly. Moreover, radiosensitivity was enhanced [254].

6.6.4   H-ras

H-ras is a further member of the ras gene family. It serves as critical component of cell-signaling pathways by serving as cell surface receptors, including control of proliferation, differentiation, and cell death [255]. In a study, the role of H-ras in tumor growth of two pancreatic cell lines was demonstrated [256]. Isis 2503 is a PS-ODN targeted to the translation initiation site of H-ras mRNA. In vitro studies showed the potential of Isis 2503 as a selective inhibitor of H-ras [257]. A clinical study phase I demonstrated that gemcitabine and Isis 2503 can be given in combination [258]. In a multicenter phase II study 48 patients with pancreatic cancer were enrolled. The combination of gemcitabine and Isis 2503 was well tolerated. However, the benefit for patients using the combination was unclear [259].

6.6.5 *VEGF*

High expression of VEGF in pancreatic cancer is associated with liver metastasis and poor prognosis [260, 261]. In a orthotopic mice model the effect of VEGF AS-ODNs on tumor growth and angiogenesis was tested. As a result, tumor growth and metastasis were reduced, while survival was improved. The authors raise the possibility that VEGF blockage by AS-ODNs may be useful in treatment of pancreatic cancer [262].

7 AS-ODNs and siRNAs in Treatment of Nonmalignant Diseases

7.1 Amyotrophic Lateral Sclerosis

Amyotrophic lateral sclerosis (ALS) is a progressive fatal neurodegenerative disease that attacks nerve cells in the brain and spinal cord. It is associated with mutations in SOD1 gene [263]. One therapeutic approach is to reduce the level of this protein. In an animal model a 20mer PS-ODN gapmer with 2′-O-(2-methoxy)ethyl substitutions of the first and last five nucleotides was tested. Both in brain and spinal cord SOD1 protein and mRNA was reduced after intraventricular administration. Additionally, a significantly slowed disease progression was observed [264].

7.2 Alzheimer's Disease

AChE is responsible for hydrolysis of acetylcholine (ACh) in the brain. Reduced levels of ACh in the brain of Alzheimer's disease (AD) patients leave a relative excess of AChE. Inhibition of AChE is suggested to be the most successful strategy to enhance cholinergic neurotransmission in AD patients [265]. In an AD model in mice AChE mRNA inhibition by a PS-ODN improves the learning and memory function, suggesting the potential for a novel approach to treat cognitive deficits in AD patients [266]. EN101 is an AS-ODN also targeting AChE mRNA. It has been successfully applied in a clinical phase Ib study for the treatment of myasthenia gravis [263] (Ester Neuroscience Ltd., www.esterneuro.com).

7.3 Human Immunodeficiency Virus Type I

Human immunodeficiency virus type I (HIV-I) belongs to the retrovirus family. It primarily infects CD4+ T lymphocytes and monocytes/macrophages. Over time, virus replication leads to a slow and progressive destruction of the immune system [267]. A PS-ODN in treatment of HIV-I infection that entered clinical trials is GEM91, targeted to the translation initiation site of Gag protein [268–270]. GEM91 was well tolerated by patients in a clinical phase I study, but lack of efficiency forced the withdrawal from later clinical trials [271]. For HIV-I no clinical trial of an AS-ODN is currently in progress. siRNAs and shRNAs have been targeted to early and late HIV-encoded RNAs in cell lines and primary hematopoietic cells [272–275]. Nevertheless, clinical application of siRNA targeted HIV-encoded RNA is a challenge because of the high viral mutation rate [276]. Therefore, inhibition of cellular cofactors required for HIV infection by siRNAs are an alternative approach [277]. Therefore Chang et al. used lentiviral insulator vectors carrying siRNA targeting multiple highly conserved HIV-I sequences. Efficient inhibition in acute infection was demonstrated against different strains of HIV-I. Efficient suppression of HIV-I replication was also shown in chronically infected cells and peripheral blood mononuclear cells [278].

7.4 Respiratory Diseases

Respirable AS-ODN therapeutics were introduced in 1997 with the publication of preclinical data on EPI-2010 [279]. Respirable formulations of naked AS-ODNs were taken up quite efficiently by lung tissue targeting adenosine A1 receptor. Asthma is a complex inflammatory disease of the lung. The transcription factor NF-kappaB, which activates various genes for proinflammatory cytokines, plays an important role in regulation of asthma. The influence of inhibition of NF-kappaB p65 subunit was investigated in a mouse model. Two successive i.v. injections of p65 AS-ODN results in a significant inhibition of all asthmatic reactions [280]. The p38 mitogen-activated protein kinase (MAPK) is another critical player in asthma. Aerosolized p38 AS-ODN reduced interleukin-4 (IL-4), IL-5, and IL-13 levels in bronchoalveolar lavage. It also reduced airway mucus hypersecretion, indicating the therapeutic potential for asthma [281]. Respiratory syncytial virus (RSV) and parainfluenza virus (PIV) are two respiratory pathogens of medical significance that exert high mortality. siRNA targeting the P protein, an essential subunit of the viral RNA–dependent RNA polymerase, reduced RSV growth in culture [282]. In mice, this siRNA was applied intranasally and prevented and treated single as well as concurrent infection with RSV and PIV [283].

7.5 Crohn's Disease

Crohn's disease is an inflammatory disease of the intestinal tract, which is characterized by immune dysregulation and leucocyte recruitment into gastrointestinal tissue [284]. Human intercellular adhesion molecule-1 (ICAM-1, CD54) is a member of the immunoglobulin superfamily. It is constitutively expressed at low levels in vascular endothelial cells and leucocytes [285–287]. ICAM-1 is upregulated in response to proinflammatory stimulators [288]. Tissue expression of ICAM-1 correlates with disease activity [289]. Alicaforsen (Isis 2302) is a 20mer PS-ODN complementary to the 3′ untranslated region of ICAM-1 mRNA. Several studies demonstrated the safety and tolerability of alicaforsen [290, 291]. In a double-blind placebo-controlled trial, clinical remission rates correlated with the patient's achieved drug exposure. Subjects with high drug exposure had significantly higher remission rates [292].

7.6 Hepatitis C Virus

Hepatitis C virus (HCV) is a serious human pathogen, which leads to chronic liver disease, including liver cirrhosis and hepatocellular carcinoma [293–295]. Isis 14803 is a PS-ODN targeting HCV internal ribosome entry site. It inhibited replication and protein in cell culture and mouse model [296, 297]. In a clinical phase I study with 24 patients with HCV genotype 1 chronic hepatitis C Isis 14803 significantly reduced the viral load in two patients; and somewhat reduced the viral load in a further nine patients [298]. In a phase I, open-label, dose-escalation trial only three of 28 patients showed HCV reduction [299]. However, further studies are required to evaluate Isis 14803 treatment. siRNA targeting NS5A RNA of HCV reduced NS5A RNA and protein in a HepG2 cell line [300].

8 Conclusion

AS-ODNs and siRNAs have demonstrated to be valid approaches in modulation of gene expression. Several AS-ODNs are in clinical phase I, II, or III and fomivirsen is the first approved AS-ODN on the market. Nevertheless, there are still significant hurdles to overcome, including systemic delivery, enhanced biostability, and low toxicity. Antisense and siRNA technology imply a more rational and possibly less expensive design of new drugs in malignant as well as nonmalignant diseases.

References

1. Crooke ST. Antisense strategies. Curr Mol Med 2004, 4:465–487.
2. Crooke ST. Progress in antisense technology. Annu Rev Med 2004, 55:61–95.
3. Walder RY, Walder JA. Role of RNase H in hybrid-arrested translation by antisense oligonucleotides. Proc Natl Acad Sci U S A 1988, 85:5011–5015.
4. Minshull J, Hunt T. The use of single-stranded DNA and RNase H to promote quantitative 'hybrid arrest of translation' of mRNA/DNA hybrids in reticulocyte lysate cell-free translations. Nucleic Acids Res 1986, 14:6433–6451.
5. Crum C, Johnson JD, Nelson A, et al. Complementary oligodeoxynucleotide mediated inhibition of tobacco mosaic virus RNA translation in vitro. Nucleic Acids Res 1988, 16:4569–4581.
6. Monia BP, Lesnik EA, Gonzalez C, et al. Evaluation of 2′-modified oligonucleotides containing 2′-deoxy gaps as antisense inhibitors of gene expression. J Biol Chem 1993, 268:14514–14522.
7. Condon TP, Bennett CF. Altered mRNA splicing and inhibition of human E-selectin expression by an antisense oligonucleotide in human umbilical vein endothelial cells. J Biol Chem 1996, 271:30398–30403.
8. Giles RV, Tidd DM. Increased specificity for antisense oligodeoxynucleotide targeting of RNA cleavage by RNase H using chimeric methylphosphonodiester/phosphodiester structures. Nucleic Acids Res 1992, 20:763–770.
9. Baker BF, Monia BP. Novel mechanisms for antisense–mediated regulation of gene expression. Biochim Biophys Acta 1999, 1489:3–18.

10. Baker BF, Lot SS, Condon TP, et al. 2′-O-(2-Methoxy)ethyl-modified anti-intercellular adhesion molecule 1 (ICAM-1) oligonucleotides selectively increase the ICAM-1 mRNA level and inhibit formation of the ICAM-1 translation initiation complex in human umbilical vein endothelial cells. J Biol Chem. 1997, 272:11994–12000.

11. Bernstein E, Denli AM, Hannon GJ. The rest is silence. RNA 2001a, 7:1509–1521.

12. Kurreck J. Antisense technologies. Improvement through novel chemical modifications. Eur J Biochem 2003, 270:1628–1644.

13. Bernstein E, Caudy AA, Hammond SM, et al. Role for a bidentate ribonuclease in the initiation step of RNA interference. Nature 2001, 409:363–366.

14. Elbashir SM, Martinez J, Patkaniowska A, et al. Functional anatomy of siRNAs for mediating efficient RNAi in *Drosophila melanogaster* embryo lysate. EMBO J 2001, 20:6877–6888.

15. Yang S, Tutton S, Pierce E, et al. Specific double-stranded RNA interference in undifferentiated mouse embryonic stem cells. Mol Cell Biol 2001, 21: 7807–7816.

16. Hammond SM, Boettcher S, Caudy AA, et al. Argonaute2, a link between genetic and biochemical analyses of RNAi. Science 2001, 293:1146–1150.

17. Liu J, Carmell MA, Rivas FV, et al. Argonaute2 is the catalytic engine of mammalian RNAi. Science 2004, 305:1437–1441.

18. Nykanen A, Haley B, Zamore PD. ATP requirements and small interfering RNA structure in the RNA interference pathway. Cell 2001, 107:309–321.

19. Elbashir SM, Lendeckel W, Tuschl T. RNA interference is mediated by 21- and 22-nucleotide RNAs. Genes Dev 2001, 15:188–200.

20. Meister G, Tuschl T. Mechanisms of gene silencing by double-stranded RNA. Nature 2004, 431:343–349.

21. Crooke ST. Antisense drug technology: principles, strategies and applications. New York: Marcel Dekker, 2001.

22. Eckstein F. Phosphorothioate oligodeoxynucleotides: what is their origin and what is unique about them? Antisense Nucleic Acid Drug Dev 2000, 10:117–121.

23. Campbell JM, Bacon TA, Wickstrom E. Oligodeoxynucleotide phosphorothioate stability in subcellular extracts, culture media, sera and cerebrospinal fluid. J Biochem Biophys Meth 1990, 20:259–267.

24. Phillips MI, Zhang YC. Basic principles of using antisense oligonucleotides in vivo. Methods Enzymol 2000, 313:46–56.

25. Kurreck J, Wyszko E, Gillen C, et al. Design of antisense oligonucleotides stabilized by locked nucleic acids. Nucleic Acids Res 2002, 30:1911–1918.

26. Crooke ST. Progress in antisense technology: the end of the beginning. Meth Enzymol 2000, 313:3–45.

27. Guvakova MA, Yakubov LA, Vlodavsky I, et al. Phosphorothioate oligodeoxynucleotides bind to basic fibroblast growth factor, inhibit its binding to cell surface receptors, and remove it from low affinity binding sites on extracellular matrix. J Biol Chem 1995, 270:2620–2627.

28. Fennewald SM, Rando RF. Inhibition of high affinity basic fibroblast growth factor binding by oligonucleotides. J Biol Chem 1995, 270:21718–21721.

29. Rockwell P, O'Connor WJ, King K, et al. Cell-surface perturbations of the epidermal growth factor and vascular endothelial growth factor receptors by phosphorothioate oligodeoxynucleotides. Proc Natl Acad Sci U S A 1997, 94:6523–6528.

30. Stein CA, Benimetskaya L, Mani S. Antisense strategies for oncogene inactivation. Semin Oncol 2005, 32:563–572.

31. Levin AA. A review of the issues in the pharmacokinetics and toxicology of phosphorothioate antisense oligonucleotides. Biochim Biophys Acta 1999, 1489:69–84.

32. Cotter FE, Waters J, Cunningham D. Human Bcl-2 antisense therapy for lymphomas. Biochim Biophys Acta 1999, 1489:97–106.

33. Raynaud FI, Orr RM, Goddard PM, et al. Pharmacokinetics of G3139, a phosphorothioate oligodeoxynucleotide antisense to bcl-2, after intravenous administration or continuous subcutaneous infusion to mice. J Pharmacol Exp Ther 1997, 281:420–427.

34. Geary RS, Henry SP, Grillone LR. Fomivirsen: clinical pharmacology and potential drug interactions. Clin Pharmacokinet 2002, 41:255–260.

35. Crooke ST, Lemonidis KM, Neilson L, et al. Kinetic characteristics of Escherichia coli RNase H1: cleavage of various antisense oligonucleotide-RNA duplexes. Biochem J 1995, 312:599–608.

36. Zamaratski E, Pradeepkumar PI, Chattopadhyaya J. A critical survey of the structure-function of the antisense oligo/RNA heteroduplex as substrate for RNase H. J Biochem Biophys Meth 2001, 48: 189–208.

37. McKay RA, Miraglia LJ, Cummins LL, et al. Characterization of a potent and specific class of antisense oligonucleotide inhibitor of human protein kinase C-alpha expression. J Biol Chem 1999, 274:1715–1722.

38. Yu RZ, Geary RS, Monteith DK, et al. Tissue disposition of 2′-O-(2-methoxy) ethyl modified antisense oligonucleotides in monkeys. J Pharm Sci 2004, 93:48–59.

39. Gleave ME, Monia BP. Antisense therapy for cancer. Nat Rev Cancer 2005, 5:468–479.

40. Nielsen PE, Egholm M, Berg RH, et al. Sequence-selective recognition of DNA by strand displacement with a thymine-substituted polyamide. Science 1991, 254:1497–1500.

41. Nielsen PE. PNA Technology. Mol Biotechnol 2004, 26:233–248.

42. Elayadi AN, Corey DR. Application of PNA and LNA oligomers to chemotherapy. Curr Opin Investig Drugs 2001, 2:558–561.

43. Braasch DA, Corey DR. Locked nucleic acid (LNA): fine-tuning the recognition of DNA and RNA. Chem Biol 2001, 8:1–7.

44. Orum H, Wengel J. Locked nucleic acids: a promising molecular family for gene-function analysis and antisense drug development. Curr Opin Mol Ther 2001, 3:239–243.

45. Vester B, Wengel J. LNA (locked nucleic acid): high-affinity targeting of complementary RNA and DNA. Biochemistry 2004, 43:13233–13241.

46. Amantana A, Iversen PL. Pharmacokinetics and biodistribution of phosphorodiamidate morpholino antisense oligomers. Curr Opin Pharmacol 2005, 5:550–555.

47. Heasman J. Morpholino oligos: making sense of antisense? Dev Biol 2002, 243:209–214.

48. McCaffrey AP, Meuse L, Karimi M, et al. A potent and specific morpholino antisense inhibitor of hepatitis C translation in mice. Hepatology 2003, 38:503–508.

49. Geller BL, Deere JD, Stein DA, et al. Inhibition of gene expression in Escherichia coli by antisense phosphorodiamidate morpholino oligomers. Antimicrob Agents Chemother 2003, 47:3233–3239.

50. Crooke ST. Antisense drug technologies: principles, strategies and applications. New York: Marcel Dekker, 2001.

51. Arora V, Knapp DC, Reddy MT, et al. Bioavailability and efficacy of antisense morpholino oligomers targeted to c-myc and cytochrome P-450 3A2 following oral administration in rats. J Pharm Sci 2002, 91:1009–1018.

52. Hudziak RM, Barofsky E, Barofsky DF, et al. Resistance of morpholino phosphorodiamidate oligomers to enzymatic degradation. Antisense Nucleic Acid Drug Dev 1996, 6:267–272.

53. Chiu YL, Rana TM. siRNA function in RNAi: a chemical modification analysis. Rna. 2003, 9:1034–1048.

54. Amarzguioui M, Holen T, Babaie E, et al. Tolerance for mutations and chemical modifications in a siRNA. Nucleic Acids Res 2003, 31:589–595.

55. Prakash TP, Allerson CR, Dande P, et al. Positional effect of chemical modifications on short interference RNA activity in mammalian cells. J Med Chem 2005, 48:4247–4253.

56. Braasch DA, Jensen S, Liu Y, et al. RNA interference in mammalian cells by chemically-modified RNA. Biochemistry 2003, 42:7967–7975.

57. Parrish S, Fleenor J, Xu S, et al. Functional anatomy of a dsRNA trigger: differential requirement for the two trigger strands in RNA interference. Mol Cell 2000, 6:1077–1087.

58. Harborth J, Elbashir SM, Vandenburgh K, et al. Sequence, chemical, and structural variation of small interfering RNAs and short hairpin RNAs and the effect on mammalian gene silencing. Antisense Nucleic Acid Drug Dev 2003, 13:83–105.

59. Czauderna F, Fechtner M, Dames S, et al. Structural variations and stabilising modifications of synthetic siRNAs in mammalian cells. Nucleic Acids Res 2003, 31:2705–2716.

60. Caplen NJ, Parrish S, Imani F, et al. Specific inhibition of gene expression by small double-stranded RNAs in invertebrate and vertebrate systems. Proc Natl Acad Sci U S A 2001, 98:9742–9747.

61. Chiu YL, Rana TM. RNAi in human cells: basic structural and functional features of small interfering RNA. Mol Cell 2002, 10:549–561.

62. Forster Y, Meye A, Krause S, et al. Antisense-mediated VEGF suppression in bladder and breast cancer cells. Cancer Lett 2004, 212:95–103.

63. Peyman A, Helsberg M, Kretzschmar G, et al. Inhibition of viral growth by antisense oligonucleotides directed against the IE110 and the UL30 mRNA of herpes simplex virus type-1. Biol Chem Hoppe Seyler 1995, 376:195–198.

64. Monia BP, Johnston JF, Geiger T, et al. Antitumor activity of a phosphorothioate antisense oligodeoxynucleotide targeted against C-raf kinase. Nat Med 1996, 2:668–675.

65. Chan JH, Lim S, Wong WS. Antisense oligonucleotides: from design to therapeutic application. Clin Exp Pharmacol Physiol 2006, 33:533–540.

66. Sczakiel G. Theoretical and experimental approaches to design effective antisense oligonucleotides. Front Biosci 2000, 5:D194–201.

67. Zuker M. Mfold web server for nucleic acid folding and hybridization prediction. Nucleic Acids Res 2003, 31:3406–3415.

68. Patzel V, Steidl U, Kronenwett R, Haas R, Sczakiel G. A theoretical approach to select effective antisense oligodeoxyribonucleotides at high statistical probability. Nucleic Acids Res 1999, 27:4328–34.

69. Lima WF, Monia BP, Ecker DJ, et al. Implication of RNA structure on antisense oligonucleotide hybridisation kinetics. Biochemistry 1992, 31:12055–12061.

70. Thierry AR, Rahman A, Dritschilo A. Overcoming multidrug resistance in human tumor cells using free and liposomally encapsulated antisense

oligodeoxynucleotides. Biochem Biophys Res Commun 1993, 190:952–960.

71. Laptev AV, Lu Z, Colige A, et al. Specific inhibition of expression of a human collagen gene (COL1A1) with modified antisense oligonucleotides. The most effective target sites are clustered in double-stranded regions of the predicted secondary structure for the mRNA. Biochemistry 1994, 33:11033–11039.

72. Forster Y, Schwenzer B. Inhibition of TF gene expression by antisense oligonucleotides in different cancer cell lines. J Exp Ther Oncol 2004, 4:281–289.

73. Matveeva OV, Tsodikov AD, Giddings M, et al. Identification of sequence motifs in oligonucleotides whose presence is correlated with antisense activity. Nucleic Acids Res 2000, 28:2862–2865.

74. Bohl M, Schwenzer B. A potent inhibitor of prothrombin gene expression as a result of standardized target site selection and design of antisense oligonucleotides. Oligonucleotides 2005, 15:172–182.

75. Matveeva OV, Mathews DH, Tsodikov AD, et al. Thermodynamic criteria for high hit rate antisense oligonucleotide design. Nucleic Acids Res 2003, 31:4989–4994.

76. Mathews DH, Burkard ME, Freier SM, et al. Predicting oligonucleotide affinity to nucleic acid targets. RNA 1999, 5:1458–1469.

77. Akhtar S. Antisense technology: selection and delivery of optimally acting antisense oligonucleotides. J Drug Target 1998, 5:225–234.

78. Sohail M, Southern EM. Selecting optimal antisense reagents. Adv Drug Deliv Rev 2000, 44:23–34.

79. Bruice TW, Lima WF. Control of complexity constraints on combinatorial screening for preferred oligonucleotide hybridization sites on structured RNA. Biochemistry 1997, 36:5004–5019.

80. Ho SP, Britton DH, Stone BA, et al. Potent antisense oligonucleotides to the human multidrug resistance-1 mRNA are rationally selected by mapping RNA-accessible sites with oligonucleotide libraries. Nucleic Acids Res 1996, 24:1901–1907.

81. Ho SP, Bao Y, Lesher T, et al. Mapping of RNA accessible sites for antisense experiments with oligonucleotide libraries. Nat Biotechnol 1998, 16:59–63.

82. Milner N, Mir KU, Southern EM. Selecting effective antisense reagents on combinatorial oligonucleotide arrays. Nat Biotechnol 1997, 15:537–541.

83. Southern EM, Case-Green SC, Elder JK, et al. Arrays of complementary oligonucleotides for analysing the hybridisation behaviour of nucleic acids. Nucleic Acids Res 1994, 22:1368–1373.

84. Holen T, Amarzguioui M, Wiiger MT, et al. Positional effects of short interfering RNAs targeting the human coagulation trigger Tissue Factor. Nucleic Acids Res 2002, 30:1757–1766.

85. Reynolds A, Leake D, Boese Q, et al. Rational siRNA design for RNA interference. Nat Biotechnol 2004, 22:326–330.

86. Ren Y, Gong W, Xu Q, et al. siRecords: an extensive database of mammalian siRNAs with efficacy ratings. Bioinformatics 2006, 22:1027–1028.

87. Chalk AM, Warfinge RE, Georgii-Hemming P, et al. siRNAdb: a database of siRNA sequences. Nucleic Acids Res 2005, 33:D131–134.

88. Smith C. Sharpening the tools of RNA interference. Nat Meth 2006, 3:475–486.

89. Ui-Tei K, Naito Y, Takahashi F, et al. Guidelines for the selection of highly effective siRNA sequences for mammalian and chick RNA interference. Nucleic Acids Res 2004, 32:936–948.

90. Yuan B, Latek R, Hossbach M, et al. siRNA Selection Server: an automated siRNA oligonucleotide prediction server. Nucleic Acids Res 2004, 32: W130–134.

91. Takasaki S, Kotani S, Konagaya A. An effective method for selecting siRNA target sequences in mammalian cells. Cell Cycle 2004, 3:790–795.

92. Jagla B, Aulner N, Kelly PD, et al. Sequence characteristics of functional siRNAs. RNA 2005, 11:864–872.

93. Chalk AM, Wahlestedt C, Sonnhammer EL. Improved and automated prediction of effective siRNA. Biochem Biophys Res Commun 2004, 319:264–274.

94. Amarzguioui M, Prydz H. An algorithm for selection of functional siRNA sequences. Biochem Biophys Res Commun 2004, 316:1050–1058.

95. Khvorova A, Reynolds A, Jayasena SD. Functional siRNAs and miRNAs exhibit strand bias. Cell 2003, 115:209–216.

96. Mittal V. Improving the efficiency of RNA interference in mammals. Nat Rev Genet 2004, 5:355–365.

97. Hsieh AC, Bo R, Manola J, et al. A library of siRNA duplexes targeting the phosphoinositide 3-kinase pathway: determinants of gene silencing for use in cell-based screens. Nucleic Acids Res 2004, 32:893–901.

98. Luo KQ, Chang DC. The gene-silencing efficiency of siRNA is strongly dependent on the local structure of mRNA at the targeted region. Biochem Biophys Res Commun 2004, 318:303–310.

99. Ding Y, Chan CY, Lawrence CE. Sfold web server for statistical folding and rational design of nucleic acids. Nucleic Acids Res 2004, 32:W135–141.

100. Kretschmer-Kazemi Far R, Sczakiel G. The activity of siRNA in mammalian cells is related to structural target accessibility: a comparison with antisense oligonucleotides. Nucleic Acids Res 2003, 31:4417–4424.

101. Schubert S, Grunweller A, Erdmann VA, et al. Local RNA target structure influences siRNA efficacy: systematic analysis of intentionally designed binding regions. J Mol Biol 2005, 348:883–893.

102. Overhoff M, Alken M, Far RK, et al. Local RNA target structure influences siRNA efficacy: a systematic global analysis. J Mol Biol 2005, 348:871–881.

103. Sheehan JP, Phan TM. Phosphorothioate oligonucleotides inhibit the intrinsic tenase complex by an allosteric mechanism. Biochemistry 2001, 40:4980–4989.

104. Henry SP, Giclas PC, Leeds J, et al. Activation of the alternative pathway of complement by a phosphorothioate oligonucleotide: potential mechanism of action. J Pharmacol Exp Ther 1997, 281:810–816.

105. Krieg AM, Stein CA. Phosphorothioate oligodeoxynucleotides: antisense or anti–protein? Antisense Res Dev 1995, 5:241.

106. Krieg AM. Mechanisms and applications of immune stimulatory CpG oligodeoxynucleotides. Biochim Biophys Acta 1999, 1489:107–116.

107. Mui B, Raney SG, Semple SC, et al. Immune stimulation by a CpG-containing oligodeoxynucleotide is enhanced when encapsulated and delivered in lipid particles. J Pharmacol Exp Ther 2001, 298:1185–1192.

108. Sledz CA, Holko M, de Veer MJ, et al. Activation of the interferon system by short-interfering RNAs. Nat Cell Biol 2003, 5:834–839.

109. Persengiev SP, Zhu X, Green MR. Nonspecific, concentration-dependent stimulation and repression of mammalian gene expression by small interfering RNAs (siRNAs). RNA 2004, 10:12–18.

110. Dorsett Y, Tuschl T. siRNAs: applications in functional genomics and potential as therapeutics. Nat Rev Drug Discov 2004, 3:318–329.

111. Bridge AJ, Pebernard S, Ducraux A, et al. Induction of an interferon response by RNAi vectors in mammalian cells. Nat Genet 2003, 34:263–264.

112. Judge AD, Sood V, Shaw JR, et al. Sequence-dependent stimulation of the mammalian innate immune response by synthetic siRNA. Nat Biotechnol 2005, 23:457–462.

113. Salcman M. Supratentorial gliomas: clinical features and surgical therapy. In: Wilkins RH, Rengachary SS, eds. Neurosurgery. New York: McGraw-Hill, 1996:777–778.

114. Louis DN, Holland EC, Cairncross JG. Glioma classification: a molecular reappraisal. Am J Pathol 2001, 159:779–786.

115. Annegers JF, Schoenberg BS, Okazaki H, et al. Epidemiologic study of primary intracranial neoplasms. Arch Neurol 1981, 38:217–219.

116. Daumas-Duport C, Scheithauer B, O'Fallon J, et al. Grading of astrocytomas. A simple and reproducible method. Cancer 1988, 62:2152–2165.

117. Haskell CM, ed. Cancer treatment. Philadelphia: Saunders, 1995.

118. Pulkkanen KJ, Yla-Herttuala S. Gene therapy for malignant glioma: current clinical status. Mol Ther 2005, 12:585–598.

119. Nishizuka Y. The molecular heterogeneity of protein kinase C and its implications for cellular regulation. Nature 1988, 334:661–665.

120. Couldwell WT, Uhm JH, Antel JP, et al. Enhanced protein kinase C activity correlates with the growth rate of malignant gliomas in vitro. Neurosurgery 1991, 29:880–886, discussion 886–887.

121. Baltuch GH, Yong VW. Signal transduction for proliferation of glioma cells in vitro occurs predominantly through a protein kinase C-mediated pathway. Brain Res 1996, 710:143–149.

122. Grossman SA, Alavi JB, Supko JG, et al. Efficacy and toxicity of the antisense oligonucleotide aprinocarsen directed against protein kinase C-alpha delivered as a 21-day continuous intravenous infusion in patients with recurrent high-grade astrocytomas. Neuro-oncology 2005, 7:32–40.

123. Lahn MM, Sundell KL, Paterson BM. The role of protein kinase C-alpha in malignancies of the nervous system and implications for the clinical development of the specific PKC-alpha inhibitor aprinocarsen. Oncol Rep 2004, 11:515–522.

124. Vansteenkiste J, Canon JL, Riska H, et al. Randomized phase II evaluation of aprinocarsen in combination with gemcitabine and cisplatin for patients with advanced/metastatic non-small cell lung cancer. Invest New Drugs 2005, 23:263–269.

125. Yuen AR, Halsey J, Fisher GA, et al. Phase I study of an antisense oligonucleotide to protein kinase C-alpha (ISIS 3521/CGP 64128A) in patients with cancer. Clin Cancer Res 1999, 5:3357–363.

126. Kuan CT, Wikstrand CJ, Bigner DD. EGF mutant receptor vIII as a molecular target in cancer therapy. Endocr Relat Cancer 2001, 8:83–96.

127. Nicholson RI, Gee JM, Harper ME. EGFR and cancer prognosis. Eur J Cancer 2001, 37(Suppl 4): S9–15.

128. Stoll SW, Benedict M, Mitra R, et al. EGF receptor signaling inhibits keratinocyte apoptosis: evidence for mediation by Bcl-XL. Oncogene 1998, 16:1493–1499.

129. Tian XX, Lam PY, Chen J, et al. Antisense epidermal growth factor receptor RNA transfection in human malignant glioma cells leads to inhibition of proliferation and induction of differentiation. Neuropathol Appl Neurobiol 1998, 24:389–396.

130. Kang CS, Zhang ZY, Jia ZF, et al. Suppression of EGFR expression by antisense or small interference RNA inhibits U251 glioma cell growth in vitro and in vivo. Cancer Gene Ther 2006, 13:530–538.

131. Zhang Y, Zhu C, Pardridge WM. Antisense gene therapy of brain cancer with an artificial virus gene delivery system. Mol Ther 2002, 6:67–72.

132. Zhang Y, Zhang YF, Bryant J, et al. Intravenous RNA interference gene therapy targeting the human epidermal growth factor receptor prolongs survival in intracranial brain cancer. Clin Cancer Res 2004, 10:3667–3677.

133. Massague J. TGF-beta signal transduction. Annu Rev Biochem 1998, 67:753–791.

134. Pasche B. Role of transforming growth factor beta in cancer. J Cell Physiol 2001, 186:153–168.

135. Gold LI. The role for transforming growth factor-beta (TGF-beta) in human cancer. Crit Rev Oncog 1999, 10:303–360.

136. Liau LM, Fakhrai H, Black KL. Prolonged survival of rats with intracranial C6 gliomas by treatment with TGF-beta antisense gene. Neurol Res 1998, 20:742–747.

137. Schlingensiepen KH, Schlingensiepen R, Steinb recher A, et al. Targeted tumor therapy with the TGF-beta2 antisense compound AP 12009. Cytokine Growth Factor Rev 2006, 17:129–139.

138. Kim SJ, Johnson M, Koterba K, et al. Reduced c-Met expression by an adenovirus expressing a c-Met ribozyme inhibits tumorigenic growth and lymph node metastases of PC3-LN4 prostate tumor cells in an orthotopic nude mouse model. Clin Cancer Res 2003, 9:5161–5170.

139. Kim SJ, Uehara H, Yazici S, et al. Simultaneous blockade of platelet-derived growth factor-receptor and epidermal growth factor-receptor signaling and systemic administration of paclitaxel as therapy for human prostate cancer metastasis in bone of nude mice. Cancer Res 2004, 64:4201–4208.

140. Patel P, Ashdown D, James N. Is gene therapy the answer for prostate cancer? Prostate Cancer Prostatic Dis 2004, 7(Suppl 1):S14–19.

141. Pinthus JH, Waks T, Malina V, et al. Adoptive immunotherapy of prostate cancer bone lesions using redirected effector lymphocytes. J Clin Invest 2004, 114:1774–1781.

142. Petrylak DP, Tangen CM, Hussain MH, et al. Docetaxel and estramustine compared with mitoxantrone and prednisone for advanced refractory prostate cancer. N Engl J Med 2004, 351:1513–1520.

143. Tannock IF, de Wit R, Berry WR, et al. Docetaxel plus prednisone or mitoxantrone plus prednisone for advanced prostate cancer. N Engl J Med 2004, 351:1502–1512.

144. Chi KN. Targeting Bcl-2 with oblimersen for patients with hormone refractory prostate cancer. World J Urol 2005, 23:33–37.

145. Apakama I, Robinson MC, Walter NM, et al. bcl-2 overexpression combined with p53 protein accumulation correlates with hormone-refractory prostate cancer. Br J Cancer 1996, 74:1258–1262.

146. Furuya Y, Krajewski S, Epstein JI, et al. Expression of bcl-2 and the progression of human and rodent prostatic cancers. Clin Cancer Res 1996, 2:389–398.

147. McDonnell TJ, Troncoso P, Brisbay SM, et al. Expression of the protooncogene bcl-2 in the prostate and its association with emergence of androgen-independent prostate cancer. Cancer Res 1992, 52:6940–6944.

148. Stattin P, Damber JE, Karlberg L, et al. Bcl-2 immunoreactivity in prostate tumorigenesis in relation to prostatic intraepithelial neoplasia, grade, hormonal status, metastatic growth and survival. Urol Res 1996, 24:257–264.

149. Lipponen P, Vesalainen S. Expression of the apoptosis suppressing protein bcl-2 in prostatic adenocarcinoma is related to tumor malignancy. Prostate 1997, 32:9–15.

150. Scherr DS, Vaughan ED Jr, Wei J, et al. BCL-2 and p53 expression in clinically localized prostate cancer predicts response to external beam radiotherapy. J Urol 1999, 162:12–16, discussion 16–17.

151. Raffo AJ, Perlman H, Chen MW, et al. Overexpression of bcl-2 protects prostate cancer cells from apoptosis in vitro and confers resistance to androgen depletion in vivo. Cancer Res 1995, 55:4438–4445.

152. Bubendorf L, Sauter G, Moch H, et al. Prognostic significance of Bcl-2 in clinically localized prostate cancer. Am J Pathol 1996, 148:1557–1565.

153. McDonnell TJ, Navone NM, Troncoso P, et al. Expression of bcl-2 oncoprotein and p53 protein accumulation in bone marrow metastases of androgen independent prostate cancer. J Urol 1997, 157:569–574.

154. Tolcher AW, Chi K, Kuhn J, et al. A phase II, pharmacokinetic, and biological correlative study of oblimersen sodium and docetaxel in patients with hormone-refractory prostate cancer. Clin Cancer Res 2005, 11:3854–3861.

155. Leung S, Miyake H, Zellweger T, et al. Synergistic chemosensitization and inhibition of progression to androgen independence by antisense Bcl-2 oligodeoxynucleotide and paclitaxel in the LNCaP prostate tumor model. Int J Cancer 2001, 91:846–850.

156. Mu Z, Hachem P, Pollack A. Antisense Bcl-2 sensitizes prostate cancer cells to radiation. Prostate 2005, 65:331–340.

157. Miyake H, Nelson C, Rennie PS, et al. Testosterone-repressed prostate message-2 is an antiapoptotic gene involved in progression to androgen independence in prostate cancer. Cancer Res 2000, 60:170–176.

158. Sensibar JA, Sutkowski DM, Raffo A, et al. Prevention of cell death induced by tumor necrosis factor alpha in LNCaP cells by overexpression of sulfated glycoprotein-2 (clusterin). Cancer Res 1995, 55:2431–2437.

159. Zellweger T, Chi K, Miyake H, et al. Enhanced radiation sensitivity in prostate cancer by inhibition of the cell survival protein clusterin. Clin Cancer Res 2002, 8:3276–3284.

160. Steinberg J, Oyasu R, Lang S, et al. Intracellular levels of SGP-2 (Clusterin) correlate with tumor grade in prostate cancer. Clin Cancer Res 1997, 3:1707–1711.

161. Miyake H, Chi KN, Gleave ME. Antisense TRPM-2 oligodeoxynucleotides chemosensitize human androgen-independent PC-3 prostate cancer cells both in vitro and in vivo. Clin Cancer Res 2000, 6:1655–1663.

162. Miyake H, Nelson C, Rennie PS, et al. Acquisition of chemoresistant phenotype by overexpression of the antiapoptotic gene testosterone-repressed prostate message-2 in prostate cancer xenograft models. Cancer Res 2000, 60:2547–2554.

163. July LV, Akbari M, Zellweger T, et al. Clusterin expression is significantly enhanced in prostate cancer cells following androgen withdrawal therapy. Prostate 2002, 50:179–188.

164. Chi KN, Eisenhauer E, Fazli L, et al. A phase I pharmacokinetic and pharmacodynamic study of OGX-011, a 2′-methoxyethyl antisense oligonucleotide to clusterin, in patients with localized prostate cancer. J Natl Cancer Inst 2005, 97:1287–1296.

165. Miyake H, Hara I, Fujisawa M, et al. The potential of clusterin inhibiting antisense oligodeoxynucleotide therapy for prostate cancer. Expert Opin Invest Drugs 2006, 15:507–517.

166. Duffy MJ. The urokinase plasminogen activator system: role in malignancy. Curr Pharm Des 2004, 10:39–49.

167. Aguirre-Ghiso JA, Estrada Y, Liu D, et al. ERK(MAPK) activity as a determinant of tumor growth and dormancy, regulation by p38(SAPK). Cancer Res 2003, 63:1684–1695.

168. Ma Z, Webb DJ, Jo M, et al. Endogenously produced urokinase-type plasminogen activator is a major determinant of the basal level of activated ERK/MAP kinase and prevents apoptosis in MDA-MB-231 breast cancer cells. J Cell Sci 2001, 114:3387–3396.

169. Hendry L, John S. Regulation of STAT signalling by proteolytic processing. Eur J Biochem 2004, 271:4613–4620.

170. Mora LB, Buettner R, Seigne J, et al. Constitutive activation of Stat3 in human prostate tumors and cell lines: direct inhibition of Stat3 signaling induces apoptosis of prostate cancer cells. Cancer Res 2002, 62:6659–6666.

171. Margheri F, D'Alessio S, Serrati S, et al. Effects of blocking urokinase receptor signaling by antisense oligonucleotides in a mouse model of experimental prostate cancer bone metastases. Gene Ther 2005, 12:702–714.

172. Pulukuri SM, Gondi CS, Lakka SS, et al. RNA interference-directed knockdown of urokinase plasminogen activator and urokinase plasminogen activator receptor inhibits prostate cancer cell invasion, survival, and tumorigenicity in vivo. J Biol Chem 2005, 280:36529–36540.

173. Liu B, Maher RJ, Hannun YA, et al. 12(S)-HETE enhancement of prostate tumor cell invasion: selective role of PKC alpha. J Natl Cancer Inst 1994, 86:1145–1151.

174. Cornford P, Evans J, Dodson A, et al. Protein kinase C isoenzyme patterns characteristically modulated in early prostate cancer. Am J Pathol 1999, 154: 137–144.

175. Tolcher AW, Reyno L, Venner PM, et al. A randomized phase II and pharmacokinetic study of the antisense oligonucleotides ISIS 3521 and ISIS 5132 in patients with hormone-refractory prostate cancer. Clin Cancer Res 2002, 8:2530–2535.

176. Beeram M, Patnaik A, Rowinsky EK. Regulation of c-Raf-1: therapeutic implications. Clin Adv Hematol Oncol 2003, 1:476–481.

177. Monia BP, Sasmor H, Johnston JF, et al. Sequence-specific antitumor activity of a phosphorothioate oligodeoxyribonucleotide targeted to human C-raf kinase supports an antisense mechanism of action in vivo. Proc Natl Acad Sci U S A 1996, 93:15481–15484.

178. Rocchi P, Jugpal P, So A, et al. Small interference RNA targeting heat-shock protein 27 inhibits the growth of prostatic cell lines and induces apoptosis via caspase-3 activation in vitro. BJU Int 2006, 98:1082–1089.

179. Rocchi P, So A, Kojima S, et al. Heat shock protein 27 increases after androgen ablation and plays a cytoprotective role in hormone-refractory prostate cancer. Cancer Res 2004, 64:6595–6602.

180. Bostwick DG. Immunohistochemical changes in prostate cancer after androgen deprivation therapy. Mol Urol 2000, 4:101–106,discussion 107.

181. Bubendorf L, Kolmer M, Kononen J, et al. Hormone therapy failure in human prostate cancer: analysis by complementary DNA and tissue microarrays. J Natl Cancer Inst 1999, 91:1758–1764.

182. Rocchi P, Beraldi E, Ettinger S, et al. Increased Hsp27 after androgen ablation facilitates androgen-independent progression in prostate cancer via signal transducers and activators of transcription 3-mediated suppression of apoptosis. Cancer Res 2005, 65:11083–11093.

183. Feldman AR, Kessler L, Myers MH, et al. The prevalence of cancer. Estimates based on the Connecticut Tumor Registry. N Engl J Med 1986, 315:1394–1397.

184. Malmstrom PU, Busch C, Norlen BJ. Recurrence, progression and survival in bladder cancer. A retrospective analysis of 232 patients with greater than or equal to 5-year follow-up. Scand J Urol Nephrol 1987, 21:185–195.

185. Wolf H, Melsen F, Pedersen SE, et al. Natural history of carcinoma in situ of the urinary bladder. Scand J Urol Nephrol Suppl 1994, 157:147–151.

186. So A, Rocchi P, Gleave M. Antisense oligonucleotide therapy in the management of bladder cancer. Curr Opin Urol 2005, 15:320–327.

187. Miyake H, Gleave M, Kamidono S, et al. Overexpression of clusterin in transitional cell carcinoma of the bladder is related to disease progression and recurrence. Urology 2002, 59:150–154.

188. Miyake H, Hara I, Kamidono S, et al. Synergistic chemsensitization and inhibition of tumor growth and metastasis by the antisense oligodeoxynucleotide targeting clusterin gene in a human bladder cancer model. Clin Cancer Res 2001, 7:4245–4252.

189. Chung J, Kwak C, Jin RJ, et al. Enhanced chemosensitivity of bladder cancer cells to cisplatin by suppression of clusterin in vitro. Cancer Lett 2004, 203:155–161.

190. Miyake H, Eto H, Hara I, et al. Synergistic antitumor activity by combined treatment with gemcitabine and antisense oligodeoxynucleotide targeting clusterin gene in an intravesical administration model against human bladder cancer kotcc-1 cells. J Urol 2004, 171:2477–24781.

191. Ciocca DR, Oesterreich S, Chamness GC, et al. Biological and clinical implications of heat shock protein 27,000 (Hsp27): a review. J Natl Cancer Inst 1993, 85:1558–1570.

192. Levine AJ, Momand J, Finlay CA. The p53 tumour suppressor gene. Nature 1991, 351:453–456.

193. Tomei LD, Kiecolt-Glaser JK, Kennedy S, et al. Psychological stress and phorbol ester inhibition of radiation-induced apoptosis in human peripheral blood leukocytes. Psychiatry Res 1990, 33:59–71.

194. Kassem H, Sangar V, Cowan R, et al. A potential role of heat shock proteins and nicotinamide N-methyl transferase in predicting response to radiation in bladder cancer. Int J Cancer 2002, 101:454–460.

195. Marme D. Tumor angiogenesis: the pivotal role of vascular endothelial growth factor. World J Urol 1996, 14:166–174.

196. Dvorak HF, Brown LF, Detmar M, Dvorak AM. Vascular permeability factor/vascular endothelial growth factor, microvascular hyperpermeability, and angiogenesis. Am J Pathol 1995, 146:1029–1039.

197. Brown LF, Detmar M, Claffey K, et al. Vascular permeability factor/vascular endothelial growth factor: a multifunctional angiogenic cytokine. Exs 1997, 79:233–269.

198. Zhang HT, Craft P, Scott PA, et al. Enhancement of tumor growth and vascular density by transfection of vascular endothelial cell growth factor into MCF-7 human breast carcinoma cells. J Natl Cancer Inst 1995, 87:213–219.

199. Sato K, Sasaki R, Ogura Y, et al. Expression of vascular endothelial growth factor gene and its receptor (flt-1) gene in urinary bladder cancer. Tohoku J Exp Med 1998, 185:173–184.

200. Luo J, Guo P, Matsuda K, et al. Pancreatic cancer cell-derived vascular endothelial growth factor is biologically active in vitro and enhances tumorigenicity in vivo. Int J Cancer 2001, 92:361–369.

201. Krause S, Forster Y, Kraemer K, et al. Vascular endothelial growth factor antisense pretreatment of bladder cancer cells significantly enhances the cytotoxicity of mitomycin C, gemcitabine and Cisplatin. J Urol 2005, 174:328–331.

202. Paz-Ares L, Douillard JY, Koralewski P, et al. Phase III study of gemcitabine and cisplatin with or without aprinocarsen, a protein kinase C-alpha antisense oligonucleotide, in patients with advanced-stage non-small–cell lung cancer. J Clin Oncol 2006, 24:1428–1434.

203. Dean NM, McKay R, Condon TP, et al. Inhibition of protein kinase C-alpha expression in human A549 cells by antisense oligonucleotides inhibits induction of intercellular adhesion molecule 1 (ICAM-1) mRNA by phorbol esters. J Biol Chem 1994, 269:16416–16424.

204. Yuen AR, Halsey J, Fisher GA, al. e. Phase I/II trial of Isis 3521, an antisense inhibitor of PKC-alpha, with carboplatin and paclitaxel in non-small cell lung cancer. Proc Am Soc Clin Oncol 2001, 20:1234.

205. Laudanski J, Niklinska W, Burzykowski T, et al. Prognostic significance of p53 and bcl-2 abnormalities in operable nonsmall cell lung cancer. Eur Respir J 2001, 17:660–666.

206. Hu Y, Bebb G, Tan S, et al. Antitumor efficacy of oblimersen Bcl-2 antisense oligonucleotide alone and in combination with vinorelbine in xenograft models of human non-small cell lung cancer. Clin Cancer Res 2004, 10:7662–7670.

207. Rudin CM, Kozloff M, Hoffman PC, et al. Phase I study of G3139, a bcl-2 antisense oligonucleotide, combined with carboplatin and etoposide in patients with small-cell lung cancer. J Clin Oncol 2004, 22:1110–1117.

208. Klapper LN, Kirschbaum MH, Sela M, et al. Biochemical and clinical implications of the ErbB/HER signaling network of growth factor receptors. Adv Cancer Res 2000, 77:25–79.

209. Andrulis IL, Bull SB, Blackstein ME, et al. neu/erbB-2 amplification identifies a poor-prognosis group of women with node-negative breast cancer. Toronto Breast Cancer Study Group. J Clin Oncol 1998, 16:1340–1349.

210. Tsutsui S, Ohno S, Murakami S, et al. Prognostic value of c-erbB2 expression in breast cancer. J Surg Oncol 2002, 79:216–223.

211. Pegram MD, Finn RS, Arzoo K, et al. The effect of HER-2/neu overexpression on chemotherapeutic drug sensitivity in human breast and ovarian cancer cells. Oncogene 1997, 15:537–547.

212. Bertram J, Killian M, Brysch W, et al. Reduction of erbB2 gene product in mamma carcinoma cell lines by erbB2 mRNA-specific and tyrosine kinase consensus phosphorothioate antisense oligonucleotides. Biochem Biophys Res Commun 1994, 200:661–667.

213. Roh H, Hirose CB, Boswell CB, et al. Synergistic antitumor effects of HER2/neu antisense oligodeoxynucleotides and conventional chemotherapeutic agents. Surgery 1999, 126:413–421.

214. Rait AS, Pirollo KF, Xiang L, et al. Tumor-targeting, systemically delivered antisense HER-2 chemosensitizes human breast cancer xenografts irrespective of HER-2 levels. Mol Med 2002, 8:475–486.

215. Hogrefe RI, Lebedev AV, Zon G, et al. Chemically modified short interfering hybrids (siHYBRIDS): nanoimmunoliposome delivery in vitro and in vivo for RNAi of HER-2. Nucleosides Nucleotides Nucleic Acids 2006, 25:889–907.

216. Olopade OI, Adeyanju MO, Safa AR, et al. Overexpression of BCL-x protein in primary breast cancer is associated with high tumor grade and nodal metastases. Cancer J Sci Am 1997, 3:230–237.

217. Reed JC. Bcl-2: prevention of apoptosis as a mechanism of drug resistance. Hematol Oncol Clin North Am 1995, 9:451–473.

218. Emi M, Kim R, Tanabe K, et al. Targeted therapy against Bcl-2-related proteins in breast cancer cells. Breast Cancer Res 2005, 7:R940–952.

219. Kim R, Tanabe K, Emi M, et al. Modulation of tamoxifen sensitivity by antisense Bcl-2 and trastuzumab in breast carcinoma cells. Cancer 2005, 103:2199–2207.

220. Hynes RO. Integrins: bidirectional, allosteric signaling machines. Cell 2002, 110:673–687.

221. Hood JD, Cheresh DA. Role of integrins in cell invasion and migration. Nat Rev Cancer 2002, 2:91–100.

222. Townsend PA, Villanova I, Uhlmann E, et al. An antisense oligonucleotide targeting the alphaV integrin gene inhibits adhesion and induces apoptosis in breast cancer cells. Eur J Cancer 2000, 36:397–409.

223. Cao Q, Cai W, Li T, et al. Combination of integrin siRNA and irradiation for breast cancer therapy. Biochem Biophys Res Commun 2006, 351:726–732.

224. Barak Y, Juven T, Haffner R, Oren M. mdm2 expression is induced by wild type p53 activity. EMBO J 1993, 12:461–468.

225. Momand J, Zambetti GP, Olson DC, et al. The mdm-2 oncogene product forms a complex with the p53 protein and inhibits p53-mediated transactivation. Cell 1992, 69:1237–1245.

226. Wu X, Bayle JH, Olson D, et al. The p53-mdm-2 autoregulatory feedback loop. Genes Dev 1993, 7:1126–1132.

227. Momand J, Jung D, Wilczynski S, et al. The MDM2 gene amplification database. Nucleic Acids Res 1998, 26:3453–3459.

228. Momand J, Wu HH, Dasgupta G. MDM2—master regulator of the p53 tumor suppressor protein. Gene 2000, 242:15–29.

229. Wang H, Oliver P, Zhang Z, et al. Chemosensitization and radiosensitization of human cancer by antisense anti-MDM2 oligonucleotides: in vitro and in vivo activities and mechanisms. Ann N Y Acad Sci 2003, 1002:217–235.

230. Zhang WH. MDM2 oncogene as a novel target for human cancer therapy. Curr Pharm Des 2000, 6:393–416.

231. Wang H, Nan L, Yu D, et al. Antisense anti-MDM2 oligonucleotides as a novel therapeutic approach to human breast cancer: in vitro and in vivo activities and mechanisms. Clin Cancer Res 2001, 7:3613–3624.

232. Zhang Z, Wang H, Prasad G, et al. Radiosensitization by antisense anti-MDM2 mixed-backbone oligonucleotide in in vitro and in vivo human cancer models. Clin Cancer Res 2004, 10:1263–1273.

233. Muller A, Homey B, Soto H, et al. Involvement of chemokine receptors in breast cancer metastasis. Nature 2001, 410:50–56.

234. Liang Z, Yoon Y, Votaw J, et al. Silencing of CXCR4 blocks breast cancer metastasis. Cancer Res 2005, 65:967–971.

235. Surmacz E, Guvakova MA, Nolan MK, et al. Type I insulin-like growth factor receptor function in breast cancer. Breast Cancer Res Treat 1998, 47:255–267.

236. Salatino M, Schillaci R, Proietti CJ, et al. Inhibition of in vivo breast cancer growth by antisense oligodeoxynucleotides to type I insulin-like growth factor receptor mRNA involves inactivation of ErbBs, PI-3K/Akt and p42/p44 MAPK signaling pathways but not modulation of progesterone receptor activity. Oncogene 2004, 23:5161–5174.

237. Jemal A, Murray T, Ward E, et al. Cancer statistics, 2005. CA Cancer J Clin 2005, 55:10–30.

238. MacKenzie MJ. Molecular therapy in pancreatic adenocarcinoma. Lancet Oncol 2004, 5:541–549.

239. Bhattacharyya M, Lemoine NR. Gene therapy developments for pancreatic cancer. Best Pract Res Clin Gastroenterol 2006, 20:285–298.

240. Hruban RH, van Mansfeld AD, Offerhaus GJ, et al. K-ras oncogene activation in adenocarcinoma of the human pancreas. A study of 82 carcinomas using a combination of mutant-enriched polymerase chain reaction analysis and allele5specific oligonucleotide hybridization. Am J Pathol 1993, 143:545–554.

241. Nakada Y, Saito S, Ohzawa K, et al. Antisense oligonucleotides specific to mutated K-ras genes inhibit invasiveness of human pancreatic cancer cell lines. Pancreatology 2001, 1:314–319.

242. Miura Y, Ohnami S, Yoshida K, et al. Intraperitoneal injection of adenovirus expressing antisense K-ras RNA suppresses peritoneal dissemination of hamster syngeneic pancreatic cancer without systemic toxicity. Cancer Lett 2005, 218:53–62.

243. Fleming JB, Shen GL, Holloway SE, et al. Molecular consequences of silencing mutant K-ras in pancreatic cancer cells: justification for K-ras-directed therapy. Mol Cancer Res 2005, 3:413–423.

244. Zhang YA, Nemunaitis J, Samuel SK, et al. Antitumor activity of an oncolytic adenovirus–delivered oncogene small interfering RNA. Cancer Res 2006, 66:9736–9743.

245. Ryan HE, Lo J, Johnson RS. HIF-1 alpha is required for solid tumor formation and embryonic vascularization. EMBO J 1998, 17:3005–3015.

246. Talks KL, Turley H, Gatter KC, et al. The expression and distribution of the hypoxia-inducible factors HIF-1alpha and HIF-2alpha in normal human tissues, cancers, and tumor-associated macrophages. Am J Pathol 2000, 157:411–421.

247. Jiang BH, Semenza GL, Bauer C, et al. Hypoxia-inducible factor 1 levels vary exponentially over a physiologically relevant range of O2 tension. Am J Physiol 1996, 271:C1172–1180.

248. Chang Q, Qin R, Huang T, et al. Effect of antisense hypoxia-inducible factor 1alpha on progression, metastasis, and chemosensitivity of pancreatic cancer. Pancreas 2006, 32:297–305.

249. Mizuno T, Nagao M, Yamada Y, et al. Small interfering RNA expression vector targeting hypoxia-inducible factor 1 alpha inhibits tumor growth in hepatobiliary and pancreatic cancers. Cancer Gene Ther 2006, 13:131–140.

250. Altieri DC. The molecular basis and potential role of survivin in cancer diagnosis and therapy. Trends Mol Med 2001, 7:542–547.

251. Chiou SK, Jones MK, Tarnawski AS. Survivin—an anti-apoptosis protein: its biological roles and implications for cancer and beyond. Med Sci Monit 2003, 9:PI25–129.

252. Kami K, Doi R, Koizumi M, et al. Survivin expression is a prognostic marker in pancreatic cancer patients. Surgery 2004, 136:443–448.

253. Kami K, Doi R, Koizumi M, et al. Downregulation of survivin by siRNA diminishes radioresistance of pancreatic cancer cells. Surgery 2005, 138:299–305.

254. Guan HT, Xue XH, Dai ZJ, et al. Down-regulation of survivin expression by small interfering RNA induces pancreatic cancer cell apoptosis and enhances its radiosensitivity. World J Gastroenterol 2006, 12:2901–2907.

255. Rebollo A, Martinez AC. Ras proteins: recent advances and new functions. Blood 1999, 94:2971–2980.

256. Seufferlein T, Van Lint J, Liptay S, et al. Transforming growth factor alpha activates Ha-Ras in human pancreatic cancer cells with Ki-ras mutations. Gastroenterology 1999, 116:1441–1452.

257. Monia BP, Johnston JF, Ecker DJ, et al. Selective inhibition of mutant Ha-ras mRNA expression by antisense oligonucleotides. J Biol Chem 1992, 267:19954–19962.

258. Adjei AA, Dy GK, Erlichman C, et al. A phase I trial of ISIS 2503, an antisense inhibitor of H-ras, in combination with gemcitabine in patients with advanced cancer. Clin Cancer Res 2003, 9:115–123.

259. Alberts SR, Schroeder M, Erlichman C, et al. Gemcitabine and ISIS-2503 for patients with locally advanced or metastatic pancreatic adenocarcinoma: a North Central Cancer Treatment Group phase II trial. J Clin Oncol 2004, 22:4944–4950.

260. Ikeda N, Adachi M, Taki T, et al. Prognostic significance of angiogenesis in human pancreatic cancer. Br J Cancer 1999, 79:1553–1563.

261. Seo Y, Baba H, Fukuda T, et al. High expression of vascular endothelial growth factor is associated with liver metastasis and a poor prognosis for

patients with ductal pancreatic adenocarcinoma. Cancer 2000, 88:2239–2245.

262. Hotz HG, Hines OJ, Masood R, et al. VEGF antisense therapy inhibits tumor growth and improves survival in experimental pancreatic cancer. Surgery 2005, 137:192–199.

263. Forte A, Cipollaro M, Cascino A, et al. Small interfering RNAs and antisense oligonucleotides for treatment of neurological diseases. Curr Drug Targets 2005, 6:21–29.

264. Smith RA, Miller TM, Yamanaka K, et al. Antisense oligonucleotide therapy for neurodegenerative disease. J Clin Invest 2006, 116:2290–2296.

265. Lane RM, Kivipelto M, Greig NH. Acetylcholinesterase and its inhibition in Alzheimer disease. Clin Neuropharmacol 2004, 27:141–149.

266. Fu AL, Zhang XM, Sun MJ. Antisense inhibition of acetylcholinesterase gene expression for treating cognition deficit in Alzheimer's disease model mice. Brain Res 2005, 1066:10–15.

267. Nielsen MH, Pedersen FS, Kjems J. Molecular strategies to inhibit HIV-1 replication. Retrovirology 2005, 2:10.

268. Lisziewicz J, Sun D, Metelev V, et al. Long-term treatment of human immunodeficiency virusinfected cells with antisense oligonucleotide phosphorothioates. Proc Natl Acad Sci U S A 1993, 90:3860–3864.

269. Lisziewicz J, Sun D, Weichold FF, et al. Antisense oligodeoxynucleotide phosphorothioate complementary to Gag mRNA blocks replication of human immunodeficiency virus type 1 in human peripheral blood cells. Proc Natl Acad Sci U S A 1994, 91:7942–7946.

270. Agrawal S, Tang JY. GEM 91—an antisense oligonucleotide phosphorothioate as a therapeutic agent for AIDS. Antisense Res Dev 1992, 2:261–266.

271. Turner JJ, Fabani M, Arzumanov AA, et al. Targeting the HIV-1 RNA leader sequence with synthetic oligonucleotides and siRNA: chemistry and cell delivery. Biochim Biophys Acta 2006, 1758:290–300.

272. Jacque JM, Triques K, Stevenson M. Modulation of HIV-1 replication by RNA interference. Nature 2002, 418:435–438.

273. Coburn GA, Cullen BR. Potent and specific inhibition of human immunodeficiency virus type 1 replication by RNA interference. J Virol 2002, 76:9225–9231.

274. Lee NS, Dohjima T, Bauer G, et al. Expression of small interfering RNAs targeted against HIV-1 rev transcripts in human cells. Nat Biotechnol 2002, 20:500–505.

275. Surabhi RM, Gaynor RB. RNA interference directed against viral and cellular targets inhibits human immunodeficiency Virus Type 1 replication. J Virol 2002, 76:12963–12973.

276. Boden D, Pusch O, Lee F, et al. Human immunodeficiency virus type 1 escape from RNA interference. J Virol 2003, 77:11531–11535.

277. Hannon GJ, Rossi JJ. Unlocking the potential of the human genome with RNA interference. Nature 2004, 431:371–378.

278. Chang LJ, Liu X, He J. Lentiviral siRNAs targeting multiple highly conserved RNA sequences of human immunodeficiency virus type 1. Gene Ther 2005, 12:1133–1144.

279. Nyce JW, Metzger WJ. DNA antisense therapy for asthma in an animal model. Nature 1997, 385: 721–725.

280. Choi IW, Kim DK, Ko HM, et al. Administration of antisense phosphorothioate oligonucleotide to the p65 subunit of NF-kappaB inhibits established asthmatic reaction in mice. Int Immunopharmacol 2004, 4:1817–1828.

281. Duan W, Chan JH, McKay K, et al. Inhaled p38alpha mitogen-activated protein kinase antisense oligonucleotide attenuates asthma in mice. Am J Respir Crit Care Med. 2005, 171:571–578.

282. Bitko V, Barik S. Phenotypic silencing of cytoplasmic genes using sequence-specific double-stranded short interfering RNA and its application in the reverse genetics of wild type negative-strand RNA viruses. BMC Microbiol 2001, 1:34.

283. Bitko V, Musiyenko A, Shulyayeva O, et al. Inhibition of respiratory viruses by nasally administered siRNA. Nat Med 2005, 11:50–55.

284. Levine DS. Clinical features and complications of Crohn's disease. In: Targan SR, Shanahan F, eds. Inflammatory bowel disease: from bench to bedside. Baltimore: Williams & Wilkins, 1993:296–316.

285. Dustin ML, Rothlein R, Bhan AK, et al. Induction by IL 1 and interferon-gamma: tissue distribution, biochemistry, and function of a natural adherence molecule (ICAM-1). J Immunol 1986, 137:245–254.

286. Rothlein R, Dustin ML, Marlin SD, et al. A human intercellular adhesion molecule (ICAM-1) distinct from LFA-1. J Immunol 1986, 137:1270–1274.

287. Simmons D, Makgoba MW, Seed B. ICAM, an adhesion ligand of LFA-1, is homologous to the neural cell adhesion molecule NCAM. Nature. 1988, 331:624–627.

288. To SS, Newman PM, Hyland VJ, et al. Regulation of adhesion molecule expression by human synovial microvascular endothelial cells in vitro. Arthritis Rheum 1996, 39:467–477.

289. Vainer B, Nielsen OH, Horn T. Comparative studies of the colonic in situ expression of intercellular adhesion molecules (ICAM-1, -2, and -3), beta2 integrins (LFA-1, Mac-1, and p150,95), and PECAM-1 in ulcerative colitis and Crohn's disease. Am J Surg Pathol 2000, 24:1115–1124.

290. Yacyshyn BR, Bowen-Yacyshyn MB, Jewell L, et al. A placebo-controlled trial of ICAM-1 antisense oligonucleotide in the treatment of Crohn's disease. Gastroenterology 1998, 114:1133–1142.

291. Glover JM, Leeds JM, Mant TG, et al. Phase I safety and pharmacokinetic profile of an intercellular adhesion molecule-1 antisense oligodeoxynucleotide (ISIS 2302). J Pharmacol Exp Ther 1997, 282:1173–1180.

292. Yacyshyn BR, Chey WY, Goff J, et al. Double blind, placebo controlled trial of the remission inducing and steroid sparing properties of an ICAM-1 antisense oligodeoxynucleotide, alicaforsen (ISIS 2302), in active steroid dependent Crohn's disease. Gut 2002, 51:30–36.

293. Di Bisceglie AM, Carithers RL Jr, Gores GJ. Hepatocellular carcinoma. Hepatology 1998, 28:1161–1165.

294. Hayashi J, Aoki H, Arakawa Y, et al. Hepatitis C virus and hepatocarcinogenesis. Intervirology 1999, 42:205–210.

295. Jeffers L. Hepatocellular carcinoma: an emerging problem with hepatitis C. J Natl Med Assoc 2000, 92:369–371.

296. Hanecak R, Brown-Driver V, Fox MC, et al. Antisense oligonucleotide inhibition of hepatitis C virus gene expression in transformed hepatocytes. J Virol 1996, 70:5203–5212.

297. Zhang H, Hanecak R, Brown-Driver V, et al. Antisense oligonucleotide inhibition of hepatitis C virus (HCV) gene expression in livers of mice infected with an HCV-vaccinia virus recombinant. Antimicrob Agents Chemother 1999, 43:347–353.

298. Soler M, McHutchison JG, Kwoh TJ, et al. Virological effects of ISIS 14803, an antisense oligonucleotide inhibitor of hepatitis C virus (HCV) internal ribosome entry site (IRES), on HCV IRES in chronic hepatitis C patients and examination of the potential role of primary and secondary HCV resistance in the outcome of treatment. Antivir Ther 2004, 9:953–968.

299. McHutchison JG, Patel K, Pockros P, et al. A phase I trial of an antisense inhibitor of hepatitis C virus (ISIS 14803), administered to chronic hepatitis C patients. J Hepatol 2006, 44:88–96.

300. Sen A, Steele R, Ghosh AK, Basu A, et al. Inhibition of hepatitis C virus protein expression by RNA interference. Virus Res 2003, 96:27–35.

301. Far RK, Nedbal W, Sczakiel G. Concepts to automate the theoretical design of effective antisense oligonucleotides. Bioinformatics 2001, 17:1058–1061.

302. Zhou D, He QS, Wang C, et al. RNA interference and potential applications. Curr Top Med Chem 2006, 6:901–911.

Part V
Sensitization Tailored to Individual Patients

Chapter 21
DNA Polymorphisms Affecting Chemosensitivity Toward Drugs

Thomas Efferth and Michael Wink

1 Introduction

1.1 Heterogeneity of Drug Response and Pharmacogenetics

Clinically it is well known that the same doses of medication cause considerable heterogeneity in efficacy and toxicity across human populations [1, 2]. This heterogeneity can lead to unpredictable life-threatening or even lethal adverse effects in patients who react hypersensitively [3, 4]. Since the inter-individual variability in drug response cannot satisfactorily be explained by factors such as renal and liver function, patients' age and co-morbidity, life style, or co-medication and compliance, molecular factors came into the center of interest.

The term *pharmacogenetics* was coined in the 1950s, as it became evident that there is an inherited basis for differences in the disposition and effects of drugs and xenobiotics in subjects deficient in glucose-6-phosphate dehydrogenase [5]. Another milestone of pharmacogenetics was the discovery of a debrisoquine metabolism in the 1970s, which causes a decline of blood pressure. Also at that time it became apparent that the oxidation of the anti-arrhythmic drug sparteine is a polymorphic trait. The underlying reason is a CYP2D6 deficiency (cytochrome P450 mono-oxygenase). The CYP proteins belong to phase I enzymes. The metabolism of xenobiotic compounds consists of three phases for the detoxification of these compounds. Phase I enyzmes oxidize, reduce, or hydrolyze lipophilic xenobiotics, to make them more water soluble. Phase II enzymes bind them to carrier molecules, and phase III proteins transport these conjugates. Today, more than 70 genetic variants are known in this gene, which influence the metabolism of over 40 drugs.

Pharmacogenetics focuses on the prediction of drug efficacy and toxicity based on a patient's or a tumor's genetic profile with routinely applicable genetic tests and easily accessible test samples, that is, tumor biopsies or peripheral blood. Pharmacogenetic biomarkers hold great promise for the individualization of therapeutic intervention to select the most appropriate medication and apply the optimal dose for each individual patient according to precise marker-assisted screening tests. The hope is to stratify individual patients based on their probability of response to a particular customized therapy.

1.2 Gene Polymorphisms

An important result of the human genome project is the high genomic variability (polymorphisms). Statistically a genetic variation occurs once in 1,200 bases. In most cases, polymorphic gene variants lead to a diminished protein function, in some cases, however, increased activities have been reported [1]. In contrast to somatic mutations, i.e., in cancer, polymorphisms in germ line cells are stable and heritable. Polymorphisms include single nucleotide polymorphisms (SNPs),

From: *Sensitization of Cancer Cells for Chemo/Immuno/Radio-therapy, 1st Edition.*
Edited by: Benjamin Bonavida © Humana Press, Totowa, NJ

and length differences in micro- and minisatellites. An SNP represents a single base exchange that may or may not cause an amino acid exchange in the encoded protein. The frequency of SNPs is >1% in a population and accounts for >90% of genetic variation in the human genome. The number of SNPs in the human genome has been estimated in a range from 1 to 10 million [6–8]. Between 50,000 and 250,000 SNPs are distributed in and around coding genes [9]. Many SNPs do not confer phenotypic alterations but are tagged to certain haplotypes. They might, therefore, be of importance as genetic markers too. Microsatellites (or tandem repeats) are multiple copies of repeated DNA sequences (2–4 nucleotides). Minisatellites consist of repeated small sequences up to 40–60 nucleotides in a length of 0.1–10 kb [10]. During the past decade an increasing number of polymorphisms have been analyzed for their association with diseases [11].

Alleles are alternate forms of a gene. The gametic haplotype is the sum of alleles at a genetic locus. If alleles at different genetic loci are nonrandomly associated, the term *linkage disequilibrium* is used. Haplotype and linkage disequilibrium mappings are used to explain drug response phenotypes of complex diseases such as cancer [12].

In contrast to germ line polymorphisms, mutations are rare changes in the DNA sequence (<1% frequency within a species) of somatic cells that can silence the activity of the encoded gene product. Mutations appear as single base substitutions (nonsense or missense mutations), deletions, insertions, or gene rearrangements. Mutations also occur in somatic cells and lead to the aging of organisms and to the development of certain diseases such as cancer. Life span correlates with the cellular capacity to repair molecular damage and withstand environmental oxidative stress to maintain physiologic functions of organisms [13, 14]. Biological aging represents a random, stochastic process occurring long after reproductive maturation. Thus, aging is unlikely to be a matter of evolutionary selection pressure. Natural selection must guarantee an organism to live long enough to reach sexual maturation and reproduce itself. There is no or only little evolutionary selection pressure to avoid DNA lesions which could be detrimental in later stages of life [15]. This has been termed "mutation accumulation theory."

2   Monogenic Diseases: Glucose-6-Phosphate Dehydrogenase Deficiency

An important example of a heritable monogenic disease, which is associated with drug hypersensitivity, represents glucose-6-phosphate dehydrogenase (G6PD) deficiency. In early observational studies, it was recognized that antimalarial drugs and certain foods (soy beans) cause hemolytic crises in patients with glucose-6-phosphate dehydrogenase (G6PD) deficiency. Deficiency in glucose-6-phosphate dehydrogenase (G6PD) is the most common hereditary enzymopathy worldwide affecting an estimated 400 million people [16]. Although single base alterations in the G6PD gene are by far the most common reason for reduced enzyme activity (>140 different variants), other rare alterations such as splice site mutations have also been discovered [17, 18]. G6PD is the first enzyme of the pentose phosphate pathway that converts α-D-glucose-6-phospate into D-glucono-1,5-lactone-6-phosphate and is involved in the generation of NADPH. As erythrocytes lack the citric acid cycle and respiratory chain, the pentose phosphate shunt is the only source of NADPH for them. NADPH is required for the generation of reduced glutathione, which is important for the protection against oxidative damage. Prolonged neonatal jaundice and hemolytic anemia are common clinical manifestations. Infections, ingestion of fava beans, and some drugs can trigger life-threatening hemolytic anemia. Relevant drugs in this context are the antimalarial primaquine, sulfonamide antibiotics, dapsone (to treat leprosy), the antidiabetic glibenclamide, the antibiotic nitrofurantoin, and others. Since glutathione represents a cellular protectant against oxidative damage, which inhibits apoptosis, it can be hypothesized that nucleated cells of G6PD-deficient patients should be more susceptible to DNA damage and apoptosis induced by oxidative stress. Peripheral mononuclear cells (PBMC) from a G6PD-deficient male (Fig. 21.1) showed significantly higher apoptotic rates upon challenge by cytostatic drugs (daunorubicin), gamma-irradiation, or glucocorticoids (dexamethasone) than PBMC from healthy males [19]. The induction of oxidative stress / generation of reactive oxygen species by all three agents has

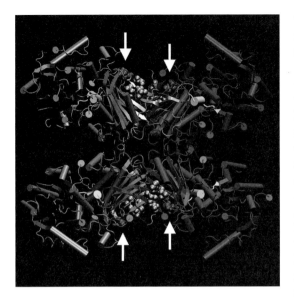

FIGURE 21.1 Model of the G6PD Aachen tetramer. This G6PD variant has originally been described by Kahn et al. [253]. The mutation in the G6PD Aachen variant has been determined by us previously [254]. A mutation 1089C>G results in a predicted amino acid change 363Asn>Lys. The 1089C>G point mutation is unique, but produces the identical amino acid change found in another variant of G6PD deficiency, G6PD Loma Linda. The 363Asn>Lys exchange in G6PD Loma Linda is caused by a 1089C>A mutation [255]. Using the available three-dimensional structure of the human G6PD tetrameric protein complex [256], the location of the point mutation of amino acid 363 in G6PD Aachen is found at the surface of a monomer in close proximity to NADP$^+$ and more than 20Å away from the glucose-6-phosphate binding site. This residue is probably involved in NADP$^+$ binding that in turn is required for tetramer stability [256]. Thus, Arg363 may be required to indirectly maintain the structural integrity of the functional unit. Replacing it with a positively charged Lys residue would lead to charge–charge repulsion between Lys363 and NADP$^+$, thus affecting NADP$^+$ binding and tetramer formation. The two pairs of dimer-forming monomers are colored in ice blue/blue and in orange/red. The mutation site (Asn363) and the cofactor NADP$^+$ are shown in van der Waals representation in purple and grey, respectively. The conserved eight-residue peptide RIDHYLGK corresponding to the substrate binding site is colored in yellow. The figure was prepared from the crystal structure 1QKI using the program VMD [257] (*See Color Plates*)

been reported [20–26]. This implies that diverse ROS-generating agents affect white blood cells of G6PD-deficient patients leading to apoptosis due to insufficient protection from oxidative damage by reduced glutathione. G6PD is essential for defense against oxidative stress using G6PD knockout mice [27, 28]. Knockout clones of embryonic stem cells were extremely sensitive to hydrogen peroxide and the sulfhydryl group oxidizing agent diamide. UV-light induced more DNA damage and more apoptosis PBMC of G6PD-deficient males than in PBMC of healthy male subjects. PBMC of heterozygote females showed intermediate rates of DNA damage and apoptosis [29]. Increased DNA damage and apoptosis may be a result of increased oxidative stress in G6PD-deficient patients. UV-radiation is known to induce DNA damage and apoptosis by the generation of reactive oxygen species [30–33]. Interestingly, UV-radiation induces hemolysis—a major complication in G6PD-deficiency—which is reversible by the addition of glutathione [34]. Recently, Mesbah-Namin et al. [35] hypothesized that failure to detoxify hydrogen peroxide in G6PD-deficient leukocytes could induce primary DNA damage. Using the comet assay, these authors found that PBMC of 36 infants suffering from the Mediterranean variant of G6PD deficiency showed a significantly higher level of DNA stand breakage than PBMC of healthy control persons.

3 Complex Diseases: Cancer

3.1 The Resistance Problem

Tracing the rise of tumor chemotherapy from the discovery of the first anticancer drug, nitrogen mustard [36] to the advent of molecular approaches sheds light on the problem of drug resistance, which has dogged oncology for the past half century. The concept of combination therapy in oncology is based on the view that cancer cells which are resistant to one drug remain susceptible to other drugs. Clinically, the success of combination treatments can be hampered by the development of broad-spectrum or multidrug resistance (MDR). Due to the modest tumor specificity of many anticancer drugs, normal tissues are also damaged. This prevents the application of sufficient high doses to eradicate less sensitive tumor cell populations. In this manner tumors develop drug resistance that leads to treatment failure and fatal consequences for patients. Novel strategies to broaden the narrow therapeutic range by separating the effective dose and toxic dose would be of great benefit for the improvement of cancer chemotherapy.

A number of strategies have been advised to combat the war against cancer, e.g., modulators of drug resistance, non–cross-resistant drug derivatives, high-dose therapy, hematopoietic stem cell transplantation, and supportive gene therapy protocols. Another way is the a priori diagnosis of drug-resistant tumors.

The question arises as to which particular cytostatic agent and which combination of substances is most suited for an individual tumor. While the statistical probability of therapeutic success is well known for larger groups of patients from clinical therapy trials, it is, however, not possible to predict how an individual tumor will respond to chemotherapy. Although clinicopathologic prognostic factors such as tumor size, lymph node, and far distance metastases are valuable for the determination of prognosis of larger cohorts, they are less helpful for the development of personalized therapy options. Biomarkers are desired to stratify patients into groups of different likelihood of the tumors' responsiveness or of toxicity in normal organs.

In the decades efforts have been undertaken to determine the resistance profiles of tumors by testing drug response in vitro [37, 38]. While the concept of individualized therapy itself traces back to the 1950s [39], the advances in cell and molecular biology have opened new avenues for the characterization of drug-resistant tumors, and the transfer from the bench to the bed is a straightforward requirement. The current progress in molecular biology gives reason to believe that molecular approaches will significantly improve individual tumor therapy.

In the past, methods to detect protein or mRNA expression were more at the center of interest. Protein detection by immunohistochemistry or flow cytometry and mRNA detection by reverse transcription-polymerase chain reaction (RT-PCR) may be more suited for clinical routine applications than more sophisticated assays such as Western blotting, Northern blotting, or RNAse protection assay. Immunohistochemistry, however, came under debate with respect to its reproducibility to detect low levels of P-glycoprotein expression in large multicentric evaluations [40, 41]. To define consensus recommendations for the standardized

of detection of resistance genes expressed in low amounts is therefore more difficult as initially estimated [42]. Although immunohistochemistry is a morphologic means with inherent limitations concerning quantitative evaluation, many semi-quantitative immunohistochemical studies provided evidence for the prognostic value of P-gp for patients' survival or a disease-free interval and corroboration has come from investigations measuring mRNA expression [43–46].

Apart from single factors with prognostic relevance, a panoply of diverse genes contribute to drug resistance and hamper the evaluation of impact of any single resistance gene. Although anticancer drugs are extremely divergent in their chemical and physical structures and biological actions, a synopsis of the relevant mechanisms, which influence drug effects allows us categorize them as to whether they act upstream of the actual drug target, at critical target sites, or downstream of them [47, 48] (Fig. 21.2).

1. Mechanisms acting upstream include transporter proteins for uptake or excretion (i.e., ATP-binding cassette transporters, reduced folate carriers, and nucleoside transporters) and drug-metabolizing enzymes that activate, inactivate, or detoxify drugs (i.e., phase I and II enzymes). Metabolizing enzymes and transporter molecules often do not exhibit specificity for certain anticancer drugs but are operative toward a wide array of different xenobiotic drugs, including anticancer agents. Drug-metabolizing enzymes may influence pharmacokinetics and dynamics.

2. Drug targets are the DNA (and DNA repair mechanisms) for alkylating agents and platinum drugs, RNA (RNA synthesis inhibitors, i.e., actinomycin D), and specific proteins such as DNA topoisomerases I and II (camptothecins, anthracyclines, and epipodophyllotoxins), tubulins (Vinca alkaloids and taxanes), or enzymes of DNA biosynthesis (antimetabolites).

3. Mechanisms downstream of the actual drug targets and at distinct intracellular locations are operative after attack by drugs has taken place. The most important downstream mechanisms are the diverse apoptosis pathways. Their deregulation may lead to drug resistance and survival of cancer cells despite target molecules

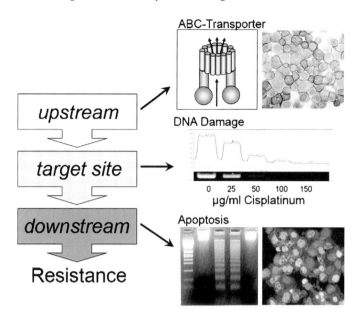

FIGURE 21.2. Multifactorial drug resistance of tumors. Molecular determinants of drug resistance are at the upstream level, target site, or downstream of it. Upstream mechanisms are ATP-binding cassette (ABC) transporters in the membrane, which extrude drug molecules out of the cell. As an example, the immunohistochemical detection of the ABC transporter P-glycoprotein in multidrug-resistant leukemia cells is shown [59]. DNA damage by cytostatic agents such as cisplatin can be measured by the PCR stop assay. Since the processivity of Taq polymerase is disturbed by DNA lesions, increasing doses of DNA-damaging agents lead to decreased PCR amplification [258, 259]. Drugs that bypass upstream and target side resistance mechanisms lethally damage the cancer cell and induce apoptosis. Apoptotic cells undergo nuclear fragmentation [49]. This figure is modified according to Efferth [260]

have been successfully targeted by anticancer drugs [49, 50]. Programmed cell death is not only regulated by the proteins directly involved in the apoptotic cascade but also by external factors, i.e., chemokines that act as "survival factors" aiding to prevent apoptosis and, hence, contributing to survival and drug resistance of tumor cells after chemotherapeutic insult [51, 52].

Genes at known upstream sites, drug target sites, and downstream sites can serve as candidates for the analysis of polymorphisms with prognostic power for the efficacy and toxicity of drugs and survival of cancer patients after chemotherapy. This is more a practical than a systematic approach. A still-unknown number of SNPs may be hidden in the human genome that are not located within or in close vicinity to known candidate genes but that also reveal prognostic significance for the response of tumors to chemotherapy and for the survival of patients. As the knowledge on the human genome expands, novel polymorphisms may be

unraveled to predict the efficacy and toxicity of anticancer drugs.

3.2 Upstream Mechanisms

3.2.1 Drug Transporters

The ATP-binding cassette (ABC) transporter family consists of 49 members. More than 10 of them are implicated in drug resistance to cancer chemotherapy [53–56]. They are important determinants of drug absorption, tissue targeting, and drug elimination. ABC transporters confer drug resistance by lowering the intracellular drug concentrations down to sublethal levels.

3.2.1.1 P-Glycoprotein *(ABCB1, MDR1)*

Cancer cells, which express P-glycoprotein, reveal a multidrug resistance phenotype to a broad range of structurally and functionally different drugs, including anthracyclines, anthracendiones, *Vinca* alkaloids, taxanes, epipodophyllotoxins, and others. P-glycoprotein is also expressed in various normal organs, such as brain vessels, adrenal gland, kidney, liver, and gastrointestinal tract [57]. P-glycoprotein contributes to the blood–brain barrier, translocates hormones, and detoxifies xenobiotics taken up along with nutrients. Although clinical therapy failure of tumors is multifaceted [58–64], its role for drug resistance is evident [65, 66]. The prognostic significance of P-glyco-protein/MDR1 as indicator for failure to chemo-therapy and worse outcome has been demonstrated in a number of clinical studies [67–69]. As of yet, 29 polymorphisms have been identified in the *ABCB1* gene, with considerable differences in their frequencies among ethnic groups [70–73]. Of them, the G2677T/A SNP in exon 21 and the C3435T SNP in exon 26 have been most inten-sively studied, because they diminish the expres-sion and function of P-glycoprotein (Fig. 21.3) [72, 74]. Whereas the C2677T/A polymorphism causes an A893S/T amino acid substitution, the functional relevance of the C3435T variant is unknown, because this is a silent SNP. It is possible that specific haplotypes of the ABCB1 (MDR1) gene might determine the efficacy and toxicity of drugs. The effect of the G2677T and C3435T SNPs on the pharmacokinetics and pharmacodynamics of drugs is still a matter of a controversy, since contrasting results have been provided [75–81]. This discrep-ancy is apparent not only for anticancer drugs, but also for cardiac glycosides, immunosuppressants, HIV protease inhibitors, tricyclic antidepressants, and others [82]. The expression of P-glycoprotein in normal tissues is thought to play an important role for the pharmacokinetics and pharmacody-namics of many drugs of different drug classes. A definitive answer on the role of *ABCB1* SNPs requires further studies.

3.2.1.2   Multidrug Resistance–Related Proteins

Multidrug resistance–related proteins (ABCC fam-ily members, MRPs) also act as efflux pumps ren-dering cells resistant to anticancer agents [83]. The cross-resistance profiles overlap that of P-glycopro-tein but are not identical. They are expressed both in drug-resistant tumors and normal tissues, i.e., lung, testis, peripheral blood mononuclear cells, etc. MRPs are located in the outer basal plasma

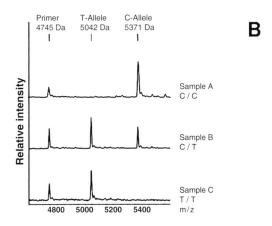

WT
5'-CAAGAGAT**C**GTGAGGGCAGCAAAGG -3'
3'-   **G**CACT CCCGTCGTTTCC –5'

C3435T
5'-GAAGAGAT**T**GTGAGGGCAGCAAAGG -3'
3'-   **A**CACT CCCGTCGTTTCC –5'

Extension primer reverse: 4745 Da
                 CACTCCCGTCGTTTCC

Extension dG and ddA:
    Extension product WT: 5371 Da
                 AGCACTCCCGTCGTTTCC
    Extension product C3435T: 5042 Da
                 ACACTCCCGTCGTTTCC

MDR1 (exon 26): SNP C3435T

FIGURE 21.3. Single nucleotide polymorphism (SNP) C3435T in the *ABCB1 (MDR1)* gene. **A.** Localization of the SNP and primers used. **B.** Detection of the C3435T genotype by means of MALDI-TOF. The DNA sequence containing the SNP is amplified by PCR. The PCR product and a reverse extension primer are used for a subsequent extension reaction with dGTP and ddATP. The extension products are purified, and the molecular weight is determined by MALDI-TOF. (Reprinted from Efferth T, Sauerbrey A, Steinbach D, et al. Analysis of single nucleotide polymorphism C3435T of the multid-rug resistance gene MDR1 in acute lymphoblastic leuke-mia. Int J Oncol 2003, 23:509–517.)

membrane as well as in intracellular vesicles and the Golgi apparatus, indicating that sequestration of drugs into vesicles takes place before drugs are released out of the cell. They transport drugs and xenobiotics as glutathione conjugates. A prognostic role for clinical tumor resistance and patients' survival has been suggested but not yet established [84]. The same applies to MRP polymorphisms, although several SNPs in the *MRP1* and *MRP2* genes are associated with the Dubin-Johnson syndrome [85–90].

3.2.1.3 Breast Cancer–Related Protein

In contrast to P-glycoprotein and MRPs, BCRP represents an ABC half-transporter, which forms homodimers in the plasma membrane. These functional dimers extrude drugs and xenobiotics out of the cell and mediate multidrug resistance. It is expressed in several normal organs and has a role for stem cell protection and regulation and also in hypoxic defense mechanisms [91]. A prognostic role of BCRP for clinical chemotherapy and survival of cancer patients has been shown for some tumor types [92, 93]. Variations in the BCRP amino acid sequence have been shown to impact substrate specificity and protein expression levels [94, 95].

3.2.2 *Drug-Metabolizing Phase I Enzymes*

3.2.2.1 Cytochrome P450 Monooxygenases

Cytochrome P450 monooxygenases (CYPs) are phase I enzymes that oxidize, reduce, or hydrolyze xenobiotics [96]. CYPs can either detoxify xenobiotics such as anticancer drugs, i.e., epipodophyllotoxins, paclitaxel, *Vinca* alkaloids, and tamoxifen, or activate inactive prodrugs, i.e., cyclophosphamide [97]. There are 55 *CYP* genes in the human genome, of which the *CYP1A, CYP2B, CYP2C*, and *CYP3A* subfamilies are involved in the metabolism of anticancer drugs. CYPs are mostly expressed in the liver and also in intestinal epithelia. Because some anticancer drugs (i.e., cyclophosphamide and paclitaxel) are metabolized by CYP isoenzymes, they contribute to drug response and/or toxicity [98]. The CYP3A isoenzymes account for approximately half of the metabolic activity of all CYPs. The two main forms of CYP3A isoenzymes are CYP3A4 and CYP3A5. The CYP3A4*1B and CYP3A5*3 variants had a statistically decreased risk of peripheral neuropathy in pediatric acute lymphoblastic leukemia patients [99]. Secondary leukemia can occur in later ages, if juvenile cancer patients are treated with epipodophyllotoxins. It has

been shown that an SNP in the 5V promotor region of CYP3A4, termed CYP3A4*V, reduces enzyme activity, and hence the metabolization of epipodophyllotoxins [100]. This is a genetic explanation for the increased risk for leukemia in pediatric oncology upon treatment with teniposide or etoposide (derivatives of podophyllotoxin). CYP2C8 metabolizes paclitaxel by 6a-hydroxylation. The CYP2C8*3 allele, which appears only in Caucasians is associated with a less efficient paclitaxel metabolism [101, 102]. The antiangiogenic drug thalidomide is activated by CYP2C19-mediated 5-hydroxylation. In prostate cancer patients, CYP2C19*2 homozygotes showed a poor metabolizing phenotype and did not respond to thalidomide treatment [103].

3.2.3 *Drug-Metabolizing Phase II Enzymes*

Phase II enzymes conjugate phase I products, other intermediates, or the parent compound for renal or biliary elimination. Among others, phase II enzymes are glutathione S-transferases (GST) and UDP-glucuronosyltransferases (UGT).

Glutathione S-transferases (GST): They conjugate glutathione to electrophilic molecules and oxidized metabolites. They play a role for mutagenesis and carcinogenesis and influence the cellular response to anticancer drugs [48, 58, 60, 104, 105]. Six subclasses of GSTs have been described: α (GSTA), π (GSTP), μ (GSTM), o (GSTO), τ (GSTT), and ζ (GSTZ). GST genes are highly polymorphic, including null variants (GSTM and GSTT1), low-activity variants (GSTP1), or variants with altered inducibility (GSTM3) [106]. The differences of frequencies of gene deletions were considerable between the different ethnic groups [107]. The null genotypes for GSTM1 or GSTT1 were associated with a reduction in risk of relapse in several tumor types treated with chemotherapy [108–114]. In addition to null phenotypes, SNPs also affect response to chemotherapy and survival of patients. The GSTT1 homozygous genotype predicted improved response rates in multiagent chemotherapy for acute lymphoblastic leukemia [115, 116]. The I105V SNP in the GSTP1 gene was of prognostic value for the patients' survival [117–119].

Patients with the 105VV homozygote allele had an improved progression-free survival. This substitution of isoleucine with valine at position

105 reduced enzyme activity against alkylating agents [120]. Other SNPs also conferred resistance to alkylating agents [121]. In colorectal cancer patients who received 5-fluorouracil/oxaliplatin–based chemotherapy, the valine homozygotes had a survival advantage compared with heterozygotes and isoleucine homozygotes [122]. The proportion of breast cancer patients surviving at 5 years was smaller among patients who had 0 or 1 GSTA1*B allele than among GSTA1*B/*B subjects [123]. In addition, GST polymorphisms may also be useful to predict drug-related toxicity. Ototoxicity is a frequent side effect of cisplatinum treatment. Patients with the GSTP3*B allele were found to have a higher risk of cisplatinum-induced ototoxicity [124]. The frequency of toxicity-related death in remission was increased in GSTT1-negative homozygotes compared with GSTT1-positive patients with acute myeloblastic leukemia [125]. The inheritance of at least 1 Val allele at GSTP1 at position 105 increased the risk of secondary, therapy-related acute myeloblastic leukemia after chemotherapy [126].

3.2.3.1   UDP-Glucuronosyltransferases

These enzymes catalyze the glucuronidation of many lipophilic xenobiotics to make them more water soluble and thereby enhance their elimination. More than 30 UGT isoforms with overlapping substrate specificity have been identified, with two major classes, UGT1 and UGT2 [127]. The prodrug irinotecan (CPT-11) (derived from camptothecin) is converted in the liver to an active metabolite, SN-38, which is a DNA topoisomerase I inhibitor for the treatment of colorectal cancer. UGT1A1 conjugates SN-38 to the inactive SN-38 glucuronide, which is excreted into bile and urine. With reduced capacity for glucuronidation, SN-38 can cause life-threatening diarrhea provoked by SN-38–mediated enteric injury [128]. Reduced glucuronidation occurs as a consequence of reduced transcription rate due to abnormal dinucleotide repeat sequences (5–8 repeats) within the TATA box of the UGT1A1 gene promoter [129]. There was an inverse relationship between the number of TA repeats and the UGT1A1 transcription rate. This promoter polymorphism was found in patients with Gilbert's syndrome (mostly as a (TA)7 repeat called UGT1A1*28 allele), a mild form of inherited

unconjugated hyperbilirubinemia, and was responsible both for the inherited disease itself as well as for severe toxicity upon CPT-11 treatment. The UGT1*28 allele was associated with both a reduced area under the curve (AUC) ratio for SN-38 and an increased total bilirubin level pointing to the relevance for the pharmacokinetics of SN-38 [130]. The usefulness of the identification of UGT1A1*28 homozygotes or heterozygotes for the prediction of severe irinotecan toxicity has been shown in clinical studies [131–135]. Font et al. [133] reported that 8 of 23 patients with non-small cell lung carcinomas (34%) with the common genotype achieved disease control (partial response or stable disease) compared with 13 of 24 patients (54%) with the variant genotypes. Furthermore, survival in patients with the variant genotypes was higher than in those with the common genotype.

3.3   Drug Target Interactions

3.3.1   DNA Biosynthesis and Metabolism

3.3.1.1   Thiopurine S-Methyltransferase

6-Mercaptopurine (6-MP) and 6-thioguanine (6-TG) need to be activated by hypoxanthine guanine phosphoribosyl transferase (HGPRT) to 6-thioguanine nucleotides (TGN), which are incorporated into the DNA. Thiopurine S-methyltransferase (TPMT) inactivates 6-MP and 6-TG by S-methylation and thereby inhibits TGN generation. Oxidation via xanthine oxidase can also inactivate the compounds. TPMT activity is found in the liver and in erythrocytes. Ten polymorphisms have been identified, of which TMPT*2 (G238C), TMPT*3A (G460A and A719G), and TMPT*3C (A719C) were found in 90% of subjects with reduced enzyme activity [136]. The three mutant variants enhanced TMPT protein degradation [137]. While 90% of the human population carries wild-type TPMT with normal enzyme activities, patients with homozygous mutant alleles (<1%) produce high TGN levels and are at high risk for severe bone marrow toxicity toward standard doses of 6-MP or 6-TG [138–145]. As hematopoietic cells lack xanthine oxidase activity, TPMT is the sole salvage enzyme against thiopurine toxicity [146]. If TPMT activity is diminished, high levels of TGN are formed with detrimental effects on hematopoietic tissues. Such patients need a dose reduction to 1/10 of the normal dose. Heterozygotes have interme-diate TPMT levels and need corresponding dose adjustment. Reduced TPMT activity increased the risk of secondary malignancies in pediatric cancer patients upon combination treatment of 6-MP with etoposide or irradiation [142–144, 147, 148]. Despite higher toxicity, leukemia patients with reduced enzyme activities had improved response rates to 6-MP therapy and a better prognosis for being cured compared with patients with wild-type TMTP [142–144, 149, 150]. Genotyping of TPMT represents a widely used approach for dose adjustment of thiopurines.

3.3.1.2 Thymidylate Synthase

The enzyme catalyzes the methylation of deoxy-uridine monophosphate (dUMP) to deoxy-thymidine monophosphate (dTMP). 5-Fluorouracil (5-FU) is converted to 5-fluoro-UMP, which binds to thymidylate synthase (TS) and blocks dTMP generation and DNA synthesis [151]. TS expression levels correlate inversely with the 5-FU sensitivity of tumors [152, 153]. TS expression can be regulated by a polymorphism in the 5V promoter enhancer region, termed thymidylate synthase 5V promotor enhancer region (TSER). Alleles with 2, 3, 4, 5, and 9 tandem repeats of a 28-bp sequence (TSER*2, TSER*3, TSER*4, TSER*5, and TSER*9) are known [154–157]. The higher the repeat number, the more TS is expressed [158]. TSER*2 homozygotes responded better to 5-FU-based chemotherapy and have a better prognosis than TSER*2/*3 heterozygotes and TSER*3 homozygotes [159–167]. The TS genotype represents an independent predictor of progression-free and overall survival [134, 135, 168].

3.3.1.3 5,10-Methylenetetrahydrofolate Reductase

This enzyme reduces 5,10-methylenetetrahyxro-folate to 5-methyltetrafolate needed for DNA synthesis. The C677T (A222V) and A1298C (E429A) polymorphisms alone or in combination caused reduced enzyme activities [169]. Since reduced 5,10-methylenetetrahydrofolate reductase (MTHFR) activity affects intracellular folate pools, antifolates can provoke increased toxicity. Indeed, the MTHFR 677TT genotype is linked with oral mucositis or bone marrow toxicity methotrexate-treated cancer patients [170–173]. The MTHFR genotype also affects the response of tumors to 5-FU– or MTX-containing chemotherapy. Patients with the 677TT allele responded better than the other genotypes [174–176]. The time to progression was longer in patients with CC genotype than in patients with heterozygous or homozygous TT genotypes [177].

3.4 DNA Repair Mechanisms

Because DNA repair enzymes are correctives for DNA damage induced by anticancer agents, it was hypothesized that SNPs in DNA repair genes may influence treatment outcome.

3.4.1 HO^6-Methylguanine-DNA Methyltransferase

Some antitumor agents can alkylate guanine at the O^6 position, resulting in GC to AT transition mutations. MGMT transfers the O^6 position of guanine to cysteine 145 of the protein. Thereby, the protein is irreversibly inactivated [178]. Tumor cells develop resistance to antitumor agents that methylate (dacarbazine, temozolomide) or chloroethyl ate (bis-chloroethylnitrosourea) DNA [179]. Various MGMT inhibitors have been developed as pseudosubstrates for MGMT. O^6-Benzylguanine and derivatives inactivate MGMT and resensitize tumor cells to O^6-alkylating drugs [180, 181]. Using the crystal structure of non-alkylated MGMT, molecular models of mutant proteins encoded by SNPs have been built (Fig. 21.4) [182]. Most of the mutations were located either within the DNA binding region (A121E, A121T, G132R, and N123V) or in the vicinity of the active C145 (I143V and G160R). The L84F variant might affect Zn^{2+} binding. W65C generates an unstable enzyme. The G160R polymorphism existing in 15% of the Japanese population [183] was less inactivated by O^6-benzylguanine than the wild-type allele [184], suggesting that a subpopulation of patients was resistant to O^6-benzylguanine. This SNP was present in the healthy noncancerous white population of the United States with a probability of <1.6% [185]. These results point to the importance of the ethnic origin of patients for the presence or absence of therapeutically relevant SNPs.

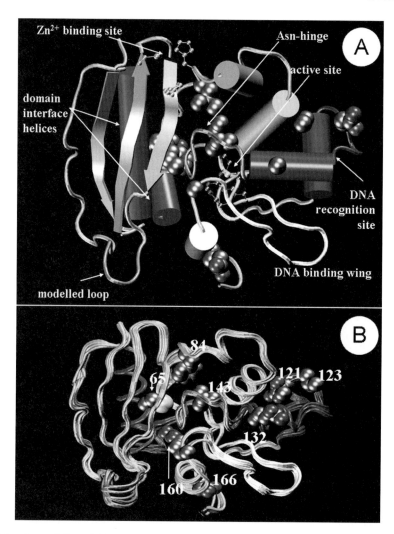

FIGURE 21.4 Molecular modeling of SNP-based variants of MGMT. **A.** Wild-type structure of MGMT. Mutation sites are shown in van der Waals representation. Color codes of helices: red, N-terminal α-helices, blue, DNA recognition site with helix-turn-helix motif. The second DNA recognition site binds to the major groove of DNA, yellow, 3–10 helix with conserved Pro-Cys-His-Arg sequence. The active Cys145 is located here. Color code of loops: orange, Asn-hinge that joins the DNA recognition helix and the active site. It also provides 40% of the contact between N-terminal and C-terminal domains, white, DNA binding wing. O6-alkylguanine lies between the binding wing and the recognition helix. Color code of side chains shown in CPK representation: per atom, conserved active Pro-Cys-Arg sequence, pink, Zn^{2+} binding site (residues C5, C24, H29, H85). Cys5 is missing, since it is not resolved in the crystal structure 1QNT. **B.** Localization of amino acid change and overlap of structures of all mutants. Mutated side chains are shown in van der Waals representations. (Reprinted from Schwarzl SM, Smith JC, Kaina B, et al. Molecular modeling of O6-methylguanine-DNA methyltransferase mutant proteins encoded by single nucleotide polymorphisms. Int J Mol Med 2005, 16:553–557.) (*See Color Plates*)

3.4.1.1 X-Ray Cross-Complementation Group 1 Gene

XRCC1 is a key enzyme in base excision repair for the removal of oxidative DNA damage and non-bulky adducts of alkylating agents. Base excision repair contributes to resistance towards mitomycin C, mafosfamide, chlorambucil, and 1,3-bis(2-chloroethyl)-1-nitrosourea [186–188]. In colorectal cancer, Gln mutant alleles of the R399Q SNP were associated with decreased response chemotherapy

[189]. Variant alleles of *XRCC1* were associated with shorter overall survival in non-small cell lung cancer [190]. The RR phenotype of *XRCC1* codon 399 was associated with an increased occurrence of radiation therapy–induced subcutaneous fibrosis [191].

3.4.1.2 Excision-Repair Cross-Complementing Genes 1 and 2

The *ERCC1* gene encodes a helicase required in the nucleotide excision repair pathway. This repair pathway is responsible for resistance to cisplatinum and chlorambucil [192–194]. Several *ERCC1* polymorphisms affecting DNA repair capacities were reported. In ovarian carcinoma cell lines, a silent C118T SNP was found [195]. The C>T transition coded in both variants for the same amino acid (asparagines), although the TT genotype may be associated with a reduction in codon usage by half and a reduction in *ERCC1* mRNA production (and subsequent protein translation too). This SNP may therefore be associated with reduced DNA repair capacity [196]. Lung cancer patients treated with cisplatinum and homozygous for the ERCC1 118 C allele had a significantly better survival [197, 198]. In colorectal carcinoma patients treated with 5-FU and oxaliplatin, the K751Q SNP in the *ERCC2* gene determined in peripheral blood lymphocytes were of prognostic relevance. KK homozygotes were more sensitive to chemotherapy and lived significantly longer than did heterozygotes or QQ homozygotes [199]. Contradictory results on the prognostic relevance of *ERCC2* variant alleles were reported for non-small cell lung cancer patients [190, 198, 200].

Human MutS homologue 2 (*hMSH2*) belongs to the DNA mismatch repair pathway, which has been associated with resistance to cisplatin, doxorubicin, etoposide, 6-mercaptopurine, 6-thioguanine, busulfan, procarbazine, and temozolomide [201–205]. Worrillow et al. [206] analyzed whether the -6 exon 13 T>C polymorphism in *hMSH2* modulated susceptibility to secondary leukemia after prior therapy with O^6-guanine-alkylating chemotherapy. Indeed, the authors found clues that this allele may confer non-disabling DNA mismatch repair defects with concomitant moderate alkylation tolerance, which predispose to the development of secondary leukemia.

3.5 Downstream Mechanisms

There are two main pathways for the programmed cell death [207, 208]. Both the extrinsic and the intrinsic pathways of apoptosis are regulated by the tumor suppressor p53 [209]. Members of the tumor necrosis factor (TNF) family trigger the receptor-mediated extrinsic pathway. FAS/CD95/APO-1 is the most prominent member of these death receptors. The mitochondrial intrinsic pathway is activated by stress signals or loss of survival signals. The *BCL-2* gene family is involved in the mitochondrial regulation of apoptosis. The different members of this gene family act as homo- or heterodimers, either in an anti-apoptotic (i.e., BCL-2, BCL-xL, BCL-w, A1, MCL-1, and BOO) or pro-apoptotic manner (i.e., BAX, BclxS, BAK, BOK, BIK, BAD, BID, HRK, and NOXA). Mitochondrial outer membrane permeabilization results in the release of caspase-activating molecules. As a consequence, initiator caspases activate effector caspases that catalyze proteolysis, ultimately leading to cell death. The tumor suppressor p53, FAS, BCL2 members, and caspases are also involved in the development of drug resistance [210–213]. Their value as prognostic factors for therapy response of tumors and survival times of patients' expression has been a matter of intense efforts over the past years [214–222].

3.5.1 Tumor Suppressor p53

Somatic mutations of the tumor suppressor gene tumor suppressor p53 (*TP53*) occur in about 50% of all human cancers, and p53 is a major player in carcinogenesis [223]. In response to DNA damage, p53 regulates cell cycle arrest, DNA repair, and apoptosis and guards, thereby the integrity of the genome. In addition to its role in carcinogenesis, *TP53* mutations render tumor cells resistant to chemotherapy [224, 225]. Interestingly, germ line SNPs have been identified in the *TP53* gene in addition to somatic mutations. A G>C base change at position 13,964 (G13964C) was observed in 3 of 42 women of Ashkenazi ancestry with hereditary breast cancer [226]. Two immortalized lymphoblastoid cell lines derived from patients with this SNP were resistant to cisplatinum treatment and showed decreased drug-induced apoptosis. Upon treatment with cisplatinum-based combination therapy, head and neck cancers with the *TP53*

72R SNP has a worse response than those with the 72P SNP [227]. Sullivan et al. [228] reported that the outcome of chemo-radiotherapy of squamous carcinomas is more favorable in cancers retaining a wild-type p53. Both overall survival and progression-free survival were significantly longer in patients whose cancers retained a wild-type p53 allele than in those lacking a wild-type allele. These clinical results correspond with in vitro data. In cell culture, the expression of 72P wild-type p53 resulted in G1 cell cycle arrest with minor induction of apoptosis after challenge with drug doses that cause extensive apoptosis in 72R-expressing variants [227].

3.5.2 BAX

A heterozygous G(-248)A polymorphism in the 5'Vuntranslated region of the BAX gene was present in 69% of stage I–IV patients with chronic lymphoblastic leukemia and 5.5% of stage 0 patients, and in 4.0% of healthy controls. It was significantly associated with reduced protein expression, tumor progression, and failure to achieve complete response [229].

4   Synopsis and Perspectives

Candidate gene-based approaches require the a priori knowledge and selection of a small number of genes for hypothesis testing. While this approach is feasible in monogenic diseases such as G6PD deficiency, complex human disorders such as cancer are caused by multiple factors. This represents an inherent problem to all single gene approaches trying to predict tumor drug response and normal tissue toxicity, because the understanding of the precise role of all participating factors is still limited. Although there is a large body of evidence that even the determination of single SNPs in cancers can produce meaningful information concerning chemosensitivity, toxicity, and prognosis, single SNPs do not cover the complexity of the drug resistance problem. As shown here, multiple factors at the upstream level, target interaction site, and downstream level all contribute to drug resistance.

Therefore, genome-wide linkage analyses may be a more systematic approach to fully discover genomic regions likely to harbor genes with relevance to

chemotherapy response of tumors and survival of patients. Watters et al. [230] mapped a quantitative trait locus influencing the cellular response to 5-FU treatment to chromosome 9q13–q22 and two quantitative trait loci influencing the cellular effects of docetaxel to chromosomes 5q11–21 and 9q13–q22.

Well-defined inbred mouse strains may also be useful for genome-wide mapping of genes affecting drug response [231]. The advantages of this approach are:

1. A reduced genetic complexity in inbred strains
2. A reduced dietary and environmental influence on drug response by standardized conditions of animal maintaining
3. A high degree of similarity between the mouse and human genomes

Generally, proteins act in complex networks. It is another argument that combinations of several polymorphisms of genes belonging to the same pathway might influence the therapeutic response rather than SNPs in single genes [232]. The metabolic pathway of 5-FU, which consists of 29 gene products [233] represents a suitable example to illustrate the magnitude of this problem. This situation is further complicated, if one tries to estimate how much a single candidate gene contributes to drug response in an individual cancer patient. The utility of genotyping of multiple SNPs in candidate genes has been shown recently [234–236].

As integrated drug pathway analyses might predict tumors' response or normal tissues' toxicity more reliable than single gene approaches, SNP genotyping should be accompanied by drug pathway profiling. The complexity of a combined approach of drug pathway profiling and multiple SNP testing necessitates the availability of both sophisticated bioinformatic tools and high throughput genotyping technology (i.e., MALDI-TOF, FP-TDI, SNP microarrays, and pyrosequencing) for large-scale generation and processing of data, to distill meaningful information from expanded data sets. The construction of comprehensive and reliable SNP and haplotype maps of the human genome will facilitate predictive genotyping in individual patients [237]. In addition to polymorphisms in nucleic DNA, sequence variants in the mitochondrial DNA may also be relevant for predicting drug effects. Peters et al. [238] found a linkage between the mitochondrial haplogroup J and cisplatinum-induced hearing impairment. Mitochondrial DNA has been neglected in pharmacogenetic research as of yet and may also be useful for the prediction of drug response.

While focusing on DNA polymorphisms, it should not be forgotten that other DNA alterations also contribute to drug resistance such as DNA methylation [239, 240] and genomic instability, that is, gain or loss of chromosomal material as well as inversions and translocations [241–244]. Sophisticated methods are available to detect therapy-related changes at a genome-wide level, e.g., methylation-sensitive microarrays, matrix comparative genomic hybridization, etc. Genome-wide scanning of genetic instabilities and gene expression profiles will achieve importance for the diagnosis of tumors with poor prognosis. While the cell's genome is rather static in nature, the expression and functions of proteins are extremely dynamic, depending on environmental conditions. Alterations in mRNA [245–248] and protein expression [61–63, 249–252] as well as metabolites are subject to transcriptome, proteome, and metabolome analyses as counterparts to genome analyses. These kinds of analyses allow the determination of dynamic alterations under chemotherapy under well-defined experimental conditions. It is not beyond the limits of expectation that combined "-omic" approaches emerge as predictors for the response of tumors toward chemotherapeutical agents. With such novel approaches of molecular diagnostics in place, there is reason to hope that individual cytostatic treatment schedules for each patient may pave new ways to defeat clinical drug resistance.

References

1. Evans WE, Relling MV. Pharmacogenomics: translating functional genomics into rational therapeutics. Science 1999, 286:487–491.
2. Fagerlund TH, Braaten O. No pain relief from codeine…? An introduction to pharmacogenomics. Acta Anaesthesiol Scand 2001, 45:140–149.
3. Rothenberg, ML, Meropol NJ, Poplin EA, et al. Mortality associated with irinotecan plus bolus fluorouracil/leucovorin: summary findings of an independent panel. J Clin Oncol 2001, 19:3801–3807.
4. Sargent DJ, Niedzwiecki D, O'Connell MJ, et al. Recommendation for caution with irinotecan, fluor-ouracil, and leucovorin for colorectal cancer. N Engl J Med 2001, 345:144–145.
5. Vogel F. Moderne Probleme der Humangenetik. Ergeb Inn Med Kinderheilkd 1959, 12:52–125.
6. Sachidanandam R, Weissman D, Schmidt SC, et al. A map of human genome sequence variation containing 1.42 million single nucleotide polymorphisms. Nature 2001, 409:928–933.
7. Botstein D, Risch N. Discovering genotypes underlying human phenotypes: past successes for Mendelian disease, future approaches for complex disease. Nat Genet 2003, 33(Suppl):228–237.
8. Carlson CS, Eberle MA, Rieder MJ, et al. Additional SNPs and linkage-disequilibrium analyses are necessary for whole-genome association studies in humans. Nat Genet 2003, 33:518–521.
9. Risch N. The genetic epidemiology of cancer: interpreting family and twin studies and their implications for molecular genetic approaches. Cancer Epidemiol Biomark Prev 2001, 10:733–741.
10. Danesi R, De Braud F, Fogli S, et al. Pharmacogenetic determinants of anti-cancer drug activity and toxicity. Trends Pharmacol Sci 2001, 22:420–426.
11. Hemminki K, Lorenzo Bernejo J, et al. The balance between heritable and environmental aetiology of human disease. Nat Rev Genet 2006, 7:958–965.
12. Nagasubramanian R, Innocenti F, Ratain MJ. Pharmacogenetics in cancer treatment. Annu Rev Med 2003, 54:437–452.
13. Martin GM, Austad SN, Johnson TE. Genetic analysis of aging: role of oxidative damage and environmental stress. Nat Genet 1996, 13:25–34.
14. Kapahi P, Boulton ME, Kirkwood TBL. Positive correlation between mammalian life span and cellular resistance to stress. Free Radic Biol Med 1999, 26:495–500.
15. Medawar PB. An unresolved problem of biology. London: Lewis, 1952.
16. Miwa S, Fujii H. Molecular basis of erythroenzymopathies associated with hereditary hemolytic anemia: tabulation of mutant enzymes. Am J Hematol 1996, 51:122–132.
17. Beutler E, Vulliamy TJ. Hematologically important mutations: glucose-6-phosphate dehydrogenase. Blood Cells Mol Dis 2002, 28:93–103.
18. Efferth T, Bachli EB, Schwarzl SM, et al. Glucose-6-phosphate dehydrogenase (G6PD) deficiency type Zurich: a splice site mutation as an uncommon mechanism producing enzyme deficiency. Blood 2004, 104:2608.
19. Efferth T, Fabry U, Glatte P, et al. Increased induction of apoptosis in mononuclear cells of a glucose-6-phosphate dehydrogenase deficient patient. J Mol Med 1995, 73:47–49.

20. Briehl MM, Baker AF, Siemankowski LM, et al. Modulation of antioxidant defenses during apoptosis. Oncol Res 1997, 9:281–285.

21. Müller I, Niethammer D, Bruchelt G. Anthracycline-derived chemotherapeutics in apoptosis and free radical cytotoxicity. Int J Mol Med 1998, 1:491–494.

22. Taatjes DJ, Fenick DJ, Gaudiano G, et al. A redox pathway leading to the alkylation of nucleic acids by doxorubicin and related anthracyclines: application to the design of antitumor drugs for resistant cancer. Curr Pharm Des 1998, 4:203–218.

23. Tome ME, Briehl MM. Thymocytes selected for resistance to hydrogen peroxide show altered antioxidant enzyme profiles and resistance to dexamethasone-induced apoptosis. Cell Death Diff 2001, 8:953–961.

24. Tome ME, Baker AF, Powis G, et al. Catalase-overexpressing thymocytes are resistant to glucocorticoid-induced apoptosis and exhibit increased net tumor growth. Cancer Res 2001, 61:2766–2773.

25. Miura Y. Oxidative stress, radiation-adaptive responses, and aging. J Radiat Res 2004, 45:357–372.

26. Spitz DR, Azzam EI, Li JJ, et al. Metabolic oxidation/reduction reactions and cellular responses to ionizing radiation: a unifying concept in stress response biology. Cancer Metastasis Rev 2004, 23:311–322.

27. Pandolfi PP, Sonati F, Rivi R, et al. Targeted disruption of the housekeeping gene encoding glucose 6-phosphate dehydrogenase (G6PD): G6PD is dispensable for pentose synthesis but essential for defense against oxidative stress. EMBO J 1995, 14:5209–5215.

28. Fico A, Paglialunga F, Cigliano L, et al. Glucose-6-phosphate dehydrogenase plays a crucial role in protection from redox-stress-induced apoptosis. Cell Death Differ 2004, 11:823–831.

29. Efferth T, Fabry U, Osieka R. DNA damage and apoptosis in mononuclear cells from glucose-6-phosphate dehydrogenase-deficient patients (G6PD Aachen variant) after UV irradiation. J Leukoc Biol 2001, 69:340–342.

30. Zhang X, Rosenstein BS, Wang Y, et al. Identification of possible reactive oxygen species involved in ultraviolet radiation-induced oxidative DNA damage. Free Radic Biol Med 1997, 23:980–985.

31. Pourzand C, Tyrrell RM. Apoptosis, the role of oxidative stress and the example of solar UV radiation. Photochem Photobiol 1999, 70:380–390.

32. Kawanishi S, Hiraku Y, Oikawa S. Mechanism of guanine-specific DNA damage by oxidative stress and its role in carcinogenesis and aging. Mutat Res 2001, 488:65–76.

33. Nishigori C, Hattori Y, Toyokuni S. Role of reactive oxygen species in skin carcinogenesis. Antioxid Redox Signal 2004, 6:561–570.

34. Godar DE, Thomas DP, Miller SA, et al. Long-wavelength UVA radiation induces oxidative stress, cytoskeletal damage and hemolysis. Photochem Photobiol 1993, 57:1018–1026.

35. Mesbah-Namin SA, Nemati A, Tiraihi T. Evaluation of DNA damage in leukocytes of G6PD-deficient Iranian newborns (Mediterranean variant) using comet assay. Mutat Res 2004, 568:179–185.

36. Rhoads CP. Report on a cooperative study of nitrogen mustard (HN2) therapy of neoplastic disease. Trans Assoc Am Physicians 1947, 60:110–117.

37. Salmon SE, Hamburger AW, Soehnlein B, et al. Quantitation of differential sensitivity of human-tumor stem cells to anticancer drugs. N Engl J Med 1978, 298:1321–1327.

38. Volm M, Drings P, Mattern J, et al. Prognostic significance of DNA patterns and resistance-predictive tests in non-small cell lung carcinoma. Cancer 1985, 56:1396–1403.

39. Cortazar P, Johnson BE. Review of the efficacy of individualized chemotherapy selected by in vitro drug sensitivity testing for patients with cancer. J Clin Oncol 1999, 17:1625–1631.

40. Beck WT, Grogan TM, Willmann CL, et al. Methods to detect P-glycoprotein-associated multidrug resistance in patients' tumors: consensus recommendations. Cancer Res 1996, 56:3010–3020.

41. Marie J-P, Huet S, Faussat A-M, et al. Multicentric evaluation of the MDR phenotype in leukemia. Leukemia 1997, 11:1086–1094.

42. Beck WT, Grogan TM. Methods to detect P-glycoprotein and implications for other drug resistance-associated proteins. Leukemia 1997, 11:1107–1109.

43. Pirker R, Wallner J, Geissler K, et al. MDR1 gene expression and treatment outcome in acute myeloid leukemia. J Natl Cancer Inst 1991, 83:708–712.

44. Wood P, Burgess R, MacGregor A, et al. P-glycoprotein expression on acute myeloid leukaemia blast cells at diagnosis predicts response to chemotherapy and survival. Br J Haematol 1994, 87:509–514.

45. Hunault M, Zhou D, Dehner A, et al. Multidrug resistance gene expression in acute myeloid leukemia, major prognosis significance for in vivo drug resistance to induction treatment. Ann Hematol 1997, 74:65–71.

46. van den Heuvel Eibrink MM, van der Holt B, te Boekhorst PA, et al. MDR1 expression is an independent prognostic factor for response and survival in de novo acute myeloid leukaemia. Br J Haematol 1997, 99:76–83.

47. Efferth T, Grassmann R. Impact of viral oncogenesis on responses to anti-cancer drugs and irradiation. Crit Rev Oncog 2000, 11:165–187.

48. Efferth T, Volm M. Pharmacogenetics for individualized cancer chemotherapy. Pharmacol Ther 2005, 107:155–176.

49. Efferth T, Fabry U, Osieka R. Apoptosis and resistance to daunorubicin in human leukemic cells. Leukemia 1997, 11:1180–1186.

50. Pommier Y, Sordet O, Antony S, et al. Apoptosis defects and chemotherapy resistance: molecular interaction maps and networks. Oncogene 2004, 23:2934–2939.

51. Lotem J, Sachs L. Control of apoptosis in hematopoiesis and leukemia by cytokines, tumor suppressor and oncogenes. Leukemia 1996, 10:925–931.

52. Efferth T, Fabry U, Osieka R. Interleukin-6 affects melphalan-induced DNA damage and repair in human multiple myeloma cells. Anticancer Res 2002, 22:231–234.

53. Efferth T. The human ATP-binding cassette transporter genes: from the bench to the bedside. Curr Mol Med 2001, 1:45–65.

54. Gottesman MM, Fojo T, Bates SE. Multidrug resistance in cancer: role of ATP-dependent transporters. Nat Rev Cancer 2002, 2:48–58.

55. Gillet JP, Efferth T, Steinbach D, et al. Microarray-based detection of multidrug resistance in human tumor cells by expression profiling of ATP-binding cassette transporter genes. Cancer Res 2004, 64:8987–8993.

56. Gillet JP, Efferth T, Remacle J. Chemotherapy-induced resistance by ATP-binding cassette transporter genes. Biochem Biophys Acta 2007, 1775:237–262.

57. Fojo AT, Ueda K, Slamon DJ, et al. Expression of a multidrug-resistance gene in human tumors and tissues. Proc Natl Acad Sci U S A 1987, 84:265–269.

58. Efferth T, Mattern J, Volm M. Immunohistochemical detection of P glycoprotein, glutathione S transferase and DNA topoisomerase II in human tumors. Oncology 1992, 49:368–375.

59. Efferth T, Volm M. Reversal of doxorubicin-resistance in sarcoma 180 tumor cells by inhibition of different resistance mechanisms. Cancer Lett 1993, 70:197–202.

60. Volm M, Kästel M, Mattern J, et al. Expression of resistance factors (P-glycoprotein, glutathione S-transferase-pi, and topoisomerase II) and their interrelationship to proto-oncogene products in renal cell carcinomas. Cancer 1993, 71:3981–3987.

61. Volm M, Koomägi R, Mattern J, et al. Expression profile of genes in non-small cell lung carcinomas from long-term surviving patients. Clin Cancer Res 2002, 8:1843–1848.

62. Volm M, Koomägi R, Mattern J, et al. Protein expression profiles indicative for drug resistance of non-small cell lung cancer. Br J Cancer 2002, 87:251–257.

63. Volm M, Koomägi R, Efferth T. Prediction of drug sensitivity and resistance of cancer by protein expression profiling. Cancer Genom Proteom 2004, 1:157–166.

64. Verrills NM, Kavallaris M. Drug resistance mechanisms in cancer cells: a proteomics perspective. Curr Opin Mol Ther 2003, 5:258–265.

65. Efferth T, Osieka R. Clinical relevance of the MDR-1 gene and its gene product, P-glycoprotein, for cancer chemotherapy: a metaanalysis. Tumor Diagn Ther 1993, 14:238–243.

66. Trock BJ, Leonessa F, Clarke R. Multidrug resistance in breast cancer: a meta-analysis of MDR1/gp170 expression and its possible functional significance. J Natl Cancer Inst 1997, 89:917–931.

67. Sauerbrey A, Zintl F, Volm M. P-glycoprotein and glutathione S-transferase pi in childhood acute lymphoblastic leukaemia. Br J Cancer 1994, 70:1144–1149.

68. Leonessa F, Clarke R. ATP binding cassette transporters and drug resistance in breast cancer. Endocr Relat Cancer 2003, 10:43–73.

69. Mahadevan D, List AF. Targeting the multidrug resistance-1 transporter in AML: molecular regulation and therapeutic strategies. Blood 2004, 104:1940–1951.

70. Hoffmeyer S, Burk O, von Richter O, et al. Functional polymorphisms of the human multidrug-resistance gene: multiple sequence variations and correlation of one allele with P-glycoprotein expression and activity in vivo. Proc Natl Acad Sci U S A 2000, 97:3473–3478.

71. Ameyaw MM, Regateiro F, Li T, et al. MDR1 pharmacogenetics: frequency of the C3435T mutation in exon 26 is significantly influenced by ethnicity. Pharmacogenetics 2001, 11:217–221.

72. Schwab M, Eichelbaum M, Fromm MF. Genetic polymorphism of the human MDR1 drug transporter. Annu Rev Pharmacol Toxicol 2003, 43:285–307.

73. Honda T, Dan Y, Koyabu N, et al. Polymorphism of MDR1 gene in healthy Japanese subjects: a novel SNP with an amino acid substitution (Glu108Lys). Drug Metab Pharmacokinet 2004, 17:479–481.

74. Sakaeda T, Nakamura T, Okumura K. Pharmacogenetics of MDR1 and its impact on the pharmacokinetics and pharmacodynamics of drugs. Pharmacogenomics 2003, 4:397–410.

75. Sakaeda T, Nakamura T, Horinouchi M, et al. MDR1 genotype-related pharmacokinetics of digoxin after single oral administration in healthy Japanese subjects. Pharm Res 2001, 18:1400–1404.

76. Goh BC, Lee SC, Wang LZ, et al. Explaining inter-individual variability of docetaxel pharmacokinetics and pharmacodynamics in Asians through phenotyping and genotyping strategies. J Clin Oncol 2002, 20:3683–3690.

77. Illmer T, Schuler US, Thiede C, et al. MDR1 gene polymorphisms affect therapy outcome in acute myeloid leukemia patients. Cancer Res 2002, 62:4955–4962.

78. Kim RB. MDR1 single nucleotide polymorphisms: multiplicity of haplotypes and functional consequences. Pharmacogenetics 2002, 12:425–427.

79. Kurata Y, Ieri I, Kimura M, et al. Role of human MDR1 gene polymorphism in bioavailability and interaction of digoxin, a substrate of P-glycoprotein. Clin Pharmacol Ther 2002, 72:209–221.

80. Efferth T, Sauerbrey A, Steinbach D, et al. Analysis of single nucleotide polymorphism C3435T of the multidrug resistance gene MDR1 in acute lymphoblastic leukemia. Int J Oncol 2003, 23:509–517.

81. Plasschaert SL, Groninger E, Boezen M, et al. Influence of functional polymorphisms of the MDR1 gene on vincristine pharmacokinetics in childhood acute lymphoblastic leukemia. Clin Pharmacol Ther 2004, 76:220–229.

82. Eichelbaum M, Fromm MF, Schwab M. Clinical aspects of the MDR1 (ABCB1) gene polymorphism. Ther Drug Monit 2004, 26:180–185.

83. Haimeur A, Conseil G, Deeley RG, et al. The MRP-related and BCRP/ABCG2 multidrug resistance proteins: biology, substrate specificity and regulation. Curr Drug Metab 2004, 5:21–53.

84. van den Heuvel-Eibrink MM, Sonneveld P, Pieters R. The prognostic significance of membrane transport-associated multidrug resistance (MDR) proteins in leukemia. Int J Clin Pharmacol Ther 2000, 38:94–110.

85. Conrad S, Kauffmann HM, Ito K, et al. Identification of human multidrug resistance protein 1 (MRP1) mutations and characterization of a G671V substitution. J Hum Genet 2001, 46:656–663.

86. Ito S, Ieiri I, Tanabe M, et al. Polymorphism of the ABC transporter genes, MDR1, MRP1 and MRP2/cMOAT, in healthy Japanese subjects. Pharmacogenetics 2001, 11:175–184.

87. Perdu J, Germain DP. Identification of novel polymorphisms in the pM5 and MRP1 (ABCC1) genes at locus 16p13.1 and exclusion of both genes as responsible for pseudoxanthoma elasticum. Hum Mutat 2001, 17:74–75.

88. Itoda M, Saito Y, Soyama A, et al. Polymorphisms in the ABCC2 (cMOAT/MRP2) gene found in 72 established cell lines derived from Japanese individuals: an association between single nucleotide polymorphisms in the 5V-untranslated region and exon 28. Drug Metab Dispos 2002, 30:363–364.

89. Suzuki H, Sugiyama Y. Single nucleotide polymorphisms in multidrug resistance associated protein 2 (MRP2/ABCC2): its impact on drug disposition. Adv Drug Del Rev 2002, 54:1311–1331.

90. Oselin K, Mrozikiewicz PM, Gaikovitch E, et al. Frequency of MRP1 genetic polymorphisms and their functional significance in Caucasians: detection of a novel mutation G816A in the human MRP1 gene. Eur J Clin Pharmacol 2003, 59:347–350.

91. Sarkadi B, Ozvegy-Laczka C, Nemet K, et al. ABCG2—a transporter for all seasons. FEBS Lett 2004, 567:116–120.

92. Sauerbrey A, Sell W, Steinbach D, et al. Expression of the BCRP gene (ABCG2/MXR/ABCP) in childhood acute lymphoblastic leukaemia. Br J Haematol 2002, 118:147–150.

93. Diestra JE, Condom E, Del Muro XG, et al. Expression of multidrug resistance proteins P-glycoprotein, multidrug resistance protein 1, breast cancer resistance protein and lung resistance related protein in locally advanced bladder cancer treated with neoadjuvant chemotherapy: biological and clinical implications. J Urol 2003, 170:1383–1387.

94. Imai Y, Nakane M, Kage K, et al. C421A polymorphism in the human breast cancer resistance protein gene is associated with low expression of Q141K protein and low-level drug resistance. Mol Cancer Ther 2002, 1:611–616.

95. Mizuarai S, Aozasa N, Kotani H. Single nucleotide polymorphisms result in impaired membrane localization and reduced atpase activity in multidrug transporter ABCG2. Int J Cancer 2004, 109:238–246.

96. Gonzalez FJ, Nebert DW. Evolution of the P450 gene superfamily: animal–plant warfare, molecular drive and human genetic differences in drug oxidation. Trends Genet 1990, 6:182–186.

97. Kivisto KT, Kroemer HK, Eichelbaum M. The role of human cytochrome P450 enzymes in the metabolism of anticancer agents: implications for drug interactions. Br J Clin Pharmacol 1995, 40:523–530.

98. van Schaik RH. Implications of cytochrome P450 genetic polymorphisms on the toxicity of antitumor agents. Ther Drug Monit 2004, 26:236–240.

99. Aplenc R, Glatfelter W, Han P, et al. CYP3A genotypes and treatment response in paediatric acute lymphoblastic leukaemia. Br J Haematol 2003, 122:240–244.

100. Felix CA, Walker AH, Lange BJ, et al. Association of CYP3A4 genotype with treatment-related leukemia. Proc Natl Acad Sci U S A 1998, 95:13176–13181.

101. Dai D, Zeldin DC, Blaisdell JA, et al. Polymorphisms in human CYP2C8 decrease metabolism of the anticancer drug paclitaxel and arachidonic acid. Pharmacogenetics 2001, 11:597–607.

102. Nagasubramanian R, Innocenti F, Ratain MJ. Pharmacogenetics in cancer treatment. Annu Rev Med 2003, 54:437–452.

103. Ando Y, Price DK, Dahut WL, et al. Pharmacogenetic associations of CYP2C19 genotype with in vivo metabolisms and pharmacological effects of tha-lidomide. Cancer Biol Ther 2002, 1:669–673.

104. Ketterer B. Protective role of glutathione and glu-tathione transferases in mutagenesis and carcino-genesis. Mutat Res 1988, 202:343–361.

105. Tew KD. Glutathione-associated enzymes in anti-cancer drug resistance. Cancer Res 1994, 54:4313–4320.

106. Peters U, Preisler-Adams S, Hebeisen A, et al. Glutathione S-transferase genetic polymorphisms and individual sensitivity to the ototoxic effect of cisplatin. Anticancer Drugs 2000, 11:639–643.

107. Chen CL, Liu Q, Relling MV. Simultaneous char-acterization of glutathione S-transferase M1 and T1 polymorphisms by polymerase chain reaction in American whites and blacks. Pharmacogenetics 1996, 6:187–191.

108. Stanulla M, Schrappe M, Brechlin AM, et al. Polymorphisms within glutathione S-transferase genes (GSTM1, GSTT1, GSTP1) and risk of relapse in childhood B-cell precursor acute lym-phoblastic leukemia: a case-control study. Blood 2000, 95:1222–1228.

109. Takanashi M, Morimoto A, Yagi T, et al. Impact of glutathione S-transferase gene deletion on early relapse in childhood B-precursor acute lymphoblas-tic leukemia. Haematologica 2003, 88:1238–1244.

110. Naoe T, Tagawa Y, Kiyoi H, et al. Prognostic signif-icance of the null genotype of glutathione S trans-ferase-T1 in patients with acute myeloid leukemia: increased early death after chemotherapy. Leukemia 2002, 16:203–208.

111. Ambrosone CB, Sweeney C, Coles BF, et al. Polymorphisms in glutathione S transferases (GSTM1 and GSTT1) and survival after treatment for breast cancer. Cancer Res 2001, 61:7130–7135.

112. Howells RE, Redman CW, Dhar KK, et al. Association of glutathione S-transferase GSTM1 and GSTT1 null genotypes with clinical outcome in epithelial ovarian cancer. Clin Cancer Res 1998, 4:2439–2445.

113. Medeiros R, Pereira D, Afonso N, et al. Platinum/paclitaxel-based chemotherapy in advanced ovar-ian carcinoma: glutathione S-transferase genetic polymorphisms as predictive biomarkers of disease outcome. Int J Clin Oncol 2003, 8:156–161.

114. Sweeney C, Ambrosone CB, Joseph L, et al. Association between a glutathione S transferase A1 promoter polymorphism and survival after breast cancer treatment. Int J Cancer 2003, 103:810–814.

115. Anderer G, Schrappe M, Brechlin AM, et al. Polymorphisms within glutathione S-transferase genes and initial response to glucocorticoids in childhood acute lymphoblastic leukaemia. Pharmacogenetics 2000, 10:715–726.

116. Meissner B, Stanulla M, Ludwig WD, et al. The GSTT1 deletion polymorphism is associated with initial response to glucocorticoids in childhood acute lymphoblastic leukemia. Leukemia 2004, 18:1920–1923.

117. Sweeney C, McClure GY, Fares MY, et al. Association between survival after treatment for breast cancer and glutathione S-transferase P1 Ile105Val polymor-phism. Cancer Res 2000, 60:5621–5624.

118. Dasgupta RK, Adamson PJ, Davies FE, et al. Polymorphic variation in GSTP1 modulates out-come following therapy for multiple myeloma. Blood 2003, 102:2345–2350.

119. Yang G, Shu XO, Ruan ZX, et al. Genetic polymor-phisms in glutathione-S-transferase genes (GSTM1, GSTT1, GSTP1) and survival after chemother-apy for invasive breast carcinoma. Cancer 2004, 103:52–58.

120. Watson MA, Stewart RK, Smith GB, et al. Human glutathione S-transferase P1 polymorphisms: rela-tionship to lung tissue enzyme activity and popula-tion frequency distribution. Carcinogenesis 1998, 19:275–280.

121. Pandya U, Srivastava SK, Singhal SS, et al. Activity of allelic variants of Pi class human glutathione S-transferase toward chlorambucil. Biochem Biophys Res Commun 2000, 278:258–262.

122. Stoehlmacher J, Park DJ, Zhang W, et al. Association between glutathione S-transferase P1, T1, and M1 genetic polymorphism and survival of patients with metastatic colorectal cancer. J Natl Cancer Inst 2002, 94:936–942.

123. Sweeney C, Nazar-Stewart V, Stapleton PL, et al. Glutathione S-transferase M1, T1, and P1 polymor-phisms and survival among lung cancer patients. Cancer Epidemiol Biomark Prev 2003, 12:527–533.

124. Peters U, Preisler-Adams S, Hebeisen A, et al. Glutathione S-transferase genetic polymorphisms and individual sensitivity to the ototoxic effect of cisplatin. Anticancer Drugs 2000, 11:639–643.

125. Davies SM, Robison LL, Buckley JD, et al. Glutathione S-transferase polymorphisms and out-come of chemotherapy in childhood acute myeloid leukemia. J Clin Oncol 2001, 19:1279–1287.

126. Allan JM, Wild CP, Rollinson S, et al. Polymorphism in glutathione S-transferase P1 is associated with susceptibility to chemotherapy-induced leukemia. Proc Natl Acad Sci U S A 2001, 98:11592–11597.

127. King CD, Rios GR, Green MD, et al. UDP glu-curonosyltransferases. Curr Drug Metab 2000, 1:143–161.

128. Gupta E, Lestingi TM, Mick R, et al. Metabolic fate of irinotecan in humans: correlation of glucuronidation with diarrhea. Cancer Res 1994, 54:3723–3725.

129. Innocenti F, Iyer L, Ratain MJ. Pharmacogenetics of anticancer agents: lessons from amonafide and irinotecan. Drug Metab Dis 2001, 29:596–600.

130. Sai K, Saeki M, Saito Y, et al. UGT1A1 haplotypes associated with reduced glucuronidation and increased serum bilirubin in irinotecan-administered Japanese patients with cancer. Clin Pharmacol Ther 2004, 75:501–515.

131. Ando Y, Saka H, Ando M, et al. Polymorphisms of UDP-glucuronosyltransferase gene and irinotecan toxicity: a pharmacogenetic analysis. Cancer Res 2000, 60:6921–6926.

132. Iyer L, Das S, Janisch L, et al. UGT1A1*28 polymorphism as a determinant of irinotecan disposition and toxicity. Pharmacogenomics J 2002, 2:43–47.

133. Font A, Sanchez JM, Taron M, et al. Weekly regimen of irinotecan/docetaxel in previously treated non-small cell lung cancer patients and correlation with uridine diphosphate glucuronosyltransferase 1A1 (UGT1A1) polymorphism. Invest New Drugs 2003, 21:435–443.

134. Marcuello E, Altes A, del Rio E, et al. Single nucleotide polymorphism in the 5V tandem repeat sequences of thymidylate synthase gene predicts for response to fluorouracil-based chemotherapy in advanced colorectal cancer patients. Int J Cancer 2004, 112:733–777.

135. Marcuello E, Altes A, Menoyo A, et al. UGT1A1 gene variations and irinotecan treatment in patients with metastatic colorectal cancer. Br J Cancer 2004, 91:678–682.

136. McLeod HL, Siva C. The thiopurine S-methyltransferase gene locus—implications for clinical pharmacogenomics. Pharmacogenomics 2002, 3:89–98.

137. Tai HL, Krynetski EY, Schuetz EG, et al. Enhanced proteolysis of thiopurine S-methyltransferase (TPMT) encoded by mutant alleles in humans/ TPMT*3A, TPMT*2): mechanisms for the genetic polymorphism of TPMT activity. Proc Natl Acad Sci USA 1997, 94:6444–6449.

138. Lennard L, Lilleyman JS, van Loon J, et al. Genetic variation in response to 6-mercaptopurine for childhood acute lymphoblastic leukaemia. Lancet 1990, 336:225–229.

139. Evans WE, Horner M, Chu YQ, et al. Altered mercaptopurine metabolism, toxic effects, and dosage requirement in a thiopurine methyltransferase deficient child with acute lymphoblastic leukaemia. J Pediatr 1991, 119:985–989.

140. Evans WE, Hon YY, Bomgaars L, et al. Preponderance of thiopurine S-methyltransferase deficiency and heterozygosity among patients intolerant to mercaptopurine or azathioprine. J Clin Oncol 2001, 19:2293–2301.

141. Alves S, Prata MJ, Ferreira F, et al. Thiopurine methyltransferase pharmacogenetics: alternative molecular diagnosis and preliminary data from Northern Portugal. Pharmacogenetics 1999, 9:257–261.

142. Relling MV, Hancock ML, Boyett JM, et al. Prognostic importance of 6-mercaptopurine dose intensity in acute lymphoblastic leukaemia. Blood 1999, 93:2817–2823.

143. Relling MV, Hancock ML, Rivera GK, et al. Mercaptopurine therapy intolerance and heterozygosity at the thiopurine S-methyltransferase gene locus. J Natl Cancer Inst 1999, 91:2001–2008.

144. Relling MV, Rubnitz JE, Rivera GK, et al. High incidence of secondary brain tumours after radiotherapy and antimetabolites. Lancet 1999, 354:34–39.

145. Coulthard SA, Rabello C, Robson J, et al. A comparison of molecular and enzyme-based assays for the detection of thiopurine methyltransferase mutations. Br J Haematol 2000, 110:599–604.

146. Krynetski EY, Evans WE. Pharmacogenetics of cancer therapy: getting personal. Am J Hum Genet 1998, 63:11–16.

147. Relling MV, Yanishevski Y, Nemec J, et al. Etoposide and antimetabolite pharmacology in patients who develop secondary acute myeloid leukemia. Leukemia 1998, 12:346–352.

148. McLeod HL, Krynetski EY, Relling MV, et al. Genetic polymorphism of thiopurine methyltransferase and its clinical relevance for childhood acute lymphoblastic leukemia. Leukemia 2000, 14:567–572.

149. McLeod HL, Relling MV, Liu Q, et al. Polymorphic thiopurine methyltransferase in erythrocytes is indicative of activity in leukemic blasts from children with acute lymphoblastic leukemia. Blood 1995, 85:1897–1902.

150. Lennard L, Lilleyman JS, van Loon J, et al. Genetic variation in response to 6-mercaptopurine for childhood acute lymphoblastic leukaemia. Lancet 1990, 336:225–229.

151. Rustum YM, Hartstrick A, Cao S, et al. Thymidylate synthase inhibitors in cancer therapy: direct and indirect inhibitors. J Clin Oncol 1997, 15:389–400.

152. Johnston PG, Lenz HJ, Leichman CG, et al. Thymidylate synthase gene and protein expression correlate and are associated with response to 5-fluorouracil in human colorectal and gastric tumors. Cancer Res 1995, 55:1407–1412.

153. Leichman CG, Lenz HJ, Leichman L, et al. Quatitation of intratumoral thymidylate synthase expression predicts for disseminated colorectal

cancer response to protracted infusion fluorouracil and weekly leucovorin. J Clin Oncol 1997, 15:3223–3229.

154. Kaneda S, Takeishi K, Ayusawa D, et al. Role in translation of a triple tandemly repeated sequence in the 5V-untranslated region of human thymidylate synthase mRNA. Nucleic Acids Res 1987, 15:1259–1270.

155. Horie N, Aiba H, Oguro K, et al. Functional analysis and DNA polymorphism of the tandemly repeated sequences in the 5V-terminal regulatory region of the human gene for thymidylate synthase. Cell Struct Funct 1995, 20:191–197.

156. Marsh S, Ameyaw MM, Githang'a J, et al. Novel thymidylate synthase enhancer region alleles in African populations. Hum Mutat 2000, 16:528.

157. Luo HR, Lu XM, Yao N, et al. Length polymorphism of thymidalate synthase regulatory region in Chinese populations and evolution of the novel alleles. Biochem Genet 2002, 40:41–51.

158. Kawakami K, Omura K, Kanehira E, et al. Polymorphic tandem repeats in the thymidylate synthase gene is associated with its protein expression in human gastrointestinal cancers. Anticancer Res 1999, 19:3249–3252.

159. Iacopetta B, Grieu F, Joseph D, et al. A polymorphism in the enhancer region of the thymidylate synthase gene influences the survival of colorectal cancer patients treated with 5-fluorouracil. Br J Cancer 2001, 85:827–830.

160. Marsh S, McLeod HL. Thymidylate synthase pharmacogenetics in colorectal cancer. Clin Colorectal Cancer 2001, 1:175–178.

161. Marsh S, McKay JA, Cassidy HL, et al. Polymorphism in the thymidylate synthase promotor enhancer region in colorectal cancer. Int J Oncol 2001, 19:383–386.

162. Pullarkat ST, Stoehlmacher J, Ghaderi V, et al. Thymidylate synthase gene polymorphism determines response and toxicity of 5-FU chemotherapy. Pharmacogenomics J 2001, 1:65–70.

163. Villafranca E, Okruzhnov Y, Dominguez MA, et al. Polymorphisms of the repeated sequences in the enhancer region of the thymidylate synthase gene promotor may predict downstaging after preoperative chemoradiation in rectal cancer. J Clin Oncol 2001, 19:1779–1786.

164. Etienne MC, Chazal M, Laurent-Puig P, et al. Prognostic value of tumoral thymidylate synthase and p53 in metastatic colorectal cancer patients receiving fluorouracil-based chemotherapy: phenotypic and genotypic analyses. J Clin Oncol 2002, 20:2832–2843.

165. Chen J, Hunter DJ, Stampfer MJ, et al. Polymorphism in the thymidylate synthase promoter enhancer region modifies the risk and survival of colorectal cancer. Cancer Epidemiol Biomark Prev 2003, 12:958–962.

166. Tsuji T, Hidaka S, Sawai T, et al. Polymorphism in the thymidylate synthase promoter enhancer region is not an efficacious marker for tumor sensitivity to 5-fluorouracil-based oral adjuvant chemotherapy in colorectal cancer. Clin Cancer Res 2003, 9:3700–3704.

167. Uchida K, Hayashi K, Kawakami K, et al. Loss of heterozygosity at the thymidylate synthase (TS) locus on chromosome 18 affects tumor response and survival in individuals heterozygous for a 28-bp polymorphism in the TS gene. Clin Cancer Res 2004, 10:433–439.

168. Lecomte T, Ferraz JM, Zinzindohoue F, et al. Thymidylate synthase gene polymorphism predicts toxicity in colorectal cancer patients receiving 5-fluorouracil-based chemotherapy. Clin Cancer Res 2004, 10:5880–5888.

169. Frosst P, Blom HJ, Milos R, et al. A candidate genetic risk factor for vascular disease: a common mutation in methylenetetrahydrofolate reductase. Nat Genet 1995, 10:111–113.

170. Toffoli G, Veronesi A, Boiocchi M, et al. MTHFR gene polymorphism and severe toxicity during adjuvant treatment of early breast cancer with cyclophosphamide, methotrexate, and fluorouracil (CMF). Ann Oncol 2000, 11:373–374.

171. Ulrich CM, Yasui Y, Storb R, et al. Pharmacogenetics of methotrexate: toxicity among marrow transplantation patients varies with the methylenetetrahydrofolate reductase C677T polymorphism. Blood 2001, 98:231–234.

172. Toffoli G, Russo A, Innocenti F, et al. Effect of methylenetetrahydrofolate reductase 677CYT polymorphism on toxicity and homocysteine plasma level after chronic methotrexate treatment of ovarian cancer patients. Int J Cancer 2003, 103:294–299.

173. Chiusolo P, Reddiconto G, Casorelli I, et al. Preponderance of methylenetetrahydrofolate reductase C677T homozygosity among leukemia patients intolerant to methotrexate. Ann Oncol 2002, 13:1915–1918.

174. Cohen V, Panet-Raymond V, Sabbaghian N, et al. Methylenetetrahydrofolate reductase polymorphism in advanced colorectal cancer: a novel genomic predictor of clinical response to fluoropyrimidine-based chemotherapy. Clin Cancer Res 2003, 9:1611–1615.

175. Sohn KJ, Croxford R, Yates Z, et al. Effect of the methylenetetrahydrofolate reductase C677T polymorphism on chemosensitivity of colon and breast cancer cells to 5-fluorouracil and methotrexate. J Natl Cancer Inst 2004, 96:134–144.

176. Taub JW, Matherly LH, Ravindranath Y, et al. Polymorphisms in methylenetetrahydrofolate reductase and methotrexate sensitivity in childhood acute lymphoblastic leukemia. Leukemia 2002, 16:764–765.

177. Alberola V, Sarries C, Rosell R, et al. Effect of the methylenetetrahydrofolate reductase C677T polymorphism on patients with cisplatin/gemcitabine-treated stage IV non-small-cell lung cancer. Clin Lung Cancer 2004, 5:360–365.

178. Mitra S, Kaina B. Regulation of repair of alkylation damage in mammalian genomes. Prog Nucleic Acid Res Mol Biol 1993, 44:109–142.

179. Glassner BJ, Weeda G, Allan JM, et al. DNA repair methyltransferase (Mgmt) knockout mice are sensitive to the lethal effects of chemotherapeutic alkylating agents. Mutagenesis 199, 14:339–347.

180. Dolan ME, Moschel RC, Pegg AE. Depletion of mammalian O6-alkylguanine-DNA alkyltransferase activity by O6-benzylguanine provides a means to evaluate the role of this protein in protection against carcinogenic and therapeutic alkylating agents. Proc Natl Acad Sci U S A 1990, 87:5368–5372.

181. Kaina B, Mühlhausen U, Piee-Staffa A, et al. Inhibition of O6-methylguanine-DNA methyltransferase by glucose-conjugated inhibitors: comparison with nonconjugated inhibitors and effect on fotemustine and temozolomide-induced cell death. J Pharmacol Exp Ther 2004, 311:585–583.

182. Schwarzl SM, Smith JC, Kaina B, et al. Molecular modeling of O6-methylguanine-DNA methyltransferase mutant proteins encoded by single nucleotide polymorphisms. Int J Mol Med 2005, 16:553–557.

183. Imai Y, Oda H, Nakatsuru Y, et al. A polymorphism at codon 160 of human O6-methylguanine-DNA methyltransferase gene in young patients with adult type cancers and functional assay. Carcinogenesis 1995, 16:2441–2445.

184. Edara S, Kanugula S, Goodtzova K, et al. Resistance of the human O6-alkylguanine-DNA alkyltransferase containing arginine at codon 160 to inactivation by O6-benzylguanine. Cancer Res 1996, 56:5571–5575.

185. Wu MH, Lohrbach KE, Olopade OI, et al. Lack of evidence for a polymorphism at codon 160 of human O6-alkylguanine-DNA alkyltransferase gene in normal tissue and cancer. Clin Cancer Res 1999, 5:209–213.

186. Allan JM, Engelward BP, Dreslin AJ, et al. Mammalian 3-methyladenine DNA glycosylase protects against toxicity and clastogenicity of certain chemotherapeutic DNA cross-linking agents. Cancer Res 1998, 58:3965–3973.

187. Ochs K, Sobol RW, Wilson SH, et al. Cells deficient in DNA polymerase beta are hypersensitive to alkylating agent-induced apoptosis and chromosomal breakage. Cancer Res 1999, 59:1544–1551.

188. Panasci L, Paiement JP, Christodoupoulos G, et al. Chlorambucil drug resistance in chronic lymphocytic leukemia: the emerging role of DNA repair. Clin Cancer Res 2001, 7:454–461.

189. Stoehlmacher J, Ghaderi V, Iobal S, et al. A polymorphism of the XRCC1 gene predicts for response to platinum based treatment in advanced colorectal cancer. Anticancer Res 2001, 21:3075–3079.

190. Gurubhagavatula S, Liu G, Park S, et al. XPD and XRCC1 genetic polymorphisms are prognostic factors in advanced non-small-cell lung cancer patients treated with platinum chemotherapy. J Clin Oncol 2004, 22:2594–2601.

191. Andreassen CN, Alsner J, Overgaard M, et al. Prediction of normal tissue radiosensitivity from polymorphisms in candidate genes. Radiother Oncol 2003, 69:127–135.

192. Reed E. Platinum-DNA adduct, nucleotide excision repair and platinum based anticancer chemotherapy. Cancer Treat Res 1998, 24:331–344.

193. Panasci L, Paiement JP, Christodoupoulos G, et al. Chlorambucil drug resistance in chronic lymphocytic leukemia: the emerging role of DNA repair. Clin Cancer Res 2001, 7:454–461.

194. Rosell R, Lord RV, Taron M, et al. DNA repair and cisplatin resistance in non-small cell lung cancer. Lung Cancer 2002, 38:217–227.

195. Yu JJ, Mu C, Lee KB, et al. A nucleotide polymorphism in ERCC1 in human ovarian cancer cell lines and tumor tissues. Mutat Res 1997, 382:13–20.

196. Yu JJ, Lee KB, Mu C, et al. Comparison of two human ovarian carcinoma cell lines (A2780/CP70 and MCAS) that are equally resistant to platinum, but differ at codon 118 of the ERCC1 gene. Int J Oncol 2000, 16:555–560.

197. Isla D, Sarries C, Rosell R, et al. Single nucleotide polymorphisms and outcome in docetaxel cisplatin-treated advanced non-small-cell lung cancer. Ann Oncol 2004, 15:1194–1203.

198. Ryu JS, Hong YC, Han HS, et al. Association between polymorphisms of ERCC1 and XPD and survival in non-small-cell lung cancer patients treated with cisplatin combination chemotherapy. Lung Cancer 2004, 44:311–316.

199. Park DJ, Stoehlmacher J, Zhang W, et al. A Xeroderma pigmentosum group D gene polymorphism predicts clinical outcome to platinum-based chemotherapy in patients with advanced colorectal cancer. Cancer Res 2001, 61:8654–8658.

200. Rosell R, Crino L, Danenberg K, et al. Targeted therapy in combination with gemcitabine in nonsmall cell lung cancer. Semin Oncol 2001, 30(Suppl 10):19–25.

201. Drummond JT, Anthoney A, Brown R, et al. Cisplatin and adriamycin resistance are associated with MutLalpha and mismatch repair deficiency in an ovarian tumor cell line. J Biol Chem 1996, 271:19645–19648.

202. Fink D, Nebel S, Aebi S, et al. The role of DNA mismatch repair in platinum drug resistance. Cancer Res 1996, 56:4881–4886.

203. Fink D, Aebi S, Howell SB. The role of DNA mismatch repair in drug resistance. Clin Cancer Res 1998, 4:1–6.

204. de las Alas MM, Aebi S, Fink D, et al. Loss of DNA mismatch repair: effects on the rate of mutation to drug resistance. J Natl Cancer Inst 1997, 89:1537–1541.

205. Vaisman A, Varchenko M, Umar A, et al. The role of hMLH1, hMSH3, and hMSH6 defects in cisplatin and oxaliplatin resistance: correlation with replicative bypass of platinum-DNA adducts. Cancer Res 2002, 58:3579–3585.

206. Worrillow LJ, Travis LB, Smith AG, et al. An intron splice acceptor polymorphism in hMSH2 and risk of leukemia after treatment with chemotherapeutic alkylating agents. Clin Cancer Res 2003, 9:3012–3020.

207. Schimmer AD, Hedley DW, Penn LZ, et al. Receptor- and mitochondrial-mediated apoptosis in acute leukemia: a translational view. Blood 2001, 98:3541–3553.

208. Green DR, Kroemer G. The pathophysiology of mitochondrial cell death. Science 2004, 305:626–629.

209. Haupt S, Berger M, Goldberg Z, et al. Apoptosis—the p53 network. J Cell Sci 2003, 116:4077–4085.

210. Volm M, Mattern J. Increased expression of Bcl-2 in drug-resistant squamous cell lung carcinomas. Int J Oncol 1995, 7:1333–1338.

211. Koomägi R, Volm M. Expression of Fas (CD95/APO-1) and Fas ligand in lung cancer, its prognostic and predictive relevance. Int J Cancer (Pred Oncol) 1999, 84:239–243.

212. Volm M, Zintl F, Sauerbrey A, et al. Expression of Fas ligand in newly diagnosed childhood acute lymphoblastic leukemia. Anticancer Res 199, 19:3399–3402.

213. Pommier Y, Sordet O, Antony S, et al. Apoptosis defects and chemotherapy resistance: molecular interaction maps and networks. Oncogene 2004, 23:2934–2949.

214. Baekelandt M, Holm R, Nesland JM, et al. Expression of apoptosis-related proteins in an independent

215. Deng X, Kornblau SM, Ruvolo PP, et al. Regulation of Bcl2 phosphorylation and potential significance for leukemic cell chemoresistance. J Natl Cancer Inst Monographs 2001, 28:30–37.

216. Ravandi F, Kantarjian HM, Talpaz M, et al. Expression of apoptosis proteins in chronic myelogenous leukemia: associations and significance. Cancer 2001, 91:1964–1967.

217. Konstantinidou AE, Korkolopoulou P, Patsouris E. Apoptotic markers for tumor recurrence: a minireview. Apoptosis 2002, 7:461–470.

218. Borresen-Dale AL. TP53 and breast cancer. Hum Mutat 2003, 21:292–300.

219. Iacopetta B. TP53 mutation in colorectal cancer. Hum Mutat 2003, 21:271–276.

220. Martin B, Paesmans M, Berghmans T, et al. Role of Bcl-2 as a prognostic factor for survival in lung cancer: a systematic review of the literature with meta-analysis. Br J Cancer 2003, 89:55–64.

221. Schimmer AD, Pedersen IM, Kitada S, et al. Functional blocks in caspase activation pathways are common in leukemia and predict patient response to induction chemotherapy. Cancer Res 2003, 63:1242–1248.

222. Botti C, Buglioni S, Benevolo M, et al. Altered expression of FAS system is related to adverse clinical outcome in stage I–II breast cancer patients treated with adjuvant anthracycline-based chemotherapy. Clin Cancer Res 2004, 10:1360–1365.

223. Hofseth LJ, Hussain SP, Harris CC. p53: 25 years after its discovery. Trends Pharmacol Sci 2004, 25:177–181.

224. Peller S. Clinical implications of p53: effect on prognosis, tumor progression and chemotherapy response. Semin Cancer Biol 1998, 8:379–387.

225. Weller M. Predicting response to cancer chemotherapy: the role of p53. Cell Tissue Res 1998, 292:435–445.

226. Lehman TA, Haffty BG, Carbone CJ, et al. Elevated frequency and functional activity of a specific germ-line p53 intron mutation in familial breast cancer. Cancer Res 2000, 60:1062–1069.

227. Bergamaschi D, Gasco M, Hiller L, et al. p53 polymorphism influences response in cancer chemotherapy via modulation of p73-dependent apoptosis. Cancer Cell 2003, 3:387–402.

228. Sullivan A, Syed N, Gasco M, et al. Polymorphism in wild-type p53 modulates response to chemotherapy in vitro and in vivo. Oncogene 2004, 23:3328–3337.

229. Saxena A, Moshynska O, Sankaran K, et al. Association of a novel single nucleotide polymorphism, G(-248)A, in the 5V-UTR of BAX

gene in chronic lymphocytic leukemia with disease progression and treatment resistance. Cancer Lett 2002, 187:199–205.

230. Watters JW, Kraja A, Meucci MA, et al. Genome-wide discovery of loci influencing chemotherapy cytotoxicity. Proc Natl Acad Sci U S A 2004, 101:11809–11814.

231. Watters JW, McLeod HL. Cancer pharmacogenomics: current and future applications. Biochim Biophys Acta 2003, 1603:99–111.

232. Ulrich CM, Robien K, McLeod HL. Cancer pharmacogenetics: polymorphisms, pathways and beyond. Nat Rev Cancer 2003, 3:912–920.

233. Marsh S, McLeod HL. Cancer pharmacogenetics. Br J Cancer 2004, 90:8–11.

234. Krajinovic M, Costea I, Chiasson S. Polymorphism of the thymidylate synthase gene and outcome of acute lymphoblastic leukaemia. Lancet 2002, 359:1033–1034.

235. Krajinovic M, Labuda D, Mathonnet G, et al. Polymorphisms in genes encoding drugs and xenobiotic metabolizing enzymes, DNA repair enzymes, and response to treatment of childhood acute lymphoblastic leukemia. Clin Cancer Res 2002, 8: 802–810.

236. Stoehlmacher J, Park DJ, Zhang W, et al. A multivariate analysis of genomic polymorphisms: prediction of clinical outcome to 5-FU/oxaliplatin combination chemotherapy in refractory colorectal cancer. Br J Cancer 2004, 91:344–354.

237. Marsh S, Kwok P, McLeod HL. SNP databases and pharmacogenetics: great start, but a long way to go. Human Mutat 2002, 20:174–179.

238. Peters U, Preisler-Adams S, Lanvers-Kaminsky C, et al. Sequence variations of mitochondrial DNA and individual sensitivity to the ototoxic effect of cisplatin. Anticancer Res 2003, 23:1249–1255.

239. Efferth T, Futscher BW, Osieka R. 5-Azacytidine modulates the response of sensitive and multidrug-resistant K562 leukemic cells to cytostatic drugs. Blood Cells Mol Dis 2001, 27:637–648.

240. Yan PS, Efferth T, Chen HL, et al. Use of CpG island microarrays to identify colorectal tumors with a high degree of concurrent methylation. Methods 2002, 27:162–169.

241. Gebhart E, Thoma K, Verdorfer I, et al. Genomic imbalances in T-cell acute lymphoblastic leukemia cell lines. Int J Oncol 2002, 21:887–894.

242. Efferth T, Verdorfer I, Miyachi H, et al. Genomic imbalances in drug-resistant T-cell acute lymphoblastic CEM leukemia cell lines. Blood Cells Mol Dis 2002, 29:1–13.

243. Efferth T, Ramirez T, Gebhart E, et al. Combination treatment of glioblastoma multiforme cell lines

with the anti-malarial artesunate and the epidermal growth factor receptor tyrosine kinase inhibitor OSI-774. Biochem Pharmacol 2004, 67:1689–1700.

244. Gebhart E, Ries J, Wiltfang J, et al. Genomic gain of the epidermal growth factor receptor harboring band 7p12 is part of a complex pattern of genomic imbalances in oral squamous cell carcinomas. Arch Med Res 2004, 35:385–394.

245. Gillet JP, Efferth T, Steinbach D, et al. Microarray-based detection of multidrug resistance in human tumor cells by expression profiling of ATP-binding cassette transporter genes. Cancer Res 2004,64:8987–8993.

246. Steinbach D, Gillet JP, Sauerbrey A, et al. ABCA3 as a possible cause of drug resistance in childhood acute myeloid leukemia. Clin Cancer Res 2006, 12.4357–4363.

247. Efferth T, Gillet JP, Sauerbrey A, et al. Expression profiling of ATP-binding cassette transporters in childhood T-cell acute lymphoblastic leukemia. Mol Cancer Ther 2006, 5:1986–1994.

248. Gillet JP, Schneider J, Bertholet V, et al. Microarray expression profiling of ABC transporters in human breast cancer. Cancer Genomics Proteomics 2006, 3:97–106.

249. Efferth T, Koomägi R, Mattern J, et al. Expression profile of proteins involved in the xenotransplantability of non-small cell lung cancers into athymic nude mice. Int J Oncol 2002, 20:391–395.

250. Volm M, Sauerbrey A, Zintl F, et al. Protein expression profile of newly diagnosed acute lymphoblastic leukemia in children developing relapses. Oncol Rep 2002, 9:965–969.

251. Volm M, Koomägi R, Efferth T, et al. Protein expression profiles of non-small cell lung carcinomas: correlation with histological subtype. Anticancer Res 2002, 22:2321–2324.

252. Volm M, Koomägi R, Mattern J, et al. Protein expression profile of primary human squamous cell lung carcinomas indicative of the incidence of metastases. Clin Exp Metastasis 2002, 19:385–390.

253. Kahn A, Esters A, Habedank. GD(-) Aachen, a new variant of deficient glucose-6-phosphate dehydrogenase. Hum Genet 1976, 32:171–180.

254. Efferth T, Osieka R, Beutler E. Molecular characterization of a German variant of glucose-6-phosphate dehydrogenase deficiency (G6PD Aachen). Blood Cells Mol Dis 2000, 26:101–104.

255. Beutler E, Kuhl W, Gelbart T, et al. DNA sequence abnormalities of human glucose-6-phosdphate dehydrogenase variants. J Biol Chem 1991, 266:4145–4150.

256. Au SWN, Gover S, Lam VNS, et al. Human glucose-6-phosphate dehydrogenase: the crystal structure

reveals a structural NADPH+ molecule and provides insight into enzyme deficiency. Structure 2000, 8:293–303.

257. Efferth T, Schwarzl SM, Smith J, et al. Role of glucose-6-phosphate dehydrogenase for oxidative stress and apoptosis. Cell Death Differ 2006, 13:527–528.

258. Efferth T, Fabry U, Osieka R. Leptin contributes to the protection of human leukemic cells from cisplatinum cytotoxicity. Anticancer Res 2000, 20:2541–2546.

259. Efferth T, Fabry U, Osieka R. Damage of the kinesin heavy chain gene contributes to the antagonism of cisplatin and paclitaxel. Anticancer Res 2000, 20:3211–3219.

260. Efferth T. Molekulare Pharmakologie und Toxikologie. Springer, Heidelberg, 2006.

Chapter 22
Pharmacogenetics in Cancer Management: Scenario for Tailored Therapy

Erika Cecchin, Massimo Libra, Calogero Cannavò, Bibiana Bruni, Alberto Fulvi, Giuseppe Toffoli, and Franca Stivala

1   Introduction

In the past two decades, clinical cancer researchers have paid increased attention to the role of genetic pathways encoding enzymes involved in metabolism, cellular transport, and mechanism of action of chemotherapeutic agents. Pharmacogenetics studies in cancer therapeutics have shown an association between specific genetic variants of drug-metabolizing enzymes (pharmacogenetic determinants of response) and adverse drug reaction (ADR) or toxicity. Primarily, it has been described that genetic abnormalities influence response to treatment [1–3]. This chapter evaluates the implications of genetic influences on the dose–response relationships of a drug and the current approaches to integrating pharmacogenetics in drug development.

Pharmacogenetic approaches in cancer therapeutics aim at identifying patients at risk of developing severe toxicity before the administration of a chemotherapeutic agent. These approaches should allow individualizing therapy through personalized dosing or using treatment modification strategies, thereby avoiding genetically altered drug metabolic pathways. It was suggested that in >50% of drugs examined in ADR studies, genetic variability, at least in one of the enzymes associated with altered metabolism, was responsible for the development of such ADR [4].

2   Drug Metabolism

Drug metabolizing enzymes are classically divided into two broad categories. Phase I enzymes include reductases, oxidases, and hydrolases. This category almost exclusively involves cytochrome P450 (CYP), which functions by insertion of one atom of atmospheric oxygen into a relatively inert substrate. At least 55 kinds of CYPs are encoded in the human genome. Their sequence homology has allowed a classification into families and subfamilies [5]. They have a dichotomist role in chemical carcinogenesis, being responsible for either detoxification or activation of exogenous molecules [6]. Phase II enzymes act on the oxygenated intermediates by conjugation with various endogenous moieties (glucuronide, glutathione, and sulfate) to produce extremely hydrophilic products that are easily extracted from the cell. Among drug-metabolizing enzymes, the family of cytochrome P450 enzymes are involved in the metabolism of the most clinically used drugs in ovarian cancer (cyclophosphamide, Taxol, anthracyclines, and etoposide). Several polymorphisms were described among the P450 subfamilies, and at present a great effort is being made to discover a correlation between genotype and phenotype.

From: *Sensitization of Cancer Cells for Chemo/Immuno/Radio-therapy, 1st Edition*.
Edited by: Benjamin Bonavida © Humana Press, Totowa, NJ

2.1 Phase I Metabolism

2.1.1 Cytochrome p450 Isoform 2C8

The cytochromes belonging to CYP2C subfamily appear to metabolize principally a number of clinically used drugs, such as nonsteroidal anti-inflammatory drugs, warfarin, omeprazole, propranolol phenytoin, and many others [7]. Four known human CYP2C enzymatic isoforms are described: CYP2C8, CYP2C9, CYP2C18, and CYP2C19, all of which are known to be polymorphic [8]. Particularly CYP2C8 is known to mediate the major pathway of Taxol metabolism in humans and human-derived in vitro systems, hydroxylating the taxane at position six and removing paclitaxel pharmacologic activity [9]. Dai et al. [10] described for the first time the existence of polymorphisms in the CYP2C8 gene. One allele, CYP2C8*3, including both a lysine to arginine and an arginine to lysine substitution in position 139 (exon 3) and 399 (exon 8), respectively, was demonstrated to be markedly defective in the metabolism of paclitaxel, with a frequency of about 15% in the White population but almost absent in Asian and Black African populations. Another allele, CYP2C8*2, with a phenylalanine to isoleucine substitution in position 269 (exon 5) characterizing only Asians resulted in a twofold lower clearance for paclitaxel than the wild-type allele [10]. Further studies in a White population confirmed these data, highlighting significantly lower paclitaxel transformation in CYP2C8*3 heterozygous human liver microsome samples compared with wild-type samples [11]. No studies on the effect of polymorphism in the in vivo metabolism of paclitaxel are currently available.

2.1.2 Cytochrome p450 Subfamily 3A

Four isozymes with very similar substrate specificity, whose genes are located on chromosome 7, are included in the CYP3A subfamily (CYP3A4, CYP3A5, CYP3A7, and CYP3A43) [12]. CYP3A4 is the most expressed CYP in the liver, comprising 25–40% of total hepatic P450 content and representing the most abundant CYP isoform in the gastrointestinal tract. CYP3A5 is commonly considered the second most important CYP3A protein in the liver, although its expression is largely variable inter-individually (10–97% of human livers) [13].

Enzymes belonging to the CYP3A subfamily (CYP3A4, CYP3A5, and CYP3A7) were also detected in tumor cells of the breast, colon, lung, and stomach, as well as in renal tumor cells, suggesting a possible role of this enzyme's family in carcinogenesis [12]. Actually the CYP3A4*1B polymorphism, discussed later, was associated with a higher risk of developing prostate cancer- and treatment-related leukemia. Enzymes belonging to the CYP3A subfamily were reported to exhibit a large inter-subject variability in hepatic and intestinal activity (40-fold in vitro, 5- to 10-fold in vivo) [6]. The biochemical bases of such differences are not yet completely understood, but they have a deep influence on drug efficacy and toxicity. Although it has been assumed that an important role could be related to polymorphic genetic variants, the existence of a polymorphism causing macroscopic variations of CYP3A4 activity in such a high percentage of people was not clearly demonstrated [14]. CYP3A4*1B is a polymorphism characterized by a A→G point mutation in a regulatory element of the 5′-flanking region of the gene, whose role was investigated by several studies without final conclusions [15]. At present, CYP3A4*1B is hypothesized to be the main CYP3A4 polymorphism correlated with inter-individual variability in irinotecan metabolism, also because of its relatively higher allelic frequency (9.6% in the White population and 52% in Africans) compared with other 2. 1. 3 CYP3A4 polymorphisms [16].

CYP3A5 expression is higher than mean values in 10–30% of livers in Whites and about 60% in Africans [5]. This variability has been associated with a single nucleotide polymorphism, CYP3A5*1 (G6986A), which abolishes a cryptic site in the intron 3, with consequent stabilization of the protein transcript. The presence of this mutation in at least one of the alleles ends up in an increase of the expression levels of the protein, with results similar to the reported peak level of CYP3A4, accounting for at least 50% of the total CYP3A hepatic content, while it usually has just a marginal role [5]. Many drugs, including anticancer agents such as irinotecan, cyclophosphamide, tamoxifen, taxanes, and doxorubicin, are substrates for the CYP3A subfamily, thus suggesting an involvement of CYP3A in the pharmacodynamics of these compounds [17, 18]. Nevertheless, not many clinical studies are currently available that elucidate the role of CYP3A family member polymorphisms and the pharmacology of these drugs. A clinical trial on 31 docetaxel-treated patients failed to correlate pharmacokinetic and pharmacodynamic profiles with CYP3A4 or CYP3A5 genotypes, but the polymorphisms were associated with the risk of developing toxicity after treatment in pediatric acute lymphoblastic leukemia [17]. A recent study of 65 cancer patients treated with irinotecan did not point out any relationship between the pharmacokinetics of the drug and CYP3A4 and CYP3A5 polymorphisms [19]. The data on this topic are still fragmentary, and further trials on larger populations are needed to establish the polymorphisms' exact role.

2.1.4 Cytochrome p450 Isoform 2B6

CYP2B6 is one of the p450 forms found in the liver. It is one of the less characterized forms probably because it was initially thought to constitute only a small fraction of total hepatic p450 [20]. Lately due to its important role in the metabolism of several drugs, particularly the alkylating agents cyclophosphamide, CYP2B6 has gained more importance [21]. Recently new tools for the investigation of CYP2B6 mRNA expression revealed a quite sharp, inter-individual variability (up to 100-fold) [22]. To explain these differences among individuals, the role of some genetic polymorphisms was investigated. Nine novel mutations have been described so far; five cause an amino acid exchange, whereas the other three are silent mutations. The 5-missense point mutations were found to describe seven different allelic variants [23]. One of them, CYP2B6*6, resulting from the combination of two different mutations in exons 4 and 5, was demonstrated to have an enhanced catalytic power in cyclophosphamide activation as well as a slightly reduced expression [24].

Presently, the role of the polymorphism has been investigated only in vitro [23, 25], but the encouraging results urge researchers to investigate further in order to verify the importance of CYP2B6*6 in clinical practice.

2.2 Phase II Metabolism

2.2.1 Uridine Diphosphate Glucuronosyltransferase

Glucuronidation is a phase II metabolic reaction that confers polarity to xenobiotics and endogenous substrates to easily eliminate them from the body

through the urine and bile. It is catalyzed by the reticular uridine 5′ diphosphate glucuronosyltransferase that utilizes uridine glucuronose-diphosphate (UDP)-glucuronic acid as a cosubstrate [26].

Nowadays at least 15 enzymatic isoforms belonging to UGT superfamily have been described, and at least five of them are polymorphic. UGT1A1 is the isoform with the highest number of polymorphic structures [27].

UGT1A1 is the isoform responsible for the glucuronidation of SN-38, the toxic metabolite of irinotecan; therefore, it has an important role of detoxification. It also glucuronidates the parental compound irinotecan [28]. The great inter-patient variability observed in response to irinotecan therapy might be associated with a reduced rate of glucuronidation of SN-38. These are normally associated with the different enzymatic activity of various UGT1A1 polymorphic structures [29], leading to accumulation of active SN-38 due to a scarce formation of the inactive form of SN-38G. The wild-type allele, UGT1A1*1, has six thymine adenine (TA) repeats in the atypical TATA box region of the UGT1A1 promoter. The variant allele, UGT1A1*28, consists of seven TA repeats in the A(TA)7TAA motif [30]. An increase in the number of repeats is associated with a reduced expression of the gene UGT1A1 and consequently with more severe toxicity. Polymorphic structures with five (UGT1A1*33) or eight (UGT1A1*34) repeats were described, especially among the African population. They are correlated with a sharper variation in enzyme expression levels and a more evident influence on irinotecan metabolism [31].

UGT1A1*28 polymorphism has been reported to be associated with increased toxicity after irinotecan chemotherapy, and the irinotecan label was modified by the FDA to indicate the role of UGT1A1*28 polymorphism in the metabolism of irinotecan and the associated increased risk of severe neutropenia, although data reported to date are conflicting in some cases. In particular, Carlini et al. investigated polymorphisms in UGT1A7 and UGT1A9 genes, also involved in CPT11 glucuronidation, highlighting that they may be predictors of response and toxicity to CPT11 therapy; but they failed to find any association between the UGT1A1*28 polymorphism and either toxicity or response to chemotherapy [32]. Mathjissen et al. extended the analysis to polymorphisms of genes involved in metabolic and transport steps, such as

ABCB1 and ABCC2 (CPT11 cell detoxification), CYP3A4 and CYP3A5 (CPT11 oxidative metabolism), and CES1 and CES2 [33]. A significant association with CPT11 pharmacokinetic parameters was found only for the ABCB1-1236C>T polymorphism, whereas no association was reported for UGT1A1*28. However, the authors noted the low prevalence of the UGT1A1*28 polymorphism in the Asian population used in the study. The correlation between CES2 polymorphisms and CPT11 toxicity and pharmacokinetics has not been fully explored. Nonetheless, recent studies evidenced a significant relationship between CES2 mRNA expression levels in lymphocytes and pharmacokinetics ($p = 0013$), and an association of CES2 mRNA expression levels with drug toxicity in 45 CRC patients treated with the FOLFIRI regimen [34]. These data encourage further studies on the possible genetic markers of protein expression.

Recently a predictive role of the polymorphism on SN38 active metabolite exposure and lower absolute neutrophil count nadir was confirmed on 86 advanced cancer patients who were refractory to other treatments and were treated with irinotecan [35]. The authors reported a stronger predictive role of UGT1A1*28 polymorphism in response to an irinotecan-containing regimen in advanced colorectal cancer patients than that in drug-related toxicity, which is significant after the first cycle of chemotherapy but less relevant over the entire course of treatment [36]. This suggests the need for careful consideration before irinotecan dose reduction in patients carrying the UGT1A1*28 polymorphism.

2.2.2 Glutathione S-Transferases

Glutathione S-transferases (GSTs) constitute a superfamily of phase II detoxifying enzymes. They play a central role in the cellular defense system, inactivating cytotoxic and mutagenic targets by conjugation to glutathione [37]. There are four cytosolic families of GSTs: GSTA (α), GSTP (π), GSTT (τ), and GSTM (μ) [38]. Polymorphisms were described for GST family members. The main GSTP1 polymorphism consists of a single nucleotide substitution (A>G) in position 313 of the gene and results in Ile105Val in the protein, substantially decreasing its enzymatic activity [39]. A GSTA1 polymorphism was described in the proximal promoter region of the gene and was linked with reduced levels of GSTA1 enzyme [40]. GSTM1 and GSTT1 exhibit quite widespread polymorphisms (about 50% and 40% of Whites, respectively), resulting in the homozygous deletion of the gene expression with a consequent lack of conjugation activity [41, 42]. A reduction or absence of GST activity correlated to all of these polymorphisms causes a greater exposition to the toxic insults of carcinogenic substances. These are substrates of the enzymes that inactivate them, particularly tobacco derivatives. Actually individuals with an absence of both genes are more disposed to develop tumors, especially those related to exposure to toxic substances such as tobacco derivatives and alcohol. Particularly GSTP1 Ile105Val was associated with oral cancer [43]; adult brain tumors [44]; and bladder, testicular, and prostate cancers; and smoking [45] with susceptibility to chemotherapy-induced leukemia [46]. GSTT1 and GSTM1 null genotypes were correlated with renal cell carcinoma [47], breast cancer with alcohol consumption [48] and acute myeloid leukemia [49]. The lung cancer risk in smokers was analyzed in relation to GSTs polymorphisms, but the results were controversial [50–52].

Several drugs constitute a substrate for GSTs (etoposide, Adriamycin, carboplatin, and cyclophosphamide) [53], and polymorphic structures of the enzymes with a role in their metabolism could be involved in a pharmaco-resistant insurgence. In vitro studies demonstrated that overexpression of GST enzymes is associated with decreased sensitivity to GST substrates such as platinum derivatives [54, 55]. Several studies also investigated the role of GST polymorphism in chemotherapy. In particular, a recent study highlighted a relationship between the polymorphisms characterizing GSTM1-T1 enzymes and disease outcome in advanced ovarian carcinoma patients, after platinum/paclitaxel–based chemotherapy [56]. In another study, GSTP1 Ile105Val polymorphism was associated with increased survival of patients with advanced colorectal cancer who were treated with oxaliplatin-5FU [57]. Concerning GSTA1, a recent report described the effect of the enzyme's polymorphism and an increased survival of patients with breast cancer after cyclophosphamide chemotherapy [40]. The polymorphic structure of the GST family's enzymes leads to reduced metabolic activity and could have a predictive value on chemotherapy with drugs that constitute a substrate for them. Specifically, polymorphism carriers for these enzymes demonstrate a main role in platinum derivatives and cyclophosphamide metabolism, and are less prone to develop pharmaco-resistance to those drugs.

3 Drug Transport

3.1 Multidrug Resistance Gene 1 and Multidrug Resistance Associated Protein 2

MDR1 (human multidrug resistance) gene and MRP2 (human multidrug resistance–associated protein) belong to ATP-binding cassette (ABC) transporter genes superfamily. They encode for integral membrane glycoproteins that function as ATP-dependent export pumps with substrate specificity. These ABC transporters can be intrinsically expressed in many human tumors; otherwise, their expression can be induced by chemotherapeutic treatment [58]. Many normal tissues with excretory function or barrier activity also express these ABC transporters. High expression levels of these proteins in tumor cells result in resistance to a variety of anticancer drugs, whereas their activity in normal, non-neoplastic cells prevents toxic side effects. Based on phylogenetic analysis, the ABC transporters were categorized into several subfamilies designated as ABCA through ABCG. Recently, genetic heterogeneity was described in a number of the ABC transporters genes [59]. The role of naturally occurring polymorphisms in the genes encoding for the ABC transporters and their potential relevance to cancer chemotherapy is under investigation. Presently, polymorphisms on the MDR1 gene (belonging to the ABC subfamily B) and MRP2 genes (belonging to the ABC subfamily C) are the most studied candidates in clinical investigations.

MDR1 is the gene chiefly responsible for pleiotropic drug resistance, and it is also the most studied among the ATP-binding cassette proteins. Located on the long arm of chromosome 7, MDR1 is formed by a promoter region and 28 exons [60]. It encodes for P-glycoprotein (P-gp) an integral membrane protein responsible for the energy

dependent excretion of several xenobiotics such as anticancer agents, immunosuppressants glucocorticoids, HIV-1 protease inhibitors, and many others [61]. The increase of this efflux-pump activity in cancer cells is associated with the development of cross-resistance of tumors to many antiblastic agents such as vinca alkaloids anthracyclines, epipodophyllotoxins, taxanes, and topoisomerase I inhibitors (irinotecan and topotecan) [62]. Conversely, the presence of the transporter in normal tissue prevents the toxic effects of antineoplastic drugs. The expression of P-gp in the apical or luminal surface of the lower gastrointestinal tract, the biliary canalicular membrane of hepatocytes, the brush-border membrane of proximal tubules of the kidney, and the blood–brain barrier influences the pharmacokinetics of the drugs interfering with intestinal absorption and biliary and urinary excretion. Intrinsic P-gp expression is generally low in untreated ovarian cancer but it increases after chemotherapy [63–65].

At least 15 polymorphisms for MDR1 gene are presently known [61]. One of the detected SNPs in exon 26 determines the replacement of a cytosine with a thymidine (C3435T). The polymorphism deeply influences P-gp expression and could have some implications in chemotherapy. This polymorphism has a considerable prevalence in the White population with a T allele frequency of 47.7%, but it is quite variable among different ethnic groups with a higher prevalence (73–84%) of the C allele among Africans [60].

Even if C>T does not cause an amino acid change, it probably defines an allele so it could be linked to other important mutations still undetected on the gene regulator zones that are responsible for the effect on the protein expression.

Laboratory investigation has defined the overexpression of the MDR1 gene as a resistance mechanism to paclitaxel [66, 67] and MDR1 gene expression may be a useful predictor of paclitaxel-based chemotherapy in ovarian cancer [68], as well as in other cancers: In 50 non-small cell lung cancer patients, a good correlation was found between PgP expression levels and response to paclitaxel-based therapy [69].

A recent study investigating the role of MDR1 polymorphism on P-gP expression and clinical outcome of patients with acute myeloid leukemia highlighted how some polymorphisms (including

C3435T) are good predictors for the clinical outcome, but currently the role of C3435T variant on the response to chemotherapy in ovarian cancer patients is not defined [70]. Sauer et al. [71] analyzed the basal expression of MDR1 in correlation with the TT genotype at the polymorphic site C3435T in 38 mammary and ovarian carcinoma cell lines, finding a weak basal expression associated with the TT genotype.

Besides C3435T variant, other MDR1 polymorphisms were investigated, including one in which an adenine substituted a guanine, in position 2995 in exon 24, and another that presented a thymine in position 2677 instead of a guanine. Most of them, for instance those cited in the preceding, affect the protein structure without influencing its expression level and could actually influence therapy outcome by additional mechanisms than PgP expression as in the case of C3435T variant [61].

MRP2, also known as cMOAT (canalicular multispecific organic anion transporter) is thought to be involved in methotrexate-, cisplatin-, and irinotecan-active metabolite glucuronide transport [72]. MRP2 gene polymorphisms were studied in association with Dubin-Johnson syndrome (DJS), an autosomal recessive disorder characterized by chronic conjugated hyperbilirubinemia due to a defect in secretion of anionic conjugates from hepatocytes into bile [73]. The existence of different mutations and polymorphism was highlighted. Two of them, one causing the substitution of a thymine with a cytosine in position 24 (C24T) located in the promoter of the gene, and one that substitutes a guanine with an adenine in position 1249 (G1249A) on exon 10, showed a considerable allelic frequency in a Jewish population (12. 3–26.1%, and 18. 3–31.2%, respectively, depending on the Jewish ethnic group) [74]. The analysis of their effect on MRP2 activity in vitro was investigated, highlighting an important role of C24T influencing the expression levels of the protein1 [75]. Correlation of the MRP2 expression and the clinical outcome of the acute myeloid leukemia patients treated with conventional chemotherapy, including cytosine arabinoside and anthracyclines highlighted a better response for the MRP2-negative patients, demonstrating their importance in the pharmaco-resistance to these drugs [76]. Some in vitro studies correlated MRP2 mRNA expression with the development of pharmaco-resistance to anticancer agents such as cisplatin. This correlation was demonstrated in particular in a set of clear cell carcinoma of the ovary cell lines [77]. Yokoyama [78] reported that MRP2 expression might be a potential predictor of the response to standard chemotherapy in ovarian cancer. Nonetheless, other in vivo studies failed to confirm this association in a population of epithelial ovarian carcinoma patients treated with combination therapies, including cisplatin [79].

4 Mechanism of Action of Drugs

4.1 Folate Pathway

4.1.1 Thymidylate Synthase and Dihydropyrimidine Dehydrogenase

Thymidylate synthase (TYMS; the major intracellular 5-FU target) and dihydropyrimidine dehydrogenase (DPYD; involved in 5-FU catabolic inactivation) could play a role in 5-fluorouracil (5-FU)–based chemotherapy [80]. Many studies indicate that low levels of DPYD or TYMS mRNA expression increase the toxic effect of 5-FU. Some polymorphisms in TYMS lead to modulation of TYMS mRNA expression, including a variable number of a 28-bp repeats (VNTR) in the promoter region, a G>C nucleotide change in the third repeat of the VNTR polymorphism, and a 3′UTR 6-bp deletion. Recent data from 187 GC patients indicate that a combination of the TYMS polymorphisms resulting in low- [2 VNTR repeats, (C) in the third VNTR repeat, 3′UTR-6bp-del] or high- [2 VNTR repeats, (G) in the third VNTR repeat, 3′UTR-6bp-ins] expressing alleles can discriminate between patients with different clinical outcomes. The patients presenting a combination of high-expressing alleles had a higher risk of relapse ($p = 0.003$, by chi-squared test), lower overall survival (OS) (HR = 2.9, 95% CI: 1.7–4.1), and lower disease-free survival (DFS) (HR = 3.5, 95% CI: 2.1–4.9) [81].

Similarly, in a population of 166 colorectal cancer (CRC) patients, the combination of 5′UTR and 3′UTR TYMS polymorphisms discriminates between high and low responders in terms of better DFS ($p = 0.049$) and OS ($p = 0.029$) for individuals carrying the low-expressing alleles [82]. Furthermore, Marcuello et al. found that in

89 metastatic CRC patients, the combination of 5′UTR VNTR and G>C polymorphisms impacted clinical response (OR = 2.9, 95%CI: 1.0–5.6), OS ($p = 0.03$, log rank test), and TTP ($p = 0.07$, log rank test) [83]. In contrast, in a population of 88 CRC patients, Jakobsen et al. reported an opposite effect of the 5′UTR VNTR polymorphism compared to the previously cited works, with a lower response rate ($p = 0.03$) among low-expressing genotype carriers [84]. The authors noted differences from other studies, including the fact that they genotyped healthy tissues, employed a different method of drug administration, and used a different treatment setting (metastatic disease versus adjuvant therapy). In any case, it is likely that a single polymorphism of TYMS is not sufficient to explain changes in the clinical benefit of 5-FU, and complex combinations of variants should be considered.

4.1.2 Methylene Tetrahydrofolate Reductase

Methylene tetrahydrofolate reductase (MTHFR) catalyses the reduction of 5,10-methylenetetrahydrofolate (5,10-methyleneTHF) to 5-methyltetrahydrofolate (5-methylTHF), which is the methyl donor for methionine synthesis from homocysteine (Hcy). MTHFR plays a central role in folate metabolism, DNA synthesis and repair. A common thermolabile genetic variant (677C>T) in the MTHFR gene, resulting in an alanine/valine amino acid substitution is associated with reduced enzyme activity, impaired remethylation of Hcy to methionine, and subsequent hyperhomocysteinemia. This variant is associated with vascular disease, spina bifida, diabetic nephropathy, and human cancers [85]. Hcy levels increase with methotrexate (MTX) therapy and this increase may be a sensitive and responsive indicator of antifolate therapy. Inhibition of methylation reactions by MTX can lead to toxicity; therefore, patients with an impaired intracellular methylation process, such as those with 677C>T mutation, could be predisposed to MTX toxicity. It has also been suggested that 677C>T MTHFR mutation can determine an impaired intracellular distribution of reduced folates. Preliminary reports indicate that MTHFR C677T polymorphism increases the toxicity and response to MTX and the level of hyperhomocysteinemia subsequent to drug administration

[86]. Our group recently demonstrated in a population of 43 ovarian cancer patients treated with MTX that the homozygous 677TT carriers were more exposed to the risk of developing G3-4 toxicity after chemotherapy and hyperhomocysteinemia than the heterozygous 677AC or wild-type 677AA carriers [87]. Although our findings should be considered exploratory because of the small sample size, they strongly suggest that the lower MTHFR enzymatic activity associated with the TT MTHFR genotype, increases the pharmacologic effects of MTX, leading to toxicity. However, our results are consistent with recent studies, indicating that patients with low activity of MTHFR (TT genotype) appear at risk of MTX toxicity [88, 89].

Hcy plasma level can represent a sensitive and responsive indicator of MTX cytotoxicity, which is also affected by MTFR activity [85, 90, 91]. Important questions are whether Hcy mediates some of the side effects caused by MTX and whether patients with the TT MTHFR genotype are at increased risk of developing hyperhomocysteinemia after MTX. Quinn et al. [92] suggested that elevation of plasma Hcy consequent to MTX administration may be important in the pathogenesis of MTX-associated neurotoxicity. Haagsma et al. [93] reported higher plasma concentrations of Hcy in rheumatoid patients with MTHFR T allele treated with low-dose chronic MTX than in patients without the allele, which may be related to the increased frequency of gastrointestinal side effects. We found that patients with the TT genotype and G3-4 toxicity had higher Hcy plasma levels compared with the other patients. Moreover, after MTX administration, the plasma Hcy levels were markedly increased and significantly higher than the basal levels only in patients with the TT genotype, suggesting that the TT genotype predisposes to the development of hyperhomocysteinemia after chronic administration of MTX.

4.2 DNA Repair Pathway

4.2.1 X-Ray Repair Cross-Complementing Isoform 1

XRCC1 belongs to the base excision repair (BER) genes family. BER proteins operate on small DNA lesions such as methylated, oxidized, or reduced bases. The damaged base is removed by a specific glycosylase, and then an endonuclease restores the abasic site. A polymerase and a ligase subsequently reconstitute normal DNA. XRCC1 acts as a scaffold for other DNA repair proteins such as DNA polymerase β and DNA ligase III [94]. Three polymorphic sites were described for human XRCC1 gene: one affecting codon 399 in exon 10, resulting in an arginine to methionine substitution; another in codon 194 in exon 6, causing an arginine to tryptophan substitution; and a last one in exon 9 codon 280, resulting in an arginine to histidine substitution. They have an allelic frequency in a White population of 0.14–0.39, 0.06–0.35, and 0.02–0.10, respectively [95, 96]. These polymorphisms code for nonconservative amino acid changes, which could reflect an important functional meaning [97]. In fact, a genotype-phenotype correlation study highlighted an important role of the polymorphism Arg399Gln in affecting XRCC1 phenotype, the polymorphism was associated to increased levels of DNA damage that may be due to reduced DNA repair functionality [98]. More recently, the XRCC1 Arg399Gln polymorphism has been linked to shorter survival in 61 CRC patients treated with 5-FU/oxaliplatin ($p = 0.038$) (99), and in 103 NSCLC patients treated with cisplatin-based chemotherapy ($p = 0.07$). This XRCC1 polymorphism is located in the poly-(ADP-ribose) polymerase [PARP] binding domain [100]. The PARP is a protein involved in DNA damage recognition and it is negatively regulated by XRCC1 [101]. Actually, for this polymorphism a possible correlation with the pathogenesis of several kinds of cancers was reported. A cell with a deficiency in DNA repair process was assumed to be more exposed to mutagenic insults and to have a higher probability of transforming into a cancer cell. Indeed, it was demonstrated that polymorphism frequency is higher in lung [102], breast [103], and head and neck [104] cancer patients. Preliminary data from our research group demonstrated the same trend also for ovarian cancer. However, the role of the polymorphism in cancer development is still controversial since an opposite trend was discovered for other kinds of tumors such as bladder cancer [105], non-melanomatous skin cancer [106], and therapy related acute myeloblastic leukemia [107]. It has been hypothesized that different tumors could be differently sensitive to gene environment. In fact, a low rate of DNA repairing could carry to extensive damage with a consequent apoptosis activation that could protect cells more efficiently than DNA repair from cancer degeneration [107]. Similarly controversial is the Arg399Gln role in chemotherapy. In vitro, high expression of DNA repair genes was demonstrated to be connected to the development of a pleiotropic resistance to DNA-interactive drugs such doxorubicin, etoposide, and mafosfamide [108]. We can assume that the same result could be extended to drugs working with similar mechanisms such as cyclophosphamide and platinum derivatives. XRCC1 high expression is connected with a pleiotropic resistance to DNA-interactive drugs (platinum derivatives, cyclophosphamide, and etoposide). A study investigating the association between resistance development to platinum-based therapy in advanced colorectal cancer and XRCC1 polymorphism reported a higher response to therapy among wild-type patients, who were previously believed to be more prone to develop resistance [95]. The authors suggested that BER mechanisms are less involved than NER in platinum damage repair; therefore, XRCC1 could in this case, play a still unknown role. Nonetheless, the population study consisted of only 61 patients; thus, the results need to be confirmed by larger studies.

4.2.2 Xeroderma Pigmentosum Group D

Xeroderma pigmentosum group D (XPD) is a member of the nucleotide excision repair (NER) family; it is also called ERCC2 (excision cross complementing group II). The NER pathway deals with the repair of bulky lesions of the DNA such as pyrimidine dimers, large chemical adducts, and cross-links. Particularly, XPD is an ATP-dependent helicase that, after DNA damage recognition by a series of different proteins, contributes, with other factors of the TFIIH complex, to DNA unwinding, allowing the removal of the damaged DNA [94]. Mutations at different sites of the XPD gene are connected with different pathologic states: xeroderma pigmentosum, Cockayne's syndrome, and trichothiodystrophy [109]. The presence of xeroderma pigmentosum results in an increase of sunlight-induced cancers >1,000-fold [110]. Polymorphic structures have been reported for XPD gene. Particularly two of them are taken into account for their impact on XPD functionality. One is located in exon 10 (G>A leading to a substitution of an aspartic with an asparagine at codon 312) with an allelic frequency of 0.33–0.44 [97], the other in exon 23 (A>C resulting in the substitution of a lysine with a glutamine at codon 751) and has an allelic frequency of 0.06–0.42 [111]. Controversial data were reported on their effect on DNA repair proficiency, though recent studies agree on the low efficiency of repair of both the polymorphic variants [112–114]. Several studies have been published recently, which tried to assess the association between XPD polymorphism and cancer development. However, they did not come to a common solution. Presently, published studies are in most cases only suggestive of a result due to small size of the populations in study. With these limitations, a correlation of the polymorphisms with melanoma [115], bladder cancer [105, 116], and lung cancer [112, 117], especially among smokers, was evidenced, whereas no association was found for colorectal carcinoma [118], breast cancer [119], squamous cell carcinoma of head and neck [103], and esophageal squamous cell carcinoma [120]. XPD has also an important role in the repair of DNA damages caused by alkylating drugs, platinum analogues, and radiation, while protein levels were demonstrated to correlate to resistance to these agents in human tumor cell lines [121]. Moreover, a clinical trial on 73 metastatic colorectal cancer patients treated with an association of 5-fluorouracil and oxaliplatin demonstrated that patients with Lys/Lys genotype at codon 751 responded better ($p = 0.015$) and had a longer survival ($p = 0.002$) than patients with at least one mutated allele [111]. The same polymorphism has been linked to shorter survival in 103 non-small cell lung carcinoma (NSCLC) patients treated with cisplatin-based therapy ($p = 0.003$) [122] and this highlights that polymorphism of XPD at codon 751 could become an important marker for predicting the clinical outcome of platinum-treated patients.

4.2.3 Excision Repair Cross Complementing Group 1

Excision repair cross complementing group 1 (ERCC1) belongs to NER system. The product of ERCC1 gene together with XPF form a complex that incises DNA at the 5′ size of a bulky adduct lesion such as those caused by chemical compounds [94]. Several in vitro studies demonstrated that cells with a low or absent ERCC1 expression resulted more sensitive to platinum derivatives or alkylating agent exposure [123, 124]. The mRNA levels of ERCC1 tend to be higher in ovarian clear cell tumors as opposed to other types of epithelial ovarian cancer. This is consistent with the long-standing observation that clear cell tumors are more likely to show de novo drug resistance against DNA-damaging agents in cancer patients [125]. The same results were confirmed also by some clinical trials in other tumors. In particular, high ERCC1 mRNA levels were correlated to poor survival and response of gastric cancer patients treated with 5FU and cisplatin [126]. Shen et al. [127] described some polymorphisms for ERCC1, particularly, a silent C-T transition at codon 118 in exon 4. Indeed, it converts a common codon usage (AAC) to an infrequent one (AAT), resulting in approximately 50% reduction in codon usage. This polymorphism has an allelic frequency of 0.14–0.39 [97] and has been associated in vitro with a reduced repair of cisplatin-DNA adduc, suggesting a reduced protein production in cells with the polymorphism [127]. Another polymorphism investigated is located in the untranslated region of the gene in 3′ (C8092A) [97]. This polymorphism was associated with a non-significant higher risk of developing squamous cell carcinoma of the head and neck that became significant if considered together with an XPD polymorphism [104]. Moreover, the 8092C>A polymorphism has been associated with increased toxicity ($p = 0.03$) in 214 NSCLC patients [128] and decreased OS (overall survival) ($p = 0.006$) in 128 NSCLC patients [129] treated with cisplatin/carboplatin-based therapy. The 118C>T polymorphism in the ERCC1 gene has been associated with improved survival in 62 NSCLC patients treated with docetaxel/cisplatin ($p = 0.04$) [130], but with decreased survival in 109 NSCLC patients treated with cisplatin combination therapy ($p = 0.0058$) [131]. It was also associated with the adult-onset of glioma [132]. These data support a functional role of these polymorphisms that should be further investigated for their pharmacologic implication since the important role of NER proteins like ERCC1 in platinum and alkylating agents–based therapies.

5   Conclusion

Pharmacogenetics represents an important emerging tool in cancer chemotherapy. The growing database of polymorphisms and improving laboratory techniques for genotyping are promising for future applications of pharmacogenetics to clinical practice. It is crucial to define the pharmacologic effect of the polymorphisms not only in vitro but also in patients. However, the clinical applications of pharmacogenetics are still limited at present and the establishment of the functional significance of a polymorphism requires a rigorous setting of clinical trials. This implies a careful evaluation of the criteria for patient inclusion in the studies, times, and drug administration, adverse effect monitoring, tumor response, and long-term follow-up. To assess the value of pharmacogenetics in clinical practice, prospective studies should be performed on genetic analysis in association with pharmacokinetic studies to assess the effect of a genotype on drug metabolism and define associations between genotypes and phenotypes. It must be considered that the overall pharmacologic effect of antineoplastic drugs is rarely a monogenetic trait, but is determined by the complex interplay of various genes determining multiple pathways of drug metabolism. Therefore, the simultaneous analysis of multiple gene polymorphisms or multiple mutations within a particular gene will provide a stronger rational basis for choosing optimal drug therapy. Normal tissue genetic traits have stable genetic features, whereas genetic characteristics of the tumor are a dynamic process and several epigenetic changes characterize the tumor cells. As a result, toxicity produced in normal cells could be predicted by the analysis of polymorphisms. Conversely, response of tumor cells to the therapy appears to be a pleiotropic phenomenon and not always adequately predicted by simple polymorphism analysis.

Ovarian cancer could be an adequate candidate for the pharmacogenetics studies. The response rate to the therapy, at least, in advanced stages is not quite different among the most active drugs currently used, whereas the toxic effect resulting from the treatment can be quite different. The analysis of polymorphisms in drug target will be useful in the identification of a specific subset of patients more prone to develop toxicity. Perhaps, in the next future pharmacogenetics will have a pivot role in the individualization of anticancer therapy, not only in the choice of the antineoplastic drug, but also for its dosage giving an alternative approach to the conventional body surface.

References

1. Weinshilboum R. Inheritance and drug response. N Engl J Med 2003, 348:529–537.
2. Evans WE, McLeod Hl. Pharmacogenomics: drug disposition, drug targets, and side effects. N Engl J Med 2003, 348:538–549.
3. Goldstein DB. Pharmacogenetics in the laboratory and the clinic. N Engl J Med 2003, 348:553–556.
4. Phillips KA, Veenstra DL, Oren E, et al. Potential role of pharmacogenomics in reducing adverse drug reactions: a systematic review. JAMA 2001, 286:2270–2279.
5. Kuehl P, Zhang J, Lin Y, et al. Sequence diversity in CYP3A promoters and characterization of the genetic basis of polymorphic CYP3A5 expression. Nat Genet 2001, 27:383–391.
6. Eiselt R, Domanski TL, Zibat A, et al. Identification and functional characterization of eight CYP3A4 protein variants. Pharmacogenetics 2001, 11:447–458.
7. Goldstein JA, de Morais SM. Biochemistry and molecular biology of the human CYP2C subfamily. Pharmacogenetics 1994, 4:285–299.
8. Goldstein JA. Clinical relevance of genetic polymorphisms in the human CYP2C subfamily. Br J Clin Pharmacol 2001, 52:349–355.
9. Kumar GN, Walle UK, Walle T. Cytochrome P450 3A-mediated human liver microsomal taxol 6 alpha-hydroxylation. J Pharmacol Exp Ther 1994, 268:1160–1165.
10. Dai D, Zeldin DC, Blaisdell JA, et al. Polymorphisms in human CYP2C8 decrease metabolism of the anticancer drug paclitaxel and arachidonic acid. Pharmacogenetics 2001, 11:597–607.
11. Bahadur N, Leathart JB, Mutch E, et al. CYP2C8 polymorphisms in Caucasians and their relationship with paclitaxel 6alpha-hydroxylase activity in human liver microsomes. Biochem Pharmacol 2002, 64:1579–1589.
12. Murray GI, McFadyen MC, Mitchell RT, et al. Cytochrome P450 CYP3A in human renal cell cancer. Br J Cancer 1999, 79:1836–1842.
13. Hustert E, Haberl M, Burk O, et al. The genetic determinants of the CYP3A5 polymorphism. Pharmacogenetics 2001, 11:773–779.

14. Ozdemir V, Kalowa W, Tang BK, et al. Evaluation of the genetic component of variability in CYP3A4 activity: a repeated drug administration method. Pharmacogenetics 2000, 10:373–388.

15. Westlind A, Lofberg L, Tindberg N, et al. Interindividual differences in hepatic expression of CYP3A4: relationship to genetic polymorphism in the 5′-upstream regulatory region. Biochem Biophys Res Commun 1999, 259:201–205.

16. Dai D, Tang J, Rose R, et al. Identification of variants of CYP3A4 and characterization of their abilities to metabolize testosterone and chlorpyrifos. J Pharmacol Exp Ther 2001, 299:825–831.

17. Goh BC, Lee SC, Wang LZ, et al. Explaining inter-individual variability of docetaxel pharmacokinetics and pharmacodynamics in Asians through phenotyping and genotyping strategies. J Clin Oncol 2002, 20:3683–3690.

18. Santos A, Zanetta S, Cresteil T, et al. Metabolism of irinotecan (CPT–11) by CYP3A4 and CYP3A5 in humans. Clin Cancer Res 200, 6:2012–2020.

19. Mathijssen RH, Marsh S, Karlsson MO, et al. Irinotecan pathway genotype analysis to predict pharmacokinetics. Clin Cancer Res 2003, 9:3246–3253.

20. Mimura M, Baba T, Yamazaki H, et al. Characterization of cytochrome P-450 2B6 in human liver microsomes. Drug Metab Dispos 1993, 21:1048–1056.

21. Chang TK, Weber GF, Crespi CL, et al. Differential activation of cyclophosphamide and iphosphamide by cytochromes P-450 2B and 3A in human liver microsomes. Cancer Res 1993, 53:5629–5637.

22. Code EL, Crespi CL, Penman, BW, et al. Human cytochrome P4502B6: interindividual hepatic expression, substrate specificity, and role in procarcinogen activation. Drug Metab Dispos 1997, 25:985–993.

23. Lang T, Klein K, Fischer J, et al. Extensive genetic polymorphism in the human CYP2B6 gene with impact on expression and function in human liver. Pharmacogenetics 2001, 11:399–415.

24. Xie HJ, Yasar U, Lundgren S, et al. Role of polymorphic human CYP2B6 in cyclophosphamide bioactivation. Pharmacogenomics J 1903, 3:53–61.

25. Ariyoshi N, Miyazaki M, Toide K, et al. A single nucleotide polymorphism of CYP2b6 found in Japanese enhances catalytic activity by autoactivation. Biochem Biophys Res Commun 2001, 281:1256–1260.

26. Tukey RH, Strassburg CP. Human UDP-glucuronosyltransferases: metabolism, expression, and disease. Annu Rev Pharmacol Toxicol 2000, 40:581–616.

27. Mackenzie PI, Miners JO, McKinnon RA. Polymorphisms in UDP glucuronosyltransferase genes: functional consequences and clinical relevance. Clin Chem Lab Med 2000, 38:889–892.

28. Ando Y, Saka H, Ando M, et al. Polymorphisms of UDP-glucuronosyltransferase gene and irinotecan toxicity: a pharmacogenetic analysis. Cancer Res 2000, 60:6921–6926.

29. Ciotti M, Basu N, Brangi M, et al. Glucuronidation of 7-ethyl-10-hydroxycamptothecin (SN-38) by the human UDP-glucuronosyltransferases encoded at the UGT1 locus. Biochem Biophys Res Commun 1999, 260:199–202.

30. Monaghan G, Ryan M, Seddon R, et al B. Genetic variation in bilirubin UPD–glucuronosyltransferase gene promoter and Gilbert's syndrome. Lancet 1996, 347:578–581.

31. Beutler E, Gelbart T, Demina A. Racial variability in the UDP-glucuronosyltransferase 1 (UGT1A1) promoter: a balanced polymorphism for regulation of bilirubin metabolism? Proc Natl Acad Sci U S A 1998, 95:8170–8174.

32. Carlini LE, Meropol NJ, Bever J, et al. UGT1A7 and UGT1A9 polymorphisms predict response and toxicity in colorectal cancer patients treated with capecitabine/irinotecan. Clin Cancer Res 2005, 11:1226–1236.

33. Mathijssen RH, Marsh S, Karlsson MO, et al. Irinotecan pathway genotype analysis to predict pharmacokinetics. Clin Cancer Res 2003, 9:3246–3253.

34. Cecchin E, Corona G, Masier S, et al. Carboxylesterase isoform 2 mRNA expression in peripheral blood mononuclear cells is a predictive marker of the irinotecan to SN38 activation step in colorectal cancer patients. Clin Cancer Res 2005, 11:6901–6907.

35. Ramchandani RP, Wang Y, Booth BP et al. The role of SN-38 exposure, UGT1A1*28 polymorphism, and baseline bilirubin level in predicting severe irinotecan toxicity. J Clin Pharmacol 2007, 47(1):78–86.

36. Toffoli G, Cecchin E, Corona G, et al. The role of UGT1A1*28 polymorphism in the pharmacodynamics and pharmacokinetics of irinotecan in patients with metastatic colorectal cancer. J Clin Oncol 2006, 24(19):3061–3068.

37. Smith CA, Smith G, Wolf CR. Genetic polymorphisms in xenobiotic metabolism. Eur J Cancer 1994, 30A:1921–1935.

38. Seidegard J, Ekstrom G. The role of human glutathione transferases and epoxide hydrolases in the metabolism of xenobiotics. Environ Health Perspect 1997, 105(Suppl 4):791–799.

39. Zimniak P, Nanduri B, Pikula S, et al. Naturally occurring human glutathione S-transferase GSTP1-1 isoforms with isoleucine and valine in position 104 differ in enzymic properties. Eur J Biochem 1994, 224:893–899.

40. Sweeney C, Ambrosone CB, Joseph L, et al. Association between a glutathione S-transferase A1

promoter polymorphism and survival after breast cancer treatment. Int J Cancer 2003, 103:810–814.

41. Pemble S, Schroeder KR, Spencer SR, et al. Human glutathione S-transferase theta (GSTT1): cDNA cloning and the characterization of a genetic polymorphism. Biochem J 1994, 300(Pt 1):271–276.

42. Seidegard J, Pero RW, Miller DG, et al. A glutathione transferase in human leukocytes as a marker for the susceptibility to lung cancer. Carcinogenesis 1986, 7:751–753.

43. Park JY, Schantz SP, Stern JC, et al. Association between glutathione S-transferase pi genetic polymorphisms and oral cancer risk. Pharmacogenetics 1999, 9:497–504.

44. De Roos AJ, Rothman N, Inskip PD, et al. Genetic polymorphisms in GSTM1, –P1, –T1, and CYP2E1 and the risk of adult brain tumors. Cancer Epidemiol Biomarkers Prev 2003, 12:14–22.

45. Harries LW, Stubbins MJ, Forman D, et al. Identification of genetic polymorphisms at the glutathione S-transferase Pi locus and association with susceptibility to bladder, testicular and prostate cancer. Carcinogenesis 1997, 18:641–644.

46. Allan JM, Wild CP, Rollinson S, et al. Polymorphism in glutathione S-transferase P1 is associated with susceptibility to chemotherapy-induced leukemia. Proc Natl Acad Sci U S A 2001, 98:11592–11597.

47. Sweeney C, Farrow DC, Schwartz SM, et al. Glutathione S-transferase M1, T1, and P1 polymorphisms as risk factors for renal cell carcinoma: a case-control study. Cancer Epidemiol Biomarkers Prev 2000, 9:449–454.

48. Park SK, Kang D, Noh DY, et al. Reproductive factors, glutathione S-transferase M1 and T1 genetic polymorphism and breast cancer risk. Breast Cancer Res Treat 2003, 78:89–96.

49. Naoe T, Tagawa Y, Kiyoi H, et al. Prognostic significance of the null genotype of glutathione S-transferase-T1 in patients with acute myeloid leukemia: increased early death after chemotherapy. Leukemia 2002, 16:203–208.

50. Miller DP, De VI, Neuberg D, et al. Association between self-reported environmental tobacco smoke exposure and lung cancer: modification by GSTP1 polymorphism. Int J Cancer 2003, 104:758–763.

51. Nazar-Stewart V, Vaughan TL, Stapleton P, et al. A population-based study of glutathione S-transferase M1, T1 and P1 genotypes and risk for lung cancer. Lung Cancer 2003, 40:247–258.

52. Wang Y, Spitz MR, Schabath MB, et al. Association between glutathione S-transferase p1 polymorphisms and lung cancer risk in Caucasians: a case-control study. Lung Cancer 2003, 40:25–32.

53. Bellincampi L, Ballerini S, Bernardini S, et al. Glutathione transferase P1 polymorphism in neuroblastoma studied by endonuclease restriction mapping. Clin Chem Lab Med 2001, 39:830–835.

54. Nishimura T, Newkirk K, Sessions RB, et al. Immunohistochemical staining for glutathione S-transferase predicts response to platinum-based chemotherapy in head and neck cancer. Clin Cancer Res 1996, 2:1859–1865.

55. Oguri T, Fujiwara Y, Katoh O, et al. Glutathione S-transferase-pi gene expression and platinum drug exposure in human lung cancer. Cancer Lett 2000, 156:93–99.

56. Medeiros R, Pereira D, Afonso N, et al. Platinum/paclitaxel-based chemotherapy in advanced ovarian carcinoma: glutathione S-transferase genetic polymorphisms as predictive biomarkers of disease outcome. Int J Clin Oncol 2003, 8:156–161.

57. Stoehlmacher J, Park DJ, Zhang W, et al. Association between glutathione S-transferase P1, T1, and M1 genetic polymorphism and survival of patients with metastatic colorectal cancer. J Natl Cancer Inst 2002, 94:936–942.

58. Mickley LA, Lee JS, Weng Z, et al. Genetic polymorphism in MDR-1: a tool for examining allelic expression in normal cells, unselected and drug-selected cell lines, and human tumors. Blood 1998, 91:1749–1756.

59. Lockhart AC, Tirona RG, Kim RB. Pharmacogenetics of ATP-binding cassette transporters in cancer and chemotherapy. Mol Cancer Ther 2003, 2:685–698.

60. Ameyaw MM, Regateiro F, Li T, et al. MDR1 pharmacogenetics: frequency of the C3435T mutation in exon 26 is significantly influenced by ethnicity. Pharmacogenetics 2001, 11:217–221.

61. Brinkmann U, Eichelbaum M. Polymorphisms in the ABC drug transporter gene MDR1. Pharmacogenomics J 1903, 1:59–64.

62. Iyer L, Ramirez J, Shepard DR, et al. Biliary transport of irinotecan and metabolites in normal and P-glycoprotein-deficient mice. Cancer Chemother Pharmacol 2002, 49:336–341.

63. Chen CJ, Chin JE, Ueda K, et al. Internal duplication and homology with bacterial transport proteins in the mdr1 (P-glycoprotein) gene from multidrug-resistant human cells. Cell 1986, 47:381–389.

64. Juliano RL, Ling V. A surface glycoprotein modulating drug permeability in Chinese hamster ovary cell mutants. Biochim Biophys Acta 1976, 455:152–162.

65. Rittierodt M, Harada K. Repetitive doxorubicin treatment of glioblastoma enhances the PGP expression—a special role for endothelial cells. Exp Toxicol Pathol 2003, 55:39–44.

66. Kemper EM, van Zandbergen AE, Cleypool C, et al. Increased penetration of paclitaxel into the brain by inhibition of P-glycoprotein. Clin Cancer Res 2003, 9:2849–2855.

67. Yusuf RZ, Duan Z, Lamendola DE, et al. Paclitaxel resistance: molecular mechanisms and pharmacologic manipulation. Curr Cancer Drug Targets 2003, 3:1–19.

68. Kamazawa S, Kigawa J, Kanamori Y, et al. Multidrug resistance gene-1 is a useful predictor of Paclitaxel-based chemotherapy for patients with ovarian cancer. Gynecol Oncol 2002, 86:171–176.

69. Yeh JJ, Hsu WH, Wang JJ, et al. Predicting chemotherapy response to paclitaxel-based therapy in advanced non-small-cell lung cancer with P-glycoprotein expression. Respiration 2003, 70:32–35.

70. Illmer T, Schuler US, Thiede C, et al. MDR1 gene polymorphisms affect therapy outcome in acute myeloid leukemia patients. Cancer Res 2002, 62:4955–4962.

71. Sauer G, Kafka A, Grundmann R, et al. Basal expression of the multidrug resistance gene 1 (MDR-1) is associated with the TT genotype at the polymorphic site C3435T in mammary and ovarian carcinoma cell lines. Cancer Lett 2002, 185:79–85.

72. Mathijssen RH, van Alphen RJ, Verweij J, et al. Clinical pharmacokinetics and metabolism of irinotecan (CPT-11). Clin Cancer Res 2001, 7:2182–2194.

73. Toh S, Wada M, Uchiumi T, et al. Genomic structure of the canalicular multispecific organic anion-transporter gene (MRP2/cMOAT) and mutations in the ATP-binding-cassette region in Dubin-Johnson syndrome. Am J Hum Genet 1999, 64:739–746.

74. Mor-Cohen R, Zivelin A, Rosenberg N, et al. Identification and functional analysis of two novel mutations in the multidrug resistance protein 2 gene in Israeli patients with Dubin-Johnson syndrome. J Biol Chem 2001, 276:36923–36930.

75. Itoda M, Saito Y, Soyama A, et al. Polymorphisms in the ABCC2 (cMOAT/MRP2) gene found in 72 established cell lines derived from Japanese individuals: an association between single nucleotide polymorphisms in the 5′-untranslated region and exon 28. Drug Metab Dispos 2002, 30:363–364.

76. Schuurhuis GJ, Broxterman HJ, Ossenkoppele GJ, et al. Functional multidrug resistance phenotype associated with combined overexpression of Pgp/MDR1 and MRP together with 1-beta-D-arabinofuranosylcytosine sensitivity may predict clinical response in acute myeloid leukemia. Clin Cancer Res 1995, 1:81–93.

77. Itamochi H, Kigawa J, Sultana H, et al. Sensitivity to anticancer agents and resistance mechanisms in clear cell carcinoma of the ovary. Jpn J Cancer Res 2002, 93:723–728.

78. Yokoyama Y, Sato S, Fukushi Y, et al. Significance of multi-drug-resistant proteins in predicting chemotherapy response and prognosis in epithelial ovarian cancer. J Obstet Gynaecol Res 1999, 25:387–394.

79. Arts HJ, Katsaros D, de Vries EG, et al. Drug resistance-associated markers P-glycoprotein, multidrug resistance-associated protein 1, multidrug resistance-associated protein 2, and lung resistance protein as prognostic factors in ovarian carcinoma. Clin Cancer Res 1999, 5:2798–2805.

80. Iacopetta B, Grieu F, Joseph D, et al. A polymorphism in the enhancer region of the thymidylate synthase promoter influences the survival of colorectal cancer patients treated with 5-fluorouracil. Br J Cancer 2001, 85:827–830.

81. Kawakami K, Graziano F, Watanabe G, et al. Prognostic role of thymidylate synthase polymorphisms in gastric cancer patients treated with surgery and adjuvant chemotherapy. Clin Cancer Res 2005, 11:3778–3783.

82. Hitre E, Budai B, Adleff V, et al. Influence of thymidylate synthase gene polymorphisms on the survival of colorectal cancer patients receiving adjuvant 5-fluorouracil. Pharmacogenet Genomics 2005, 15:723–730.

83. Marcuello E, Altes A, del Rio E, et al. Single nucleotide polymorphism in the 5′ tandem repeat sequences of thymidylate synthase gene predicts for response to fluorouracil-based chemotherapy in advanced colorectal cancer patients. Int J Cancer 2004, 112:733–737.

84. Jakobsen A, Nielsen JN, Gyldenkerne N, et al. Thymidylate synthase and methylenetetrahydrofolate reductase gene polymorphism in normal tissue as predictors of fluorouracil sensitivity. J Clin Oncol 2005, 23:1365–1369.

85. Ueland PM, Hustad S, Schneede J, et al. Biological and clinical implications of the MTHFR C677T polymorphism. Trends Pharmacol Sci 2001, 22:195–201.

86. Chiusolo P, Reddiconto G, Casorelli I, et al. Preponderance of methylenetetrahydrofolate reductase C677T homozygosity among leukemia patients intolerant to methotrexate. Ann Oncol 2002, 13:1915–1918.

87. Toffoli G, Russo A, Innocenti F, et al. Effect of methylenetetrahydrofolate reductase 677C→T polymorphism on toxicity and homocysteine plasma level after chronic methotrexate treatment of ovarian cancer patients. Int J Cancer 2003, 103:294–299.

88. Ulrich CM, Yasui Y, Storb R, et al. Pharmacogenetics of methotrexate: toxicity among marrow transplantation patients varies with the methylenetetrahydrofolate reductase C677T polymorphism. Blood 2001, 98:231–234.

89. Urano W, Taniguchi A, Yamanaka H, et al. Polymorphisms in the methylenetetrahydrofolate reductase gene were associated with both the efficacy and the toxicity of methotrexate used for the treatment of rheumatoid arthritis, as evidenced by single locus and haplotype analyses. Pharmacogenetics 2002, 12:183–190.

90. Chiang PK, Gordon RK, Tal J, et al. S-Adenosylmethionine and methylation. FASEB J 1996, 10:471–480.

91. Calvert H. Folate status and the safety profile of antifolates. Semin Oncol 2002, 29:3–7.

92. Quinn CT, Griener JC, Bottiglieri T, et al. Elevation of homocysteine and excitatory amino acid neurotransmitters in the CSF of children who receive methotrexate for the treatment of cancer. J Clin Oncol 1997, 15:2800–2806.

93. Haagsma CJ, Blom HJ, van Riel PL, et al. Influence of sulphasalazine, methotrexate, and the combination of both on plasma homocysteine concentrations in patients with rheumatoid arthritis. Ann Rheum Dis 1999, 58:79–84.

94. Goode EL, Ulrich CM, Potter JD. Polymorphisms in DNA repair genes and associations with cancer risk. Cancer Epidemiol. Biomarkers Prev 2002, 11:1513–1530.

95. Stoehlmacher J, Ghaderi V, Iobal S, et al. A polymorphism of the XRCC1 gene predicts for response to platinum based treatment in advanced colorectal cancer. Anticancer Res 2001, 21:3075–3079.

96. Hu JJ, Smith TR, Miller MS, et al. Amino acid substitution variants of APE1 and XRCC1 genes associated with ionizing radiation sensitivity. Carcinogenesis 2001, 22:917–922.

97. Shen MR, Jones IM, Mohrenweiser H. Nonconservative amino acid substitution variants exist at polymorphic frequency in DNA repair genes in healthy humans. Cancer Res 1998, 58:604–608.

98. Wang Y, Spitz MR, Zhu Y, et al. From genotype to phenotype: correlating XRCC1 polymorphisms with mutagen sensitivity. DNA Repair (Amst) 2003, 2:901–908.

99. Stoehlmacher J, Ghaderi V, Iobal S, et al. A polymorphism of the XRCC1 gene predicts for response to platinum based treatment in advanced colorectal cancer. Anticancer Res 2001, 21:3075–3079.

100. Masson M, Niedergang C, Schreiber V, et al. XRCC1 is specifically associated with poly(ADP-ribose) polymerase and negatively regulates its activity following DNA damage. Mol Cell Biol 1998, 18:3563–3571.

101. Schreiber V, Ame JC, Dolle P, et al. Poly(ADP-ribose) polymerase-2 (PARP-2) is required for efficient base excision DNA repair in association with PARP-1 and XRCC1. J Biol Chem 2002, 277:23028–23036.

102. Park JY, Lee SY, Jeon HS, et al. Polymorphism of the DNA repair gene XRCC1 and risk of primary lung cancer. Cancer Epidemiol Biomarkers Prev 2002, 11:23–27.

103. Smith TR, Miller MS, Lohman K, et al. Polymorphisms of XRCC1 and XRCC3 genes and susceptibility to breast cancer. Cancer Lett 2003, 190:183–190.

104. Sturgis EM, Dahlstrom KR, Spitz MR, et al. DNA repair gene ERCC1 and ERCC2/XPD polymorphisms and risk of squamous cell carcinoma of the head and neck. Arch Otolaryngol Head Neck Surg 2002, 128:1084–1088.

105. Matullo G, Guarrera S, Carturan S, et al. DNA repair gene polymorphisms, bulky DNA adducts in white blood cells and bladder cancer in a case-control study. Int J Cancer 2001, 92:562–567.

106. Nelson HH, Kelsey KT, Mott LA, et al. The XRCC1 Arg399Gln polymorphism, sunburn, and non-melanoma skin cancer: evidence of gene-environment interaction. Cancer Res 2002, 62:152–155.

107. Seedhouse C, Bainton R, Lewis M, et al. The genotype distribution of the XRCC1 gene indicates a role for base excision repair in the development of therapy-related acute myeloblastic leukemia. Blood 2002, 100:3761–3766.

108. Levenson VV, Davidovich IA, Roninson IB. Pleiotropic resistance to DNA-interactive drugs is associated with increased expression of genes involved in DNA replication, repair, and stress response. Cancer Res 2000, 60:5027–5030.

109. Kobayashi T, Uchiyama M, Fukuro S, et al. Mutations in the XPD gene in xeroderma pigmentosum group D cell strains: confirmation of genotype-phenotype correlation. Am J Med Genet 2002, 110:248–252.

110. Eveno E, Bourre F, Quilliet X, et al. Different removal of ultraviolet photoproducts in genetically related xeroderma pigmentosum and trichothiodystrophy diseases. Cancer Res 1995, 55:4325–4332.

111. Park DJ, Stoehlmacher J, Zhang W, et al. A Xeroderma pigmentosum group D gene polymorphism predicts clinical outcome to platinum-based chemotherapy in patients with advanced colorectal cancer. Cancer Res 2001, 61:8654–8658.

112. Hou SM, Falt S, Angelini S, et al. The XPD variant alleles are associated with increased aromatic DNA adduct level and lung cancer risk. Carcinogenesis 2002, 23:599–603.

113. Lunn RM, Helzlsouer KJ, Parshad R, et al. XPD polymorphisms: effects on DNA repair proficiency. Carcinogenesis 2000, 21:551–555.

114. Qiao Y, Spitz MR, Shen H, et al. Modulation of repair of ultraviolet damage in the host-cell reactivation assay by polymorphic XPC and XPD/ERCC2 genotypes. Carcinogenesis 2002, 23:295–299.

115. Tomescu D, Kavanagh G, Ha T, et al. Nucleotide excision repair gene XPD polymorphisms and genetic predisposition to melanoma. Carcinogenesis 2001, 22:403–408.

116. Stern MC, Johnson LR, Bell DA, et al. XPD codon 751 polymorphism, metabolism genes, smoking, and bladder cancer risk. Cancer Epidemiol Biomarkers Prev 2002, 11:1004–1011.

117. Xing D, Tan W, Wei Q, et al. Polymorphisms of the DNA repair gene XPD and risk of lung cancer in a Chinese population. Lung Cancer 2002, 38:123–129.

118. Mort R, Mo L, McEwan C, et al. Lack of involvement of nucleotide excision repair gene polymorphisms in colorectal cancer. Br J Cancer 2003, 89:333–337.

119. Tang D, Cho S, Rundle A, et al. Polymorphisms in the DNA repair enzyme XPD are associated with increased levels of PAH-DNA adducts in a case-control study of breast cancer. Breast Cancer Res Treat 2002, 75:159–166.

120. Xing D, Qi J, Miao X, et al. Polymorphisms of DNA repair genes XRCC1 and XPD and their associations with risk of esophageal squamous cell carcinoma in a Chinese population. Int J Cancer 2002, 100:600–605.

121. Chen ZP, Malapetsa A, Monks A, et al. Nucleotide excision repair protein levels vis-a-vis anticancer drug resistance in 60 human tumor cell lines. Ai Zheng 2002, 21:233–239.

122. Gurubhagavatula S, Liu G, Park S, et al. XPD and XRCC1 genetic polymorphisms are prognostic factors in advanced non-small-cell lung cancer patients treated with platinum chemotherapy. J Clin Oncol 2004, 22:2594–2601.

123. Dabholkar M, Vionnet J, Bostick-Bruton F, et al. Messenger RNA levels of XPAC and ERCC1 in ovarian cancer tissue correlate with response to

platinum-based chemotherapy. J Clin Invest 1994, 94:703–708.

124. Guichard S, Arnould S, Hennebelle I, et al. Combination of oxaliplatin and irinotecan on human colon cancer cell lines: activity in vitro and in vivo. Anticancer Drugs 2001, 12:741–751.

125. Reed E, Yu JJ, Davies A, et al. Clear cell tumors have higher mRNA levels of ERCC1 and XPB than other histological types of epithelial ovarian cancer. Clin Cancer Res 2003, 9:5299–5305.

126. Metzger R, Leichman CG, Danenberg KD, et al. ERCC1 mRNA levels complement thymidylate synthase mRNA levels in predicting response and survival for gastric cancer patients receiving combination cisplatin and fluorouracil chemotherapy. J Clin Oncol 1998, 16:309–316.

127. Yu JJ, Mu C, Lee KB, et al. A nucleotide polymorphism in ERCC1 in human ovarian cancer cell lines and tumor tissues. Mutat Res 1997, 382:13–20.

128. Suk R, Gurubhagavatula S, Park S, et al. Polymorphisms in ERCC1 and grade 3 or 4 toxicity in non-small cell lung cancer patients. Clin Cancer Res 2005, 11:1534–1538.

129. Zhou W, Gurubhagavatula S, Liu G, et al. Excision repair cross-complementation group 1 polymorphism predicts overall survival in advanced non-small cell lung cancer patients treated with platinum-based chemotherapy. Clin Cancer Res 2004, 10:4939–4943.

130. Isla D, Sarries C, Rosell R, et al. Single nucleotide polymorphisms and outcome in docetaxel-cisplatin-treated advanced non-small-cell lung cancer. Ann Oncol 2004, 15:1194–1203.

131. Ryu JS, Hong YC, Han HS, et al. Association between polymorphisms of ERCC1 and XPD and survival in non-small-cell lung cancer patients treated with cisplatin combination chemotherapy. Lung Cancer 2004, 44:311–316.

132. Chen P, Wiencke J, Aldape K, et al. Association of an ERCC1 polymorphism with adult-onset glioma. Cancer Epidemiol Biomarkers Prev 2000, 9:843–847.

Index

Printed in the United States of America